SYSTEMS ANALYSIS AND DESIGN

Eighth Edition: Video Enhanced

Gary B. Shelly

Harry J. Rosenblatt

Shelly Cashman Series®
An imprint of Course Technology, Cengage Learning

COURSE TECHNOLOGY
CENGAGE Learning™

Australia • Brazil • Japan • Korea • Mexico • Singapore • Spain • United Kingdom • United States

Systems Analysis and Design, Eighth Edition:
Video Enhanced
Gary B. Shelly
Harry J. Rosenblatt

Executive Editor: Kathleen McMahon

Associate Acquisitions Editor: Reed Curry

Associate Product Manager: Jon Farnham

Editorial Assistant: Lauren Brody

Director of Marketing: Cheryl Costantini

Marketing Manager: Tristen Kendall

Marketing Coordinator: Stacey Leasca

Print Buyer: Julio Esperas

Content Project Manager: Lisa Weidenfeld

Development Editor: Deb Kaufmann

Proofreaders: Karen Annett, Kim Kosmatka

Art Director: Marissa Falco

Interior and Text Design: Joel Sadagursky

Cover Photos: Jon Chomitz

Compositor: GEX Publishing Services

Indexer: Rich Carlson

ISBN-13: 978-0-538-47988-2

ISBN-10: 0-538-47988-4

Course Technology
20 Channel Center Street
Boston, MA 02210
USA

Cengage Learning is a leading provider of customized learning solutions with office locations around the globe, including Singapore, the United Kingdom, Australia, Mexico, Brazil, and Japan. Locate your local office at:
www.cengage.com/global

Cengage Learning products are represented in Canada by Nelson Education, Ltd.

To learn more about Course Technology, visit **www.cengage.com/coursetechnology**

To learn more about Cengage Learning, visit **www.cengage.com**.

Purchase any of our products at your local college store or at our preferred online store **www.CengageBrain.com**

Printed in the United States of America
1 2 3 4 5 6 15 14 13 12 11 10

BRIEF CONTENTS

TABLE OF CONTENTS

PHASE 1: SYSTEMS PLANNING

Chapter 3

Managing Systems Projects

PHASE 2: SYSTEMS ANALYSIS

Chapter 4

Requirements Modeling

Chapter 5

Data and Process Modeling

Chapter 6

Object Modeling

Chapter 7

Development Strategies

PHASE 3: SYSTEMS DESIGN

Chapter 8

Output and User Interface Design

Chapter 9

Data Design

Chapter 10

System Architecture

PHASE 4: SYSTEMS IMPLEMENTATION

Chapter 11

Managing Systems Implementation

PHASE 5: SYSTEMS SUPPORT AND SECURITY

Chapter 12

Managing Systems Support and Security

THE SYSTEMS ANALYST'S TOOLKIT

Toolkit 1

Communication Tools

Toolkit 2

CASE Tools

Toolkit 3

Financial Analysis Tools

Toolkit 4

Internet Resource Tools

PREFACE

The Shelly Cashman Series® offers the finest textbooks in computer education. We are proud that our previous editions of *Systems Analysis and Design* have been so well received by instructors and students. *Systems Analysis and Design, Eighth Edition: Video Enhanced* continues with the innovation, quality, and reliability you have come to expect from the Shelly Cashman Series.

Overview

Systems Analysis and Design, Eighth Edition: Video Enhanced includes exciting new Video Learning Sessions, developed to maximize the learning experience. The Video Learning Sessions combined with the text present a practical, visually appealing approach to information systems development. Many two- and four-year colleges and schools use this book in information systems, computer science, and e-commerce curriculums. The textbook emphasizes the role of the systems analyst in a dynamic, business-related environment.

Facing a challenging global marketplace, companies need strong IT resources to survive and compete effectively. Many of today's students will become the systems analysts, managers, and IT professionals of tomorrow. This textbook will help prepare students for those roles.

Using this book, students learn how to translate business requirements into information systems that support a company's short- and long-term objectives. Case studies and assignments teach analytical and problem-solving skills. Students learn about traditional structured analysis, object-oriented concepts, and agile methods. Extensive end-of-chapter exercises emphasize critical-thinking skills.

Features new to the *Eighth Edition* include a new chapter on project management, a new IT ethics feature, greater emphasis on IT security, and an overall update that explains new systems development methods and trends.

Objectives of This Textbook

Systems Analysis and Design, Eighth Edition: Video Enhanced is intended for a three credit-hour introductory systems analysis and design course. This textbook is designed to:

- Explain systems analysis and design using an appealing full-color format, numerous screen shots and illustrations, and an easy-to-read style that invites students to learn.

- Introduce project management concepts early in the systems development process, with a new chapter that explains project management tools and techniques.

- Challenge students with a Question of Ethics mini-case in each chapter that asks them to respond to real-life ethical issues in an IT environment.

- Provide multi-method coverage, including a comparison of structured, object-oriented, and agile systems development methods.

- Emphasize the importance of planning, implementing, and managing an effective IT security program.

- Explain how IT supports business requirements in today's intensely competitive environment, and describe major IT developments and trends.

- Provide case studies and exercises that promote critical-thinking skills and encourage students to apply their skills and knowledge.

- Describe a systems analyst's job in a typical business organization, and show students how to use various tools and techniques to improve their skills and manage their careers.

- Provide students with a comprehensive Systems Analyst's Toolkit that highlights four major cross-functional tools, including: Communications Tools, CASE Tools, Financial Analysis Tools, and Internet Resource Tools.

New Video Learning Sessions

Fourteen Video Learning Sessions enhance the textbook and describe important systems analysis skills and concepts. The sessions provide step-by-step explanations that are easy to follow and understand.

- Topics include DFDs, data normalization, entity-relationship diagrams, decision tables, financial tools, and project management.
- A **Your Turn** feature in every Video Learning Session challenges students to apply their skills and check their work against sample answers. This hands-on practice can help students better handle actual assignments and tasks.
- The Video Learning Sessions offer a self-paced multimedia format that students can review at their convenience.
- Instructors may use the Video Learning Sessions as classroom presentations, distance-education support, student review tools, and exam preparation.

New and Updated Features in This Text

Systems Analysis and Design, Eighth Edition: Video Enhanced offers these exciting new and expanded features:

- Each development phase opens with an eye-catching Dilbert© cartoon and a multicolor Gantt chart that provides a visual roadmap for students.
- New Project Management chapter adds coverage of project management tools, techniques, and a full set of practice tasks and activities. A link is provided to Open Workbench, open-source project management software that students can download and install. Each Chapter Capstone case now includes a Gantt chart example, showing concurrent and dependent tasks, and each Chapter Opener case also includes a Gantt chart.
- New Question of Ethics mini-case in each chapter challenges students with real-life ethical issues in an IT environment.
- Multi-method coverage provides comparison of structured, object-oriented, and agile development methods, starting in Chapter 1. New material on agile methods includes examples of extreme programming, scrum, spiral models, and related topics.
- New coverage of risk management, both in a project management context and as a key element of IT security planning.
- Extensive update of networking coverage, including new material on switches, routers, and multistation access units. New coverage of wireless networks, including wireless standards, topologies, and trends.
- Major expansion of IT security material, including risk management, fault management, backup and recovery, wireless security issues, and a six-level security framework.
- Expanded coverage of IT trends, including cloud computing, Web 2.0, RFID, wireless networks, mobile computing, offshore outsourcing, e-business, ERP, Web hosting, client/server architecture, network concepts, Webinars, podcasts, RSS feeds, Web-based applications, and others.

- Revised Systems Analyst's Toolkit teaches students IT support skills in four cross-functional areas, including Communication Tools, CASE Tools, Financial Analysis Tools, and Internet Resource Tools.

Organization of This Textbook

Systems Analysis and Design, Eighth Edition: Video Enhanced contains 16 learning units in twelve chapters and a four-part Systems Analyst's Toolkit that teaches valuable cross-functional skills.

Chapter 1 – Introduction to Systems Analysis and Design Chapter 1 provides an up-to-date overview of IT issues, major trends, and various systems development approaches, including structured, object-oriented, and agile methods. The chapter emphasizes the important role of systems analysis and design in supporting business objectives.

Chapter 2 – Analyzing the Business Case Chapter 2 offers a business-related starting point for successful systems analysis. Topics include strategic planning, review of systems requests, how to conduct a feasibility study, and the steps in a preliminary investigation.

Chapter 3 – Managing Systems Projects Chapter 3 explains project management, cost estimating, and change control for information systems. This chapter includes hands-on skills that systems analysts can use to create Gantt charts and PERT charts.

Chapter 4 – Requirements Modeling Chapter 4 describes fact-finding techniques and team-based modeling methods, including JAD and RAD, that systems analysts use to model and document a new system.

Chapter 5 – Data and Process Modeling Chapter 5 explains how systems analysts create a logical model for the new system by using data flow diagrams and process description tools, including structured English, decision tables, and decision trees.

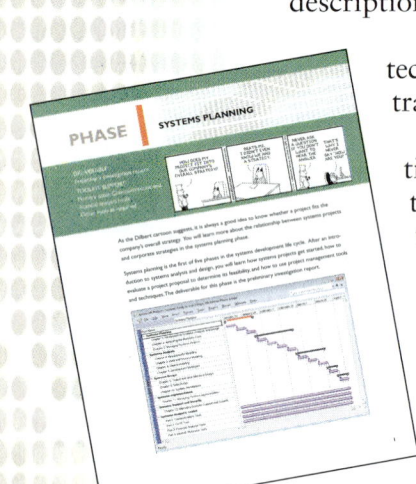

Chapter 6 – Object Modeling Chapter 6 explains object-oriented tools and techniques, including use case diagrams, class diagrams, sequence diagrams, state-transition diagrams, activity diagrams, and the Unified Modeling Language.

Chapter 7 – Development Strategies Chapter 7 focuses on software acquisition options, including outsourcing and offshore outsourcing options, application service providers, and other trends that view software as a service rather than a product.

Chapter 8 – Output and User Interface Design Chapter 8 highlights output and report design, the interaction between humans and computers, including usability issues, graphical screen design, input issues, and data entry guidelines.

Chapter 9 – Data Design Chapter 9 describes data design terms, concepts, and skills including entity-relationship diagrams, cardinality, data normalization rules, data warehousing, data mining, a comparison of logical and physical records, and data control measures.

Chapter 10 – System Architecture Chapter 10 explains the elements of system architecture, with emphasis on RFID, ERP, supply chain management, client/server architecture, and network topology, including wireless networking standards and trends.

Chapter 11 – Managing Systems Implementation Chapter 11 includes coverage of application development and implementation topics, including structure charts, documentation techniques, system testing, user training, data conversion, changeover methods, and post-implementation evaluation.

Chapter 12 – Managing Systems Support and Security Chapter 12 describes user support, maintenance techniques, and factors that indicate the end of a system's useful life. This chapter explains IT security concepts, techniques, and tools, and specifically addresses six security levels: physical, network, application, file, user, and procedural security. Chapter 12 also describes risk management, data backup and disaster recovery, and explains future challenges and opportunities that IT professionals will face in a dynamic workplace.

Toolkit Part 1 – Communication Tools Part 1 of the Toolkit describes oral and written communication tools that can make a systems analyst more effective. Topics include guidelines for successful communications, tips for better readability, how to organize and plan a presentation, effective speaking techniques, and managing communication skills.

Toolkit Part 2 – CASE Tools Part 2 of the Toolkit focuses on computer-aided software engineering (CASE) tools that systems analysts use to document, model, and develop information systems. Examples of several popular CASE tools are provided, along with sample screens that show CASE tool features.

Toolkit Part 3 – Financial Analysis Tools Part 3 of the Toolkit explains various tools that systems analysts use to determine feasibility and evaluate the costs and benefits of an information system. Specific tools include payback analysis, return on investment (ROI), and net present value (NPV).

Toolkit Part 4 – Internet Resource Tools Part 4 of the Toolkit explains Internet-based information gathering strategies. Topics include search engines, subject directories, the invisible Web, advanced search techniques, Boolean logic and Venn diagrams. This Toolkit Part also discusses newsgroups, newsletters, blogs, podcasts, RSS feeds, Webinars, mailing lists, Web-based discussion groups, chat rooms, instant messaging, and online learning opportunities.

FOR THE STUDENT

The Shelly Cashman Series wants you to have a valuable learning experience that will provide the knowledge and skills you need to be successful. With that goal in mind, we have included many activities, games, and learning tools, that we hope you will find interesting, challenging, and enjoyable. For example, because a picture is worth a thousand words, each systems development phase will begin with an eye-catching Dilbert© cartoon and a multi-color Gantt chart that provides a *"You are Here"* roadmap.

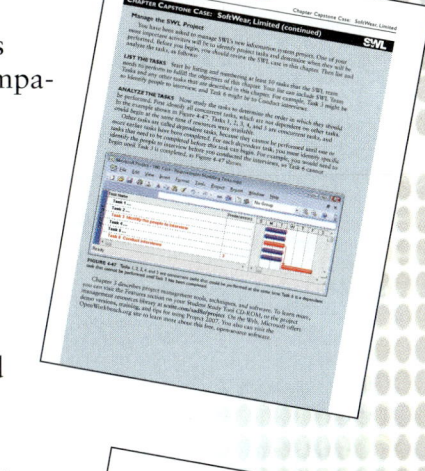

The following sections describe features at the beginning, learning tools within, and exercises at the end of each chapter. Other support tools accompanying this textbook also are described.

Chapter Opening Features

Each chapter contains the following features to help you get started:

- **Chapter Introduction** Read the Chapter Introduction for a brief overview of the chapter.

- **Chapter Objectives** The Chapter Objectives lists the main skills and knowledge you will have when you finish the chapter.

- **Chapter Introduction Case: Mountain View College Bookstore** The Mountain View College Bookstore case is a continuing case study that introduces each chapter and provides a real-world overview of the topics that will be covered in the chapter. As you work through the textbook, you will see how the Mountain View IT team discusses the issues, identifies the key points, and creates specific task lists.

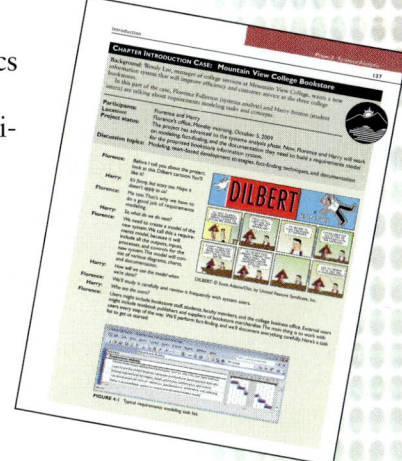

Learning Tools within the Chapter

As you work through each chapter, you will find these helpful tools and features:

- **A Question of Ethics** A mini-case in each chapter will challenge you with real-life ethical issues in an IT environment.

- **Case in Point** This exciting feature provides four embedded mini-case opportunities for you to analyze and apply the skills and concepts you are learning in the chapter.

- **Toolkit Time** The Systems Analyst's Toolkit explains skills that you can apply at any point in the textbook. Toolkit Time marginal notes remind you about the Toolkit, where to find it, and how it might help you address the issues or material in the chapter.

- **On the Web** Learn more about a topic by visiting the suggested Web sites and exploring the links we have provided.

End-of-Chapter Exercises

The following exercises are in every text chapter:

- **Learn It Online** Each chapter features a Learn It Online page that includes six exercises. These exercises utilize the Web to offer chapter-related reinforcement activities that will help you gain confidence in systems analysis and design. These exercises include True/False, Multiple Choice, Short Answer, Flash Cards, Practice Test, and several learning games.

- **CASE SIM: SCR Associates** This is an interactive Web-based case study, with a work session at the end of each chapter. Visit SCR's Web site and log on to the company's intranet to read e-mail messages addressed to you, listen to voice mail messages, and perform assigned tasks in a realistic corporate setting. In this simulation you report to Jesse Baker, but you e-mail your completed assignments to your instructor. Detailed instructions on how to use this case are available at www.scsite.com/sad8e/scr. To log on to the SCR intranet, you must use the password *sad8*. When you log on to the SCR intranet, you also will be asked to enter your first and last name so your e-mail can be addressed to you correctly.

- **Chapter Exercises** In this section, you will find 10 Review Questions, four Discussion Topics, and four Projects. These exercises allow you to apply your understanding of the material and will help to prepare you for tests and assessments.

- **Apply Your Knowledge** This section includes four mini-cases per chapter. Each mini-case requires you to use the knowledge and skills you learned in the chapter.

- **Case Studies** Case studies provide practical experience and allow you to practice specific skills learned in the chapter. Each chapter contains several case studies, two of which (New Century Health Clinic and Personal Trainer, Inc.) continue throughout the textbook. You can complete your assignments using Microsoft Word and Excel forms, available at www.scsite.com/sad8e/forms.

- **Chapter Capstone Case: SoftWear, Limited** SoftWear, Limited (SWL) is a continuing case study where students act as members of the SWL systems development team and perform various assignments in each chapter, including a set of project management tasks and a sample Gantt chart.

Additional Support Tools

These additional tools are provided to enhance your learning experience:

GLOSSARY/INDEX This edition of the textbook includes a glossary/index feature. Check your understanding of key terms and phrases, or use the glossary as a quick reference tool.

STUDENT STUDY TOOL This CD-ROM, provided in the back of your book, contains:

- Detailed outlines of every chapter that highlight key topics covered and can be used as a guide when reviewing for an exam
- Chapter glossaries that allow you to look up all key terms in one place. They also provide page references where key terms can be found if you need more information
- Web Links, Figures, and Test Yourself questions that provide additional reinforcement of chapter concepts
- User guide for Open Workbench (a free, open-source project management program), and links to download and install a trial version of Microsoft Project and a full version of Open Workbench

ONLINE COMPANION Broaden your learning experience and enhance your understanding of the material in each chapter with the Online Companion Web site. Visit scsite.com/sad8e for access to:

- Video Learning Sessions and Your Turn exercises
- On the Web links
- Learn It Online exercises, including True/False, Multiple Choice, Short Answer, Flash Cards, Practice Test, and several learning games
- SCR Associates Internet and intranet sites
- Forms Library
- Project Management Resources

FOR THE INSTRUCTOR

The Shelly Cashman Series is dedicated to providing you all of the tools you need to make your class a success. Information on all supplementary materials is available through your Course Technology representative or by calling one of the following telephone numbers: Colleges, Universities, Continuing Education Departments, Post-Secondary Vocational Schools, Career Colleges, Business, Industry, Government, Trade, Retailer, Wholesaler, Library, and Resellers, call Cengage Learning at 800-354-9706; K-12 Schools, Secondary and Vocational Schools, Adult Education, and School Districts, call Cengage Learning at 800-354-9706. In Canada, call Nelson Cengage Learning at 800-268-2222.

Instructor Resources Disc

The Instructor Resources disc (0-324-59767-3) for this textbook includes both teaching and testing aids. The contents of the disc are listed below:

- **Instructor's Manual** Includes lecture notes summarizing the chapter sections, figures and boxed elements found in every chapter, teacher tips, classroom activities, lab activities, and quick quizzes in Microsoft Word files.
- **Syllabi** Easily customizable sample syllabi that cover policies, assignments, exams, and other course information. Also included is a Microsoft Project file used to create the five Phase Opener Gantt charts. An instructor can use this project file to create a visual syllabus that could include additional tasks, quizzes, and projects. The project file also can be used to track class progress through the course. Instructors are welcome to distribute this file to students, and show them how to manage tasks, resources, and deadlines for team projects that might be assigned.

- **PowerPoint Presentations** A multimedia lecture presentation system provides slides for each chapter, based on chapter objectives.
- **Figure Files** Illustrations for every figure in the textbook in electronic form.
- **Solutions to Exercises** Includes solutions for end-of-chapter exercises, chapter reinforcement exercises, and extra case studies.
- **Test Bank & Test Engine** Test Banks include 112 questions for every chapter, and feature objective-based and critical thinking question types, page number references, and figure references when appropriate. The ExamView test engine is the ultimate tool for your testing needs.
- **Additional Activities for Students** The forms that students can use to complete the Case Studies are included. Two additional case studies are also provided for every chapter, to be assigned as homework, extra credit, or assessment tools. Also included are Chapter Reinforcement Exercises, which are true/false, multiple-choice, and short answer questions that help students gain confidence in the material learned.
- **Additional Faculty Files** A copy of the powerful CASE tool, Visible Analyst — Student Edition, is provided for your evaluation. Several sample solutions to case study tasks also are included. To install this program, you follow a simple registration process that entitles you to use the software and obtain support. Detailed instructions are provided on the Instructor Resources disc. Also included are Word document versions of the e-mail and voice mail messages posted for students on the SCR Web site and the Interview Summaries for the New Century Case Study.

Content for Online Learning

Course Technology has partnered with the leading distance learning solution providers and class-management platforms today. To access this material, visit www.cengage.com/webtutor and search for your title. Instructor resources include the following: additional case projects, sample syllabi, PowerPoint presentations, and more. For students to access this material, they must have purchased a WebTutor PIN-code specific to this title and your campus platform. The resources for students might include (based on instructor preferences): topic reviews, review questions, practice tests, and more. For additional information, please contact your sales representative.

SOFTWARE BUNDLING OPPORTUNITIES *Systems Analysis and Design, Eighth Edition: Video Enhanced* can be bundled with several popular software programs:

- **Visible Analyst, Student Edition** Whether you are designing e-business applications, developing a data warehouse, or integrating your legacy systems with new enterprise applications, Visible Analyst has all the power you need. Educating tomorrow's developers today, Visible Analyst helps students become more marketable with advanced, affordable, application development and training tools. This Student Edition of Visible Analyst is a separate product with a unique goal of educating the future application development workforce. A concurrent user network version, called the Visible Analyst University Edition, is available to colleges and universities for installation in computer labs.
- **Microsoft Visio®** Create business and technical diagrams that document and organize complex ideas, processes, and systems. Diagrams created in Visio enable you to visualize and communicate information clearly, concisely, and effectively in ways that text and numbers cannot. Visio also automates data visualization by

synchronizing directly with data sources to provide up-to-date diagrams, and it can be customized to meet the needs of your organization.

- **Microsoft Office Project** Project managers everywhere rely on Microsoft Office Project to plan and manage their projects. With Microsoft Office Project, efficiently organize and track tasks and resources to keep your projects on time and within budget. Extensive help resources and printing assistance make Project easy to learn, so that you can be productive quickly. Project is an integral part of the Microsoft Office System, so you can smoothly use products like Microsoft Office PowerPoint® and Microsoft Office Visio® to present project status effectively. And, as a user of the leading desktop project management program, support is readily available through a broad community of solution providers and user groups.

ACKNOWLEDGMENTS

Special thanks go to Deb Kaufmann, development editor; Raymond Enger, author of the current Student Study Tool and to a review team that included Barry Andrews, Mt. San Antonio Community College; Lennie Alice Cooper, Miami Dade College; Makoto Nakayama, DePaul University; Michael J. Nicholas, Davenport University; Paul Weaver, Bossier Parish Community College. Each reviewer taught a class using the prior edition of *Systems Analysis and Design*, and provided real-time feedback that was extremely important in shaping the *Eighth Edition*. We would also like to express our thanks to the students at College of the Albemarle who provided valuable input, and to Aaron Sawyer for his insight and helpful suggestions about networking, wireless issues, agile methods, and IT security topics.

ABOUT OUR COVERS

Learning styles of students have changed, but the Shelly Cashman Series' dedication to their success has remained steadfast for over 30 years. We are committed to continually updating our approach and content to reflect the way today's students learn and experience new technology.

This focus on the user is reflected in our bold cover design, which features photographs of real students using the Shelly Cashman Series in their courses. Each book features a different user, reflecting the many ages, experiences, and backgrounds of all of the students learning with our books. When you use the Shelly Cashman Series, you can be assured that you are learning computer skills using the most effective courseware available.

We would like to thank the administration and faculty at the participating schools for their help in making our vision a reality. Most of all, we'd like to thank the wonderful students from all over the world who learn from our texts and appear on our covers.

PHASE | SYSTEMS PLANNING

As the Dilbert cartoon suggests, it is always a good idea to know whether a project fits the company's overall strategy. You will learn more about the relationship between systems projects and corporate strategies in the systems planning phase.

Systems planning is the first of five phases in the systems development life cycle. After an introduction to systems analysis and design, you will learn how systems projects get started, how to evaluate a project proposal to determine its feasibility, and how to use project management tools and techniques. The deliverable for this phase is the preliminary investigation report.

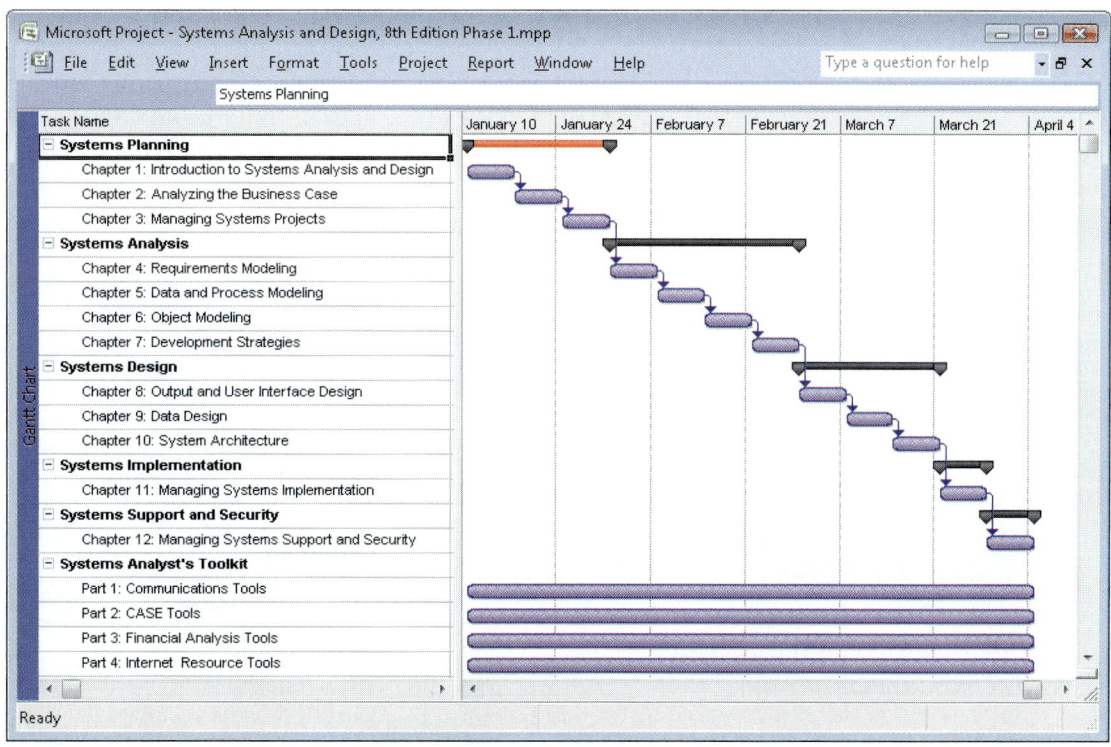

1

CHAPTER

Introduction to Systems Analysis and Design

Chapter I is the first of three chapters in the systems planning phase. This chapter describes the role of information technology in today's dynamic business environment. In this chapter, you will learn about the development of information systems, systems analysis and design concepts, and various systems development methods. This chapter also describes the role of the information technology department and its people.

INTRODUCTION

OBJECTIVES

When you finish this chapter, you will be able to:

- Describe the impact of information technology on business strategy and success
- Define an information system and describe its components
- Explain how profiles and models can represent business functions and operations
- Explain how the Internet has affected business strategies and relationships
- Identify various types of information systems and explain who uses them
- Distinguish between structured analysis, object-oriented analysis, and agile methods
- Compare the traditional waterfall model with agile methods and models
- Discuss the role of the information technology department and the systems analysts who work there

The headlines in Figure 1-1 offer dramatic examples of how information technology affects our society. Companies use information as a weapon in the battle to increase productivity, deliver quality products and services, maintain customer loyalty, and make sound decisions. In a global economy with intense competition, information technology can mean the difference between success and failure.

FIGURE I-I   These headlines show the enormous impact of information technology in the twenty-first century.

CHAPTER INTRODUCTION CASE: Mountain View College Bookstore

Background: Mountain View College is located in a large, southwestern city. The school has grown rapidly and now has 8,000 students at three campuses, each with a branch bookstore. Wendy Lee, manager of college services, is responsible for all bookstore operations. Wendy wants a new information system that will increase efficiency and improve customer service.

As the case begins, Florence Fullerton, a systems analyst in the college's Information Technology Department, is talking with Harry Boston. Harry is majoring in information systems at Mountain View College and is earning credit toward his degree by working part-time as a student intern.

Participants:	Florence and Harry
Location:	Florence's office, 10 a.m. Monday morning, August 24, 2009
Project status:	Initial discussion
Discussion topics:	Basic systems analysis and design concepts

Florence: Welcome aboard, Harry.

Harry: *I'm glad to be here. What's on the agenda?*

Florence: Well, there's been some talk about a new bookstore information system. Wendy says nothing is definite yet, but she suggested that we should get ready.

Harry: *So we start by learning about the bookstore business?*

Florence: Yes, the best system in the world isn't worth much unless it supports business and information needs. But let's not get ahead of ourselves. First, we need to talk about business information systems in general. Then we'll build a business model so we can understand the specific operations and processes at the bookstore. We'll also discuss systems analysis and design tools and techniques. Let's start with an overview of information systems and their characteristics.

Harry: *That makes sense. What about the basic systems analysis techniques you mentioned? Can you tell me a bit more?*

Florence: On this project, we'll use what's called a structured method, which is based on the concept of a systems development life cycle, or SDLC for short. I'll also explain object-oriented and agile methods to you. You'll also learn about waterfall models and spiral models.

Harry: *How does the SDLC work?*

Florence: The SDLC is like constructing a building. First, you would list specific objectives for the project. Then, you might hire an architect to create a set of drawings to show the building concept. Later, you'd need detailed blueprints for the workers who would do the actual construction. When the building is done, you would check everything, turn it over to the new occupants, and make sure they're happy with the results.

Harry: *And that's how we'll develop new information systems?*

Florence: It sure is. We'll use a program called Microsoft Project to create a list of tasks we can work on.

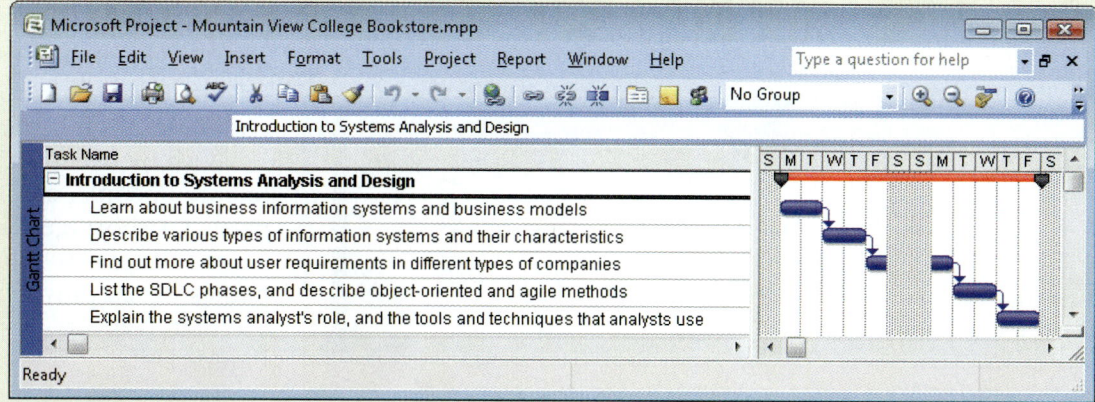

FIGURE I-2 Typical introductory tasks for systems projects

THE IMPACT OF INFORMATION TECHNOLOGY

Information technology (IT) refers to the combination of hardware and software products and services that people use to manage, access, communicate, and share information. Successful firms treat information as a vital asset that must be used effectively, updated constantly, and safeguarded carefully. Although fictitious, the headlines in Figure 1-1 on page 2 show the impact of IT on businesses, large and small.

The Future of IT

> **ON THE WEB**
>
> For more information about the future of IT, visit **scsite.com/sad8e/more**, locate Chapter 1, and then click The Future of IT link.

More than ever, business success depends on information technology. IT is driving a new economy, where advances in hardware, software, and connectivity are providing enormous benefits to businesses and individuals around the world. As the table shown in Figure 1-3 indicates, the global online population skyrocketed between 2000 and 2007, with well over a billion users worldwide!

As IBM stated in its 2007 annual report, "The basic computing model has changed. The PC model of the 1980s has receded in importance to clients, and has been replaced by a new paradigm, based on openness, networks, powerful new technology and the integration of digital intelligence into the fabric of work and life."

Demand for IT jobs also is expected to remain strong. According to a December 2007 report by the U.S. Department of Labor, many IT occupations will see robust growth for at least a decade. The greatest need will be for systems analysts, network administrators, data communications analysts, and software engineers. In the IT sector overall, over a million new jobs are expected by 2016.

Although economic trends affect IT spending levels, most businesses give IT budgets a relatively high priority, in good times or bad. The reason is simple — during periods of growth, companies cannot afford to lag behind the IT curve. Conversely, when the economy slows down, firms often use IT to reduce operating costs and improve efficiency.

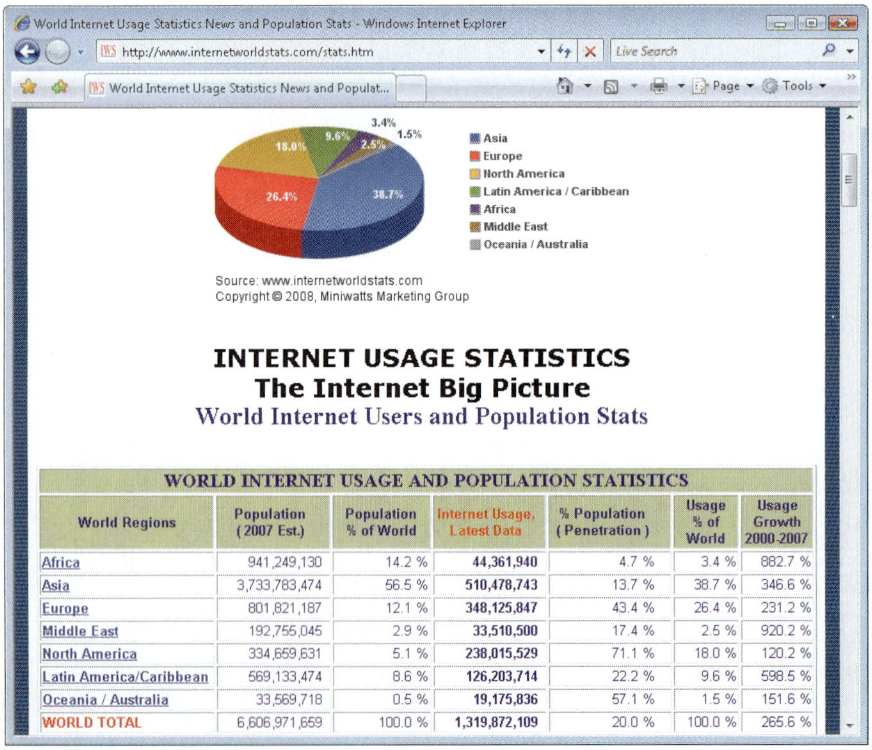

WORLD INTERNET USAGE AND POPULATION STATISTICS

World Regions	Population (2007 Est.)	Population % of World	Internet Usage, Latest Data	% Population (Penetration)	Usage % of World	Usage Growth 2000-2007
Africa	941,249,130	14.2 %	**44,361,940**	4.7 %	3.4 %	882.7 %
Asia	3,733,783,474	56.5 %	**510,478,743**	13.7 %	38.7 %	346.6 %
Europe	801,821,187	12.1 %	**348,125,847**	43.4 %	26.4 %	231.2 %
Middle East	192,755,045	2.9 %	**33,510,500**	17.4 %	2.5 %	920.2 %
North America	334,659,631	5.1 %	**238,015,529**	71.1 %	18.0 %	120.2 %
Latin America/Caribbean	569,133,474	8.6 %	**126,203,714**	22.2 %	9.6 %	598.5 %
Oceania / Australia	33,569,718	0.5 %	**19,175,836**	57.1 %	1.5 %	151.6 %
WORLD TOTAL	6,606,971,659	100.0 %	**1,319,872,109**	20.0 %	100.0 %	265.6 %

FIGURE 1-3 Internet World Stats provides usage and online population data. Although total numbers are not large in some regions, notice the huge percentage increases in Internet usage.

The Role of Systems Analysis and Design

Systems analysis and design is a step-by-step process for developing high-quality information systems. An **information system** combines information technology, people, and data to support business requirements. For example, information systems handle daily business transactions, improve company productivity, and help managers make sound decisions. The IT department team includes **systems analysts** who plan, develop, and maintain information systems.

With increasing demand for talented people, employment experts predict a shortage of qualified applicants to fill IT positions. Many companies list employment opportunities on their Web sites, as shown in Figure 1-4.

Who Develops Information Systems?

Traditionally, a company either developed its own information systems, called **in-house applications**, or purchased systems called **software packages** from outside vendors. Today, the choice is much more complex. Options include Internet-based application services, outsourcing, custom solutions from IT consultants, and enterprise-wide software strategies.

Regardless of the development method, launching a new information system involves risks as well as benefits. The greatest risk occurs when a company tries to decide *how* the system will be implemented before determining *what* the system is supposed to do. Instead of putting the cart before the horse, a company must begin by outlining its business needs and identifying possible IT solutions. Typically, this important work is performed by systems analysts and other IT professionals. A firm should not consider implementation options until it has a clear set of objectives. Later on, as the system is developed, a systems analyst's role will vary depending on the implementation option selected.

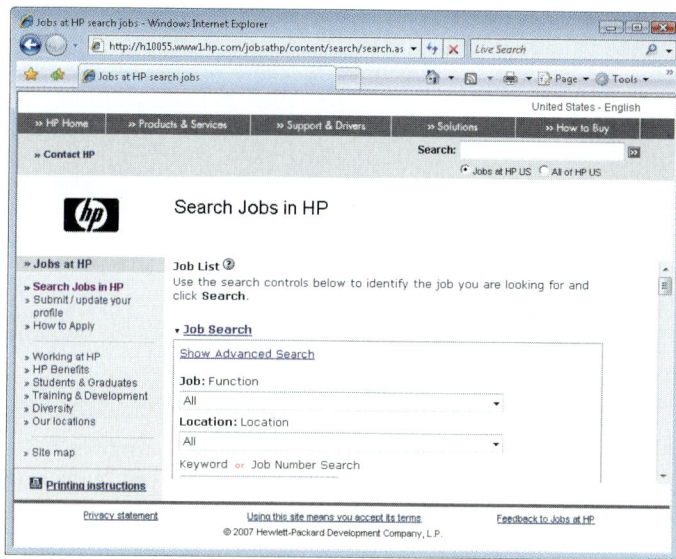

FIGURE 1-4 HP offers a search feature that allows potential applicants to search by location, type of job, and other criteria.

INFORMATION SYSTEM COMPONENTS

A **system** is a set of related components that produces specific results. For example, specialized systems route Internet traffic, manufacture microchips, and control complex entities like the International Space Station shown in Figure 1-5. A **mission-critical system** is one that is vital to a company's operations. An order processing system, for example, is mission-critical because the company cannot do business without it.

Every system requires input data. For example, your computer receives data when you press a key or click a menu command. In an information system, **data** consists of basic facts that are the system's raw material. **Information** is data that has been transformed into output that is valuable to users. For example, Figure 1-6 on the next page shows an order processing system that displays an order form. When a sales representative enters data (customer number, product code, and quantity ordered), the system creates a customer order with all the necessary information. Large businesses with thousands or millions of sales transactions require company-wide information systems and powerful servers, such as those shown in Figure 1-7 on the next page.

An information system has five key components, as shown in Figure 1-8 on the next page: hardware, software, data, processes, and people.

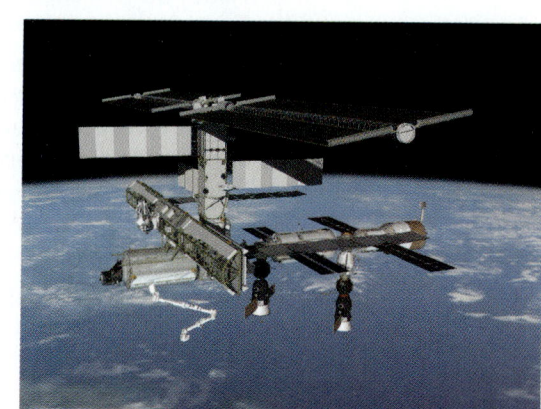

FIGURE 1-5 Imagine the complexity of the systems used to launch and operate the International Space Station.

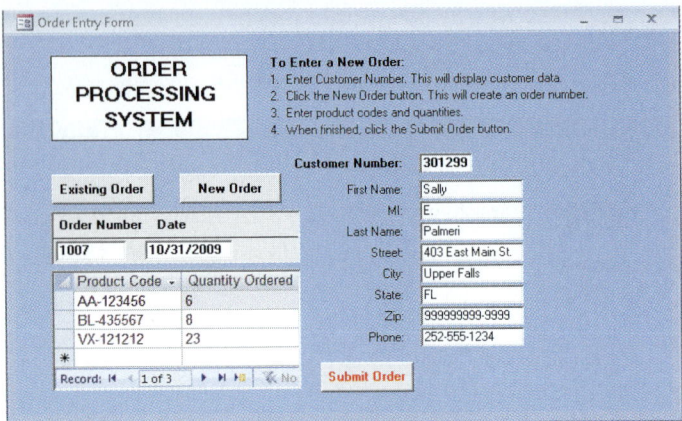

FIGURE 1-6 After a sales representative enters a customer number and product details, the system supplies the rest of the data and creates a sales order.

FIGURE 1-7 Multiple servers provide the power and speed that modern IT systems require.

ON THE WEB

For more information about Moore's Law, visit **scsite.com/ sad8e/more**, locate Chapter 1, and then click the Moore's Law link.

Hardware

Hardware consists of everything in the physical layer of the information system. For example, hardware can include servers, workstations, networks, telecommunications equipment, fiber-optic cables, handheld computers, scanners, digital capture devices, and other technology-based infrastructure. As new technologies emerge, manufacturers race to market the innovations and reap the rewards.

Hardware purchasers today face a wide array of technology choices and decisions. In 1965, Gordon Moore, a cofounder of Intel, predicted that the number of transistors on an integrated circuit would double about every 24 months. Figure 1-9 shows that his concept, called Moore's Law, has remained valid for more than 40 years. Fortunately, as hardware became more powerful, it also became less expensive.

Software

Software refers to the programs that control the hardware and produce the desired information or results. Software consists of system software and application software.

System software manages the hardware components, which can include a single workstation or a global network with many thousands of clients. Either the hardware manufacturer supplies the system software or a company purchases it from a vendor. Examples of system software include the operating system, security software that protects the computer from intrusion, device drivers that communicate with hardware such as printers, and utility programs that handle specific tasks such as data backup and disk management. System software also controls the flow of data, provides data security, and manages network operations. In today's interconnected business world, network software is vitally important.

Application software consists of programs that support day-to-day business functions and provide users with the information they require. Application software can serve one user or thousands of users throughout the organization. Examples of company-wide applications, called **enterprise applications**, include order processing systems, payroll systems, and company communications networks. On a smaller scale, individual users increase their productivity with tools such as spreadsheets, word processors, and database management systems.

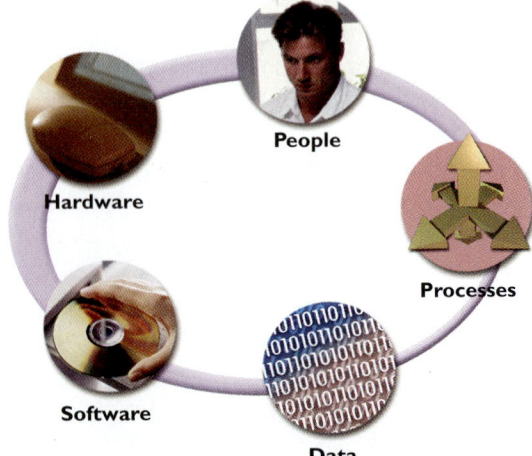

FIGURE 1-8 The five main components of an information system.

Application software includes horizontal and vertical systems. A **horizontal system** is a system, such as an inventory or payroll application, that can be adapted for use in many different types of companies. A **vertical system** is designed to meet the unique requirements of a specific business or industry, such as a Web-based retailer, a medical practice, or a video chain.

Most companies use a combination of software that is acquired at various times. When planning an information system, a company must consider how a new system will interface with older systems, which are called **legacy systems**. For example, a new human resources system might need to exchange data with an older payroll application.

Data

Data is the raw material that an information system transforms into useful information. An information system can store data in various locations, called tables. By linking the tables, the system can extract specific information. Figure 1-10 shows a payroll system that stores data in four separate tables. At the end of a pay period, the payroll system produces a paycheck that accurately reflects the employee's hours worked, gross pay, current deductions, and net pay.

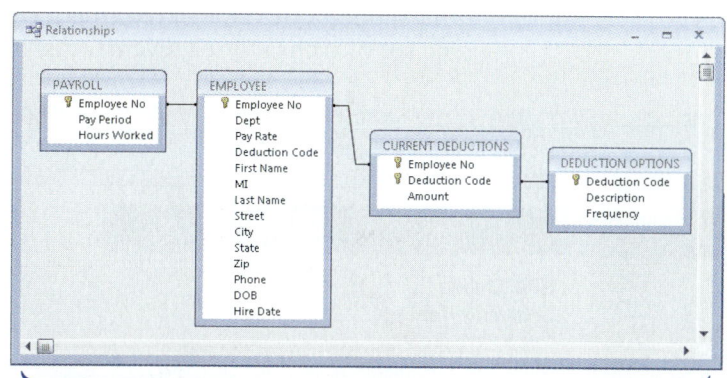

FIGURE 1-9 Moore's Law has remained valid for more than 40 years.

Processes

Processes describe the tasks and business functions that users, managers, and IT staff members perform to achieve specific results. Processes are the building blocks of an information system because they represent actual day-to-day business operations. To build a successful information system, analysts must understand business processes and document them carefully.

People

People who have an interest in an information system are called **stakeholders**. Stakeholders include the management group responsible for the system, the **users** (sometimes called **end users**) inside and outside the company who will interact with the system, and IT staff members, such as systems analysts, programmers, and network administrators who develop and support the system.

Each stakeholder group has a vital interest in the information system, but most experienced IT professionals agree that the success or failure of a system usually depends on whether it meets the needs of its users. For that reason, it is essential to understand user requirements and expectations throughout the development process.

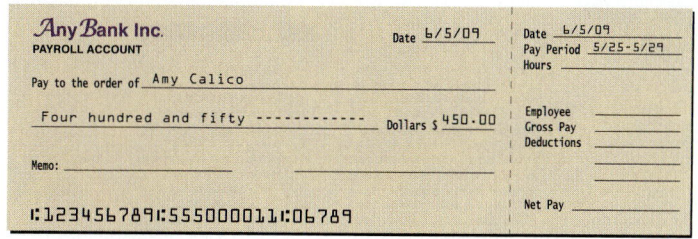

FIGURE 1-10 By linking several tables, a payroll system can extract specific information to produce a paycheck that accurately reflects the employee's hours worked, gross pay, current deductions, and net pay.

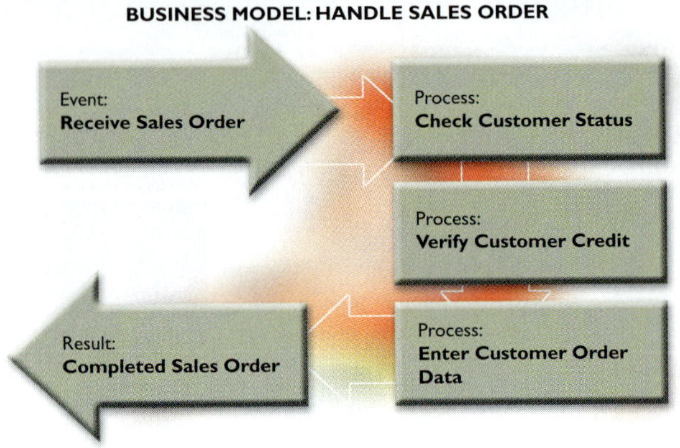

BUSINESS MODEL: HANDLE SALES ORDER

FIGURE 1-11 A simple business model might consist of an event, three processes, and a result.

UNDERSTANDING THE BUSINESS

IT professionals must understand a company's business operations to design successful systems. Each business situation is different. For example, a retail store, an Internet auction site, and a hotel chain all have unique information systems requirements. Systems analysts use a process called **business process modeling** to represent a company's operations and information needs. Business process modeling requires a business profile and a series of models that document various business processes.

As the business world changes, systems analysts can expect to work in new kinds of companies that require innovative IT solutions, including Web-based systems that serve customers and carry out online transactions with other businesses.

Business Profile

A **business profile** is an overview that describes a company's overall functions, processes, organization, products, services, customers, suppliers, competitors, constraints, and future direction. To develop a business profile, a systems analyst investigates a company's products, services, and Internet opportunities. The analyst also studies interactivity among the firm's information systems, specialized information needs, and future growth projections. Armed with a business profile, the analyst then creates a series of business models.

Business Models

Business models make it easier for managers and systems analysts to understand day-to-day business operations. A **business model** is a graphical representation of one or more business processes that a company performs, such as accepting an airline reservation, selling a ticket, or crediting a customer account. A **business process** describes a specific set of transactions, events, tasks, and results. For example, Figure 1-11 shows a business model called HANDLE SALES ORDER. Notice that the model represents an event, three separate business processes, and a result. Complex business operations require a series of linked models to show the overall picture.

When companies attempt to simplify operations or reduce costs, a popular strategy is to have managers and systems analysts perform **business process reengineering (BPR)**. ProSci's BPR Online Learning Center shown in Figure 1-12 offers comprehensive resources for business process reengineering, including articles, tutorials, and information on reengineering toolkits and templates.

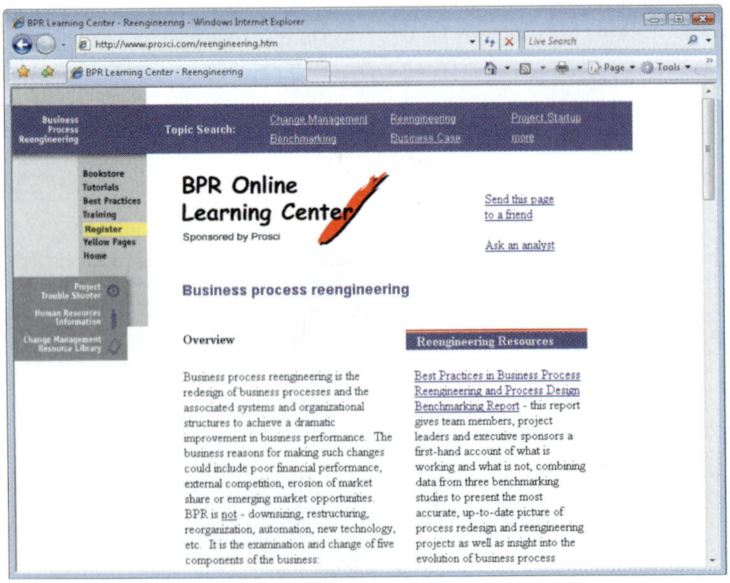

FIGURE 1-12 ProSci's BPR Online Learning Center offers many resources for business process reengineering.

New Kinds of Companies

Traditionally, IT companies were identified as product-oriented or service-oriented. **Product-oriented** firms manufactured computers, routers, or the microchips shown in Figure 1-13, whereas **service-oriented** companies included resellers and providers of information and various IT services.

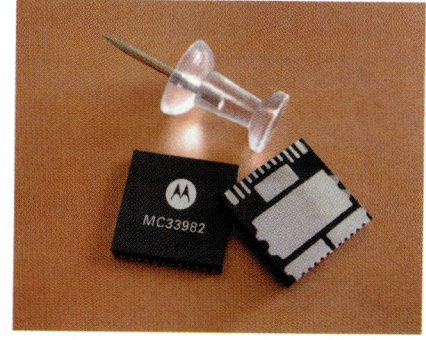

FIGURE 1-13 Motorola is an example of a product-oriented company that manufactures technology products, such as the microchip shown here.

Today, those distinctions are much less meaningful. Most successful IT companies offer a mix of products, technical and financial services, consulting, and customer support. Many firms believe that long-term profitability lies in value-added services rather than hardware, which customers sometimes view as a commodity. In a striking example of this trend, IBM stated in its 2007 annual report that software and services produced 77% of total revenues, whereas hardware accounted for only 23% of sales.

The newest company category is the **Internet-dependent** firm, often described as a **dot-com (.com)** company because its primary business depends on the Internet rather than a traditional business channel. Google, Yahoo!, AOL, and eBay are examples of pure dot-com companies. At the other end of the spectrum are more traditional companies, sometimes called **brick-and-mortar** companies because they conduct business primarily from physical locations. Today, that distinction no longer exists. Most successful brick-and-mortar firms—such as Lowe's, Target, and Wal-Mart—have added Web-based marketing channels to increase sales and serve customers more effectively. This has allowed them to combine the convenience of online shopping and the alternative of hands-on purchasing for customers who prefer that option.

In recent years, some Internet-based companies have enjoyed spectacular growth, while others have fallen by the wayside. As competition heats up for the online consumer, dot-com companies will need to work hard to survive and grow in a dynamic marketplace.

Rising fuel prices are also a factor in the success of Web-based firms. The Motley Fool, a well-known financial advisory company, recently commented that "The winners in this are the Internet companies." Citing online companies such as Netflix, the article pointed out that in the old days, e-commerce was attractive to shoppers who wanted to avoid a crowded parking lot. Now, the real question is "Do you want to drive at all?"

CASE IN POINT 1.1: CLOUD NINE FINANCIAL ADVISORS

Cloud Nine provides its clients with a monthly newsletter that offers recommendations about stocks to buy or sell. Doug Layton, Cloud Nine's president, has asked your opinion on whether dot-com stocks might be good investments for the future. He specifically mentioned Google, eBay, Amazon.com, and Yahoo!, but he said you could suggest other companies. Doug wants you to do some Internet research to learn more about these Web-based companies and their future prospects. You can use a search engine, or start by visiting the Web sites of publications such as *Forbes*, *Fortune Magazine*, *Business Week*, or *The Wall Street Journal*, among others.

IMPACT OF THE INTERNET

Internet-based commerce is called **e-commerce (electronic commerce)** or **I-commerce (Internet commerce)**. E-commerce includes two main sectors: B2C **(business-to-consumer)** and B2B **(business-to-business)**.

B2C (Business-to-Consumer)

ON THE WEB

For more information about electronic commerce, visit **scsite.com/ sad8e/ more**, locate Chapter 1, and then click the Electronic Commerce link.

Using the Internet, consumers can go online to purchase an enormous variety of products and services. This new shopping environment allows customers to do research, compare prices and features, check availability, arrange delivery, and choose payment methods in a single convenient session. Many companies, such as airlines, offer incentives for online transactions because Web-based processing costs are lower than traditional methods. By making flight information available online to last-minute travelers, some airlines also offer special discounts on seats that might otherwise go unfilled.

B2C commerce is changing traditional business models and creating new ones. For example, a common business model is a retail store that sells a product to a customer. To carry out that same transaction on the Internet, the company must develop an online store and deal with a totally different set of marketing, advertising, and profitability issues. Some companies have found new ways to use established business models. For example, eBay.com has transformed a traditional auction concept into a new, popular, and successful method of buying goods and services.

In recent years, B2C transactions accounted for a small portion of total retail sales, but B2C activity is expected to grow significantly. The surge in B2C marketing has created strong competition among Web designers to create attractive sites that increase online sales. The B2C trend also means more demand for systems analysts and programmers who can develop Web-based information systems and applications.

B2B (Business-to-Business)

Although the business-to-consumer (B2C) sector is more familiar to retail customers, the volume of business-to-business (B2B) transactions is many times greater. Industry observers predict that B2B sales will increase sharply in the future as more firms use advanced technology to improve efficiency and lower their acquisition costs.

Online trading marketplaces initially were developed as company-to-company data-sharing arrangements called **electronic data interchange (EDI)**. EDI enabled computer-to-computer transfer of data between companies, usually over private telecommunications networks. Firms used EDI to plan production, adjust inventory levels, or stock up on raw materials using data from another company's information system. As B2B volume soared, the development of **extensible markup language (XML)** enabled company-to-company traffic to migrate to the Internet, which offered standard protocols, universal availability, and low communication costs. XML is a flexible data description language that allows Web-based communication between different hardware and software environments.

ON THE WEB

For more information about XML, visit **scsite.com/ sad8e/ more**, locate Chapter 1, and then click the Extensible Markup Language link.

Because it allows companies to access the global marketplace, B2B is especially important to firms under pressure to reduce costs. B2B enables smaller suppliers to contact large customers, and allows purchasers to obtain instant information about market prices and availability. On an industry-wide scale, many B2B sites exist where buyers, sellers, distributors, and manufacturers can offer products, submit specifications, and transact business. This popular form of online B2B interaction is called **supplier relationship management (SRM)**. Figure 1-14 shows the site of Infor, a software firm that offers SRM solutions designed to reduce supply chain costs.

Web-Based System Development

Internet-based systems development is changing rapidly, as software industry giants compete in market for overall software services, rather than individual products. These services include powerful Web-development environments and software solutions. For example, IBM claims that its **WebSphere** strategy is best, while Microsoft counters with a broad vision called **.NET** that redefines that company's approach to Web-based application development. These alternatives are discussed in more detail in Chapter 7, Development Strategies. Web-based databases are discussed in Chapter 9, Data Design, and Chapter 10, System Architecture. In addition, many firms offer **Web services,** which are Internet-based support programs that can be executed as an integral part of an information system. For example, a real estate brokerage Web site might offer instant mortgage calculations, which are performed by a Web service provided by a third-party company.

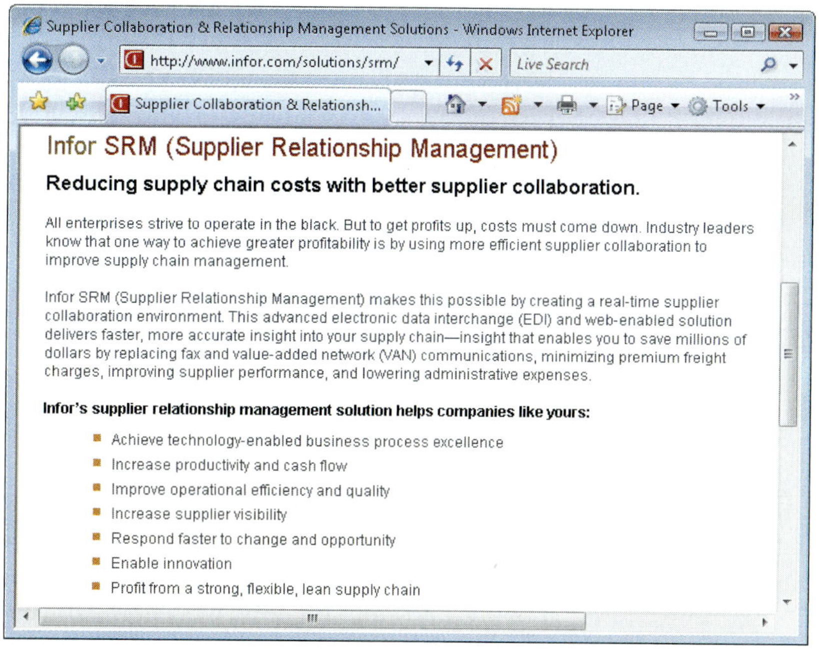

Internet-based systems involve various hardware and software designs, but a simple model is a series of Web pages that provides a user interface, which communicates with one or more levels of data management software and a Web-based database server. As companies build more Internet-based systems, career opportunities will expand for IT professionals, including Web designers, database developers, and systems analysts. The surge in demand will come from dot-com companies and mainstream retailers who have worldwide brand recognition. In the e-commerce battles, the real winners will be online consumers, who will have access to more information, better choices, and an enhanced shopping experience. For example, in addition to the traditional offerings, the Lowe's site shown in Figure 1-15 includes a gift advisor, buying guides, how-to clinics, and interactive design tools.

FIGURE 1-14 Infor is a software provider that offers SRM solutions based on real-time supplier collaboration.

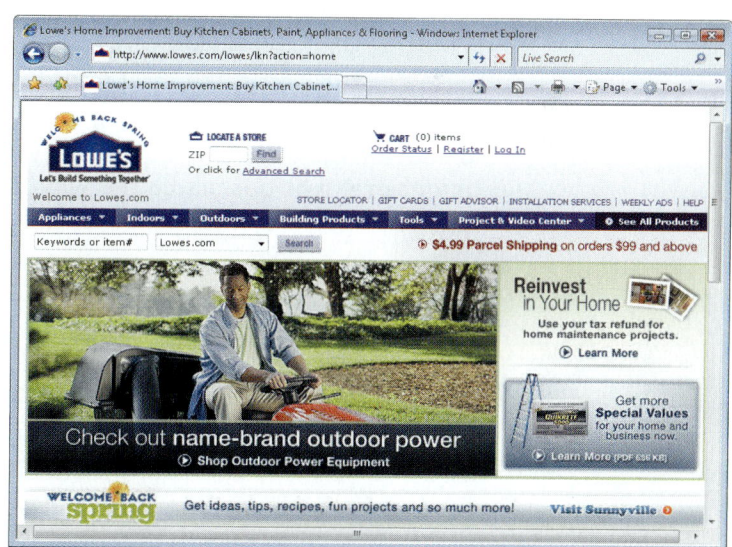

FIGURE 1-15 Lowe's is an example of a mainstream retailer that effectively combines traditional and online marketing strategies.

How Business Uses Information Systems

In the past, IT managers divided systems into categories based on the user group the system served. Categories and users included office systems (administrative staff), operational systems (operational personnel), decision support systems (middle-managers and knowledge workers), and executive information systems (top managers).

Today, traditional labels no longer apply. For example, all employees, including top managers, use office productivity systems. Similarly, operational users often require decision support systems. As business changes, information use also changes in most companies. Today, it makes more sense to identify a system by its functions and features, rather than by its users. A new set of system definitions includes enterprise computing systems, transaction processing systems, business support systems, knowledge management systems, and user productivity systems.

Enterprise Computing Systems

Enterprise computing refers to information systems that support company-wide operations and data management requirements. Wal-Mart's inventory control system, Boeing's production control system, and Hilton Hotels' reservation system are examples of enterprise computing systems. The main objective of enterprise computing is to integrate a company's primary functions (such as production, sales, services, inventory control, and accounting) to improve efficiency, reduce costs, and help managers make key decisions. Enterprise computing also improves data security and reliability by imposing a company-wide framework for data access and storage.

In many large companies, applications called **enterprise resource planning (ERP)** systems provide cost-effective support for users and managers throughout the company. For example, a car rental company can use ERP to forecast customer demand for rental cars at hundreds of locations.

By providing a company-wide computing environment, many firms have been able to achieve dramatic cost reductions. Other companies have been disappointed in the time, money, and commitment necessary to implement ERP successfully. A potential disadvantage of ERP is that ERP systems generally impose an overall structure that might or might not match the way a company operates. ERP is described in more detail in Chapter 7, which discusses system development strategies.

ON THE WEB

For more information about enterprise resource planning, visit **scsite.com/sad8e/ more**, locate Chapter 1, and then click the Enterprise Resource Planning link.

Because of its growth and potential, many hardware and software vendors target the enterprise computing market and offer a wide array of products and services. Figure 1-16 shows an IBM Web site that is dedicated to marketing enterprise computing software and solutions.

Transaction Processing Systems

Transaction processing (TP) systems process data generated by day-to-day business operations. Examples of TP systems include customer order processing, accounts receivable, and warranty claim processing.

TP systems perform a series of tasks whenever a specific transaction occurs. In the example shown in Figure 1-17, a TP system verifies customer data, checks the customer's

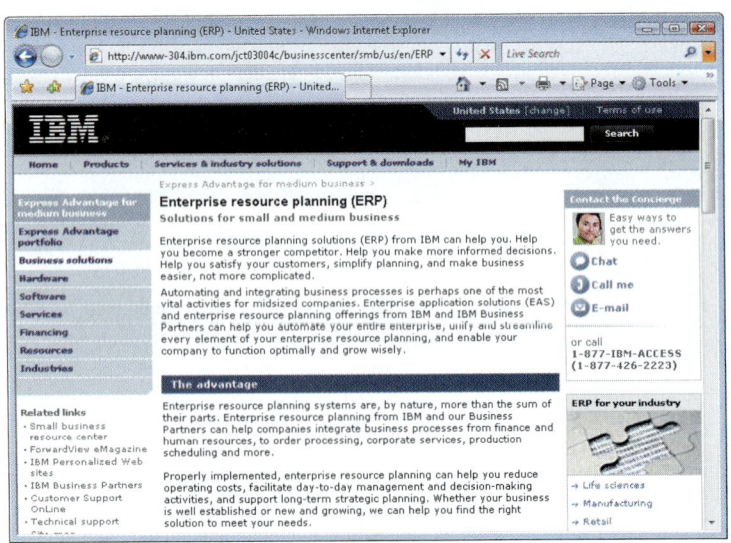

FIGURE 1-16 IBM maintains a site dedicated to enterprise computing.

credit status, posts the invoice to the accounts receivable system, checks to ensure that the item is in stock, adjusts inventory data to reflect a sale, and updates the sales activity file. TP systems typically involve large amounts of data and are mission-critical systems because the enterprise cannot function without them.

TP systems are efficient because they process a set of transaction-related commands as a group rather than individually. To protect data integrity, however, TP systems ensure that if any single element of a transaction fails, the system does not process the rest of the transaction.

Business Support Systems

Business support systems provide job-related information support to users at all levels of a company. These systems can analyze transactional data, generate information needed to manage and control business processes, and provide information that leads to better decision-making.

The earliest business computer systems replaced manual tasks, such as payroll processing. Companies soon realized that computers also could produce valuable information. The new systems were called **management information systems (MIS)** because managers were the primary users. Today, e12mployees at *all* levels need information to perform their jobs, and they rely on information systems for that support.

A business support system can work hand in hand with a TP system. For example, when a company sells merchandise to a customer, a TP system records the sale, updates the customer's balance, and makes a deduction from inventory. A related business support system highlights slow- or fast-moving items, customers with past due balances, and inventory levels that need adjustment.

To compete effectively, firms must collect production, sales, and shipping data and update the company-wide business support system immediately. The newest development in data acquisition is called **radio frequency identification (RFID)** technology, which uses high-frequency radio waves to track physical objects, such as the item shown in Figure 1-18. RFID is expected to grow dramatically as the U.S. Department of Defense and companies such as Wal-Mart begin to require suppliers to add RFID tags to their goods.

An important feature of a business support system is decision support capability. Decision support helps users make decisions by creating a computer model and applying a set of variables. For example, a truck fleet dispatcher might run a series of **what-if** scenarios to determine the impact of increased shipments or bad weather. Alternatively, a retailer might use what-if analysis to determine the price it must charge to increase profits by a specific amount while volume and costs remain unchanged.

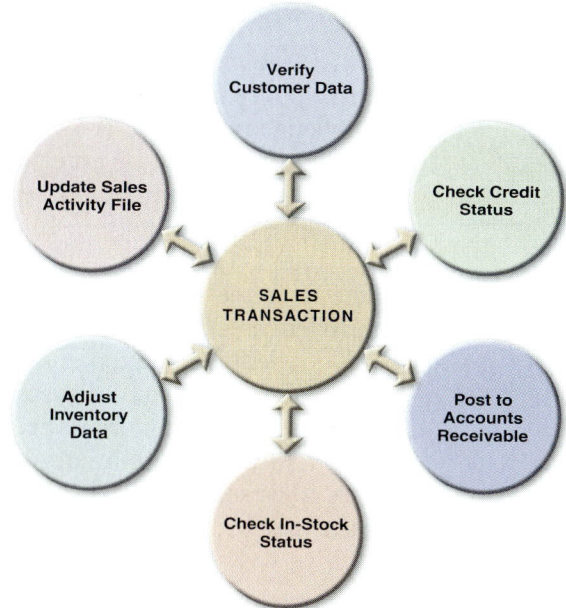

FIGURE 1-17 A single sales transaction consists of six separate tasks, which the TP system processes as a group.

ON THE WEB

For more information about RFID, visit **scsite.com/ sad8e/more**, locate Chapter 1, and then click the RFID link.

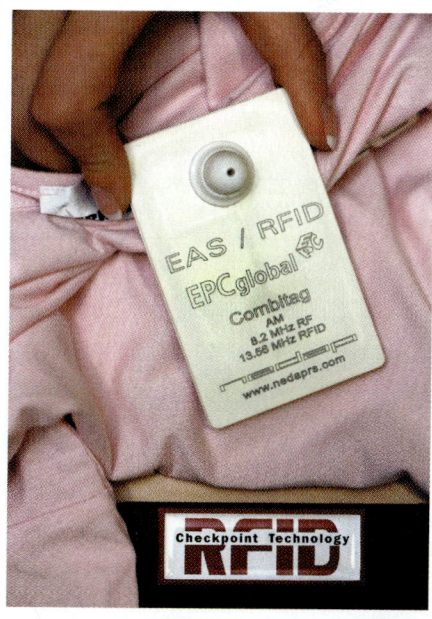

FIGURE 1-18 Retailers use RFID tags for security and inventory control.

Knowledge Management Systems

ON THE WEB

For more information about knowledge management systems, visit **scsite.com/sad8e/more**, locate Chapter 1, and then click the Knowledge Management Systems link.

Knowledge management systems are called **expert systems** because they simulate human reasoning by combining a knowledge base and inference rules that determine how the knowledge is applied. A **knowledge base** consists of a large database that allows users to find information by entering keywords or questions in normal English phrases. A knowledge management system uses **inference rules**, which are logical rules that identify data patterns and relationships.

Figure 1-19 shows a knowledge management system that 3Com maintains for its customers and users. After a user enters a symptom, problem, or question, 3Com's Knowledgebase searches for a solution and displays the results.

Knowledge management systems do not use strict logical rules. Instead, many knowledge management systems use a technique called **fuzzy logic** that allows inferences to be drawn from imprecise relationships. Using fuzzy logic, values need not be black and white, like binary logic, but can be many shades of gray. The results of a fuzzy logic search will display in priority order, with the most relevant results at the top of the list.

User Productivity Systems

Companies provide employees at all levels with technology that improves productivity. Examples of **user productivity systems** include e-mail, voice mail, fax, video conferencing, word processing, automated calendars, database management, spreadsheets, desktop publishing, presentation graphics, company intranets, and high-speed Internet access. User productivity systems also include groupware. **Groupware** programs run on a company intranet and enable users to share data, collaborate on projects, and work in teams. GroupWise, offered by Novell, is a popular example of groupware.

When companies first installed word processing systems, managers expected to reduce the number of employees as office efficiency increased. That did not happen, primarily because the basic nature of clerical work changed. With computers performing most of the repetitive work, managers realized that office personnel could handle tasks that required more judgment, decision-making, and access to information.

Computer-based office work expanded rapidly as companies assigned more responsibility to employees at lower organizational levels. Relatively inexpensive hardware, powerful networks, corporate downsizing, and a move toward employee empowerment also contributed to this trend. Today, administrative assistants and company presidents alike are networked, use computer workstations, and need to share corporate data to perform their jobs.

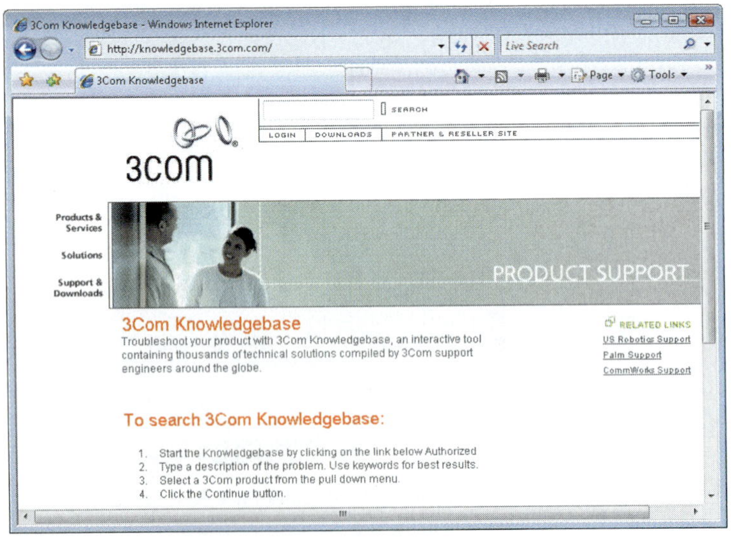

Information Systems Integration

Most large companies require systems that combine transaction processing, business support, knowledge management, and user productivity features. For example, suppose an international customer has a problem with a product and makes a warranty claim. A customer service representative enters the claim into a TP system. The transaction updates two other systems: a knowledge management system that tracks product problems and warranty activity, and a quality control system with decision support capabilities.

FIGURE 1-19 The interactive 3Com Knowledgebase allows users to search for solutions.

A quality control engineer uses what-if analysis to determine if it would be advantageous to make product design changes to reduce warranty claims. In this example, a TP system is integrated with a knowledge management system and a business support system with decision support features.

INFORMATION SYSTEM USERS AND THEIR NEEDS

Corporate organizational structure has changed considerably in recent years. As part of downsizing and business process reengineering, many companies reduced the number of management levels and delegated responsibility to operational personnel. Although modern organization charts tend to be flatter, an organizational hierarchy still exists in most companies.

A typical organizational model identifies business functions and organizational levels, as shown in Figure 1-20. Within the functional areas, operational personnel report to supervisors and team leaders. The next level includes middle managers and knowledge workers, who, in turn, report to top managers. In a corporate structure, the top managers report to a board of directors elected by the company's shareholders.

A systems analyst must understand the company's organizational model to recognize who is responsible for specific processes and decisions and to be aware of what information is required by whom.

Top Managers

Top managers develop long-range plans, called **strategic plans**, which define the company's overall mission and goals. To plot a future course, top managers ask questions such as "How much should the company invest in information technology?" or "How much will Internet sales grow in the next five years?" or "Should the company build new factories or contract out the production functions?"

Strategic planning affects the company's future survival and growth, including long-term IT plans. Top managers focus on the overall business enterprise and use IT to set the company's course and direction. To develop a strategic plan, top managers also need information from outside the company, such as economic forecasts, technology trends, competitive threats, and governmental issues.

Middle Managers and Knowledge Workers

Just below the top management level, most companies have a layer of middle managers and knowledge workers. Middle managers provide direction, necessary resources, and performance feedback to supervisors and team leaders. Because they focus on a somewhat

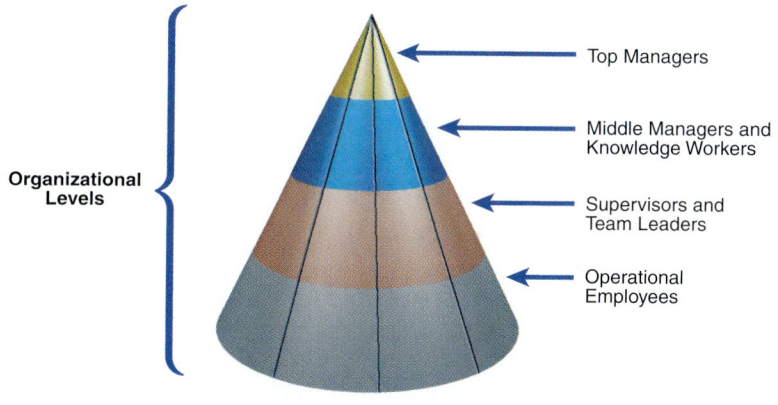

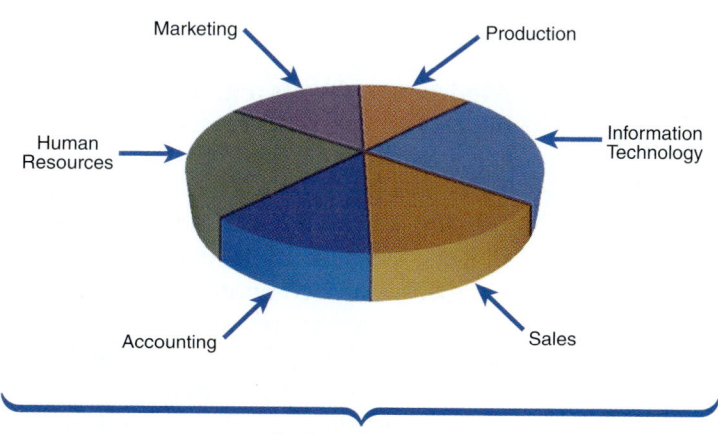

FIGURE 1-20 A typical organizational model identifies business functions and organizational levels.

shorter time frame, middle managers need more detailed information than top managers, but somewhat less than supervisors who oversee day-to-day operations. For example, a middle manager might review a weekly sales summary for a three-state area, whereas a local sales team leader would need a daily report on customer sales at a single location.

In addition to middle managers, every company has people called knowledge workers. **Knowledge workers** include professional staff members such as systems analysts, programmers, accountants, researchers, trainers, and human resource specialists. Knowledge workers also use business support systems, knowledge management systems, and user productivity systems. Knowledge workers provide support for the organization's basic functions. Just as a military unit requires logistical support, a successful company needs knowledge workers to carry out its mission.

Supervisors and Team Leaders

Supervisors, often called team leaders, oversee operational employees and carry out day-to-day functions. They coordinate operational tasks and people, make necessary decisions, and ensure that the right tools, materials, and training are available. Like other managers, supervisors and team leaders need decision support information, knowledge management systems, and user productivity systems to carry out their responsibilities.

Operational Employees

Operational employees include users who rely on TP systems to enter and receive data they need to perform their jobs. In many companies, operational users also need information to handle tasks and make decisions that were assigned previously to supervisors. This trend, called **empowerment**, gives employees more responsibility and accountability. Many companies find that empowerment improves employee motivation and increases customer satisfaction.

SYSTEMS DEVELOPMENT TOOLS

In addition to understanding business operations, systems analysts must know how to use a variety of techniques, such as modeling, prototyping, and computer-aided systems engineering tools to plan, design, and implement information systems. Systems analysts work with these tools in a team environment, where input from users, managers, and IT staff contributes to the system design.

Modeling

Modeling produces a graphical representation of a concept or process that systems developers can analyze, test, and modify. A systems analyst can describe and simplify an information system by using a set of business, data, object, network, and process models.

A **business model**, or **requirements model**, describes the information that a system must provide. A **data model** describes data structures and design. An **object model** describes objects, which combine data and processes. A **network model** describes the design and protocols of telecommunications links. A **process model** describes the logic that programmers use to write code modules. Although the models might appear to overlap, they actually work together to describe the same environment from different points of view.

System developers often use multipurpose charting tools such as Microsoft Visio 2007 to display business-related models. Visio is a popular tool that systems analysts can use to create business process diagrams, flowcharts, organization charts, network diagrams, floor plans, project timelines, and work flow diagrams, among others.

TOOLKIT TIME

The CASE tools in Part 2 of the Systems Analyst's Toolkit can help you develop and maintain complex information systems. To learn more about these tools, turn to Part 2 of the four-part Toolkit that follows Chapter 12.

Figure 1-21 shows how you can drag and drop various symbols from the left pane into the drawing on the right, and connect them to show a business process.

Modeling involves various techniques, including data flow diagrams and entity-relationship diagrams (described in Chapters 5 and 9), and unified modeling language diagrams (described in Chapters 4 and 6).

Prototyping

Prototyping tests system concepts and provides an opportunity to examine input, output, and user interfaces before final decisions are made. A **prototype** is an early working version of an information system. Just as an aircraft manufacturer tests a new design in a wind tunnel, systems analysts construct and study information system prototypes. A prototype can serve as an initial model that is used as a benchmark to evaluate the finished system, or the prototype itself can develop into the final version of the system. Either way, prototyping speeds up the development process significantly.

A possible disadvantage of prototyping is that important decisions might be made too early, before business or IT issues are understood thoroughly. A prototype based on careful fact-finding and modeling techniques, however, can be an extremely valuable tool.

Computer-Aided Systems Engineering (CASE) Tools

Computer-aided systems engineering (CASE), also called **computer-aided software engineering**, is a technique that uses powerful software, called **CASE tools**, to help systems analysts develop and maintain information systems. CASE tools provide an overall framework for systems development and support a wide variety of design methodologies, including structured analysis and object-oriented analysis.

Because CASE tools make it easier to build an information system, they boost IT productivity and improve the quality of the finished product. Part 2 of the Systems Analyst's Toolkit explains how analysts use CASE tools to create business profiles, build business models, and document complex processes. After developing a model, many CASE tools can generate program code, which speeds the implementation process. Figure 1-22 shows the Web site for Visible Systems Corporation, a leading vendor of CASE tools.

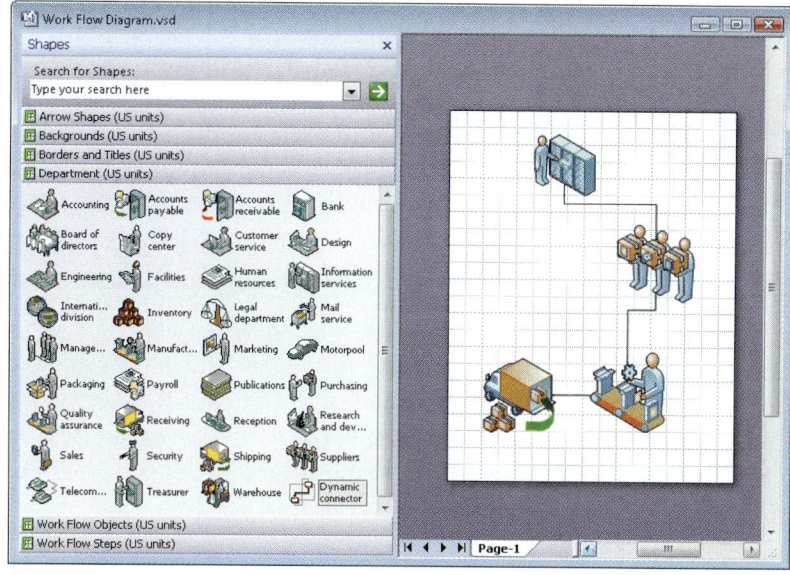

FIGURE 1-21 Microsoft Visio allows you to drag and drop various symbols and connect them to show a business process.

ON THE WEB

For more information about CASE Tools, visit **scsite.com/ sad8e/more**, locate Chapter 1, and then click the CASE Tools link.

FIGURE 1-22 Visible Systems Corporation offers a wide array of software engineering tools, including Visible Analyst, a popular CASE tool.

OVERVIEW OF SYSTEMS DEVELOPMENT METHODS

Many options exist for developing information systems, but the most popular alternatives are **structured analysis**, which is a traditional method that still is widely used, **object-oriented analysis (O-O)**, which is a more recent approach that many analysts prefer, and **agile methods**, also called **adaptive methods**, which include the latest trends in software development. Figure 1-23 provides an overview of the three methods, which are discussed in the following sections.

	STRUCTURED ANALYSIS	OBJECT-ORIENTED ANALYSIS	AGILE/ADAPTIVE METHODS
Description	Represents the system in terms of data and the processes that act upon that data. System development is organized into phases, with deliverables and milestones to measure progress. The SDLC waterfall model typically consists of five phases. Iteration is possible among the phases, as shown in Figure 1-25.	Views the system in terms of objects that combine data and processes. The objects represent actual people, things, transactions, and events, as shown in Figure 1-26. Compared to structured analysis, O-O phases tend to be more interactive. Can use the waterfall model or the model that stresses greater iteration, as shown in Figure 1-27.	Stresses intense team-based effort, as shown in Figures 1-28 and 1-29. Breaks development process down into cycles, or iterations that add functionality. Each iteration is designed, built, and tested in an ongoing process. Attempts to reduce major risks by incremental steps in short time intervals. Typically uses a spiral model, as shown in Figure 1-30.
Modeling tools	Data flow diagrams (DFDs) and process descriptions, which are described in Chapter 5.	Various object-oriented diagrams depict system actors, methods, and messages, which are described in Chapter 6.	Uses tools that facilitate team communication, such as collaborative software, interactive presentations, traditional whiteboards and face-to-face contact.
Pros	Traditional method, which has been very popular over time. Relies heavily on written documentation. Frequent phase iteration can provide flexibility comparable with other methods. Well-suited to project management tools and techniques.	Integrates easily with object-oriented programming languages. Code is modular and reusable, which can reduce cost and development time. Easy to maintain and expand as new objects can be cloned using inherited properties.	Very flexible and efficient in dealing with change. Stresses team interaction and reflects a set of community-based values. Frequent deliverables constantly validate the project and reduce risk.
Cons	Changes can be costly, especially in later phases. Requirements are defined early, and can change during development. Users might not be able to describe their needs until they can see examples of features and functions.	Somewhat newer method might be less familiar to development team members. Interaction of objects and classes can be complex in larger systems.	Team members need a high level of technical and communications skills. Lack of structure and documentation can introduce risk factors. Overall project might be subject to scope change as user requirements change.

FIGURE 1-23   Comparison of structured, object-oriented, and agile/adaptive development methods.

Although most projects utilize one of these approaches, it is not unusual for system developers to mix and match methods to gain a better perspective. In addition to these three main development methods, some organizations choose to develop their own in-house approaches or use techniques offered by software suppliers, CASE tool vendors, or consultants. Many alternatives exist, and most IT experts agree that no one system development method is best in all cases. An approach that works well for one project might have major disadvantages or risks in another situation. The important thing is for a systems analyst to understand the various methods and the strengths and weaknesses of each approach.

Regardless of the development strategy, people, tasks, timetables, and costs must be managed effectively. Complex projects can involve dozens of people, hundreds of tasks, and many thousands of dollars. **Project management** is the process of planning, scheduling, monitoring, controlling, and reporting upon the development of an information system. Chapter 3 describes project management tools and techniques in detail.

Structured Analysis

Structured analysis is a traditional systems development technique that is time-tested and easy to understand. Structured analysis uses a series of phases, called the **systems development life cycle (SDLC)**, to plan, analyze, design, implement, and support an information system. Although structured analysis evolved many years ago, it remains a popular systems development method. Structured analysis is based on an overall plan, similar to a blueprint for constructing a building, so it is called a **predictive** approach.

Structured analysis uses a set of process models to describe a system graphically. Because it focuses on processes that transform data into useful information, structured analysis is called a **process-centered** technique. In addition to modeling the processes, structured analysis also addresses data organization and structure, relational database design, and user interface issues.

Process modeling identifies the data flowing into a process, the business rules that transform the data, and the resulting output data flow. Figure 1-24 on the next page shows a simple process model that represents a school registration process with related input and output.

Structured analysis uses the SDLC to plan and manage the systems development process. The SDLC describes activities and functions that all systems developers perform, regardless of which approach they use. In the **waterfall model**, the result of each phase is called a **deliverable**, or **end product**, which flows into the next phase.

Some analysts see a disadvantage in the built-in structure of the SDLC, because the waterfall model does not emphasize interactivity among the phases. This criticism can be valid if the SDLC phases are followed too rigidly. However, adjacent phases usually interact, as shown by the dotted lines in Figure 1-25 on the next page, and interaction among several phases is not uncommon. Other analysts regard the waterfall model as a two-way *water flow* model, with emphasis on iteration and user input. Used in this manner, the traditional model is not as different from agile methods as it might appear to be.

The SDLC model usually includes five steps, which are described in the following sections: systems planning, systems analysis, systems design, systems implementation, and systems support and security.

SYSTEMS PLANNING   The **systems planning phase** usually begins with a formal request to the IT department, called a **systems request**, which describes problems or desired changes in an information system or a business process. In many companies, IT systems planning is an integral part of overall business planning. When managers and users develop their business plans, they usually include IT requirements that generate systems requests. A systems request can come from a top manager, a planning team, a department head, or the IT department itself. The request can be very significant or

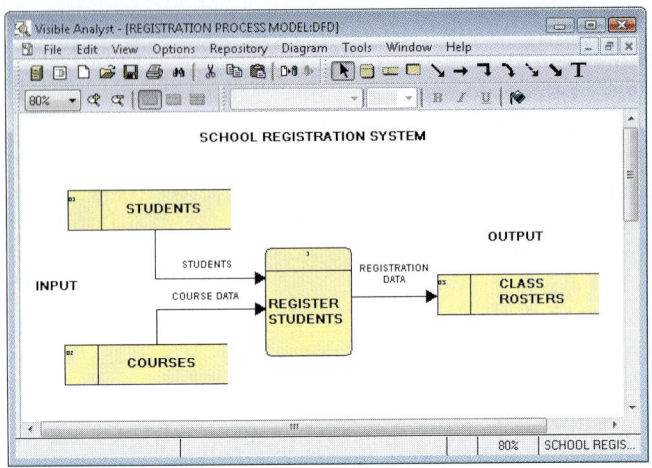

FIGURE 1-24 This Visible Analyst screen shows a process model for a school registration system. The REGISTER STUDENTS process accepts input data from two sources and transforms it into output data.

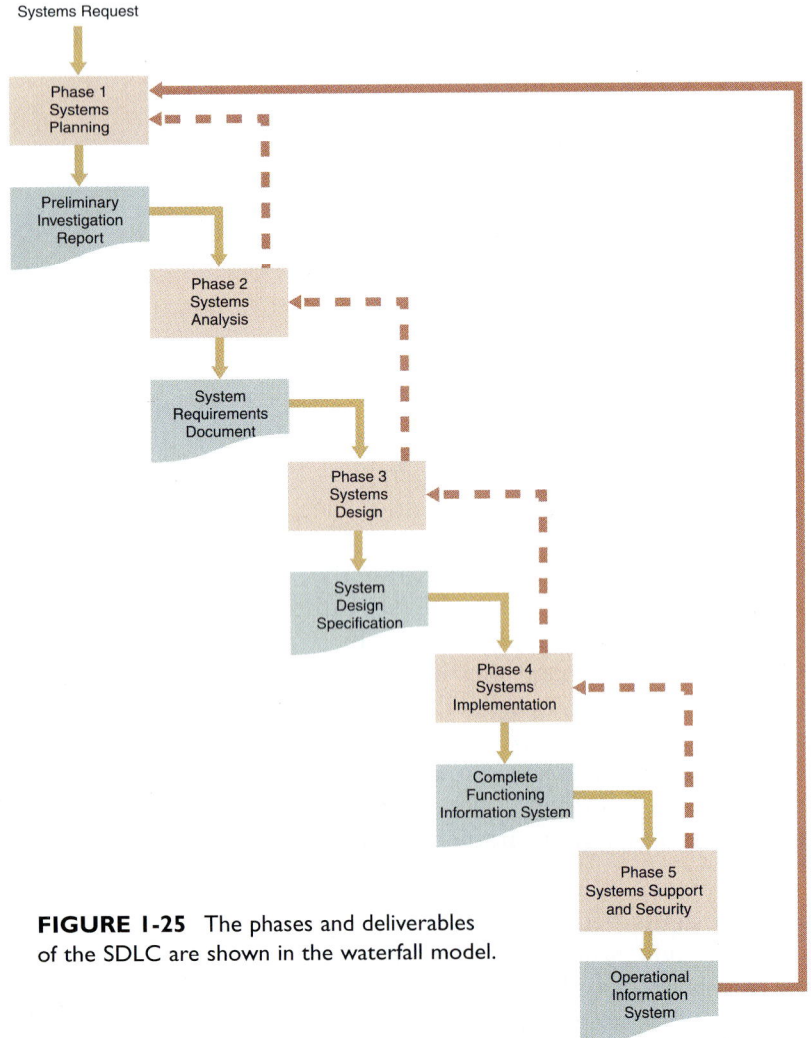

FIGURE 1-25 The phases and deliverables of the SDLC are shown in the waterfall model.

relatively minor. A major request might involve a new information system or the upgrading of an existing system. In contrast, a minor request might ask for a new feature or a change to the user interface.

The purpose of this phase is to perform a **preliminary investigation** to evaluate an IT-related business opportunity or problem. The preliminary investigation is a critical step because the outcome will affect the entire development process. A key part of the preliminary investigation is a **feasibility study** that reviews anticipated costs and benefits and recommends a course of action based on operational, technical, economic, and time factors.

Suppose you are a systems analyst and you receive a request for a system change or improvement. Your first step is to determine whether it makes sense to launch a preliminary investigation at all. Often you will need to learn more about business operations before you can reach a conclusion. After an investigation, you might find that the information system functions properly, but users need more training. In some situations, you might recommend a business process review, rather than an IT solution. In other cases, you might conclude that a full-scale systems review is necessary. If the development process continues, the next step is the systems analysis phase.

SYSTEMS ANALYSIS The purpose of the **systems analysis phase** is to build a logical model of the new system. The first step is **requirements modeling**, where you investigate business processes and document what the new system must do to satisfy users. Requirements modeling continues the investigation that began during the systems planning phase. To understand the system, you perform fact-finding using techniques such as interviews, surveys, document review, observation, and sampling. You use the fact-finding results to build business models, data and process models, and object models.

The deliverable for the systems analysis phase is the **system requirements document**. The system requirements document describes management and user requirements, costs and benefits, and outlines alternative development strategies.

SYSTEMS DESIGN The purpose of the **systems design phase** is to create a physical model that will satisfy all documented requirements for the system. At this stage, you design the user interface and identify necessary outputs, inputs, and processes. In addition, you design internal and external controls, including computer-based and manual features to guarantee that the system will be reliable, accurate, maintainable, and secure. During the systems design phase, you also determine the application architecture, which programmers will use to transform the logical design into program modules and code.

The deliverable for this phase is the **system design specification**, which is presented to management and users for review and approval. Management and user involvement is critical to avoid any misunderstanding about what the new system will do, how it will do it, and what it will cost.

SYSTEMS IMPLEMENTATION During the **systems implementation phase**, the new system is constructed. Whether the developers use structured analysis or O-O methods, the procedure is the same — programs are written, tested, and documented, and the system is installed. If the system was purchased as a package, systems analysts configure the software and perform any necessary modifications. The objective of the systems implementation phase is to deliver a completely functioning and documented information system. At the conclusion of this phase, the system is ready for use. Final preparations include converting data to the new system's files, training users, and performing the actual transition to the new system.

The systems implementation phase also includes an assessment, called a **systems evaluation**, to determine whether the system operates properly and if costs and benefits are within expectations.

SYSTEMS SUPPORT AND SECURITY During the **systems support and security phase**, the IT staff maintains, enhances, and protects the system. Maintenance changes correct errors and adapt to changes in the environment, such as new tax rates. Enhancements provide new features and benefits. The objective during this phase is to maximize return on the IT investment. Security controls safeguard the system from both external and internal threats. A well-designed system must be secure, reliable, maintainable, and scalable. A **scalable** design can expand to meet new business requirements and volumes. Information systems development is always a work in progress. Business processes change rapidly, and most information systems need to be updated significantly or replaced after several years of operation.

Object-Oriented Analysis

Whereas structured analysis treats processes and data as separate components, object-oriented analysis combines data and the processes that act on the data into things called **objects**. Systems analysts use O-O to model real-world business processes and operations. The result is a set of software objects that represent actual people, things, transactions, and events. Using an O-O programming language, a programmer then writes the code that creates the objects.

An object is a member of a **class**, which is a collection of similar objects. Objects possess characteristics called **properties**, which the object inherits from its class or possesses on its own. As shown in Figure 1-26, the class called PERSON includes INSTRUCTOR and STUDENT. Because the PERSON class has a property

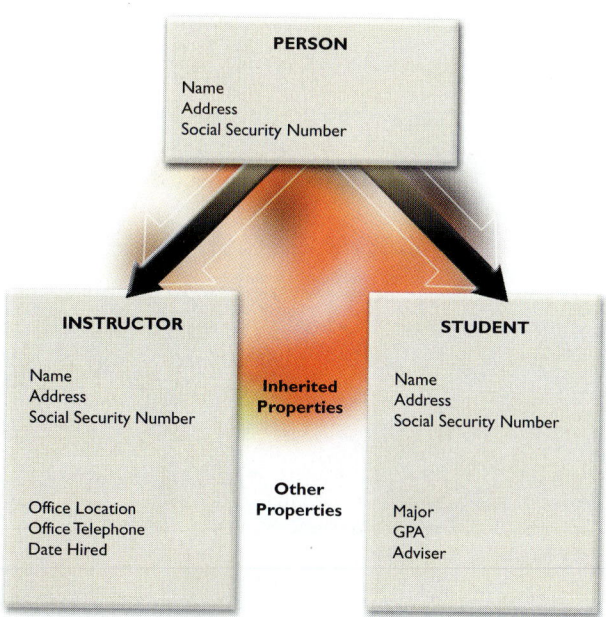

FIGURE 1-26 The PERSON class includes INSTRUCTOR and STUDENT objects, which have their own properties and inherited properties.

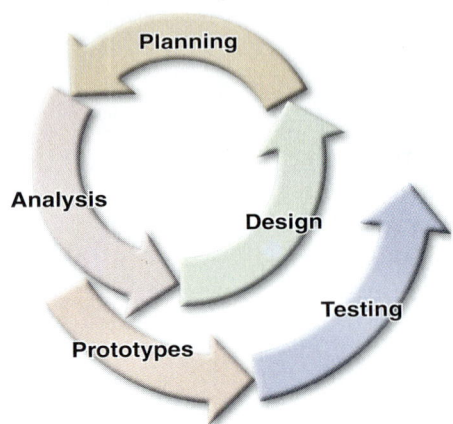

FIGURE 1-27 In this model, planning, analysis, and design tasks interact continuously. Interactive models often are used with O-O development methods.

called Address, a STUDENT inherits the Address property. A STUDENT also has a property called Major that is not shared by other members of the PERSON class.

In O-O design, built-in processes called **methods** can change an object's properties. For example, in a Web-based catalog store, an ORDER object might have a property called STATUS that changes when a CUSTOMER object clicks to place, confirm, or cancel the order.

One object can send information to another object by using a message. A **message** requests specific behavior or information from another object. For example, an ORDER object might send a message to a CUSTOMER object that requests a shipping address. When it receives the message, the CUSTOMER object supplies the information. The ORDER object has the capability to send the message, and the CUSTOMER object knows what actions to perform when it receives the message. O-O analysis uses object models to represent data and behavior, and to show how objects affect other objects. By describing the objects and methods needed to support a business operation, a system developer can design reusable components that speed up system implementation and reduce development cost.

Object-oriented methods usually follow a series of analysis and design phases that are similar to the SDLC, although there is less agreement on the number of phases and their names. In an O-O model, the phases tend to be more interactive. Figure 1-27 shows a system development model where planning, analysis, and design tasks interact continuously to produce prototypes that can be tested and implemented. The result is an **interactive model** that can accurately depict real-world business processes.

O-O methodology is popular because it provides an easy transition to O-O programming languages such as Java, Smalltalk, C++, and Perl. Chapter 6 covers O-O analysis and design, with a detailed description of O-O terms, concepts, tools, and techniques.

Agile Methods

Development techniques change over time. For example, structured analysis is a traditional approach, and agile methods are the newest development. Structured analysis builds an overall plan for the information system, just as a contractor might use a blueprint for constructing a building. Agile methods, in contrast, attempt to develop a system incrementally, by building a series of prototypes and constantly adjusting them to user requirements. As the agile process continues, developers revise, extend, and merge earlier versions into the final product. An agile approach emphasizes continuous feedback, and each incremental step is affected by what was learned in the prior steps.

Although relatively new to software development, the notion of **iterative** development can be traced back about 20 years to Japanese auto firms that were able to boost productivity by using a more flexible manufacturing system, where team-based effort and short-term milestones helped keep quality up and costs down. Agile methods have attracted a wide following and an entire community of users, as shown in Figure 1-28. Because it is stresses a team-based culture, the agile community has published the **Agile Manifesto**, which is the set of principles shown in Figure 1-29.

ON THE WEB

For more information about agile systems development methods, visit **scsite.com/sad8e/more**, locate Chapter 1, and then click the Agile Methods link.

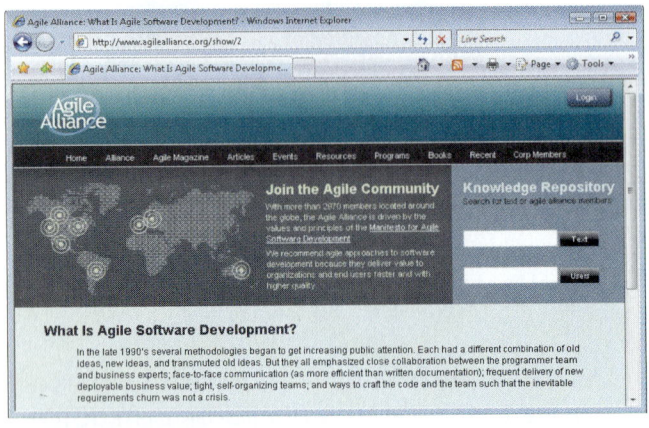

FIGURE 1-28 Agile methods have attracted a wide following and an entire community of users.

Agile methods typically use a **spiral model**, which represents a series of iterations, or revisions, based on user feedback, as shown in Figure 1-30. As the process continues, the final product gradually evolves. An agile approach requires intense interactivity between developers and individual users, and does not begin with an overall objective. Instead, the agile process determines the end result. Proponents of the spiral model believe that this approach reduces risks and speeds up software development.

Spiral models initially were suggested in the 1990s by Barry Boehm, a noted software engineering professor. He stated that each iteration, or phase, of the model must have a specific goal that is accepted, rejected, or changed by the user, or client. Thus, each iteration produces feedback and enhancements, which enable the team to reach the overall project goal. Typically, each iteration in a spiral model includes planning, risk analysis, engineering, and evaluation, as shown in the table in Figure 1-31 on the next page. The repeated iterations produce a series of prototypes, which evolve into the finished system. Notice that these phases resemble SDLC tasks, which also can be iterative.

Numerous other adaptive variations and related methods exist, and most IT developers expect this trend to continue in the future. Two examples are Scrum and Extreme Programming (XP). **Scrum**, which actually is a rugby term, is a popular process with agile developers, and refers to a powerful effort to achieve short-term goals. In Scrum, team members play specific roles and interact in intense sessions. According to Wikipedia, the term initially was used in a 1986 article by professors Hirotaka Takeuchi and Ikujiro Nonaka, who suggested a new product development method where phases overlap and the entire process is performed by one cross-functional team. They stated that the process was more like rugby, where the whole team goes downfield while passing the ball back and forth, compared with a relay race, where only one team member performs at a time.

THE AGILE MANIFESTO

We follow these principles:

- Our highest priority is to satisfy the customer through early and continuous delivery of valuable software.
- Welcome changing requirements, even late in development. Agile processes harness change for the customer's competitive advantage.
- Deliver working software frequently, from a couple of weeks to a couple of months, with a preference to the shorter timescale.
- Business people and developers must work together daily throughout the project.
- Build projects around motivated individuals. Give them the environment and support they need, and trust them to get the job done.
- The most efficient and effective method of conveying information to and within a development team is face-to-face conversation.
- Working software is the primary measure of progress.
- Agile processes promote sustainable development. The sponsors, developers, and users should be able to maintain a constant pace indefinitely.
- Continuous attention to technical excellence and good design enhances agility.
- Simplicity—the art of maximizing the amount of work not done—is essential.
- The best architectures, requirements, and designs emerge from self-organizing teams.
- At regular intervals, the team reflects on how to become more effective, then tunes and adjusts its behavior accordingly.

Source: The Agile Manifesto, www.agilemanifesto.org/principles

FIGURE 1-29 The Agile Manifesto is a set of team-based principles published by the agile community.

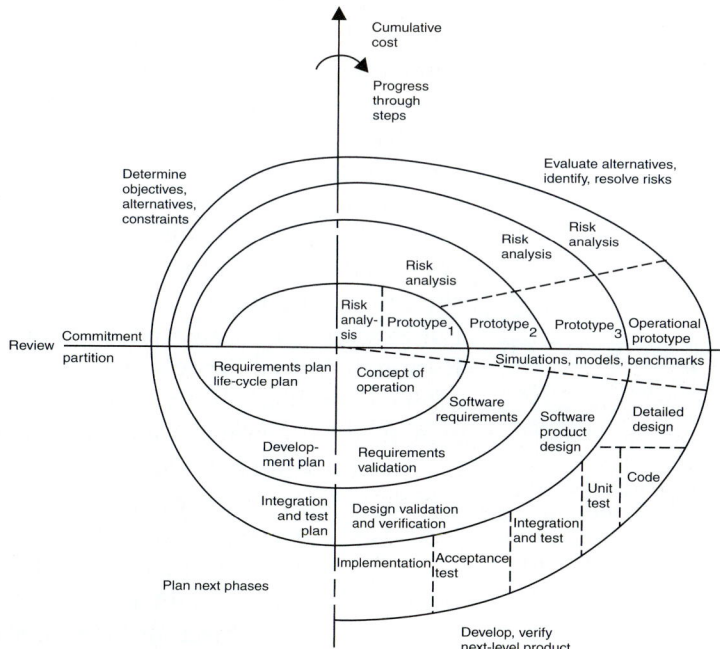

FIGURE 1-30 Agile methods typically use a spiral model, which represents a series of iterations, or versions, based on user feedback.

PHASE	TASKS
Planning	Define objectives, constraints, and deliverables
Risk analysis	Identify risks and develop acceptable resolutions
Engineering	Develop a prototype that includes all deliverables
Evaluation	Perform assessment and testing to develop objectives for next iteration

FIGURE 1-31 Typical phases and tasks in a spiral model.

Extreme Programming (XP) is another adaptive process that focuses on forceful interaction between developers and users to define and achieve project goals. XP, like agile methods generally, stresses certain key values, such as communication, simplicity, feedback, courage, and respect among team members. When properly implemented, its proponents believe that Extreme Programming can speed up development, reduce costs, and improve software quality. Time will tell whether this innovative approach will be widely accepted.

Although agile methods are becoming popular, analysts should recognize that these approaches have advantages and disadvantages. By their nature, agile methods can allow developers to be much more flexible and responsive, but can be riskier than more traditional methods. For example, without a detailed set of system requirements, certain features requested by some users might not be consistent with the company's larger game plan.

Other potential disadvantages of agile methods can include weak documentation, blurred lines of accountability, and too little emphasis on the larger business picture. Also, unless properly implemented, a long series of iterations might actually add to project cost and development time. The bottom line is that systems analysts should understand the pros and cons of any approach before selecting a development method for a specific project.

Other Development Methods

Although agile methods are relatively new, IT departments have long sought to avoid systems that were developed without sufficient input from users. Over time, many companies discovered that systems development teams composed of IT staff, users, and managers could complete their work more rapidly and produce better results. Two methodologies became popular: **joint application development (JAD)** and **rapid application development (RAD)**.

Both JAD and RAD use teams composed of users, managers, and IT staff. The difference is that JAD focuses on team-based fact-finding, which is only one phase of the development process, whereas RAD is more like a compressed version of the entire process. JAD, RAD, and agile methods are described in more detail in Chapter 4.

In addition to the methods described in this chapter, you might encounter other systems development techniques. If a systems analyst wants additional choices, he or she can choose from an entire industry of IT software companies and consulting firms. For example, a popular approach offered by the Rational group at IBM is called the **Rational Unified Process (RUP®)**, as shown in Figure 1-32. According to IBM, RUP® offers a flexible, iterative process for managing software development projects that can minimize risk, ensure predictable results, and deliver high-quality software on time.

Another option is what Microsoft calls **Microsoft Solutions Framework (MSF)**, which documents the experience of its own software development teams. Although the Microsoft process differs from the SDLC phase-oriented approach, MSF developers perform the same kind of planning, ask the same kinds of fact-finding questions, deal with the same kinds of design and implementation issues, and resolve the same kinds of problems. Using this approach, MSF examines a broader business and organizational context that surrounds the development of an information system.

ON THE WEB

For more information about Microsoft Solutions Framework, visit **scsite.com/ sad8e/more**, locate Chapter 1, and then click the Microsoft Solutions Framework link.

Companies often choose to follow their own methodology. Using CASE tools, an IT team can apply a variety of techniques rather than being bound to a single, rigid methodology. As shown in Part 2 of the Systems Analyst's Toolkit, many CASE tools offer a complete set of analysis and modeling tools that support various methods and strategies. Regardless of the development model, it will be necessary to manage people, tasks, timetables, and expenses by using various project management tools and techniques.

SYSTEMS DEVELOPMENT GUIDELINES

With experience as a systems analyst, you will develop your own style and techniques. Although each project is different, you should consider some basic guidelines as you build an information system.

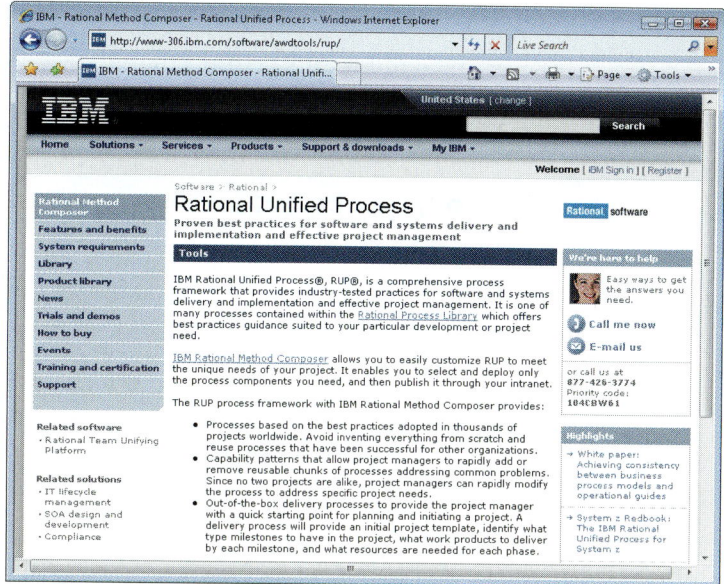

FIGURE 1-32 IBM's Rational Group offers a development method called the Rational Unified Process®.

Develop a Project Plan

Prepare an overall project plan and stick to it. If you use the SDLC as a framework for systems development, complete the phases in sequence. If you use an O-O methodology, follow a logical series of steps as you define the components. If you use agile methods, set the ground rules and be sure they are understood clearly.

Involve Users and Listen Carefully to Them

Ensure that users are involved in the development process, especially when identifying and modeling system requirements. Modeling and prototyping can help you understand user needs and develop a better system. When you interact with users, put aside any preconceived notions and listen closely to what they are saying. Chapter 4 describes the interview process and contains many tips about getting the most out of face-to-face communication.

Use Project Management Tools to Identify Tasks and Milestones

Regardless of the development methodology, the systems analyst must keep the project on track and avoid surprises. Create a reasonable number of checkpoints — too many can be burdensome, but too few will not provide adequate control. An example of a checkpoint might be the completion of interviews conducted during a preliminary investigation.

In Chapter 3, you will learn how to use Microsoft Project 2007 to help you define tasks, manage resources, monitor progress, and create reports on systems development projects. This powerful project management tool includes a Project Guide, as shown in Figure 1-33, that offers step-by-step instructions and guidance.

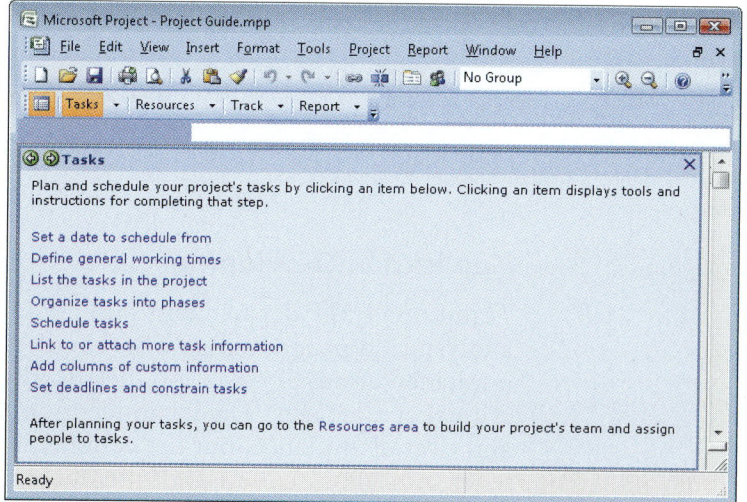

FIGURE 1-33 The Project Guide in Microsoft Project 2007 offers step-by-step instructions and guidance.

Develop Accurate Cost and Benefit Information

Provide accurate and reliable cost and benefit information. Managers need to know the cost of developing and operating a system. At the start of each phase, provide specific estimates, and update these as necessary.

Remain Flexible

Be flexible within the framework of your plan. Systems development is a dynamic process, and overlap often exists between the phases of systems planning, analysis, design, and implementation. For example, when you investigate a systems request, you begin a fact-finding process that often carries over into the next phase. Similarly, you often start building process models before fact-finding is complete. The ability to overlap phases is especially important when you are working on a system that must be developed rapidly.

INFORMATION TECHNOLOGY DEPARTMENT

The information technology (IT) department develops and maintains a company's information systems. The structure of the IT department varies among companies, as does its name and placement within the organization. In a small firm, one person might handle all computer support activities and services, whereas a large corporation might require many people with specialized skills to provide information systems support. Figure 1-34 shows a typical IT organization in a company that has networked PCs, enterprise-wide databases, centralized processing, and Web-based operations.

The IT group provides **technical support**, which includes six main functions: application development, systems support and security, user support, database administration, network administration, and Web support. These functions overlap considerably and often have different names in different companies.

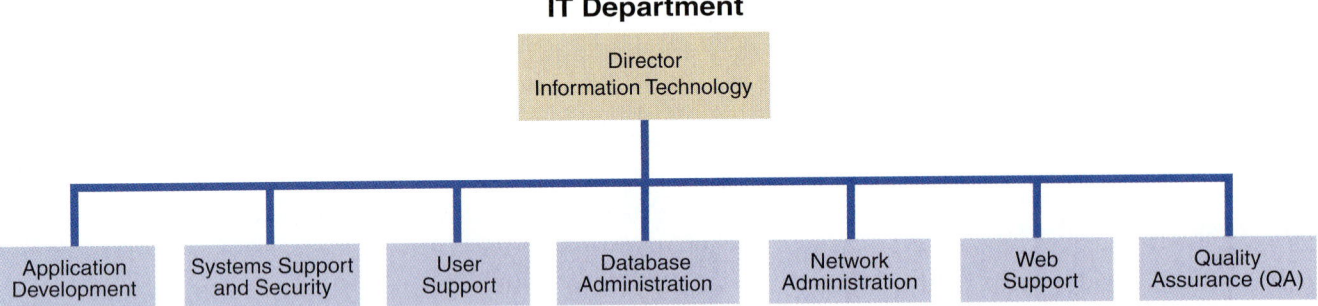

FIGURE 1-34 Depending on its size, an IT department might have separate organizational units for these functions, or they might be combined into a smaller number of teams.

Application Development

Traditionally, IT departments had an **application development** group composed of systems analysts and programmers who handled information system design, development, and implementation. Today, regardless of the development method, user involvement is seen as critical at all stages. The IT application development group typically provides leadership and overall guidance, but the systems themselves are developed by teams consisting of users, managers, and IT staff members. A popular model for information systems development is a project-oriented team using RAD or JAD, with IT professionals providing overall coordination, guidance, and technical support.

CASE IN POINT 1.2: GLOBAL HOTELS AND MOMMA'S MOTELS

Suppose you work in the IT department of Global Hotels, a multinational hotel chain. Global Hotels runs several specialized business support systems, including a guest reservations system that was developed in-house to meet the requirements of a large company with worldwide operations. Guests can make one-stop online reservations by visiting Global's Web site, which has links to all major travel industry sites.

Global Hotels just acquired Momma's, a regional chain of 20 motels in western Canada. Momma's uses a vertical reservations package suitable for small- to medium-sized businesses, and a generic accounting and finance package. Should Momma's use Global Hotels' information systems or continue with its own? In your answer, consider issues such as business profiles, business processes, system interactivity, EDI, XML, e-commerce, and the characteristics of both information systems. What additional information would be helpful to you in making a recommendation?

Systems Support and Security

Systems support and security provides vital protection and maintenance services for system hardware and software, including enterprise computing systems, networks, transaction processing systems, and corporate IT infrastructure. The systems support and security group implements and monitors physical and electronic security hardware, software, and procedures. This group also installs and supports operating systems, telecommunications software, and centralized database management systems. In addition, systems support and security technicians provide technical assistance to other groups in the IT department. If a site has a large number of remote clients, the systems support group often includes a **deployment team** that installs and configures the workstations.

User Support

User support provides users with technical information, training, and productivity support. The user support function usually is called a **help desk** or **information center (IC)**. A help desk's staff trains users and managers on application software such as e-mail, word processors, spreadsheets, and graphics packages. User support specialists answer questions, troubleshoot problems, and serve as a clearinghouse for user problems and solutions.

In many companies, the user support team also installs and configures software applications that are used within the organization. Although user support specialists coordinate with other technical support areas, their primary focus is user productivity and support for user business processes.

Database Administration

Database administration involves database design, management, security, backup, and user access. In small- and medium-sized companies, an IT support person performs those roles in addition to other duties. Regardless of company size, mission-critical database applications require continuous attention and technical support.

Network Administration

Business operations depend on telecommunication networks that enable company-wide information systems. **Network administration** includes hardware and software mainte- nance, support, and security. In addition to controlling user access, network administra- tors install, configure, manage, monitor, and maintain network applications. Network administration is discussed in more detail in Chapter 10.

Web Support

Web support is a vital technical support function. Web support specialists, often called **webmasters**, support a company's Internet and intranet operations. Web support involves design and construction of Web pages, monitoring traffic, managing hardware and software, and linking Web-based applications to the company's existing informa- tion systems. Reliable, high-quality Web support is especially critical for companies engaged in e-commerce.

Quality Assurance (QA)

Many large IT departments also use a **quality assurance (QA)** team that reviews and tests all applications and systems changes to verify specifications and software quality standards. The QA team usually is a separate unit that reports directly to IT management.

CASE IN POINT 1.3:   WHAT SHOULD LISA DO?

Lisa Jameson has two job offers. One is from Pembroke Boats, a boat manufacturer that employs 200 people in a small Ohio town. Pembroke does not have an IT department and wants her to create one. The job position is called information coordinator, but she would be the only IT person.

   The other offer, which pays about $7,500 more annually, is from Albemarle Express, a nationwide trucking firm located in Detroit. At Albemarle Express, Lisa would be a program- mer-analyst, with the promise that if she does well in her position, she eventually will move into a systems analyst position and work on new systems development. Lisa has heard a rumor that another company might acquire Albemarle Express, but that rumor has occurred before and nothing has ever happened. What should Lisa do, and why?

THE SYSTEMS ANALYST POSITION

A systems analyst investigates, analyzes, designs, develops, installs, evaluates, and maintains a company's information systems. To perform those tasks, a systems analyst constantly interacts with users and managers within and outside the company. On large projects, the analyst works as a member of an IT department team; on smaller assignments, he or she might work alone.

   Most companies assign systems analysts to the IT department, but analysts also can report to a specific user area such as marketing, sales, or accounting. As a member of a functional team, an analyst is better able to understand the needs of that group and how information systems support the department's mission. Smaller companies often use consultants to perform systems analysis work on an as-needed basis.

Responsibilities

The systems analyst's job overlaps business and technical issues. Analysts help translate business requirements into IT projects. When assigned to a systems development team, an analyst might help document business profiles, review business processes, select hardware and software packages, design information systems, train users, and plan e-commerce Web sites.

A systems analyst plans projects, develops schedules, and estimates costs. To keep managers and users informed, an analyst conducts meetings, delivers presentations, and writes memos, reports, and documentation. The Systems Analyst's Toolkit that follows Chapter 12 includes various tools to help you with each of those important skills.

⚡ TOOLKIT TIME

The Communications Tools in Part I of the Systems Analyst's Toolkit can help you develop better reports and presentations. To learn more about these tools, turn to Part I of the four-part Toolkit that follows Chapter 12.

Required Skills and Background

A systems analyst needs solid technical knowledge, strong oral and written communication skills, good analytical ability, and an understanding of business operations and processes. Companies typically require that systems analysts have a college degree in information systems, computer science, business, or a closely related field, and some IT experience usually is required. For higher-level positions, many companies require a master of science degree and additional experience.

A systems analyst needs good interpersonal skills to interact with people at all levels, from operational staff to senior executives, including people outside the company, such as software and hardware vendors, customers, and government officials.

Often an analyst must lead an IT development team. As a team leader, an analyst plans, estimates, and manages the project, and uses leadership and team-building skills to coach and motivate team members.

State-of-the-art knowledge is extremely important in a rapidly changing business and technical environment. The Internet offers numerous opportunities to update technical knowledge and skills. Many sites, such as the TechRepublic site shown in Figure 1-35, offer free subscriptions and enable IT professionals to learn about the latest technical developments, exchange experiences, and ask questions.

Analysts also maintain their skills by attending training courses, both on-site and online. Networking with colleagues is another way to keep up with new developments, and membership in professional associations also is important. A systems analyst, like any other professional, needs to manage his or her own career by developing knowledge and skills that are valuable and expected in the marketplace.

FIGURE 1-35 The TechRepublic Web site offers support for IT professionals. Features include newsletters, forums, product information, and a searchable knowledge base.

Certification

Many hardware and software companies offer certification for IT professionals. **Certification** verifies that an individual demonstrated a certain level of knowledge and skill on a standardized test. Certification is an excellent way for IT professionals to learn new skills and gain recognition for their efforts. Although certification alone does not guarantee competence or ability, many companies regard certification as an important credential for hiring or promotion. You can learn more about certification by visiting the Web sites of individual companies such as Microsoft, Cisco Systems, Sun Microsystems, and Novell.

Career Opportunities

The demand for systems analysts is expected to remain strong well into the twenty-first century. Companies will need systems analysts to apply new information technology, and the explosion in e-commerce will fuel IT job growth. The systems analyst position is a challenging and rewarding one that can lead to a top management position. With an understanding of technical and business issues, a systems analyst has an unlimited horizon. Many companies have presidents and senior managers who started in IT departments as systems analysts.

The responsibilities of a systems analyst at a small firm are different from those at a large corporation. Would you be better off at a small or large company? Where will you find the best opportunity for experience and professional growth? Each person looks for different rewards in a job. What will be important to you?

JOB TITLES   First, do not rely on job titles alone. Some positions are called systems analysts, but involve only programming or technical support. In other cases, systems analyst responsibilities are found in positions titled computer specialist, programmer, programmer/analyst, systems designer, software engineer, and various others. Be sure the responsibilities of the job are stated clearly when you consider a position.

COMPANY ORGANIZATION   Find out all you can about the company and where the IT department fits in the organization chart. Where are IT functions performed, and by whom? A firm might have a central IT group, but decentralize the systems development function. This situation sometimes occurs in large conglomerates, where the parent company consolidates information that actually is developed and managed at the subsidiary level. Where would you rather work?

TOOLKIT TIME

The Internet Resource Tools in Part 4 of the Systems Analyst's Toolkit can help you obtain technical data, advance your career, and network with other IT professionals. To learn more about these tools, turn to Part 4 of the four-part Toolkit that follows Chapter 12.

COMPANY SIZE   If you like more variety, a smaller firm might suit you best. If you want to specialize, however, then consider a larger company with state-of-the-art systems. Although you might have more responsibility in a smaller company, the promotional opportunities and financial rewards often are greater in larger companies. You also might want to consider working as an independent consultant, either on your own or with others. Many consulting firms have been successful in offering their services to smaller business enterprises that do not have the expertise to handle systems development on their own.

CORPORATE CULTURE   In addition to having goals, methods, and information systems requirements, every firm has an underlying **corporate culture**. A corporate culture is the set of beliefs, rules, traditions, values, and attitudes that define a company and influence its way of doing business. To be successful, a systems analyst must understand the corporate culture and how it affects the way information is managed. Companies sometimes include statements about corporate culture in their mission statements, which are explained in Chapter 2.

SALARY, LOCATION, AND FUTURE GROWTH   Finally, consider salary, location, and the company's prospects for future growth and success. Think about your impressions of the company and the people you met during your interviews. Most important, review your short- and long-term goals very carefully before deciding which position is best for you.

CASE IN POINT 1.4: JUST-IN-TIME AIRFREIGHT, INC.

Suppose you are the IT director at Just-in-Time Airfreight, and you have received authorization to hire another systems analyst. This will be an entry-level position, and the person will assist senior systems analysts on various projects involving the reservations and the human resources systems. Using the information in this chapter, draft an ad that would appear in *The Wall Street Journal*, local newspapers, and online. You can get some ideas by visiting monster.com, or a similar site. In your ad, be sure to list desired skills, experience, and educational requirements.

A QUESTION OF ETHICS

You are enjoying your job as a summer intern in the IT department of a local company. At lunch yesterday, several people were discussing ethical issues. You learned that some of them belong to IT organizations that have ethical codes to guide members and set professional standards. For example, Ann, your supervisor, belongs to the Association for Computing Machinery (ACM), which has over 82,000 members and a Web site at acm.org. Ann said that the ACM code of ethics is important to her, and would definitely influence her views. On the other hand, Jack, a senior programmer, believes that his own personal standards would be sufficient to guide him if ethical questions were to arise.

Because you are excited about your career as an IT professional, you decide to examine the ACM code of ethics and make up your own mind. After you do so, would you tend to agree more with Ann or with Jack?

CHAPTER SUMMARY

In this chapter, you learned that information technology (IT) refers to the combination of hardware and software resources that companies use to manage, access, communicate, and share information. IT supports business operations, improves productivity, and helps managers make decisions. Systems analysis and design is the process of developing information systems that transform data into useful information.

Traditionally, companies either developed in-house applications or purchased software packages from vendors. Today, the choice is much more complex, but it is always important for companies to plan the system carefully before considering implementation options.

The essential components of an information system are hardware, software, data, processes, and people. Hardware consists of everything in the physical layer of the information system. Software consists of system software, which manages the hardware components, and application software, which supports day-to-day business operations. Data is the raw material that an information system transforms into useful information. Processes describe the tasks and functions that users, managers, and IT staff members perform. People who interact with a system include users, from both within and outside the company.

A systems analyst starts with a business profile, which is an overview of company functions, and then he or she creates a series of business models that represent business processes, which describe specific transactions, events, tasks, and results. Companies engage in business process reengineering to simplify operations or reduce costs.

Most successful companies offer a mix of products, technical and financial services, consulting, and customer support. A rapidly growing business category is the Internet-dependent (dot-com) firm, which relies solely on Internet-based operations. E-commerce includes business-to-consumer (B2C) sales, and business-to-business (B2B) transactions that use Internet-based digital marketplaces or private electronic data interchange (EDI) systems.

Based on their functions and features, business information systems are identified as enterprise computing systems, transaction processing systems, business support systems, knowledge management systems, or user productivity systems. In most companies, significant overlap and integration exists among the various types of information systems.

A typical organization structure includes top managers, middle managers and knowledge workers, supervisors and team leaders, and operational employees. Top managers develop strategic plans, which define an overall mission and goals. Middle managers provide direction, resources, and feedback to supervisors and team leaders. Knowledge workers include various professionals who function as support staff. Supervisors and team leaders oversee operational employees. Each organizational level has a different set of responsibilities and information needs.

Systems analysts use modeling, prototyping, and computer-aided systems engineering (CASE) tools. Modeling produces a graphical representation of a concept or process, whereas prototyping involves the creation of an early working model of the information or its components. A systems analyst uses CASE tools to perform various systems development tasks.

Three popular system development approaches are structured analysis, which is a traditional method that still is widely used, object-oriented analysis (O-O), which is a more recent approach that many analysts prefer, and agile methods, also called adaptive methods, which include the latest trends in software development.

Structured analysis uses a series of phases, called the systems development life cycle (SDLC) that usually is shown as a waterfall model. Structured analysis uses an overall plan, similar to a blueprint for constructing a building, so it is called a predictive approach. This method uses a set of process models to describe a system graphically, and also addresses data organization and structure, relational database design, and user interface issues.

Object-oriented analysis combines data and the processes that act on the data into things called objects that represent people, things, transactions, and events. Objects have characteristics called properties, built-in processes called methods, and can send information to other objects by using messages. Using an O-O programming language, a programmer then writes the code that creates the objects. Object-oriented methods usually follow a series of analysis and design phases similar to the SDLC, but the phases are more interactive.

Agile methods are the newest development approach, and attempt to develop a system incrementally by building a series of prototypes and constantly adjusting them to user requirements. Agile methods typically use a spiral model, which represents a series of iterations, or revisions, based on user feedback. The repeated iterations produce a series of prototypes, which evolve into the finished system.

Regardless of the development strategy, people, tasks, timetables, and costs must be managed effectively using project management tools and techniques, which are described in detail in Chapter 3.

Some firms choose to develop their own in-house methods or adopt techniques offered by software suppliers, CASE tool vendors, or consultants. Examples include IBM's Rational Unified Process (RUP®) and Microsoft Solutions Framework (MSF). Using CASE tools, an IT team can apply a variety of techniques rather than being bound to a single methodology. Companies also use team-based strategies called joint application development (JAD) and rapid application development (RAD). JAD focuses on team-based fact-finding, whereas RAD is more like a compressed version of the entire process. JAD and RAD are described in more detail in Chapter 4.

The IT department develops, maintains, and operates a company's information systems. IT staff members provide technical support, including application development, systems support, user support, database administration, network administration, and Web support. These functions overlap considerably and often have different names in different companies.

A systems analyst investigates, analyzes, designs, develops, installs, evaluates, and maintains information systems. Systems analysts need a combination of technical and business knowledge, analytical ability, and communication skills. In addition to education and experience, various certifications are available to systems analysts. A systems analyst's responsibilities depend on a company's organization, size, and culture. Systems analysts need to consider salary, location, and future growth potential when making a career decision.

Key Terms and Phrases

adaptive methods 18
Agile Manifesto 22
agile methods 18
application development 26
application software 6
B2B (business-to-business) 9
B2C (business-to-consumer) 9
brick-and-mortar 9
business model 8, 16
business process 8
business process modeling 8
business process reengineering (BPR) 8
business profile 8
business support systems 13
CASE tools 17
certification 29
class 19
computer-aided software engineering 17
computer-aided systems engineering (CASE) 17
corporate culture 30
data 5
data model 16
database administration 27
deliverable 19
deployment team 27
dot-com (.com) 9
e-commerce (electronic commerce) 9
electronic data interchange (EDI) 10
empowerment 16
end product 19
end users 7
enterprise applications 6
enterprise computing 12
enterprise resource planning (ERP) 12
expert systems 14
extensible markup language (XML) 10
extreme programming (XP) 24
feasibility study 20
fuzzy logic 14
groupware 14
hardware 6
help desk 27

horizontal system 7
I-commerce (Internet commerce) 9
inference rules 14
information 5
information center (IC) 27
information system 4
information technology (IT) 4
in-house applications 5
interactive model 20
Internet-dependent 9
iterative 22
joint application development (JAD) 24
knowledge base 14
knowledge management systems 14
knowledge workers 16
legacy systems 7
management information systems (MIS) 13
message 22
methods 22
Microsoft Solutions Framework (MSF) 24
mission-critical system 5
modeling 16
Moore's Law 6
.NET 11
network administration 28
network model 16
object-oriented analysis (O-O) 18
object model 16
objects 21
predictive 19
preliminary investigation 20
process model 16
process-centered 19
processes 7
product-oriented 9
project management 18
properties 21
prototype 17
quality assurance (QA) 28
radio frequency identification (RFID) 13
rapid application development (RAD) 24

Rational Unified Process (RUP®) 24
requirements model 16
requirements modeling 20
scalable 21
Scrum 23
service-oriented 9
software 6
software packages 5
spiral model 23
stakeholder 7
strategic plans 15
structured analysis 18
supplier relationship management (SRM) 10
system 5
system design specification 21
system requirements document 20
system software 6
systems analysis and design 4
systems analysis phase 20
systems analysts 4
systems design phase 21
systems development life cycle (SDLC) 19
systems evaluation 21
systems implementation phase 21
systems planning phase 19
systems request 19
systems support and security 27
systems support and security phase 21
technical support 26
transaction processing (TP) systems 12
user productivity systems 14
user support 27
users 7
vertical system 7
waterfall model 19
Web services 11
Web support 28
webmasters 28
WebSphere 11
what-if 13

Learn It Online

Instructions: To complete the Learn It Online exercises, start your browser, click the Address bar, and then enter the Web address **scsite.com/sad8e/learn**. When the Systems Analysis and Design Learn It Online page is displayed, follow the instructions in the exercises below. Each exercise has instructions for saving your results, either for your own records or for submission to your instructor.

1 Chapter Reinforcement

TF, MC, and SA

Below SAD Chapter 1, click the Chapter Reinforcement link. Print the quiz by clicking Print on the File menu for each page. Answer each question.

2 Flash Cards

Below SAD Chapter 1, click the Flash Cards link and read the instructions. Type 20 (or a number specified by your instructor) in the Number of playing cards text box, type your name in the Enter your Name text box, and then click the Flip Card button. When the flash card is displayed, read the question and then click the ANSWER box arrow to select an answer. Flip through the Flash Cards. If your score is 15 (75%) correct or greater, click Print on the File menu to print your results. If your score is less than 15 (75%) correct, then redo this exercise by clicking the Replay button.

3 Practice Test

Below SAD Chapter 1, click the Practice Test link. Answer each question, enter your first and last name at the bottom of the page, and then click the Grade Test button. When the graded practice test is displayed on your screen, click Print on the File menu to print a hard copy. Continue to take practice tests until you score 80% or better.

4 Who Wants To Be a Computer Genius?

Below SAD Chapter 1, click the Computer Genius link. Read the instructions, enter your first and last name at the bottom of the page, and then click the Play button. When your score is displayed, click the PRINT RESULTS link to print a hard copy.

5 Wheel of Terms

Below SAD Chapter 1, click the Wheel of Terms link. Read the instructions, and then enter your first and last name and your school name. Click the PLAY button. When your score is displayed on the screen, right-click the score and then click Print on the shortcut menu to print a hard copy.

6 Crossword Puzzle Challenge

Below SAD Chapter 1, click the Crossword Puzzle Challenge link. Read the instructions, and then enter your first and last name. Click the SUBMIT button. Work the crossword puzzle. When you are finished, click the Submit button. When the crossword puzzle is redisplayed, click the Print Puzzle button to print a hard copy.

Case-Sim: SCR Associates

Background

SCR Associates is a consulting firm that offers IT solutions and training. SCR needs an information system to manage operations at its new training center. The new system will be called TIMS (Training Information Management System). As a newly hired systems analyst, you will report to Jesse Baker, systems group manager. You will work on various tasks and practice the skills you learned in this chapter.

Using the Case

The SCR Associates case study is a Web-based simulation that allows you to practice your skills in a real-world environment. The case study transports you to SCR's company intranet, where you can complete 12 work sessions, each aligning with a chapter. As you work on the case, you will receive e-mail and voice mail messages, obtain information from SCR's online libraries, and perform various tasks. The first time you enter the SCR Case, you should go to the starting page at **scsite.com/sad8e/scr** for detailed instructions.

Preview: Session 1

This is your second day on the job as a systems analyst at SCR Associates. You spent most of yesterday filling out personnel forms and learning your way around the office. This morning, you sit at your desk and examine SCR's Internet site. You explore the entire site, which reflects SCR's history, purpose, and values. You especially are impressed with the emphasis SCR puts on its relationships with clients. When you finish examining the site, you are more convinced than ever that SCR will be a great career opportunity for you. You are excited about your new job, and eager to get started.

To start a work session, you log on to the SCR intranet. When you enter your name and password, an opening screen displays links to the work sessions, and you select Session 1. At this point, you check your e-mail and voice mail messages carefully. You then begin working on your task list, which includes the following items:

Tasks: Introduction

Jesse wants me to learn as much as possible about SCR, and she gave me a checklist to get me started:

1. *Investigate SCR's Internet site and learn about the company's history, purpose, and values. Send Jesse a brief memo with suggestions to expand or improve these sections.*

2. *On the SCR intranet, visit the data, forms, and resource libraries and review a sample of the information in each library.*

3. *Using the SCR functions and organization listed in the data library, create an organization chart using Microsoft Word, Visio, or a drawing program.*

4. *Jesse says that SCR has plenty of competition in the IT consulting field. Get on the Internet and find three other IT consulting firms. She wants a brief description of each firm and the services it offers.*

FIGURE 1-36   Task list: Session 1.

Chapter Exercises

Review Questions

1. What is information technology, and why is it important to a business?
2. Define business profiles, business models, and business processes.
3. Identify the main components of an information system, and describe the system's stakeholders.
4. Explain the difference between vertical and horizontal systems packages.
5. How do companies use EDI? What are some advantages of using XML?
6. Describe five types of information systems, and give an example of each.
7. Describe four organizational levels of a typical business and their information requirements.
8. Describe the phases of the systems development life cycle, and compare the SDLC waterfall model with the spiral model.
9. Explain the use of models, prototypes, and CASE tools in the systems development process. Also explain the pros and cons of agile development methods.
10. What is object-oriented analysis, and how does it differ from structured analysis?

Discussion Topics

1. Some experts believe that the growth in e-commerce will cause states and local governments to lose a significant amount of sales tax revenue, unless Internet transactions are subject to sales tax. Do you agree? Why or why not?
2. Present an argument for and against the following proposition: Because IT managers must understand all phases of the business, a company should fill top management vacancies by promoting IT managers.
3. The head of the IT group in a company often is called the chief information officer (CIO) or chief technology officer (CTO). Should the CIO or CTO report to the company president, to the finance department, where many of the information systems are used, or to someone or somewhere else? Why would it matter?
4. Computers perform many jobs that previously were performed by people. Will computer-based transactions and expanded e-commerce eventually replace person-to-person contact? From a customer's point of view, is this better? Why or why not?

Projects

1. Contact at least three people at your school or a nearby company who use information systems. List the systems, the position titles of the users, and the business functions that the systems support.
2. Research newspaper, business magazine articles, or the Web to find computer companies whose stock is traded publicly. Choose a company and pretend to buy $1,000 of its stock. What is the current price per share? Why did you choose that company? Report each week to your class on how your stock is doing.
3. Do a search on the Web to learn more about agile system development approaches and spiral models. Prepare a summary of the results and a list of the sites you visited.
4. In recent years, technology stocks have climbed to staggering heights, only to fall sharply. Imagine that you are a financial advisor, and an investor has come to you for advice. Perform Internet research to learn more about investing in technology stocks. Would it be important to know something about the investor, including his or her goals and risk tolerance? How would you decide what companies to invest in?

Apply Your Knowledge

The Apply Your Knowledge section contains four mini-cases. Each case describes a situation, explains your role in the case, and asks you to respond to questions. You can answer the questions by applying knowledge you learned in the chapter.

1 Low-Voltage Components

Situation:

You are the IT manager at Low-Voltage Components, a medium-sized firm that makes specialized circuit boards. Low-Voltage's largest customer, TX Industries, recently installed a computerized purchasing system. If Low-Voltage connects to the TX system, TX will be able to submit purchase orders electronically. Although Low-Voltage has a computerized accounting system, that system is not capable of handling EDI.

1. Should Low-Voltage develop a system to connect with TX Industries' purchasing system? Why or why not?
2. What terms or concepts describe the proposed computer-to-computer relationship between Low-Voltage and TX Industries?
3. Is Low-Voltage's proposed new system a transaction processing system? Why or why not?
4. Before Low-Voltage makes a final decision, should the company consider an ERP system? Why or why not?

2 Systems Analyst Salaries

Situation:

As part of your job search, you decide to find out more about salaries and qualifications for systems analysts in the area where you would like to work. To increase your knowledge, search the Internet to perform the following research:

1. Find information about a career as a systems analyst.
2. Using the Internet, determine whether the Federal Bureau of Labor Statistics lists salary information for systems analysts. If so, summarize the information you find.
3. Find at least two online ads for systems analysts and list the employers, the qualifications, and the salaries, if mentioned.
4. Find at least one ad for an IT position that specifically mentions e-commerce.

3   MultiTech Interview

Situation:

You have an interview for an IT position with MultiTech, a large telecommunications company, and you want to learn more about the firm and its organizational structure. To prepare for the interview, you decide to review your knowledge about corporations, including the following questions:

1. What are the four organizational levels in a typical company?
2. How can you classify companies based on their mix of products and services?
3. What is empowerment?
4. What types of information systems might a large company use?

4   Rainbow's End Interview

Situation:

Your MultiTech interview seemed to go well, but you did not get the job. During the meeting, the interviewer mentioned that MultiTech uses structured analysis and relies heavily on modeling, prototyping, and CASE tools. Thinking back, you realize that you did not fully understand those terms. As you prepare for an interview with Rainbow's End, a large retail chain, you decide to review some IT terms and concepts. You want to be ready for the following questions:

1. What are the main differences between structured, O-O, and agile development methods?
2. What is a CASE tool and what does it do?
3. What is modeling and how is it done?
4. What is prototyping and why is it important?

Case Studies

Case studies allow you to practice specific skills learned in the chapter. Each chapter contains several case studies that continue throughout the textbook, and a chapter capstone case.

New Century Health Clinic

New Century Health Clinic offers preventive medicine and traditional medical care. In your role as an IT consultant, you will help New Century develop a new information system.

Background

Five years ago, cardiologists Timothy Jones and Dolores Garcia decided to combine their individual practices in Brea, California, to form New Century Health Clinic. They wanted to concentrate on preventive medicine by helping patients maintain health and fitness and by providing traditional medical care. Dr. Jones recently asked you to work with him as an IT consultant. He wants you to help New Century develop an information system that will support the clinic's operations and future growth. At your initial meeting, he provided you with some background information and asked for your suggestions about how to begin.

At your desk, you begin to review New Century's situation. The clinic is located near a new shopping mall in a busy section of the city. New Century's staff includes four doctors, three registered nurses, four physical therapists, and six office staff workers. The clinic currently has a patient base of 3,500 patients from 275 different employers, many of which provide insurance coverage for employee wellness and health maintenance. Currently, New Century accepts 34 different insurance policies.

Anita Davenport, who has been with New Century since its inception, is the office manager. She supervises the staff, including Fred Brown, Susan Gifford, Tom Capaletti, Lisa Sung, and Carla Herrera. Fred Brown handles office payroll, tax reporting, and profit distribution among the associates. Susan Gifford is responsible for the maintenance of patient records. Tom Capaletti handles most of the paperwork concerning insurance reporting and accounting. Lisa Sung has the primary responsibility for the appointment book, and her duties include making reminder calls to patients and preparing daily appointment lists. Carla Herrera is concerned primarily with ordering and organizing office and clinic supplies.

Each of the six office staff people has one or more primary responsibilities; however, all members of the staff help out whenever necessary with patient records, insurance processing, and appointment processing. In addition to their regular responsibilities, all six office workers are involved in the preparation of patient statements at the end of each month.

Using this information, you begin to prepare for your next meeting with Dr. Jones.

Assignments

1. Create an organization chart of the office staff using Microsoft Word or a similar program, or you can draw it by hand. In Word 2003, click Insert Diagram or Organization Chart on the Drawing toolbar; in Word 2007, on the Insert tab click SmartArt then Organization Chart.
2. Identify at least three business processes that New Century performs, and explain who is responsible for the specific tasks.
3. Explain how New Century might use a transaction processing system, a business support system, and a user productivity system. For each type of system, provide a specific example, and explain how the system would benefit the clinic.
4. During the systems development process, should New Century consider any of the following: B2B, vertical and horizontal system packages, or Internet-based solutions? Explain your answers.

PERSONAL TRAINER, INC.

Personal Trainer, Inc. owns and operates fitness centers in a dozen midwestern cities. The centers have done well, and the company is planning an international expansion by opening a new "supercenter" in the Toronto area.

Background

Cassia Umi, president, heads Personal Trainer's management team. Three managers report to her at the firm's Chicago headquarters: Janet McDonald, manager, finance; Tai Tranh, manager, sales and marketing; and Reed Cotter, manager, operations. The managers who run the 12 existing centers all report to Reed.

Cassia wants the new supercenter to emphasize a wide variety of personal services and special programs for members. If the supercenter approach is successful, it will become the model for Personal Trainer's future growth. Cassia personally selected Gray Lewis, a manager with three years of fitness center experience, to run the new facility.

The new supercenter will feature a large exercise area with state-of-the-art equipment, a swimming pool, a sporting goods shop, a health food store, and a snack bar. In addition, the center will offer child care with special programs for various ages, a teen center, and a computer café. Cassia also wants members to have online access to customized training programs and progress reports.

Personal Trainer currently uses BumbleBee, a popular accounting package, to manage its receivables, payables, and general ledger. Membership lists and word processing are handled with Microsoft Office products.

Cassia believes the new supercenter will require additional data management capability, and she decided to hire Patterson and Wilder, an IT consulting firm, to help Personal Trainer develop an information system for the new operation. The firm assigned Susan Park, an experienced consultant, to work with the Personal Trainer team.

Susan's first task was to learn more about business operations at the new center, so she requested a meeting with Gray. After some small talk, the discussion went like this:

Susan: Tell me about your plans for the new operation. I'm especially interested in what kind of information management you'll need.

Gray: *Cassia thinks that we'll need more information support because of the size and complexity of the new operation. To tell the truth, I'm not so sure. We've had no problem with BumbleBee at the other centers, and I don't really want to reinvent the wheel.*

Susan: Maybe we should start by looking at the similarities — and the differences — between the new center and the existing ones.

Gray: *Okay, let's do that. First of all, we offer the same basic services everywhere. That includes the exercise equipment, a pool, and, in most centers, a snack bar. Some centers also sell sporting goods, and one offers child care — but not child-fitness programs. It is true that we've never put all this together under one roof. And, I admit, we've never offered online access. To be honest, I'm not absolutely sure what Cassia and Zachary have in mind when they talk about 24/7 Web-based access. One more feature — we plan to set up two levels of membership — let's call them silver and gold for now. Silver members can use all the basic services, but will pay additional fees for some special programs, such as child fitness. Gold members will have unlimited use of all services.*

Susan: So, with all this going on, wouldn't an overall system make your job easier?

Gray: *Yes, but I don't know where to start.*

(continued)

Susan: Gray, that's why I'm here. I'll work with you and the rest of the team to come up with a solution that supports your business.

Gray: *Sounds good to me. When can we start?*

Susan: Let's get together first thing tomorrow. Bring along an organization chart and think about how you plan to run the new facility. We'll try to build a model of the new operation so we can identify the business functions. When we know what the functions are, we can figure out what kind of information is needed or generated by each function. That will be our starting point.

Assignments

1. Develop a business profile for Personal Trainer, based on the facts provided. List at least three of Personal Trainer's business processes.

2. Create an organization chart for Personal Trainer using Microsoft Word or a similar program, or you can draw it by hand. In Word 2003, click Insert Diagram or Organization Chart on the Drawing toolbar; in Word 2007, on the Insert tab click SmartArt then Organization Chart.

3. Review the conversation between Susan and Gray. In your opinion, is Gray totally supportive of the new system? Why or why not? Do you agree with the way that Susan responds to Gray's comments? Why or why not?

4. Should Personal Trainer consider any of the following systems: enterprise computing, transaction processing, business support, knowledge management, or user productivity? Why or why not? What opportunities might Personal Trainer have for Web-based B2C transactions in the future? What about B2B?

Original Kayak Adventures

Original Kayak Adventures (OKA) offers guided eco-tours and kayak rentals along the Hudson River.

Background

John and Edie Caputo, who are avid kayakers and amateur naturalists, founded OKA two years ago. The Caputos spent many weekends and vacations exploring the Hudson's numerous creeks and tributaries. John was a sales representative and Edie worked for a Web design firm. Two years ago, John's division was purchased by a rival company, which announced plans to move operations to another state. Rather than relocate, the Caputos decided to launch OKA. They reasoned that Edie could leave her job and work as a freelance Web designer, which would provide some income while John tried to build OKA into a profitable business. John and Edie are convinced that the ecotourism market will expand greatly, and they look forward to sharing their experience and knowledge with others who enjoy nature and kayaking.

Original Kayak Adventures advertises in regional magazines and maintains a Web site, which Edie designed. Customers say that the site is attractive and informative, but the Caputos are not sure of its effectiveness in attracting new business. At this time, no other kayak rental firms operate within 20 miles of OKA's location.

So far, the Caputos' plan is working out well. OKA rents space at a nearby marina, where Edie runs the office and operates her Web design business. She also handles rentals when John is giving lessons or busy with a tour group. On summer weekends and holidays, Janet Jacobs, a local college student, handles telephone inquiries and reservations.

OKA's inventory includes 16 rental kayaks of various types, eight car-top carriers, and a large assortment of accessories and safety equipment. Based on customer requests, Edie is considering adding a selection of books and videos about kayaking and ecotourism.

OKA has three main business segments: rentals, instruction, and guided tours. Most customers make advance reservations for scheduled tours and instruction sessions, but sometimes space is available for last-minute customers. Rentals are split evenly between reservations and walk-in customers.

Reservations are entered in a loose-leaf binder, with separate tabs for each business activity. Edie also created a Microsoft Access database to record reservations. When she has time, she enters the reservation date, the reservation details and kayak type, and the customer information into a table, which is sorted by reservation date. Each day, she prints a reservation list. For quick reference, Edie also displays kayak availability on a wall-mounted board with color-coded magnets that show the available or reserved status of each rental kayak. In addition to the database, Edie uses an inexpensive accounting package to keep OKA's books.

Although the OKA database handles the basic information, the Caputos have noticed some drawbacks. For example, reservations for guided tours or instruction sessions sometimes conflict with John's or Edie's availability. The Caputos also would like to get more information about rental patterns, customer profiles, advertising effectiveness, and future business opportunities. John and Edie have talked about updating the system, but they have been too busy to do so.

Assignments

1. Develop a business profile for Original Kayak Adventures. The profile should include information about OKA's business activities, organization, resources, customers, and potential opportunity to engage in e-commerce.

2. List OKA's main functions and business processes. Draw a model of an OKA business process, including possible events, processes, and results.

3. What types of information systems does OKA use? Do these systems support its current and future business objectives? Why or why not?

4. From an object-oriented viewpoint, OKA treats reservations as a class. Based on the background information provided, what are some properties of reservation objects?

CHAPTER CAPSTONE CASE: SoftWear, Limited

SoftWear, Limited (SWL) is a continuing case study that illustrates the knowledge and skills described in each chapter. In this case study, the student acts as a member of the SWL systems development team and performs various tasks.

Background

SoftWear, Limited, manufactures and sells casual and recreational clothing for men and women. SWL was formed about 10 years ago when a national firm sold the division during a corporate downsizing. A group of managers obtained financing and became owners of the company. With clever marketing, competitive pricing, and efficient production, SWL has grown to more than 450 employees, including the corporate headquarters and manufacturing plants. Last year, SWL had sales of $700 million.

The company employs 90 people at its Raleigh, North Carolina, headquarters, including officers, managers, and support staff. Another 30 salaried and 340 hourly people are employed at production facilities in Haskell, California, and Florence, Texas. The company also is considering new factories in Canada and Australia.

SWL maintains a Web site with information about the company and its products. SWL's Web site features text, graphics, and audio and allows customers to send e-mail, order products from the SoftWear catalog, and request special promotional items, including beach umbrellas, hats, and T-shirts customized with the purchaser's logo. SWL also is studying other ways to use the Internet to boost product sales and expand its marketing efforts, including a special European promotion designed to increase awareness of SWL's Web site.

Organization

SWL's headquarters includes the executive, operations, marketing, finance, and human resources departments. Figure 1-37 shows the organization chart of the management positions within SWL. Notice that the director of information technology, Ann Hon, reports to Michael Jeremy, vice president of finance. The director of the payroll department, Amy Calico, also reports to Mr. Jeremy.

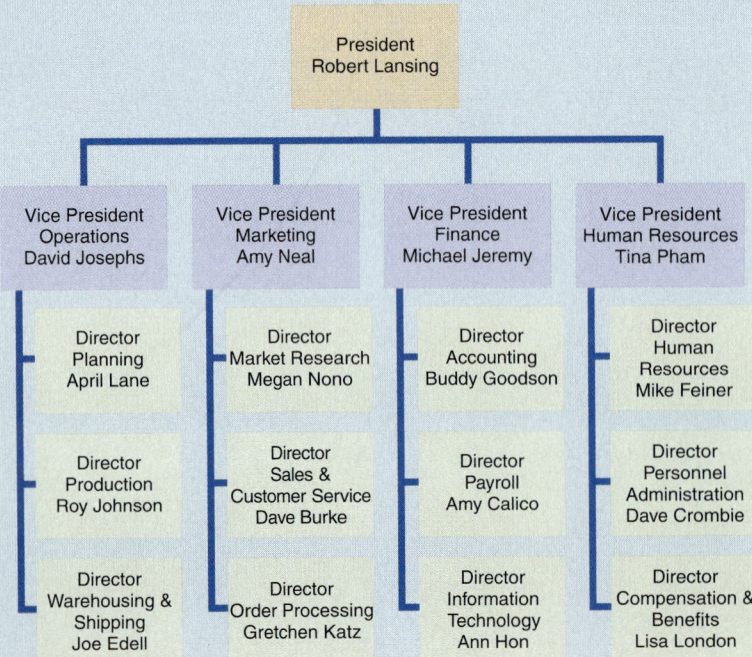

FIGURE 1-37 Organization chart of SoftWear, Limited.

CHAPTER CAPSTONE CASE: SoftWear, Limited (continued)

The IT department includes Ann Hon, the director; Jane Rossman, the systems support manager; Zachary Ridgefield, the user support manager; and Ella Allen, the Web support manager. Figure 1-38 shows the organization of the IT department. At SWL, the systems support group also handles new systems development, network administration, and database administration.

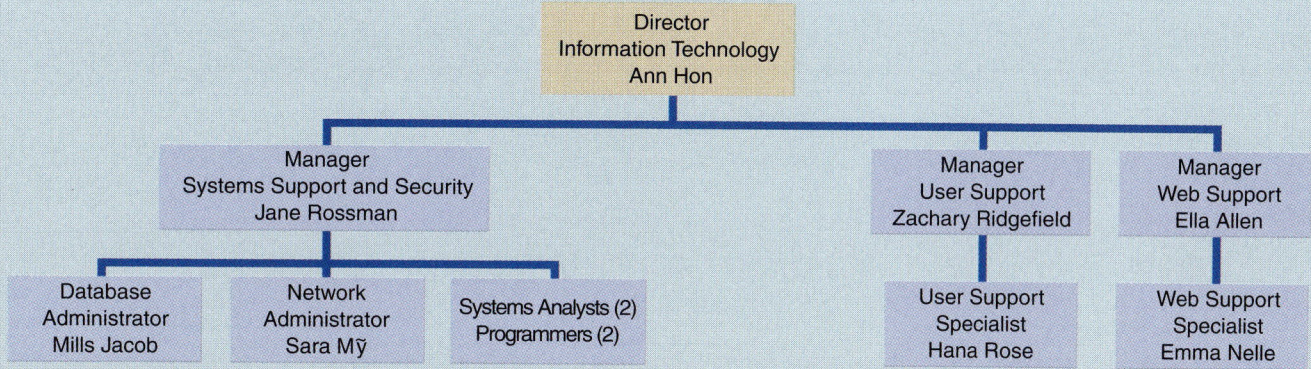

FIGURE 1-38 Organization chart of the IT department of SoftWear, Limited.

Systems analysts and programmers report to Jane Rossman, systems support manager. Systems analysts primarily analyze and design information systems. Programmers primarily develop, test, and implement code necessary for systems development, enhancements, and maintenance. In addition to the current staff, SWL is planning to hire a programmer-analyst who will divide his or her time between systems analysis and programming duties.

The technical support staff members are responsible for the system software on all SWL computers. They also provide technical advice and guidance to the other groups within the IT department.

The operations staff is responsible for centralized IT functions, including SWL's mainframe computer, and provides network and database administration.

Current Systems

SWL uses a manufacturing and inventory control system at its factories, but the system does not exchange data with SWL's suppliers at this time. The company's sales processing system handles online and catalog transactions, and produces sales reports. The marketing staff, however, wants even more information about sales trends and marketing analysis data. A company intranet connects employees at all locations, and provides e-mail, shared calendars, and a document library. Most administrative employees have workstations with Microsoft Office applications, but SWL has not provided company-wide training or help desk support.

CHAPTER CAPSTONE CASE: SoftWear, Limited (continued)

SWL Team Tasks

1. Write an employment advertisement for a new systems analyst position at SWL. Perform Internet research to locate examples of advertisements for systems analysts, and consider SWL's business profile when you write the advertisement.

2. Should SWL consider any of the following systems: ERP, business support, or knowledge management? Why or why not?

3. What opportunities might SWL have for Web-based B2B transactions in the future?

4. Should SWL consider ways to increase a sense of empowerment among its employees? Why or why not? Could user productivity software play a role in that effort? How?

Manage the SWL Project

You have been asked to manage SWL's new information system project. One of your most important activities will be to identify project tasks and determine when they will be performed. Before you begin, you should review the SWL case in this chapter. Then list and analyze the tasks, as follows:

LIST THE TASKS Start by listing and numbering at least 10 tasks that the SWL team needs to perform to fulfill the objectives of this chapter. Your list can include SWL Team Tasks and any other tasks that are described in this chapter. For example, Task 3 might be to Draw an SWL organization chart, and Task 6 might be to Identify the various levels of SWL management.

ANALYZE THE TASKS Now study the tasks to determine the order in which they should be performed. First identify all concurrent tasks, which are not dependent on other tasks. In the example shown in Figure 1-39, Tasks 1, 2, 3, 4, and 5 are concurrent tasks, and could begin at the same time if resources were available.

Other tasks are called dependent tasks, because they cannot be performed until one or more earlier tasks have been completed. For each dependent task, you must identify specific tasks that need to be completed before this task can begin. For example, you would want an organization chart to help you identify the management levels, so Task 6 cannot begin until Task 3 is completed, as Figure 1-39 shows.

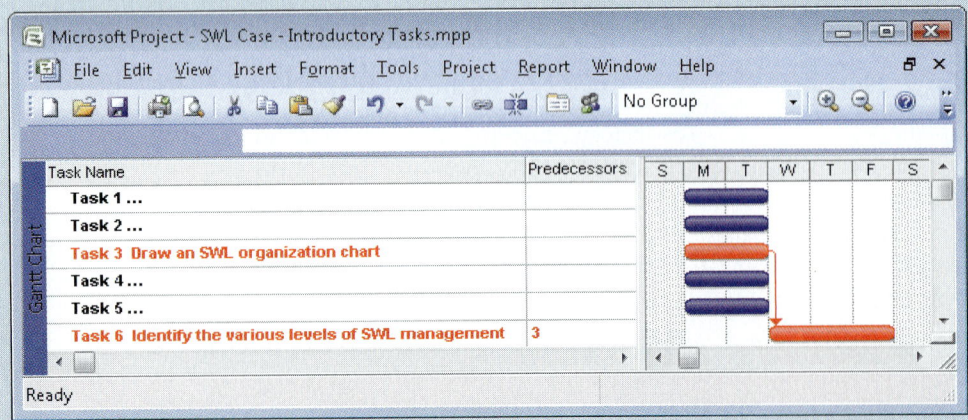

FIGURE 1-39 Tasks 1, 2, 3, 4, and 5 are concurrent tasks that could be performed at the same time. Task 6 is a dependent task that cannot be performed until Task 3 has been completed.

Chapter 3 describes project management tools, techniques, and software. To learn more, you can visit the Features section on your Student Study Tool CD-ROM, or the project management resources library at **scsite.com/sad8e/project**. On the Web, Microsoft offers demo versions, training, and tips for using Project 2007. You also can visit the OpenWorkbench.org site to learn more about this free, open-source software.

CHAPTER 2 **Analyzing the Business Case**

Chapter 2 explains how to understand and analyze a business case. This chapter also explains why it is important to understand business operations and requirements, how IT projects support a company's overall strategic plan, how systems projects get started, and how systems analysts conduct a preliminary investigation and feasibility study.

INTRODUCTION

OBJECTIVES

When you finish this chapter, you will be able to:

- Explain the concept of a business case and how a business case affects an IT project

- Describe the strategic planning process and why it is important to the IT team

- Conduct a SWOT analysis and describe the four factors involved

- Explain the purpose of a mission statement

- Describe the SDLC, and explain how it serves as a framework for systems development and business modeling

- List the reasons for information systems projects and the factors that affect such projects

- Explain the initial review of systems requests and the role of the systems review committee

- Define operational feasibility, technical feasibility, economic feasibility, and schedule feasibility

- Describe the steps in a preliminary investigation and the end product of an investigation

During the systems planning phase, the IT team reviews a proposal to determine if it presents a strong business case. The term **business case** refers to the reasons, or justification, for a proposal. A strong business case suggests that the company should pursue the alternative, above other options, because it would be in the firm's best interest to do so. To analyze the business case for a specific proposal, the analyst must consider the company's overall mission, objectives, and IT needs.

This chapter begins with a discussion of strategic planning, because the IT team must understand, support, and help plan long-term strategic goals. Along with financial, marketing, and human resources, companies need information technology to achieve growth and success.

Systems development typically starts with a systems request, followed by a preliminary investigation, which includes a feasibility study. You will learn how systems requests originate, how they are evaluated, and how to conduct a preliminary investigation. You also will learn about fact-finding techniques that begin at this point and carry over into later development phases. Finally, you will examine the report to management, which concludes the systems planning phase.

CHAPTER INTRODUCTION CASE: Mountain View College Bookstore

Background: Wendy Lee, manager of college services at Mountain View College, wants a new information system that will improve efficiency and customer service at the three college bookstores.

In this part of the case, Florence Fullerton (systems analyst) and Harry Boston (student intern) are talking about justification for the new system and the project's feasibility.

Participants:	Florence and Harry
Location:	Mountain View College cafeteria, Tuesday afternoon, September 8, 2009
Project status:	Florence has received a systems request from Wendy Lee for a new bookstore information system.
Discussion topics:	Analysis of business justification and project feasibility

Florence:	Hi, Harry. Are you ready to get started?
Harry:	*Sure. What's our next step?*
Florence:	Well, when we analyze a specific systems request, we need to see how the proposal fits into the overall picture at the college. In other words, we have to analyze the business case for the request.
Harry:	*What's a business case?*
Florence:	A business case is the justification for a project. A strong business case means that a proposal will add substantial value to the organization and support our strategic plan.
Harry:	*What's a strategic plan?*
Florence:	A strategic plan is like a road map for the future. Without a long-range plan, it's hard to know if you're heading in the right direction. Our plan starts with a mission statement, which reflects our purpose, our vision, and our values.
Harry:	*I see what you mean. I read the mission statement this morning. It says that we will strive to be an efficient, customer-friendly bookstore that uses a mix of interpersonal skills and technology to serve our students and support the overall objectives of the college. That says a lot in just one sentence.*
Florence:	It sure does. Now, let's get to the specifics. I just received a systems request from the college business manager. She wants us to develop a new information system for the bookstore.
Harry:	*Do we have a green light to get started?*
Florence:	Yes and no. Mountain View College doesn't have a formal procedure for evaluating IT requests, and we don't have a systems review committee. Maybe that's something we should consider for the future. Meanwhile, we need to conduct a preliminary investigation to see whether this request is feasible.
Harry:	*What do you mean by "feasible"?*
Florence:	To see if a systems request is feasible, we have to look at four separate yardsticks: operational feasibility, technical feasibility, economic feasibility, and schedule feasibility. If the request passes all the tests, we continue working on the system. If not, we stop.
Harry:	*How will we know if the request passes the tests?*
Florence:	That's our next step. Here's a task list to get us started:

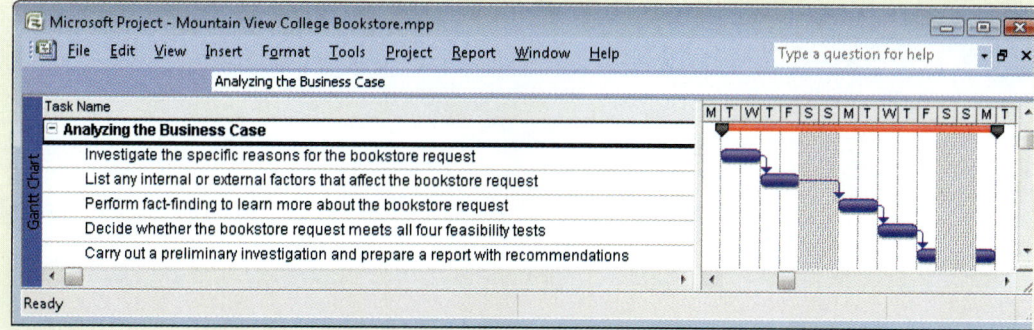

FIGURE 2-1 Typical business case analysis task list.

STRATEGIC PLANNING — A FRAMEWORK FOR IT SYSTEMS DEVELOPMENT

ON THE WEB

For more information about strategic planning, visit **scsite.com/sad8e/more**, locate Chapter 2, and then click the Strategic Planning link.

Companies develop and maintain IT systems to support their current and future business operations. Some IT needs are immediate, such as fixing a logic problem in a payroll system. Other needs might be further on the horizon, such as planning IT support for a new factory, a future merger, or a corporate restructuring. In most companies, the IT team reviews each IT-related proposal, project, and systems request to determine if it presents a strong business case, or justification.

Most successful IT managers engage in long-range planning, even as they handle day-to-day maintenance and support. To carry out this task effectively, they must understand and participate in the firm's strategic planning process. **Strategic planning** is the process of identifying long-term organizational goals, strategies, and resources. Strategic planning looks beyond day-to-day activities and focuses on a horizon that is 3, 5, or even 10 years in the future.

Strategic Planning Overview

Why does a systems analyst need to know about strategic planning? The answer might be found in an old story about two stonecutters who were hard at work when a passerby asked them what they were doing. "I am cutting stones," said the first worker. The second worker replied, "I am building a cathedral." So it is with information technology: One analyst might say, "I am using a CASE tool," whereas another might say, "I am helping the company succeed in a major new business venture." Systems analysts should focus on the larger, strategic role of IT as they carry out their day-to-day responsibilities.

During strategic planning, top managers ask a series of questions that is called a **SWOT analysis** because it examines a company's strengths (S), weaknesses (W), opportunities (O), and threats (T). Each question might lead to an IT-related issue, which in turn requires more review, analysis, and planning. For example:

- What are our major strengths, and how can we maximize them in the future? What must we do to strengthen our IT function, including our people and technology infrastructure?

- What are our major weaknesses, and how can we overcome them? How should we address weaknesses in IT resources and capability?

- What are our major opportunities, and how can we take full advantage of them? What IT plans do we have to support business opportunities?

- What major threats do we face, and what can we do about them? What can we do to deal with potential threats to IT success?

A SWOT analysis contributes to the strategic planning process by identifying technical, human, and financial resources. For example, suppose that a typical company assessed its IT resources by performing a SWOT analysis. Figure 2-2 shows some examples of strengths, weaknesses, opportunities, and threats.

From Strategic Plans to Business Results

Figure 2-3 shows the strategic planning process. A company develops a mission statement based on the firm's purpose, vision, and values. The mission statement is the foundation for the company's major goals, shorter-term objectives, and day-to-day business operations.

A **mission statement** describes a company for its stakeholders and briefly states the company's overall purpose, products, services, and values. Stakeholders include anyone affected by the company's operations, such as customers, employees, suppliers, stockholders, and members of the community. Figure 2-4 on the next page shows examples of mission statements from three organizations.

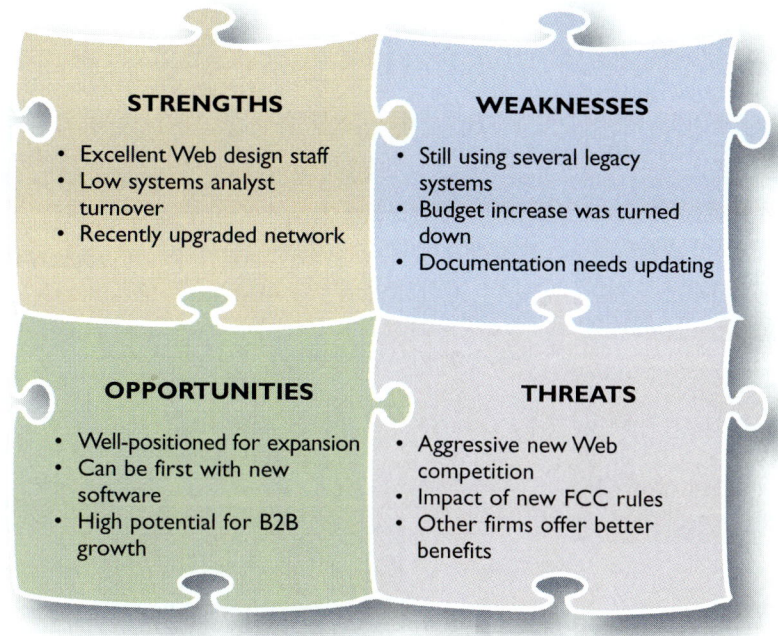

STRENGTHS
- Excellent Web design staff
- Low systems analyst turnover
- Recently upgraded network

WEAKNESSES
- Still using several legacy systems
- Budget increase was turned down
- Documentation needs updating

OPPORTUNITIES
- Well-positioned for expansion
- Can be first with new software
- High potential for B2B growth

THREATS
- Aggressive new Web competition
- Impact of new FCC rules
- Other firms offer better benefits

FIGURE 2-2 A SWOT analysis might produce results similar to those shown here.

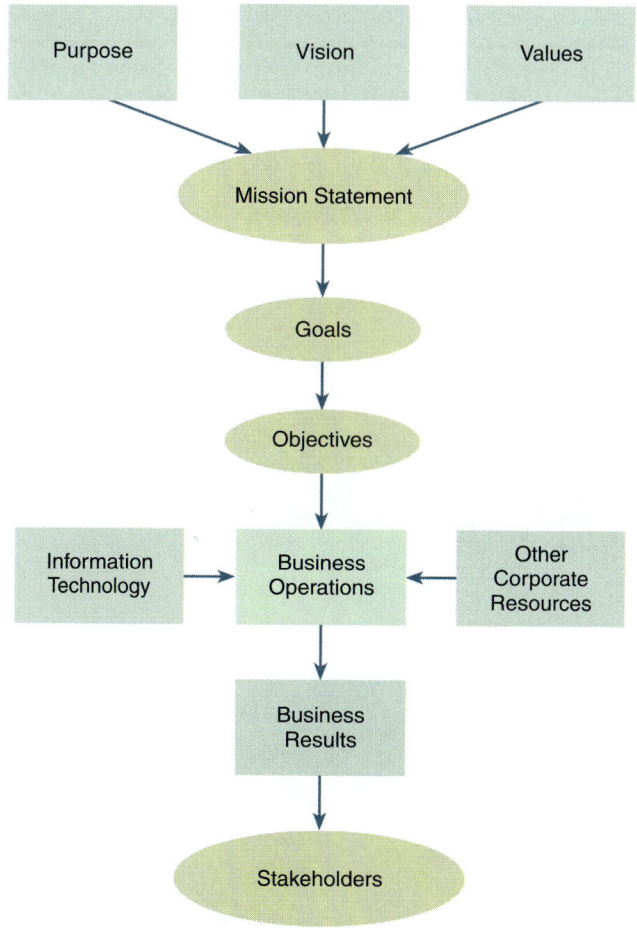

FIGURE 2-3 In the strategic planning process, a company's purpose, vision, and values shape its mission statement, which in turn leads to goals, objectives, business operations, and business results that affect company stakeholders.

FIGURE 2-4 Three examples of mission statements.

A mission statement is just the starting point. Next, the company identifies a set of **goals** that will accomplish the mission. For example, the company might establish one-year, three-year, and five-year goals for expanding market share. To achieve those goals, the company develops a list of shorter-term **objectives**. For example, if a goal is to increase Web-based orders by 30% next year, a company might set quarterly objectives with monthly milestones. Objectives also might include tactical plans, such as creating a new Web site and training a special customer support group to answer e-mail inquiries. Finally, the objectives translate into day-to-day business operations, supported by IT and other corporate resources. The outcome is a set of business results that affect company stakeholders.

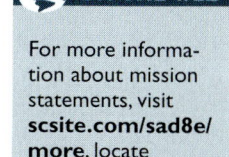

ON THE WEB

For more information about mission statements, visit **scsite.com/sad8e/more**, locate Chapter 2, and then click the Mission Statements link.

CASE IN POINT 2.1: LO CARB MEALS

Lo Carb is a successful new company that has published several cookbooks, and marketed its own line of low-carbohydrate meals. Joe Turner, Lo Carb's president, has asked your opinion. He wants to know whether a mission statement really is necessary. After you review the chapter material, write a brief memo with your views. It might be a good idea to include online references that illustrate good (or not-so-good) examples of mission statements.

A CASE Tool Example

You are a systems analyst working for Sally, the IT manager for a large hotel chain. Sally is working with top management to develop a strategic plan, and she asked you to assist her. The plan will guide future company goals and objectives, including IT projects.

Sally has experience with the Visible Analyst CASE tool, but she has never used it for strategic planning, so she asked you to do some research. You use the program's Help feature to locate the Strategic Planning – Overview section shown at the top of Figure 2-5. After reviewing that information, you navigate to the Planning Phase section, which explains how senior-level management captures and documents a business vision and a strategic business plan. You also learn that the input to this process is a series of planning statements, as shown at the bottom of Figure 2-5.

As you examine the list, you learn that planning statements can include assumptions, goals, objectives, and critical success factors, and many other types of statements. You also see that planning statements can document the strengths, weaknesses, opportunities, and threats found in a SWOT analysis.

When you present your results to Sally, she seems pleased. Because the term is new to you, you ask her what critical success factors are, and she replies that **critical success factors** are vital objectives that must be achieved for the company to fulfill its mission.

The Role of the IT Department in Project Evaluation

Management leadership and information technology are linked closely, and remarkable changes have occurred in both areas. Ten years ago, a typical IT department handled all aspects of systems development and consulted users only when, and if, the department wanted user input. Today, systems development is much more team-oriented. New approaches to systems development, such as joint application development (JAD) and rapid application development (RAD), typically involve groups of users, managers, and IT staff working together right from the start.

Although team-oriented development is the norm, some companies see the role of the IT department as a gatekeeper, responsible for screening and evaluating systems requests. Should the IT department perform the initial evaluation, or should a cross-functional team do it? The answer depends on the company's size and organization, and whether IT is tightly integrated into business operations. In smaller companies or firms where only one person has IT skills, that person acts as a coordinator and consults closely with users and managers to evaluate systems requests. Larger firms are more likely to use an evaluation team or systems review committee.

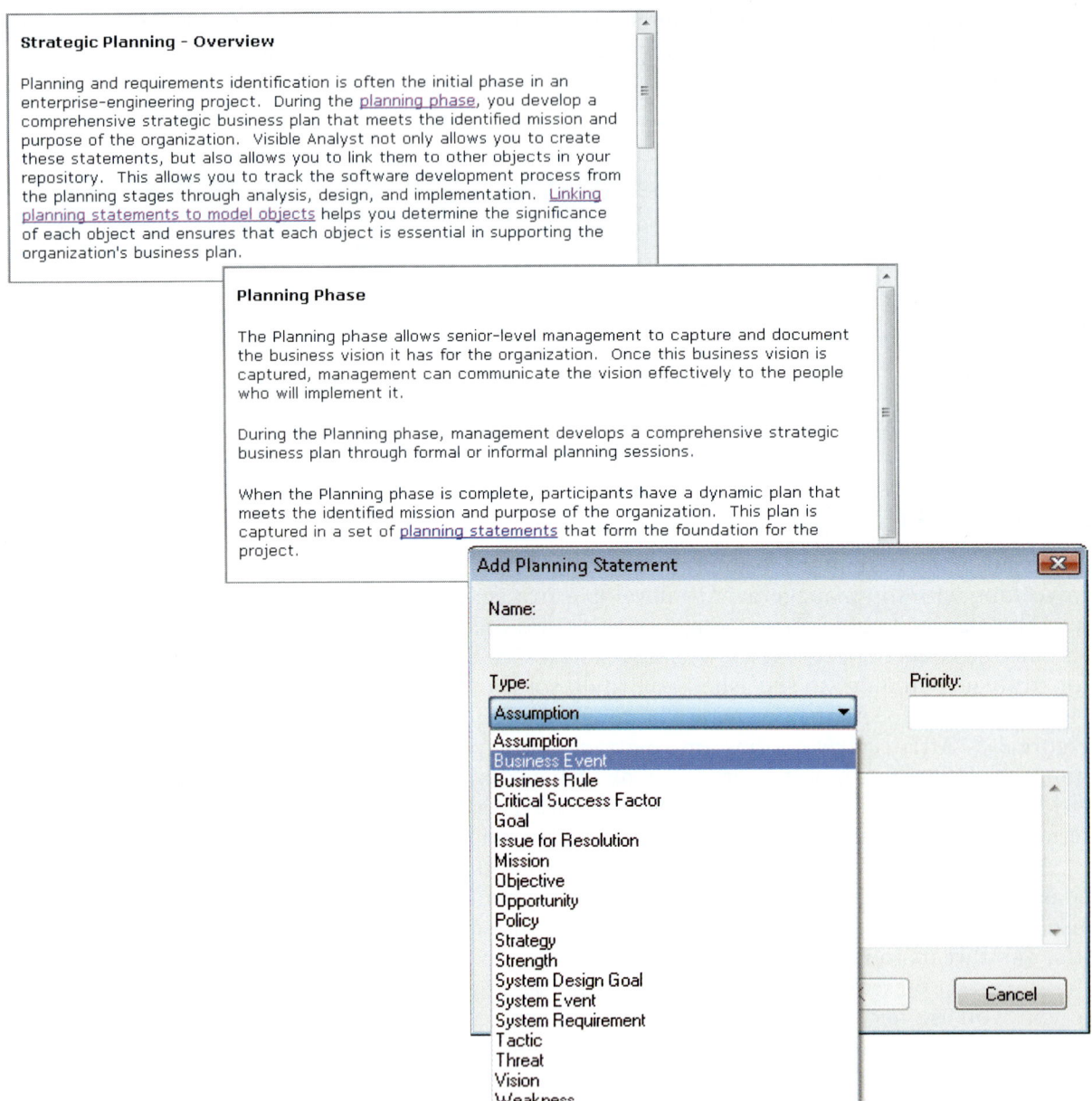

FIGURE 2-5 The Visible Analyst CASE tool can be used to develop and document a company's strategic business plan.

The Future

If you could look into the future, here is what you might see: new industries, products, and services emerging from amazing advances in information technology, customers who expect world-class IT support, a surge in Internet-based commerce, and a global business environment that is dynamic and incredibly challenging. To some firms, these changes will be threatening; other companies will see opportunities and take advantage of them by creating and following a strategic plan.

CASE IN POINT 2.2: ATTAWAY AIRLINES, PART ONE

You are the IT director at Attaway Airlines, a small regional air carrier. You chair the company's systems review committee, and you currently are dealing with strong disagreements about two key projects. Dan Esposito, the marketing manager, says it is vital to have a new computerized reservation system that can provide better customer service and reduce operational costs. Molly Kinnon, vice president of finance, is equally adamant that a new accounting system is needed immediately, because it will be very expensive to adjust the current system to new federal reporting requirements. Molly outranks Dan, and she is your boss. The next meeting, which promises to be a real showdown, is set for 9:00 a.m. tomorrow. How will you prepare for the meeting? What questions and issues should be discussed?

WHAT IS A BUSINESS CASE?

As mentioned earlier, the term *business case* refers to the reasons, or justification, for a proposal. A business case should be comprehensive, yet easy to understand. It should describe the project clearly, provide the justification to proceed, and estimate the project's financial impact. ProSci's BPR Online Learning Center, as shown in Figure 2-6, offers a Business Case Tutorial Series. According to ProSci, the business case should answer questions such as the following:

- Why are we doing this project?
- What is the project about?
- How does this solution address key business issues?
- How much will it cost and how long will it take?
- Will we suffer a productivity loss during the transition?
- What is the return on investment and payback period?
- What are the risks of doing the project? What are the risks of *not* doing the project?
- How will we measure success?
- What alternatives do we have?

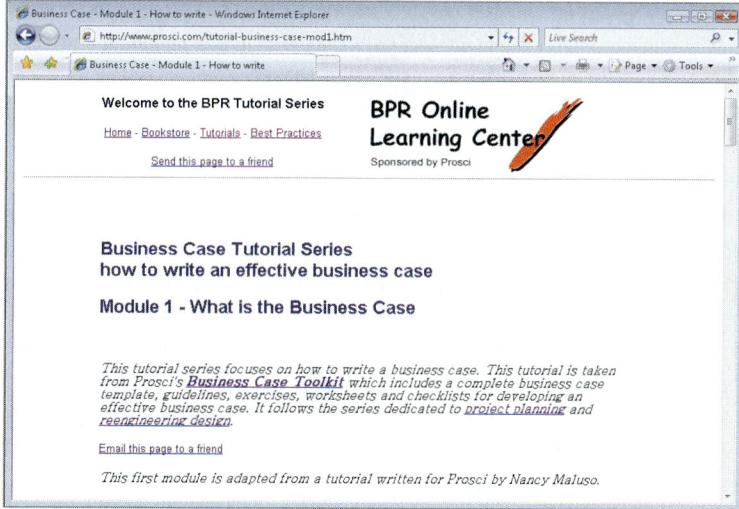

FIGURE 2-6 ProSci's BPR Online Learning Center offers a Business Case Tutorial Series that focuses on how to write a business case.

INFORMATION SYSTEMS PROJECTS

This section discusses reasons for systems projects and the internal and external factors that affect systems projects. The section also includes a preview of project management, which is discussed in detail in Chapter 3.

Main Reasons for Systems Projects

The starting point for most projects is called a **systems request,** which is a formal way of asking for IT support. A systems request might propose enhancements for an existing system, the correction of problems, the replacement of an older system, or the development of an entirely new information system that is needed to support a company's current and future business needs.

As Figure 2-7 shows, the main reasons for systems requests are improved service to customers, better performance, support for new products and services, more information, stronger controls, and reduced cost.

IMPROVED SERVICE Systems requests often are aimed at improving service to customers or users within the company. Allowing mutual fund investors to check their account balances on a Web site, storing data on rental car customer preferences, or creating an online college registration system are examples that provide valuable services and increased customer satisfaction.

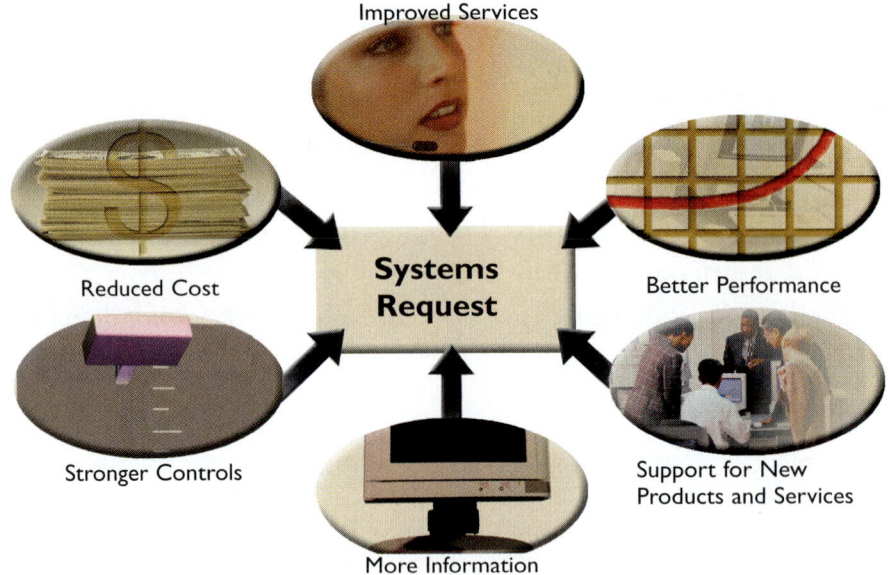

FIGURE 2-7 Six main reasons for systems requests.

SUPPORT FOR NEW PRODUCTS AND SERVICES New products and services often require new types or levels of IT support. For example, a software vendor might offer an automatic upgrade service for subscribers; or a package delivery company might add a special service for RFID-tagged shipments. In situations like these, it is most likely that additional IT support will be required. At the other end of the spectrum, product obsolescence can also be an important factor in IT planning. As new products enter the marketplace, vendors often announce that they will no longer provide support for older versions. A lack of vendor support would be an important consideration in deciding whether or not to upgrade.

BETTER PERFORMANCE The current system might not meet performance requirements. For example, it might respond slowly to data inquiries at certain times, or it might be unable to support company growth. Performance limitations also result when a system that was designed for a specific hardware configuration becomes obsolete when new hardware is introduced.

**ON THE WEB**

For more information about biometric devices, visit **scsite.com/sad8e/ more**, locate Chapter 2, and then click the Biometric Devices link.

MORE INFORMATION The system might produce information that is insufficient, incomplete, or unable to support the company's changing information needs. For example, a system that tracks customer orders might not be capable of analyzing and predicting marketing trends. In the face of intense competition and rapid product development cycles, managers need the best possible information to make major decisions on planning, designing, and marketing new products and services.

STRONGER CONTROLS A system must have effective controls to ensure that data is secure and accurate. Some common security controls include passwords, various levels of user access, and **encryption**, or coding of data to keep it safe from unauthorized users. Hardware-based security controls include **biometric devices** that can identify a person by a retina scan or by mapping a facial pattern. A new biometric tool scans hands, rather than faces. The technology uses infrared scanners that create images with thousands of measurements of hand and finger characteristics, as shown in Figure 2-8.

FIGURE 2-8 Students at West Virginia University use a hand scanning device to identify themselves.

In addition to being secure, data also must be accurate. Controls should minimize data entry errors whenever possible. For example, if a user enters an invalid customer number, the order processing system should reject the entry immediately and prompt the user to enter a valid number. Data entry controls must be effective without being excessive. If a system requires users to confirm every item with an "Are you sure? Y/N" message, internal users and customers might complain that the system is not user-friendly.

REDUCED COST The current system could be expensive to operate or maintain as a result of technical problems, design weaknesses, or the changing demands of the business. It might be possible to adapt the system to newer technology or upgrade it. On the other hand, cost-benefit analysis might show that a new system would be more cost effective and provide better support for long-term objectives.

CASE IN POINT 2.3: TRENT COLLEGE

Trent College is a private school in a small Maryland town. The college has outgrown its computerized registration system and is considering a new system. Althea Riddick, the college president, has asked you to list the reasons for systems projects, which are described on pages 56–58, and assign a relative weight to each reason, using a scale of 1 – 10, low to high. She said to use your best judgment, and support your conclusions in a brief memo to her. She also wants you to create a Microsoft Excel spreadsheet that will calculate the weighted values automatically for each reason.

Factors that Affect Systems Projects

Internal and external factors affect every business decision that a company makes, and IT systems projects are no exception. Figure 2-9 shows the main internal and external factors.

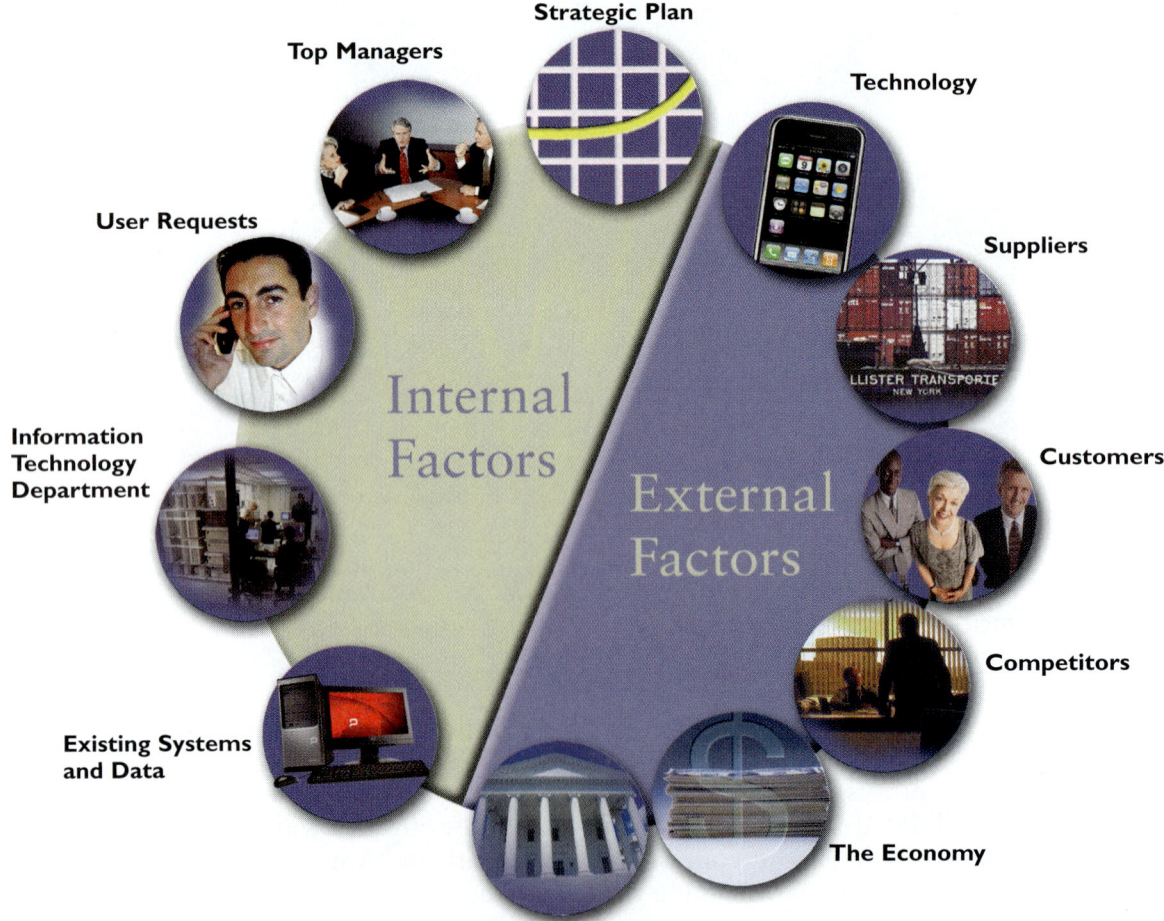

FIGURE 2-9 Internal and external factors that affect IT systems projects.

Internal Factors

Internal factors include the strategic plan, top managers, user requests, information technology department, and existing systems and data.

STRATEGIC PLAN A company's strategic plan sets the overall direction for the firm and has an important impact on IT projects. Company goals and objectives that need IT support will generate systems requests and influence IT priorities. A strategic plan that stresses technology tends to create a favorable climate for IT projects that extends throughout the organization.

TOP MANAGERS Directives from top managers are a prime source of large-scale systems projects. Those directives often result from strategic business decisions that require new IT systems, more information for decision making, or better support for mission-critical information systems.

USER REQUESTS As users rely more heavily on information systems to perform their jobs, they are likely to request even more IT services and support. For example, sales reps might request improvements to the company's Web site, a more powerful sales analysis report, a network to link all sales locations, or an online system that allows customers to obtain the status of their orders instantly. Or, users might not be satisfied with the current system because it is difficult to learn or lacks flexibility. They might want information systems support for business requirements that did not even exist when the system was developed.

INFORMATION TECHNOLOGY DEPARTMENT Many systems project requests come from the IT department. IT staff members often make recommendations based on their knowledge of business operations and technology trends. IT proposals might be strictly technical matters, such as replacement of certain network components, or suggestions might be more business oriented, such as proposing a new reporting or data collection system.

EXISTING SYSTEMS AND DATA Errors or problems in existing systems can trigger requests for systems projects. When dealing with older systems, analysts sometimes spend too much time reacting to day-to-day problems without looking at underlying causes. This approach can turn an information system into a patchwork of corrections and changes that cannot support the company's overall business needs. This problem typically occurs with legacy systems, which are older systems that are less technologically advanced. When migrating to a new system, IT planners must plan the conversion of existing data, which is described in detail in Chapter 11, Systems Implementation.

External Factors

External factors include technology, suppliers, customers, competitors, the economy, and government.

TECHNOLOGY Changing technology is a major force affecting business and society in general. For example, the rapid growth of telecommunications has created entire new industries and technologies. Technology also dramatically reshapes existing business operations. The success of scanner technology resulted in universal bar coding that now affects virtually all products.

Some industry experts predict that bar code technology will be overshadowed in the future by **electronic product code (EPC)** technology that uses RFID tags to identify and monitor the movement of each individual product, from the factory floor to the retail checkout counter.

ON THE WEB

For more information about JIT systems, visit **scsite.com/sad8e/more**, locate Chapter 2, and then click the JIT Systems link.

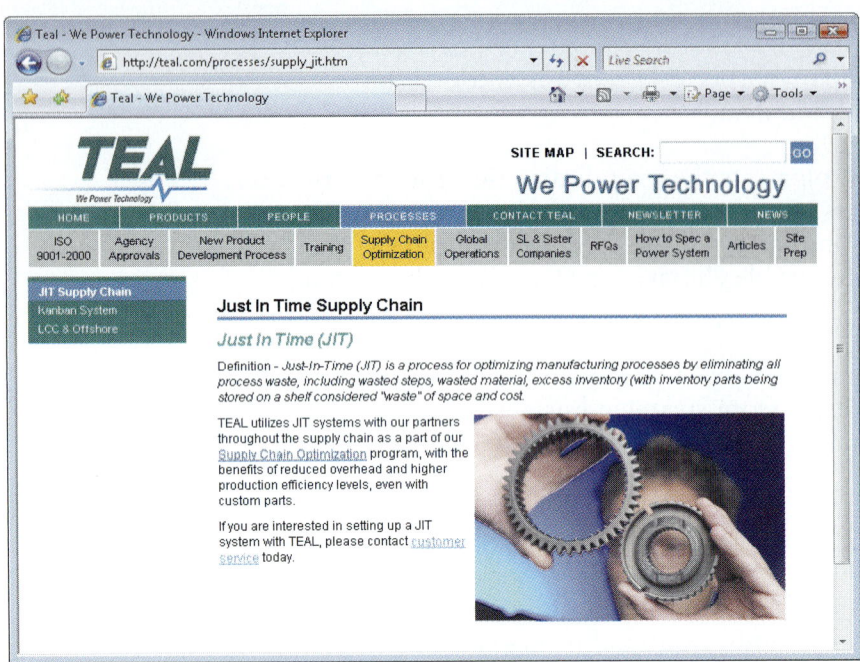

FIGURE 2-10 Just-in-time (JIT) inventory systems rely on computer-to-computer data exchange to minimize unnecessary inventory.

SUPPLIERS With the growth of electronic data interchange (EDI), relationships with suppliers are critically important. For example, an automobile company might require that suppliers code their parts in a certain manner to match the auto company's inventory control system. EDI also enables **just-in-time (JIT)** inventory systems, as shown in Figure 2-10, which rely on computer-to-computer data exchange to minimize unnecessary inventory. The purpose of a JIT system is to provide the right product at the right place at the right time.

CUSTOMERS Customers are vitally important to any business. Information systems that interact with customers usually receive top priority. Many companies implement **customer relationship management (CRM)** systems that integrate all customer-related events and transactions, including marketing, sales, and customer service activities. Vendor-oriented CRM systems often interconnect with supplier relationship management (SRM) systems, which were discussed in Chapter 1. CRM components can provide automated responses to sales inquiries, Web-based order processing, and online inventory tracking. Because an efficient warehouse is just as important as a successful Web site, suppliers use *smart* forklifts that can read RFID tags or UPC numbers and transmit data to a CRM system, as shown in Figure 2-11.

One of the newest RFID applications is called **electronic proof of delivery (EPOD)**. Using EPOD, a supplier uses RFID tags on each crate, case, or shipping unit to create a digital shipping list. The customer receives the list and scans the incoming shipment. If a discrepancy is detected, it is reported and adjusted automatically. Because they would be expensive to investigate manually, small shipping inconsistencies might not otherwise be traced. This is an example of technology-related cost control.

COMPETITORS Competition drives many information systems decisions. For example, if one cellular telephone provider offers a new type of digital service, other firms must match the plan in order to remain competitive. New product research and development, marketing, sales, and service all require IT support.

THE ECONOMY Economic activity has a powerful influence on corporate information management. In a period of economic expansion, firms need to be ready with scalable systems that can handle additional volume and growth. Predicting the business cycle is not an exact science, and careful research and planning is critically important.

GOVERNMENT Federal, state, and local government regulations affect the design of corporate information systems. For example, income tax reporting requirements must be designed into a payroll

ON THE WEB

For more information about CRM systems, visit **scsite.com/sad8e/more**, locate Chapter 2, and then click the CRM Systems link.

FIGURE 2-11 In an efficient warehouse, *smart* forklifts can read RFID tags or UPC numbers and transmit data to a CRM system.

package. The debate about Internet sales tax issues could profoundly affect e-commerce, as well as traditional retail businesses.

Project Management

As mentioned earlier, business case analysis involves consideration of project reasons, costs, benefits, and risks. At the end of the preliminary investigation, if the project is approved, it can be planned, scheduled, monitored and controlled, and reported upon. Individual analysts or IT staff members often handle small projects, but companies usually designate a project manager to coordinate the overall effort for complex projects.

In Chapter 3, you will study project management concepts, skills, tools, and techniques. You also will learn about project risk management, and how to perform the following tasks:

- Develop a project risk management plan
- Identify the risks
- Analyze the risks
- Create a risk response plan
- Monitor and respond to risks

Microsoft Project is a popular project management tool, as shown in Figure 2-12. Using this program, a project manager can define project tasks, list activities and participants, plan the sequence of work, estimate project milestone dates, and track costs.

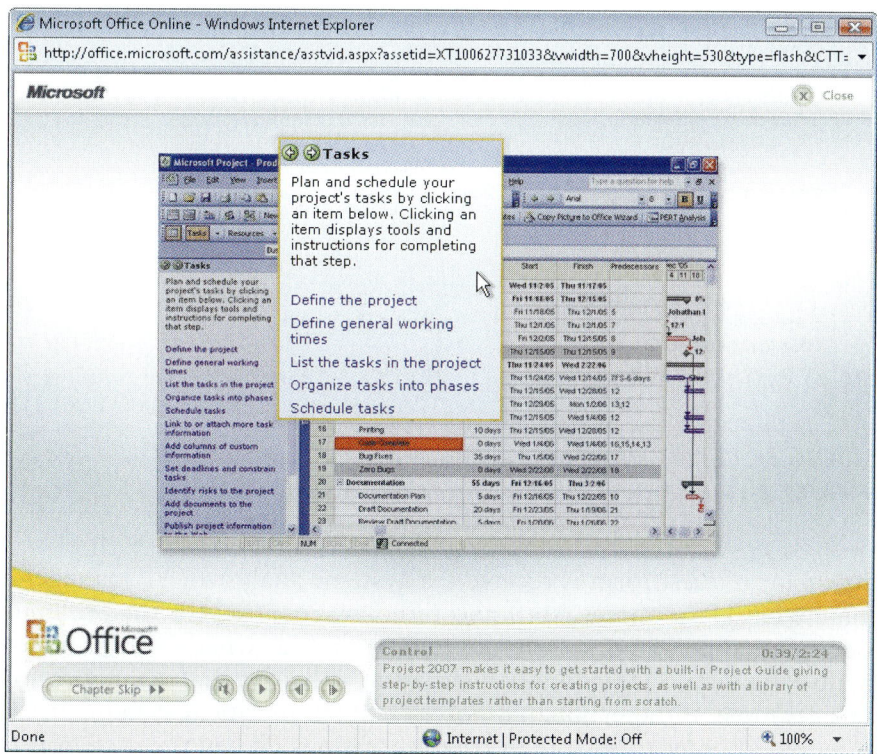

FIGURE 2-12 Microsoft has a Web site with demo versions, training, and tips for working with Microsoft Project software.

EVALUATION OF SYSTEMS REQUESTS

In most organizations, the IT department receives more systems requests than it can handle. Many organizations assign responsibility for evaluating systems requests to a group of key managers and users. Many companies call this group a **systems review committee** or a **computer resources committee**. Regardless of the name, the objective is to use the combined judgment and experience of several managers to evaluate systems projects.

Systems Request Forms

Many organizations use a special form for systems requests, similar to the online sample shown in Figure 2-13. A properly designed form streamlines the request process and ensures consistency. The form must be easy to understand and include clear instructions. It should include enough space for all required information and should indicate what supporting documents are needed. Many companies use online systems request forms that can be filled in and submitted electronically.

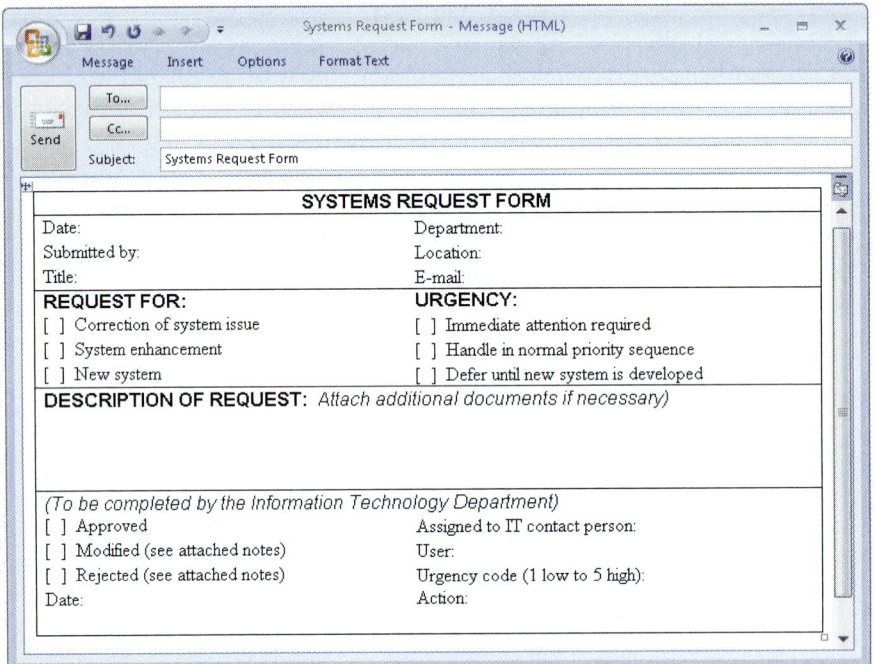

FIGURE 2-13 Example of an online systems request form.

When a systems request form is received, a systems analyst or IT manager examines it to determine what IT resources (staff and time) are required for the preliminary investigation. A designated person or a committee then decides whether to proceed with a preliminary investigation. Occasionally a situation will arise that requires an immediate response. For example, if the problem involves a mission-critical system, an IT maintenance team would attempt to restore normal operations. When the system is functioning properly, the team conducts a review and prepares a systems request to cover the work that was performed.

Systems Review Committee

Most large companies use a systems review committee to evaluate systems requests. Instead of relying on a single individual, a committee approach provides a variety of experience and knowledge. With a broader viewpoint, a committee can establish priorities more effectively than an individual, and one person's bias is less likely to affect the decisions. A typical committee consists of the IT director and several managers from other departments. The IT director usually serves as a technical consultant to ensure that committee members are aware of crucial issues, problems, and opportunities.

Although a committee offers many advantages, some disadvantages exist. For example, action on requests must wait until the committee meets. To avoid delay, committee members typically use e-mail and teleconferencing to communicate. Another potential disadvantage of a committee is that members might favor projects requested by their own departments, and internal political differences could delay important decisions.

Many smaller companies rely on one person to evaluate system requests instead of a committee. If only one person has the necessary IT skills and experience, that person must consult closely with users and managers throughout the company to ensure that business and operational needs are considered carefully.

Whether one person or a committee is responsible, the goal is to evaluate the requests and set priorities. Suppose four requests must be reviewed: a request from the marketing group to analyze current customer spending habits and forecast future trends; a request from the technical support group for a cellular link so service representatives can download technical data instantly; a request from the accounting department to redesign customer statements and allow Internet access; and a request from the production staff for an inventory control system that can exchange data with major suppliers. Which of those projects should the firm pursue? What criteria should be applied? How should priorities be determined? To answer those questions, the individual or the committee must assess the feasibility of each systems request.

OVERVIEW OF FEASIBILITY

As you learned in Chapter 1, a systems request must pass several tests, called a feasibility study, to see whether it is worthwhile to proceed further. As shown in Figure 2-14, a feasibility study uses four main yardsticks to measure a proposal: operational feasibility, technical feasibility, economic feasibility, and schedule feasibility.

Sometimes a feasibility study is quite simple and can be done in a few hours. If the request involves a new system or a major change, however, extensive fact-finding and investigation is required.

How much effort needs to go into a feasibility study? That depends on the request. For example, if a department wants an existing report sorted in a different order, the analyst can decide quickly whether the request is feasible. On the other hand, a proposal by the marketing department for a new market research system to predict sales trends requires more effort. In both cases, the systems analyst asks these important questions:

- Is the proposal desirable in an operational sense? Is it a practical approach that will solve a problem or take advantage of an opportunity to achieve company goals?

- Is the proposal technically feasible? Are the necessary technical resources and people available for the project?

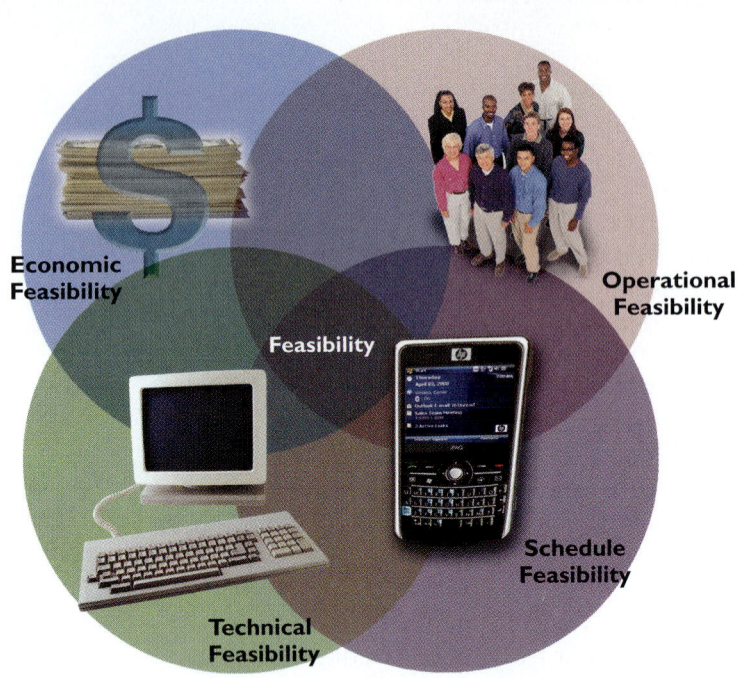

FIGURE 2-14 A feasibility study includes tests for operational, technical, economic, and schedule feasibility.

- Is the proposal economically desirable? What are the projected savings and costs? Are other intangible factors involved, such as customer satisfaction or company image? Is the problem worth solving, and will the request result in a sound business investment?

- Can the proposal be accomplished within an acceptable time frame?

To obtain more information about a systems request, you might perform initial fact-finding by studying organization charts, performing interviews, reviewing current documentation, observing operations, and surveying users. If the systems request is approved, more intensive fact-finding will continue during the systems analysis phase.

Operational Feasibility

Operational feasibility means that a proposed system will be used effectively after it has been developed. If users have difficulty with a new system, it will not produce the expected benefits. Operational feasibility depends on several vital issues. For example, consider the following questions:

- Does management support the project? Do users support the project? Is the current system well liked and effectively used? Do users see the need for change?

- Will the new system result in a workforce reduction? If so, what will happen to affected employees?

- Will the new system require training for users? If so, is the company prepared to provide the necessary resources for training current employees?

- Will users be involved in planning the new system right from the start?

- Will the new system place any new demands on users or require any operating changes? For example, will any information be less accessible or produced less frequently? Will performance decline in any way? If so, will an overall gain to the organization outweigh individual losses?

- Will customers experience adverse effects in any way, either temporarily or permanently?

- Will any risk to the company's image or goodwill result?

- Does the development schedule conflict with other company priorities?

- Do legal or ethical issues need to be considered?

Technical Feasibility

Technical feasibility refers to the technical resources needed to develop, purchase, install, or operate the system. When assessing technical feasibility, an analyst must consider the following points:

- Does the company have the necessary hardware, software, and network resources? If not, can those resources be acquired without difficulty?

- Does the company have the needed technical expertise? If not, can it be acquired?

- Does the proposed platform have sufficient capacity for future needs? If not, can it be expanded?

- Will a prototype be required?

- Will the hardware and software environment be reliable? Will it integrate with other company information systems, both now and in the future? Will it interface properly with external systems operated by customers and suppliers?

- Will the combination of hardware and software supply adequate performance? Do clear expectations and performance specifications exist?

- Will the system be able to handle future transaction volume and company growth?

Economic Feasibility

Economic feasibility means that the projected benefits of the proposed system outweigh the estimated costs usually considered the **total cost of ownership** (**TCO**), which includes ongoing support and maintenance costs, as well as acquisition costs. To determine TCO, the analyst must estimate costs in each of the following areas:

- People, including IT staff and users

- Hardware and equipment

- Software, including in-house development as well as purchases from vendors

- Formal and informal training

- Licenses and fees

- Consulting expenses

- Facility costs

- The estimated cost of not developing the system or postponing the project

In addition to costs, you need to assess tangible and intangible benefits to the company. The systems review committee will use those figures, along with your cost estimates, to decide whether to pursue the project beyond the preliminary investigation phase.

Tangible benefits are benefits that can be measured in dollars. Tangible benefits result from a decrease in expenses, an increase in revenues, or both. Examples of tangible benefits include the following:

- A new scheduling system that reduces overtime

- An online package tracking system that improves service and decreases the need for clerical staff

- A sophisticated inventory control system that cuts excess inventory and eliminates production delays

Intangible benefits are advantages that are difficult to measure in dollars but are important to the company. Examples of intangible benefits include the following:

- A user-friendly system that improves employee job satisfaction

- A sales tracking system that supplies better information for marketing decisions

- A new Web site that enhances the company's image

You also must consider the development timetable, because some benefits might occur as soon as the system is operational, but others might not take place until later.

ON THE WEB

For more information about TCO, visit **scsite.com/sad8e/more**, locate Chapter 2, and then click the TCO link.

TOOLKIT TIME

The Financial Analysis tools in Part 3 of the Systems Analyst's Toolkit can help you analyze project costs, benefits, and economic feasibility. To learn more about these tools, turn to Part 3 of the four-part Toolkit that follows Chapter 12.

Schedule Feasibility

Schedule feasibility means that a project can be implemented in an acceptable time frame. When assessing schedule feasibility, a systems analyst must consider the interaction between time and costs. For example, speeding up a project schedule might make a project feasible, but much more expensive.

Other issues that relate to schedule feasibility include the following:

- Can the company or the IT team control the factors that affect schedule feasibility?
- Has management established a firm timetable for the project?
- What conditions must be satisfied during the development of the system?
- Will an accelerated schedule pose any risks? If so, are the risks acceptable?
- Will project management techniques be available to coordinate and control the project?
- Will a project manager be appointed?

Chapter 3 describes various project management tools and techniques.

EVALUATING FEASIBILITY

The first step in evaluating feasibility is to identify and weed out systems requests that are not feasible. For example, a request would not be feasible if it required hardware or software that the company already had rejected.

Even if the request is feasible, it might not be necessary. For example, a request for multiple versions of a report could require considerable design and programming effort. A better alternative might be to download the server data to a personal computer-based software package and show users how to produce their own reports. In this case, training users would be a better investment than producing reports for them.

Also keep in mind that systems requests that are not currently feasible can be resubmitted as new hardware, software, or expertise becomes available. Development costs might decrease, or the value of benefits might increase enough that a systems request eventually becomes feasible. Conversely, an initially feasible project can be rejected later. As the project progresses, conditions often change. Acquisition costs might increase, and the project might become more expensive than anticipated. In addition, managers and users sometimes lose confidence in a project. For all those reasons, feasibility analysis is an ongoing task that must be performed throughout the systems development process.

SETTING PRIORITIES

After rejecting systems requests that are not feasible, the systems review committee must establish priorities for the remaining items. The highest priority goes to projects that provide the greatest benefit, at the lowest cost, in the shortest period of time. Many factors, however, influence project evaluation.

Factors that Affect Priority

When assessing a project's priority, a systems analyst should consider the following:

- Will the proposed system reduce costs? Where? When? How? How much?
- Will the system increase revenue for the company? Where? When? How? How much?
- Will the systems project result in more information or produce better results? How? Are the results measurable?
- Will the system serve customers better?
- Will the system serve the organization better?
- Can the project be implemented in a reasonable time period? How long will the results last?
- Are the necessary financial, human, and technical resources available?

Very few projects will score high in all areas. Some proposed systems might not reduce costs but will provide important new features. Other systems might reduce operating costs substantially but require the purchase or lease of additional hardware. Some systems might be very desirable but require several years of development before producing significant benefits.

Whenever possible, the analyst should evaluate a proposed project based on tangible costs and benefits that represent actual (or approximate) dollar values. For example, a reduction of $8,000 in network maintenance is an example of a tangible benefit.

Often, the evaluation involves intangible costs or benefits, as described in the section on economic feasibility. In contrast to tangible benefits, such as the network cost reduction example, it is more difficult to assign dollar values to intangible benefits such as enhancing the organization's image, raising employee morale, or improving customer service. Intangible costs and benefits often influence systems decisions and priorities and must be considered carefully.

Discretionary and Nondiscretionary Projects

Is the project absolutely necessary? Projects where management has a choice in implementing them are called **discretionary projects**. Projects where no choice exists are called **nondiscretionary projects**. Creating a new report for a user is an example of a discretionary project; adding a report required by a new federal law is an example of a nondiscretionary project.

If a particular project is not discretionary, is it really necessary for the systems review committee to evaluate it? Some people believe that waiting for committee approval delays critical nondiscretionary projects unnecessarily. Others believe that by submitting all systems requests to the systems review committee, the committee is kept aware of all projects that compete for the resources of the IT department. As a result, the committee assesses the priority of discretionary projects and can schedule them more realistically. Additionally, the committee might need to prioritize nondiscretionary projects when funds or staff are limited.

Many nondiscretionary projects are predictable. Examples include annual updates to payroll, tax percentages, or quarterly changes in reporting requirements for an insurance processing system. By planning ahead for predictable projects, the IT department manages its resources better and keeps the systems review committee fully informed without needing prior approval in every case.

CASE IN POINT 2.4:   ATTAWAY AIRLINES, PART TWO

Back at Attaway Airlines, the morning meeting ended with no agreement between Dan Esposito and Molly Kinnon. In fact, a new issue arose. Molly now says that the new accounting system is entitled to the highest priority because the federal government soon will require the reporting of certain types of company-paid health insurance premiums. Because the current system will not handle this report, she insists that the entire accounting system is a nondiscretionary project. As you might expect, Dan is upset. Can part of a project be nondiscretionary? What issues need to be discussed? The committee meets again tomorrow, and the members will look to you, as the IT director, for guidance.

PRELIMINARY INVESTIGATION OVERVIEW

A systems analyst conducts a **preliminary investigation** to study the systems request and recommend specific action. After obtaining an authorization to proceed, the analyst interacts with managers and users, as shown in the model in Figure 2-15. The analyst gathers facts about the problem or opportunity, project scope and constraints, project benefits, and estimated development time and costs. The end product of the preliminary investigation is a report to management.

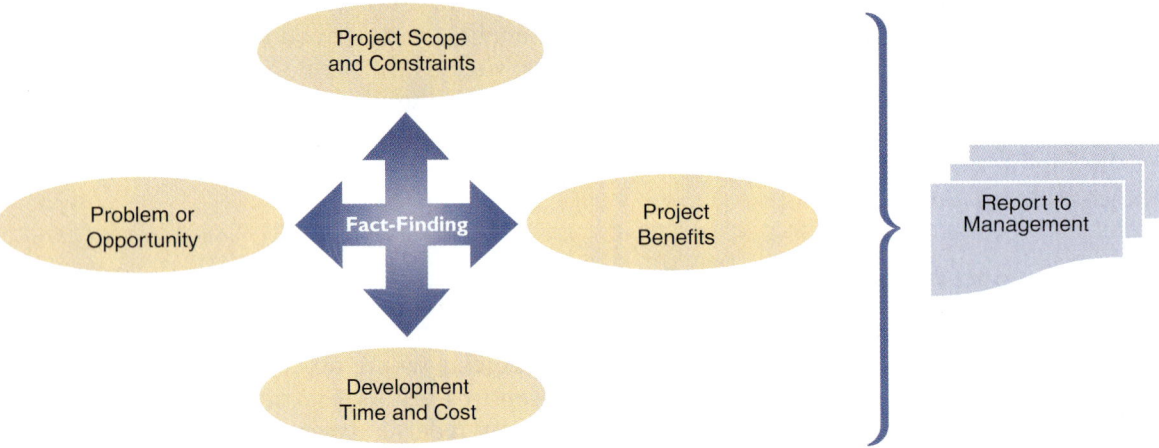

FIGURE 2-15   Model of a preliminary investigation.

Interaction with Managers and Users

Before beginning a preliminary investigation, a memo or an e-mail message should let people know about the investigation and explain your role. You should meet with key managers, users, and IT staff to describe the project, explain your responsibilities, answer questions, and invite comments. This starts an important dialogue with users that will continue throughout the entire development process.

A systems project often produces significant changes in company operations. Employees may be curious, concerned, or even opposed to those changes. It is not surprising to encounter some user resistance during a preliminary investigation. Employee attitudes and reactions are important and must be considered.

When interacting with users, you should be careful in your use of the word *problem*, because generally it has a negative meaning. When you ask users about *problems*, some will stress current system limitations rather than desirable new features or enhancements. Instead of focusing on difficulties, you should question users about additional capability they would like to have. Using this approach, you highlight ways to improve the user's job, you get a better understanding of operations, and you build better, more positive relationships with users.

Planning the Preliminary Investigation

During a preliminary investigation, a systems analyst typically follows a series of steps, as shown in Figure 2-16. The exact procedure depends on the nature of the request, the size of the project, and the degree of urgency.

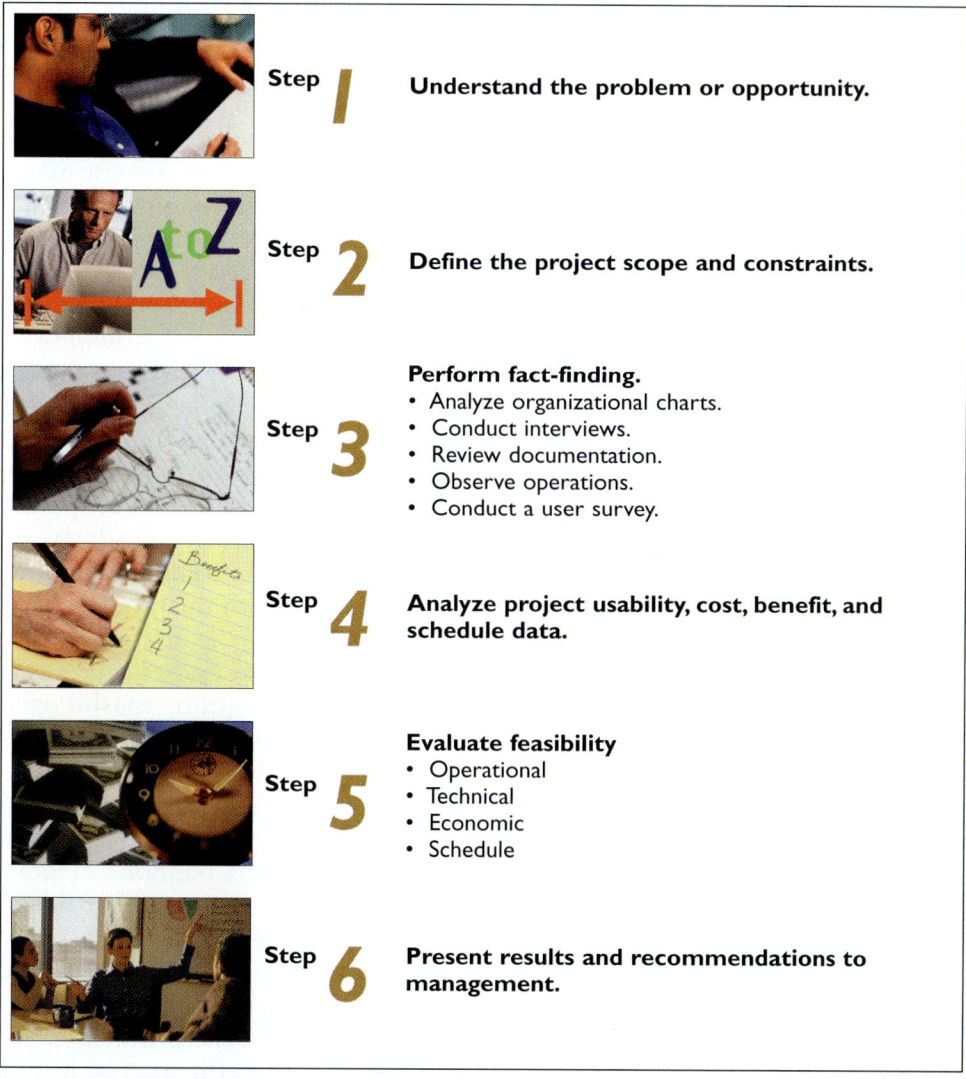

Step 1 Understand the problem or opportunity.

Step 2 Define the project scope and constraints.

Step 3 Perform fact-finding.
- Analyze organizational charts.
- Conduct interviews.
- Review documentation.
- Observe operations.
- Conduct a user survey.

Step 4 Analyze project usability, cost, benefit, and schedule data.

Step 5 Evaluate feasibility
- Operational
- Technical
- Economic
- Schedule

Step 6 Present results and recommendations to management.

FIGURE 2-16 Six steps in a preliminary investigation.

Figure 2-17 on the next page shows how a systems analyst might use Microsoft Project to plan and manage the preliminary investigation. Notice that the analyst has listed the tasks, estimated the duration of each task, and designated a specific order in which the tasks must be performed.

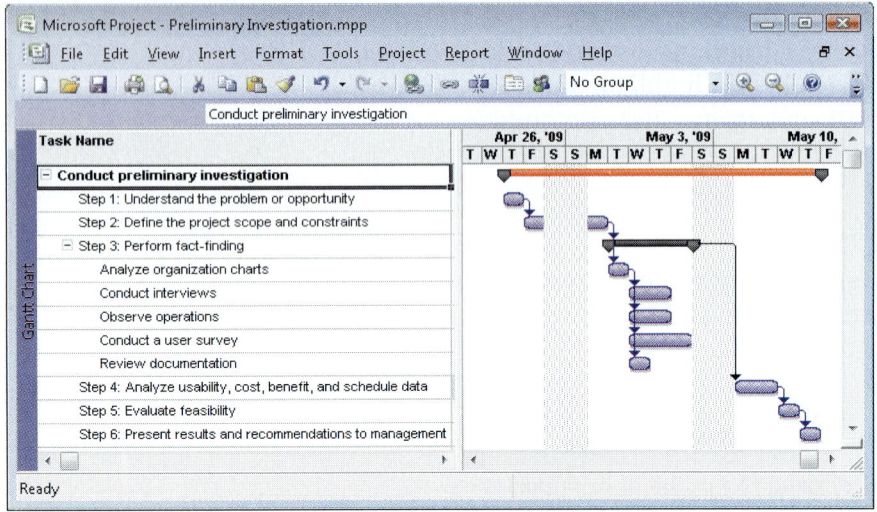

FIGURE 2-17 A systems analyst can use Microsoft Project to plan, schedule, monitor, and report on a preliminary investigation.

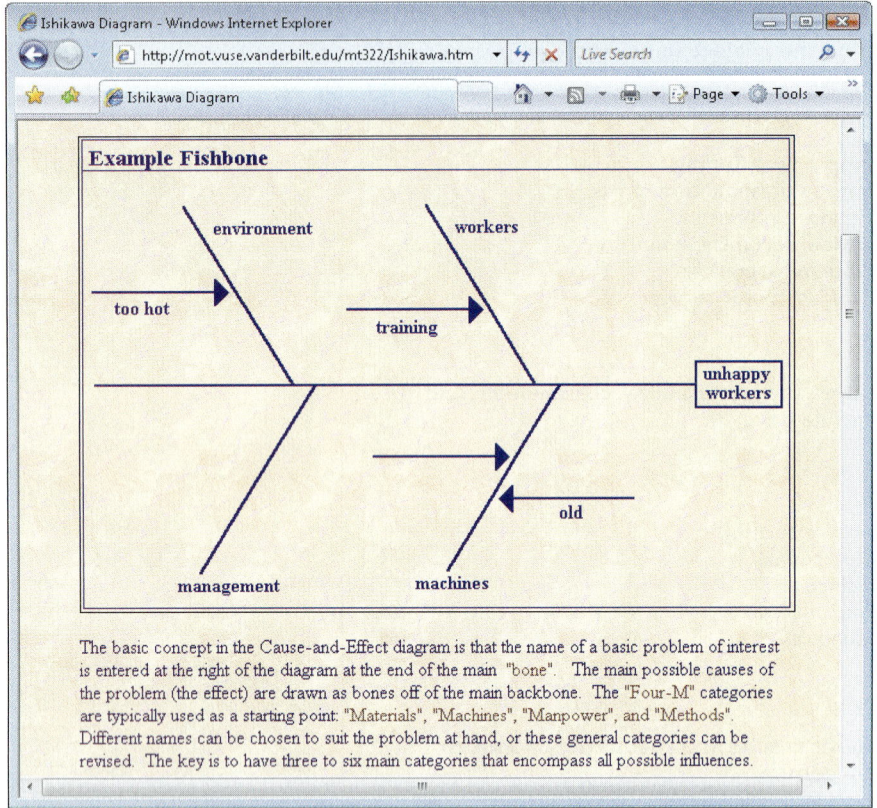

FIGURE 2-18 A fishbone diagram is an analysis tool that represents the possible causes of a problem as a graphical outline.

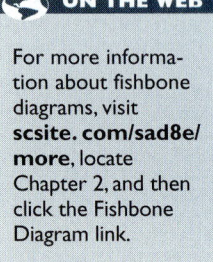

ON THE WEB

For more information about fishbone diagrams, visit **scsite. com/sad8e/ more**, locate Chapter 2, and then click the Fishbone Diagram link.

Step 1: Understand the Problem or Opportunity

If the systems request involves a new information system or a substantial change in an existing system, systems analysts might need to develop a business profile that describes business processes and functions, as explained in Chapter 1. Even where the request involves relatively minor changes or enhancements, you need to understand how those modifications will affect business operations and other information systems. Often a change in one system has an unexpected effect on another system. When you analyze a systems request, you need to determine which departments, users, and business processes are involved.

In many cases, the systems request does not reveal the underlying problem, but only a symptom. For example, a request to investigate mainframe processing delays might reveal improper scheduling practices rather than hardware problems. Similarly, a request for analysis of customer complaints might disclose a lack of sales representative training, rather than problems with the product.

A popular technique for investigating causes and effects is called a **fishbone diagram**, or **Ishikawa diagram**, as shown in Figure 2-18. A fishbone diagram is an analysis tool that represents the possible causes of a problem as a graphical outline. When using a fishbone diagram, an analyst first states the problem and draws a main bone with sub-bones that represent possible causes of the problem. In the example shown in Figure 2-18, the problem is *unhappy workers*, and the analyst has identified four areas to investigate: *environment, workers, management,* and *machines.* In each area, the analyst identifies possible causes and draws them as horizontal sub-bones. For example, *too hot* is a possible cause in the *environment* bone. For each cause, the analyst must dig deeper and ask the question: What could be causing *this* symptom to occur? For example, *why* is it too hot? If the answer is insufficient air conditioning capacity, the analyst indicates this as a sub-bone to the *too hot* cause. In this manner, the analyst adds additional sub-bones to the diagram, until he or she uncovers root causes of a problem, rather than just the symptoms.

The **Pareto chart** is another widely used tool for visualizing and prioritizing issues that need attention. Named for a nineteenth century economist, a Pareto chart is drawn as a vertical bar graph, as shown in Figure 2-19. The bars, which represent various causes of a problem, are arranged in descending order, so the team can focus on the most important causes. In the example shown, a systems analyst might use a Pareto chart to learn more about the causes of inventory system problems, so that necessary improvements can be made. Creating Pareto charts with Excel is a simple process.

FIGURE 2-19   A Pareto chart can display and prioritize causes of a problem.

Step 2: Define the Project Scope and Constraints

Determining the **project scope** means defining the specific boundaries, or extent, of the project. For example, a statement that, *payroll is not being produced accurately* is very general, compared with the statement *overtime pay is not being calculated correctly for production workers on the second shift at the Yorktown plant*. Similarly, the statement, *the project scope is to modify the accounts receivable system*, is not as specific as the statement, *the project scope is to allow customers to inquire online about account balances and recent transactions*.

Some analysts find it helpful to define project scope by creating a list with sections called *Must Do, Should Do, Could Do,* and *Won't Do*. This list can be reviewed later, during the systems analysis phase, when the systems requirements document is developed.

Projects with very general scope definitions are at risk of expanding gradually, without specific authorization, in a process called **project creep**. To avoid this problem, you should define project scope as clearly as possible. You might want to use a graphical model that shows the systems, people, and business processes that will be affected. The scope of the project also establishes the boundaries of the preliminary investigation itself. A systems analyst should limit the focus to the problem at hand and avoid unnecessary expenditure of time and money.

Along with defining the scope of the project, you need to identify any constraints on the system. A **constraint** is a requirement or condition that the system must satisfy or an outcome that the system must achieve. A constraint can involve hardware, software, time, policy, law, or cost. System constraints also define project scope. For example, if the system must operate with existing hardware, that is a constraint that affects potential solutions. Other examples of constraints are: The order entry system must accept input from 15 remote sites; the human resources information system must produce statistics on hiring practices; and the new Web site must be operational by March 1. When examining constraints, you should identify their characteristics.

PRESENT VERSUS FUTURE Is the constraint something that must be met as soon as the system is developed or modified, or is the constraint necessary at some future time?

INTERNAL VERSUS EXTERNAL Is the constraint due to a requirement within the organization or does some external force, such as government regulation, impose it?

MANDATORY VERSUS DESIRABLE Is the constraint mandatory? Is it absolutely essential to meet the constraint, or is it merely desirable?

Figure 2-20 shows five examples of constraints. Notice that each constraint has three characteristics, which are indicated by its position in the figure and by the symbol that represents the constraint. The constraint in Example A is present, external, and mandatory. The constraint in Example B is future, external, and mandatory. The constraint in Example C is present, internal, and desirable. The constraint in Example D is present, internal, and mandatory. The constraint in Example E is future, internal, and desirable.

Regardless of the type, all constraints should be identified as early as possible to avoid future problems and surprises. A clear definition of project scope and constraints avoids misunderstandings that arise when managers assume that the system will have a certain feature or support for a project, but later find that the feature is not included.

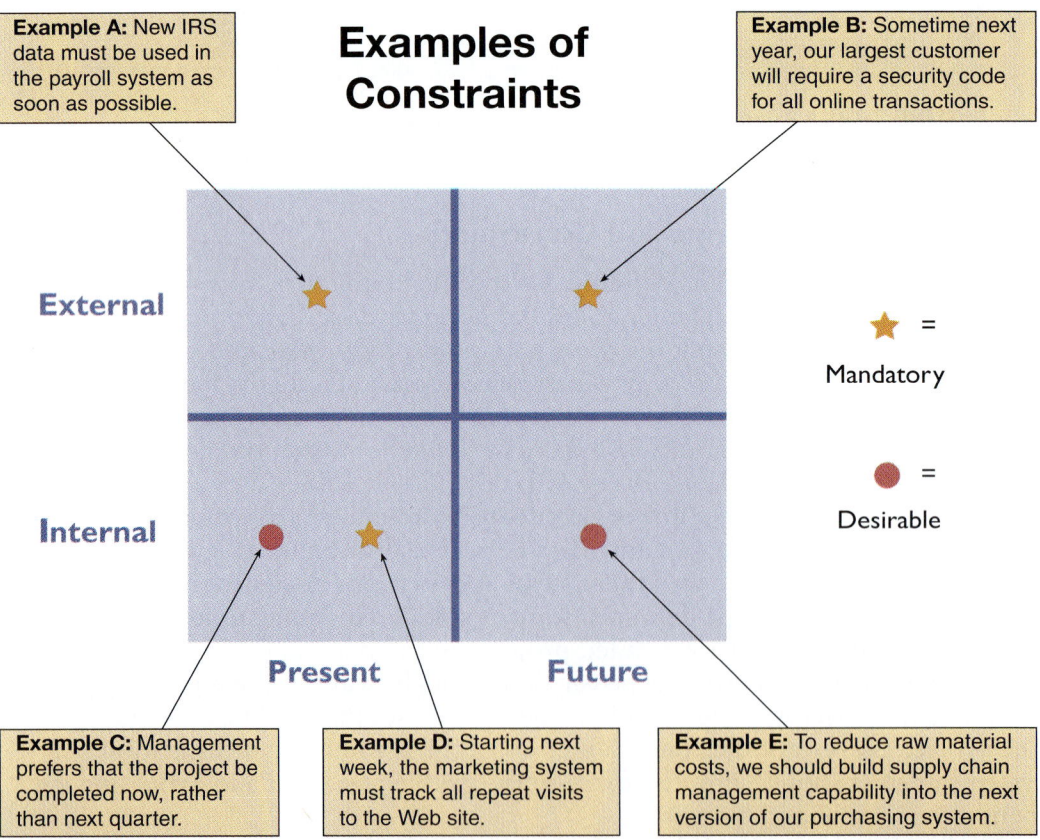

FIGURE 2-20 Examples of various types of constraints. The constraint in Example A is present, external, and mandatory. The constraint in Example B is future, external, and mandatory. The constraint in Example C is present, internal, and desirable. The constraint in Example D is present, internal, and mandatory. The constraint in Example E is future, internal, and desirable.

Step 3: Perform Fact-Finding

The objective of fact-finding is to gather data about project usability, costs, benefits, and schedules. Fact-finding involves various techniques, which are described below. Depending on what information is needed to investigate the systems request, fact-finding might consume several hours, days, or weeks. For example, a change in a report format or data entry screen might require a single telephone call or e-mail message to a user, whereas a new inventory system would involve a series of interviews. During fact-finding, you might analyze organization charts, conduct interviews, review current documentation, observe operations, and carry out a user survey.

ANALYZE ORGANIZATION CHARTS In many instances, you will not know the organizational structure of departments involved in the study. You should obtain organization charts to understand how the department functions and identify individuals you might want to interview. Organization charts often can be obtained from the company's human resources department. If such charts are unavailable, you should obtain the necessary information directly from department personnel and then construct your own charts, as shown in Figure 2-21.

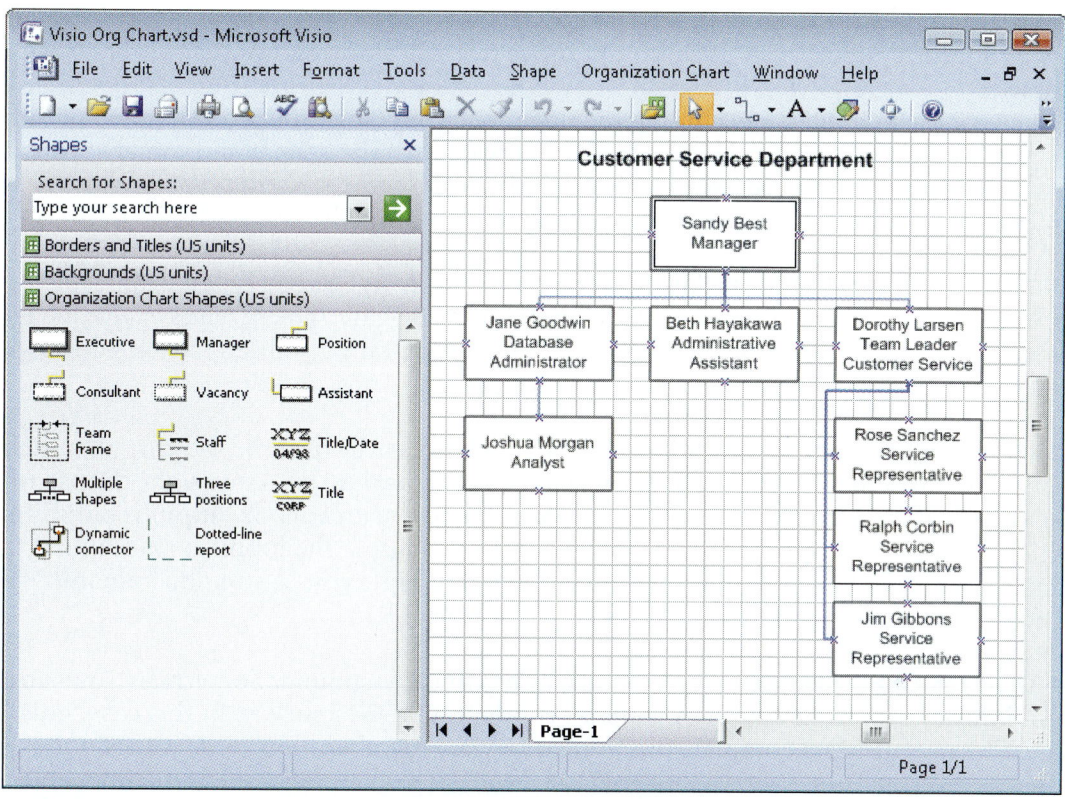

FIGURE 2-21 Microsoft Visio includes an organization chart drawing tool that is powerful and easy to use.

When organization charts are available, you should verify their accuracy. Keep in mind that organization charts show formal reporting relationships but not the informal alignment of a group, which also is important.

CONDUCT INTERVIEWS The primary method of obtaining information during the preliminary investigation is the interview, as shown in Figure 2-22. The interviewing process involves a series of steps:

1. Determine the people to interview.
2. Establish objectives for the interview.

FIGURE 2-22 The interview is the primary method of obtaining information.

3. Develop interview questions.

4. Prepare for the interview.

5. Conduct the interview.

6. Document the interview.

7. Evaluate the interview.

These seven steps are discussed in detail in Chapter 4, which describes fact-finding techniques that occur during the systems analysis phase of the SDLC.

Remember that the purpose of the interview, and of the preliminary investigation itself, is to uncover facts, not to convince others that the project is justified. Your primary role in an interview is to ask effective questions and listen carefully. If you plan to talk to several people about the same topic, you should prepare a standard set of questions for all the interviews. Also be sure to include open-ended questions, such as "What else do you think I should know about the system?" or "Is there any other relevant information that we have not discussed?"

When conducting interviews during the preliminary investigation, you should interview managers and supervisors who have a broad knowledge of the system and can give you an overview of the business processes involved. Depending on the situation, you might talk to operational personnel to learn how the system functions on a day-to-day basis.

REVIEW DOCUMENTATION   Although interviews are an extremely important method of obtaining information, you also might want to investigate the current system documentation. The documentation might not be up to date, so you should check with users to confirm that you are receiving accurate and complete information.

OBSERVE OPERATIONS   Another fact-finding method is to observe the current system in operation. You might see how workers carry out typical tasks. You might choose to trace or follow the actual paths taken by input source documents or output reports. In addition to observing operations, you might want to sample the inputs or outputs of the system. Using sampling techniques described in Chapter 4, you can obtain valuable information about the nature and frequency of the problem.

CONDUCT A USER SURVEY   Interviews can be time consuming. Sometimes you can obtain information from a larger group by conducting a user survey. In this case, you design a form that users complete and return to you for tabulation. A survey is not as flexible as a series of interviews, but it is less expensive, generally takes less time, and can involve a broad cross-section of people.

Step 4: Analyze Project Usability, Cost, Benefit, and Schedule Data

During fact-finding, you gathered data about the project's predicted costs, anticipated benefits, and schedule issues that could affect implementation. Before you can evaluate feasibility, you must analyze this data carefully. If you conducted interviews or used surveys, you should tabulate the data to make it easier to understand. If you observed current operations, you should review the results and highlight key facts that will be useful in the feasibility analysis. If you gathered cost and benefit data, you should be able to prepare financial analysis and impact statements using spreadsheets and other decision support tools.

Also, you should develop time and cost estimates for the requirements modeling tasks for the next SDLC phase, systems analysis. Specifically, you should consider the following:

- What information must you obtain, and how will you gather and analyze the information?

- Will you conduct interviews? How many people will you interview, and how much time will you need to meet with the people and summarize their responses?

- Will you conduct a survey? Who will be involved? How much time will it take people to complete it? How much time will it take to tabulate the results?

- How much will it cost to analyze the information and prepare a report with findings and recommendations?

Step 5: Evaluate Feasibility

You have analyzed the problem or opportunity, defined the project scope and constraints, and performed fact-finding to evaluate project usability, costs, benefits, and time constraints. Now you are ready to evaluate the project's feasibility. You should start by reviewing the answers to the questions listed on pages 63–66. Also consider the following guidelines:

OPERATIONAL FEASIBILITY Your fact-finding should have included a review of user needs, requirements, and expectations. When you analyze this data, you should look for areas that might present problems for system users and how they might be resolved. Because operational feasibility means that a system will be used effectively, this is a vital area of concern.

TECHNICAL FEASIBILITY The fact-finding data should identify the hardware, software, and network resources needed to develop, install, and operate the system. With this data, you can develop a checklist that will highlight technical costs and concerns, if any.

ECONOMIC FEASIBILITY Using the fact-finding data, you can apply the financial analysis tools described in Part 3 of the Systems Analyst's Toolkit to assess feasibility. The cost-benefit data will be an important factor for management to consider. Also, a cost estimate for the project development team will be built into the project management plan.

SCHEDULE FEASIBILITY The fact-finding data should include stakeholder expectations regarding acceptable timing and completion dates. As mentioned previously, often a trade-off exists between a project's schedule and its costs. For example, compressing a project schedule might be possible, but only if the budget is increased accordingly. The schedule data will be incorporated into the project plan in the form of task durations and milestones.

Step 6: Present Results and Recommendations to Management

At this stage, you have several alternatives. You might find that no action is necessary or that some other strategy, such as additional training, is needed. To solve a minor problem, you might implement a simple solution without performing further analysis. In other situations, you will recommend that the project proceed to the next development phase, which is systems analysis.

The final task in the preliminary investigation is to prepare a report to management, and possibly deliver a presentation, as shown in Figure 2-23. The report includes an evaluation of the systems request, an estimate of costs and benefits, and a **case for action**, which is a summary of the project request and a specific recommendation.

The format of a preliminary investigation report varies from one company to another. A typical report might consist of the following sections:

FIGURE 2-23 Oral presentations often are required during systems development, and systems analysts need to develop strong presentation skills.

- *Introduction* — the first section is an overview of the report. The introduction contains a brief description of the system, the name of the person or group who performed the investigation, and the name of the person or group who initiated the investigation.

- *Systems Request Summary* — the summary describes the basis of the systems request.

- *Findings* — the findings section contains the results of your preliminary investigation, including a description of the project's scope, constraints, and feasibility.

- *Case for Action* — a summary of the project request and a specific recommendation. Management will make the final decision, but the IT department's input is an important factor.

- *Project Roles* — this section lists the people who will participate in the project, and describes each person's role.

- *Time and Cost Estimates* — this section describes the cost of acquiring and installing the system, and the total cost of ownership during the system's useful life.

- *Expected Benefits* — this section includes anticipated tangible and intangible benefits and a timetable that shows when they are to occur.

- *Appendix* — an appendix is included in the report if you need to attach supporting information. For example, you might list the interviews you conducted, the documentation you reviewed, and other sources for the information you obtained. You do not need detailed reports of the interviews or other lengthy documentation. It is critical that you retain those documents to support your findings and for future reference.

A QUESTION OF ETHICS

As a new systems analyst at Premier Financial Services, you are getting quite an education. You report to Mary, the IT manager, who also chairs the systems review committee. Several months ago, the committee rejected a request from Jack, the finance director, for an expensive new accounts payable system, because the benefits did not appear to outweigh the costs.

Yesterday, Mary's boss called her in and asked her to reconsider Jack's request, and to persuade the other members to approve it. Mary wanted to discuss the merits of the request, but he cut her off rather abruptly. Mary happens to know that Jack and her boss are longtime friends.

Mary has confided in you. She is very uncomfortable about the meeting with her boss, and she believes that his request would undermine the integrity of the systems review process. Mary feels it would be unethical to grant preferred treatment just because a friendship is involved. She is thinking of submitting a request to step down as review committee chair, even though that might harm her career at the company.

Is this an ethical question, or just a matter of office politics? What would you say to Mary?

CHAPTER SUMMARY

Systems planning is the first phase of the systems development life cycle. Effective information systems help an organization support its business processes, carry out its mission, and serve its stakeholders. Strategic planning allows a company to examine its purpose, vision, and values and develops a mission statement, which leads to goals, objectives, day-to-day operations, and business results that affect company stakeholders.

During the systems planning phase, an analyst reviews the business case, which is the basis, or reason, for a proposed system. A business case should describe the project clearly, provide the justification to proceed, and estimate the project's financial impact.

Systems projects are initiated to improve performance, provide more information, reduce costs, strengthen controls, or provide better service. Various internal and external factors affect systems projects, such as user requests, top management directives, existing systems, the IT department, software and hardware vendors, technology, customers, competitors, the economy, and government.

During the preliminary investigation, the analyst evaluates the systems request and determines whether the project is feasible from an operation, technical, economic, and schedule standpoint. Analysts evaluate systems requests on the basis of their expected costs and benefits, both tangible and intangible.

The steps in the preliminary investigation are to understand the problem or opportunity; define the project scope and constraints; perform fact-finding; analyze project usability, cost, benefit, and schedule data; evaluate feasibility; and present results and recommendations to management. During the preliminary investigation, analysts often use investigative tools such as fishbone or Ishikawa diagrams and Pareto charts. The last task in a preliminary investigation is to prepare a report to management. The report must include an estimate of time, staffing requirements, costs, benefits, and expected results for the next phase of the SDLC.

Key Terms and Phrases

biometric devices 57
business case 48
case for action 76
computer resources committee 62
constraint 71
critical success factors 53
customer relationship management (CRM) 60
discretionary projects 67
economic feasibility 65
electronic product code (EPC) 59
electronic proof of delivery (EPOD) 60
fishbone diagram 70
goals 53
intangible benefits 65
Ishikawa diagram 70
just-in-time (JIT) 60

mission statement 51
nondiscretionary projects 67
objectives 53
operational feasibility 64
Pareto chart 71
preliminary investigation 68
project creep 71
project scope 71
schedule feasibility 66
strategic planning 50
SWOT analysis 50
systems request 56
systems review committee 62
tangible benefits 65
technical feasibility 64
total cost of ownership (TCO) 65

Learn It Online

Instructions: To complete the Learn It Online exercises, start your browser, click the Address bar, and then enter the Web address **scsite.com/sad8e/learn**. When the Systems Analysis and Design Learn It Online page is displayed, follow the instructions in the exercises below. Each exercise has instructions for saving your results, either for your own records or for submission to your instructor.

1 Chapter Reinforcement

TF, MC, and SA

Below SAD Chapter 2, click the Chapter Reinforcement link. Print the quiz by clicking Print on the File menu for each page. Answer each question.

2 Flash Cards

Below SAD Chapter 2, click the Flash Cards link and read the instructions. Type 20 (or a number specified by your instructor) in the Number of playing cards text box, type your name in the Enter your Name text box, and then click the Flip Card button. When the flash card is displayed, read the question and then click the ANSWER box arrow to select an answer. Flip through the Flash Cards. If your score is 15 (75%) correct or greater, click Print on the File menu to print your results. If your score is less than 15 (75%) correct, then redo this exercise by clicking the Replay button.

3 Practice Test

Below SAD Chapter 2, click the Practice Test link. Answer each question, enter your first and last name at the bottom of the page, and then click the Grade Test button. When the graded practice test is displayed on your screen, click Print on the File menu to print a hard copy. Continue to take practice tests until you score 80% or better.

4 Who Wants To Be a Computer Genius?

Below SAD Chapter 2, click the Computer Genius link. Read the instructions, enter your first and last name at the bottom of the page, and then click the Play button. When your score is displayed, click the PRINT RESULTS link to print a hard copy.

5 Wheel of Terms

Below SAD Chapter 2, click the Wheel of Terms link. Read the instructions, and then enter your first and last name and your school name. Click the PLAY button. When your score is displayed on the screen, right-click the score and then click Print on the shortcut menu to print a hard copy.

6 Crossword Puzzle Challenge

Below SAD Chapter 2, click the Crossword Puzzle Challenge link. Read the instructions, and then enter your first and last name. Click the SUBMIT button. Work the crossword puzzle. When you are finished, click the Submit button. When the crossword puzzle is displayed, click the Print Puzzle button to print a hard copy.

Case-Sim: SCR Associates

Background

SCR Associates is a consulting firm that offers IT solutions and training. SCR needs an information system to manage operations at its new training center. The new system will be called TIMS (Training Information Management System). As a newly hired systems analyst, you will report to Jesse Baker, systems group manager. You will work on various tasks and practice the skills you learned in this chapter.

Using the Case

The SCR Associates case study is a Web-based simulation that allows you to practice your skills in a real-world environment. The case study transports you to SCR's company intranet, where you can complete 12 work sessions, each aligning with a chapter. As you work on the case, you will receive e-mail and voice mail messages, obtain information from SCR's online libraries, and perform various tasks. The first time you enter the SCR Case, you should go to the starting page at **scsite.com/sad8e/scr** for detailed instructions.

Preview: Session 2

During your orientation, you found your way around the office and had a chance to explore the SCR Internet site. Now, after a week on the job, your supervisor, Jesse Baker, has explained the new TIMS system and asked you to lead the systems development effort. She suggested that you review SCR's mission statement, think about a systems review committee, draft a project scope statement, and prepare to interview people to learn more about the new system.

To start a work session, you log on to the SCR intranet. When you enter your name and password, an opening screen displays links to the work sessions, and you select Session 2. At this point, you check your e-mail and voice mail messages carefully. You then begin working on your task list, which includes the following items:

Tasks: Analyzing the Business Case

1. *We need a corporate goal for SCR that refers to our new training activity. Prepare a draft to show Jesse.*

2. *Jesse wants my opinion on whether or not SCR needs a system review committee. Need to prepare a recommendation and reasons.*

3. *Draft a project scope statement for the TIMS system and describe the constraints. She said be specific.*

4. *Need to identify the people I want to interview to learn more about the new training activity, and prepare a list of the questions I will ask.*

FIGURE 2-24 Task list: Session 2.

Chapter Exercises

Review Questions

1. What is a business case? How does a business case affect an IT project?
2. What is a SWOT analysis and why is it important?
3. What are five common reasons for systems projects?
4. What are some internal and external factors that affect systems projects?
5. What are some advantages and disadvantages of a systems review committee?
6. What is feasibility? List and briefly discuss four feasibility tests.
7. How do tangible benefits differ from intangible benefits?
8. What are the steps in a preliminary investigation?
9. What is project scope? What is a constraint? In what three ways are constraints classified?
10. What are three fact-finding techniques that systems analysts use during the preliminary investigation?

Discussion Topics

1. Directives from top management often trigger IT projects. Suppose that the vice president of marketing tells you to write a program to create mailing labels for a one-time advertising promotion. As the IT manager, you know that the labels can be prepared more efficiently by simply exporting the data to a word processing program with a mail merge feature. How would you handle this situation?
2. The vice president of accounting says to you, the IT director, "This systems development life cycle stuff takes too long." She tells you that her people know what they are doing and that all systems requests coming from her department are necessary and important to the organization. She suggests that the IT department bypass the initial steps for any accounting department request and immediately get to work at the solution. What would you say to her?
3. One of your coworkers says, "Mission statements are nice, but they really don't change things down here where the work gets done." How would you reply?
4. Would you continue to work for a company if you disagreed with the firm's mission statement? Why or why not?

Projects

1. Use the Internet to find an example of a corporate mission statement.
2. Many articles have been written on how to develop, understand, and evaluate a business case. Visit the Web sites for TechRepublic, CIO, or another IT magazine, and find one or more articles that might be of interest to your class. For more information, you can visit the Resources Library at the online SCR Associates case, which lists more than a dozen IT news sources. To view these sources, go to **scsite.com/sad8e/scr**, log on to the SCR intranet, and navigate to the library. When your research is done, write a brief summary of what you learned.
3. Suppose you own a travel agency in a large city. You have many corporate clients, but growth has slowed somewhat. Some long-term employees are getting discouraged, but you feel that there might be a way to make technology work in your favor. Use your imagination and suggest at least one strength, weakness, opportunity, and threat that your business faces.
4. Write a mission statement and at least three goals for the travel agency described in Project 3.

Apply Your Knowledge

The section contains four mini-cases. Each case describes a situation, explains your role in the case, and asks you to respond to questions. You can answer the questions by applying knowledge you learned in the chapter.

 Last Chance Securities

Situation:

The IT director opened the department staff meeting today by saying "I've got some good news and some bad news. The good news is that management approved the payroll system project this morning. The new system will reduce clerical time and errors, improve morale in the payroll department, and avoid possible fines and penalties for noncompliance. The bad news is that the system must be installed by the end of December in order to meet new federal reporting rules, costs must be within the budgeted amount, the new system must interact with existing systems, and the vice president of finance insists on approving the final design."

1. Name the constraints and indicate whether each is present, future, internal, external, mandatory, or desirable.
2. Explain why it is important to define the payroll project's scope. Explain how to define project scope.
3. Identify tangible and intangible benefits of the new payroll system.
4. What topics should be included in a report to management at the end of the preliminary investigation?

2 Way Out Bikes

Situation:

The owner of Way Out Bikes asked you for advice about acquiring an information system for her business. The company specializes in helping customers select exactly the right bicycle for their needs and lifestyles. Way Out cannot compete on price with mass merchandisers, but it seeks to offer value and expertise for which customers are willing to pay. You ask the owner whether she has long-range plans for the company, and she replies that she has not really thought beyond a one-year time frame.

1. Explain the concept of strategic planning to Way Out's owner.
2. Decide what else you might want to know about Way Out. Consider the internal and external factors described on pages 58 to 61, and make a list of questions to ask the owner.
3. Draft a mission statement for Way Out.
4. Make a list of Way Out's stakeholders.

3 The Monday IT Department Staff Meeting

Situation:

Your boss, the IT manager, was ready to explode. "Why can't we get our priorities straight?" he fumed. "Here we go again, working on a low-value project, just because it's a favorite of the marketing group. I wish we could get away from departmental politics! I want you to draft a memo that proposes a systems review committee for this company. Explain the advantages, but don't step on anyone's toes!"

1. Write a draft of the proposal, as your boss requested.
2. Write a memo to your boss explaining potential disadvantages of the committee approach.
3. Draft a set of ground rules for committee meetings. Try to suggest rules that will minimize political differences and focus on the overall benefit to the company.
4. Most people serve on a committee at some point in their lives. Write a brief memo describing your committee experiences, good or bad.

4 The Friday IT Department Staff Meeting

Situation:

By the end of the week, things quieted down. The IT staff discussed how to prioritize IT project requests, taking into account technical, operational, economic, and schedule feasibility. The IT manager asked for suggestions from the group.

1. Provide three examples of why a project might lack technical feasibility.
2. Provide three examples of why a project might lack operational feasibility.
3. Provide three examples of why a project might lack economic feasibility.
4. Provide three examples of why a project might lack schedule feasibility.

Case Studies

Case studies allow you to practice specific skills learned in the chapter. Each chapter contains several case studies that continue throughout the textbook, and a chapter capstone case.

NEW CENTURY HEALTH CLINIC

New Century Health Clinic offers preventive medicine and traditional medical care. In your role as an IT consultant, you will help New Century develop a new information system.

Background

New Century Health Clinic's office manager, Anita Davenport, recently asked permission to hire an additional office clerk because she feels the current staff can no longer handle the growing workload. The associates discussed Anita's request during a recent meeting. They were not surprised that the office staff was feeling overwhelmed by the constantly growing workload.

Because the clinic was busier and more profitable than ever, they all agreed that New Century could afford to hire another office worker. Dr. Jones then came up with another idea. He suggested that they investigate the possibility of computerizing New Century's office systems. Dr. Jones said that a computerized system could keep track of patients, appointments, charges, and insurance claim processing and reduce paperwork. All the associates were enthusiastic about the possibilities and voted to follow up on the suggestion. Dr. Jones agreed to direct the project.

Because no member of the staff had computer experience, Dr. Jones decided to hire a consultant to study the current office systems and recommend a course of action. Several friends recommended you as a person who has considerable experience with computerized business applications.

Assignments

1. Dr. Jones has arranged an introductory meeting between the associates of New Century Health Clinic and you to determine if mutual interest exists in pursuing the project. What should the associates try to learn about you? What should you try to learn in this meeting?
2. Does the proposed system present a strong business case? Why or why not?
3. For each type of feasibility, prepare at least two questions that will help you reach a feasibility determination.
4. You begin the preliminary investigation. What information is needed? From whom will you obtain it? What techniques will you use in your fact-finding?

PERSONAL TRAINER, INC.

Personal Trainer, Inc., owns and operates fitness centers in a dozen midwestern cities. The centers have done well, and the company is planning an international expansion by opening a new "supercenter" in the Toronto area. Personal Trainer's president, Cassia Umi, hired an IT consultant, Susan Park, to help develop an information system for the new facility. During the project, Susan will work closely with Gray Lewis, who will manage the new operation.

Background

At their initial meeting, Susan and Gray discussed some initial steps in planning a new information system for the new facility. The next morning, they worked together on a business profile, drew an organization chart, discussed feasibility issues, and talked about various types of information systems that would provide the best support for the supercenter's operations. Their main objective was to carry out a preliminary investigation of the new system and report their recommendations to Personal Trainer's top managers.

After the working session with Gray, Susan returned to her office and reviewed her notes. She knew that Personal Trainer's president, Cassia Umi, wanted the supercenter to become a model for the company's future growth, but she did not remember any mention of an overall strategic plan for the company. Susan also wondered whether the firm had done a SWOT analysis or analyzed the internal and external factors that might affect an information system for the supercenter.

Because the new operation would be so important to the company, Susan believed that Personal Trainer should consider an enterprise resource planning strategy that could provide a company-wide framework for information management. After she finished compiling her notes, Susan listed several topics that might need more study and called Gray to arrange another meeting the following day.

Assignments

1. Based on the background facts described in Chapter 1, draft a mission statement for Personal Trainer. Consider the firm's overall direction, and the services, products, and experiences the company might want to offer its customers in the future. In your statement, consider all the stakeholders affected by Personal Trainer's operations.
2. Susan and Gray probably will need more information about the proposed system. Make a list of people whom they might want to interview. Also, suggest other fact-finding techniques they should consider.
3. Consider the internal and external factors that affect information systems. Which factors, in your opinion, will have the greatest impact on the system proposed for the new supercenter? Explain your answer.
4. At the conclusion of the preliminary investigation, Susan and Gray will deliver a written summary of the results and deliver a brief presentation to Personal Trainer's management team. Prepare a list of recommendations that will help make their written and oral communications more effective. Put your list in priority order, starting with what you consider to be the most important suggestions. Before you complete this task, you should review Part 1 of the Systems Analyst's Toolkit, which provides suggestions for oral and written presentations.

ORIGINAL KAYAK ADVENTURES

Original Kayak Adventures (OKA) offers guided eco-tours and kayak rentals along the Hudson River.

Background

In Chapter 1, you learned that John and Edie Caputo founded OKA two years ago. Now John and Edie are thinking about replacing their current system, which is a mix of manual and computer-based techniques, with a new information system that would meet their current and future needs. Before you answer the following questions, you should review the fact statement in Chapter 1.

Assignments

1. Does a strong business case exist for developing an information system to support the Caputos' business? Explain your answer.
2. In a small- to medium-sized business, such as OKA, is it really important to use a structured approach for information systems development? Why or why not?
3. Based on the facts provided, draft a mission statement for OKA. In your statement, consider all the stakeholders who might be affected by OKA operations.
4. What internal and external factors might affect OKA's business success?

TOWN OF EDEN BAY

The town of Eden Bay owns and maintains a fleet of vehicles. You are a systems analyst reporting to Dawn, the town's IT manager.

Background

Eden Bay is a medium-sized municipality. The town has grown rapidly, and so has the demand for town services. Eden Bay currently owns 90 vehicles, which the town's equipment department maintains. The fleet includes police cars, sanitation trucks, fire trucks, and other vehicles assigned to town employees. The maintenance budget has risen sharply in recent years, and people are asking whether the town should continue to perform its own maintenance or outsource it to private firms.

This morning, Dawn called you into her office to discuss the situation. A summary of her comments follows.

> **Dawn (IT manager):** When I came here two years ago, I was told that Eden Bay had a computerized information system for vehicle maintenance. What I found was a spreadsheet application designed by a part-time employee as a quick answer to a much more complex problem. It's probably better than no system at all, but I can't justify spending any time on it. The system should never have been designed as a spreadsheet in the first place.
>
> I've discussed the situation with the equipment department people. Rather than tinker with the current system, I think we should press for a new information system project, and I've developed an initial proposal. I've code-named the new system RAVE, which stands for Repair Analysis for Vehicular Equipment. I know that commercial fleet maintenance packages exist, but they are very expensive.
>
> I did some fact-finding, and I want you to start by reading the interview summaries I prepared.

Before You Begin ...

Review the following interview summaries from Marie (town manager), Martin (equipment department manager), Phil (maintenance supervisor), Alice (maintenance clerk), and Joe (mechanic).

> **Marie (town manager):** Maintenance costs have risen 14 to 16% annually. I'm not sure that we have any real control over these costs. Some members of the town council think we should get out of the maintenance business and contract it out to a private firm. That might mean laying off current employees, and I'm not sure whether outsourcing is the right way to go.
>
> Both the equipment department manager and the IT manager tell me that our current record-keeping system is outdated, and I wonder if a new information system would give us a better handle on the problem. My own view is that if there's a way we can become more efficient, we should continue to perform our own maintenance.
>
> Dawn, our IT manager, tells me that she has developed a proposal for a maintenance information system. I plan to bring it up at the next council meeting.

> **Martin (equipment department manager):** I hear a lot of criticism about the maintenance budget, but I'm doing the best I can. We operate from one budget year to the next, without a long-term plan. I belong to a professional association of fleet maintenance managers, and I know that we should be developing a strategic plan instead of juggling annual budget figures.
>
> I'd like to build this department into a first-class organization. Our people are great, but they could use more technical training. Our shop and equipment are generally adequate for what we do, but we haven't kept up with some of the newer diagnostic equipment. We have a real problem in record keeping. Instead of a short-term solution, Eden Bay should have developed a maintenance information system years ago. Prior to taking this position, I was assistant maintenance manager in a medium-sized city, and they had developed a system that handled scheduling and cost analysis, in addition to day-to-day maintenance operations.

Phil (maintenance supervisor): I'm in the middle — I get pressure from above to cut costs, and I get complaints from below that management doesn't know what it's doing. One thing for sure — short-term solutions are not the answer. I hope they don't ask me to cut back on preventive maintenance. The last time we did that, we extended routine oil changes and servicing, and we ended up with even more repairs than we had previously.

My mechanics are capable people, and they're doing the best they can. One problem I see is that it's hard to pull up a history for a particular vehicle. We keep the data on a computer, but different people used different codes and procedures over the years, and the system probably needs a good overhaul.

Alice (maintenance clerk): I'm in charge of maintenance record keeping. We use a spreadsheet system that was designed by a part-time employee who is no longer around. Because we work on a monthly budget, the spreadsheet has a separate page for each month. When the year is over, we start a new set of monthly pages. The spreadsheet is supposed to record labor and parts used, and assign the cost to a specific vehicle, but it doesn't always work out that way.

I also use a notebook to keep track of vehicle mileage and scheduled service intervals, so I can let the department heads know when a vehicle needs to come in for service. I write up work orders for scheduled service or necessary repairs, but often a mechanic finds other problems and has to write up an additional charges form.

Each time a vehicle comes into the shop, I start a new row on the spreadsheet. I enter the vehicle number, mileage, and date. Then I enter the rest of the data into the columns for parts, labor hours, job code, shop supplies, and miscellaneous charges. At the end of the month, I calculate total costs from the spreadsheet, and we compare these with actual payroll and parts vouchers for the month. If the totals are close, everyone is happy. If not, we try to figure out what work didn't get reported and entered into the spreadsheet.

The labor codes also are a problem. Specific codes are assigned for certain types of shop labor, but these were changed three years ago when the new Director arrived. Also, about half the labor can be coded, but the rest has to be entered manually — and there are no standards. Two mechanics might do the same job, and one records four specific tasks, while the other calls it a tune-up.

I know the mechanics don't like paperwork, but what can I do? I asked the IT manager if she could do anything to help, but she says that it isn't worthwhile to update the current system. She says she has heard some talk about developing a new information system specifically designed for vehicle fleet maintenance. It can't be soon enough for me.

Joe (mechanic): I love my job, but I hate the paperwork. We get a work order from the clerk for all scheduled maintenance, but if we find other problems, we have to handwrite an additional work ticket. Personally, I think some of these vehicles should be retired before they get too expensive to maintain.

I would hate to see the town contract out the maintenance. I've put in 17 years here, and I don't want to lose my job, but I know that some specialized repairs would be less expensive on the outside. Most of the mechanics realize this, but let management figure it out — they're the ones with the fancy computer system.

Assignments

1. Upon investigation, you learn that the town does not have a strategic plan or a mission statement. In your view, does this affect the current situation? Why or why not?

2. Based on the fact statements provided, summarize the maintenance department's most important strengths, weaknesses, opportunities, and threats.

3. Describe the specific steps you will follow during a preliminary investigation, including any fact-finding techniques you will use.

4. Of the four tests of feasibility — operational, technical, economic, and schedule — which would you perform first to measure the system project's feasibility? Why?

CHAPTER CAPSTONE CASE: SoftWear, Limited

SoftWear, Limited (SWL), is a continuing case study that illustrates the knowledge and skills described in each chapter. In this case study, the student acts as a member of the SWL systems development team and performs various tasks.

Background

SWL outsources the company's payroll processing to an outside firm called Business Information Systems (BIS). SWL's payroll department submits data to BIS, which uses its own hardware and software to produce employee paychecks and generate payroll reports. BIS performs payroll processing for dozens of companies. Contractual agreements between BIS and its customers identify specific information processing services and prices.

SWL's information technology department is located at the company headquarters in Raleigh and reports to the vice president of finance. The IT staff is responsible for SWL's mainframe computer and supports the company's Web site and the inventory, marketing, customer order entry, and accounting systems.

Robert Lansing, SWL's president, believes that IT support is vital to the company's strategic long-range plans and has approved increased IT budgets and expansion of the IT staff. In addition to the mainframe, the company networked personal computers in all offices and many shop floor locations and implemented a company intranet linking all SWL locations.

Even though it could handle its own payroll processing, SWL continues to use BIS for payroll services because BIS does a good job at a reasonable cost, and it relieves SWL of this responsibility. Recently, problems with the payroll system developed, and SWL's payroll department employees had to work overtime to correct errors involving employee deductions.

SWL employees can make two types of voluntary payroll deductions. Starting in 2004, employees could contribute to the newly formed SWL credit union. To enroll or make changes, an employee must complete a deduction form. In 2007, the company gave employees an opportunity to purchase SWL company stock through payroll deductions. Employees enroll in the stock purchase plan or change their deductions by visiting the human resources department, which then sends a weekly list of transactions to SWL's payroll department.

In addition to the credit union and stock purchase deductions, SWL employees soon may have other savings and investment choices. SWL's top management, with strong support from the vice president of human resources, may consider a new Employee Savings and Investment Plan (ESIP) that allows employees to purchase mutual funds, stocks, and other investments through regular payroll deductions. Under this new 401(k) plan, an outside investment firm, Court Street Securities, manages tax-sheltered deductions and services the individual accounts. Each employee maintains direct control over his or her investments using a 24-hour toll-free number or accessing the Court Street Securities Web site. Management expects to make a final decision about the new ESIP in several months.

Request for Information Technology Services

Tina Pham, vice president of human resources, learned that a number of SWL employees had complained about improper paycheck deductions, and she became concerned about employee morale. She decided to discuss the subject with Michael Jeremy, vice president of finance. At their meeting, he listened carefully and promised to look into the matter further.

That afternoon, Mr. Jeremy met with Amy Calico, director of payroll, to ask her about the problem as well as a recent increase in overtime pay in her group. Amy stated that the overtime became necessary because payroll operations recently required more time and effort. She also noted that, because this workload increase came about recently, she lacked the money in her budget to hire any additional people. She did not provide any specific explanation for the payroll deduction errors.

CHAPTER CAPSTONE CASE: SoftWear, Limited (continued)

Mr. Jeremy then decided to ask the IT department to investigate the payroll system. He prepared a systems request, as shown in Figure 2-25, and sent it to the IT department for action. In the request, he mentioned problems with the payroll system and requested help but did not identify the causes of the problems or propose a solution.

Jane Rossman, manager of applications, normally receives systems requests and does an initial review to see whether a preliminary investigation is warranted. After a quick look at Mr. Jeremy's request, Jane decided to contact her boss, Ann Hon, director of information technology. Because SWL does not have a formal systems review committee, Ann normally makes the initial decision on most systems requests. She always consults with other managers if the proposal is significant or could affect their areas. After discussing the proposal, Jane and Ann decided that a preliminary investigation should start right away. Given that the system was eight years old and had never received a major update, it seemed likely that they would find some problems that warranted attention.

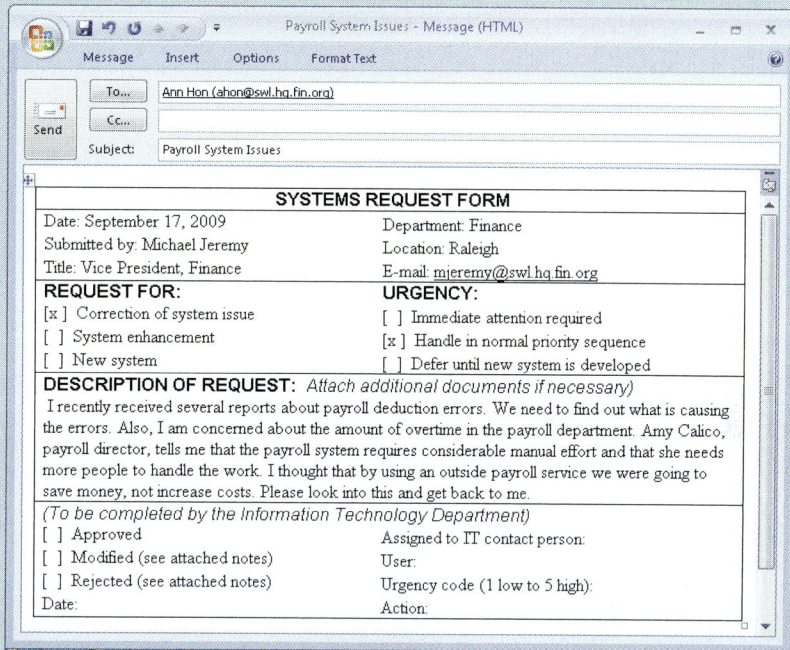

FIGURE 2-25 Michael Jeremy's systems request.

Jane assigned Rick Williams, a systems analyst, to conduct a preliminary investigation of the payroll system. Ann sent the e-mail message shown in Figure 2-26 to Mr. Jeremy so he knew that Rick would start the preliminary investigation the following week.

Because the information technology department reports to him, Mr. Jeremy sent the e-mail shown in Figure 2-27 to all SWL departments. Although the message gives few details, it explains that Rick Williams has been authorized to conduct a preliminary investigation and requests everyone's cooperation.

Payroll Department Organization

To begin his investigation, Rick met with Tina Pham, vice president of human resources. She gave Rick copies of job descriptions for all payroll department positions but did not have a current organization chart for that group.

After reviewing the descriptions, Rick visited Amy Calico, director of payroll. She explained how the payroll department was organized. She explained that two people report directly to her: Nelson White, payroll manager, and Nancy Farmer, administrative assistant. Two payroll technicians, Britton Ellis and Debra Williams, report to Nelson White.

Interviews

Rick next decided to interview Michael Jeremy, Amy Calico, and Mike Feiner, director of human resources.

CHAPTER CAPSTONE CASE: SoftWear, Limited (continued)

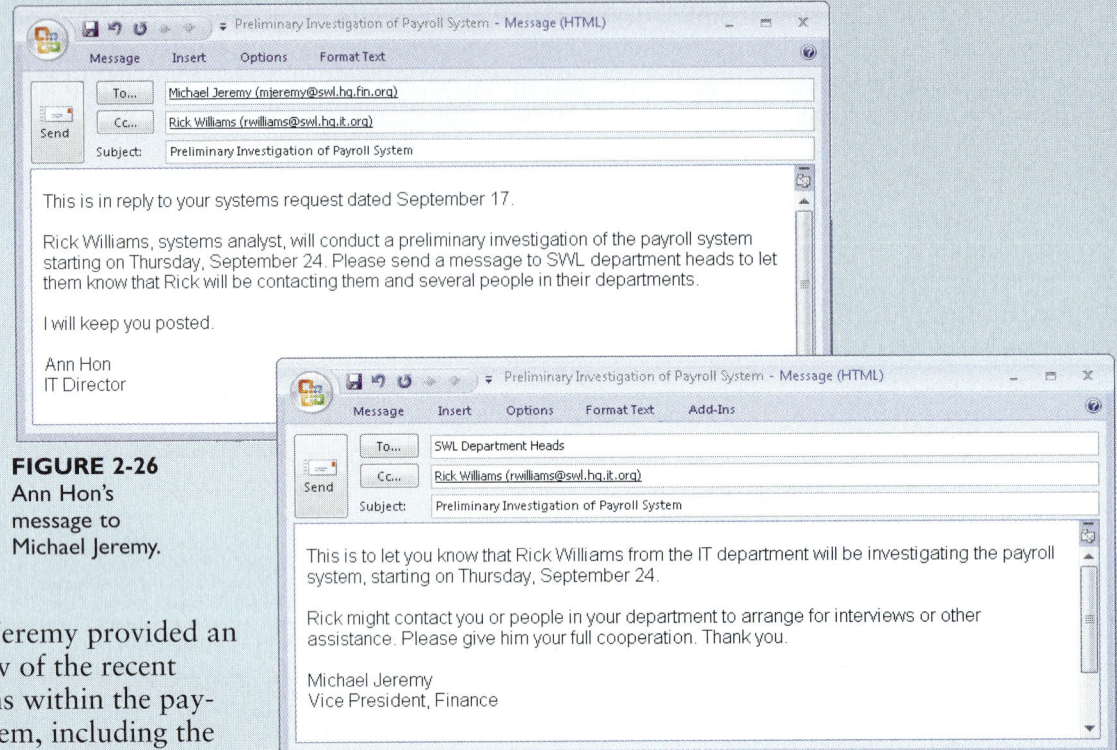

FIGURE 2-26
Ann Hon's
message to
Michael Jeremy.

FIGURE 2-27 Michael Jeremy's message announcing the start of the payroll system investigation.

Mr. Jeremy provided an overview of the recent problems within the payroll system, including the costs of the current system. He had no specific data, but he believed, from what he had heard, that the majority of the errors involved stock purchases rather than credit union deductions.

Later that day, in his meeting with Mike Feiner, Rick found out more about the reported deduction errors. He learned that stock purchase enrollments and changes are handled differently from credit union deductions. For legal reasons, Mike explained, employees must complete a special form for stock purchase plan transactions. When enrolling or making changes, an employee visits the human resources department for a brochure and an information package called a prospectus, which also includes the form required to enroll. At the end of each week, the human resources department prepares a summary of deduction requests and sends it to the payroll department. Payroll clerks then file the changes with the employee's master record.

The next morning, Rick again met with Amy Calico. In the interview, Amy told Rick that some problems with deductions existed, but she did not feel that the payroll clerks were at fault. She suggested that he look elsewhere for the source of the problem. Amy stated that the payroll process generally works well, although it requires a substantial amount of manual effort. She said that if she could hire two additional clerks, it would resolve any remaining problems. During the course of the meeting, Rick began to feel that Amy's opinion might be somewhat biased. As payroll director, she might not want to call attention to problems in her department, and, Rick guessed, it involved other potential issues — such as her wanting more reports and wanting to expand her department. He made a mental note of those possibilities so he could factor them in when considering her comments and assessment of the problem.

Current Documentation

After completing the three interviews, Rick reviewed his notes and decided to find out more about the actual sequence of operations in the current system. He studied the documentation and found that it provided step-by-step procedures for preparing the payroll. When he asked the payroll clerks about those procedures, he learned that some sections were outdated. The actual sequence of events is shown in Figure 2-28.

Step 1: A new SWL employee completes an employee master sheet and a W-4 form. The human resources department then enters the employee's status and pay rate. Copies of these forms are sent to the payroll department. The payroll department updates the employee master sheet whenever changes are received from the employee or the human resources department. Updates are made with various forms, including forms for credit union and employee stock purchase plan enrollment and changes.

Step 2: On the last day of a weekly pay period, the payroll department prepares and distributes time sheets to all SWL departments. The time sheets list each employee, with codes for various status items such as regular pay, overtime, sick leave, vacation, jury duty, and personal leave.

Step 3: Department heads complete the time sheets on the first business day after the end of a pay period. The sheets then go to the payroll department, where they are reviewed. A payroll clerk enters pay rates and deduction information and forwards the time sheets to the BIS service bureau.

Step 4: BIS enters and processes the time sheet data, prints SWL paychecks, and prepares a payroll register.

Step 5: The checks, time sheets, and payroll register are returned to SWL. The payroll department distributes checks to each department, creates reports for credit union and stock purchase plan deductions, and then transfers necessary funds.

FIGURE 2-28   Sequence of events in payroll processing at SoftWear, Limited.

Rick also discovered that the payroll department never sees a copy of the form that an employee fills out in the human resources department when joining the stock purchase plan or changing deductions. Rick obtained a copy of the SWL stock purchase form from the human resources department and copies of several forms from the payroll department — including employee master sheets, employee time sheets, and credit union deduction forms. Rick put them in a file for later review.

During the preliminary investigation, Rick did not show concern with the detailed information on each form. He would review that information only after management authorized the IT department to continue with the systems analysis phase.

Presentation to Management

After Rick finished his investigation, he analyzed his findings, prepared a preliminary investigation report, and met with Jane and Ann to plan the presentation to management. Ann sent an advance copy of the report to Mr. Jeremy with an e-mail that announced the time and location of the presentation.

Figure 2-29 shows the cover message and the preliminary investigation report. Following the presentation to SWL's top managers and department heads, a question-and-answer session took place. The management group discussed the findings and recommendations and decided that the payroll system needed further analysis. The group also wanted to know if

CHAPTER CAPSTONE CASE: SoftWear, Limited (continued)

the BIS service bureau could handle the ESIP using their current arrangement. Ann replied that no clear answer could be given, and everyone agreed that the project scope should be broadened to include that question.

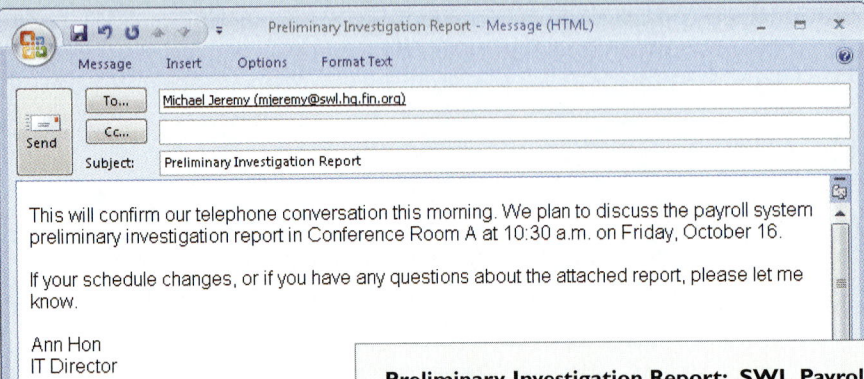

Preliminary Investigation Report: SWL Payroll System   **October 8, 2009**

Introduction
The IT department completed a preliminary investigation of the payroll system on October 8. This investigation was the result of a systems request submitted by Michael Jeremy, vice president, finance, on September 17.

Systems Request Summary
Two problems were mentioned in the request: incorrect deductions from employee paychecks, and excessive payroll department overtime to perform manual processing tasks and make corrections.

Preliminary Investigation Findings
1.   The human resources department sends a summary of employee stock purchase deductions to the payroll department. It is likely that data errors occur during this process. Although the errors are corrected, we believe that incorrect payroll information adversely affects employee morale.

2.   The payroll processing arrangement with Business Information Systems (BIS) requires considerable manual effort. BIS does not provide summary reports that SWL needs to verify and apply credit union and stock purchase deductions. Currently, the payroll department handles these tasks manually at the end of each pay period.

3.   Payroll department overtime averages about eight hours per week, plus an additional eight hours at the end of the month, when stock purchase deductions are applied. Total annual overtime is about 512 hours. The average hourly base rate for payroll staff is $16.00, with an overtime rate of $24.00 per hour. The additional expense is about $12,288 per year.

4.   SWL developed its current payroll procedures 10 years ago, when the company had only 75 employees. At that time, the only payroll deductions were legally required tax items. Today, the payroll system handles over 450 people and many deduction options that must be verified and applied manually.

Recommendations
The current problems will intensify as SWL continues to grow. At this point, it is unclear whether the current system can be modified to handle tasks that are being done manually. Accordingly, the IT department recommends a full analysis of the current system and possible solutions. The project should focus on two main areas: manual processing at SWL and computer-based payroll processing at BIS.

Time and Cost Estimates
We can perform a study during a two-week period. In addition to the time spent by IT staff, we will conduct about 20 hours of interviews with people outside the IT department. The following is a rough estimate of costs through the systems analysis phase:

Systems analyst	2.0 weeks @ $1,400 per week	$2,800
Other SWL staff	0.5 weeks @ $1,000 per week (average)	500
	Total:	$3,300

If the project continues beyond the systems analysis phase, total cost will depend on what development strategy is followed. If the current system can be modified, we estimate a total project effort of $20,000 to $30,000 over a four-month period. If modification is not feasible, a revised cost estimate will be submitted.

Expected Benefits
A sharp reduction in overtime costs and processing errors will avoid unnecessary expense and improve employee morale. During the systems analysis phase, the IT department will investigate various strategies and solutions to address current problems and strengthen SWL's ability to handle payroll-related IT issues in the future.

FIGURE 2-29 Ann Hon's e-mail to Michael Jeremy, with preliminary investigation report attached.

CHAPTER CAPSTONE CASE: SoftWear, Limited (continued)

SWL Team Tasks

1. You have been assigned to write a formal mission statement for SWL. Start by reviewing SWL's background in Chapter 1, then do Internet research to find mission statements that seem clear, focused, and easy to understand. Pay special attention to Web-based and catalog retail firms to see how they approach the issue.
2. Review the preliminary investigation report to see whether all four feasibility tests were discussed in the report. Write a brief summary of your findings.
3. Review the payroll department organization information on page 86. Using this information, prepare an organization chart for this group. In Word 2003, click Insert Diagram or Organization Chart on the Drawing toolbar; in Word 2007, on the Insert tab click SmartArt then Organization Chart.
4. Rick asked you to investigate other firms that offer payroll processing services. Perform an Internet search using the term "payroll processing services." Try your search both with and without placing quotes around the phrase and notice what happens. Based on your search results, select an example of a payroll processing firm and write a brief report to Rick. Include the firm's name, Web address, and services offered.

Manage the SWL Project

You have been asked to manage SWL's new information system project. One of your most important activities will be to identify project tasks and determine when they will be performed. Before you begin, you should review the SWL case in this chapter. Then list and analyze the tasks, as follows:

LIST THE TASKS Start by listing and numbering at least 10 tasks that the SWL team needs to perform to fulfill the objectives of this chapter. Your list can include SWL Team Tasks and any other tasks that are described in this chapter. For example, Task 3 might be to Prepare a payroll department organization chart, and Task 6 might be to Review payroll department job descriptions.

(continues)

CHAPTER CAPSTONE CASE: SoftWear, Limited (continued)

ANALYZE THE TASKS Now study the tasks to determine the order in which they should be performed. First identify all concurrent tasks, which are not dependent on other tasks. In the example shown in Figure 2-30, Tasks 1, 2, 3, 4, and 5 are concurrent tasks, and could begin at the same time if resources were available.

Other tasks are called dependent tasks, because they cannot be performed until one or more earlier tasks have been completed. For each dependent task, you must identify specific tasks that need to be completed before this task can begin. For example, you would want an organization chart to help you identify the payroll department positions, so Task 6 cannot begin until Task 3 is completed, as Figure 2-30 shows.

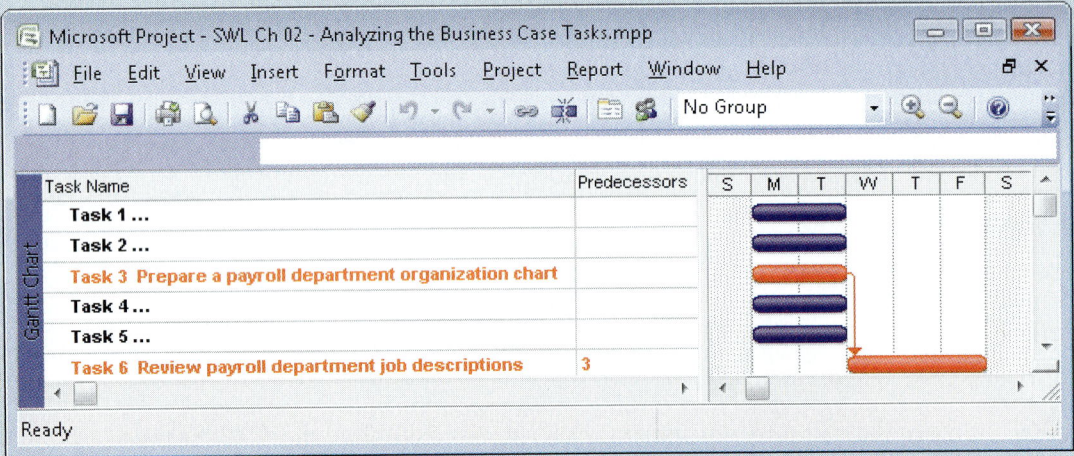

FIGURE 2-30 Tasks 1, 2, 3, 4, and 5 are concurrent tasks that could be performed at the same time. Task 6 is a dependent task that cannot be performed until Task 3 has been completed.

Chapter 3 describes project management tools, techniques, and software. To learn more, you can visit the Features section on your Student Study Tool CD-ROM, or the project management resources library at **scsite.com/sad8e/project**. On the Web, Microsoft offers demo versions, training, and tips for using Project 2007. You also can visit the OpenWorkbench.org site to learn more about this free, open-source software.

CHAPTER 3 Managing Systems Projects

Chapter 3 is the final chapter in the systems planning phase of the SDLC. In this chapter, you will learn about project management and how to plan, schedule, monitor, and report on IT projects.

INTRODUCTION

OBJECTIVES

When you finish this chapter, you will be able to:

- Describe project management tools and how they are used

- Explain project planning, scheduling, monitoring and controlling, and reporting

- Discuss the importance of project risk management

- Explain techniques for estimating task completion times and costs

- Describe various scheduling tools, including Gantt charts and PERT/CPM charts

- Analyze task dependencies, durations, start dates, and end dates

- Identify examples of project management software and explain how these programs can assist you in project planning, estimating, scheduling, monitoring, and reporting

- Explain software change control

- Understand why projects sometimes fail

Chapter 3 explains project management, cost estimating, and change control for information systems projects. You will learn about project planning, scheduling, monitoring, reporting, and the use of project management software. You will learn how to use Gantt charts and PERT/CPM techniques to schedule and monitor projects. You also will learn how to control and manage project changes as they occur.

In addition to the project management material in this chapter, you can visit the Features section on your Student Study Tool CD-ROM, where you can learn more about Microsoft Project and Open Workbench, an open-source project management program that you can download and install. You can also visit **scsite.com/sad8e/swlproject** and explore links in the SWL project management resources library.

CHAPTER INTRODUCTION CASE: Mountain View College Bookstore

Background: Wendy Lee, manager of college services at Mountain View College, wants a new information system that will improve efficiency and customer service at the three college bookstores.

In this part of the case, Florence Fullerton, systems analyst, and Harry Boston, student intern, are talking about project management tools and techniques.

Participants:	Florence and Harry
Location:	Mountain View College Cafeteria, Monday afternoon, September 21, 2009
Discussion topics:	Project planning and estimating, Gantt charts, PERT/CPM charts, Microsoft Project and Open Workbench software, project monitoring, and software change control.

Florence: Hi, Harry. Glad I ran into you. I'd like to talk with you about project management, which we'll be using as we plan and execute the bookstore information system project.

Harry: *Sure. I've read a little about project management, but I don't know the specifics.*

Florence: Well, we manage business and personal projects every day, but we don't always give it much thought. To manage large-scale IT projects, you need specific tools and techniques. You also need a project manager, who is responsible for planning, leading, organizing, and controlling all the tasks.

Harry: *I guess that's you?*

Florence: Sure is. No matter which tools you use, the idea is to break the project down into individual tasks, determine the order in which the tasks need to be performed, and figure out how long each task will take. With this information, you can use Gantt charts or PERT/CPM charts to schedule and manage the work.

Harry: *I've seen Gantt charts — they're the ones that look like horizontal bar charts?*

Florence: Right. In addition to Gantt charts, we'll use PERT/CPM charts, which look like network diagrams that show all the tasks, patterns, and calculations that we'll need. We'll learn how to create PERT/CPM charts manually, and we'll also experiment with Microsoft Project and Open Workbench, which are powerful project management tools.

Harry: *Anything else we need to know?*

Florence: Yes. After we have a specific plan, we need to monitor it carefully, report the progress, and employ a process called software change control. If you are ready, here's a task list to get us started:

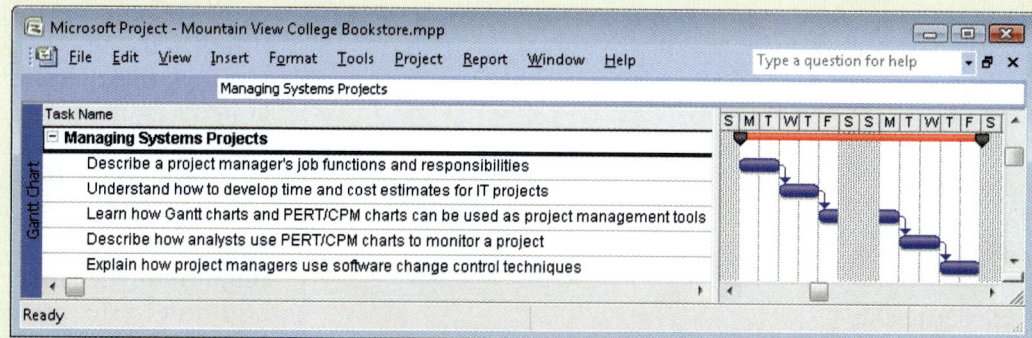

FIGURE 3-1   Typical project management tasks.

FIGURE 3-2 Building construction and systems development projects both need careful management and monitoring.

PROJECT MANAGEMENT OVERVIEW

Whether you are developing an information system or constructing a building like the one in Figure 3-2, the process is similar. **Project management** is the process of planning, scheduling, monitoring and controlling, and reporting upon the development of an information system.

A successful project must be completed on time, within budget, and deliver a quality product that satisfies users and meets requirements. Project management techniques can be used throughout the SDLC. System developers can initiate a formal project as early as the preliminary investigation stage, or later on, as analysis, design, and implementation activities occur.

There is an old story about a sign in an auto repair shop that reads: *Cheap Prices, Fast Service, Quality Work — You Can Choose Any Two Out of Three.* The same concept applies to systems development, except that four factors exist, as shown in Figure 3-3, including project scope, budget, time constraints, and quality standards. As long as everything is in balance, the project will be successful. However, if one factor changes, adjustments must be made to maintain the balance. Because the factors interact constantly, a project manager must be able to respond quickly. For example, if an extremely time-critical project starts to slip, the project manager might have to trim some features, seek approval for a budget increase, simplify the testing plan, or perform some combination of all three actions.

Unfortunately, many systems projects fail. A 2006 report published by The Standish Group concluded that only 35% of software development projects were successful, in the sense that they met budget, schedule, scope, and quality constraints. Even so, the report pointed out that this statistic was a significant improvement over a 1994 study that found a success rate of 16%. Standish chairman Jim Johnson said that several factors were responsible for the improvement, including better project management techniques, more iterative development methods, and utilization of the Web in facilitating communication and feedback between developers and users.

Whether a project involves a new office building or an information system, good leadership is essential. In a systems project, the **project manager**, or **project leader**, usually is a senior systems analyst or an IT department manager if the project is large. An analyst or a programmer/analyst might manage smaller projects. In addition to the project manager, most large projects have a project coordinator. The **project coordinator** handles administrative responsibilities for the development team and negotiates with users who might have conflicting requirements or want changes that would require additional time or expense.

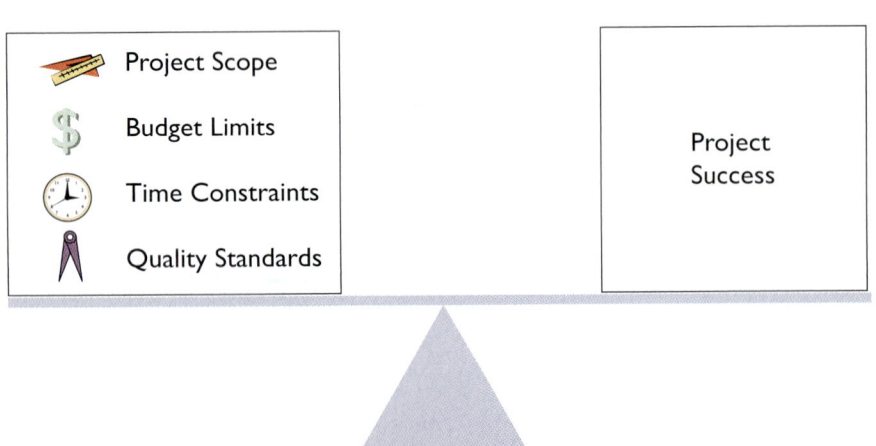

Project Scope

Budget Limits

Time Constraints

Quality Standards

Project Success

FIGURE 3-3 If one factor changes, adjustments must be made to keep things in balance.

Project managers typically perform four main tasks: project planning, scheduling, monitoring and controlling, and reporting:

- **Project planning** includes identifying project tasks and estimating completion time and costs.

- **Project scheduling** involves the creation of a specific timetable, usually in the form of charts that show tasks, task dependencies, and critical tasks that might delay the project. Scheduling also involves selecting and staffing the project team and assigning specific tasks to team members. Project scheduling uses Gantt charts and PERT/CPM charts, which are explained in the following sections.

- **Project monitoring and controlling** requires guiding, supervising, and coordinating the project team's workload. The project manager must monitor the project's progress, evaluate the results, and take corrective action when necessary to control the project and stay on target.

- **Project reporting** tasks include regular progress reports to management, users, and the project team itself. Effective reporting requires strong communication skills and a sense of what others want and need to know about the project.

ON THE WEB

For more information about project management, visit **scsite.com/sad8e/ more**, locate Chapter 3, and then click the Project Management link.

CASE IN POINT 3.1: SPRING FORWARD PRODUCTS

After three years with the company, you recently were asked to manage several IT projects. You are confident that you have the technical skills you need, but you are concerned about morale at the company. There has been some downsizing, and many employees are worried about the future.

As a longtime fan of the Dilbert cartoon strip, you know that maintaining morale can be a real challenge. Your current project involves a team of a dozen people, several of whom remind you of Dilbert and his coworkers. What are some techniques that you might use to motivate the team and inspire its members? What are some things you might *not* want to do?

PROJECT PLANNING

The project plan provides an overall framework for managing costs and schedules. Project planning takes place at the beginning and end of each SDLC phase to develop a plan and schedule for the phases that follow.

The planning process starts with a list of tasks or activities. A **task**, or **activity**, is any work that has a beginning and an end and requires the use of company resources such as people, time, or money. Examples of tasks include conducting interviews, designing a report, selecting software, waiting for the delivery of equipment, or training users. Tasks are basic units of work that the project manager plans, monitors, and tracks, so tasks should be relatively small and manageable.

In addition to tasks, every project has events, or milestones. An **event**, or **milestone**, is a recognizable reference point that can be used to monitor progress and manage the project. For example, an event might be the start of user training, conversion of system data, or completion of interviews. A milestone such as *Complete 50 percent of program testing* would not be useful information unless you could determine exactly when that event should occur.

Project managers must begin by identifying all tasks and developing time and cost estimates for each task. The next step is to determine the order in which the tasks must be performed and assign tasks to specific members of the project team. As the work is performed, the project manager leads and coordinates the team, monitors events, and

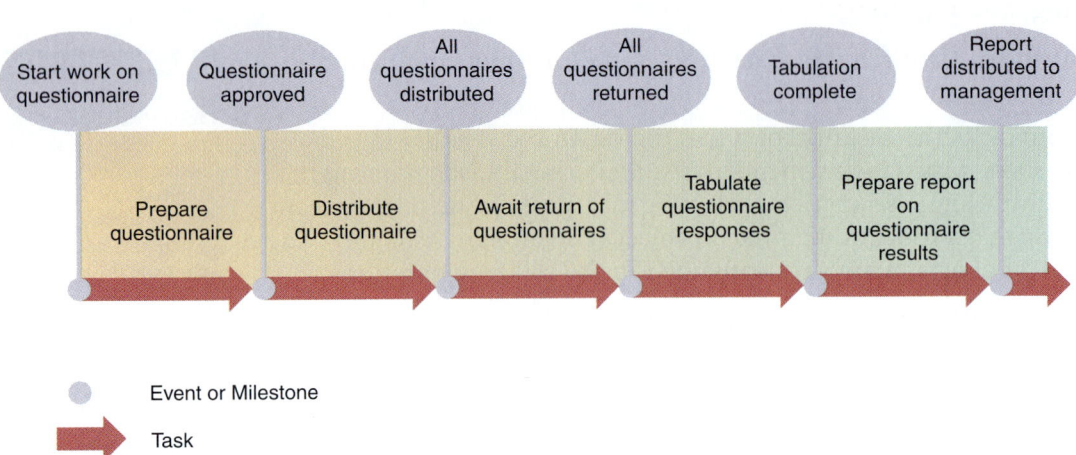

FIGURE 3-4 Using a questionnaire requires a series of tasks and events to track the progress. The illustration shows the relationship between the tasks and the events, or milestones, that mark the beginning and end of each task.

reports on progress. Figure 3-4 shows an example of tasks and events that might be involved in the creation, distribution, and tabulation of a questionnaire. Notice that the beginning and end of each task is marked by a recognizable event.

If you tried to manage a project as one large task, it would be impossible. Instead, you break the project down into a series of smaller tasks, called a **work breakdown structure (WBS)**. The first step in creating a WBS is to identify all tasks.

Identifying Tasks

When identifying tasks, a project manager considers many factors. One important variable is the project size, because the amount of work increases dramatically as project scope increases. For example, consider Figure 3-5, which shows Project A and Project B. Each project contains symbols that represent team members and programs, with connecting lines to show possible interactions.

Figure 3-5 shows that a project that is twice as large will be much more than twice as complex. For example, only one interaction among team members exists in Project A. Project B, however, has a four-member team, so as many as six different interactions can take place. Unless carefully coordinated, multiple interactions can lead to misunderstandings and delays.

As you learned in Chapter 2, projects with general scope definitions are risky, because they tend to expand gradually, without specific authorization, in a process called **project creep**. However, even when a project is clearly described, it must be managed constantly.

Also notice the interfaces among programs. Project A has three programs, so only three possible interfaces exist. In contrast, Project B has six programs, so it has 15 possible interfaces, each with its own set of specifications, requirements, and potential problems.

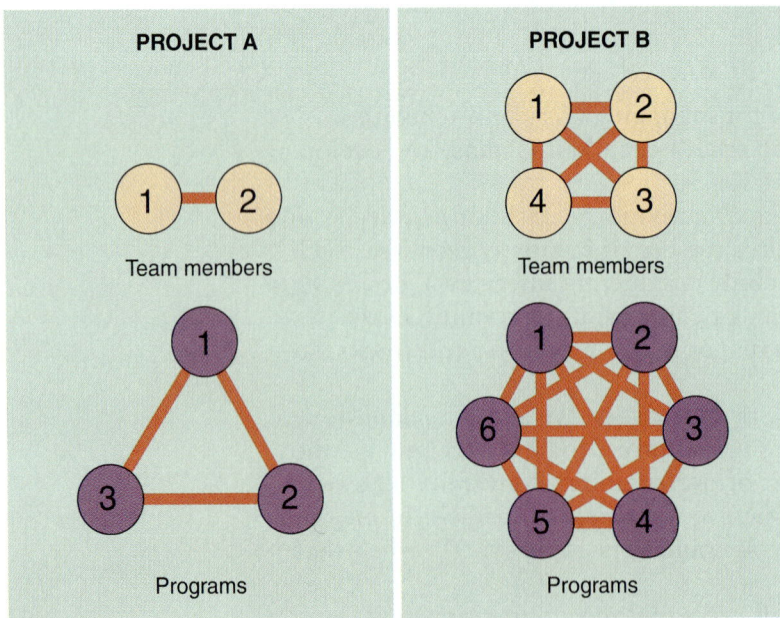

FIGURE 3-5 Project B has twice as many components as Project A, but three times as many team interactions, and five times as many program interfaces.

Figure 3-6 shows the relationship between project resources and project size. Notice that the dashed line indicates a linear forecast. Most analysts would agree that the solid line is a better forecast.

The capabilities of project team members also affect time requirements. A less experienced analyst will need more time to complete a task than an experienced team member. Other factors can affect project time requirements, including the attitudes of users, the degree of management support, and the priority of the project compared with other projects within the organization.

Scheduling people and tasks also can be affected by a principle called **Brooks' Law**. This interesting concept was stated by Frederick Brooks, Jr., an IBM engineer, who observed that adding manpower to a late software project only makes it later. Brooks reached this conclusion when he saw that new workers on a project first had to be educated and instructed by existing employees whose own productivity was reduced accordingly.

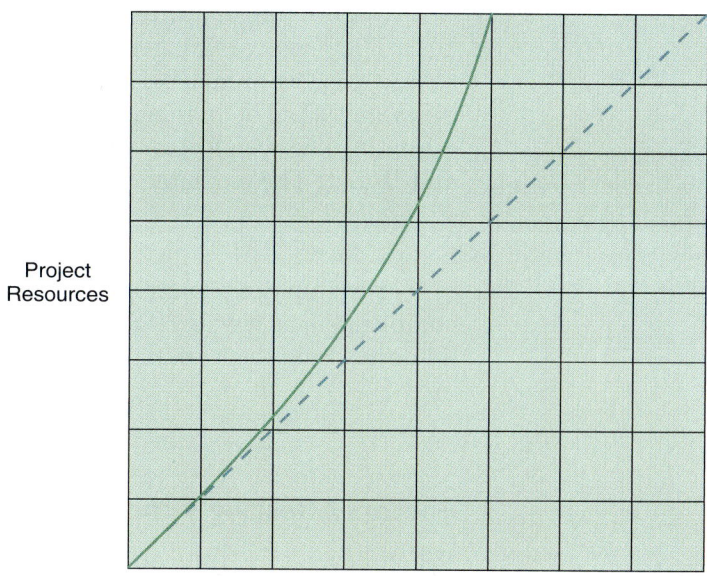

FIGURE 3-6 As the size of a project grows, the resources needed to develop it grow even faster. Notice the difference between the linear forecast (dashed line) and the actual forecast (solid line).

CASE IN POINT 3.2: PARALLEL SERVICES

The project management team at Parallel Services is having a debate about how to define tasks in the work breakdown structure (WBS). Ann, the project manager, wants to break tasks down into the smallest possible units. For example, she objected to a broad task statement called *Develop a training schedule*. Instead, she suggested three subtasks: (1) *Determine availability of training room*, (2) *Determine availability of attendees*, and (3) *Select specific dates and training times*.

Karen, another project team member, disagrees. She feels that the broader task statement is better, because it allows more flexibility and will produce the same result. Karen says that if you break tasks into pieces that are too small, you risk overmanaging the work and spending more time on monitoring than actually performing the tasks. As a member of the team, would you tend to agree more with Ann or Karen? What are the pros and cons of each approach?

Estimating Task Completion Time and Cost

Task completion times and related cost estimates usually are expressed in **person-days** that represent the amount of work that one person can complete in one day. This approach, however, can present some problems. For example, if it will take one person 20 days to perform a particular task, it might not be true that two people could complete the same task in 10 days or that 10 people could perform the task in two days.

Some tasks can be divided evenly so it is possible to use different combinations of time and people, up to a point. For instance, if it takes two person-days to install the cables for a new local area network, one person might do the task in two days, two people in one day, or four people in half a day. In most systems analysis tasks, however, time and people are not interchangeable. If one analyst needs two hours to interview a user, two analysts also will need two hours to do the same interview.

Project managers often use a weighted formula for estimating the duration of each task. The project manager first makes three time estimates for each task: an optimistic, or **best-case estimate** (B), a **probable-case estimate** (P), and a pessimistic, or **worst-case estimate** (W). The manager then assigns a **weight**, which is an importance value, to each estimate. The weight can vary, but a common approach is to use a ratio of B = 1, P = 4, and W = 1. The expected task duration is calculated as follows:

$$\frac{(B+4P+W)}{6}$$

For example, a project manager might estimate that a file-conversion task could be completed in as few as 20 days or could take as many as 34 days, but most likely will require 24 days. Using the formula, the expected task duration is 25 days, calculated as follows:

$$\frac{(20+(4*24)+34)}{6} = 25$$

Factors Affecting Time and Cost Estimates

When developing time and cost estimates, project managers must consider four main factors:

- Project size and scope
- IT resources
- Prior experience with similar projects or systems
- Applicable constraints

PROJECT SIZE AND SCOPE You learned in Chapter 1 that information systems have various characteristics that affect their complexity and cost. In addition to considering those factors, a project manager must estimate the time required to complete each project phase. To develop accurate estimates, a project manager must identify all project tasks, from initial fact-finding to system implementation. Regardless of the systems development methodology used, the project manager must determine how much time will be needed to perform each task. In developing an estimate, the project manager must allow time for meetings, project reviews, training, and any other factors that could affect the productivity of the development team.

IT RESOURCES Companies must invest heavily in cutting-edge technology and Web-based systems to remain competitive in a connected world. In many areas, skilled IT professionals are in great demand, and firms must work hard to attract and retain the talent they need. A project manager must assemble and guide a development team that has the skill and experience to handle the project. If necessary, additional systems analysts or programmers must be hired or trained, and this must be accomplished within a specific time frame. After a project gets under way, the project manager must deal with turnover, job vacancies, and escalating salaries in the technology sector — all of which can affect whether the project can be completed on time and within budget.

PRIOR EXPERIENCE WITH SIMILAR PROJECTS OR SYSTEMS A project manager can develop time and cost estimates based on the resources used for similar, previously developed information systems. The experience method works best for small- or medium-sized projects where the two systems are similar in size, basic content, and operating environment. In large systems with more variables, the estimates are less reliable.

In addition, you might not be able to use experience from projects that were developed in a different environment. For example, when you use a new Web-based database application, you might not have previous experience to measure in this environment. In this situation, you can design a prototype or pilot system to gain technical and cost estimating experience. A pilot system is a small system that is developed as a basis for understanding a new environment. The concept is similar to pilot operation, which is one of the four system changeover methods that you will learn about in Chapter 11, but it takes place much earlier in the systems development life cycle.

CONSTRAINTS You learned in Chapter 2 that constraints must be defined as part of the preliminary investigation of a project. A constraint is a condition, restriction, or requirement that the system must satisfy. For example, a constraint might involve maximums for one or more resources, such as time, dollars, or people. Given those limitations, the project manager must define the system requirements that can be achieved realistically within the required constraints. In the absence of constraints, the project manager calculates the resources needed. In contrast, if constraints are present, the project manager either must adjust other resources or must change the scope of the project. This approach is similar to the what-if analysis that is described in Chapter 12.

CASE IN POINT 3.3: SUNRISE SOFTWARE

A lively discussion is under way at Sunrise Software, where you are a project manager. The main question is whether the person-days concept has limitations. In other words, if a task will require 100 person-days, does it matter whether the work is performed by two people in 50 days, five people in 20 days, 10 people in 10 days, or some other combination that adds up to 100?

Programmers Paula and Ethan seem to think it doesn't matter. On the other hand, Hector, a systems analyst, says it is ridiculous to think that any combination would work. To support his point, he offers this extreme example: Could a task estimated at 100 person-days be accomplished by 100 people in one day?

Is Hector correct? If so, what are the limits in the people versus days equation? Taking the concept a step farther, is there an *optimum* number of people to be assigned to a task? If so, how would that number be determined? You need to offer some guidance at the next project team meeting. What will you say?

PROJECT SCHEDULING

A project schedule is a specific timetable, usually in the form of charts that show tasks, task dependencies, and critical tasks that might delay the project. Project scheduling also involves selecting and staffing the project team, assigning specific tasks to team members, and arranging for other necessary resources.

When scheduling a project, the project manager must know the duration of each task, the order in which the activities will be performed, the start and end times for each task, and the person(s) assigned to each specific task.

Once the duration for each task is estimated, the project manager determines whether or not the task is dependent on other tasks. For example, you cannot tabulate questionnaires until they have been developed, tested, approved, distributed, and returned. After the project manager identifies all the task dependencies, he or she arranges the tasks in a logical sequence.

The next step is to set starting and ending times for each task. A task cannot start until all preceding activities on which it depends are completed. The ending time for a task is its start time plus whatever time it takes to complete the task.

When scheduling a project, project managers decide how they will assign people to the work. Assignments should not overload or underutilize team members, and alternate periods of inactivity followed by intense effort can cause problems and should be avoided. Although scheduling can be a difficult task, a project manager must balance task time estimates, sequences, and personnel assignments to achieve a workable schedule.

Several graphical planning aids can help a project manager in the scheduling process. We will examine two of those tools: Gantt charts and PERT/CPM charts.

GANTT CHARTS

ON THE WEB

For more information about Gantt charts, visit **scsite.com/sad8e/more**, locate Chapter 3, and then click the Gantt Charts link.

Gantt charts were developed almost 100 years ago by Henry L. Gantt, a mechanical engineer and management consultant. His goal was to design a chart that could show planned and actual progress on a complex project.

A **Gantt chart** is a horizontal bar chart that represents a series of tasks. The Gantt chart shown in Figure 3-7 shows 11 tasks in an IT project. The chart displays time on the horizontal axis and arranges the tasks vertically, from top to bottom. The position of the bar shows the planned start and end of the task, and the length of the bar indicates its duration. On the horizontal axis, time can be shown as elapsed time from a fixed starting point, or as actual calendar dates.

A project can have dozens of tasks, and larger projects might have hundreds or thousands of tasks and activities. A Gantt chart for a very large project would be complex and might be hard to understand. To simplify the chart, a project manager can combine related activities. For example, in Figure 3-7, the activities labeled Programming, Write User Manual, and Program Testing are **task groups**, where each group might represent several activities that can be shown in a more detailed chart.

A Gantt chart can show task status in several ways, as shown in Figure 3-8. In each case, the current date is represented by a vertical line. In the top two examples, different colors or arrowheads show how much of the task has been completed and whether or not the work is on schedule. The bottom example shows a Microsoft Project screen that displays similar information, using contrasting color bars to show task progress. In all three examples, notice that Task 1 was completed ahead of schedule, Task 2 is only about 80 percent done and is running behind schedule, Task 3 should have started, but no work has been done,

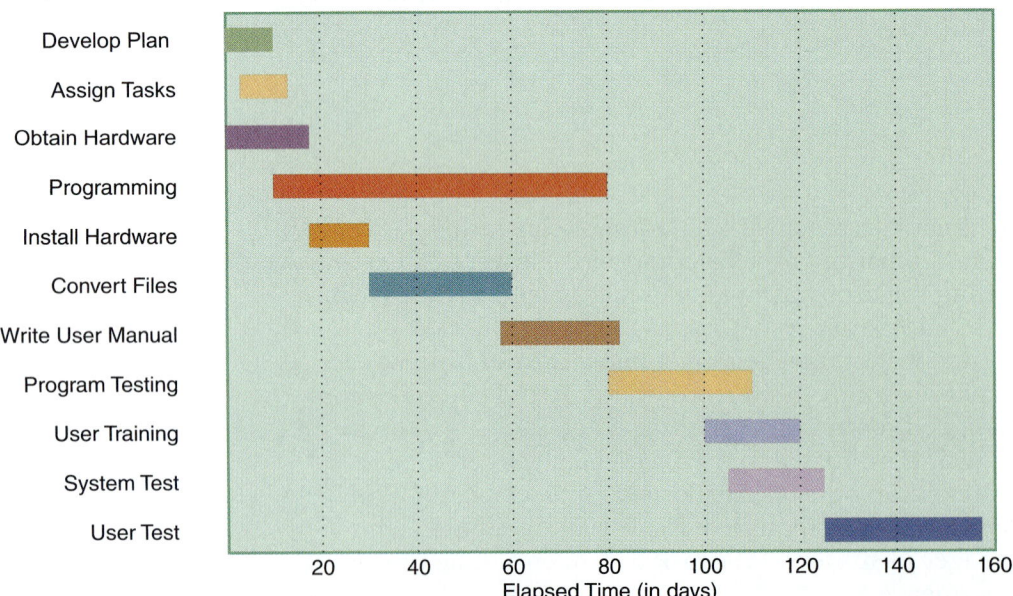

FIGURE 3-7 A Gantt chart for the implementation phase of a project. The chart shows 11 activities on the vertical axis and the elapsed time on the horizontal axis.

Task 4 actually is running ahead of schedule, and Task 5 will begin in several weeks.

Gantt charts can present an overview of the project's status, but they do not provide detailed information that is necessary when managing a complex project. Most project managers find that PERT/CPM charts, which are discussed in the following section, are better tools for managing large projects.

PERT/CPM Charts

The **Program Evaluation Review Technique (PERT)** was developed by the U.S. Navy to manage very complex projects, such as the construction of nuclear submarines. At approximately the same time, the **Critical Path Method (CPM)** was developed by private industry to meet similar project management needs. The distinction between the two methods has disappeared over time, and today the technique is called either PERT, CPM, or **PERT/CPM**. PERT/CPM is called a **bottom-up technique**, because it analyzes a large, complex project as a series of individual tasks, called **project tasks**.

To create a PERT/CPM chart, you first identify all the project tasks and estimate how much time each task will take to perform. Next, you must determine the logical order in which the tasks must be performed. For example, some tasks cannot start until other tasks have been completed. In other situations, several tasks can be performed at the same time.

Once you know the tasks, their durations, and the order in which they must be performed, you can calculate the time that it will take to complete the project. You also will be able to identify the specific tasks that will be critical to the project's on-time completion. These techniques are explained in the following sections.

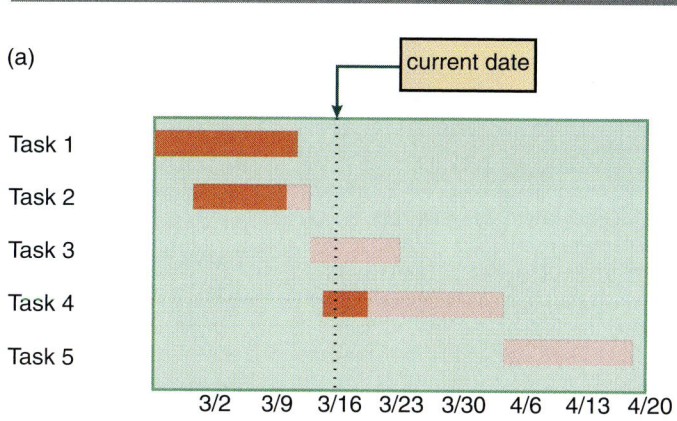

(a)

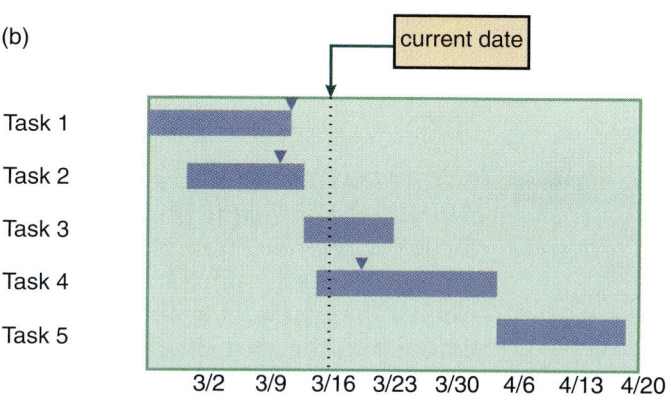

(b)

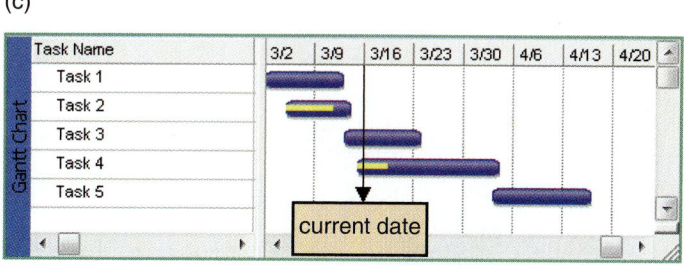

(c)

FIGURE 3-8   Three different ways to show that Task 1 was completed ahead of schedule, Task 2 is about 80 percent done and is behind schedule, Task 3 is late starting, Task 4 is ahead of schedule, and Task 5 will begin in several weeks.

TASK BOX FORMAT

Task Name	
Start Day/Date	Task ID
Finish Day/Date	Task Duration

FIGURE 3-9 Each section of the task box contains important information about the task, including the Task Name, Task ID, Task Duration, Start Day/Date, and Finish Day/Date.

PERT/CPM Tasks

In a PERT/CPM chart, project tasks are shown as rectangular boxes, arranged in the sequence in which they must be performed. Each rectangular box, called a **task box**, has five sections, as shown in Figure 3-9. Each section of the task box contains important information about the task, including the Task Name, Task ID, Task Duration, Start Day/Date, and Finish Day/Date.

TASK NAME The **task name** should be brief and descriptive, but it does not have to be unique in the project. For example, a task named *Conduct Interviews* might occur in several phases of the project.

TASK ID The **task ID** can be a number or code that provides unique identification.

TASK DURATION The **duration** is the amount of time it will take to complete a task. All tasks must use the same time units, which can be hours, days, weeks, or months, depending on the project. An actual project starts on a specific date, but can also be measured from a point in time, such as *Day 1*.

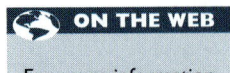

ON THE WEB

For more information about PERT/CPM, visit **scsite.com/ sad8e/more**, locate Chapter 3, and then click the PERT/ CPM link.

START DAY/DATE The **start** is the time that a task is scheduled to begin. For example, suppose that a simple project has two tasks: Task 1 and Task 2. Also suppose that Task 2 cannot begin until Task 1 is finished. An analogy might be that you cannot run a program until you turn on your computer. If Task 1 begins on Day 1 and has a duration of three days, it will finish on Day 3. Because Task 2 cannot begin until Task 1 is completed, the start time for Task 2 is Day 4, the day after Task 1 is finished.

FINISH DAY/DATE The **finish** is the time that a task is scheduled to be completed. To calculate the finish day or date, you add the duration to the start day or date. When you do this, you must be very careful not to add too many days. For example, if a task starts on Day 10 and has a duration of 5 days, then the finish would be on Day 14 — *not* Day 15.

Task Patterns

In any project, large or small, tasks depend on each other and must be performed in a sequence, not unlike the commands in a software program. **Task patterns** can involve dependent tasks, multiple successor tasks, and multiple predecessor tasks. In larger projects, these patterns can become quite complex, and an analyst must study the logical flow carefully.

DEPENDENT TASKS When tasks must be completed one after another, like the relay race shown in Figure 3-10, they are called **dependent tasks**, because one depends on the other. For example, Figure 3-11 shows that Task 2 depends on Task 1, because Task 2 cannot start until Task 1 is completed. In this example, the finish time of Task 1, Day 5, controls the start date of Task 2, which is Day 6.

FIGURE 3-10 In a relay race, each runner is dependent on the preceding runner and cannot start until the earlier runner finishes.

MULTIPLE SUCCESSOR TASKS

When several tasks can start at the same time, each is called a **concurrent task**. Often, two or more concurrent tasks depend on a single prior task, which is called a **predecessor task**. In this situation, each concurrent task is called a **successor task**. In the example shown in Figure 3-12, successor Tasks 2 and 3 both can begin as soon as Task 1 is finished. Notice that the finish time for Task 1 determines the start time for both Tasks 2 and 3. In other words, the earliest that Task 1 can finish is day 30, so day 31 is the earliest that Tasks 2 and 3 can start.

MULTIPLE PREDECESSOR TASKS

Suppose that a task requires two or more prior tasks to be completed before it can start. Figure 3-13 shows an example of this situation, where Task 3 cannot begin until Tasks 1 and 2 are both completed. Because the two tasks might not finish at the same time, the longest (latest) predecessor task becomes the controlling factor. Notice that the start for Task 3 is 16, not 6. Why is this so? Because Task 3 depends on two predecessor tasks, Tasks 1 and 2, Task 3 cannot begin until the *later* of those tasks is complete. Therefore, the start time for a successor task must be the latest (largest) finish time for any of its preceding tasks. In the example shown, Task 1 ends on Day 15, while Task 2 ends on Day 5, so Task 1 controls the start time for Task 3.

Complex Task Patterns

When various task patterns combine, you must study the facts carefully in order to understand the logical sequence. A project schedule will not be accurate unless the underlying task pattern is logically correct. For example, consider the following three fact statements and the task patterns they represent, which are shown in Figure 3-14 on the next page.

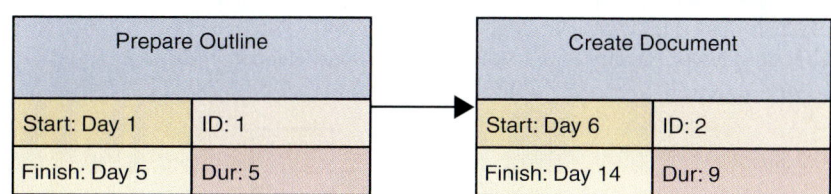

EXAMPLE OF A DEPENDENT TASK

FIGURE 3-11 This example of a dependent task shows that the finish time of Task 1, Day 5, controls the start date of Task 2, which is Day 6.

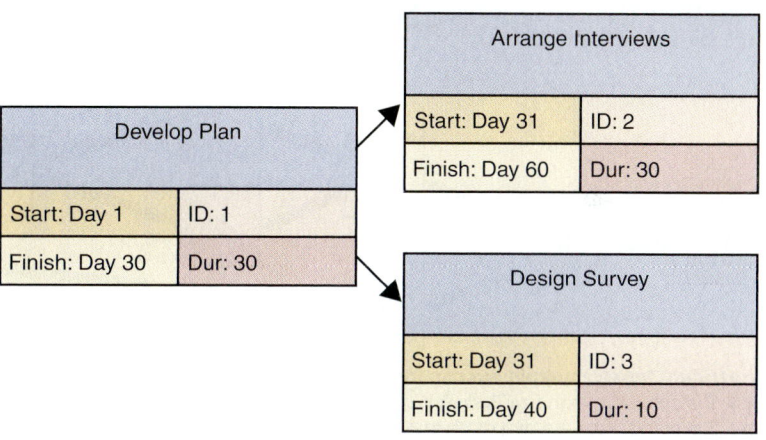

EXAMPLE OF MULTIPLE SUCCESSOR TASKS

FIGURE 3-12 This example of multiple successor tasks shows that the finish time for Task 1 determines the start time for both Tasks 2 and 3.

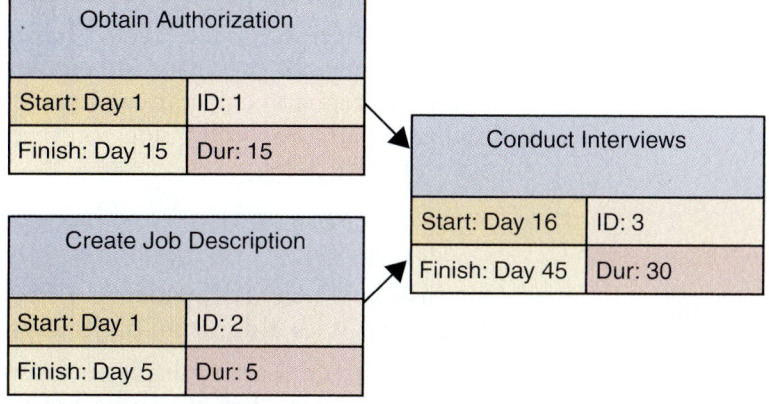

EXAMPLE OF MULTIPLE PREDECESSOR TASKS

FIGURE 3-13 This example of multiple predecessor tasks shows that the start time for a successor task must be the latest (largest) finish time for any of its preceding tasks. In the example shown, Task 1 ends on Day 15, while Task 2 ends on Day 5, so Task 1 controls the start time for Task 3.

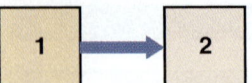

Dependent tasks: Perform Task 1. When Task 1 is complete, perform Task 2.

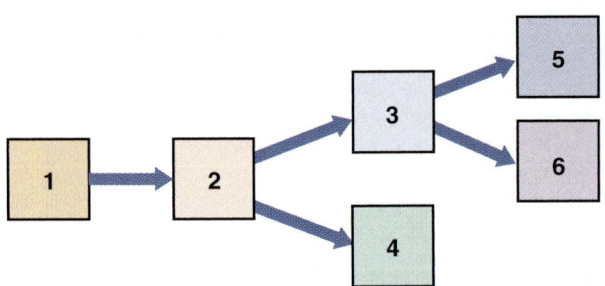

Dependent tasks and multiple successor tasks: Perform Task 1. When Task 1 is complete, perform Task 2. When Task 2 is finished, start two tasks: Task 3 and Task 4. When Task 3 is complete, start two more tasks: Task 5 and Task 6.

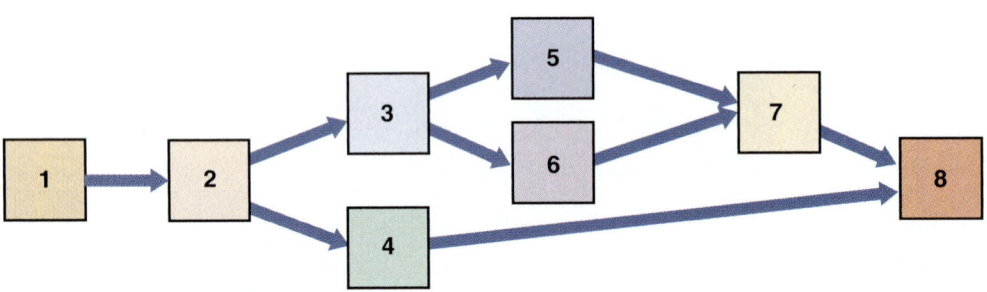

Dependent tasks, multiple successor tasks, and multiple predecessor tasks: Perform Task 1. When Task 1 is complete, perform Task 2. When Task 2 is finished, start two Tasks: Task 3 and Task 4. When Task 3 is complete, start two more tasks: Task 5 and Task 6. When Tasks 5 and 6 are done, start Task 7. Then, when Tasks 4 and 7 are finished, perform Task 8.

FIGURE 3-14

DEPENDENT TASKS Perform Task 1. When Task 1 is complete, perform Task 2.

DEPENDENT TASKS AND MULTIPLE SUCCESSOR TASKS Perform Task 1. When Task 1 is complete, perform Task 2. When Task 2 is finished, start two tasks: Task 3 and Task 4. When Task 3 is complete, start two more tasks: Task 5 and Task 6.

DEPENDENT TASKS, MULTIPLE SUCCESSOR TASKS, AND MULTIPLE PREDECESSOR TASKS Perform Task 1. When Task 1 is complete, perform Task 2. When Task 2 is finished, start two Tasks: Task 3 and Task 4. When Task 3 is complete, start two more tasks: Task 5 and Task 6. When Tasks 5 and 6 are done, start Task 7. Then, when Tasks 4 and 7 are finished, perform Task 8.

A PERT/CPM Example with Five Tasks

Figure 3-15 shows a PERT/CPM chart with five tasks. Task 2 is a dependent task that has multiple successor tasks. Task 5 has multiple predecessor tasks. In this figure, the analyst has arranged the tasks and entered task names, IDs, and durations.

The next step is to calculate and enter start and finish times for each task. The following explanation will guide you through the calculations, and the results are shown in Figure 3-16.

- Task 1 starts on Day 1 and has a duration of 10 days, so the finish date is Day 10.

- Task 2, which is dependent on Task 1, can start on Day 11 — the day after Task 1 ends. With a duration of 30 days, Task 2 will end on Day 40.

- Tasks 3 and 4 are multiple successor tasks, and depend on Task 2. Task 2 ends on Day 40, so Tasks 3 and 4 both can start on Day 41. Task 3 has a duration of 5 days and will end on Day 45. Task 4 has a duration of 25 days, and will not end until Day 65.

- Task 5 depends on Tasks 3 and 4, which are multiple predecessors. Because Task 5 depends on both tasks, it cannot start until the later of the two tasks is complete. In this example, Task 3 ends earlier, but Task 4 will not be completed until Day 65, so Task 5 must start on Day 66.

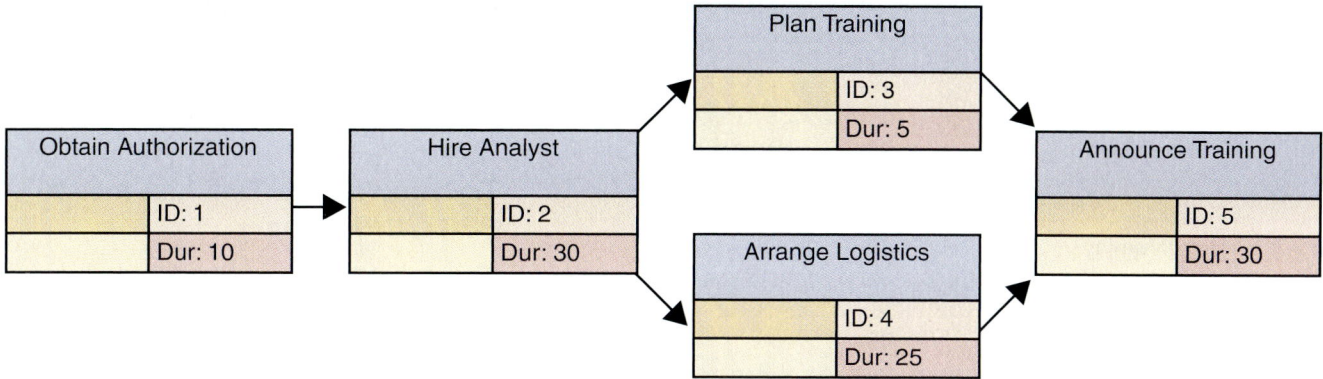

FIGURE 3-15 Example of a PERT/CPM chart with five tasks. Task 2 is a dependent task that has multiple successor tasks. Task 5 has multiple predecessor tasks. In this figure, the analyst has arranged the tasks and entered task names, IDs, and durations.

Critical Path

Once the tasks and relationships have been defined as shown in the previous examples, you can determine the critical path. The best way to describe the critical path concept is to give a familiar example. Suppose that you invite Joan and Jim and to your home for dinner. Joan arrives on time, but Jim is 30 minutes late, which delays the meal by 30 minutes because you do not want to start without him.

Similarly, as you can see in Figure 3-16, Task 5 cannot begin until Tasks 3 and 4 both are completed. In this case, Task 4 is the controlling factor, because Task 4 finishes on Day 65, which is 20 days later than Task 3, which is completed on Day 45.

Tasks 1, 2, 4, and 5 represent the critical path, which is highlighted in the figure. A **critical path** is a series of tasks which, if delayed, would affect the final completion date of the overall project. In other words, tasks on the critical path have no slack time. **Slack time** is the amount of time that the task could be late without pushing back the completion date of the entire project.

In this example, Task 2 could be as much as 10 days late before it would have an impact on the overall project completion date. In Figure 3-16, 20 days of slack time exist for Task 3, so Task 3 could start up to 20 days late and still not affect the overall project completion date.

If any task along the critical path falls behind schedule, the entire project is delayed. As the name implies, a critical path includes all tasks that are vital to the project schedule. Project managers always must be aware of the critical path, so they can monitor the vital tasks and keep the project on track. Microsoft Project and other project management

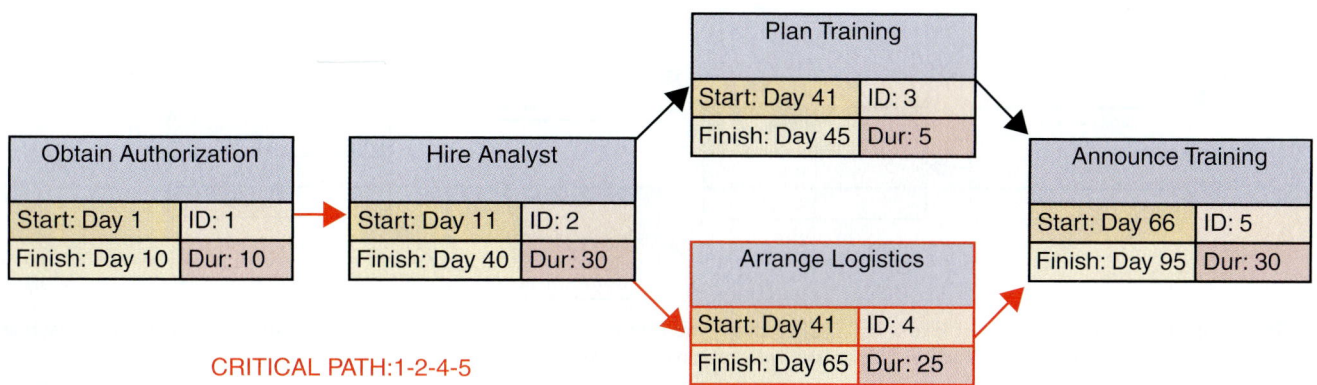

CRITICAL PATH:1-2-4-5

FIGURE 3-16 Now the analyst has entered the start and finish times, using the rules explained in this section. Notice that the overall project has a duration of 95 days.

TASK	DESCRIPTION	DURATION (DAYS)	PREDECESSOR TASKS
1	Develop Plan	1	-
2	Assign Tasks	4	1
3	Obtain Hardware	17	1
4	Programming	70	2
5	Install Hardware	10	3
6	Program Test	30	4
7	Write User Manual	25	5
8	Convert Files	20	5
9	System Test	25	6
10	User Training	20	7, 8
11	User Test	25	9,10

FIGURE 3-17 Example of a table listing 11 tasks, together with their descriptions, durations, and predecessor tasks.

software can display critical path information. If necessary, a project manager can reassign resources to keep the project on schedule.

Transforming a Task List into a PERT/CPM Chart

Figure 3-17 shows a list of 11 tasks. Notice that each task has an ID, a description, a duration, and a reference to predecessor tasks, if any, which must be completed before the task can begin. Also notice that dependent tasks can have one predecessor task, or several. You construct a PERT/CPM chart from this task list in a two-step process:

STEP 1: CREATE THE WORK BREAKDOWN STRUCTURE In the first step, as shown in Figure 3-18, you identify the tasks, determine task dependencies, and enter the name, ID, and duration for each task. Notice that this example includes dependent tasks, tasks with multiple successor tasks, and tasks with multiple predecessor tasks.

STEP 2: ENTER START AND FINISH TIMES In the second step, as shown in Figure 3-19, you enter the start and finish times by applying the guidelines in this section. For example, Task 1 has a one-day duration, so you enter the start and finish times for Task 1 as Day 1. Then you enter Day 2 as the start time for successor Tasks 2 and 3. Continuing from left

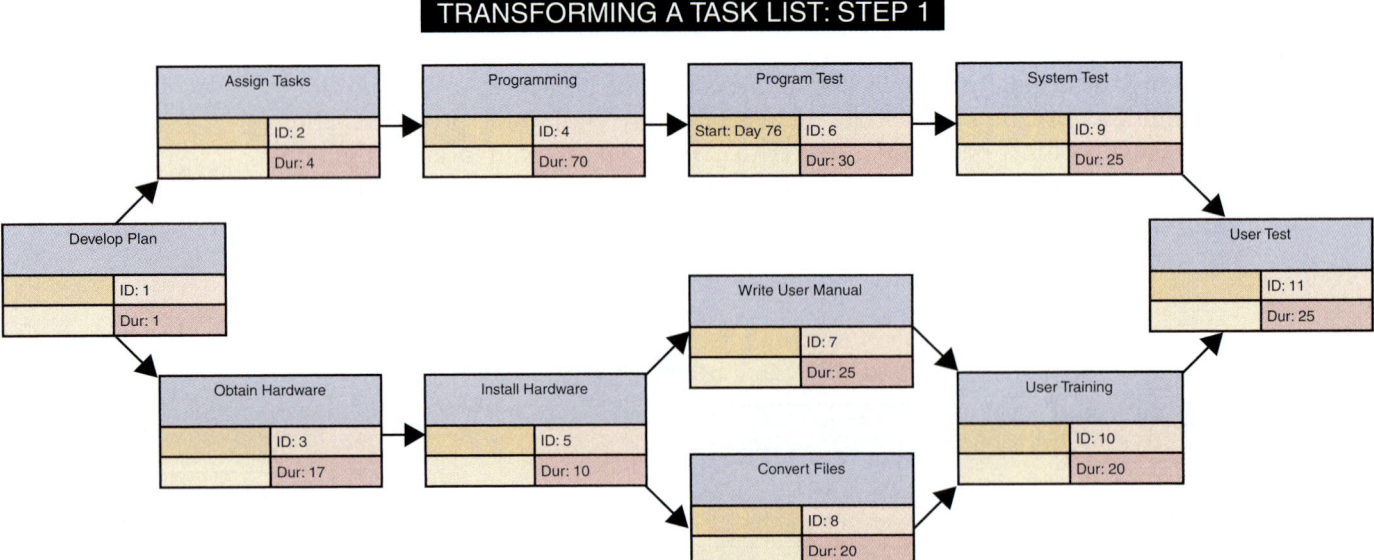

TRANSFORMING A TASK LIST: STEP 1

FIGURE 3-18 To transform a task list into a PERT/CPM chart, you first enter the task name, ID, duration, and predecessors for each task. Notice that this example includes dependent tasks, tasks with multiple successors, and tasks with multiple predecessors.

to right, you add the task duration for each task to its start time to determine its finish time. As you proceed, there are three important rules you must keep in mind:

- If a successor task has more than one predecessor task, use the *latest* finish time of the predecessor tasks to determine the start time for the successor task.

- If a predecessor task has more than one successor task, use the predecessor task's finish time to determine the start time for *all* successor tasks.

- Continuing from left to right, add the task duration for each task to its start time to determine and enter its finish time. Again, be very careful not to add too many days. For example, if a task starts on Day 10 and has a duration of 5 days, then the finish would be Day 14 — *not* Day 15.

When you have entered all the start and finish times, you determine that the project will be completed on Day 155. Also, you note that Tasks 1, 2, 4, 6, 9, and 11 represent the critical path, as shown by the red arrows in Figure 3-19.

Comparing Gantt Charts and PERT/CPM Charts

Although a Gantt chart offers a rapid overview that graphically displays the timing, duration, and progress of each task, many project managers find PERT/CPM charts more helpful for scheduling, monitoring, and controlling projects. A project manager can convert task start and finish times to actual dates by laying out the entire project on a calendar. Then, on any given day, the manager can compare what should be happening with what is taking place, and react accordingly.

Also, a PERT/CPM chart displays complex task patterns and relationships. This information is valuable to a manager who is trying to address the highest priority issues. PERT/CPM and Gantt charts are not mutually exclusive techniques. Project managers often use both methods.

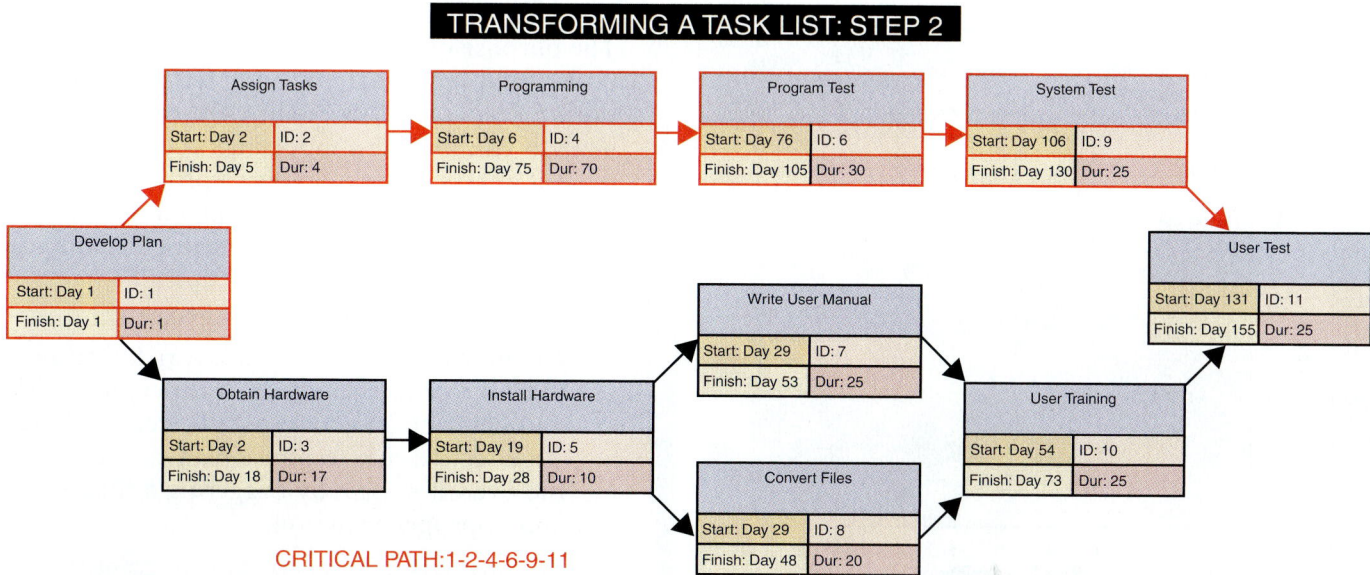

FIGURE 3-19 To complete the PERT/CPM chart, you apply the guidelines explained in this section. For example, Task 1 has a one-day duration, so you enter the start and finish for Task 1 as Day 1. Then you enter Day 2 as the start for successor Tasks 2 and 3.

PROJECT RISK MANAGEMENT

Every IT project involves risks that systems analysts and project managers must address. A **risk** is an event that could affect the project negatively. **Risk management** is the process of identifying, analyzing, anticipating, and monitoring risks to minimize their impact on the project.

Steps in Risk Management

The first step in risk management is to develop a specific plan. Although project management experts differ with regard to the number of steps or phases, a basic list would include the following tasks:

- *Develop a risk management plan.* A **risk management plan** includes a review of the project's scope, stakeholders, budget, schedule, and any other internal or external factors that might affect the project. The plan should define project roles and responsibilities, risk management methods and procedures, categories of risks, and contingency plans.

- *Identify the risks.* **Risk identification** lists each risk and assesses the likelihood that it could affect the project. The details would depend on the specific project, but most lists would include a means of identification, and a brief description of the risk, what might cause it to occur, who would be responsible for responding, and the potential impact of the risk.

- *Analyze the risks.* This typically is a two-step process: Qualitative risk analysis and quantitative risk analysis. **Qualitative risk analysis** evaluates each risk by estimating the probability that it will occur and the degree of impact. Project managers can use a formula to weigh risk and impact values, or they can display the results in a two-axis grid. For example, a Microsoft Excel XY chart can be used to display the matrix, as shown in Figure 3-20. In the chart, notice the various combinations of risk and impact ratings for the five sample values. This tool can help a project manager focus on the most critical areas, where risk probability and potential impact are high. The purpose of **quantitative risk analysis** is to understand the actual impact in terms of dollars, time, project scope, or quality. Quantitative risk analysis can involve a modeling process called what-if analysis, which allows a project manager to vary one or more element(s) in a model to measure the effect on other elements. This topic is discussed in more detail in Chapter 12, Managing Systems Support and Security.

- *Create a risk response plan.* A **risk response plan** is a proactive effort to anticipate a risk and describe an action plan to deal with it. An effective risk response plan can reduce the overall impact by triggering a timely and appropriate action.

- *Monitor risks.* This activity is ongoing throughout the risk management process. It is important to conduct a continuous tracking process that can identify new risks, notice changes in existing risks, and update any other areas of the risk management plan.

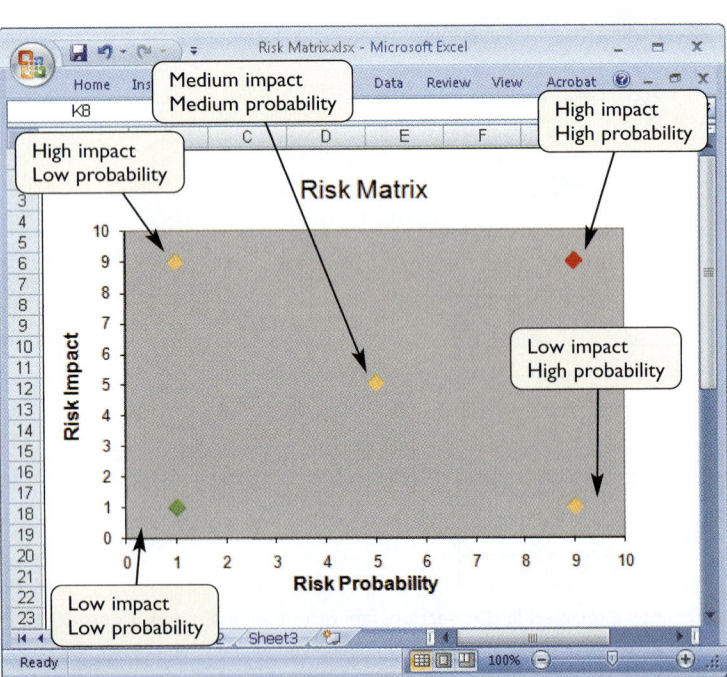

FIGURE 3-20 You can use a Microsoft Excel XY chart type to display a risk matrix that shows risk probability and potential impact.

Risk Management Software Tools

Most project management software programs, such as Microsoft Project, contain various tools that a project manager can use. For example, he or she can assign specific dates as constraints, align task dependencies, note external factors that might affect a task, track task progress, and display tasks that are behind schedule. In addition, some vendors offer risk management add-ons, such as the one shown in Figure 3-21.

In addition to offering Microsoft Office Project 2007 Standard and Project 2007 Professional versions, Microsoft sells an enterprise edition called Microsoft Office Project Server 2007. This version has a built-in risk management capability that can be used for large, corporate-wide projects. Microsoft claims that the software can link risks with specific tasks and projects, specify probability and impact, assign ownership, and track progress to manage projects more efficiently.

FIGURE 3-21 Some vendors offer risk management software that can enhance Microsoft Project's capabilities.

Microsoft's risk management model includes the following factors:

- Probability, which represents the likelihood that the risk will happen, expressed as a percentage
- Impact, which indicates the degree of adverse effect should the risk occur, on a scale of 1 to 10
- Cost, which indicates the potential financial impact of the risk
- Category, which specifies the risk type
- Description, which specifies the nature of the risk
- Mitigation plan, which identifies plans to control or limit the risk
- Contingency plan, which specifies actions to be taken if the risk occurs
- Trigger, which identifies a condition that would initiate the contingency plan

Armed with this information, the IT team can make a recommendation regarding the risks associated with the project. Depending on the nature and magnitude of the risks, the final decision might be made by management.

PROJECT MONITORING AND CONTROL

A project must be planned and scheduled before the work actually starts. After the project tasks begin, the project manager concentrates on monitoring and controlling the project.

Monitoring and Control Techniques

Regardless of whether the project was planned and scheduled with project management software or in some other manner, the project manager must keep track of the tasks and progress of team members, compare actual progress with the project plan, verify the completion of project milestones, and set standards and ensure that they are followed.

To help ensure that quality standards are met, many project managers institute structured walk-throughs. A **structured walk-through** is a review of a project team member's work by other members of the team. Generally, systems analysts review the work of other systems analysts, and programmers review the work of other programmers, as a form of peer review. Structured walk-throughs take place throughout the SDLC and are called **design reviews**, **code reviews**, or **testing reviews**, depending on the phase in which they occur.

Maintaining a Schedule

Maintaining a project schedule can be a challenging task, and most projects run into at least some problems or delays. By monitoring and controlling the work, the project manager tries to anticipate problems, avoid them or minimize their impact, identify potential solutions, and select the best way to solve the problem.

The better the original plan, the easier it will be to control the project. If clear, verifiable milestones exist, it will be simple to determine if and when those targets are achieved. If enough milestones and frequent checkpoints exist, problems will be detected rapidly.

A project that is planned and scheduled with PERT/CPM can be tracked and controlled using these same techniques. As work continues, the project manager revises the plan to record actual times for completed tasks and revises times for tasks that are not yet finished.

Project managers often spend most of their time tracking the tasks along the critical path, because delays in those tasks have the greatest potential to delay or jeopardize the project. Other tasks cannot be ignored, however. For example, suppose that a task not on the critical path takes too long and depletes the allotted slack time. At that point, the task actually becomes part of the critical path, and any further delay will push back the overall project.

PROJECT REPORTING

Members of the project team regularly report their progress to the project manager, who in turn reports to management and users. As shown in Figure 3-22, the project manager collects, verifies, organizes, and evaluates the information he or she receives from the team. Then the manager decides which information needs to be passed along, prepares a summary that can be understood easily, adds comments and explanations if needed, and submits it to management and users.

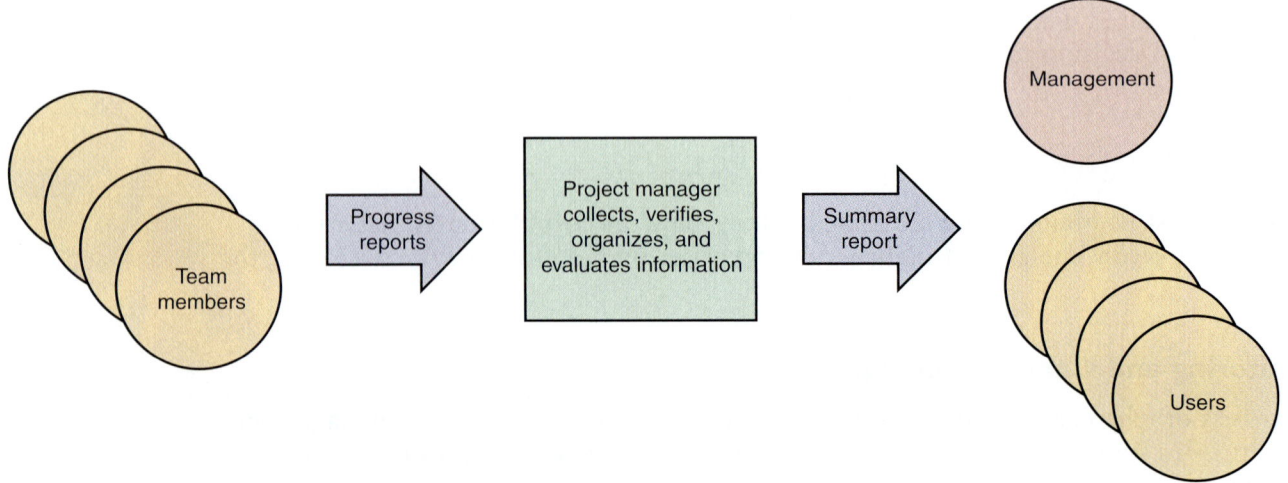

FIGURE 3-22 Members of the project team regularly report their progress to the project manager, who in turn reports to management and users.

Project Status Meetings

Project managers like the one shown in Figure 3-23 schedule regular meetings to update the team and discuss project status, issues, problems, and opportunities. Although meetings can be time consuming, most project managers believe they are worth the effort. The sessions give team members an opportunity to share information, discuss common problems, and explain new techniques. The meetings also give the project manager an opportunity to seek input and conduct brainstorming sessions.

Project Status Reports

A project manager must report regularly to his or her immediate supervisor, upper management, and users. Although a progress report might be given verbally to an immediate supervisor, reports to management and users usually are written. Gantt charts often are included in progress reports to show project status graphically.

Deciding how to handle potential problems can be difficult. At what point should you inform management about the possibility of cost overruns, schedule delays, or technical problems? At one extreme is the overly cautious project manager who alerts management to every potential snag and slight delay. The danger here is that the manager loses credibility over a period of time, and management might ignore potentially serious situations. At the other extreme is the project manager who tries to handle all situations single-handedly and does not alert management until a problem is serious. By the time management learns of the problem, little time might remain in which to react or devise a solution.

A project manager's best course of action lies somewhere between the two extremes, but is probably closer to the first. If you are unsure of the consequences, you should be cautious and warn management about the possibility of a problem. When you report the situation, you also should explain what you are doing to handle and monitor the problem. If you believe the situation is beyond your control, you might want to suggest possible actions that management can take to resolve the situation. Most managers recognize that problems do occur on most projects; it is better to alert management sooner rather than later.

FIGURE 3-23 Project managers schedule regular meetings to update the project team and discuss project status, issues, problems, and opportunities.

TOOLKIT TIME

The Communication Tools in Part I of the Systems Analyst's Toolkit can help you develop better reports and presentations. To learn more about these tools, turn to Part I of the four-part Toolkit that follows Chapter 12.

PROJECT MANAGEMENT SOFTWARE

Earlier in this chapter, you learned about project scheduling, and how task patterns and dependencies create a critical path for the project. Understanding these concepts will make it easier to learn how to use project management software to plan and manage your projects.

Project managers, like the one shown in Figure 3-24, can use **project management software** to help plan, estimate, schedule, monitor, and report on a project. Most programs offer features such as PERT/CPM, Gantt charts, resource scheduling, project calendars, and cost tracking.

FIGURE 3-24 Project managers can use powerful software to help them plan, estimate, schedule, monitor, and report on complex projects.

FIGURE 3-25   Open Workbench is a free, open-source project management program with powerful features and capabilities.

ON THE WEB

For more information about project management software, visit **scsite.com/sad8e/ more**, locate Chapter 3, and then click the Project Management Software link.

Project Management Software Examples

Microsoft Office Project 2007 is a powerful, full-featured program that holds the dominant share of the project management software market. On the Web, Microsoft offers demo versions, training, and tips for using Project 2007. Although Microsoft is the industry leader, many other vendors offer project management software, and you can explore these options by searching on the Web. One interesting product, **Open Workbench,** is available as free software, complete with manuals and sample projects, as shown in Figure 3-25. You can download this program from the Open Workbench site at *openworkbench*.org, or you can use the download link in the Features section of the Student Study Tool CD-ROM, which also contains a user manual for Open Workbench.

As the Web site explains, Open Workbench is **open-source software** that is supported by a large group of users and developers. Support options include community forums that are open to all users, various training packages, and third-party support. For many small to medium-sized projects, Open Workbench would be a cost-effective alternative that would compare favorably to Microsoft Project. Open Workbench also can exchange files with Microsoft Project by importing and exporting the data in XML file format.

A Sample Project Using Microsoft Project and Open Workbench

The first step in project management is to identify the tasks, task duration, and task dependencies. Typically, an analyst must review the results of planning and fact-finding in order to extract this information. For example, suppose you receive the task summary shown in Figure 3-26 from your IT manager.

Your manager would like you to create a Gantt chart and a PERT chart that show all tasks, dependencies, dates, and total project duration. Your first step is to create a Gantt chart showing the necessary information. You decide to use Microsoft Project to construct the chart as shown in Figure 3-27. As you enter each of the 12 tasks, you also enter the task duration and the predecessor tasks that must be completed before each task can begin.

Please study the following task summary:

- First, we will review the systems request. That will take three days.
- Then, two tasks can begin at once: We can review the documentation, which will take three days, and review the Internet access delays, which will take two days.
- When the documentation and the Internet access delays have been analyzed, we can contact managers about the interviews, which will take two days.
- After we contact the managers, we can plan the interview schedule, which will take two days.
- Next, we can prepare the preliminary investigation report, which will take two days.
- When the report is ready, we can deliver our presentation to the committee, which will take two days.
- After the presentation, three tasks can begin at once: We plan the interview questions, which will take one day; contact the interviewees, which will take one day; and send out the questionnaire, which will be returned in five days.
- When the interview questions are ready and the interviewees have been contacted, we can conduct the interviews, which will take three days.
- Finally, when the interviews have been conducted and the questionnaire results are back, we can tabulate all results, which will take one day.

FIGURE 3-26   A sample task summary.

When you are finished entering the tasks, the program displays the Gantt chart, which consists of 12 horizontal bars, connected with arrows that indicate the task dependencies. Notice that Saturdays and Sundays are shown as shaded columns, because no work will be performed on those days. The program makes these adjustments automatically. For example, Task 2, which has a duration of three days, starts on a Friday and ends on a Tuesday.

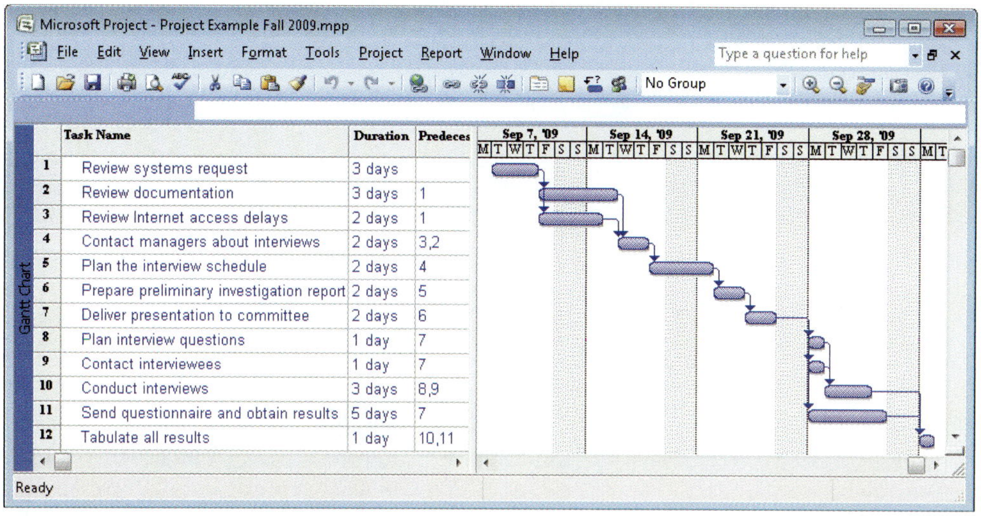

After you complete the Gantt chart, you decide to view the data in the form of a Microsoft Project network diagram, which is similar to a PERT chart. When you select the Network Diagram option on the View menu, you can see the project tasks and dependencies, as shown in Figure 3-28. You study the diagram and see that the program has calculated a start and finish date for each task.

Notice that the diagram displays the same information as the Gantt chart, including task dependencies, and also includes a red line that indicates the project's critical path. According to the diagram, if the project remains on schedule, the last task will be completed on Monday, October 5, 2009. Notice that the task boxes in Microsoft Project

FIGURE 3-27 This Microsoft Project screen shows a systems development project in the form of a Gantt chart.

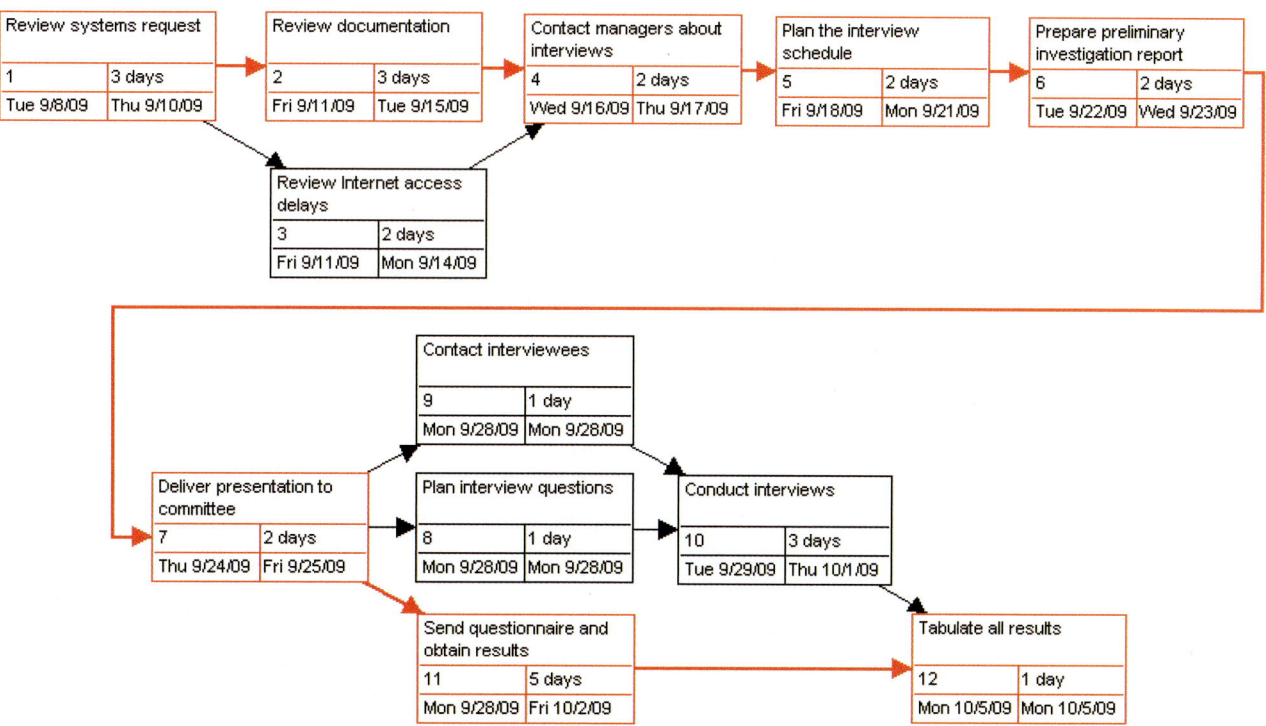

FIGURE 3-28 Using Microsoft Project, you can display a network diagram, which is similar to a PERT chart. Notice that the critical path appears as a red line.

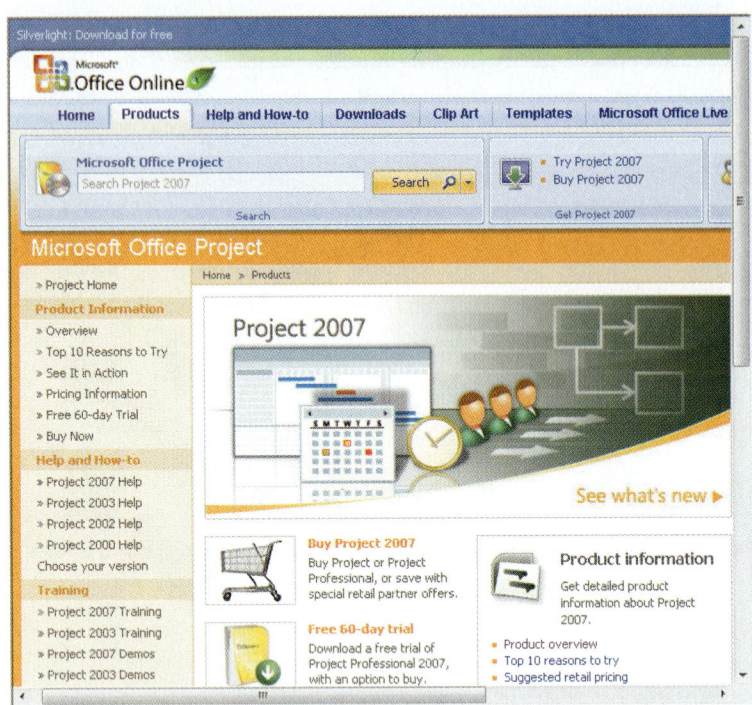

FIGURE 3-29 Microsoft Office Project 2007 is the latest version of this popular program, and includes many new features and benefits.

are similar to PERT/CPM task boxes. Using Microsoft Project, you can assign each task to one or more people, assign budget targets, produce progress reports, and readjust schedules and deadlines as necessary.

The latest version of Project is Office Project 2007, as shown in Figure 3-29. The new release is offered in a Standard version, a Professional version, and a Server version that includes support for large, enterprise-wide projects. In addition to providing a full description, demos, and training on its Web site, Microsoft also offers a free 60-day trial version that allows you to install, use, and evaluate the program.

Suppose that you did not have access to Microsoft Project, and you decided to use the Open Workbench program, which is free. Figure 3-30 shows examples of Open Workbench screens based on the same sample project that was shown in Figures 3-27 and 3-28. In Open Workbench, you create tasks and durations, indicate dependencies, and assign resources, just as you would in Microsoft Project. Notice that the critical path is highlighted, both in the Gantt chart and the network diagram.

Regardless of which software you use, you can see from these examples that project schedules, task estimates, and personnel assignments all are interrelated. Therefore, project planning is a dynamic task and involves constant change. One significant advantage of integrated interactive project management software is that it allows the project manager to adjust schedules, estimates, and resource assignments rapidly to develop a workable plan.

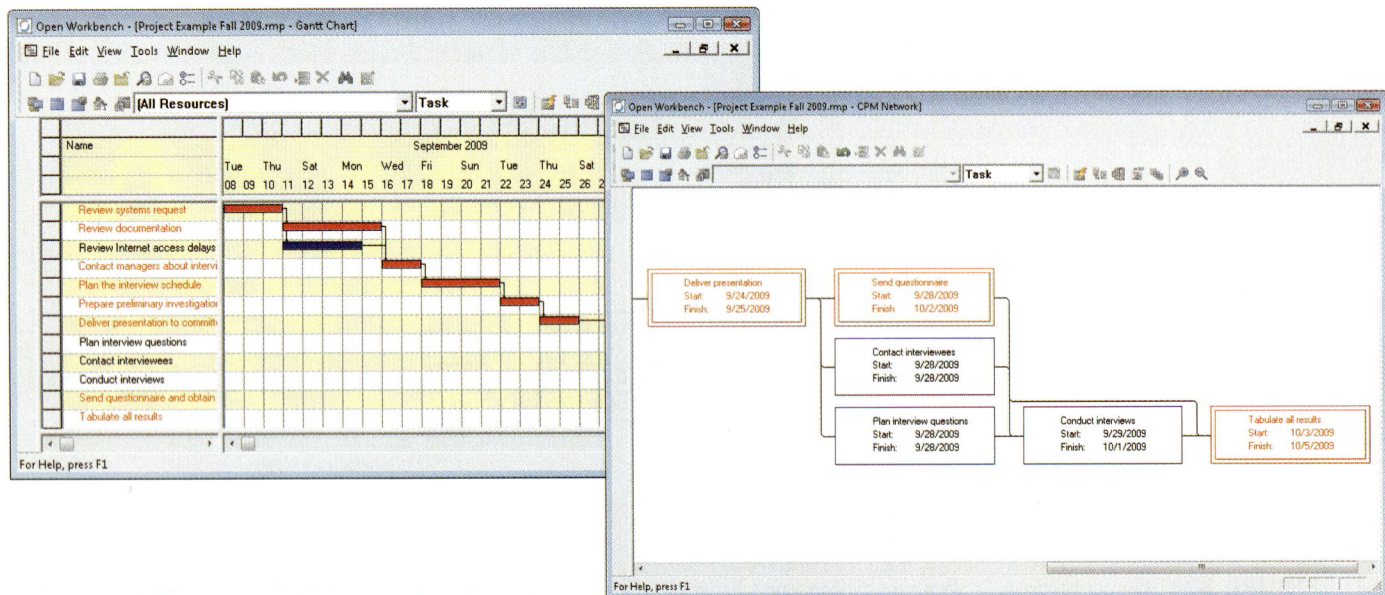

FIGURE 3-30 Open Workbench can show the sample project as a Gantt chart, or as a PERT chart that includes tasks, durations, dependencies, and a highlighted critical path.

CASE IN POINT 3.4: CENSUS 2010

In April 2008, the U.S. Commerce Department canceled a plan to acquire 500,000 handheld computers to tabulate data during the 2010 census. According to Commerce Secretary Carlos Gutierrez, costs had skyrocketed. He blamed the problem on "a lack of effective communications with one of our major suppliers."

Apparently, there was plenty of blame to go around. Secretary Gutierrez noted that the Census Bureau had submitted numerous technical changes to the vendor, Harris Corporation. This greatly increased the cost and the complexity of the devices. Gutierrez stated, "The Census Bureau was unaccustomed to working with an outside vendor on such a large contract." He also pointed out that the vendor had submitted an initial estimate of $36 million to operate a help desk to assist census-takers, but that figure had jumped to $217 million. "It was a bad estimate. I can't think of a better way to say it. Harris gave us the number. We accepted it. It was totally underestimated."

What can be learned from the failure of this project, and could it have been prevented? Suppose you were asked to head up a similar project. What would you do to prevent a similar outcome?

SOFTWARE CHANGE CONTROL

Software change control is the process of managing and controlling changes requested after the system requirements document has been submitted and accepted. Software change control can be a real problem because the development process involves many compromises, and users are never entirely satisfied with the results. Changes to an information system's requirements are inevitable. The issue, therefore, is how to create an effective process for controlling changes that protects the overall project, but allows those changes that are necessary and desirable.

The project coordinator, rather than the project manager, has primary responsibility for change control because requests for change most often are initiated by someone outside the information systems department. A specific process must be in place for handling requested changes. The process must be formal, but flexible enough to incorporate desired changes promptly with minimal impact to the overall project.

A procedure for processing requests for changes to an information system's requirements consists of four steps: Complete a change request form, take initial action on the request, analyze the impact of the requested change, and determine the disposition of the requested change.

1. Complete a change request form. The person requesting the change completes a System Requirements Change Request Form similar to the one shown in Figure 3-31. On the form, which can be stored in an online network library, the requester describes and

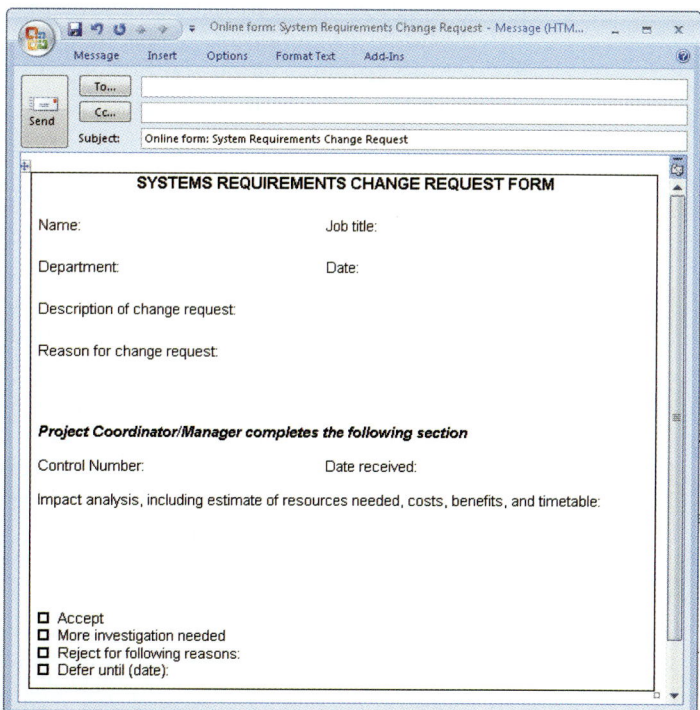

FIGURE 3-31 Sample of a Systems Requirements Change Request Form that can be stored in an online library. Notice the Impact analysis section; if a system has a high number of changes, what does that indicate?

justifies the desired changes. The requester attaches helpful documents and pertinent information, such as new calculations, copies of government regulations, and memos from executives specifying new strategies and directions.

2. Take initial action on the request. The project coordinator enters a sequential control number and the date on the change request form, reviews the specific change, and then determines if the change should be accepted, deferred to a later date, rejected for specific reasons, or investigated further. If the request is deferred or rejected, the project coordinator sends a copy of the request back to the requester. If the change is to be investigated further, then the request is reviewed for impact by the project manager or a systems analyst.

3. Analyze the impact of the requested change. The project manager or a systems analyst must review the request and determine the impact of incorporating the change into the information system's requirements. Then, the manager or analyst prepares an impact analysis that describes the effect of the change on the information system's requirements and on costs and schedules. The analysis should address the impact of incorporating the change immediately versus incorporating the change after the currently configured information system has been implemented.

4. Determine the disposition of the requested change. Based on the impact analysis and the project coordinator's recommendation, the change might be accepted, deferred, or rejected. In each of the three cases, the project coordinator informs the requester of the action taken.

ON THE WEB

For more information about software change control, visit **scsite.com/sad8e/ more**, locate Chapter 3, and then click the Software Change Control link.

KEYS TO PROJECT SUCCESS

To be successful, an information system must satisfy business requirements, stay within its budget, be completed on time, and — most important of all — be managed effectively. When a project develops problems, the reasons typically involve business, budget, or schedule issues, as explained in the following sections. In addition to planning and managing the project, a project manager must be able to recognize problems and deal with them effectively.

Business Issues

The major objective of every system is to provide a solution to a business problem or opportunity. If the system does not do this, then it is a failure — regardless of positive reaction from users, acceptable budget performance, or timely delivery. When the information system does not meet business requirements, causes might include unidentified or unclear requirements, inadequately defined scope, imprecise targets, shortcuts or sloppy work during systems analysis, poor design choices, insufficient testing or inadequate testing procedures, and lack of change control procedures. Systems also fail because of changes in the organization's culture, funding, or objectives. A system that falls short of business needs also produces problems for users and reduces employee morale and productivity.

Budget Issues

Cost overruns typically result from one or more of the following:

- Unrealistic estimates that either are too optimistic or are based on incomplete definitions of the work to be done
- Failure to develop an accurate TCO forecast that considered all cost elements
- Poor monitoring of progress and inadequate reaction to early signs of problems

- Schedule delays due to unanticipated factors
- Human resource factors, including turnover, inadequate training, and motivation issues

Schedule Issues

Problems with timetables and project milestones can indicate a failure to recognize task dependencies, confusion between effort and progress, poor monitoring and control methods, personality conflicts among team members, or turnover of project personnel.

The failure of an IT project also can be caused by poor project management techniques. If the project manager fails to plan, staff, organize, supervise, communicate, motivate, evaluate, direct, and control properly, then the project is certain to fail. Even when factors outside his or her control contribute to the failure, the project manager is responsible for recognizing the early warning signs and handling them effectively.

Successful Project Management

Project management is a challenging task. Project managers must be alert, technically competent, and highly resourceful. They also must be good communicators with strong human resource skills. A project manager rightly can be proud when he or she handles a successful project that helps the company achieve its business objectives. Unfortunately, projects can and do get derailed for a wide variety of reasons. When problems occur, the project manager's ability to handle the situation becomes the critical factor.

When a project manager first recognizes that a project is in trouble, what options are available? Alternatives can include trimming the project requirements, adding to the project resources, delaying the project deadline, and improving management controls and procedures.

Sometimes, when a project experiences delays or cost overruns, the system still can be delivered on time and within budget if several less critical requirements are trimmed. The system can be delivered to satisfy the most necessary requirements, and additional features can be added later as a part of a maintenance or enhancement project.

If a project is in trouble because of a lack of resources or organizational support, management might be willing to give the project more commitment and higher priority. For example, management might agree to add more people to a project that is behind schedule. Adding staff, however, will reduce the project's completion time only if the additional people can be integrated effectively into the development team.

If team members lack experience with certain aspects of the required technology, temporary help might be obtained from IT consultants or part-time staff. Adding staff can mean training and orienting the new people, however. In some situations, adding more people to a project actually might increase the time necessary to complete the project. More staff also means increased costs and the potential for exceeding budget limits.

When a project is behind schedule, a typical response is to push back the completion date. This is an option only if the original target date is flexible and the extension will not create excessive costs or other problems.

A QUESTION OF ETHICS

"Better blow the whistle," says Roy, your friend and project teammate at Final Four Industries. "The project is out of control, and you know it!" "Maybe so," you respond, "But that's not my call — I'm not the project manager." What you don't say is that Stephanie, the project manager, feels like her career is on the line and she is reluctant to bring bad news to management at this time. She honestly believes that the project can catch up, and says that a bad report on a major project could result in bad publicity for the firm and frighten potential customers.

To be fair, the next management progress report is scheduled in three weeks. It is possible that the team could catch up, but you doubt it. You wonder if there is an ethical question here: Even though the report isn't due yet, should a significant problem be reported to management as soon as possible? You are concerned about the issue, and you decide to discuss it with Stephanie. What will you say to her?

CHAPTER SUMMARY

Project management is the process of planning, scheduling, monitoring and controlling, and reporting upon the development of an information system. Project management is important throughout the SDLC, but is especially vital during the implementation phase of a project. The primary objective of project management is to deliver a system that meets all requirements on time and within budget. Although the project manager can use a variety of software tools that make the job easier, he or she must have a clear understanding of project management concepts and techniques.

Project management begins with identifying and planning all specific tasks or activities. Projects also have events or milestones that provide reference points to monitor progress. After identifying the tasks, project managers develop a work schedule that assigns specific tasks to project team members.

Time and cost estimates for tasks usually are made in person-days. A person-day represents the work that one person can accomplish in one day. Estimating the time for project activities is more difficult with larger systems. Project managers must consider the project size and scope, IT resources, prior experience with similar projects or systems, and applicable constraints.

In project scheduling, the project manager develops a specific time for each task, based on available resources and whether or not the task is dependent on other activities being accomplished earlier. The project manager can use graphical tools such as Gantt charts and PERT/CPM charts to assist in the scheduling process.

A Gantt chart is a horizontal bar chart that represents the project schedule with time on the horizontal axis and tasks arranged vertically. It shows individual tasks and task groups, which include several tasks. In a Gantt chart, the length of the bar indicates the duration of the tasks. A Gantt chart can display progress, but does not show task dependency details or resource assignment unless the chart was created with a project management program that supports dependency linking and the entry of other information.

A PERT/CPM chart shows the project as a network diagram with tasks connected by arrows. Using a prescribed calculation method, the project manager uses a PERT/CPM chart to determine the overall duration of the project and provide specific information for each task, including the task IDs, their durations, start and finish times, and the order in which they must be performed. With this information, the manager can determine the critical path, which is the sequence of tasks that have no slack time and must be performed on schedule in order to meet the overall project deadline. Project managers are responsible for risk management, which is the process of identifying, analyzing, anticipating, and monitoring risks to minimize their impact on the project. A project manager uses structured walk-throughs, which are reviews of a team member's work by other team members, to ensure that quality standards are met. A project manager also keeps team members and others up to date with regular reports and periodic meetings.

Project management software such as Microsoft Project and Open Workbench can assist you in project planning, estimating, scheduling, monitoring, and reporting. Software change control is concerned with change requests that arise after the system requirements document has been approved, and most companies establish a specific procedure for managing such requests. A typical change control procedure consists of four steps: completion of a change request form, initial action, impact analysis, and final disposition.

In the end, every successful information system must support business requirements, stay within budget, and be completed and available to users on time. Sound project management involves the same skills as any type of management. The project manager must be perceptive, analytical, well-organized, and a good communicator. If the project manager senses that the project is off-track, he or she must take immediate steps to diagnose and solve the problem.

Key Terms and Phrases

activity *99*
best-case estimate *102*
bottom-up technique *105*
Brooks' Law *101*
code review *114*
concurrent tasks *107*
critical path *109*
Critical Path Method (CPM) *105*
dependent task *106*
design review *114*
duration *106*
event *99*
finish day/date *106*
Gantt chart *104*
Microsoft Office Project 2007 *116*
milestone *99*
network diagram *117*
Open Workbench *116*
open-source software *116*
person-days *101*
PERT/CPM *105*
predecessor task *107*
probable-case estimate *102*
Program Evaluation Review Technique (PERT) *105*
project coordinator *98*
project creep *100*
project leader *98*
project management *98*
project management software *115*

project manager *98*
project monitoring and controlling *99*
project planning *99*
project reporting *99*
project scheduling *99*
project tasks *105*
qualitative risk analysis *112*
quantitative risk analysis *112*
risk *112*
risk identification *112*
risk management *112*
risk management plan *112*
risk response plan *112*
slack time *109*
software change control *119*
start day/date *106*
structured walk-through *114*
successor task *107*
task *99*
task box *106*
task group *104*
task ID *106*
task name *106*
task pattern *106*
testing review *114*
weight *102*
work breakdown structure (WBS) *100*
worst-case estimate *102*

Learn It Online

Instructions: To complete the Learn It Online exercises, start your browser, click the Address bar, and then enter the Web address **scsite.com/sad8e/learn**. When the Systems Analysis and Design Learn It Online page is displayed, follow the instructions in the exercises below. Each exercise has instructions for saving your results, either for your own records or for submission to your instructor.

1 Chapter Reinforcement

TF, MC, and SA

Below SAD Chapter 3, click the Chapter Reinforcement link. Print the quiz by clicking Print on the File menu for each page. Answer each question.

2 Flash Cards

Below SAD Chapter 3, click the Flash Cards link and read the instructions. Type 20 (or a number specified by your instructor) in the Number of playing cards text box, type your name in the Enter your name text box, and then click the Flip Card button. When the flash card is displayed, read the question and then click the ANSWER box arrow to select an answer. Flip through the Flash Cards. If your score is 15 (75%) correct or greater, click Print on the File menu to print your results. If your score is less than 15 (75%) correct, then redo this exercise by clicking the Replay button.

3 Practice Test

Below SAD Chapter 3, click the Practice Test link. Answer each question, enter your first and last name at the bottom of the page, and then click the Grade Test button. When the graded practice test is displayed on your screen, click Print on the File menu to print a hard copy. Continue to take practice tests until you score 80% or better.

4 Who Wants To Be a Computer Genius?

Below SAD Chapter 3, click the Computer Genius link. Read the instructions, enter your first and last name at the bottom of the page, and then click the Play button. When your score is displayed, click the PRINT RESULTS link to print a hard copy.

5 Wheel of Terms

Below SAD Chapter 3, click the Wheel of Terms link. Read the instructions, and then enter your first and last name and your school name. Click the PLAY button. When your score is displayed on the screen, right-click the score and then click Print on the shortcut menu to print a hard copy.

6 Crossword Puzzle Challenge

Below SAD Chapter 3, click the Crossword Puzzle Challenge link. Read the instructions, and then enter your first and last name. Click the SUBMIT button. Work the crossword puzzle. When you are finished, click the SUBMIT button. When the crossword puzzle is redisplayed, click the Print Puzzle button to print a hard copy.

Case-Sim: SCR Associates

Background

SCR Associates is a consulting firm that offers IT solutions and training. SCR needs an information system to manage operations at its new training center. The new system will be called TIMS (Training Information Management System). As a newly hired systems analyst, you will report to Jesse Baker, systems group manager. You will work on various tasks and practice the skills you learned in this chapter.

Using the Case

The SCR Associates case study is a Web-based simulation that allows you to practice your skills in a real-world environment. The case study transports you to SCR's company intranet, where you can complete 12 work sessions, each aligning with a chapter. As you work on the case, you will receive e-mail and voice mail messages, obtain information from SCR's online libraries, and perform various tasks. The first time you enter the SCR case, you should go to the starting page at **scsite.com/sad8e/scr** for detailed instructions.

Preview: Session 3

Now that TIMS has been approved, Jesse Baker has asked you to help her manage the project. She suggested that you review project management concepts and practice your skills, so that you will be able to handle a future project on your own.

To start a work session, you log on to the SCR intranet. When you enter your name and password, an opening screen displays links to the work sessions, and you select Session 3. At this point, you check your e-mail and voice mail messages carefully. You then begin working on your task list, which includes the following items:

Tasks: Managing Information Systems Projects

1. Jesse wants me to investigate Open Workbench software to determine whether it would be suitable for SCR. She asked me to prepare a summary of pros and cons, and a sample of screen shots and information.

2. Jesse likes the idea of using task completion estimates with best-case, probable-case, and worst-case estimates. She said that I should use typical formulas and weight values to create a Microsoft Excel spreadsheet that would make it easier to calculate expected task durations.

3. To practice my skills, Jesse asked me to create an imaginary project with 10 tasks, which include dependent, multiple predecessor, and multiple successor tasks. She wants me to create a list showing the tasks and dependencies, and then lay it out on paper to show the logical flow, and the duration, start, and finish for each task.

4. I'm excited to be part of the project team, and Jesse wants me to prepare a brief handout for the other team members with some do's and don'ts regarding project management. She said to make it look like a checklist of keys to project success.

FIGURE 3-32 Task list: Session 3.

Chapter Exercises

Review Questions

1. What is project management, and what are its main objectives?
2. What is the relationship between tasks and events, or milestones?
3. If Project A has twice as many resources as Project B, will Project A be twice as complex as Project B? Why or why not?
4. What is the difference between dependent and concurrent tasks?
5. Compare the advantages and disadvantages of Gantt and PERT/CPM charts.
6. Define the following terms: best-case estimate, probable-case estimate, and worst-case estimate, and describe how project managers use these concepts.
7. How does a project manager calculate start and finish times?
8. What is a critical path, and why is it important to project managers?
9. What are some project reporting and communication techniques?
10. What is software change control, and what are the four steps typically involved?

Discussion Topics

1. In *Poor Richard's Almanac*, Benjamin Franklin penned the familiar lines: "For the want of a nail the shoe was lost, for the want of a shoe the horse was lost, for the want of a horse the rider was lost, for the want of a rider the battle was lost, for the want of a battle the kingdom was lost — and all for the want of a horseshoe nail." Looking at the outcome in hindsight, could project management concepts have avoided the loss of the kingdom? Explain your answers.
2. Microsoft Project is an example of software that is very powerful, but quite expensive. As a project manager, how would you justify the purchase of this software? Also, would you consider using Open Workbench? Why or why not?
3. Suppose you want to manage a relatively small project, but you have no access to project management software of any kind. How could you use a spreadsheet program or a database program to manage the project? Share your ideas with the class.
4. Many managers claim to have "seat of the pants" intuition when it comes to project management. In your view, does this kind of intuition actually exist? Can you think of examples to support your views?

Projects

1. Think of all the tasks that you perform when you purchase a car. Include any research, decisions, or financial issues that relate to the purchase. Draw a Gantt chart that shows all the tasks and the estimated duration of each.
2. Perform Internet research to learn more about project risk management, and write a summary of the results. Be sure to search for a book titled *Waltzing with Bears: Managing Risk on Software Projects*, by Tom Demarco and Timothy Lister.
3. Go to Microsoft's Web site and navigate to the Download and Trials area. Select Microsoft Project Professional 2007, download the program, and install it. Then create a project based on the five tasks shown in Figure 3-16 on page 109. When the project is complete, click View, then click Network Diagram. Do the tasks resemble Figure 3-16 on page 109? Is the critical path the same?
4. Describe three personal experiences where a project management approach would have been helpful.

Apply Your Knowledge

The Apply Your Knowledge section contains four mini-cases. Each case describes a situation, explains your role in the case, and asks you to respond to questions. You can answer the questions by applying knowledge you learned in the chapter.

1 Countywide Construction

At Countywide Construction, you are trying to convince your boss that he should consider modern project management techniques to manage a complex project. Your boss says that he doesn't need anything fancy, and that he can guess the total time by the seat of his pants.

To prove your point, you decide to use a very simple example of a commercial construction project, with eight tasks. You create a hypothetical work breakdown structure, as follows:

- Prepare the foundation (10 days). Then assemble the building (4 days).
- When the building is assembled, start two tasks at once: Finish the interior work (4 days) and set up an appointment for the final building inspection (30 days).
- When the interior work is done, start two more tasks at once: landscaping (5 days) and driveway paving (2 days).
- When the landscaping and driveway are done, do the painting (5 days).
- Finally, when the painting is done and the final inspection has occurred, arrange the sale (3 days).

Now you ask your boss to estimate the total time and write his answer on a piece of paper. You look at the paper and see that his guess is wrong.

1. What is the correct answer? *Hint*: You might want to review the examples in Figure 3-14 on page 108 before you do the calculations.
2. What is the critical path?
3. Create a Gantt chart that shows the WBS.
4. Create a PERT/CPM chart.

2 Pleasantville High School Class

The computer science instructor at Pleasantville High School has asked you to visit her class and give a presentation about project management. You have just a few days to prepare, and you need to develop a presentation that briefly describes project management tools and techniques. You can be creative, and you might want to include examples of actual projects that you know about. In any case, try to describe how projects are planned, scheduled, monitored, and reported upon. Your presentation can be in the form of a Microsoft Word outline with notes, or as a set of PowerPoint slides.

1. Prepare opening comments that give the class an overview of project management.
2. Provide the class with a glossary of the most important project management terms and definitions.
3. Think of a common event like buying a new home, and show the class how a project manager might handle the matter.
4. Create a short scenario with 4 – 6 tasks, some of which depend on each other. You can use the two preceding cases as a model. Develop a sample answer that you will show the students after you give them a chance to analyze the tasks.

3   Lightfoot Industries

You have been asked to lead a training session for new employees at Lightfoot Industries. You must develop a specific schedule for the tasks listed below (the estimated task duration for each is shown in parentheses):

- First, you need to contact the participants and explain their roles (1 day). Then you must obtain approval from their department managers (5 days).
- After you obtain the approval, two tasks can begin at the same time: You can arrange the meeting room (4 days) and prepare an agenda for the initial session (11 days).
- When the agenda is ready, you can start two more concurrent tasks: Prepare the information packets (4 days) and create visual aids (8 days).
- When the meeting room is arranged and the information packets are ready, you can send out an e-mail to participants (1 day).
- Finally, after the e-mail is sent to participants and the visual aids are ready, you can conduct the JAD sessions (5 days).
  1. Prepare a list showing all tasks and their durations.
  2. Analyze the fact situation carefully to determine which tasks are concurrent and which ones are dependent on other tasks.
  3. Using PERT/CPM techniques, develop a chart that shows the project. Use a format similar to Figure 3-19 on page 111. If project management software is available, use it to develop the chart.
  4. What is the critical path for this project? How do you know?

4   Riverside Financial

At Riverside Financial, where you work as a project manager, you have been asked to conduct user training sessions during the implementation phase for a new information system. You must develop a specific schedule for the tasks (the estimated task duration for each is shown in parentheses):

- First, you need to send an e-mail message to all department managers announcing the training sessions (1 day).
- After the e-mail message goes out, two tasks can begin at the same time: You can develop the training material (4 days) and confirm arrangements for the training facility you plan to use (11 days).
- As soon as the training material is complete, you can work on two tasks at once: Arrange to have copies of handout material printed (3 days) and develop a set of PowerPoint slides (4 days).
- When the PowerPoint slides are ready, you conduct a practice training session with the instructor who will assist you (1 day).
- Finally, when the practice session is over, the handout material is ready, and the training facility is confirmed, you conduct the user training sessions (3 days).
  1. Prepare a list showing all tasks and their durations.
  2. Analyze the fact situation carefully to determine which tasks are concurrent and which ones are dependant on other tasks.
  3. Using PERT/CPM techniques, develop a chart that shows the project. Use a format similar to Figure 3-19 on page 111. If project management software is available, use it to develop the chart.
  4. What is the critical path for this project? How do you know?

Case Studies

Case studies allow you to practice specific skills learned in the chapter. Each chapter contains several case studies that continue throughout the textbook, and a chapter capstone case.

NEW CENTURY HEALTH CLINIC

New Century Health Clinic offers preventive medicine and traditional medical care. In your role as an IT consultant, you will help New Century develop a new information system.

Background

To ensure the quality, cost, and timeliness of the new information system, New Century is considering a project management approach. To obtain a better understanding of project management, Dr. Jones contacted Precision Planning, a consulting firm that specializes in managing projects of this type. He invited the company to deliver a brief presentation on project management concepts and advantages, and to submit a proposal for project management consulting services.

You joined Precision Planning two years ago as a project assistant, after working two summers as a student intern. Your supervisor, Charlie West, asked you to develop the presentation for New Century and you are excited about the opportunity. Charlie said that the main objective is to provide a clear, informative presentation.

Charlie wants you to include the following topics in your presentation: an overview of project management and its history, a description of the process, and an explanation of the most important terms and concepts. Charlie also wants you to describe task identification, various types of relationships among tasks, and schedule development. He says you should show how Gantt and PERT/CPM charts are developed, and how they can be used to plan, track, and control projects. Charlie also said that your presentation should include a specific example to illustrate all the main points.

Assignments

1. Create a Microsoft PowerPoint presentation that will meet the requirements that Charlie outlined to you.
2. Create a Microsoft Word handout that will meet the requirements that Charlie outlined to you.
3. Create a project management example with at least six tasks. Assign durations and task dependencies. At least three of the tasks should be dependent on other tasks. Use this example to display a Gantt chart.
4. Use the same data as Assignment 3 to display a PERT/CPM chart.

PERSONAL TRAINER, INC.

Personal Trainer, Inc. owns and operates fitness centers in a dozen Midwestern cities. The centers have done well, and the company is planning an international expansion by opening a new "supercenter" in the Toronto area. Personal Trainer's president, Cassia Umi, hired an IT consultant, Susan Park, to help develop an information system for the new facility. During the project, Susan will work closely with Gray Lewis, who will manage the new operation.

Background

You are enjoying your job as a student intern at Personal Trainer. Last week, Susan asked you to help her plan the new information system project. Susan knows that you have completed several information systems courses at the local college, and that you have studied project management tools and techniques.

Specifically, she wants you to get ready for the next set of systems development tasks, which will be requirements modeling for the new system. Yesterday, Susan called you into her office to discuss the specific tasks she wants you to perform. After meeting with Susan, you sit down and review your notes. She wants you to treat the set of tasks as a project, and to use project management skills to plan the tasks.

Here is what she suggested to you as a work breakdown structure, including the duration she estimated for each task:

- First, you need to meet with fitness center managers at other Personal Trainer locations (10 days).

- After these meetings, you can conduct a series of interviews (8 days).

- When the interviews are complete, two tasks can begin at the same time: You can review company records (2 days) and observe business operations (7 days).

- When you have reviewed the records and observed business operations, you can analyze the BumbleBee accounting software (3 days) and study a sample of sales and billing transactions (1 day).

You are excited about the opportunity to practice your skills, and you start to work on the following list.

Assignments

1. Create a table listing all tasks separately, with their duration.
2. Identify all dependent tasks, and indicate what predecessor tasks are required.
3. Construct a PERT/CPM chart similar to the one in Figure 3-19 on page 111. If you have access to Microsoft Project or other project management software, you can use it to help you create the chart.
4. Determine the overall duration of the project, and identify the critical path.

CHAPTER CAPSTONE CASE: SoftWear, Limited

SoftWear, Limited (SWL) is a continuing case study that illustrates the knowledge and skills described in each chapter. In this case study, the student acts as a member of the SWL systems development team and performs various tasks.

Background

At a recent management meeting, SWL's president, Robert Lansing, announced a major effort to control costs and improve quality. To help achieve this goal, Mr. Lansing stated that SWL would use project management tools and techniques to plan and manage all major corporate projects. He named several people who would work as an interdepartmental team to coordinate SWL's project management efforts. Team members included April Lane, director of planning; Mike Feiner, director of human resources; and Ann Hon, director of information technology.

The Interdepartmental Team

At their first meeting, the team came up with three main goals: Establish a company-wide understanding of project management concepts, identify suitable project management software, and develop comprehensive training for all SWL managers. Since Ann Hon had the most experience with project management, she agreed to serve as team leader. She also agreed to develop a list of concepts that the team could use as a starting point.

Project Management Concepts

The team met again a week later, and Ann distributed a list of 10 key questions:
1. What is a project?
2. What are project characteristics, constraints, and risks?
3. What is a project stakeholder?
4. What is the role of a project manager?
5. What is project planning?
6. What is project scheduling?
7. What is project monitoring and controlling?
8. What is project reporting?
9. What is project risk management?
10. What are the indications of project success or failure?

As the team members reviewed the list, Ann said that a set of working definitions would be a good first step in developing a company-wide approach for managing projects. She suggested that the answers were available from various sources, including a considerable body of literature and numerous online links. She also pointed out that the answers would be a key part of the proposed training program for SWL managers. The team decided to split up the research tasks and share the results at the next meeting.

Project Management Software

Ann made arrangements for the other team members to obtain a copy of Microsoft Office Project, which is the leading project management program. She also suggested that each of them try the brief Project 2007 training courses that are available on the Microsoft Web site. She then walked them though a two-hour session that demonstrated the software. She showed examples of Gantt charts, PERT charts, milestones, task dependencies, and resource assignments.

CHAPTER CAPSTONE CASE: SoftWear, Limited (continued)

Ann also pointed out that other software alternatives exist, including free, open-source programs, such as Open Workbench, which is supported by a large user group. For now, the team agreed to obtain pricing and licensing information for Microsoft Project, and to look into other alternatives to determine whether the other programs could exchange data with Microsoft Project.

Project Management Training

Ann suggested that the team compare the pros and cons of in-house training versus vendor-supplied training options. Again, Ann suggested the Microsoft Web site as a good starting point to evaluate third-party solutions. Using information on the site, the team was able to identify three training providers. After contacting these firms, the team had some realistic time and cost estimates for outside training solutions.

Ann suggested that the team should also consider a train-the-trainer approach where she would instruct an initial group from all SWL departments, and the training team would then provide training sessions within their respective departments. Meanwhile, Mike Feiner wondered whether any current SWL employees had listed project management experience and skills in their applications or résumés.

SWL Team Tasks

1. Using the material in this chapter and other reference material if necessary, develop a set of answers to the 10 questions that Ann presented to the team.
2. Suppose that Ann asked you to create an outline for her two-hour demo session. You can use Microsoft Project if it is available to you, or you can download a free demo version from the Microsoft Web site. In your outline, try to mention the basic information that a user would need to get started with a simple project. For some ideas, you can click the Microsoft Project 2007 View button to display and review the Project Guide.
3. Visit the Web site for Open Workbench and write a description of the product. Try to include as many features as possible, and list the pros and cons of the program. Determine whether the program can exchange information with Microsoft Project, and whether any special techniques are necessary to accomplish the transfer.
4. Microsoft has launched MPUG, which stands for Microsoft Project User Group. MPUG's stated mission is to deliver Microsoft Office Project content, resources, opportunities, and community networking worldwide. Explore the site at mpug.com and note the various levels of membership. Should SWL encourage IT staff members to join this group? Write up a recommendation with your reasons.

Manage the SWL Project

You have been asked to manage SWL's new information system project. One of your most important activities will be to identify project tasks and determine when they will be performed. Before you begin, you should review the SWL case in this chapter. Then list and analyze the tasks, as follows:

LIST THE TASKS Start by listing and numbering at least 10 tasks that the SWL team needs to perform to fulfill the objectives of this chapter. Your list can include SWL Team Tasks and any other tasks that are described in this chapter. For example, Task 3 might be to Identify the project tasks and Task 6 might be to Analyze task relationships.

(continues)

CHAPTER CAPSTONE CASE: SoftWear, Limited (continued)

ANALYZE THE TASKS   Now study the tasks to determine the order in which they should be performed. First identify all concurrent tasks, which are not dependent on other tasks. In the example shown in Figure 3-33, Tasks 1, 2, 3, 4, and 5 are concurrent tasks, and could begin at the same time if resources are available.

Other tasks are dependent tasks, because they cannot be performed until one or more earlier tasks have been completed. For each dependent task, you must identify specific tasks that need to be completed before these tasks can begin. For example, you need to identify the project tasks before you can analyze the task relationships, so Task 6 cannot begin until Task 3 is completed, as Figure 3-33 shows.

This chapter describes project management tools, techniques, and software. To learn more, you can visit the Features section on your Student Study Tool CD-ROM, or the project management resources library at **scsite.com/sad8e/project**. On the Web, Microsoft offers demo versions, training, and tips for using Project 2007. You also can visit the OpenWorkbench.org site to learn more about this free, open-source software.

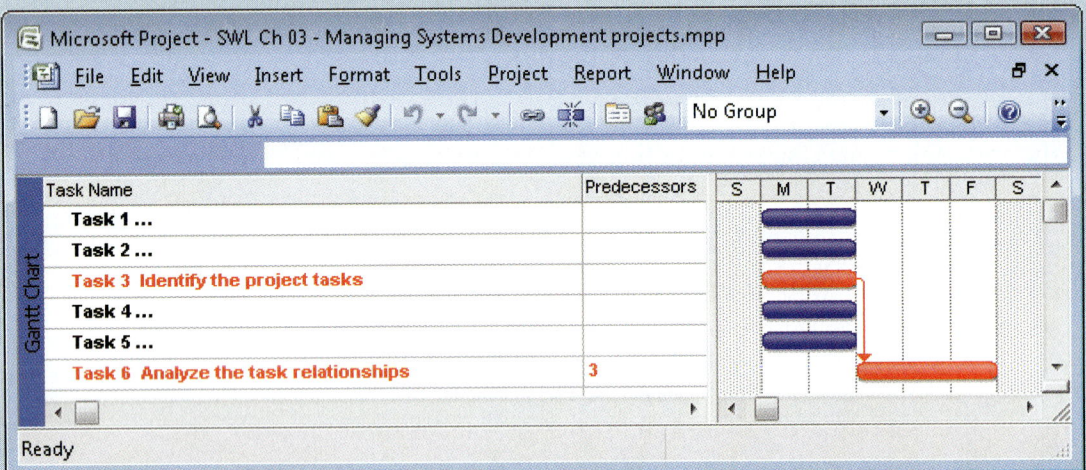

FIGURE 3-33   Tasks 1, 2, 3, 4, and 5 are concurrent tasks that could be performed at the same time. Task 6 is a dependent task that cannot be performed until Task 3 has been completed.

PHASE 2 SYSTEMS ANALYSIS

DELIVERABLE
System requirements document

TOOLKIT SUPPORT
Primary tools: Communication, CASE, and financial analysis tools
Other tools as required

As the Dilbert cartoon suggests, a successful project manager must determine the requirements before starting the design process, not the other way around. You will learn more about fact-finding and modeling system requirements in the systems analysis phase.

Systems analysis is the second of five phases in the systems development life cycle. In the previous phase, systems planning, you conducted a preliminary investigation to determine the project's feasibility. Now you will use requirements modeling, data and process modeling, and object modeling techniques to represent the new system. You also will consider various development strategies for the new system, and plan for the transition to systems design tasks. The deliverable for this phase is the system requirements document.

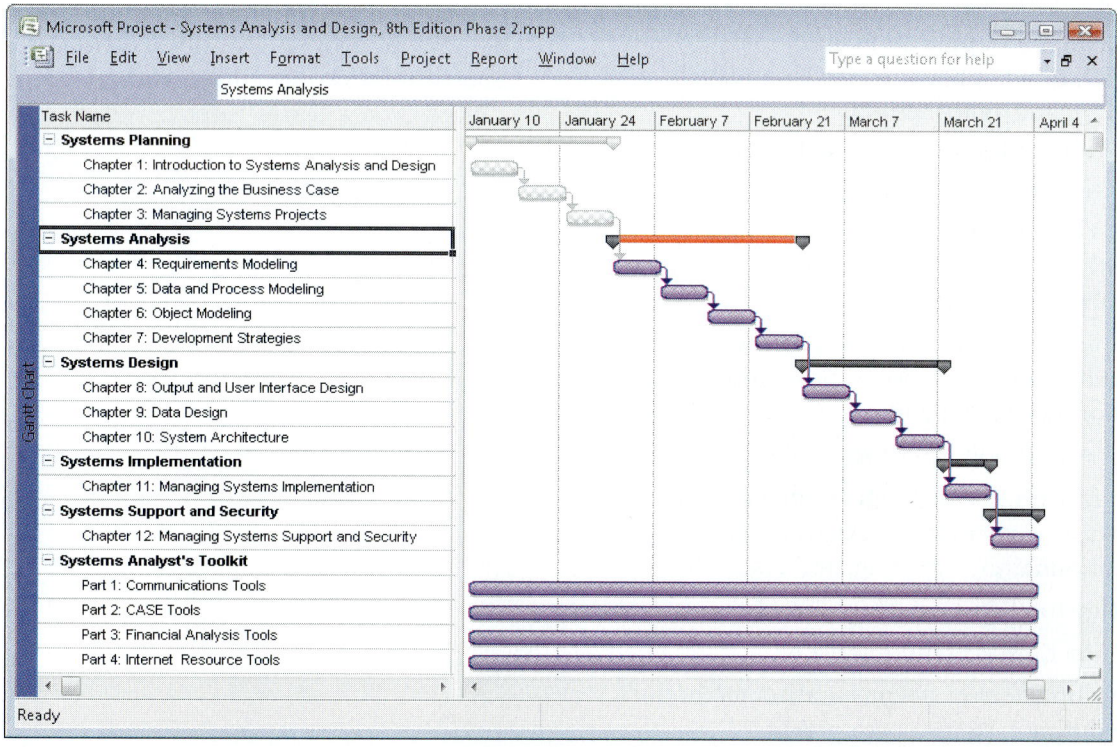

CHAPTER 4 Requirements Modeling

Chapter 4 is the first of four chapters in the systems analysis phase. This chapter describes the process of gathering facts about a systems project, preparing documentation, and creating models that will be used to design and develop the system.

INTRODUCTION

OBJECTIVES

When you finish this chapter, you will be able to:

- Describe systems analysis phase activities and the end product of the systems analysis phase
- Explain joint application development (JAD), rapid application development (RAD), and agile methods
- Understand how systems analysts use a functional decomposition diagram (FDD)
- Describe the Unified Modeling Language (UML) and explain use case diagrams and sequence diagrams
- List and describe system requirements, including outputs, inputs, processes, performance, and controls
- Explain the concept of scalability
- Use fact-finding techniques, including interviews, documentation review, observation, questionnaires, sampling, and research
- Define total cost of ownership (TCO)
- Conduct a successful interview
- Develop effective documentation methods to use during systems development

After an overview of the systems analysis phase, this chapter describes requirements modeling techniques and team-based methods that systems analysts use to visualize and document new systems. The chapter then discusses system requirements and fact-finding techniques, which include interviewing, documentation review, observation, surveys and questionnaires, sampling, and research.

CHAPTER INTRODUCTION CASE: Mountain View College Bookstore

Background: Wendy Lee, manager of college services at Mountain View College, wants a new information system that will improve efficiency and customer service at the three college bookstores.

In this part of the case, Florence Fullerton (systems analyst) and Harry Boston (student intern) are talking about requirements modeling tasks and concepts.

Participants:	Florence and Harry
Location:	Florence's office, Monday morning, October 5, 2009
Project status:	The project has advanced to the systems analysis phase. Now, Florence and Harry will work on modeling, fact-finding, and the documentation they need to build a requirements model for the proposed bookstore information system.
Discussion topics:	Modeling, team-based development strategies, fact-finding techniques, and documentation

Florence: Before I tell you about the project, look at this Dilbert cartoon. You'll like it!

Harry: *It's funny, but scary too. Hope it doesn't apply to us!*

Florence: Me too. That's why we have to do a good job of requirements modeling.

Harry: *So, what do we do next?*

Florence: We need to create a model of the new system. We call this a requirements model, because it will include all the outputs, inputs, processes, and controls for the new system. The model will consist of various diagrams, charts, and documentation.

Harry: *How will we use the model when we're done?*

Florence: We'll study it carefully and review it frequently with system users.

Harry: *Who are the users?*

DILBERT: © Scott Adams/Dist. by United Feature Syndicate, Inc.

Florence: Users might include bookstore staff, students, faculty members, and the college business office. External users might include textbook publishers and suppliers of bookstore merchandise. The main thing is to work with users every step of the way. We'll perform fact-finding, and we'll document everything carefully. Here's a task list to get us started:

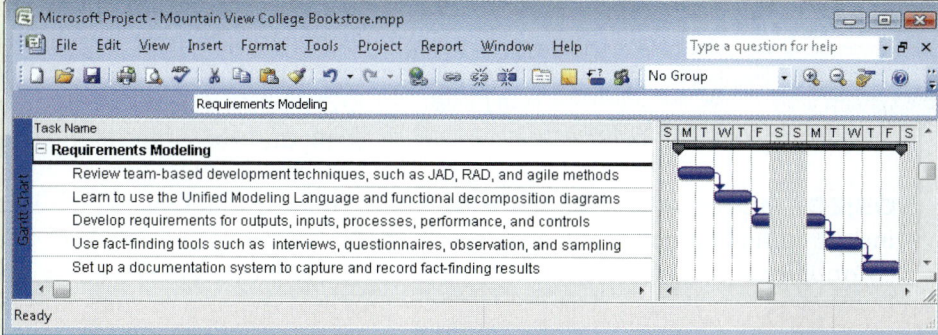

FIGURE 4-1 Typical requirements modeling task list.

SYSTEMS ANALYSIS PHASE OVERVIEW

The overall objective of the systems analysis phase is to understand the proposed project, ensure that it will support business requirements, and build a solid foundation for system development. In this phase, you use models and other documentation tools to visualize and describe the proposed system.

Systems Analysis Activities

The systems analysis phase includes the four main activities shown in Figure 4-2: requirements modeling, data and process modeling, object modeling, and consideration of development strategies.

Although the waterfall model shows sequential SDLC phases, it is not uncommon for several phases (or certain tasks within a phase) to interact during the development process, just as they would in an adaptive model. For example, this occurs whenever new facts are learned or system requirements change during the modeling process. Figure 4-2 shows typical interaction among the three modeling tasks: requirements modeling, data and process modeling, and object modeling.

Systems Analysis Phase Tasks

FIGURE 4-2   The systems analysis phase consists of requirements modeling, data and process modeling, object modeling, and consideration of development strategies. Notice that the systems analysis tasks are interactive, even though the waterfall model generally depicts sequential development.

REQUIREMENTS MODELING   This chapter describes **requirements modeling**, which involves fact-finding to describe the current system and identification of the requirements for the new system, such as outputs, inputs, processes, performance, and security. **Outputs** refer to electronic or printed information produced by the system. **Inputs** refer to necessary data that enters the system, either manually or in an automated manner. **Processes** refer to the logical rules that are applied to transform the data into meaningful information. **Performance** refers to system characteristics such as speed, volume, capacity, availability, and reliability. **Security** refers to hardware, software, and procedural controls that safeguard and protect the system and its data from internal or external threats.

DATA AND PROCESS MODELING   In Chapter 5, Data and Process Modeling, you will continue the modeling process by learning how to represent graphically system data and processes using traditional structured analysis techniques. As you learned in Chapter 1, structured analysis identifies the data flowing into a process, the business rules that transform the data, and the resulting output data flow.

OBJECT MODELING   Chapter 6 discusses object modeling, which is another popular modeling technique. While structured analysis treats processes and data as separate components, object-oriented analysis (O-O) combines data and the processes that act on the data into things called objects. These objects represent actual people, things, transactions, and events that affect the system. During the system development process, analysts often use both modeling methods to gain as much information as possible.

DEVELOPMENT STRATEGIES In Chapter 7, Development Strategies, you will consider various development options and prepare for the transition to the systems design phase of the SDLC. You will learn about software trends, acquisition and development alternatives, outsourcing, and formally documenting requirements for the new system.

The deliverable, or end product, of the systems analysis phase is a **system requirements document**, which is an overall design for the new system. In addition, each activity within the systems analysis phase has an end product and one or more milestones. As you learned in Chapter 3, project managers use various tools and techniques to coordinate people, tasks, timetables, and budgets.

Systems Analysis Skills

You will need strong analytical and interpersonal skills to build an accurate model of the new system. **Analytical skills** enable you to identify a problem, evaluate the key elements, and develop a useful solution. **Interpersonal skills** are especially valuable to a systems analyst who must work with people at all organizational levels, balance conflicting needs of users, and communicate effectively.

Because information systems affect people throughout the company, you should consider team-oriented strategies as you begin the systems analysis phase.

ON THE WEB

For more information about interpersonal skills, visit **scsite.com/ sad8e/more**, locate Chapter 4, and then click the Interpersonal Skills link.

Team-Based Techniques: JAD, RAD, and Agile Methods

The IT department's goal is to deliver the best possible information system, at the lowest possible cost, in the shortest possible time. To achieve the best results, system developers view users as partners in the development process. Greater user involvement usually results in better communication, faster development times, and more satisfied users.

The traditional model for systems development was an IT department that used structured analysis and consulted users only when their input or approval was needed. Although the IT staff still has a central role, and structured analysis remains a popular method of systems development, most IT managers invite system users to participate actively in various development tasks.

As you learned in Chapter 1, team-based approaches have been around for some time. A popular example is **joint application development (JAD)**, which is a user-oriented technique for fact-finding and requirements modeling. Because it is not linked to a specific development methodology, systems developers use JAD whenever group input and interaction are desired.

Another popular user-oriented method is **rapid application development (RAD)**. RAD resembles a condensed version of the entire SDLC, with users involved every step of the way. While JAD typically focuses only on fact-finding and requirements determination, RAD provides a fast-track approach to a full spectrum of system development tasks, including planning, design, construction, and implementation.

Finally, as you learned in Chapter 1, **agile methods** represent a recent trend that stresses intense interaction between system developers and users. JAD, RAD, and agile methods are discussed in the following sections.

JOINT APPLICATION DEVELOPMENT

Joint application development (JAD) is a popular fact-finding technique that brings users into the development process as active participants.

ON THE WEB

For more information about JAD, visit **scsite.com/sad8e/ more**, locate Chapter 4, and then click the JAD link.

User Involvement

Users have a vital stake in an information system, and they should participate fully in the development process. Until recent years, the IT department usually had sole responsibility for systems development, and users had a relatively passive role. During the development process, the IT staff would collect information from users, define system requirements, and construct the new system. At various stages of the process, the IT staff might ask users to review the design, offer comments, and submit changes.

Today, users typically have a much more active role in systems development. IT professionals now recognize that successful systems must be user-oriented, and users need to be involved, formally or informally, at every stage of system development.

One popular strategy for user involvement is a JAD team approach, which involves a task force of users, managers, and IT professionals that works together to gather information, discuss business needs, and define the new system requirements.

JAD Participants and Roles

A JAD team usually meets over a period of days or weeks in a special conference room or at an off-site location. Either way, JAD participants should be insulated from the distraction of day-to-day operations. The objective is to analyze the existing system, obtain user input and expectations, and document user requirements for the new system.

The JAD group usually has a project leader, who needs strong interpersonal and organizational skills, and one or more members who document and record the results and decisions. Figure 4-3 describes typical JAD participants and their roles. IT staff members often serve as JAD project leaders, but that is not always the case. Systems analysts on the JAD team participate in discussions, ask questions, take notes, and provide support to the team. If CASE tools are available, analysts can develop models and enter documentation from the JAD session directly into the CASE tool.

A typical JAD session agenda is shown in Figure 4-4. The JAD process involves intensive effort by all team members. Because of the wide range of input and constant interaction among the participants, many companies believe that a JAD group produces the best possible definition of the new system.

JAD PARTICIPANT	ROLE
JAD project leader	Develops an agenda, acts as a facilitator, and leads the JAD session
Top management	Provides enterprise-level authorization and support for the project
Managers	Provide department-level support for the project and understanding of how the project must support business functions and requirements
Users	Provide operational-level input on current operations, desired changes, input and output requirements, user interface issues, and how the project will support day-to-day tasks
Systems analysts and other IT staff members	Provide technical assistance and resources for JAD team members on issues such as security, backup, hardware, software, and network capability
Recorder	Documents results of JAD sessions and works with systems analysts to build system models and develop CASE tool documentation

FIGURE 4-3 Typical JAD participants and roles.

Project leader	• Introduce all JAD team members • Discuss ground rules, goals, and objectives for the JAD sessions • Explain methods of documentation and use of CASE tools, if any
Top management (sometimes called the project owner or sponsor)	• Explain the reason for the project and express top management authorization and support
Project leader	• Provide overview of the current system and proposed project scope and constraints • Present outline of specific topics and issues to be investigated
Open discussion session, moderated by project leader	• Review the main business processes, tasks, user roles, input, and output • Identify specific areas of agreement or disagreement • Break team into smaller groups to study specific issues and assign group leaders
JAD team members working in smaller group sessions, supported by IT staff	• Discuss and document all system requirements • Develop models and prototypes
Group leaders	• Report on results and assigned tasks and topics • Present issues that should be addressed by the overall JAD team
Open discussion session, moderated by project leader	• Review reports from small group sessions • Reach consensus on main issues • Document all topics
Project leader	• Present overall recap of JAD session • Prepare report that will be sent to JAD team members

FIGURE 4-4 Typical agenda for a JAD session.

JAD Advantages and Disadvantages

Compared with traditional methods, JAD is more expensive and can be cumbersome if the group is too large relative to the size of the project. Many companies find, however, that JAD allows key users to participate effectively in the requirements modeling process. When users participate in the systems development process, they are more likely to feel a sense of ownership in the results, and support for the new system. When properly used, JAD can result in a more accurate statement of system requirements, a better understanding of common goals, and a stronger commitment to the success of the new system.

RAPID APPLICATION DEVELOPMENT

Rapid application development (RAD) is a team-based technique that speeds up information systems development and produces a functioning information system. Like JAD, RAD uses a group approach, but goes much further. While the end product of JAD is a requirements model, the end product of RAD is the new information system. RAD is a complete methodology, with a four-phase life cycle that parallels the traditional SDLC phases. Companies use RAD to reduce cost and development time, and increase the probability of success.

RAD relies heavily on prototyping and user involvement. The RAD process allows users to examine a working model as early as possible, determine if it meets their needs, and suggest necessary changes. Based on user input, the prototype is modified and the interactive process continues until the system is completely developed and users are satisfied. The project team uses CASE tools to build the prototypes and create a continuous stream of documentation.

RAD Phases and Activities

The RAD model consists of four phases: requirements planning, user design, construction, and cutover, as shown in Figure 4-5. Notice the continuous interaction between the user design and construction phases.

For more information about RAD, visit **scsite.com/sad8e/more**, locate Chapter 4, and then click the RAD link.

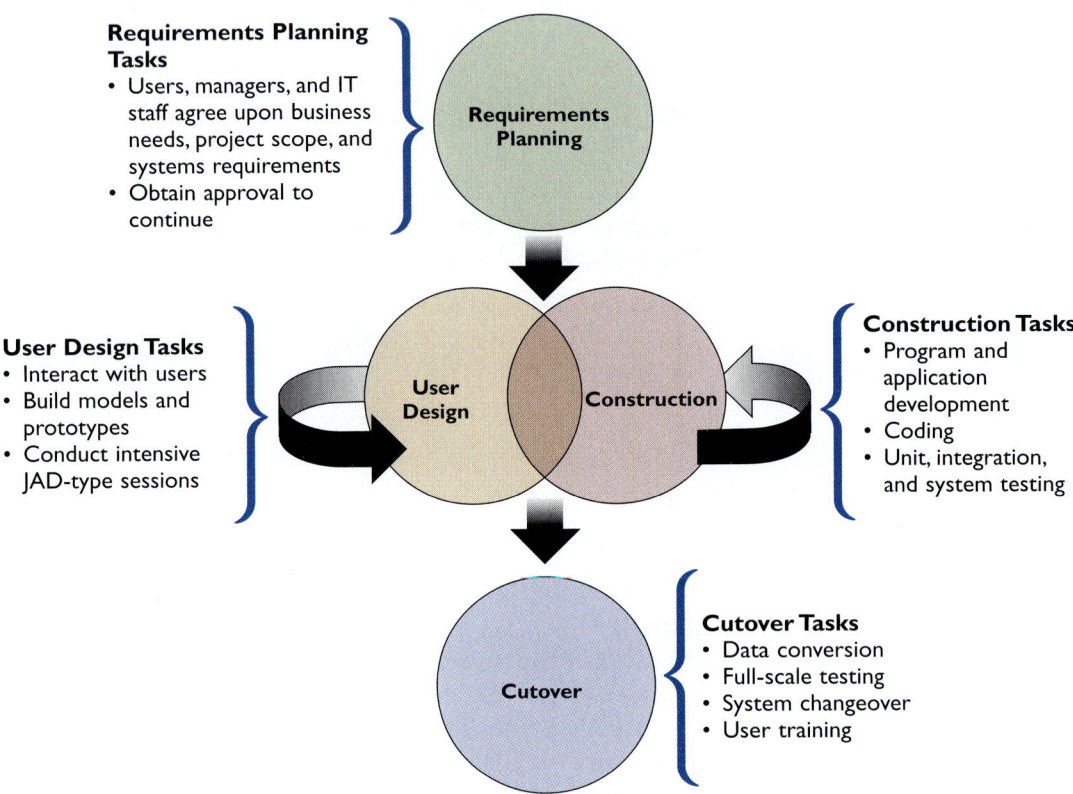

FIGURE 4-5 The four phases of the RAD model are requirements planning, user design, construction, and cutover. Notice the continuous interaction between the user design and construction phases.

REQUIREMENTS PLANNING The **requirements planning phase** combines elements of the systems planning and systems analysis phases of the SDLC. Users, managers, and IT staff members discuss and agree on business needs, project scope, constraints, and system requirements. The requirements planning phase ends when the team agrees on the key issues and obtains management authorization to continue.

USER DESIGN During the **user design phase**, users interact with systems analysts and develop models and prototypes that represent all system processes, outputs, and inputs. The RAD group or subgroups typically use a combination of JAD techniques and CASE tools to translate user needs into working models. User design is a continuous, interactive process that allows users to understand, modify, and eventually approve a working model of the system that meets their needs.

CONSTRUCTION The **construction phase** focuses on program and application development tasks similar to the SDLC. In RAD, however, users continue to participate and still can suggest changes or improvements as actual screens or reports are developed.

CUTOVER The **cutover phase** resembles the final tasks in the SDLC implementation phase, including data conversion, testing, changeover to the new system, and user training. Compared with traditional methods, the entire process is compressed. As a result, the new system is built, delivered, and placed in operation much sooner.

RAD Objectives

The main objective of all RAD approaches is to cut development time and expense by involving users in every phase of systems development. Because it is a continuous process, RAD allows the development team to make necessary modifications quickly, as the design evolves. In times of tight corporate budgets, it is especially important to limit the cost of changes that typically occur in a long, drawn-out development schedule.

In addition to user involvement, a successful RAD team must have IT resources, skills, and management support. Because it is a dynamic, user-driven process, RAD is especially valuable when a company needs an information system to support a new business function. By obtaining user input from the beginning, RAD also helps a development team design a system that requires a highly interactive or complex user interface.

RAD Advantages and Disadvantages

RAD has advantages and disadvantages compared with traditional structured analysis methods. The primary advantage is that systems can be developed more quickly with significant cost savings. A disadvantage is that RAD stresses the mechanics of the system itself and does not emphasize the company's strategic business needs. The risk is that a system might work well in the short term, but the corporate and long-term objectives for the system might not be met. Another potential disadvantage is that the accelerated time cycle might allow less time to develop quality, consistency, and design standards. RAD can be an attractive alternative, however, if an organization understands the possible risks.

AGILE METHODS

In Chapter 1, you learned that agile methods attempt to develop a system incrementally, by building a series of prototypes and constantly adjusting them to user requirements. As the agile process continues, developers revise, extend, and merge earlier versions into the final product. An agile approach emphasizes continuous feedback, and each incremental step is affected by what was learned in the prior steps.

As agile methods become more popular, a large community of agile-related software and services has evolved. For example, Visual Paradigm offers Agilian, which includes a set of agile modeling tools, as shown in Figure 4-6 on the next page. The Agilian modeling toolset includes support for many modeling tools, such as the Unified Modeling Language, entity-relationship diagrams, data flow diagrams, and business process modeling, among others.

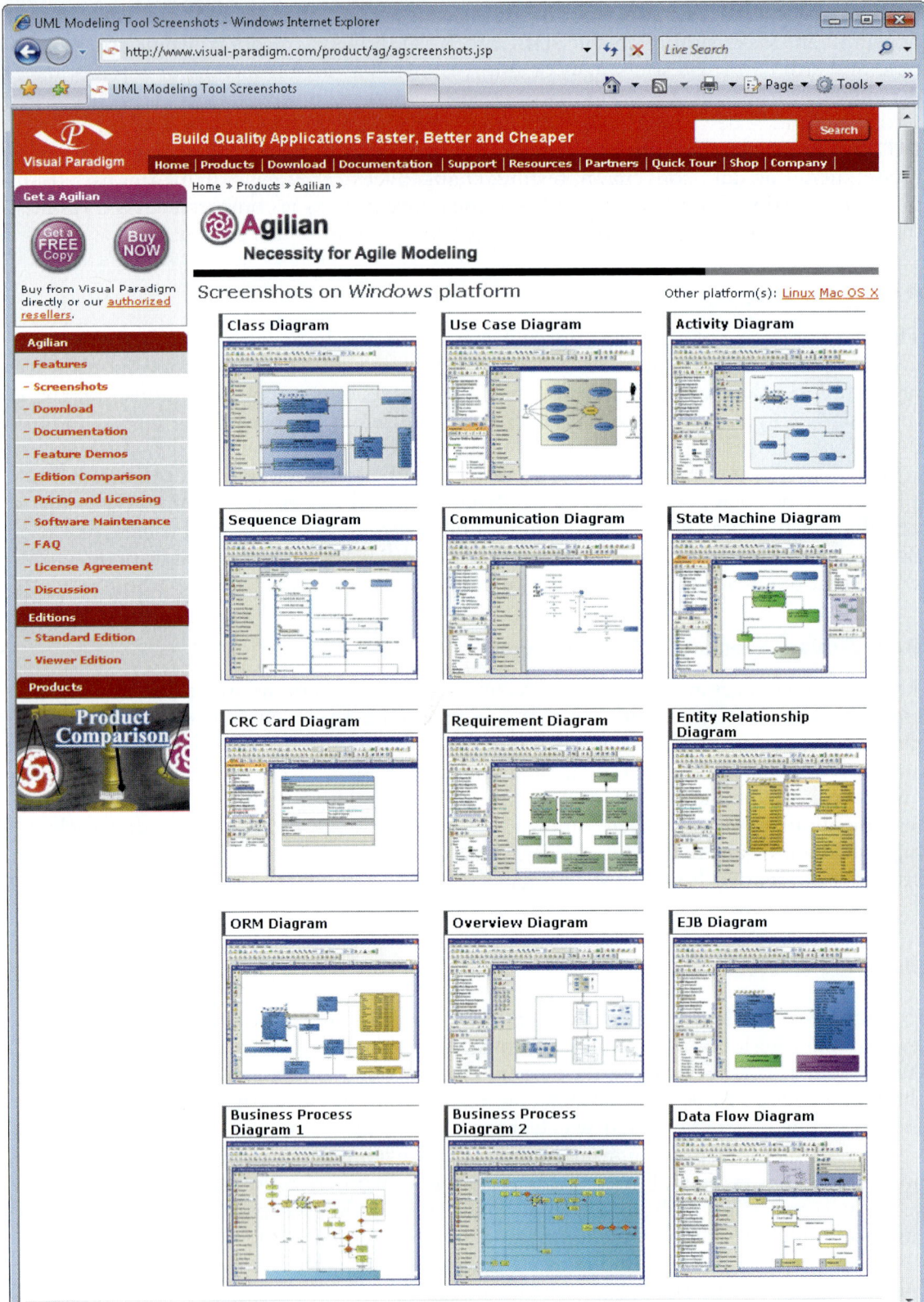

FIGURE 4-6 Visual Paradigm's Agilian includes many types of agile modeling tools.

Some agile developers prefer not to use CASE tools at all, and rely instead on whiteboard displays and arrangements of movable sticky notes. This approach, they believe, reinforces the agile strategy: simple, rapid, flexible, and user-oriented.

In Chapter 1, you also learned that *scrum* is a rugby term, as shown in Figure 4-7, where team members prepare to lunge at each other to achieve their objectives. The systems development version of Scrum involves the same intense interaction, though more mental than physical. In a Scrum session, agile team members play specific roles, including colorful designations as *pigs* or *chickens*. These roles are based on the old joke about the pig and chicken who discuss a restaurant where ham and eggs would be served. However, the pig declines, because that role would require a total commitment, while for the chicken, it would only be a contribution.

ON THE WEB

For more information about agile methods, visit **scsite.com /sad8e/more**, locate Chapter 4, and then click the Agile Methods link.

FIGURE 4-7 In a rugby scrum, team members prepare to lunge at each other to achieve their objectives.

In the agile world, the *pigs* include the product owner, the facilitator, and the development team; while the *chickens* include users, other stakeholders, and managers. Scrum sessions have specific guidelines that emphasize time blocks, interaction, and team-based activities that result in deliverable software.

Agile Method Advantages and Disadvantages

Agile, or adaptive, methods are very flexible and efficient in dealing with change. They are popular because they stress team interaction and reflect a set of community-based values. Also, frequent deliverables constantly validate the project and reduce risk.

However, some potential problems exist. For example, team members need a high level of technical and interpersonal skills. Also, a lack of structure and documentation can introduce risk factors. Finally, the overall project may be subject to significant change in scope as user requirements continue to evolve during the project.

CASE IN POINT 4.1: NORTH HILLS COLLEGE

North Hills College has decided to implement a new registration system that will allow students to register online, as well as in person. As IT manager, you decide to set up a JAD session to help define the requirements for the new system. The North Hills organization is fairly typical, with administrative staff that includes a registrar, a student support and services team, a business office, an IT group, and a number of academic departments. Using this information, you start work on a plan to carry out the JAD session. Who would you invite to the session, and why? What would be your agenda for the session, and what would take place at each stage of the session?

MODELING TOOLS AND TECHNIQUES

TOOLKIT TIME

The CASE Tools in Part 2 of the Systems Analyst's Toolkit can help you document business functions and processes, develop graphical models, and provide an overall framework for information system development. To learn more about these tools, turn to Part 2 of the four-part Toolkit that follows Chapter 12.

Models help users, managers, and IT professionals understand the design of a system. Modeling involves graphical methods and nontechnical language that represent the system at various stages of development. During requirements modeling, you can use various tools to describe business processes, requirements, and user interaction with the system.

CASE Tools

In Chapter 1, you learned that CASE tools can offer powerful modeling features. For example, using the Visible Analyst CASE tool, a systems analyst can create a model of order processing system tasks, as shown in Figure 4-8. CASE tool modeling is described in more detail in Part 2 of the Systems Analyst's Toolkit.

Systems analysts use modeling and fact-finding interactively — first they build fact-finding results into models, then they study the models to determine whether additional fact-finding is needed. To help them understand system requirements, systems analysts often use functional decomposition diagrams, which provide a business-oriented overview, and the Unified Modeling Language, which shows how people interact with the system.

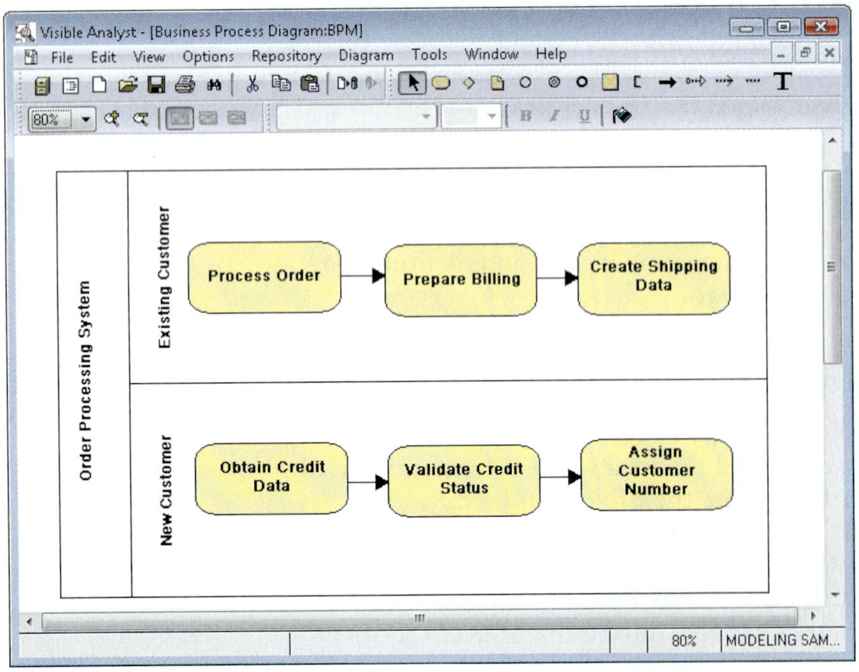

Functional Decomposition Diagrams

A **functional decomposition diagram** (FDD) is a top-down representation of a function or process. FDDs also are called **structure charts**. Using an FDD, an analyst can show business functions and break them down into lower-level functions and processes. Creating an FDD is similar to drawing an organization chart — you start at the top and work your way down. Figure 4-9 shows a four-level FDD of a library system drawn with the Visible Analyst CASE tool. FDDs can be used at several stages of systems

FIGURE 4-8 Using the Visible Analyst CASE tool, a systems analyst can create a business process diagram.

development. During requirements modeling, analysts use FDDs to model business functions and show how they are organized into lower-level processes. Those processes translate into program modules during application development.

Data Flow Diagrams

Working from a functional decomposition diagram, analysts can create **data flow diagrams (DFDs)** to show how the system stores, processes, and transforms data. The DFD in Figure 4-10 describes adding and removing books, which is a function shown in the Library Management diagram in Figure 4-9. Notice that the two shapes in the DFD represent proces154 ses, each with various inputs and outputs. Additional levels of information and detail are depicted in other, related DFDs. Data and process modeling is described in detail in Chapter 5.

Unified Modeling Language

The **Unified Modeling Language (UML)** is a widely used method of visualizing and documenting software systems design. UML uses object-oriented design concepts, but it is independent of any specific programming language and can be used to describe business processes and requirements generally.

UML provides various graphical tools, such as use case diagrams and sequence diagrams. During requirements modeling, a systems analyst can utilize the UML to represent the information system from a user's viewpoint. Use case diagrams, sequence diagrams, and other UML concepts are discussed in more detail in Chapter 6, along with other object-oriented analysis concepts. A brief description of each technique follows.

USE CASE DIAGRAMS During requirements modeling, systems analysts and users work together to document requirements and model system functions. A **use case diagram** visually represents the interaction between users and the information system. In a use case diagram, the user becomes an **actor**, with a specific role that describes how he or she interacts with the system. Systems analysts can draw use case diagrams freehand or use CASE tools that integrate the use cases into the overall system design.

Figure 4-11 on the next page shows a simple use case diagram for a sales system where the actor is a customer and the use case involves a credit card validation that is performed by the system. Because use cases depict the system through the eyes of a user,

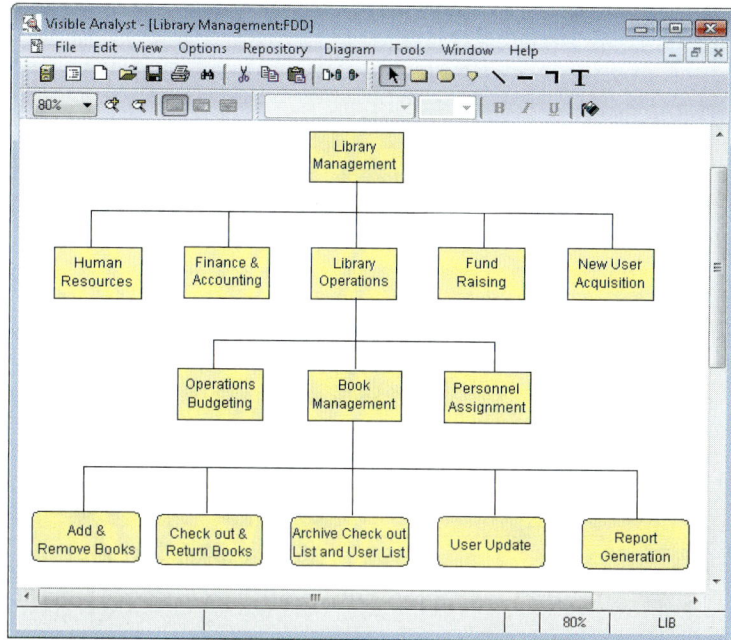

FIGURE 4-9 Four-level functional decomposition diagram (FDD) of a library system.

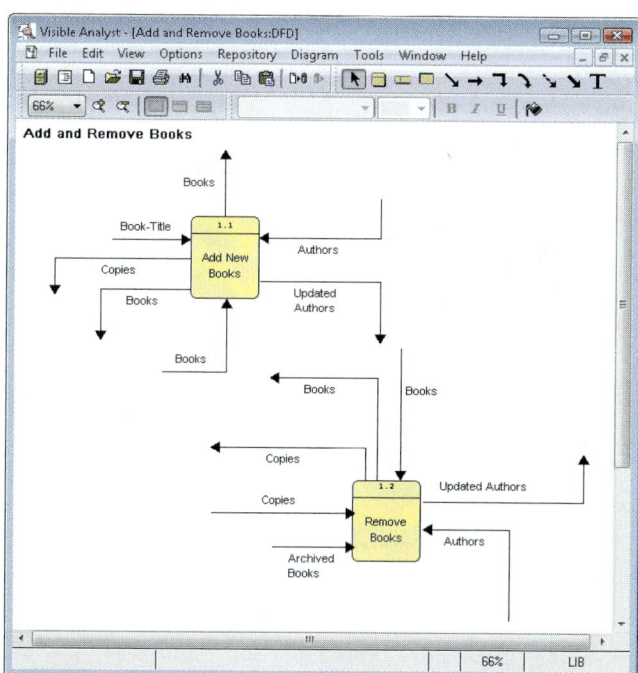

FIGURE 4-10 A library system DFD shows how books are added and removed.

For more information about the Unified Modeling Language, visit **scsite.com/sad8e/more**, locate Chapter 4, and then click The Unified Modeling Language link.

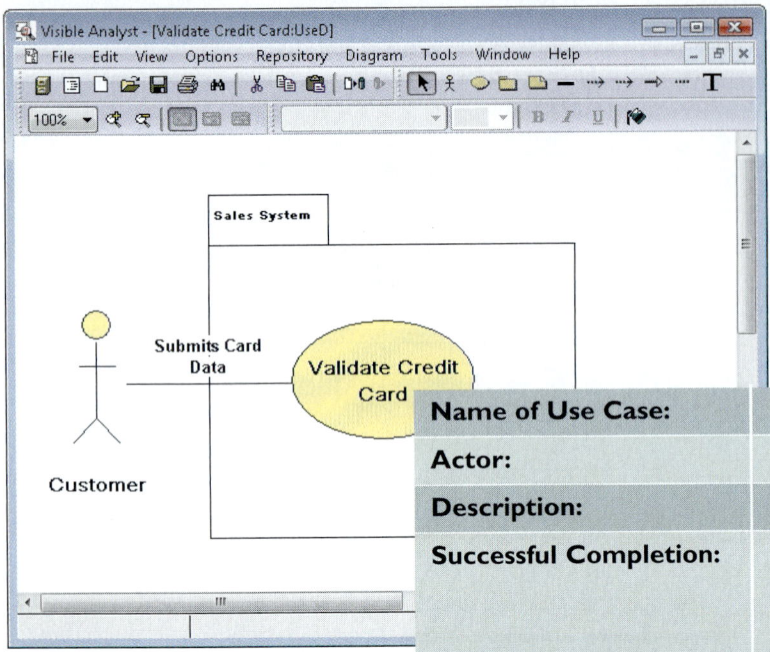

Visible Analyst - [Validate Credit Card:UseD]

File   Edit   View   Options   Repository   Diagram   Tools   Window   Help

100%

Sales System

Submits Card
Data

Validate Credit
Card

Customer

FIGURE 4-11 Use case diagram of a sales system, where the actor is a customer and the use case involves a credit card validation.

Name of Use Case:	Credit card validation process
Actor:	Customer
Description:	Describes the credit card validation process
Successful Completion:	1. Customer clicks the input selector and enters credit card number and expiration date 2. System verifies card 3. System sends authorization message
Alternative:	1. Customer clicks the input selector and enters credit card number and expiration date 2. System rejects card 3. System sends rejection message
Precondition:	Customer has selected at least one item and has proceeded to checkout area
Postcondition:	Credit card information has been validated Customer can continue with order
Assumptions:	None

FIGURE 4-12 A table documents the credit card validation use case shown in Figure 4-11.

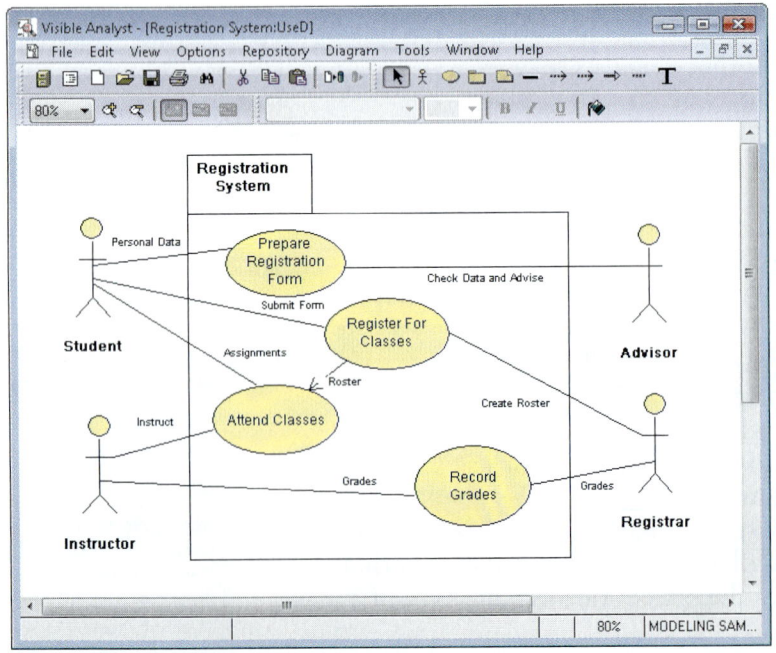

Visible Analyst - [Registration System:UseD]

File   Edit   View   Options   Repository   Diagram   Tools   Window   Help

80%

Registration
System

Personal Data

Prepare
Registration
Form

Check Data and Advise

Submit Form

Register For
Classes

Student

Assignments

Roster

Advisor

Create Roster

Instruct

Attend Classes

Grades

Record
Grades

Grades

Registrar

Instructor

80%   MODELING SAM...

FIGURE 4-13 Use case diagram of a student records system.

common business language can be used to describe the transactions. For example, Figure 4-12 shows a table that documents the credit card validation use case, and Figure 4-13 shows a student records system, with several use cases and actors.

SEQUENCE DIAGRAMS A sequence **diagram** shows the timing of interactions between objects as they occur. A systems analyst might use a sequence diagram to show all possible outcomes, or focus on a single scenario. Figure 4-14 shows a simple sequence diagram of a successful credit card validation. The interaction proceeds from top to bottom along a vertical timeline, while the horizontal arrows represent messages from one object to another.

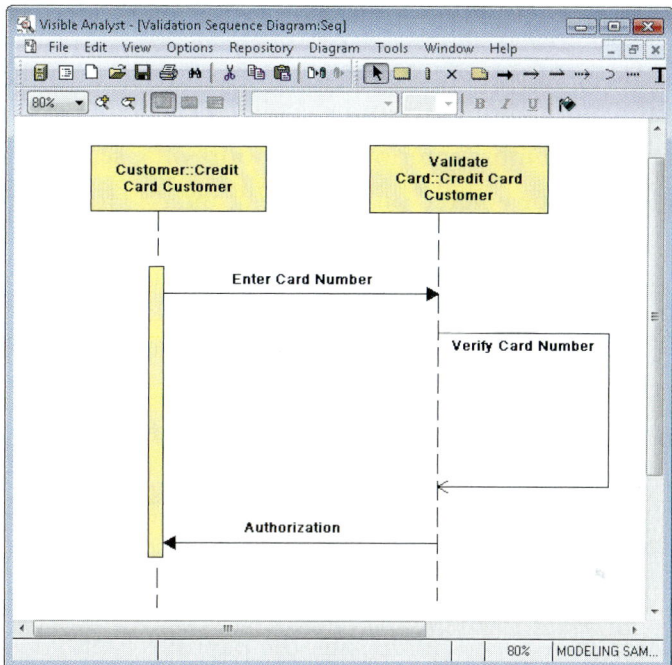

FIGURE 4-14 Sequence diagram of a successful credit card validation showing how the transaction proceeds from top to bottom, along a vertical timeline.

SYSTEM REQUIREMENTS CHECKLIST

During requirements modeling, systems developers must identify and describe all system requirements. A **system requirement** is a characteristic or feature that must be included in an information system to satisfy business requirements and be acceptable to users. System requirements serve as benchmarks to measure the overall acceptability of the finished system.

System requirements fall into five general categories: outputs, inputs, processes, performance, and controls. Typical examples of system requirements for each category are listed below.

Output Examples

- The Web site must report online volume statistics every four hours, and hourly during peak periods.

- The inventory system must produce a daily report showing the part number, description, quantity on hand, quantity allocated, quantity available, and unit cost of all sorted by part number.

- The contact management system must generate a daily reminder list for all sales reps.

- The purchasing system must provide suppliers with up-to-date specifications.

- The sales tracking system must produce a daily fast-moving-item report, listing all products that exceed the forecasted sales volume grouped by style, color, size, and reorder status.

- The customer analysis system must produce a quarterly report that identifies changes in ordering patterns or trends with statistical comparisons to the previous four quarters.

Input Examples

- Manufacturing employees must swipe their ID cards into online data collection terminals that record labor costs and calculate production efficiency.
- The department head must enter overtime hours on a separate screen.
- Student grades must be entered on machine-scannable forms prepared by the instructor.
- Each input form must include date, time, product code, customer number, and quantity.
- Data entry screens must be uniform, except for background color, which can be changed by the user.
- A data entry person at the medical group must input patient services into the billing system.

Process Examples

- The student records system must calculate the GPA at the end of each semester.
- As the final step in year-end processing, the payroll system must update employee salaries, bonuses, and benefits and produce tax data required by the IRS.
- The warehouse distribution system must analyze daily orders and create a routing pattern for delivery trucks that maximizes efficiency and reduces unnecessary mileage.
- The human resources system must interface properly with the existing payroll system.
- The video rental system must not execute new rental transactions for customers who have overdue tapes.
- The prescription system must automatically generate an insurance claim form.

Performance Examples

- The system must support 25 users online simultaneously.
- Response time must not exceed four seconds.
- The system must be operational seven days a week, 365 days a year.
- The accounts receivable system must prepare customer statements by the third business day of the following month.
- The student records system must produce class lists within five hours after the end of registration.
- The online inventory control system must flag all low-stock items within one hour after the quantity falls below a predetermined minimum.

Control Examples

- The system must provide logon security at the operating system level and at the application level.
- An employee record must be added, changed, or deleted only by a member of the human resources department.
- The system must maintain separate levels of security for users and the system administrator.

- All transactions must have audit trails.
- The manager of the sales department must approve orders that exceed a customer's credit limit.
- The system must create an error log file that includes the error type, description, and time.

FUTURE GROWTH, COSTS, AND BENEFITS

In addition to the system requirements, systems analysts must consider scalability, which determines how a system will handle future growth and demands, and the total cost of ownership, which includes all future operational and support costs.

Scalability

Scalability refers to a system's ability to handle increased business volume and transactions in the future. Because it will have a longer useful life, a scalable system offers a better return on the initial investment.

To evaluate scalability, you need information about projected future volume for all outputs, inputs, and processes. For example, for a Web-based order processing system, you would need to know the maximum projected number of concurrent users, the periods of peak online activity, the number and types of data items required for each transaction, and the method of accessing and updating customer files.

Even to print customer statements, you need to know the number of active accounts and have a forecast for one, two, or five years, because that information affects future hardware decisions. In addition, with realistic volume projections, you can provide reliable cost estimates for related expenses, such as postage and online charges.

Similarly, to ensure that a Web-based hotel reservation system is sufficiently scalable, you would need to project activity levels for several years of operation. For example, you might forecast the frequency of online queries about room availability and estimate the time required for each query and the average response time. With that information, you could estimate server transaction volume and network requirements.

Transaction volume has a significant impact on operating costs. When volume exceeds a system's limitations, maintenance costs increase sharply. Volume can change dramatically if a company expands or enters a new line of business. For example, a new Internet-based marketing effort might require an additional server and 24-hour technical support.

Data storage also is an important scalability issue. You need to determine how much data storage is required currently and predict future needs based on system activity and growth. Those requirements affect hardware, software, and network bandwidth needed to maintain system performance. You also must consider data retention requirements and determine whether data can be deleted or archived on a specific timetable.

Total Cost of Ownership

In addition to direct costs, systems developers must identify and document indirect expenses that contribute to the **total cost of ownership (TCO)**. TCO is especially important if the development team is assessing several alternatives. After considering the indirect costs, which are not always apparent, a system that seems inexpensive initially might actually turn out to be the most costly choice. One problem is that cost estimates tend to understate indirect costs such as user support and downtime productivity losses. Even if accurate figures are unavailable, systems analysts should try to identify indirect costs and include them in TCO estimates.

> **⚙ TOOLKIT TIME**
>
> The Financial Analysis tools in Part 3 of the Systems Analyst's Toolkit can help you analyze project costs, benefits, and economic feasibility. To learn more about these tools, turn to Part 3 of the four-part Toolkit that follows Chapter 12.

ON THE WEB

For more information about the REJ, visit **scsite.com/sad8e/more**, locate Chapter 4, and then click the REJ link.

Microsoft has developed a method for measuring total costs and benefits, called **Rapid Economic Justification (REJ)**, which is described in Figure 4-15. According to Microsoft, REJ is a framework to help IT professionals analyze and optimize IT investments. Notice that the primary emphasis is on business improvement, rather than operational efficiency. As the Web site points out, the strategic role of IT investments should be included, even when the specific benefits are difficult to quantify.

FIGURE 4-15   Microsoft developed Rapid Economic Justification (REJ) as a framework to help IT professionals analyze and optimize IT investments.

FACT-FINDING

Now that you understand the categories of system requirements, scalability, and TCO, the next step is to begin collecting information. Whether you are working on your own or as a member of a JAD team, during requirements modeling you will use various fact-finding techniques, including interviews, document review, observation, surveys and questionnaires, sampling, and research.

Fact-Finding Overview

Although software can help you to gather and analyze facts, no program actually performs fact-finding for you. First, you must identify the information you need. Typically, you begin by asking a series of questions, such as these:

- What business functions are supported by the current system?
- What strategic objectives and business requirements must be supported by the new system?
- What are the benefits and TCO of the proposed system?
- What transactions will the system process?
- What information do users and managers need from the system?
- Must the new system interface with legacy systems?

- What procedures could be eliminated by business process reengineering?
- What security issues exist?
- What risks are acceptable?
- What budget and timetable constraints will affect system development?

To obtain answers to these questions, you develop a fact-finding plan, which can involve another series of questions (*who*, *what*, *where*, *when*, and *how*), or use a more structured approach such as the Zachman Framework, which is explained in a following section. Either way, you will develop a strategy, carry out fact-finding techniques, document the results, and prepare a system requirements document, which is presented to management.

Who, What, Where, When, How, and Why?

Fact-finding involves answers to five familiar questions: *who*, *what*, *where*, *when*, and *how*. For each of those questions, you also must ask another very important question: *why*. Some examples of these questions are:

1. *Who?* Who performs each of the procedures within the system? Why? Are the correct people performing the activity? Could other people perform the tasks more effectively?

2. *What?* What is being done? What procedures are being followed? Why is that process necessary? Often, procedures are followed for many years and no one knows why. You should question why a procedure is being followed at all.

3. *Where?* Where are operations being performed? Why? Where could they be performed? Could they be performed more efficiently elsewhere?

4. *When?* When is a procedure performed? Why is it being performed at this time? Is this the best time?

5. *How?* How is a procedure performed? Why is it performed in that manner? Could it be performed better, more efficiently, or less expensively in some other manner?

There is a difference between asking what *is* being done and what *could* or *should* be done. The systems analyst first must understand the current situation. Only then can he or she tackle the question of what *should* be done. Figure 4-16 lists the basic questions and when they should be asked. Notice that the first two columns relate to the current system, but the third column focuses on the proposed system.

CURRENT SYSTEM		PROPOSED SYSTEM
Who does it?	Why does this person do it?	Who should do it?
What is done?	Why is it done?	What should be done?
Where is it done?	Why is it done there?	Where should it be done?
When is it done?	Why is it done then?	When should it be done?
How is it done?	Why is it done this way?	How should it be done?

FIGURE 4-16 Sample questions during requirements modeling as the focus shifts from the current system to the proposed system.

ON THE WEB

For more information about the Zachman Framework, visit **scsite.com/sad8e/more**, locate Chapter 4, and then click The Zachman Framework link.

The Zachman Framework

In the 1980s, John Zachman observed how industries such as architecture and construction handled complex projects, and he suggested that the same ideas could be applied to information systems development. His concept, the **Zachman Framework for Enterprise Architecture**, is a model that asks the traditional fact-finding questions in a systems development context, as shown in Figure 4-17. The Zachman Framework is a popular approach, and the Visible Analyst CASE tool now includes a Zachman Framework interface that allows users to view a systems project from different perspectives and levels of detail. The Zachman Framework helps managers and users understand the model and ensures that overall business goals translate into successful IT projects.

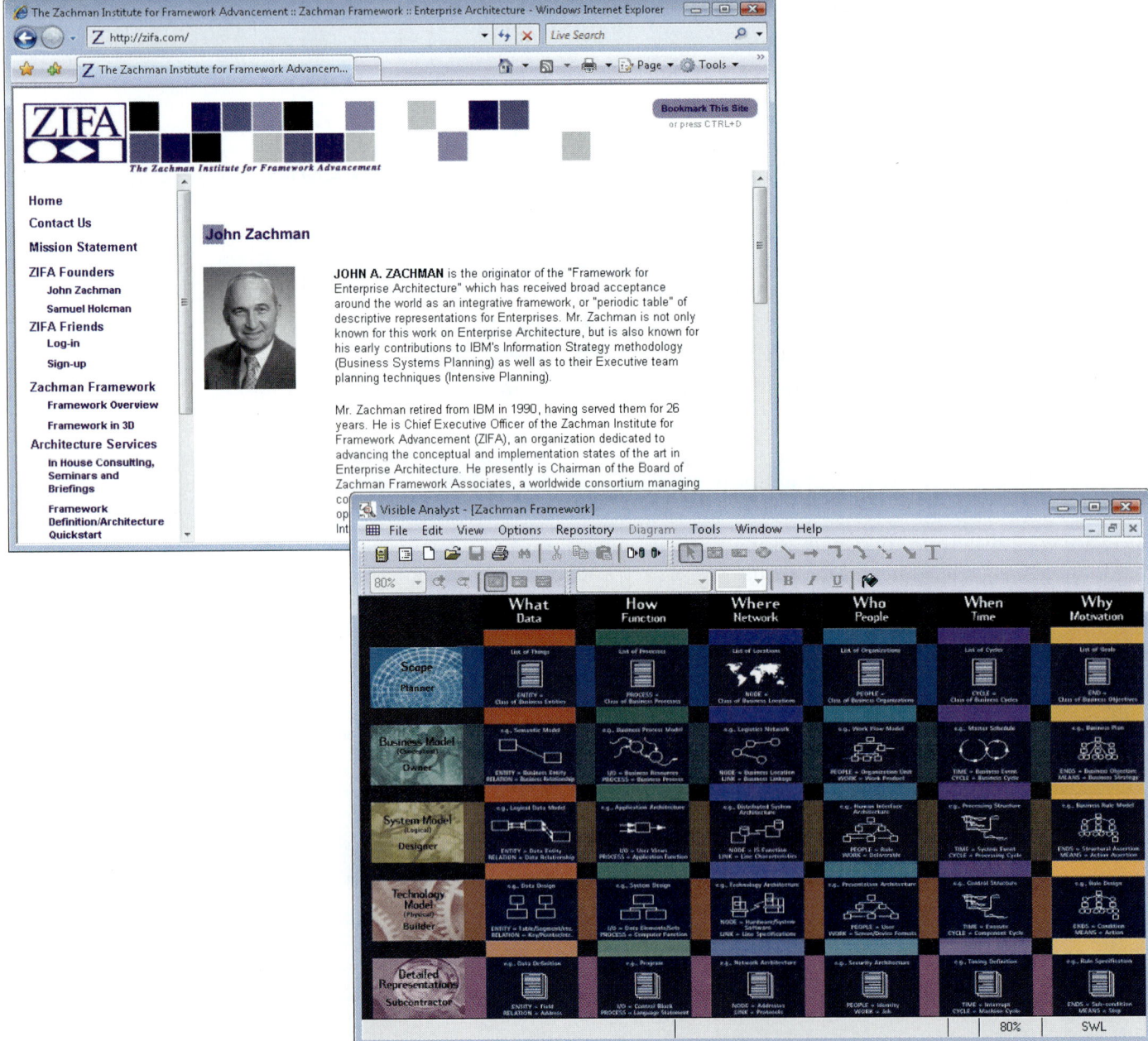

FIGURE 4-17 Visible Analyst uses the Zachman Framework for Enterprise Architecture. The Zachman concept presents traditional fact-finding questions in a systems development context.

INTERVIEWS

Interviewing is an important fact-finding tool during the systems analysis phase. An **interview** is a planned meeting during which you obtain information from another person. You must have the skills needed to plan, conduct, document, and evaluate interviews successfully.

After you identify the information you need, as described earlier in the chapter, you can begin the interviewing process, which consists of seven steps for each interview:

1. Determine the people to interview.
2. Establish objectives for the interview.
3. Develop interview questions.
4. Prepare for the interview.
5. Conduct the interview.
6. Document the interview.
7. Evaluate the interview.

Step 1: Determine the People to Interview

To get an accurate picture, you must select the right people to interview and ask them the right questions. During the preliminary investigation, you talked mainly to middle managers or department heads. Now, during the systems analysis phase, you might need to interview people from all levels of the organization.

Although you can select your interview candidates from the formal organization charts that you reviewed earlier, you also must consider any informal structures that exist in the organization. Informal structures usually are based on interpersonal relationships and can develop from previous work assignments, physical proximity, unofficial procedures, or personal relationships such as the informal gathering shown in Figure 4-18. In an **informal structure**, some people have more influence or knowledge than appears on an organization chart. Your knowledge of the company's formal and informal structures helps you determine the people to interview during the systems analysis phase.

Should you interview several people at the same time? Group interviews can save time and provide an opportunity to observe interaction among the participants. Group interviews also can present problems. One person might dominate the conversation, even when questions are addressed specifically to others. Organization level also can present a problem, as the presence of senior managers in an interview might prevent lower-level employees from expressing themselves candidly.

Step 2: Establish Objectives for the Interview

After deciding on the people to interview, you must establish objectives for the session. First, you should determine the general areas to be discussed, and then list the facts you want to gather. You also should try to solicit ideas, suggestions, and opinions during the interview.

FIGURE 4-18 An analyst must consider informal structures in the organization when selecting interview candidates.

The objectives of an interview depend on the role of the person being interviewed. Upper-level managers can provide the big picture and help you to understand the system as a whole. Specific details about operations and business processes are best learned from people who actually work with the system on a daily basis.

In the early stages of systems analysis, interviews usually are general. As the fact-finding process continues, however, the interviews focus more on specific topics. Interview objectives also vary at different stages of the investigation. By setting specific objectives, you create a framework that helps you decide what questions to ask and how to phrase the questions.

Step 3: Develop Interview Questions

Creating a standard list of interview questions helps to keep you on track and avoid unnecessary tangents. Also, if you interview several people who perform the same job, a standard question list allows you to compare their answers. Although you have a list of specific questions, you might decide to depart from it because an answer to one question leads to another topic that you want to pursue. That question or topic then should be included in a revised set of questions used to conduct future interviews. If the question proves to be extremely important, you may need to return to a previous interviewee to query him or her on the topic.

The interview should consist of several different kinds of questions: open-ended, closed-ended, or questions with a range of responses. When you phrase your questions, you should avoid **leading questions** that suggest or favor a particular reply. For example, rather than asking, "What advantages do you see in the proposed system?" you might ask, "Do you see any advantages in the proposed system?"

OPEN-ENDED QUESTIONS   **Open-ended questions** encourage spontaneous and unstructured responses. Such questions are useful when you want to understand a larger process or draw out the interviewee's opinions, attitudes, or suggestions. Here are some examples of open-ended questions: What are users saying about the new system? How is this task performed? Why do you perform the task that way? How are the checks reconciled? What added features would you like to have in the new billing system? Also, you can use an open-ended question to probe further by asking: Is there anything else you can tell me about this topic?

CLOSED-ENDED QUESTIONS   **Closed-ended questions** limit or restrict the response. You use closed-ended questions when you want information that is more specific or when you need to verify facts. Examples of closed-ended questions include the following: How many personal computers do you have in this department? Do you review the reports before they are sent out? How many hours of training does a clerk receive? Is the calculation procedure described in the manual? How many customers ordered products from the Web site last month?

RANGE-OF-RESPONSE QUESTIONS   **Range-of-response questions** are closed-ended questions that ask the person to evaluate something by providing limited answers to specific responses or on a numeric scale. This method makes it easier to tabulate the answers and interpret the results. Range-of-response questions might include these: On a scale of 1 to 10, with 1 the lowest and 10 the highest, how effective was your training? How would you rate the severity of the problem: low, medium, or high? Is the system shutdown something that occurs never, sometimes, often, usually, or always?

Step 4: Prepare for the Interview

After setting the objectives and developing the questions, you must prepare for the interview. Careful preparation is essential because an interview is an important meeting and not just a casual chat. When you schedule the interview, suggest a specific day and time and let the interviewee know how long you expect the meeting to last. It is also a good idea to send an e-mail or place a reminder call the day before the interview.

Remember that the interview is an interruption of the other person's routine, so you should limit the interview to no more than one hour. If business pressures force a post-ponement of the meeting, you should schedule another appointment as soon as it is convenient. Remember to keep department managers informed of your meetings with their staff members. Sending a message to each department manager listing your planned appointments is a good way to keep them informed. Figure 4-19 is an example of such a message.

You should send a list of topics to an interviewee several days before the meeting, especially when detailed information is needed, so the person can prepare for the interview and minimize the need for a follow-up meeting. Figure 4-20 shows a sample message that lists specific questions and confirms the date, time, location, purpose, and anticipated duration of the interview.

If you have questions about documents, ask the interviewee to have samples available at the meeting. Your advance memo should include a list of the documents you want to discuss, if you know what they are. Also, you can make a general request for documents, as the analyst did in her e-mail shown in Figure 4-20.

Two schools of thought exist about the best location for an interview. Some analysts believe that interviews should take place in the interviewee's office, whereas other analysts feel that a neutral location such as a conference room is better.

Supporters of interviews in the interviewee's office believe that is the best location because it makes the interviewee feel comfortable during the meeting. A second

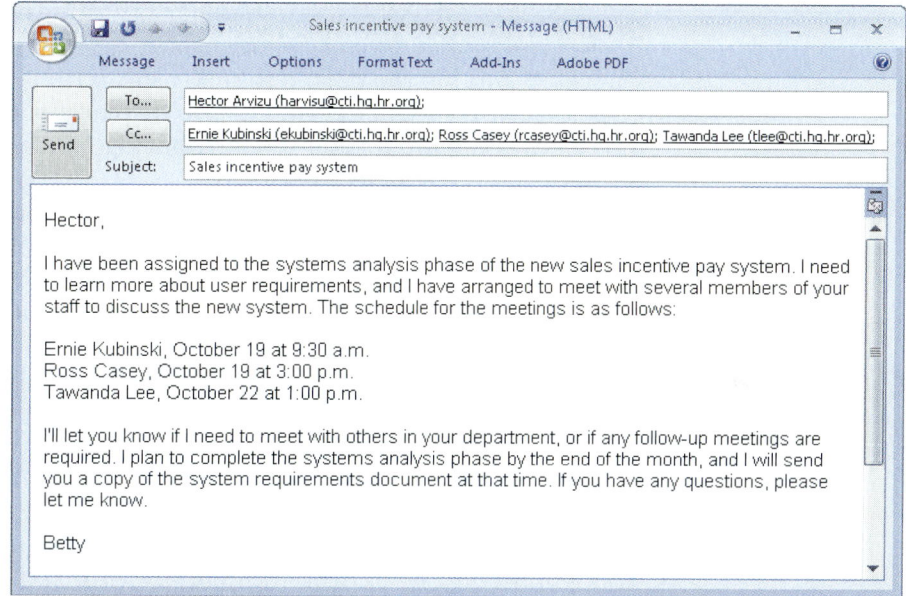

FIGURE 4-19 Sample message to a department head about interviews with people in his group.

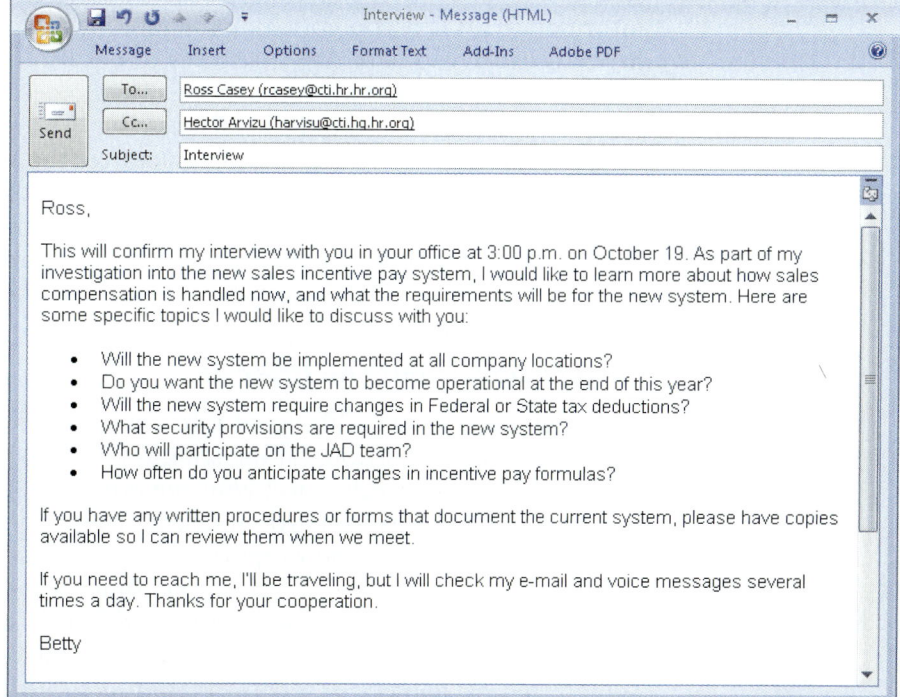

FIGURE 4-20 Sample message to confirm a planned meeting and provide advance information about topics to be discussed.

argument in favor of the interviewee's office is that the office is where he or she has the easiest access to supporting material that might be needed during the discussion. If you provide a complete list of topics in advance, however, the interviewee can bring the necessary items to a conference room or other location.

Supporters of neutral locations stress the importance of keeping interruptions to a minimum so both people can concentrate fully. In addition, an interview that is free of interruptions takes less time. If the meeting does take place in the interviewee's office, you should suggest tactfully that all calls be held until the conclusion of the interview.

Step 5: Conduct the Interview

After determining the people to interview, setting your objectives, and preparing the questions, you should develop a specific plan for the meeting. When conducting an interview, you should begin by introducing yourself, describing the project, and explaining your interview objectives.

During the interview, ask questions in the order in which you prepared them, and give the interviewee sufficient time to provide thoughtful answers. Establishing a good rapport with the interviewee is important, especially if this is your first meeting. If the other person feels comfortable and at ease, you probably will receive more complete and candid answers. Your primary responsibility during an interview is to listen carefully to the answers. Analysts sometimes hear only what they expect to hear. You must concentrate on what is said and notice any nonverbal communication that takes place. This process is called **engaged listening**.

After asking a question, allow the person enough time to think about the question and arrive at an answer. Studies have shown that the maximum pause during a conversation is usually three to five seconds. After that interval, one person will begin talking. You will need to be patient and practice your skills in many actual interview situations to be successful.

When you finish asking your questions, summarize the main points covered in the interview and explain the next course of action. For example, mention that you will send a follow-up memo or that the interviewee should get back to you with certain information. When you conclude the interview, thank the person and encourage him or her to contact you with any questions or additional comments. Also, when the interview ends, it is a good idea to ask the interviewee whether he or she can suggest any additional topics that should be discussed.

After an interview, you should summarize the session and seek a confirmation from the other person. By stating your understanding of the discussion, the interviewee can respond and correct you, if necessary. One approach is to rephrase the interviewee's answers. For example, you can say, "If I understand you correctly, you are saying that …" and then reiterate the information given by the interviewee.

Step 6: Document the Interview

Although taking notes during an interview has both advantages and disadvantages, the accepted view is that note taking should be kept to a minimum. Although you should write down a few notes to jog your memory after the interview, you should avoid writing everything that is said. Too much writing distracts the other person and makes it harder to establish a good rapport.

After conducting the interview, you must record the information quickly. You should set aside time right after the meeting to record the facts and evaluate the information. For that reason, try not to schedule back-to-back interviews. Studies have shown that 50 percent of a conversation is forgotten within 30 minutes. You, therefore, should use your notes to record the facts immediately so you will not forget them. You can summarize the facts by preparing a narrative describing what took place or by recording the answers you received next to each question on your prepared question list.

Tape recorders are effective tools for an interview; however, many people feel uncomfortable when recorders are present. Before using a recorder, you should discuss its use with the interviewee. Assure the interviewee that you will erase the tape after you transcribe your notes and that you will stop and rewind the tape anytime during the interview at his or her request. If you ask sensitive questions or the interviewee wants to answer a question without being recorded, explain that you will turn off the tape for a period of time during the interview.

Even with a tape recorder in use, you should listen carefully to the interviewee's responses so you can ask good follow-up questions. Otherwise, you might have to return for a second visit to ask the questions you missed the first time. Also, remember that each recorded interview takes twice the amount of time, because you must listen to or view the recorded meeting again after conducting the interview itself.

After the interview, send a memo to the interviewee expressing your appreciation for his or her time and cooperation. In the memo, you should note the date, time, location, purpose of the interview, and the main points you discussed so the interviewee has a written summary and can offer additions or corrections.

Step 7: Evaluate the Interview

In addition to recording the facts obtained in an interview, try to identify any possible biases. For example, an interviewee who tries to protect his or her own area or function might give incomplete answers or refrain from volunteering information. Or, an interviewee with strong opinions about the current or future system might distort the facts. Some interviewees might answer your questions in an attempt to be helpful even though they do not have the necessary experience to provide accurate information.

CASE IN POINT 4.2: DEEP RIVER COLLEGE

Deep River College is a two-year school in Southern California. Twice a year, the fund-raising office at Deep River mails requests for donations to the alumni. The staff uses a word processing program and a personal information database to create personalized letters. Data on past contributions and other alumni information, however, is stored manually. The dean, Alexandra Ali, recently submitted a systems request asking the college's IT department to develop a computerized alumni information system. The school does not have a formal systems review committee, and each department has an individual budget for information services.

Eddie Bateman, a systems analyst, performed a preliminary investigation and he concluded that the system met all the feasibility tests. After reading his report, Alexandra asked him to proceed with the systems analysis phase. Eddie has scheduled an interview with her, and he has asked you to help him prepare for the meeting. Specifically, he wants you to list all the topics he should cover during the interview. Eddie also wants you to prepare a list of specific questions that he should ask. Be sure to include open-ended, closed-ended, and range-of-response questions.

Unsuccessful Interviews

No matter how well you prepare for interviews, some are not successful. One of the main reasons could be that you and the interviewee did not get along well. Such a situation can be caused by several factors. For example, a misunderstanding or personality conflict could affect the interview negatively, or the interviewee might be afraid that the new system will eliminate or change his or her job.

In other cases, the interviewee might give only short or incomplete responses to your open-ended questions. If so, you should switch to closed-ended questions or questions with a range of responses, or try rephrasing your open-ended questions into those types of questions. If that still does not help, you should find a tactful way to conclude the meeting.

Continuing an unproductive interview is difficult. The interviewee could be more cooperative later, or you might find the information you seek elsewhere. If failure to obtain specific information will jeopardize the success of the project, inform your supervisor, who can help you decide what action to take. Your supervisor might contact the interviewee's supervisor, ask another systems analyst to interview the person, or find some other way to get the needed information.

CASE IN POINT 4.3:   FastPak Overnight Package System

FastPak, the nation's fourth-largest overnight package system, is headquartered in Los Angeles, California. Jesse Evans is a systems analyst on an IT team that is studying ways to update FastPak's package tracking system. Jesse prepared well for her interview with Jason Tanya, FastPak's executive vice president. Mr. Tanya did not ask his assistant to hold his calls during the meeting, however. After several interruptions, Jesse tactfully suggested that she could come back another time, or perhaps that Mr. Tanya might ask his assistant to hold his calls. "No way," he replied. "I'm a very busy man and we'll just have to fit this in as we can, even if it takes all day." Jesse was unprepared for his response. What are her options? Is an analyst always in control of this kind of situation? Why or why not?

OTHER FACT-FINDING TECHNIQUES

In addition to interviewing, systems analysts use other fact-finding techniques, including document review, observation, questionnaires and surveys, sampling, and research. Such techniques are used before interviewing begins to obtain a good overview and to help develop better interview questions.

Document Review

Document review can help you understand how the current system is supposed to work. Remember that system documentation sometimes is out of date. Forms can change or be discontinued, and documented procedures often are modified or eliminated. You should obtain copies of actual forms and operating documents currently in use. You also should review blank copies of forms, as well as samples of actual completed forms. You usually can obtain document samples during interviews with the people who perform that procedure. If the system uses a software package, you should review the documentation for that software.

Observation

The **observation** of current operating procedures is another fact-finding technique. Seeing the system in action gives you additional perspective and a better understanding of system procedures. Personal observation also allows you to verify statements made in interviews and determine whether procedures really operate as they are described.

Through observation, you might discover that neither the system documentation nor the interview statements are accurate.

Personal observation also can provide important advantages as the development process continues. For example, recommendations often are better accepted when they are based on personal observation of actual operations. Observation also can provide the knowledge needed to test or install future changes and can help build relationships with the users who will work with the new system.

Plan your observations in advance by preparing a checklist of specific tasks you want to observe and questions you want to ask. Consider the following issues when you prepare your list:

1. Ask sufficient questions to ensure that you have a complete understanding of the present system operation. A primary goal is to identify the methods of handling situations that are not covered by standard operating procedures. For example, what happens in a payroll system if an employee loses a time card? What is the procedure if an employee starts a shift 10 minutes late but then works 20 minutes overtime? Often, the rules for exceptions such as these are not written or formalized; therefore, you must try to document any procedures for handling exceptions.

2. Observe all the steps in a transaction and note the documents, inputs, outputs, and processes involved.

3. Examine each form, record, and report. Determine the purpose each item of information serves.

4. Consider each user who works with the system and the following questions: What information does that person receive from other people? What information does this person generate? How is the information communicated? How often do interruptions occur? How much downtime occurs? How much support does the user require, and who provides it?

5. Talk to the people who receive current reports to see whether the reports are complete, timely, accurate, and in a useful form. Ask whether information can be eliminated or improved and whether people would like to receive additional information.

As you observe people at work, as shown in Figure 4-21, consider a factor called the **Hawthorne Effect**. The name comes from a well-known study performed in the Hawthorne plant of the Western Electric Company in the 1920s. The purpose of the study was to determine how various changes in the work environment would affect employee productivity. The surprising result was that productivity improved during observation whether the conditions were made better or worse. Researchers concluded that productivity seemed to improve whenever the workers knew they were being observed.

Thus, as you observe users, remember that normal operations might not always run as smoothly as your observations indicate. Operations also might run less smoothly because workers might be nervous during

FIGURE 4-21 The Hawthorne study suggested that worker productivity improves during observation. Always consider the Hawthorne Effect when observing the operation of an existing system.

the observation. If possible, meet with workers and their supervisors to discuss your plans and objectives to help establish a good working relationship. In some situations, you might even participate in the work yourself to gain a personal understanding of the task or the environment.

Questionnaires and Surveys

In projects where it is desirable to obtain input from a large number of people, a questionnaire can be a valuable tool. A **questionnaire**, also called a **survey**, is a document containing a number of standard questions that can be sent to many individuals.

Questionnaires can be used to obtain information about a wide range of topics, including workloads, reports received, volumes of transactions handled, job duties, difficulties, and opinions of how the job could be performed better or more efficiently. Figure 4-22 shows a sample questionnaire that includes several different question and response formats.

PURCHASE REQUISITION QUESTIONNAIRE

Pat Kline, Vice President, Finance, has asked us to investigate the purchase requisition process to see if it can be improved. Your input concerning this requisition process will be very valuable. We would greatly appreciate it if you could complete the following questionnaire and return it by March 10 to Dana Juarez in information technology. If you have any questions, please call Dana at x2561.

A.   YOUR OBSERVATIONS
Please answer each question by checking one box.

1.   How many purchase requisitions did you process in the past five working days? _____

2.   What percentage of your time is spent processing requisitions?
[ ] under 20%   [ ] 60–79%
[ ] 21–39%   [ ] 80% or more
[ ] 40–59%

3.   Do you believe too many errors exist on requisitions?
[ ] yes
[ ] no

4.   Out of every 100 requisitions you process, how many contain errors?
[ ] fewer than 5   [ ] 20 to 29
[ ] 5 to 9   [ ] 30 to 39
[ ] 10 to 14   [ ] 40 to 49
[ ] 15 to 19   [ ] 50 or more

5.   What errors do you see most often on requisitions? (Place a 1 next to the most common error, place a 2 next to the second, etc.)
[ ] incorrect charge number   [ ] missing authorization
[ ] missing charge information   [ ] other (please explain) _____
[ ] arithmetic errors
[ ] incorrect discount percent used

B.   YOUR SUGGESTIONS
Please be specific, and give examples if possible.

1.   If the currently used purchase requisition form were to be redesigned, what changes to the form would you recommend?

_____
_____
_____
_____
_____

(If necessary, please attach another sheet)

2.   Would you be interested in meeting with an information technology representative to discuss your ideas further? If so, please complete the following information:

Name _____   Department _____

Telephone _____   E-mail address _____

FIGURE 4-22 Sample questionnaire.

A typical questionnaire starts with a heading, which includes a title, a brief statement of purpose, the name and telephone number of the contact person, the deadline date for completion, and how and where to return the form. The heading usually is followed by general instructions that provide clear guidance on how to answer the questions. Headings also are used to introduce each main section or portion of the survey and include instructions when the type of question or response changes. A long questionnaire might end with a conclusion that thanks the participants and reminds them how to return the form.

What about the issue of anonymity? Should people be asked to sign the questionnaire, or is it better to allow anonymous responses? The answer depends on two questions. First, does an analyst really need to know who the respondents are in order to match or correlate information? For example, it might be important to know what percentage of users need a certain software feature, but specific usernames might not be relevant. Second, does the questionnaire include any sensitive or controversial topics? Many people do not want to be identified when answering a question such as "How well has your supervisor explained the system to you?" In such cases, anonymous responses might provide better information.

When designing a questionnaire, the most important rule of all is to make sure that your questions collect the right data in a form that you can use to further your fact-finding. Here are some additional ideas to keep in mind when designing your questionnaire:

- Keep the questionnaire brief and user-friendly.

- Provide clear instructions that will answer all anticipated questions.

- Arrange the questions in a logical order, going from simple to more complex topics.

- Phrase questions to avoid misunderstandings; use simple terms and wording.

- Try not to lead the response or use questions that give clues to expected answers.

- Limit the use of open-ended questions that are difficult to tabulate.

- Limit the use of questions that can raise concerns about job security or other negative issues.

- Include a section at the end of the questionnaire for general comments.

- Test the questionnaire whenever possible on a small test group before finalizing it and distributing to a large group.

A questionnaire can be a traditional paper form, or you can create a **fill-in form** and collect data on the Internet or a company intranet. For example, you can use Microsoft Word, as shown in Figure 4-23, to create form fields, including text boxes, date pickers, and drop-down lists where users can click selections. Before you publish the form, you should protect it so users can fill it in but cannot change the layout or design. Forms also can be automated, so if a user answers *no* to question three, he or she goes directly to question eight, where the form-filling resumes.

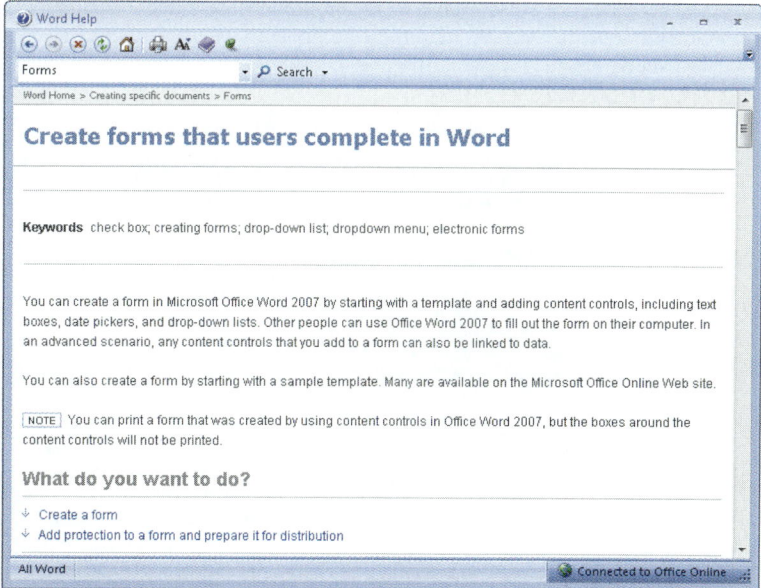

Sampling

When studying an information system, you should collect examples of actual documents using a process called **sampling**. The samples might include records, reports, operational logs,

FIGURE 4-23 Using Microsoft Word, you can create a fill-in form with text boxes, date pickers, and drop-down lists.

data entry documents, complaint summaries, work requests, and various types of forms. Sampling techniques include systematic sampling, stratified sampling, and random sampling.

Suppose you have a list of 200 customers who complained about errors in their statements, and you want to review a representative sample of 20 customers. A **systematic sample** would select every tenth customer for review. If you want to ensure that the sample is balanced geographically, however, you could use a **stratified sample** to select five customers from each of four zip codes. Another example of stratified sampling is to select a certain percentage of transactions from each zip code, rather than a fixed number. Finally, a **random sample** selects any 20 customers.

The main objective of a sample is to ensure that it represents the overall population accurately. If you are analyzing inventory transactions, for example, you should select a sample of transactions that are typical of actual inventory operations and do not include unusual or unrelated examples. For instance, if a company performs special processing on the last business day of the month, that day is not a good time to sample *typical* daily operations. To be useful, a sample must be large enough to provide a fair representation of the overall data.

You also should consider sampling when using interviews or questionnaires. Rather than interviewing everyone or sending a questionnaire to the entire group, you can use a sample of participants. You must use sound sampling techniques to reflect the overall population and obtain an accurate picture.

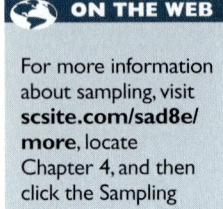

ON THE WEB

For more information about sampling, visit **scsite.com/sad8e/ more**, locate Chapter 4, and then click the Sampling link.

Research

Research is another important fact-finding technique. Your research can include the Internet, IT magazines, and books to obtain background information, technical material, and news about industry trends and developments. In addition, you can attend professional meetings, seminars, and discussions with other IT professionals, which can be very helpful in problem solving.

The Internet is an extremely valuable resource. Part 4 of the Systems Analyst's Toolkit describes a variety of Internet resource tools. Using the Internet, you also can access information from federal and state governments, as well as from publishers, universities, and libraries around the world. Online forums and newsgroups are good resources for exchanging information with other professionals, seeking answers to questions, and monitoring discussions that are of interest to you.

All major hardware and software vendors maintain sites on the Web where you can obtain information about products and services offered by the company and send e-mail with specific questions to company representatives. In addition to contacting specific firms, you can access Web sites maintained by publishers and independent firms that provide links to hundreds of hardware and software vendors, as shown in Figure 4-24. Such sites are one-stop information centers where IT

FIGURE 4-24 InfoWorld's Web site offers many resources for IT professionals.

professionals can find information, share ideas, and keep posted on developments in technology.

Research also can involve a visit to a physical location, called a **site visit**, where the objective is to observe a system in use at another location. If you are studying your firm's human resources information system, for example, you might want to see how another company's system works. Site visits also are important when considering the purchase of a software package. If the software vendor suggests possible sites to visit, be aware that such sites might constitute a biased sample. A single site visit seldom gives you true pictures, so you should try to visit more than one installation.

Before a site visit such as the one shown in Figure 4-25, prepare just as you would for an interview. Contact the appropriate manager and explain the purpose of your visit. Decide what questions you will ask and what processes you will observe. During your visit, observe how the system works and note any problems or limitations. You also will want to learn about the support provided by the vendor, the quality of the system documentation, and so on.

FIGURE 4-25 A site visit provides an opportunity to observe a system in use.

Interviews versus Questionnaires

When you seek input from a large group, a questionnaire is a very useful tool. On the other hand, if you require detailed information from only a few people, then you probably should interview each person individually. Is it better to interview or use a questionnaire? Each situation is different, and you must consider the type of information, time constraints, and expense factors.

The interview is more familiar and personal than a questionnaire. People who are unwilling to put critical or controversial comments in writing might talk more freely in person. Moreover, during a face-to-face interview, you can react immediately to anything the interviewee says. If surprising or confusing statements are made, you can pursue the topic with additional questions. In addition, during a personal interview, you can watch for clues to help you determine if responses are knowledgeable and unbiased. Participation in interviews also can affect user attitudes, because people who are asked for their opinions often view the project more favorably.

Interviewing, however, is a costly and time-consuming process. In addition to the meeting itself, both people must prepare, and the interviewer has to do follow-up work. When a number of interviews are planned, the total cost can be quite substantial. The personal interview usually is the most expensive fact-finding technique.

In contrast, a questionnaire gives many people the opportunity to provide input and suggestions. Questionnaire recipients can answer the questions at their convenience and do not have to set aside a block of time for an interview. If the questionnaire allows anonymous responses, people might offer more candid responses than they would in an interview.

Preparing a good questionnaire, however, like a good interview, requires skill and time. If a question is misinterpreted, you cannot clarify the meaning as you can in a face-to-face interview. Furthermore, unless questionnaires are designed well, recipients might view them as intrusive, time-consuming, and impersonal. As an analyst, you should select the technique that will work best in a particular situation.

Another popular method of obtaining input is called **brainstorming**, which refers to a small group discussion of a specific problem, opportunity, or issue. This technique encourages new ideas, allows team participation, and enables participants to build on each other's inputs and thoughts. Brainstorming can be structured or unstructured. In **structured brainstorming**, each participant speaks when it is his or her turn, or passes. In **unstructured brainstorming**, anyone can speak at any time. At some point, the results are recorded and made part of the fact-finding documentation process.

Ann Ellis is a systems analyst at CyberStuff, a large company that sells computer hardware and software via telephone, mail order, and the Internet. CyberStuff processes several thousand transactions per week on a three-shift operation and employs 50 full-time and 125 part-time employees. Lately, the billing department has experienced an increase in the number of customer complaints about incorrect bills. During the preliminary investigation, Ann learned that some CyberStuff representatives did not follow established order entry procedures. She feels that with more information, she might find a pattern and identify a solution for the problem.

Ann is not sure how to proceed. She came to you, her supervisor, with two separate questions. First, is a questionnaire the best approach, or would interviews be better? Second, whether she uses interviews, a questionnaire, or both techniques, should she select the participants at random, include an equal number of people from each shift, or use some other approach? As Ann's supervisor, what would you suggest, and why?

DOCUMENTATION

Keeping accurate records of interviews, facts, ideas, and observations is essential to successful systems development. The ability to manage information is the mark of a successful systems analyst and an important skill for all IT professionals.

The Need for Recording the Facts

As you gather information, the importance of a single item can be overlooked or complex system details can be forgotten. The basic rule is to write it down. You should document your work according to the following principles:

- Record information as soon as you obtain it.
- Use the simplest recording method possible.
- Record your findings in such a way that they can be understood by someone else.
- Organize your documentation so related material is located easily.

Often, systems analysts use special forms for describing a system, recording interviews, and summarizing documents. One type of documentation is a narrative list with simple statements about what is occurring, apparent problems, and suggestions for improvement. Other forms of documentation that are described in Chapter 4 include data flow diagrams, flowcharts, sample forms, and screen captures.

Software Tools

Many software programs are available to help you record and document information. Some examples are described here.

CASE TOOLS   You can use CASE tools at every stage of systems development. This chapter contains several examples of CASE tools. Part 2 of the Systems Analyst's Toolkit describes other features and capabilities of CASE tools.

PRODUCTIVITY SOFTWARE   Productivity software includes word processing, spreadsheet, database management, and presentation graphics programs. Although Microsoft Office is the best-known set of productivity software programs, other vendors offer products in each of these categories.

Using word processing software such as Microsoft Word, Corel WordPerfect, or OpenOffice.org Writer, you can create reports, summaries, tables, and forms. In addition to standard document preparation, the program can help you organize a presentation with templates, bookmarks, annotations, revision control, and an index. You can consult the program's Help system for more information about those and other features. You also can create fill-in forms to conduct surveys and questionnaires, as described earlier in this chapter.

Spreadsheet software, such as Microsoft Excel, Corel Quattro Pro, or OpenOffice.org Calc, can help you track and manage numerical data or financial information. You also can generate graphs and charts that display the data and show possible patterns, and you can use the statistical functions in a spreadsheet to tabulate and analyze questionnaire data. A graphical format often is used in quality control analysis because it highlights problems and their possible causes, and it is effective when presenting results to management. A common tool for showing the distribution of questionnaire or sampling results is a vertical bar chart called a **histogram**. Most spreadsheet programs can create histograms and other charts that can display data you have collected. Figure 4-26 displays a typical histogram that might have resulted from the questionnaire shown in Figure 4-22 on page 162.

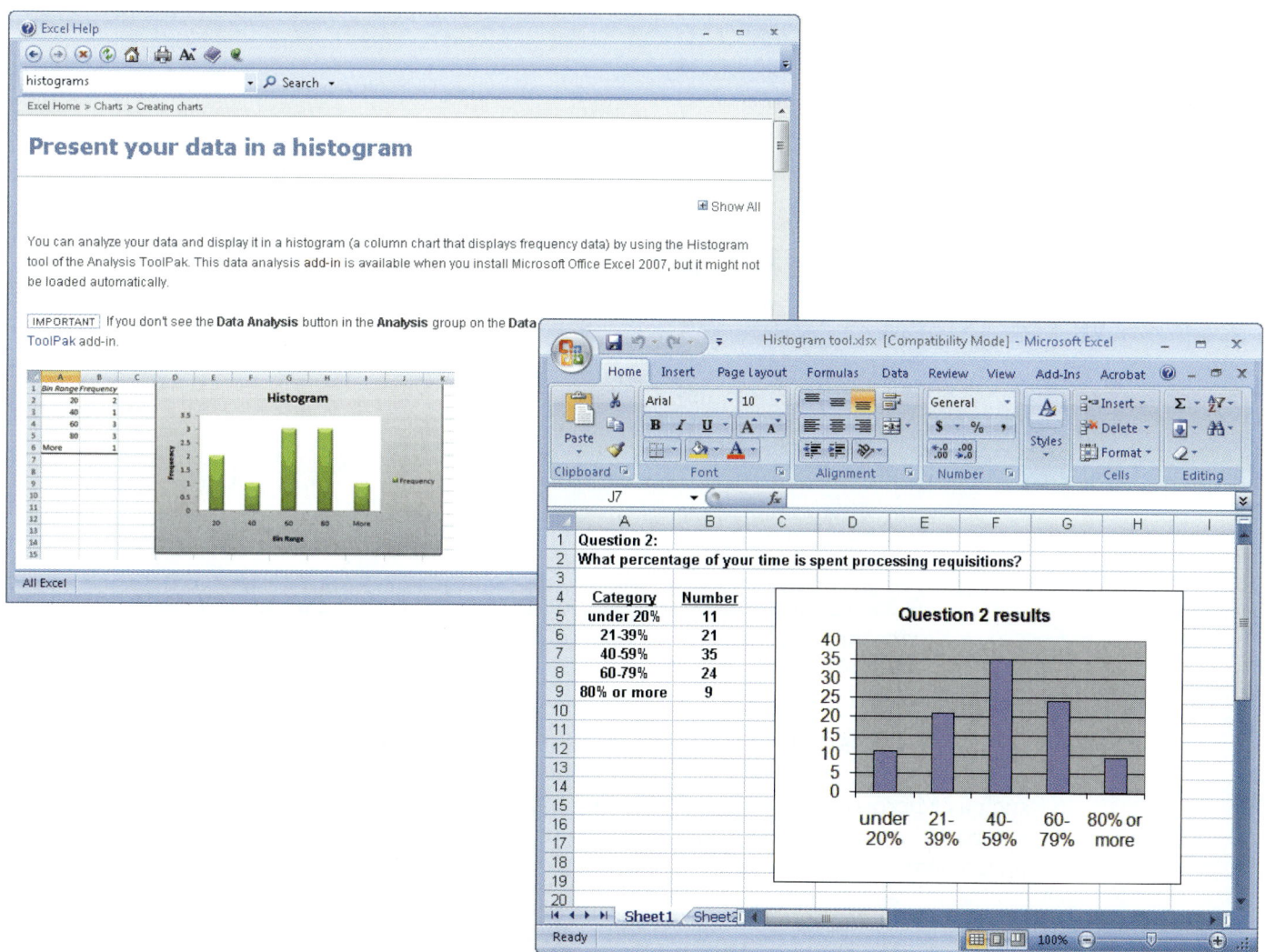

FIGURE 4-26 A Microsoft Excel histogram displays results from a questionnaire.

Database management software allows you to document and organize fact-finding results such as events, observations, and data samples. You can use a database program such as Microsoft Access to manage the details of a complex project, create queries to retrieve specific information, and generate custom reports.

Presentation graphics software, such as Microsoft PowerPoint, Apple Keynote, or OpenOffice.org Impress, is a powerful tool for organizing and developing your formal presentation. Presentation graphics programs enable you to create organization charts that can be used in a preliminary investigation and later during requirements modeling. These high-quality charts also can be included in written reports and management presentations.

GRAPHIC MODELING SOFTWARE Microsoft Visio is a popular graphic modeling tool that can produce a wide range of charts and diagrams. Visio includes a library of templates, stencils, and shapes. An analyst can use Visio to create many types of visual models, including business processes, flowcharts, network diagrams, organization charts, and Web site maps, such as the one shown in Figure 4-27.

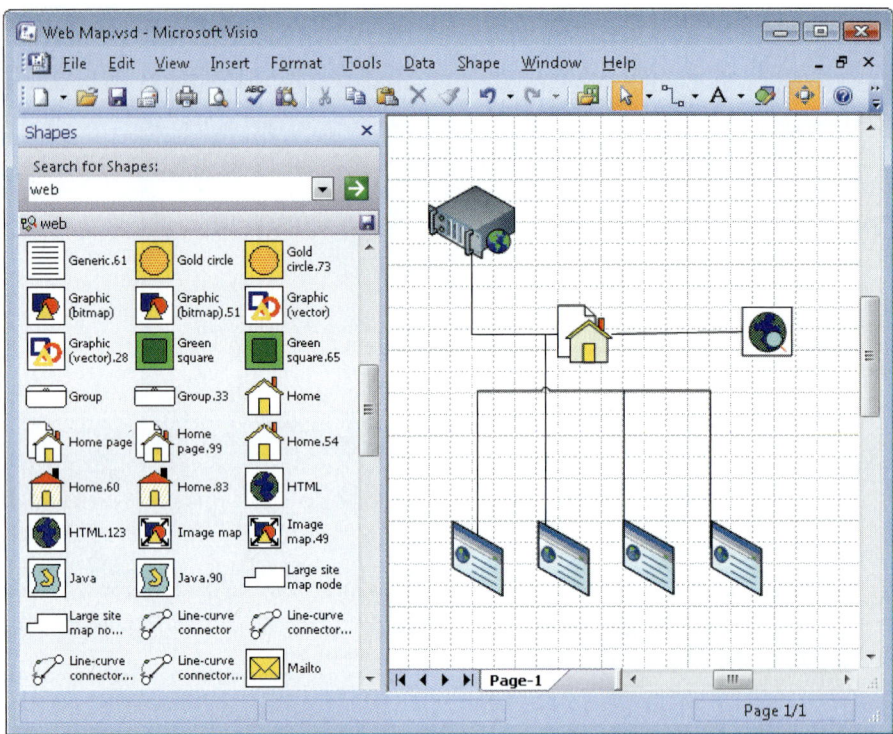

FIGURE 4-27 This Microsoft Visio screen shows shapes that can be used to create a Web site map.

PERSONAL INFORMATION MANAGERS A busy analyst needs to keep track of meetings, interviews, appointments, and deadlines. A **personal information manager (PIM)**, such as Microsoft Outlook or Lotus Organizer, can help manage those tasks and provide a personal calendar and a to-do list, with priorities and the capability to check off completed items.

In addition to desktop-based organizers, handheld computers are enormously popular. Some handheld computers, also called **personal digital assistants (PDAs)**, accept handwritten input, while others have small keyboards. These devices can handle calendars, schedules, appointments, telephone lists, and calculations. A PDA can be stand-alone, bluetooth-capable to synchronize with a desktop, or fully wireless-enabled, such as the Palm® T|X shown in Figure 4-28.

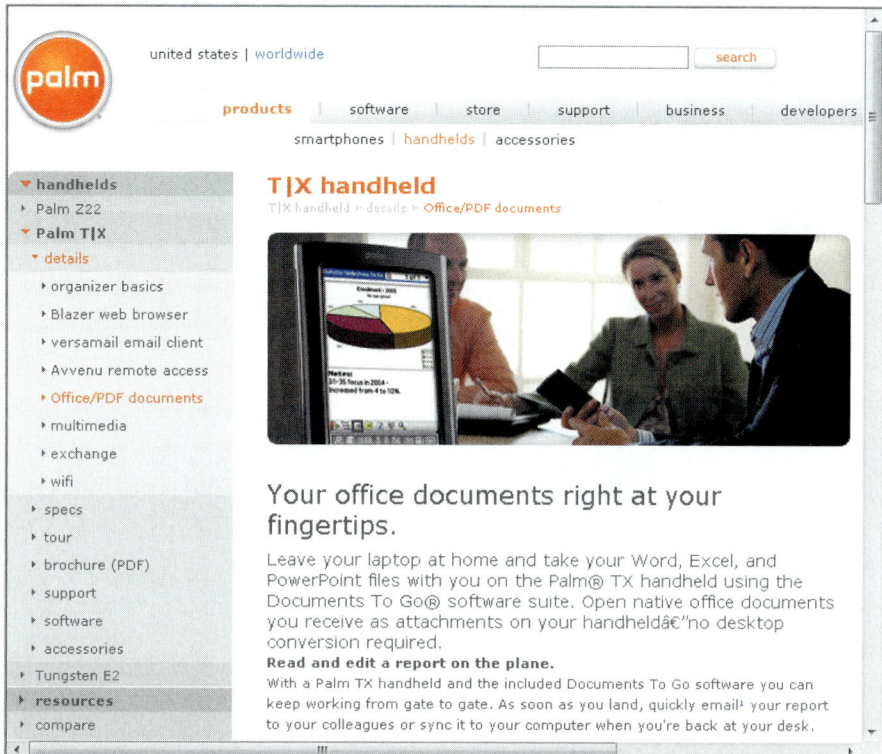

FIGURE 4-28 In addition to handling schedules, appointments, and telephone lists, handheld PDAs such as the Palm TX can run Microsoft Office applications. The Web-enabled TX can exchange data with desktop computers and corporate networks.

WIRELESS COMMUNICATION DEVICES Wireless technology is shaping the future of mobile computing, and wireless communication devices are extremely powerful tools in an analyst's arsenal.

Wireless capability allows Web access, e-mail, easy file exchange, and synchronization with desktop computers and office networks. Popular examples of these all-in-one devices include Research in Motion's BlackBerry and Motorola's Moto Q. To compete in this rapidly expanding market, cell phone makers have enhanced their products with powerful new features. For example, Kyocera's 7135 smartphone combines a Web-capable phone with a handheld PDA, and Apple's iPhone offers a striking look and feel, together with Internet access and powerful wireless features. Examples of these devices are shown in Figure 4-29 on the next page.

As wireless communication expands in the future, PDA, cell phone, and all-in-one device technologies will merge. This means that business and professional users will be able to choose from a wide array of wireless tools to enhance communication and productivity.

FIGURE 4-29 Four popular examples of wireless devices.

PREVIEW OF LOGICAL MODELING

At the conclusion of requirements modeling, systems developers should have a clear understanding of business processes and system requirements. The next step is to construct a logical model of the system.

Data and process modeling, which is described in Chapter 5, uses a structured analysis approach. Structured analysis is a popular, traditional technique that describes the system in terms of data and the processes that act on that data.

An alternative to structured analysis modeling is object modeling, which is described in Chapter 6. Object modeling is a methodology that combines data and processes into things called objects that represent actual people, things, transactions, and events. Systems analysts use object models to visualize and document real-world business processes and operations.

IT professionals have differing views about systems development methodologies, and no universally accepted approach exists. By studying both structured analysis and object-oriented methods, you gain valuable knowledge, skills, and perspective. You then can use that information to determine what method, or combination of methods, is best for the different situations you will face in your career.

A QUESTION OF ETHICS

Your supervisor manages the corporate office where you work as a systems analyst. Several weeks ago, after hearing rumors of employee dissatisfaction, he asked you to create a survey for all IT employees. After the responses were returned and tabulated, he was disappointed to learn that many employees assigned low ratings to morale and management policies.

This morning he called you into his office and asked whether you could identify the departments that submitted the lowest ratings. No names were used on the individual survey forms. However, with a little analysis, you probably could identify the departments, because several questions were department-related.

Now you are not sure how to respond. The expectation was that the survey would be anonymous. Even though no individuals would be identified, would it be ethical to reveal which departments sent in the low ratings? Would your supervisor's motives for wanting this information matter?

CHAPTER SUMMARY

The systems analysis phase includes three activities: requirements modeling, data and process modeling, and consideration of development strategies. The main objective is to understand the proposed project, ensure that it will support business requirements, and build a solid foundation for the systems design phase.

During requirements modeling, you identify the business-related requirements for the new information system, including outputs, inputs, processes, performance, and controls. You consider scalability to ensure that the system can support future growth and expansion. You also estimate total cost of ownership (TCO) to identify all costs, including indirect costs.

Popular team-based approaches include JAD, RAD, and agile methods. Joint application development (JAD) is a popular, team-based approach to fact-finding and requirements modeling. JAD involves an interactive group of users, managers, and IT professionals who participate in requirements modeling and develop a greater commitment to the project and to their common goals.

Rapid application development (RAD) is a team-based technique that speeds up information systems development and produces a functioning information system. RAD is a complete methodology, with a four-phase life cycle that parallels the traditional SDLC phases.

Agile methods attempt to develop a system incrementally, by building a series of prototypes and constantly adjusting them to user requirements.

Systems analysts use various tools and techniques to model system requirements. Unified Modeling Language (UML) is a widely used method of visualizing and documenting software design through the eyes of the business user. UML tools include use case diagrams and sequence diagrams to represent actors, their roles, and the sequence of transactions that occurs. A functional decomposition diagram (FDD) is used to represent business functions and processes.

The fact-finding process includes interviewing, document review, observation, questionnaires, sampling, and research. Successful interviewing requires good planning and strong interpersonal and communication skills. The systems analyst must decide on the people to interview, set interview objectives, and prepare for, conduct, and analyze interviews. The analyst also might find it helpful to use one or more software tools during fact-finding.

Systems analysts should carefully record and document factual information as it is collected, and various software tools can help an analyst visualize and describe an information system. The chapter concluded with a preview of logical modeling. Data and process modeling is a structured analysis approach that views the system in terms of data and the processes that act on that data. Object modeling is an approach that views the system in terms of data and the processes that act on that data.

Key Terms and Phrases

actor *147*

agile methods *139*

analytical skills *139*

brainstorming *165*

closed-ended questions *156*

construction phase *143*

cutover phase *143*

data flow diagram (DFD) *147*

document review *160*

engaged listening *158*

fill-in form *163*

functional decomposition diagram (FDD) *146*

Hawthorne Effect *161*

histogram *167*

informal structure *155*

inputs *138*

interpersonal skills *139*

interview *155*

joint application development (JAD) *139*

leading questions *156*

observation *160*

open-ended questions *156*

outputs *138*

performance *138*

personal digital assistants (PDAs) *169*

personal information manager (PIM) *168*

processes *138*

productivity software *166*

questionnaire *162*

random sample *164*

range-of-response questions *156*

rapid application development (RAD) *139*

Rapid Economic Justification (REJ) *152*

requirements modeling *138*

requirements planning phase *142*

research *164*

sampling *163*

scalability *151*

security *138*

sequence diagram *148*

site visit *165*

stratified sample *164*

structure chart *146*

structured brainstorming *165*

survey *162*

system requirement *149*

system requirements document *139*

systematic sample *164*

total cost of ownership (TCO) *151*

Unified Modeling Language (UML) *147*

unstructured brainstorming *165*

use case diagram *147*

user design phase *142*

Zachman Framework for Enterprise
  Architecture *154*

Learn It Online

Instructions: To complete the Learn It Online exercises, start your browser, click the Address bar, and then enter **scsite.com/sad8e/learn**. When the Systems Analysis and Design Learn It Online page is displayed, follow the instructions in the exercises below. Each exercise has instructions for saving your results, either for your own records or for submission to your instructor.

1 Chapter Reinforcement

TF, MC, and SA

Below SAD Chapter 4, click one of the Chapter Reinforcement links for Multiple Choice, True/False, or Short Answer. Answer each question and submit to your instructor.

2 Flash Cards

Below SAD Chapter 4, click the Flash Cards link and read the instructions. Type 20 (or a number specified by your instructor) in the Number of playing cards text box, type your name in the Enter your Name text box, and then click the Flip Card button. When the flash card is displayed, read the question and then click the ANSWER box arrow to select an answer. Flip through the Flash Cards. If your score is 15 (75%) correct or greater, click Print on the File menu to print your results. If your score is less than 15 (75%) correct, then redo this exercise by clicking the Replay button.

3 Practice Test

Below SAD Chapter 4, click the Practice Test link. Answer each question, enter your first and last name at the bottom of the page, and then click the Grade Test button. When the graded practice test is displayed on your screen, click Print on the File menu to print a hard copy. Continue to take practice tests until you score 80% or better.

4 Who Wants To Be a Computer Genius?

Below SAD Chapter 4, click the Computer Genius link. Read the instructions, enter your first and last name at the bottom of the page, and then click the Play button. When your score is displayed, click the PRINT RESULTS link to print a hard copy.

5 Wheel of Terms

Below SAD Chapter 4, click the Wheel of Terms link. Read the instructions, and then enter your first and last name and your school name. Click the PLAY button. When your score is displayed on the screen, right-click the score and then click Print on the shortcut menu to print a hard copy.

6 Crossword Puzzle Challenge

Below SAD Chapter 4, click the Crossword Puzzle Challenge link. Read the instructions, and then click the Continue button. Work the crossword puzzle. When you are finished, click the Submit button. When the crossword puzzle is redisplayed, submit it to your instructor.

Case-Sim: SCR Associates

Background

SCR Associates is a consulting firm that offers IT solutions and training. SCR needs an information system to manage operations at its new training center. The new system will be called TIMS (Training Information Management System). As a newly hired systems analyst, you will report to Jesse Baker, systems group manager. You will work on various tasks and practice the skills you learned in this chapter.

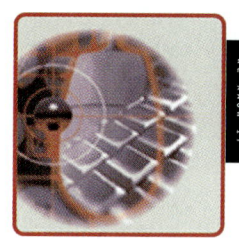

Using the Case

The SCR Associates case study is a Web-based simulation that allows you to practice your skills in a real-world environment. The case study transports you to SCR's company intranet, where you can complete 12 work sessions, each aligning with a chapter. As you work on the case, you will receive e-mail and voice mail messages, obtain information from SCR's online libraries, and perform various tasks. The first time you enter the SCR Case, you should go to the starting page at **scsite.com/sad8e/scr** for detailed instructions.

Preview: Session 4

As you begin the requirements modeling process, you receive specific directions from your supervisor, Jesse Baker. She wants you to conduct a survey of former and prospective students, lead a JAD group session, and draft a list of system requirements based on the results of the JAD session. She also wants to see a functional decomposition diagram showing the main TIMS functions.

To start a work session, you log on to the SCR intranet. When you enter your name and password, an opening screen displays links to the work sessions, and you select Session 4. At this point, you check your e-mail and voice mail messages carefully. Then you begin working on your task list, which includes the following items:

> **Tasks: Requirements Modeling**
>
> 1. Group managers said it was OK for their people to attend a three-day JAD session next week. Send a message to the JAD team members, with a brief explanation of JAD methods and a proposed agenda.
>
> 2. Design a questionnaire for former and potential students in SCR's training classes. Also, reply to Jesse's message about sampling. Give her a recommendation and reasons.
>
> 3. Read the JAD session summary in the Data Library and put together a list of system requirements, including outputs, inputs, processes, performance, and controls.
>
> 4. Draw an FDD of the main functions for TIMS and send it to Jesse.

FIGURE 4-30 Task list: Session 4.

Chapter Exercises

Review Questions

1. What are the five questions typically used in fact-finding? What additional question can be asked during this process?
2. What is a systems requirement, and how are systems requirements classified?
3. What is JAD, how does it differ from traditional methods of fact-finding, and what are some advantages and potential disadvantages of using JAD?
4. What is total cost of ownership (TCO), and why is it important?
5. What are the three different types of questions? How do those different questions affect the answers given?
6. What are three types of sampling, and why would you use them?
7. What is the Hawthorne Effect? Why is it significant?
8. What is RAD, what are the RAD phases, and what occurs in each phase?
9. What are agile methods, and what are some advantages and disadvantages of this approach?
10. To what three different audiences might you have to give a presentation? How would the presentation differ for each?

TOOLKIT TIME

Answer question 10 after you complete the presentations section in Part 1 of the four-part Systems Analyst's Toolkit that follows Chapter 12.

Discussion Topics

1. A group meeting sometimes is suggested as a useful compromise between interviews and questionnaires. In such a group meeting, one systems analyst meets with and asks questions of a number of users at one time. Discuss the advantages and disadvantages of such a group meeting.
2. JAD requires strong interpersonal and communication skills on the part of the systems analyst. Are those skills different from the ones that an analyst needs when conducting one-to-one interviews? Explain your answer.
3. Research the Internet, magazines, or textbooks to find examples of each of the following types of visual aids: bar chart, pie chart, line chart, table, diagram, and bulleted list of key points. How effective do you think each aid is? Find at least one example that you feel could be improved. Discuss its shortcomings and prepare an improved version.
4. Review the presentations section in Part 1 of the Systems Analyst's Toolkit, then attend a speech or presentation and analyze its effectiveness. Consider the speaker's delivery and how he or she organized the material, used visual aids, and handled audience questions. Describe specifically how the speech or presentation was most effective, as well as how it could have been improved.

Projects

1. Design a questionnaire to learn more about the registration process at your school or how customers place orders at a local business. Apply the guidelines you learned in this chapter.
2. Use Microsoft Word or another word processing program to design a simple form, using the program's form-filling feature.
3. Use the Internet to do research about JAD, and present the information to your class.
4. Use the Internet to find a Web site that contains current IT industry news, information, and links. Bookmark the site and print a copy of the initial screen.

Apply Your Knowledge

The Apply Your Knowledge section contains four mini-cases. Each case describes a situation, explains your role in the case, and asks you to respond to questions. You can answer the questions by applying knowledge you learned in the chapter.

1 Elmwood College

Situation:

The school is considering a new system that will speed up the registration process. As a systems analyst, you are asked to develop a plan for fact-finding.

1. List all the possible techniques that you might use.
2. Describe an advantage for each technique.
3. Suppose the development budget is tight. How might that affect the fact-finding process?
4. What are five important questions to use during fact-finding?

2 JAD Session 1

Situation:

You are an IT advisor to a JAD team that is studying a new inventory system. The proposed system will provide more information and faster updates, and automatically monitor fast- or slow-moving items. Some controversy exists about whether to use an on-site or off-site location for the JAD sessions.

1. How would you advise the project leader?
2. Who should be on the JAD team, and what would be their roles as team members?
3. The JAD project leader asked for advice about how to get the first session started. How would you reply?
4. You invited the senior vice president to the opening JAD session, but she says she is quite busy and might not be able to attend unless it is really important. What would you say to her?

3 **JAD Session 2**

Situation:

The JAD team wants you to draw up a checklist of requirements for the new system.

1. List the five main categories of system requirements.
2. Use your imagination and provide at least one example per category of a system requirement that might be appropriate for an inventory system.
3. The project leader wants you to explain the concept of scalability to the team. How will you do that?
4. Several managers on the team have heard of TCO but are not quite sure what it is. How will you explain it to them?

4 **Better Hardware Marketing System**

Situation:

Your boss, the IT director, wants you to explain the UML to a group of company managers and users who will serve on a systems development team for the new marketing system.

1. Describe the Unified Modeling Language (UML) and how it can be used during systems development.
2. Explain use case diagrams to the group, and provide a simple example.
3. Explain sequence diagrams to the group, and provide a simple example.
4. During the meeting, a manager asks you to explain why it is desirable to describe the system through the eyes of a user. How would you answer?

Case Studies

Case studies allow you to practice specific skills learned in the chapter. Each chapter contains several case studies that continue throughout the textbook, and a chapter capstone case.

NEW CENTURY HEALTH CLINIC

New Century Health Clinic offers preventive medicine and traditional medical care. In your role as an IT consultant, you will help New Century develop a new information system.

Background

New Century Health Clinic has decided to computerize its office systems. The associates hired you, a local computer consultant, to perform a preliminary investigation. You had several meetings with Dr. Tim Jones to discuss the various office records and accounting systems. Anita Davenport, New Century's office manager, participated in those meetings.

In a report to the associates at the end of your investigation, you recommended conducting a detailed analysis of the patient record system, the patient and insurance billing systems, and the patient scheduling system. You believe that New Century would benefit most from implementing those three systems. Although the systems could be developed independently, you recommended analyzing all three systems together because of the significant interaction among them.

You presented your findings and recommendations at a late afternoon meeting of the associates. After answering several questions, you left the meeting so they could discuss the matter privately. Dr. Jones began the discussion by stating that he was impressed with your knowledge and professionalism, as well as your report and presentation.

Dr. Jones recommended accepting your proposal and hiring you immediately to conduct the systems analysis phase. Dr. Garcia, however, was not as enthusiastic and pointed out that such a study would certainly disrupt office procedures. The staff already had more work than they could handle, she argued, and taking time to answer your questions would only make the situation worse. Dr. Jones countered that the office workload was going to increase in any event, and that it was important to find a long-term solution to the problem. After some additional discussion, Dr. Garcia finally agreed with Dr. Jones's assessment. The next morning, Dr. Jones called you and asked you to go ahead with the systems analysis phase of the project.

Assignments

1. Review the office organization chart you prepared in Chapter 1 for New Century.
2. List the individuals you would like to interview during the systems analysis phase.
3. Prepare a list of objectives for each of the interviews you will conduct.
4. Prepare a list of specific questions for each individual you will interview.
5. Conduct the interviews. (Consult your instructor regarding how to accomplish this. One possibility is through role-playing.)
6. Prepare a written summary of the information gained from each of the interviews. (Your instructor may want you to use a standard set of interview results.)
7. Design a questionnaire that will go to a sample of New Century patients to find out if they were satisfied with current insurance and scheduling procedures. Your questionnaire should follow the suggestions in this chapter. Also, decide what sampling method you will use and explain the reason for your choice.

PERSONAL TRAINER, INC.

Personal Trainer, Inc., owns and operates fitness centers in a dozen Midwestern cities. The centers have done well, and the company is planning an international expansion by opening a new "supercenter" in the Toronto area. Personal Trainer's president, Cassia Umi, hired an IT consultant, Susan Park, to help develop an information system for the new facility. During the project, Susan will work closely with Gray Lewis, who will manage the new operation.

Background

During requirements modeling for the new system, Susan Park met with fitness center managers at several Personal Trainer locations. She conducted a series of interviews, reviewed company records, observed business operations, analyzed the BumbleBee accounting software, and studied a sample of sales and billing transactions. Susan's objective was to develop a list of system requirements for the proposed system.

Fact-Finding Summary

- A typical center has 300–500 members, with two membership levels: full and limited. Full members have access to all activities. Limited members are restricted to activities they have selected, but they can participate in other activities by paying a usage fee. All members have charge privileges. Charges for merchandise and services are recorded on a charge slip, which is signed by the member. At the end of each day, cash sales and charges are entered into the BumbleBee accounting software, which runs on a computer workstation at each location. Daily cash receipts are deposited in a local bank and credited to the corporate Personal Trainer account. The BumbleBee program produces a daily activity report with a listing of all sales transactions. At the end of the month, the local manager uses BumbleBee to transmit an accounts receivable summary to the Personal Trainer headquarters in Chicago, where member statements are prepared and mailed. Members mail their payments to the Personal Trainer headquarters, where the payment is applied to the member account.

- The BumbleBee program stores basic member information, but does not include information about member preferences, activities, and history.

- Currently, the BumbleBee program produces one local report (the daily activity report) and three reports that are prepared at the headquarters location: a monthly member sales report, an exception report for inactive members and late payers, and a quarterly profit-and-loss report that shows a breakdown of revenue and costs for each separate activity.

During the interviews, Susan received a number of "wish list" comments from local managers and staff members. For example, many managers wanted more analytical features so they could spot trends and experiment with what-if scenarios for special promotions and discounts. The most frequent complaint was that managers wanted more frequent information about the profitability of the business activities at their centers.

To enhance their business, managers wanted to offer a computerized activity and wellness log, a personal coach service, and e-mail communication with members. Managers also wanted better ways to manage information about part-time instructors and staff. Several staff members suggested a redesign for the charge slips or scannable ID cards.

Assignments

1. List the system requirements, with examples for each category. Review the information that Susan gathered, and assume that she will add her own ideas to achieve more effective outputs, inputs, processes, performance, and controls.
2. Are there scalability issues that Susan should consider? What are they?
3. If Susan wants to conduct a survey of current or prospective members to obtain their input, what type of sampling should she use? Why?
4. Draw an FDD that shows the main operations described in the fact statement.

BAXTER COMMUNITY COLLEGE

Baxter Community College is a two-year school in New Jersey. Twice a year, the records office at Baxter mails requests for donations to the alumni. The staff uses a word processing merge file to create personalized letters, but the data on past contributions and other alumni information is stored manually. The registrar, Mary Louise, recently submitted a systems request asking the college's IT department to develop a computerized alumni information system. The school does not have a formal systems review committee, and each department head has an individual budget for routine information services.

Todd Wagner, a systems analyst, was assigned to perform a preliminary investigation. After reading his report, Mary asked him to proceed with the systems analysis phase, saying that a formal presentation was unnecessary. Todd has scheduled an interview tomorrow with her, and he asked you to help him prepare for the meeting.

Assignments

1. Make a list of the topics that you think Todd should cover during the interview.
2. Prepare a list of specific questions that Todd should ask. Include open-ended, closed-ended, and range-of-response questions.
3. Conduct student-to-student interviews, with half the students assuming Todd's role and the other half playing the registrar.
4. Document the information covered during the interviews.

TOWN OF EDEN BAY

The town of Eden Bay owns and maintains a fleet of vehicles. You are a systems analyst reporting to Dawn, the town's IT manager.

Background

In Chapter 2, you learned that the town's maintenance budget has risen sharply in recent years. Based on a preliminary investigation, the town has decided to develop a new information system to manage maintenance information and costs more effectively. The new system will be named RAVE, which stands for Repair Analysis for Vehicular Equipment.

Dawn has asked you to perform additional fact-finding to document the requirements for the new system.

Assignments

1. Review the interview summaries in Chapter 2. For each person (Marie, Martin, Phil, Alice, and Joe), develop three additional questions: an open-ended question, a closed-ended question, and a range-of-response question.
2. Based on what you know so far, list the system requirements for the new system. You can use your imagination if the facts are insufficient. Consider outputs, inputs, processes, performance, and controls. Include at least two examples for each category.
3. You decide to analyze a sample of vehicle records. What sampling methods are available to you? Which one should you use, and why?
4. Dawn thinks it would be a good idea to conduct a JAD session to perform additional fact-finding. Draft a message to the participants, with a brief explanation of JAD methods and a proposed agenda.

CHAPTER CAPSTONE CASE: SoftWear, Limited

SoftWear, Limited (SWL), is a continuing case study that illustrates the knowledge and skills described in each chapter. In this case study, the student acts as a member of the SWL systems development team and performs various tasks.

Background

In Chapter 2, you learned that SWL's vice president of finance, Michael Jeremy, submitted a request for information systems services to investigate problems with the company's payroll system. Jane Rossman, the manager of applications, assigned systems analyst Rick Williams to conduct a preliminary investigation to study the payroll system's problems.

Rick's investigation revealed several problems, including input errors and a need for manual preparation of various reports. The payroll department often is working overtime to correct those errors and produce the required reports.

The IT department recommended conducting an analysis to investigate the problem areas in the payroll system, and Mr. Jeremy approved the study. Now, as the systems analysis phase begins, the next step is requirements modeling.

Human Resources Department Interview

During the preliminary investigation phase, Rick prepared the organization chart of the human resources department shown in Figure 4-31.

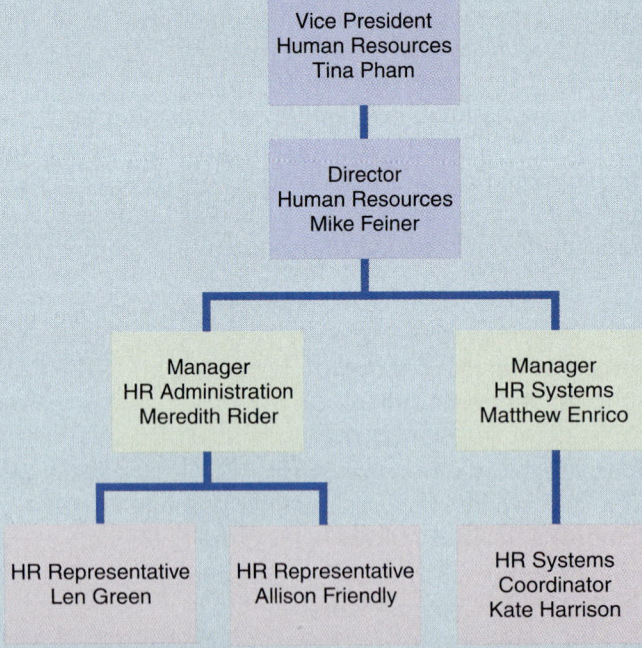

FIGURE 4-31 Human resources department organization chart.

Rick learned that some errors occurred in employee stock purchase deductions, so he decided to study that process. He knew that the human resources department initiates stock purchase deductions. He reviewed the organization chart and decided to interview Meredith Rider, manager of human resources administration. Meredith is responsible for completing the personnel records of newly hired employees and sending the forms to the payroll department.

Chapter Capstone Case: SoftWear, Limited (continued)   SWL

Before arranging any interviews, Rick sent the memo shown in Figure 4-32 to the human resources director, Mike Feiner, to keep him posted. Then Rick called Meredith to make an appointment and sent her the confirmation message shown in Figure 4-33 that described the topics and requested copies of related forms.

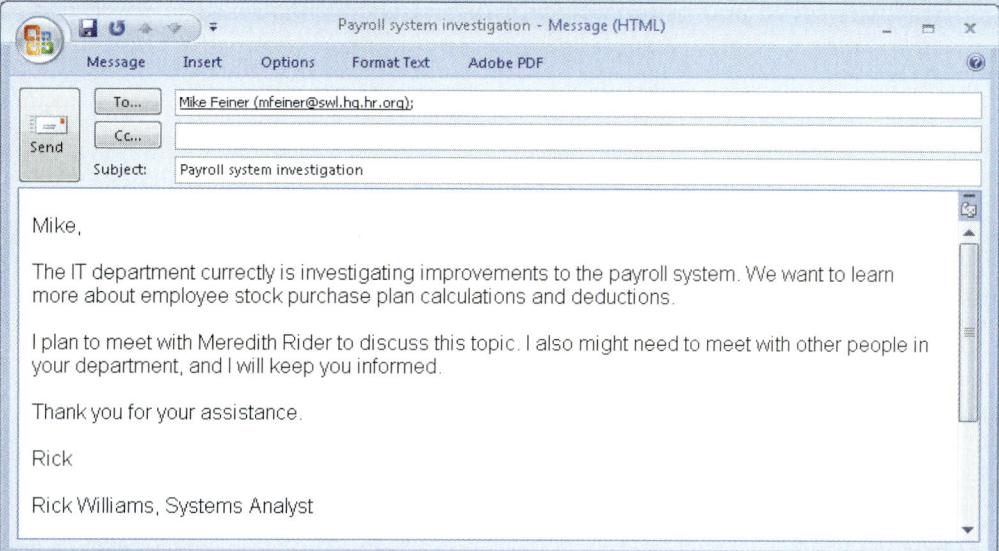

FIGURE 4-32   Rick Williams's message to Mike Feiner regarding the payroll system investigation.

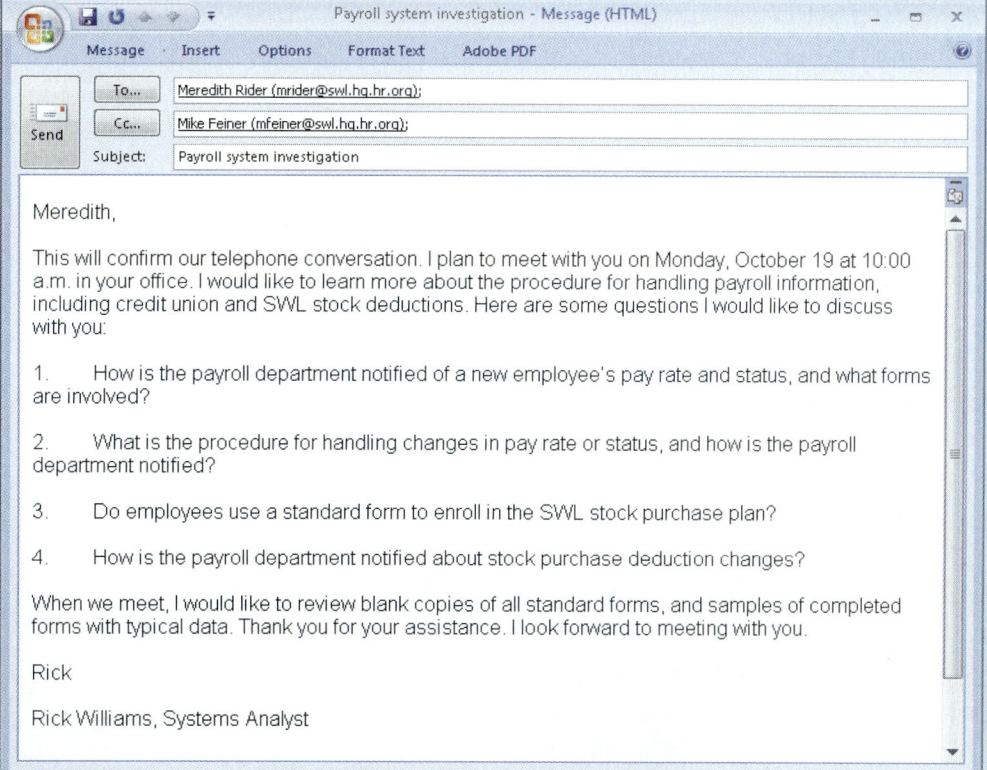

FIGURE 4-33   Rick Williams's message to Meredith Rider regarding preparation for the interview.

CHAPTER CAPSTONE CASE: SoftWear, Limited (continued)

In the interview, Meredith explained that when employees are hired, they fill in the top portion of a Payroll Master Record Form (Form PR-1) that includes personal data and other required information. The human resources department then completes the form by adding pay rate and other data and sends a copy of the PR-1 form to the payroll department. Meredith showed Rick a blank copy of an online PR-1 form shown in Figure 4-34. She explained that because payroll and personnel information is confidential, she could not give Rick a completed form.

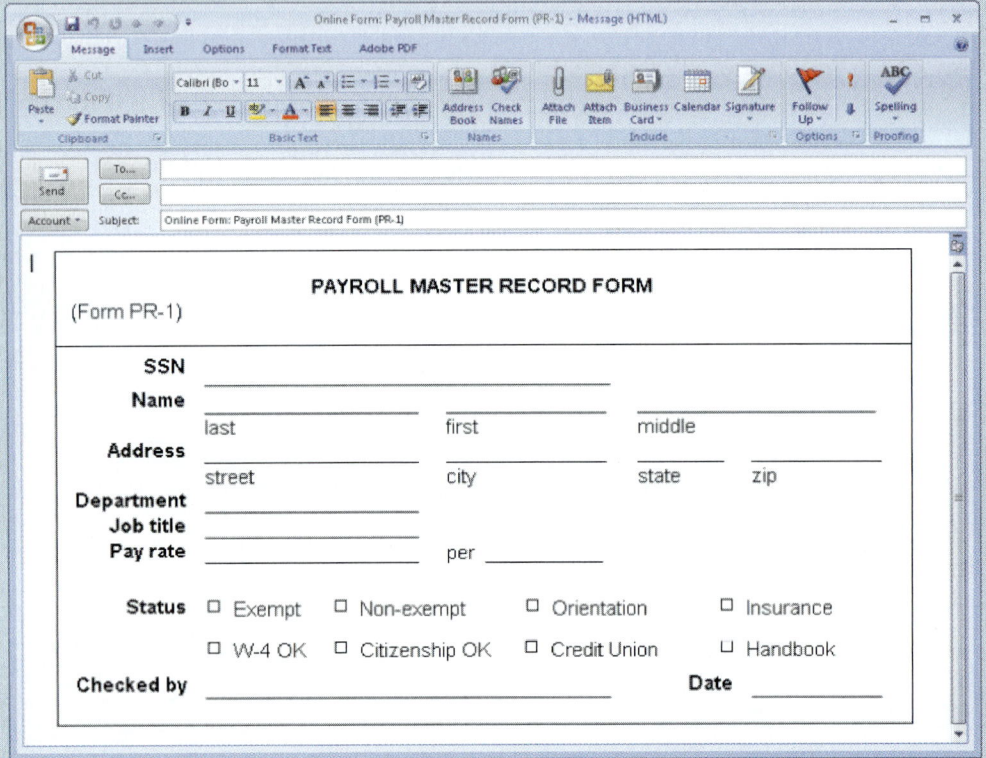

FIGURE 4-34 Payroll Master Record Form (Form PR-1).

When an employee's pay rate or status changes, the human resources department completes the online Payroll Status Change Form (Form PR-2) shown in Figure 4-35 and sends a copy to the payroll department. The payroll department files that form with the employee's PR-1.

CHAPTER CAPSTONE CASE: SoftWear, Limited (continued)

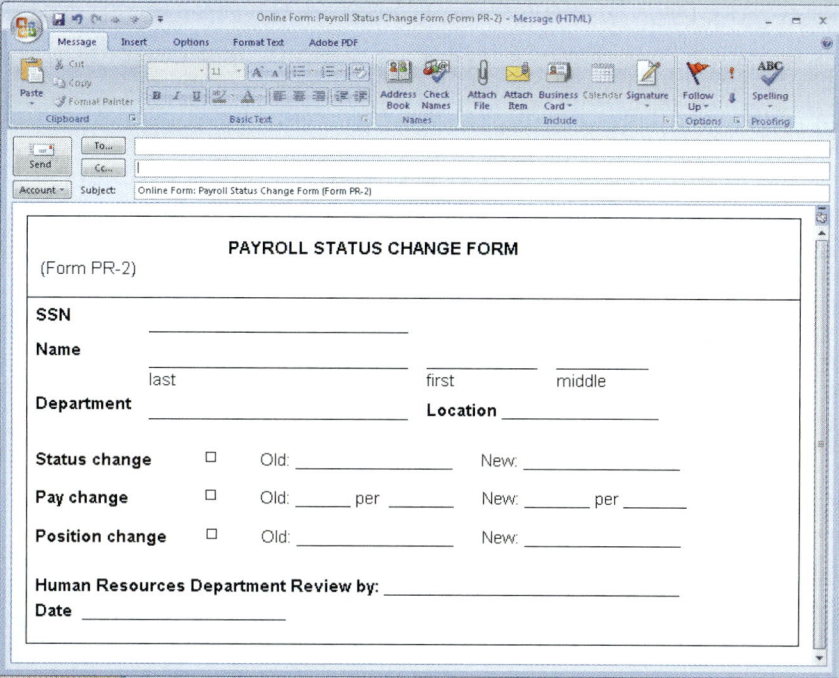

FIGURE 4-35 Payroll Status Change Form (Form PR-2).

Meredith also explained that after completing a 90-day probationary period, employees are allowed to participate in the SWL Credit Union. An employee submits the Payroll Deduction Change Form (Form PR-3) shown in Figure 4-36 to the human resources department, which forwards it to the payroll department.

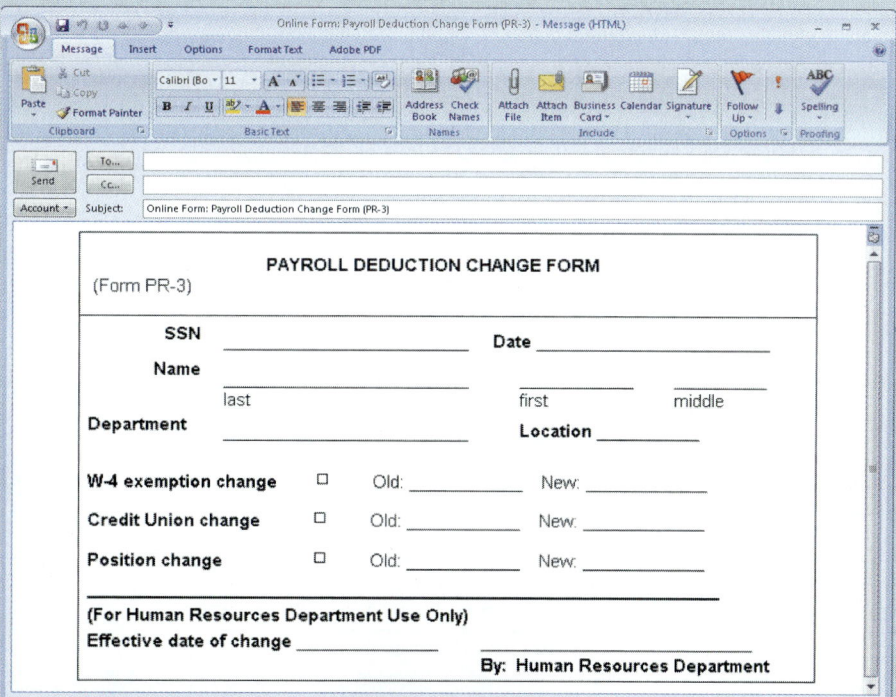

FIGURE 4-36 Payroll Deduction Change Form (Form PR-3).

CHAPTER CAPSTONE CASE: SoftWear, Limited (continued)

SWL also has an Employee Stock Purchase Plan. An individual must be employed for 180 days to be eligible for the plan. The employee receives a brochure and prospectus, and then he or she completes the printed Employee Stock Purchase Plan Enrollment and Change Form (Form PR-4) shown in Figure 4-37 to enroll. The human resources department completes the weekly report of all stock plan enrollments and changes on the Employee Stock Purchase Plan Weekly Deduction Summary Report (Form PR-5) shown in Figure 4-38 and then sends a copy to the payroll department. The payroll department records the information on a card that is filed with the employee's master record.

SWL   **EMPLOYEE STOCK PURCHASE PLAN**
Enrollment and Change Form

To be completed by Employee
(Please print clearly)

I, _____ , hereby acknowledge that I have received a brochure and prospectus on the common stock of SoftWear, Limited (SWL) and that I understand the terms and conditions by which SWL stock is offered to employees.

I understand that an account will be established in my name, and stock will be purchased through payroll deductions. I also understand that my ownership rights in SWL stock purchased for this account are subject to the provisions of the Stock Ownership Plan (the Plan), and I agree to the terms thereof.

I understand that I may change or discontinue my contributions at any time, and that I am entitled to a return of my contributions with thirty days written notice.

I wish to contribute a total of $ _____ per week to the Plan. I understand that deductions will be invested monthly on a pro rata basis pursuant to the SWL systems and procedures manual.

_____ SSN _____ Date _____
     (Employee signature)

(Form PR-4)

FIGURE 4-37 Employee Stock Purchase Plan Enrollment and Change Form (Form PR-4).

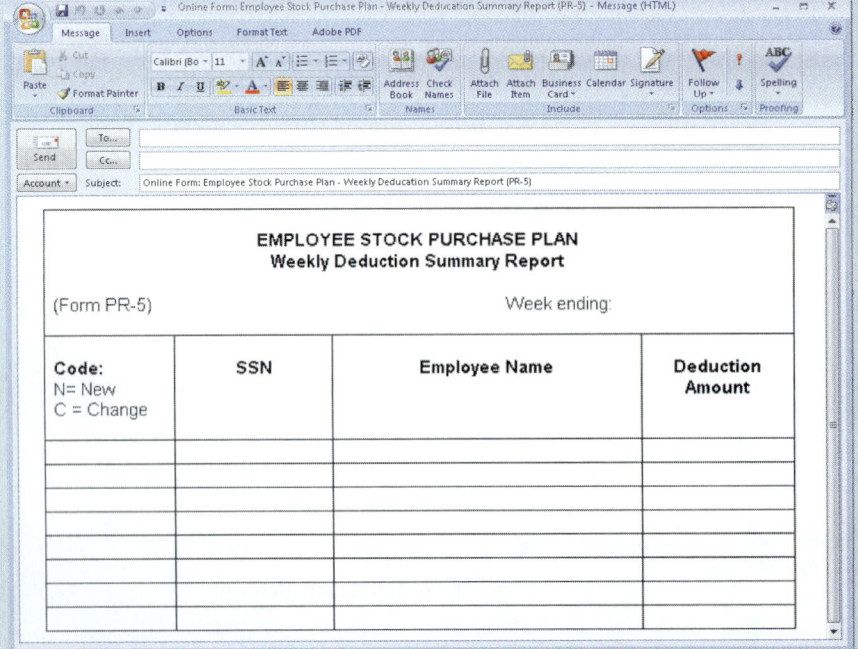

FIGURE 4-38 Employee Stock Purchase Plan Weekly Deduction Summary Report (Form PR-5).

CHAPTER CAPSTONE CASE: SoftWear, Limited (continued) SWL

After the interview with Meredith, Rick sent the follow-up message shown in Figure 4-39 and attached a copy of the interview documentation shown in Figure 4-40.

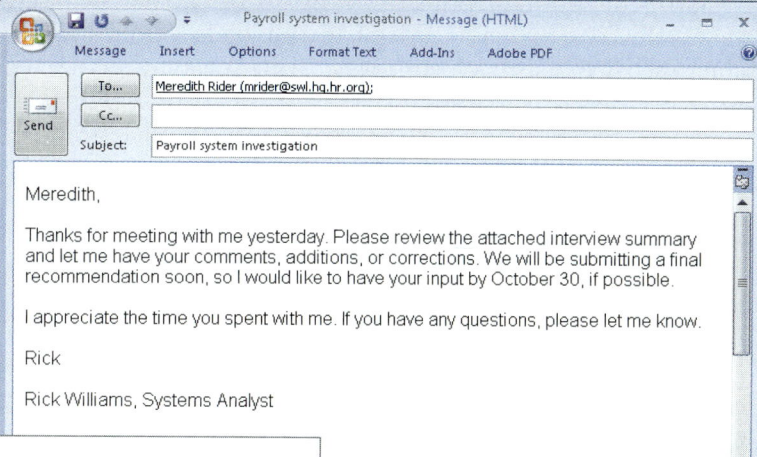

FIGURE 4-39 Follow-up message from Rick Williams to Meredith Rider, with a request for her comments on the interview summary.

Name of System: Payroll
Date: October 22, 2009
Prepared by: Rick Williams
Title: Systems Analyst
Purpose: Interview Summary: Meredith Rider, Manager of Human Resources Administration
Location: Raleigh

Five basic forms are used by the human resources department that relate to the payroll system:
1. Payroll Master Record Form (Form PR-1)
2. Payroll Status Change Form (Form PR-2)
3. Payroll Deduction Change Form (Form PR-3)
4. Employee Stock Purchase Plan Enrollment and Change Form (Form PR-4)
5. Employee Stock Purchase Plan Weekly Deduction Summary Report (Form PR-5)

When an employee is hired, the following takes place:
1. The human resources department prepares a Payroll Master Record Form (Form PR-1) with employee data, including Social Security number, name, address, telephone, emergency contact, and information about the position, title, and initial pay rate.
2. A copy of this form is sent to the payroll department, where it is filed and maintained.
3. Subsequent pay rate or status changes are submitted by the human resources department to the payroll department on a Payroll Status Change Form (Form PR-2). Payroll then files these change forms with the employee's PR-1 form.

After 90 days of employment, the employee is eligible to join the SWL Credit Union.
1. To enroll, or to make changes in existing deductions, the employee goes to the human resources department and completes a Payroll Deduction Change Form (Form PR-3). The human resources department sends the form to payroll, where it is filed with the employee's Payroll Master Record Form (Form PR-1).

After 180 days of employment, the employee is eligible to enroll in the SWL Stock Purchase Plan.
1. To enroll, an employee completes an Employee Stock Purchase Plan Enrollment and Change Form (Form PR-4).
2. The human resources department prepares an Employee Stock Purchase Plan Weekly Deduction Summary Report Form (Form PR-5) and sends it to the payroll department with copies of the PR-4 forms, which then are filed with the employee's PR-1 form.

Changes in employee status that affect payroll involve the following forms:
1. Pay rate   PR-2
2. Status (exempt vs. nonexempt)   PR-2
3. Federal tax exemptions   PR-3
4. Credit Union deductions   PR-3
5. Employee Stock Purchase Plan deductions   PR-4

When an employee changes Credit Union deductions or federal tax exemptions:
1. The employee completes a Payroll Deduction Change Form (Form PR-3).
2. The form is forwarded to the payroll department.
3. The form is filed with the employee's PR-1 form.

I have identified several problems with the current procedures:
1. Data errors can occur when the human resources staff prepares the weekly summary of employee stock purchase deductions, and no system verification takes place until incorrect deductions are reported.
2. The system performs no verification of employment dates, and it is possible that the 90- and 180-day eligibility periods are applied incorrectly.
3. The filing of the PR-2, PR-3, and PR-4 forms with the Payroll Master Record Forms in the payroll department could lead to problems. If any of the forms are lost or misfiled, incorrect data is entered into the system.

FIGURE 4-40 Documentation of the interview with Meredith Rider.

CHAPTER CAPSTONE CASE: SoftWear, Limited (continued)   SWL

Payroll Department Interview

Rick's next interview was with the lead payroll clerk, Nelson White. During the interview, Nelson confirmed that when an employee is hired, a PR-1 form is completed in the human resources department. This form then is forwarded to payroll, where it is filed. He explained that each week the payroll department sends a time sheet to every SWL department manager. The time sheet lists each employee, with space to record regular hours, vacation, sick leave, jury duty, and other codes for accounting purposes.

After each pay period, SWL managers complete their departmental time sheets and return them to the payroll department. Payroll then enters the pay rate and deduction information and delivers the sheets to Business Information Systems (BIS), the service bureau that prepares SWL's payroll.

After the payroll is run, a BIS employee returns the time sheets, paychecks, and the payroll register to SWL. The director of payroll, Amy Calico, sends the paychecks to SWL department heads for distribution to employees.

Nelson uses the weekly payroll register to prepare a report of credit union deductions and a check to the credit union for the total amount deducted. Stock purchases, on the other hand, are processed monthly, based on the stock's closing price on the last business day of the month. Using the weekly payroll registers, Nelson manually prepares a monthly report of employee stock purchases and forwards a copy of the report and a funds transfer authorization to Carolina National Bank, which is SWL's stock transfer agent.

Rick asked Nelson why BIS did not produce a report on employee stock purchase deductions. Nelson replied that although the payroll is run weekly, the stock deductions are invested only once a month. Because the two cycles do not match, the BIS system could not handle the task.

Nelson then referred Rick to the SWL Systems and Procedures Manual page that describes how monthly Employee Stock Purchase Plan investment amounts are calculated, as shown in Figure 4-41. After blanking out the employee's name and

---

SoftWear, Limited   Payroll
Systems and Procedures Manual   Page 29

VII.   Employee Stock Purchase Plan

         The human resources department will notify the payroll department of the weekly deduction that the employee has specified on the Employee Stock Purchase Plan Enrollment and Change Form and send a copy of the Employee Stock Purchase Plan Enrollment and Change Form (Form PR-4) and the Employee Stock Purchase Plan Weekly Deduction Summary Report (Form PR-5) to the payroll department.

         Deductions will be made weekly and then invested on a monthly basis. The payroll department will calculate the proper monthly investment amount on a pro rata basis, as follows:

A.   A nominal per diem deduction rate will be established by dividing the weekly deduction by 7, rounded to 3 decimal places.

B.   The monthly Plan investment will be the number of calendar days in the month times the nominal per diem rate, rounded to 2 decimal places. For example:

         Weekly deduction: $20.00 / 7 = 2.8571 = $2.857 per diem
         Month of January = 31 times 2.857 = 88.567 = $88.57

C.   At the end of each month, the payroll department will prepare a monthly deduction register (Form PR-6) that shows individual employee deductions by a weekly and monthly total.

FIGURE 4-41 Sample page from SWL Systems and Procedures Manual.

CHAPTER CAPSTONE CASE: SoftWear, Limited (continued)

Social Security number, Nelson also gave Rick a sample of two monthly deduction registers, as shown in Figure 4-42.

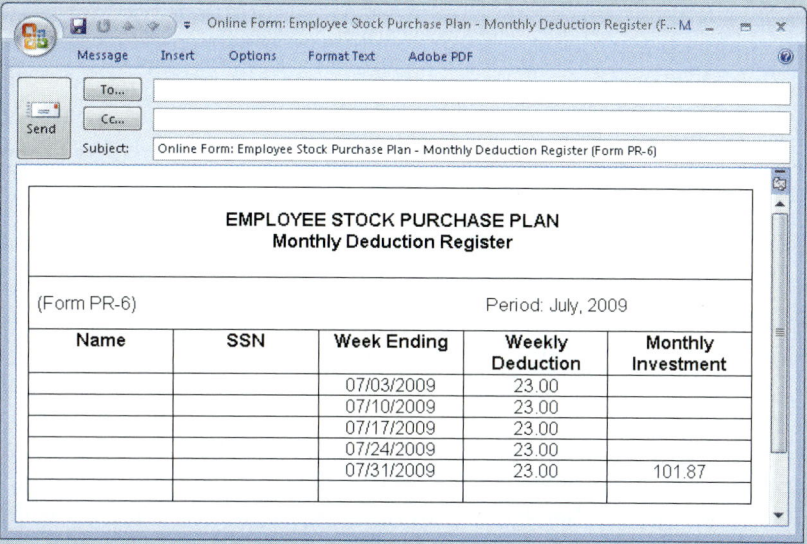

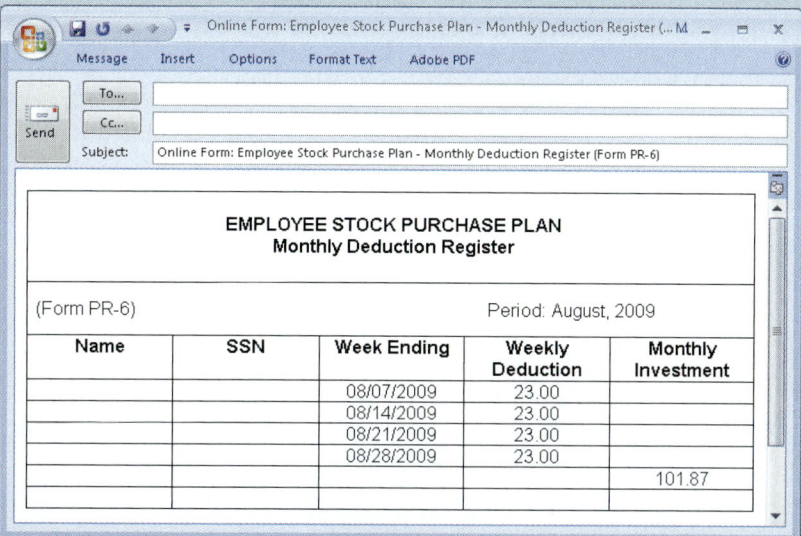

FIGURE 4-42 Sample of the ESIP Monthly Deduction Register for July and August, 2009.

Rick began to see why it was taking so much effort to prepare the reports. The process that Nelson described provided much more detail than the general description that Rick had received during the preliminary investigation from Amy Calico, director of payroll.

BIS Interview

Rick decided that he should talk with someone at the BIS service bureau to find out more about its operations. He learned from Nelson that Linda DeMarco was BIS's customer relations manager, so he scheduled an appointment with her.

When Rick arrived at BIS, Linda greeted him warmly. She explained that she had planned to meet with members of SWL's payroll department within the next month or two to discuss the latest developments. Because Rick now was working on SWL's payroll system, however, this meeting would save her a trip. Rick temporarily abandoned his interview plan and asked Linda what she had in mind.

"The payroll system that your company is using, which we call GAPP, for Generalized Automated Payroll Program, originally was developed here at BIS about eight years ago," Linda began. "In fact, SoftWear, Limited was one of our very first customers. We've worked together for a long time, and we are very committed to your firm. As you know, GAPP was modified and updated many times. But let's face it, even with the patches, GAPP is an antique! Anyway, I have some exciting news. A few months ago, our company decided to develop a new, state-of-the-art payroll system. We are going to call it CHIPS, for Comprehensive High-powered Interactive Payroll System. I am really looking forward to working with your company when you switch over to CHIPS," Linda said.

Rick took a few moments to consider this surprising development. He then asked what would happen with GAPP. Linda stated that GAPP would be available to customers for another year or two, but that BIS would make no further enhancements to that system. Using BIS resources to maintain an obsolete system would not make sense, she explained.

Before this meeting, Rick had hoped that BIS could make some minor changes to solve SWL's payroll problems. He now realized that was impossible, so he decided to learn more about CHIPS.

Rick described the problem with the mismatched deduction cycles and asked if CHIPS would handle that. Linda said that she already had looked into the matter. She pointed out that SWL was their only customer with more than one deduction application cycle. From BIS's point of view, programming CHIPS to handle multiple cycle reports did not make sense. Linda suggested that perhaps a special add-on module could be written, once CHIPS was up and running. BIS could do that kind of job on a contract basis, she added.

Rick then asked when the new system would be available and what the cost would be. Linda stated that current plans were to begin offering CHIPS sometime in the following year. She explained that the system was still in development, and she could not be more specific about timetables and costs. She was sure, however, that the monthly fee for CHIPS would not increase more than 30 percent above the current GAPP charges.

As Rick was preparing to leave, Linda urged him to keep in touch. In the next few months, she explained, plans for CHIPS would become more specific, and she would be able to answer all his questions.

New Developments

When Rick returned from his meeting with Linda, he immediately went to his manager, Jane Rossman. After he described his visit to BIS, Jane telephoned Ann Hon, director of information technology. Within the hour, Jane and Rick held a meeting with Ann in her office. Rick repeated the details of his visit, and Ann asked for his opinion on how the developments at BIS would affect SWL's current systems analysis.

Rick explained that one of the problems — possible input errors when transferring data from the human resources summary list — might be solved easily by developing a new form or procedure. Nevertheless, he saw no obvious solutions for the stock purchase deduction problems, except to change the scope of the payroll project.

Jane, Rick, and Ann then analyzed the situation. They all agreed that because of the upcoming changes at BIS, the current payroll system project would produce very limited results and should be expanded in scope. They totaled the costs of the SWL project to that point and prepared estimates for a detailed investigation of the entire payroll system in order to meet SWL's current and future needs.

CHAPTER CAPSTONE CASE: SoftWear, Limited (continued)

Later that week, Ann met with Michael Jeremy, vice president of finance, to discuss the situation and present her proposal to go forward with an expanded analysis. Before she even started, however, Mr. Jeremy filled her in on the latest announcement from SWL's top management: The company had decided to move forward with the new Employee Savings and Investment Plan (ESIP) under consideration. He said that in December, Robert Lansing, SWL's president, would announce a target date of April 1, 2010, for the new ESIP plan. Mr. Jeremy explained that the new plan would be a 401(k) plan with tax advantages for employees.

Facing the new constraints on top of the existing payroll system problems, it looked like SWL would need a new payroll system after all.

The Revised Project

Jane Rossman assigned Carla Moore, a programmer-analyst, to work with Rick Williams on the revised system project. Because they now had to determine the requirements for the complete payroll system, Rick and Carla conducted follow-up interviews with Nelson White and Meredith Rider, as well as Allison Friendly, a human resources representative, and both payroll clerks, Britton Ellis and Debra Williams. During the payroll department interviews, the payroll staff prepared samples of all the existing payroll reports. At the end of the fact-finding process, Rick and Carla decided to prepare the functional decomposition diagram shown in Figure 4-43. The diagram shows the main functions identified during the interviews.

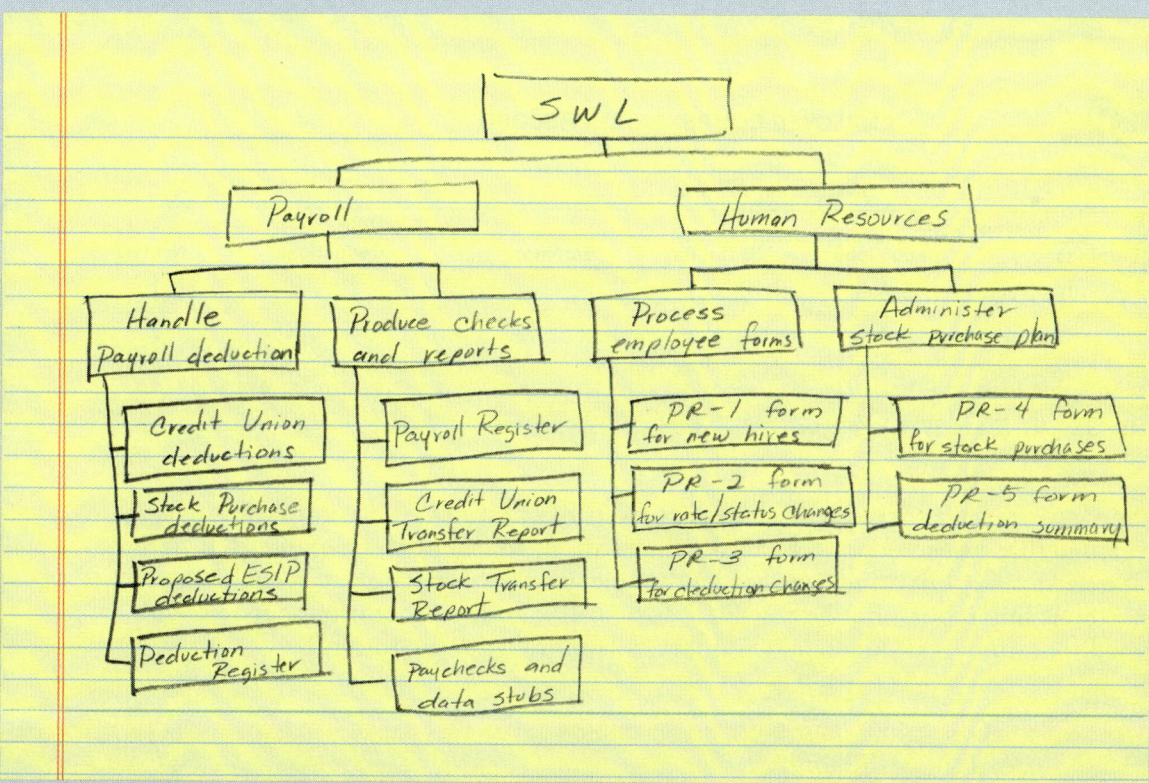

FIGURE 4-43 A functional decomposition diagram (FDD) shows the main functions that were identified during the interviews.

CHAPTER CAPSTONE CASE: SoftWear, Limited (continued)

The Payroll Register report is shown in Figure 4-44. On the report, each employee is listed on a separate line, along with his or her earnings, deductions, and net pay. BIS creates three copies of this report each week. One copy is sent to Michael Jeremy, and one copy goes to Amy Calico. The third copy is used by the payroll department for determining SWL's obligation for tax withholding and FICA payments and for applying credit union and stock purchase plan deductions. BIS also prints three copies of the Employee Compensation Record shown in Figure 4-45, which shows year-to-date payroll information for each employee.

PAYROLL REGISTER

Week Ending _____   *Page 1*

Employee Data		Earnings			Deductions					Net Pay	
Name	SSN	Regular Pay	Overtime Pay	Total Pay	Federal Tax	State Tax	FICA	Credit Union	Stock Plan	Net Amount	Check Number

FIGURE 4-44 Sample page of SWL Payroll Register report.

EMPLOYEE COMPENSATION RECORD

Name _____   SSN _____

	Weekly Payroll										Year to Date									
	Earnings			Deductions					Net Pay		Earnings			Deductions					Net Pay
Week Ending	Reg. Pay	OT Pay	Total Pay	Fed. Tax	State Tax	FICA	Credit Union	Stock Plan	Net Pay	Check No.	Reg. Pay	OT Pay	Total Pay	Fed. Tax	State Tax	FICA	Credit Union	Stock Plan	Net Amount
08/07/2009	352.00		352.00	45.00	7.40	22.40	10.00	9.20	258.00	011917	10,912.00		10,912.00	1,395.00	229.40	694.40	310.00	285.20	7,998.00
08/14/2009	352.00		352.00	45.00	7.40	22.40	10.00	9.20	258.00	016175	11,264.00		11,264.00	1,440.00	236.80	716.80	320.00	294.40	8,256.00
08/21/2009	352.00		352.00	45.00	7.40	22.40	10.00	9.20	258.00	020342	11,616.00		11,616.00	1,485.00	244.20	739.20	330.00	303.60	8,514.00
08/31/2009	352.00		352.00	45.00	7.40	22.40	10.00	9.20	258.00	030919	11,968.00		11,968.00	1,530.00	251.60	761.60	340.00	312.80	8,772.00

FIGURE 4-45 Sample page of SWL Employee Compensation Record report.

Mr. Jeremy receives a weekly overtime report from BIS that lists every employee who worked overtime that week. When Carla asked him about that report, he stated that he consulted it occasionally but admitted that he did not need the report every week. He also receives an accounting report, but he routinely forwards it to the accounting department. He mentioned that an overall financial summary was more valuable to him.

Another key output of the payroll system is the payroll paycheck and stub shown in Figure 4-46 that is distributed weekly to employees. In addition to the check itself, the stub lists current and year-to-date totals for regular pay, overtime pay, total pay, deductions, and net pay.

SWL SOFTWEAR, LIMITED 999 Technology Plaza Raleigh, NC 29991	55-555/5555 1234567	No. _____ Date _____
Pay to the Order of _____		$ _____
_____ (Not Negotiable)		Dollars
Carolina Bank 999 Ninth Street Raleigh, NC 29999	1234" 567" 8888	

Week Ending _____

	This Period	Year-to-Date Totals
EARNINGS		
Regular Pay		
Overtime Pay		
Total Pay		
DEDUCTIONS		
Federal Tax		
State Tax		
FICA		
Credit Union		
Stock Plan		
Net Pay		

FIGURE 4-46   Sample SWL employee paycheck and stub.

SWL Team Tasks

1. When Rick Williams met with Meredith Rider in the human resources department, he asked for copies of actual reports and forms that contained confidential information, but Meredith declined to provide them. Rick has asked you to suggest a reasonable compromise between confidentiality requirements and the need for analysts to review actual records, instead of fictitious data. Think about this, and write a message to Rick with your views.

2. Assume that you were with Rick at the meeting with Linda DeMarco. Review the fact statement, then write an interview summary that documents the main topics that Rick and Linda discussed.

3. Rick asked you to design a questionnaire that would measure employee satisfaction with the current payroll deduction system. Review the sample questionnaire in the chapter, and prepare a draft for Rick. Rick also wants you to suggest various sampling methods so he can make a choice. Include a brief description of various methods, and be sure to include your recommendation and reasons.

4. Rick wants you to interview several employees to learn more about their levels of satisfaction with the current system. Prepare a set of interview questions, and be sure to include at least examples of open-ended, closed-ended, and range-of-response questions. If possible, conduct role-play interviews with other students.

CHAPTER CAPSTONE CASE: SoftWear, Limited (continued)

Manage the SWL Project

You have been asked to manage SWL's new information system project. One of your most important activities will be to identify project tasks and determine when they will be performed. Before you begin, you should review the SWL case in this chapter. Then list and analyze the tasks, as follows:

LIST THE TASKS   Start by listing and numbering at least 10 tasks that the SWL team needs to perform to fulfill the objectives of this chapter. Your list can include SWL Team Tasks and any other tasks that are described in this chapter. For example, Task 3 might be to Identify people to interview, and Task 6 might be to Conduct interviews.

ANALYZE THE TASKS   Now study the tasks to determine the order in which they should be performed. First identify all concurrent tasks, which are not dependent on other tasks. In the example shown in Figure 4-47, Tasks 1, 2, 3, 4, and 5 are concurrent tasks, and could begin at the same time if resources were available.

Other tasks are called dependent tasks, because they cannot be performed until one or more earlier tasks have been completed. For each dependent task, you must identify specific tasks that need to be completed before this task can begin. For example, you would need to identify the people to interview before you conducted the interviews, so Task 6 cannot begin until Task 3 is completed, as Figure 4-47 shows.

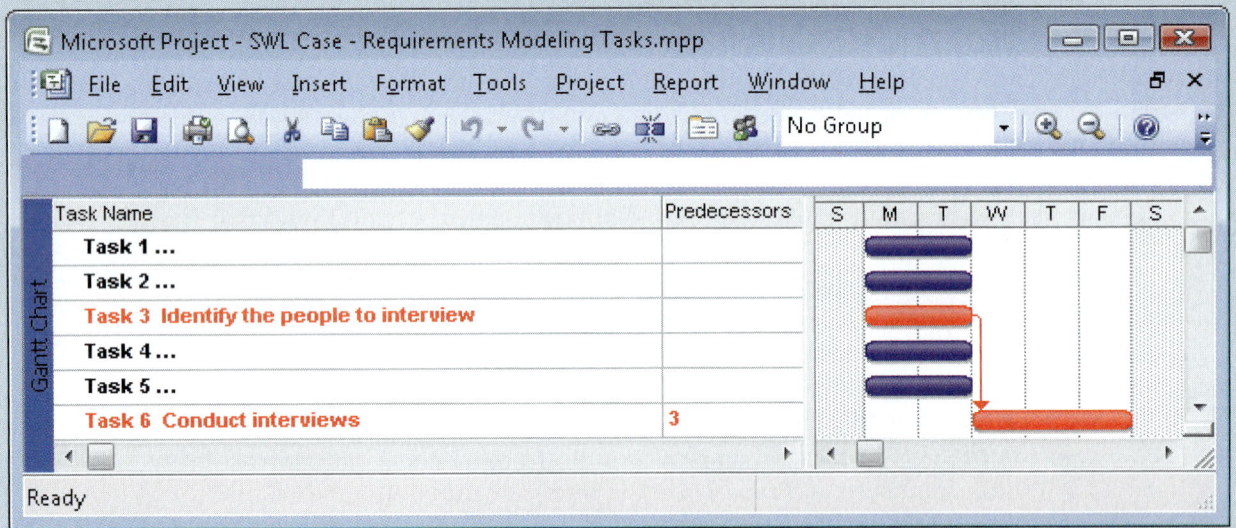

FIGURE 4-47   Tasks 1, 2, 3, 4, and 5 are concurrent tasks that could be performed at the same time. Task 6 is a dependent task that cannot be performed until Task 3 has been completed.

Chapter 3 describes project management tools, techniques, and software. To learn more, you can visit the Features section on your Student Study Tool CD-ROM, or the project management resources library at **scsite.com/sad8e/project**. On the Web, Microsoft offers demo versions, training, and tips for using Project 2007. You also can visit the OpenWorkbench.org site to learn more about this free, open-source software.

CHAPTER **5** # Data and Process Modeling

Chapter 5 is the second of four chapters in the systems analysis phase of the SDLC. This chapter discusses data and process modeling techniques that analysts use to show how the system transforms data into useful information. The deliverable, or end product, of data and process modeling is a logical model that will support business operations and meet user needs.

INTRODUCTION

OBJECTIVES

When you finish this chapter, you will be able to:

- Describe data and process modeling concepts and tools, including data flow diagrams, a data dictionary, and process descriptions

- Describe the symbols used in data flow diagrams and explain the rules for their use

- Draw data flow diagrams in a sequence, from general to specific

- Explain how to level and balance a set of data flow diagrams

- Describe how a data dictionary is used and what it contains

- Use process description tools, including structured English, decision tables, and decision trees

- Describe the relationship between logical and physical models

During the requirements modeling process described in Chapter 4, you used fact-finding techniques to investigate the current system and identify user requirements. Now, in Chapters 5 and 6 you will use that information to develop a logical model of the proposed system and document the system requirements. A **logical model** shows *what* the system must do, regardless of how it will be implemented physically. Later, in the systems design phase, you build a **physical model** that describes *how* the system will be constructed. Data and process modeling involves three main tools: data flow diagrams, a data dictionary, and process descriptions.

In Chapter 6, you will learn how to use object-oriented modeling tools and techniques. In the final stage of systems analysis, which is explained in Chapter 7, you will learn how to evaluate various development strategies, create a system requirements proposal, and prepare for the systems design phase of the SDLC.

CHAPTER INTRODUCTION CASE: Mountain View College Bookstore

Background: Wendy Lee, manager of college services at Mountain View College, wants a new information system that will improve efficiency and customer service at the three college bookstores.

In this part of the case, Florence Fullerton (systems analyst) and Harry Boston (student intern) are talking about data and process modeling tasks and concepts.

Participants:	Florence and Harry
Location:	Florence's office, Monday afternoon, October 19, 2009
Project status:	Florence and Harry have completed fact-finding for the new system and are ready to develop a requirements model using various diagrams and a data dictionary that will describe and document the proposed system.
Discussion topics:	Data flow diagrams, data dictionaries, and process description tools

Florence:	Hi, Harry. Any questions about the fact-finding we did?
Harry:	*Well, I found out that fact-finding is hard work.*
Florence:	Yes, but it was worth it. Look at what we learned — now we understand how the current system operates, and we know what users expect in the new system. This information will help us build a requirements model that we can present to Wendy and her staff.
Harry:	*What's the next step?*
Florence:	We need to draw a set of data flow diagrams, or DFDs for short.
Harry:	*Do we use a CASE tool to draw the DFDs?*
Florence:	We can draw the initial versions by hand. We'll use a CASE tool to prepare the final version of the diagrams.
Harry:	*What goes into a DFD?*
Florence:	DFDs use four basic symbols that represent processes, data flows, data stores, and entities. You'll learn about these as we go along. I'll also show you how we use techniques called leveling and balancing to develop accurate, consistent DFDs.
Harry:	*Apart from the diagrams, do we need to develop any other documentation?*
Florence:	Yes, we need to create a data dictionary and process descriptions. The data dictionary is an overall storehouse of information about the system, and serves as a central clearinghouse for all documentation. We use process descriptions to explain the logical steps that each process performs. To create these descriptions, we use three tools: structured English statements, decision tables, and decision trees.
Harry:	*Sounds like a lot to do. Where do we begin?*
Florence:	Here's a task list to get us started:

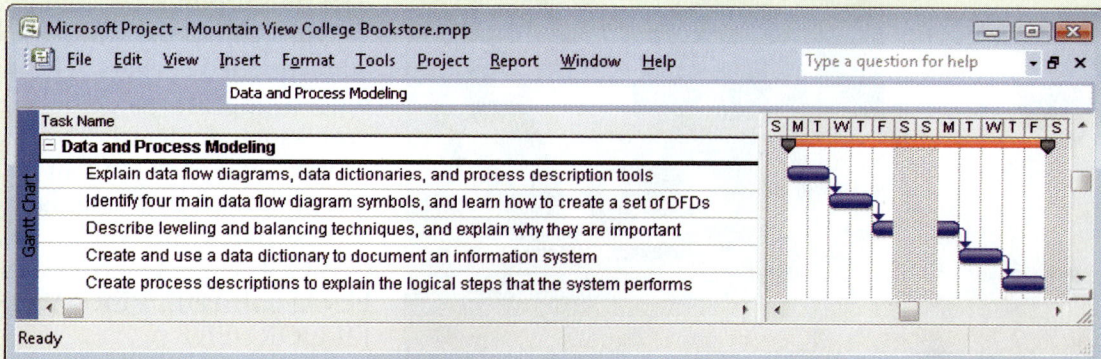

FIGURE 5-1   Typical data and process modeling task list.

OVERVIEW OF DATA AND PROCESS MODELING TOOLS

Systems analysts use many graphical techniques to describe an information system. One popular method is to draw a set of data flow diagrams. A **data flow diagram** (DFD) uses various symbols to show how the system transforms input data into useful information. Other graphical tools include object models, which are explained in Chapter 6 (Object Modeling), and entity-relationship diagrams, which are described in Chapter 9 (Data Design).

DATA FLOW DIAGRAMS

In Part 1 of the Systems Analyst's Toolkit, you learn how to use visual aids to help explain a concept, as shown in Figure 5-2. Similarly, during the systems analysis phase, you learn how to create a visual model of the information system using a set of data flow diagrams.

A data flow diagram (DFD) shows how data moves through an information system but does not show program logic or processing steps. A set of DFDs provides a logical model that shows *what* the system does, not *how* it does it. That distinction is important because focusing on implementation issues at this point would restrict your search for the most effective system design.

DFD Symbols

DFDs use four basic symbols that represent processes, data flows, data stores, and entities. Several different versions of DFD symbols exist, but they all serve the same purpose. DFD examples in this textbook use the **Gane and Sarson** symbol set. Another popular symbol set is the **Yourdon** symbol set. Figure 5-3 shows examples of both versions. Symbols are referenced by using all capital letters for the symbol name.

PROCESS SYMBOL A **process** receives input data and produces output that has a different content, form, or both. For instance, the process for calculating pay uses two inputs (pay rate and hours worked) to produce one output (total pay). Processes can be very simple or quite complex. In a typical company, processes might include calculating sales trends, filing online insurance claims, ordering inventory from a supplier's system, or verifying e-mail addresses for Web customers. Processes contain the **business logic**, also called **business rules**, that transform the data and produce the required results.

The symbol for a process is a rectangle with rounded corners. The name of the process appears inside the rectangle. The process name identifies a specific function and consists of a verb (and an adjective, if necessary) followed by a singular noun. Examples of process names are APPLY RENT PAYMENT, CALCULATE COMMISSION, ASSIGN FINAL GRADE, VERIFY ORDER, and FILL ORDER.

Processing details are not shown in a DFD. For example, you might have a process named DEPOSIT PAYMENT. The process symbol

FIGURE 5-2 Systems analysts often use visual aids during presentations.

does not reveal the business logic for the DEPOSIT PAYMENT process. To document the logic, you create a process description, which is explained later in this chapter.

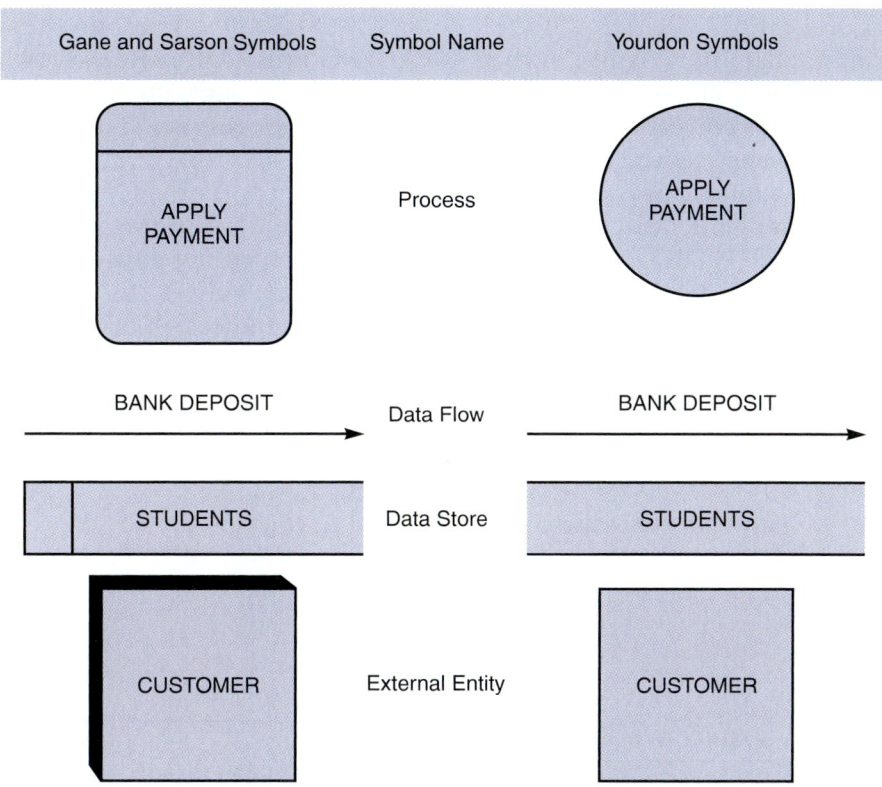

FIGURE 5-3 Data flow diagram symbols, symbol names, and examples of the Gane and Sarson and Yourdon symbol sets.

In DFDs, a process symbol can be referred to as a **black box**, because the inputs, outputs, and general functions of the process are known, but the underlying details and logic of the process are hidden. By showing processes as black boxes, an analyst can create DFDs that show how the system functions, but avoid unnecessary detail and clutter. When the analyst wishes to show additional levels of detail, he or she can zoom in on a process symbol and create a more in-depth DFD that shows the process's internal workings — which might reveal even more processes, data flows, and data stores. In this manner, the information system can be modeled as a series of increasingly detailed pictures.

The network router shown in Figure 5-4 is an example of a black box. An observer can see cables that carry data into and out of the router, but the router's internal operations are not revealed — only the results are apparent.

DATA FLOW SYMBOL A **data flow** is a path for data to move from one part of the information system to another. A data flow in a DFD represents one or more data items. For example, a data flow could consist of a single data item (such as a student ID number) or it could include a set of data (such as a class roster with student ID numbers, names, and registration dates for a specific class). Although the DFD does not show the detailed contents of a data flow, that information is included in the data dictionary, which is described later in this chapter.

FIGURE 5-4 Networks use various devices that act like black boxes. Cables carry data in and out, but internal operations are hidden inside the case.

The symbol for a data flow is a line with a single or double arrowhead. The data flow name appears above, below, or alongside the line. A data flow name consists of a singular noun and an adjective, if needed. Examples of data flow names are DEPOSIT, INVOICE PAYMENT, STUDENT GRADE, ORDER, and COMMISSION. Exceptions to the singular name rule are data flow names, such as GRADING PARAMETERS, where a singular name could mislead you into thinking a single parameter or single item of data exists.

Figure 5-5 shows correct examples of data flow and process symbol connections. Because a process changes the data's content or form, at least one data flow must enter and one data flow must exit each process symbol, as they do in the CREATE INVOICE process. A process symbol can have more than one outgoing data flow, as shown in the GRADE STUDENT WORK process, or more than one incoming data flow, as shown in the CALCULATE GROSS PAY process. A process also can connect to any other symbol, including another process symbol, as shown by the connection between VERIFY ORDER and ASSEMBLE ORDER in Figure 5-5. A data flow, therefore, *must* have a process symbol on at least one end.

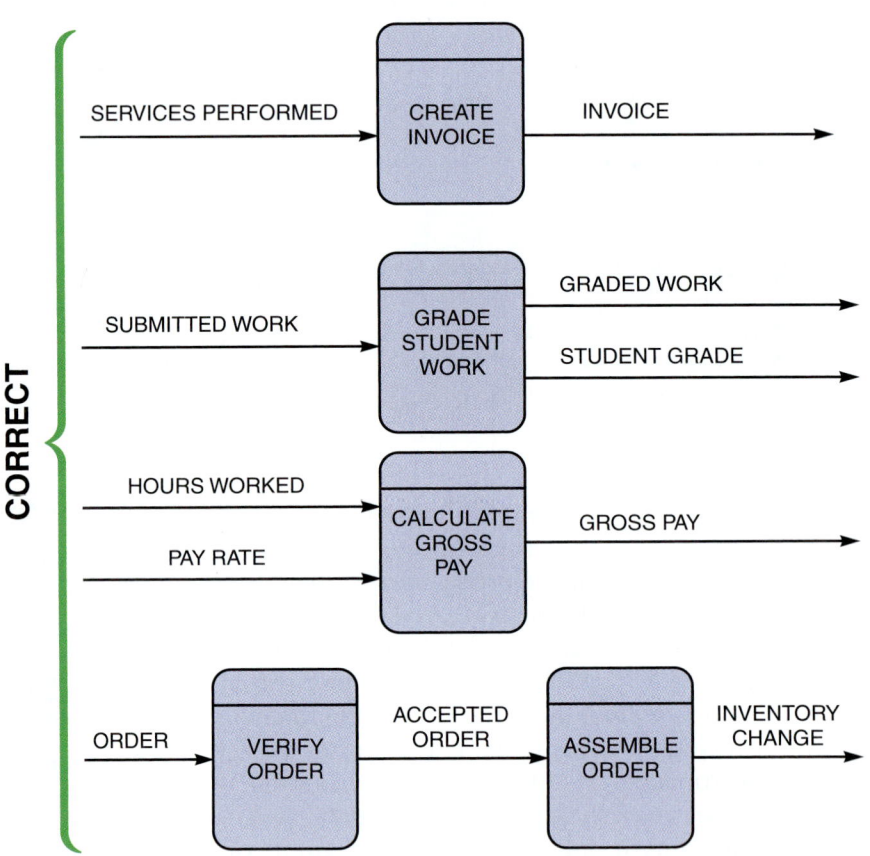

FIGURE 5-5 Examples of correct combinations of data flow and process symbols.

Figure 5-6 shows three data flow and process combinations that you must avoid:

- **Spontaneous generation.** The APPLY INSURANCE PREMIUM process, for instance, produces output, but has no input data flow. Because it has no input, the process is called a spontaneous generation process.
- **Black hole.** The CALCULATE GROSS PAY is called a black hole process, which is a process that has input, but produces no output.

- **Gray hole.** A gray hole is a process that has at least one input and one output, but the input obviously is insufficient to generate the output shown. For example, a date of birth input is not sufficient to produce a final grade output in the CALCULATE GRADE process.

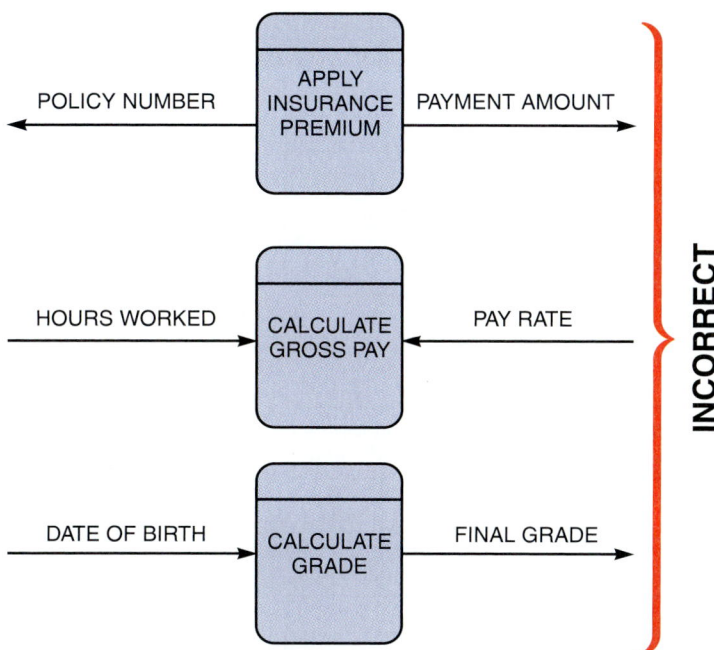

FIGURE 5-6   Examples of incorrect combinations of data flow and process symbols. APPLY INSURANCE PREMIUM has no input and is called a spontaneous generation process. CALCULATE GROSS PAY has no outputs and is called a black hole process. CALCULATE GRADE has an input that is obviously unable to produce the output. This process is called a gray hole.

Spontaneous generation, black holes, and gray holes are impossible logically in a DFD because a process must act on input, shown by an incoming data flow, and produce output, represented by an outgoing data flow.

DATA STORE SYMBOL   A **data store** is used in a DFD to represent data that the system stores because one or more processes need to use the data at a later time. For instance, instructors need to store student scores on tests and assignments during the semester so they can assign final grades at the end of the term. Similarly, a company stores employee salary and deduction data during the year in order to print W-2 forms with total earnings and deductions at the end of the year. A DFD does not show the detailed contents of a data store — the specific structure and data elements are defined in the data dictionary, which is discussed later in this chapter.

The physical characteristics of a data store are unimportant because you are concerned only with a logical model. Also, the length of time that the data is stored is unimportant — it can be a matter of seconds while a transaction is processed or a period of months while data is accumulated for year-end processing. What is important is that a process needs access to the data at some later time.

In a DFD, the Gane and Sarson symbol for a data store is a flat rectangle that is open on the right side and closed on the left side. The name of the data store appears between the lines and identifies the data it contains. A data store name is a plural name consisting of a noun and adjectives, if needed. Examples of data store names are STUDENTS, ACCOUNTS RECEIVABLE, PRODUCTS, DAILY PAYMENTS,

PURCHASE ORDERS, OUTSTANDING CHECKS, INSURANCE POLICIES, and EMPLOYEES. Exceptions to the plural name rule are collective nouns that represent multiple occurrences of objects. For example, GRADEBOOK represents a group of students and their scores.

A data store must be connected to a process with a data flow. Figure 5-7 illustrates typical examples of data stores. In each case, the data store has at least one incoming and one outgoing data flow and is connected to a process symbol with a data flow.

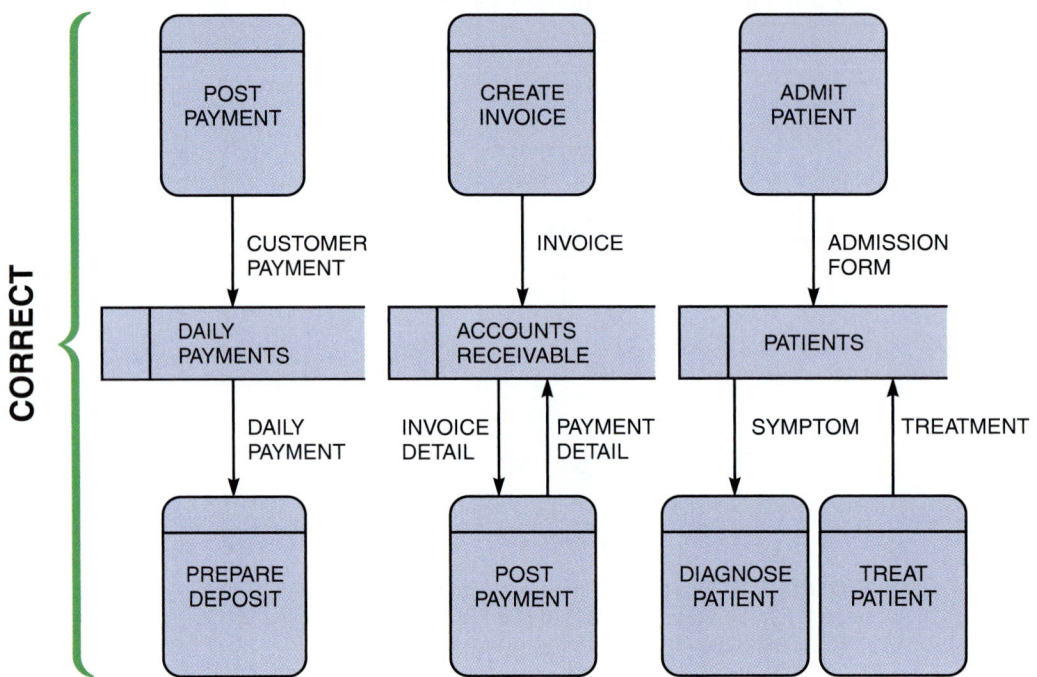

FIGURE 5-7 Examples of correct uses of data store symbols in a data flow diagram.

Violations of the rule that a data store must have at least one incoming and one outgoing data flow are shown in Figure 5-8. In the first example, two data stores are connected incorrectly because no process is between them. Also, COURSES has no incoming data flow and STUDENTS has no outgoing data flow. In the second and third examples, the data stores lack either an outgoing or incoming data flow.

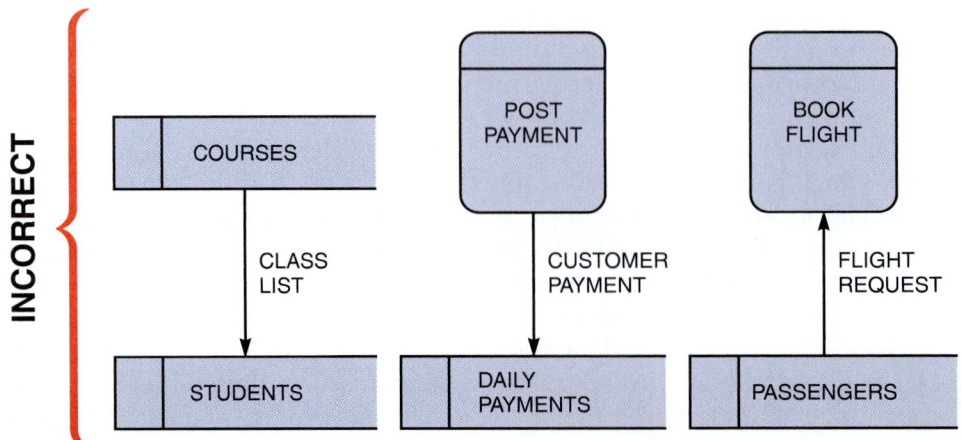

FIGURE 5-8 Examples of incorrect uses of data store symbols: Two data stores cannot be connected by a data flow without an intervening process, and each data store should have an outgoing and incoming data flow.

There is an exception to the requirement that a data store must have at least one incoming and one outgoing data flow. In some situations, a data store has no input data flow because it contains fixed reference data that is not updated by the system. For example, consider a data store called TAX TABLE, which contains withholding tax data that a company downloads from the Internal Revenue Service. When the company runs its payroll, the CALCULATE WITHHOLDING process accesses data from this data store. On a DFD, this would be represented as a one-way outgoing data flow from the TAX TABLE data store into the CALCULATE WITHHOLDING process.

ENTITY SYMBOL The symbol for an **entity** is a rectangle, which may be shaded to make it look three-dimensional. The name of the entity appears inside the symbol.

A DFD shows only external entities that provide data to the system or receive output from the system. A DFD shows the boundaries of the system and how the system interfaces with the outside world. For example, a customer entity submits an order to an order processing system. Other examples of entities include a patient who supplies data to a medical records system, a homeowner who receives a bill from a city property tax system, or an accounts payable system that receives data from the company's purchasing system.

DFD entities also are called **terminators**, because they are data origins or final destinations. Systems analysts call an entity that supplies data to the system a **source**, and an entity that receives data from the system a **sink**. An entity name is the singular form of a department, outside organization, other information system, or person. An external entity can be a source or a sink or both, but each entity must be connected to a process by a data flow. Figures 5-9 and 5-10 show correct and incorrect examples of this rule.

With an understanding of the proper use of DFD symbols, you are ready to construct diagrams that use these symbols. Figure 5-11 on the next page shows a summary of the rules for using DFD symbols.

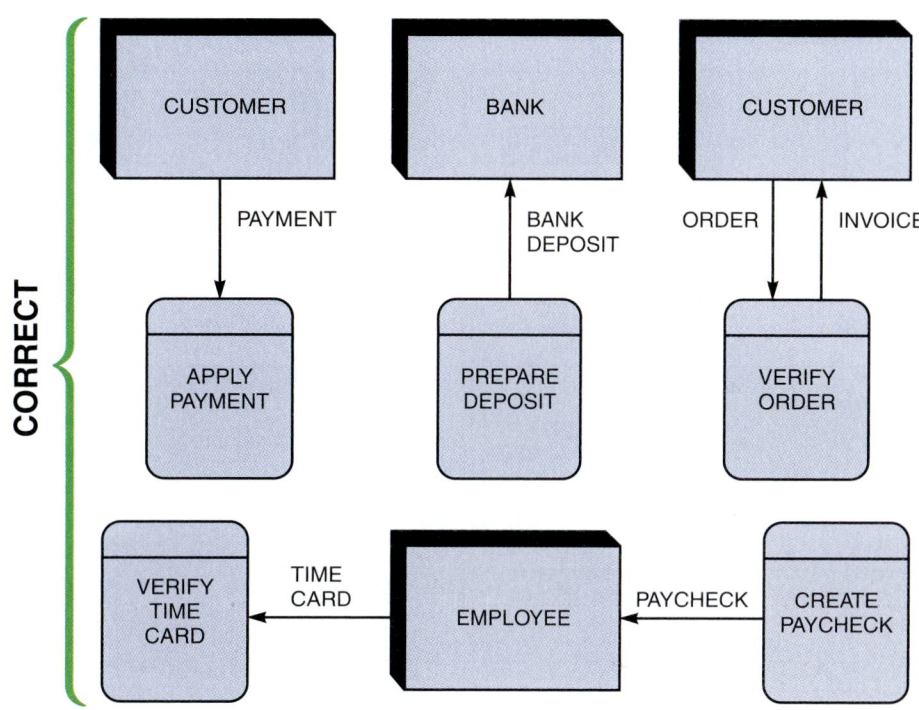

FIGURE 5-9 Examples of correct uses of external entities in a data flow diagram.

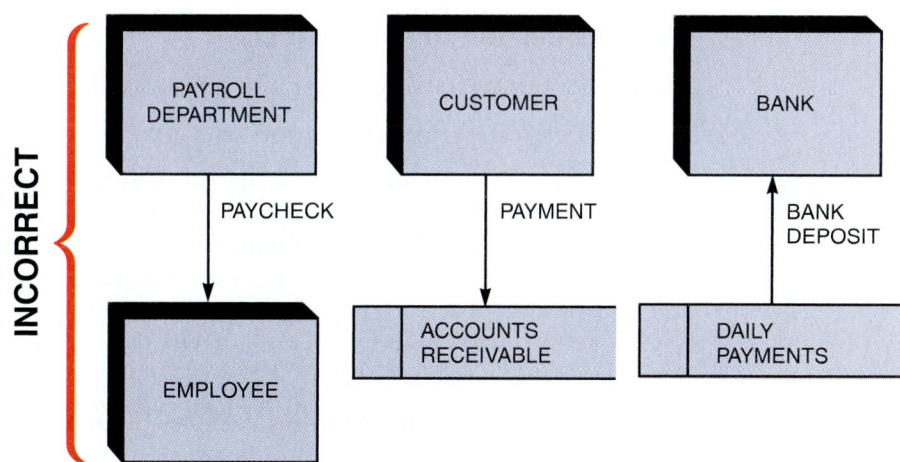

FIGURE 5-10 Examples of incorrect uses of external entities. An external entity must be connected by a data flow to a process, and not directly to a data store or to another external entity.

FIGURE 5-11 Examples of correct and incorrect uses of data flows.

CREATING A SET OF DFDs

During requirements modeling, you used interviews, questionnaires, and other techniques to gather facts about the system, and you learned how the various people, departments, data, and processes fit together to support business operations. Now you are ready to create a graphical model of the information system based on your fact-finding results.

To learn how to construct DFDs, you will use examples of two information systems. The first example is a grading system that instructors use to assign final grades based on the scores that students receive during the term. The second example is an order system that a company uses to enter orders and apply payments against a customer's balance. First, you will review a set of guidelines for drawing DFDs. Then you will learn how to apply these guidelines and create a set of DFDs using a three-step process:

Step 1: Draw a context diagram
Step 2: Draw a diagram 0 DFD
Step 3: Draw the lower-level diagrams

Guidelines for Drawing DFDs

When you draw a context diagram and other DFDs, you should follow several guidelines:

- Draw the context diagram so it fits on one page.

- Use the name of the information system as the process name in the context diagram. For example, the process name in Figure 5-12 is GRADING SYSTEM. Notice that the process name is the same as the system name. This is because the context diagram shows the entire information system as if it were a single process. For processes in lower-level DFDs, you would use a verb followed by a descriptive noun, such as ESTABLISH GRADEBOOK, ASSIGN FINAL GRADE, or PRODUCE GRADE REPORT.

- Use unique names within each set of symbols. For instance, the diagram in Figure 5-12 shows only one entity named STUDENT and only one data flow named FINAL GRADE. Whenever you see the entity STUDENT on any other DFD in the grading system, you know that you are dealing with the same entity. Whenever the FINAL GRADE data flow appears, you know that you are dealing with the same data flow. The naming convention also applies to data stores.

- Do not cross lines. One way to achieve that goal is to restrict the number of symbols in any DFD. On lower-level diagrams with multiple processes, you should not have more than nine process symbols. Including more than nine symbols usually is a signal that your diagram is too complex and that you should reconsider your analysis. Another way to avoid crossing lines is to duplicate an entity or data store. When duplicating a symbol on a diagram, make sure to document the duplication to avoid possible confusion. A special notation, such as an asterisk, next to the symbol name and inside the duplicated symbols signifies that they are duplicated on the diagram.

- Provide a unique name and reference number for each process. Because it is the highest-level DFD, the context diagram contains process 0, which represents the entire information system, but does not show the internal workings. To describe the next level of detail inside process 0, you must create a DFD named diagram 0, which will reveal additional processes that must be named and numbered. As you continue to create lower-level DFDs, you assign unique names and reference numbers to all processes, until you complete the logical model.

- Obtain as much user input and feedback as possible. Your main objective is to ensure that the model is accurate, easy to understand, and meets the needs of its users.

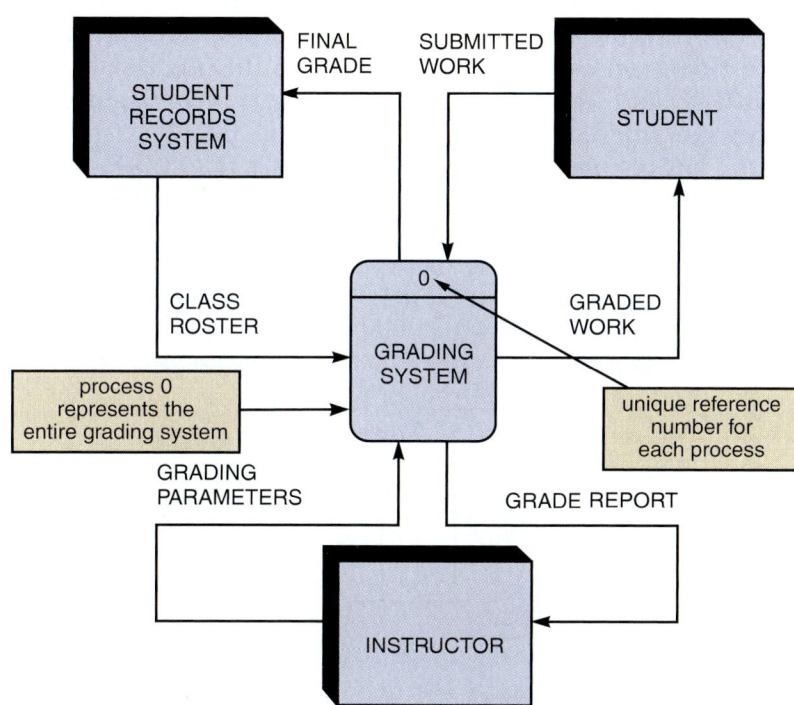

FIGURE 5-12 Context diagram DFD for a grading system.

Step 1: Draw a Context Diagram

The first step in constructing a set of DFDs is to draw a context diagram. A **context diagram** is a top-level view of an information system that shows the system's boundaries and scope. To draw a context diagram, you start by placing a single process symbol in the center of the page. The symbol represents the entire information system, and you identify it as **process 0** (the numeral zero, and not the letter O). Then you place the system entities around the perimeter of the page and use data flows to connect the entities to the central process. Data stores are not shown in the context diagram because they are contained within the system and remain hidden until more detailed diagrams are created.

How do you know which entities and data flows to place in the context diagram? You begin by reviewing the system requirements to identify all external data sources and destinations. During that process, you identify the entities, the name and content of the data flows, and the direction of the data flows. If you do that carefully, and you did a good job of fact-finding in the previous stage, you should have no difficulty drawing the context diagram. Now review the following context diagram examples.

EXAMPLE: CONTEXT DIAGRAM FOR A GRADING SYSTEM The context diagram for a grading system is shown in Figure 5-12 on the previous page. The GRADING SYSTEM process is at the center of the diagram. The three entities (STUDENT RECORDS SYSTEM, STUDENT, and INSTRUCTOR) are placed around the central process. Interaction among the central process and the entities involves six different data flows. The STUDENT RECORDS SYSTEM entity supplies data through the CLASS ROSTER data flow and receives data through the FINAL GRADE data flow. The STUDENT entity supplies data through the SUBMITTED WORK data flow and receives data through the GRADED WORK data flow. Finally, the INSTRUCTOR entity supplies data through the GRADING PARAMETERS data flow and receives data through the GRADE REPORT data flow.

EXAMPLE: CONTEXT DIAGRAM FOR AN ORDER SYSTEM The context diagram for an order system is shown in Figure 5-13. Notice that the ORDER SYSTEM process is at the center of the diagram and five entities surround the process. Three of the entities, SALES REP, BANK, and ACCOUNTING, have single incoming data flows for COMMISSION, BANK DEPOSIT, and CASH RECEIPTS ENTRY, respectively. The WAREHOUSE entity has one incoming data flow — PICKING LIST — that is, a report that shows the items ordered and their quantity, location, and sequence to pick from the warehouse. The WAREHOUSE entity has one outgoing data flow, COMPLETED ORDER. Finally, the CUSTOMER entity has two outgoing data flows, ORDER and PAYMENT, and two incoming data flows, ORDER REJECT NOTICE and INVOICE.

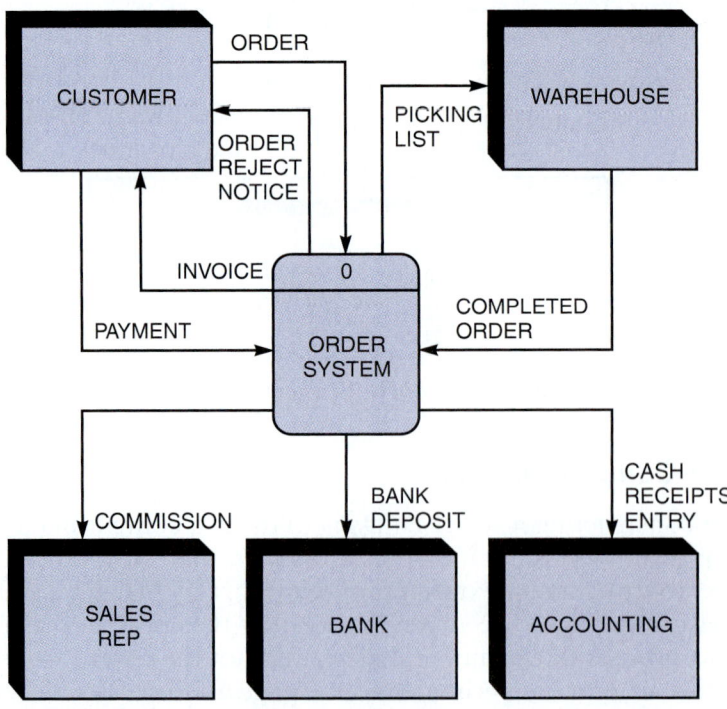

FIGURE 5-13 Context diagram DFD for an order system.

The context diagram for the order system appears more complex than the grading system because it has two more entities and three more data flows. What makes one system more complex than another is the number of components, the number of levels, and the degree of interaction among its processes, entities, data stores, and data flows.

Step 2: Draw a Diagram 0 DFD

In the previous step, you learned that a context diagram provides the most general view of an information system and contains a single process symbol, which is like a black box. To show the detail inside the black box, you create DFD diagram 0. **Diagram 0** (the numeral zero, and not the letter O) zooms in on the system and shows major internal processes, data flows, and data stores. Diagram 0 also repeats the entities and data flows that appear in the context diagram. When you expand the context diagram into DFD diagram 0, you must retain all the connections that flow into and out of process 0.

The real-life scene in Figure 5-14 represents a complex manufacturing system with many interactive processes and data. In a large system such as this, each process in diagram 0 could represent a separate system such as inventory, production control, and scheduling. Diagram 0 provides an overview of all the components that interact to form the overall system. Now review the following diagram 0 examples.

FIGURE 5-14 Complex manufacturing systems require many interactive processes and data sources.

EXAMPLE: DIAGRAM 0 DFD FOR A GRADING SYSTEM

Figure 5-15 on the next page shows a context diagram at the top and diagram 0 beneath it. Notice that diagram 0 is an expansion of process 0. Also notice that the three same entities (STUDENT RECORDS SYSTEM, STUDENT, and INSTRUCTOR) and the same six data flows (FINAL GRADE, CLASS ROSTER, SUBMITTED WORK, GRADED WORK, GRADING PARAMETERS, and GRADE REPORT) appear in both diagrams. In addition, diagram 0 expands process 0 to reveal four internal processes, one data store, and five additional data flows.

Notice that each process in diagram 0 has a reference number: ESTABLISH GRADEBOOK is 1, ASSIGN FINAL GRADE is 2, GRADE STUDENT WORK is 3, and PRODUCE GRADE REPORT is 4. These reference numbers are important because they identify a series of DFDs. If more detail were needed for ESTABLISH GRADEBOOK, for example, you would draw a diagram 1, because ESTABLISH GRADEBOOK is process 1.

The process numbers do not suggest that the processes are accomplished in a sequential order. Each process always is considered to be available, active, and awaiting data to be processed. If processes must be performed in a specific sequence, you document the information in the process descriptions (discussed later in this chapter), not in the DFD.

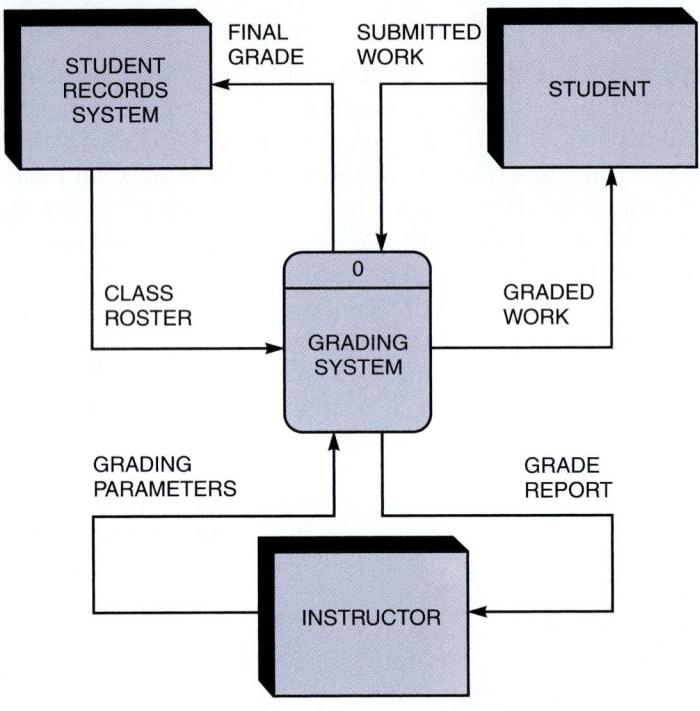

CONTEXT DIAGRAM FOR GRADING SYSTEM

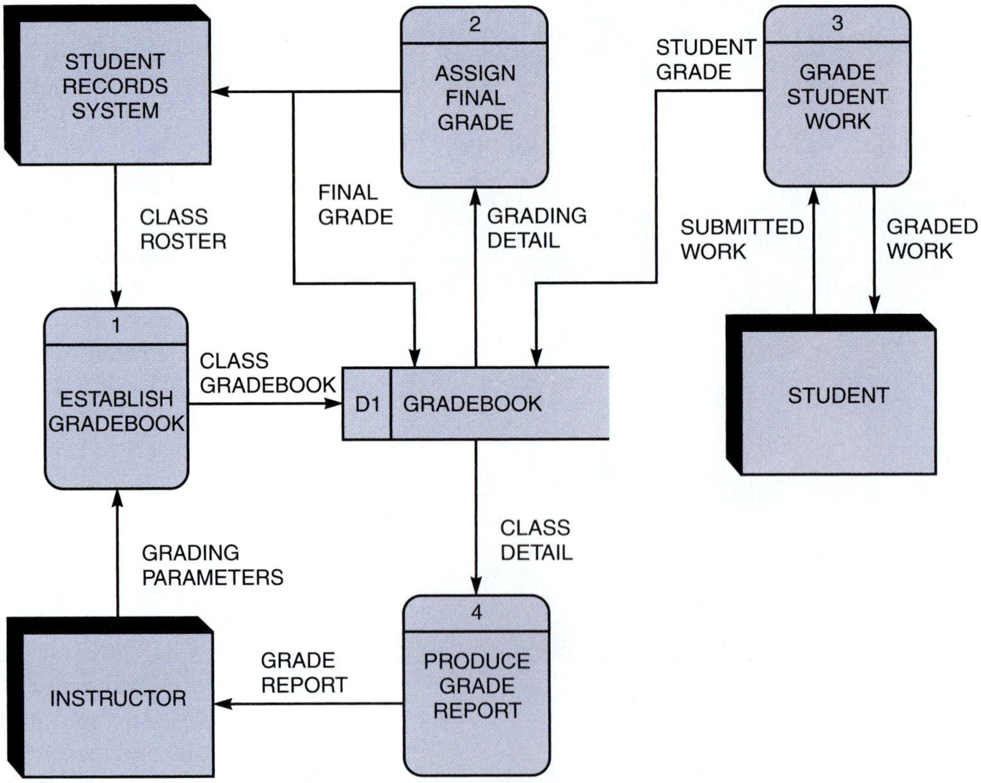

DIAGRAM 0 FOR GRADING SYSTEM

FIGURE 5-15 Context diagram and diagram 0 for the grading system.

The FINAL GRADE data flow output from the ASSIGN FINAL GRADE process is a diverging data flow that becomes an input to the STUDENT RECORDS SYSTEM entity and to the GRADEBOOK data store. A **diverging data flow** is a data flow in which the same data travels to two or more different locations. In that situation, a diverging data flow is the best way to show the flow rather than showing two identical data flows, which could be misleading.

If the same data flows in both directions, you can use a double-headed arrow to connect the symbols. To identify specific data flows into and out of a symbol, however, you use separate data flow symbols with single arrowheads. For example, in Figure 5-15 on the previous page, the separate data flows (SUBMITTED WORK and GRADED WORK) go into and out of the GRADE STUDENT WORK process.

Because diagram 0 is an exploded version of process 0, it shows considerably more detail than the context diagram. You also can refer to diagram 0 as a partitioned or decomposed view of process 0. When you explode a DFD, the higher-level diagram is called the **parent diagram**, and the lower-level diagram is referred to as the **child diagram**. The grading system is simple enough that you do not need any additional DFDs to model the system. At that point, the four processes, the one data store, and the 10 data flows can be documented in the data dictionary.

When you create a set of DFDs for a system, you break the processing logic down into smaller units, called functional primitives, that programmers will use to develop code. A **functional primitive** is a process that consists of a single function that is not exploded further. For example, each of the four processes shown in the lower portion of Figure 5-15 is a functional primitive. You document the logic for a functional primitive by writing a process description in the data dictionary. Later, when the logical design is implemented as a physical system, programmers will transform each functional primitive into program code and modules that carry out the required steps. Deciding whether to explode a process further or determine that it is a functional primitive is a matter of experience, judgment, and interaction with programmers who must translate the logical design into code.

EXAMPLE: DIAGRAM 0 DFD FOR AN ORDER SYSTEM   Figure 5-16 on the next page shows the diagram 0 for an order system. Process 0 on the order system's context diagram is exploded to reveal three processes (FILL ORDER, CREATE INVOICE, and APPLY PAYMENT), one data store (ACCOUNTS RECEIVABLE), two additional data flows (INVOICE DETAIL and PAYMENT DETAIL), and one diverging data flow (INVOICE).

The following walkthrough explains the DFD shown in Figure 5-16:

1. A CUSTOMER submits an ORDER. Depending on the processing logic, the FILL ORDER process either sends an ORDER REJECT NOTICE back to the customer or sends a PICKING LIST to the WAREHOUSE.

2. A COMPLETED ORDER from the WAREHOUSE is input to the CREATE INVOICE process, which outputs an INVOICE to both the CUSTOMER process and the ACCOUNTS RECEIVABLE data store.

3. A CUSTOMER makes a PAYMENT that is processed by APPLY PAYMENT. APPLY PAYMENT requires INVOICE DETAIL input from the ACCOUNTS RECEIVABLE data store along with the PAYMENT. APPLY PAYMENT also outputs PAYMENT DETAIL back to the ACCOUNTS RECEIVABLE data store and outputs COMMISSION to the SALES DEPT, BANK DEPOSIT to the BANK, and CASH RECEIPTS ENTRY to ACCOUNTING.

The walkthrough of diagram 0 illustrates the basic requirements of the order system. To learn more, you would examine the detailed description of each separate process.

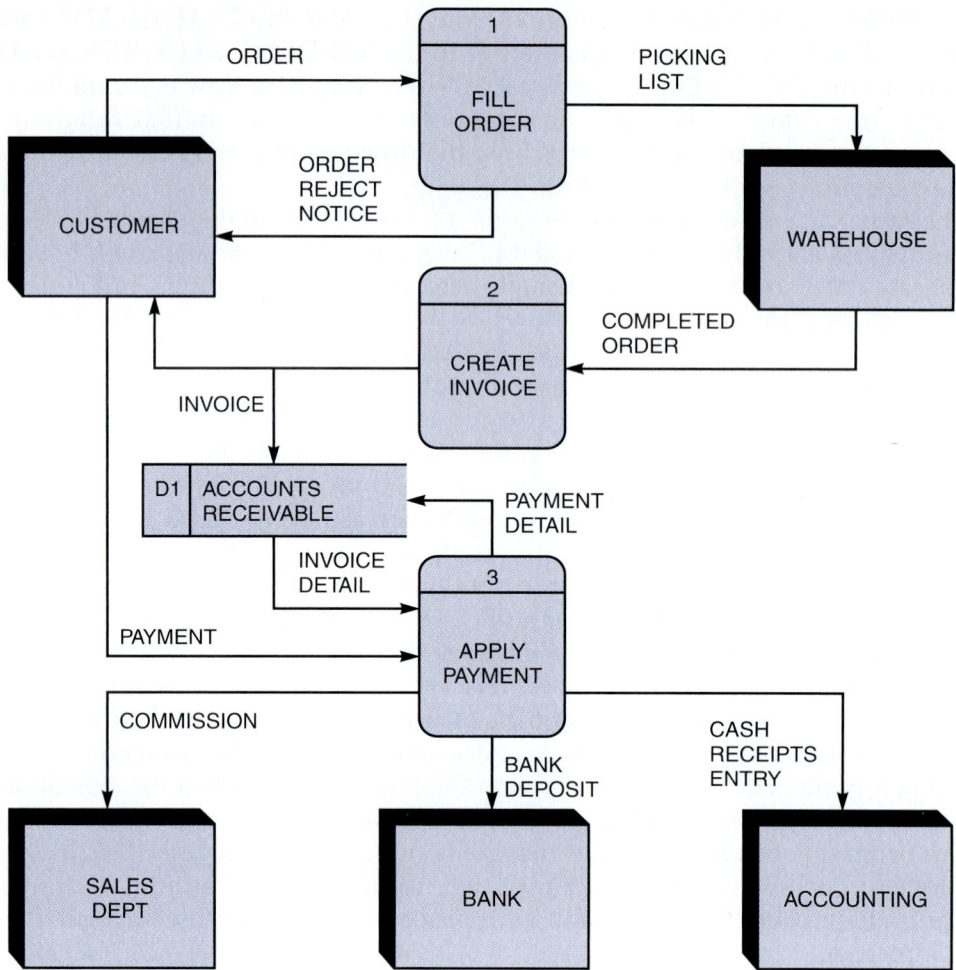

FIGURE 5-16 Diagram 0 DFD for the order system.

Step 3: Draw the Lower-Level Diagrams

This set of lower-level DFDs is based on the order system. To create lower-level diagrams, you must use leveling and balancing techniques. **Leveling** is the process of drawing a series of increasingly detailed diagrams, until all functional primitives are identified. **Balancing** maintains consistency among a set of DFDs by ensuring that input and output data flows align properly. Leveling and balancing are described in more detail in the following sections.

LEVELING EXAMPLES Leveling uses a series of increasingly detailed DFDs to describe an information system. For example, a system might consist of dozens, or even hundreds, of separate processes. Using leveling, an analyst starts with an overall view, which is a context diagram with a single process symbol. Next, the analyst creates diagram 0, which shows more detail. The analyst continues to create lower-level DFDs until all processes are identified as functional primitives, which represent single processing functions. More complex systems have more processes, and analysts must work through many levels to identify the functional primitives. Leveling also is called **exploding, partitioning,** or **decomposing.**

Figure 5-16 and Figure 5-17 provide an example of leveling. Figure 5-16 shows diagram 0 for an order system, with the FILL ORDER process labeled as process 1. Now consider Figure 5-17, which provides an exploded view of the FILL ORDER process. Notice that FILL ORDER (process 1) actually consists of three processes: VERIFY ORDER (process 1.1), PREPARE REJECT NOTICE (process 1.2), and ASSEMBLE ORDER (process 1.3).

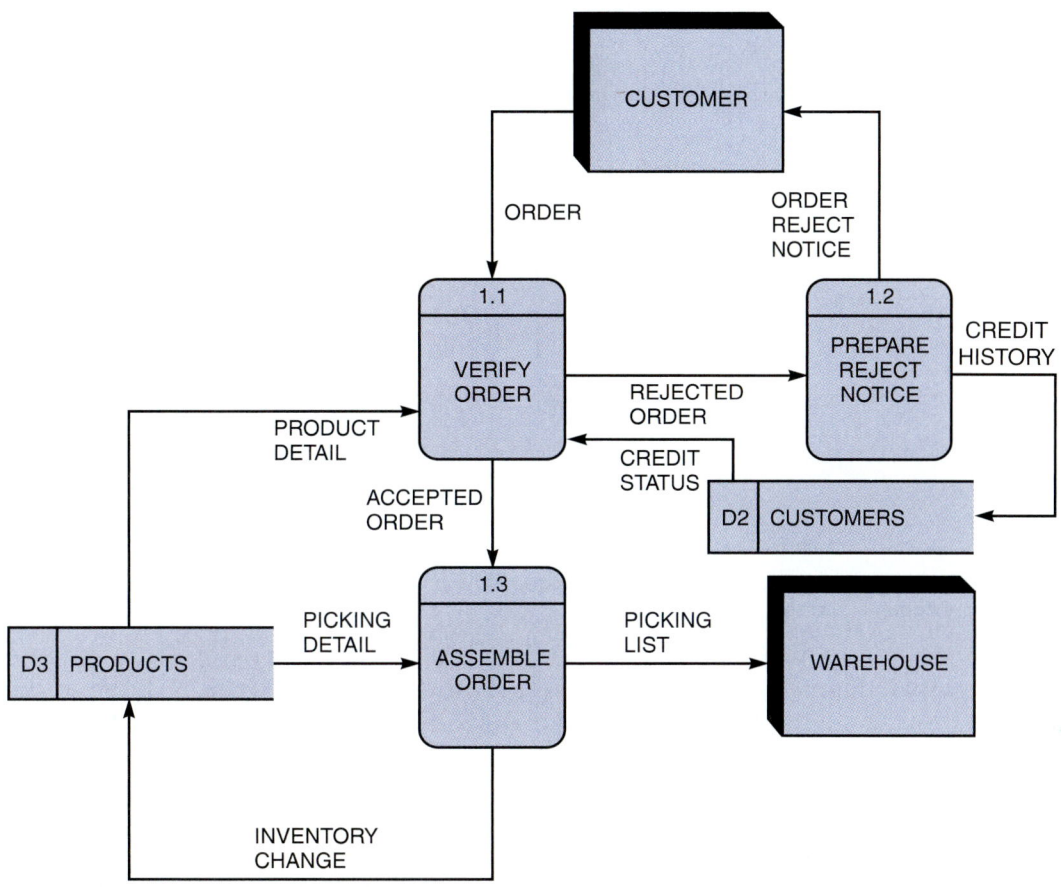

FIGURE 5-17 Diagram I DFD shows details of the FILL ORDER process in the order system.

As Figure 5-17 shows, all processes are numbered using a decimal notation consisting of the parent's reference number, a decimal point, and a sequence number within the new diagram. In Figure 5-17, the parent process of diagram 1 is process 1, so the processes in diagram 1 have reference numbers of 1.1, 1.2, and 1.3. If process 1.3, ASSEMBLE ORDER, is decomposed further, then it would appear in diagram 1.3 and the processes in diagram 1.3 would be numbered as 1.3.1, 1.3.2, 1.3.3, and so on. This numbering technique makes it easy to integrate and identify all DFDs.

When you compare Figures 5-16 and 5-17, you will notice that Figure 5-17 (the exploded FILL ORDER process) shows two data stores (CUSTOMERS and PRODUCTS) that do not appear on Figure 5-16, which is the parent DFD. Why not? The answer is based on a simple rule: When drawing DFDs, you show a data store only when two or more processes use that data store. The CUSTOMERS and PRODUCTS data stores were internal to the FILL ORDER process, so the analyst did not show them on diagram 0, which is the parent. When you explode the FILL ORDER process into diagram 1 DFD, however, you see that three processes (1.1, 1.2, and 1.3) interact with the two data stores, which now are shown.

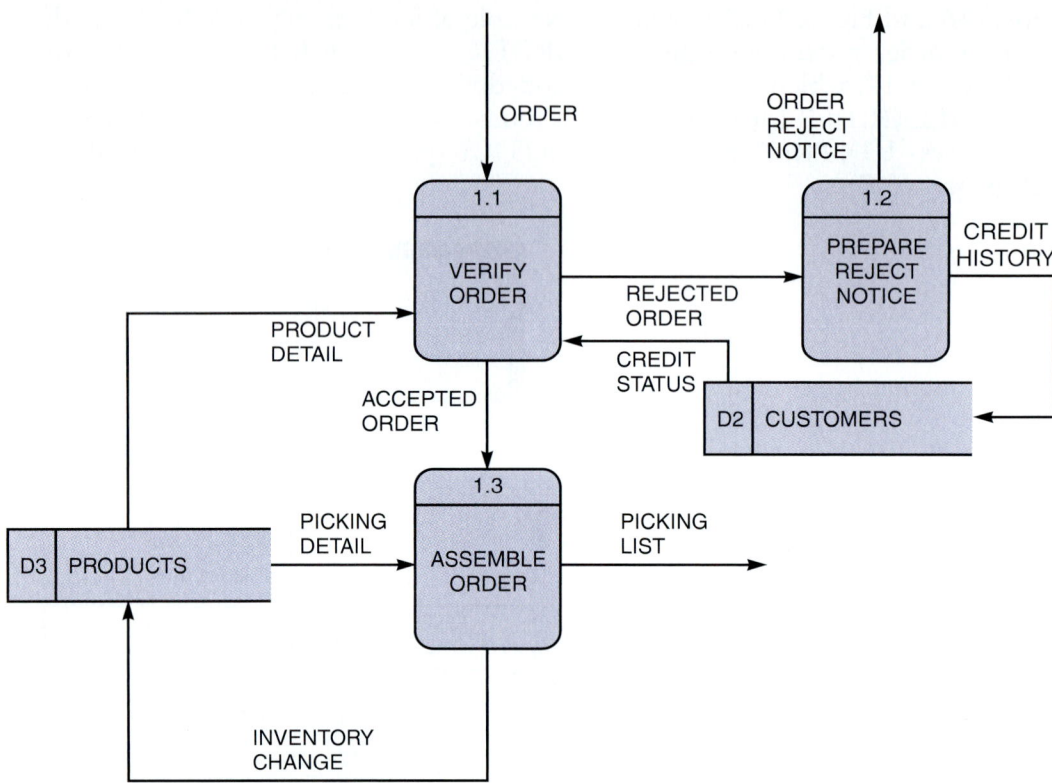

FIGURE 5-18 This diagram does not show the symbols that connect to data flows entering or leaving FILL ORDER on the context diagram.

Now compare Figure 5-17 and Figure 5-18. Notice that that Figure 5-18 shows the same data flows as Figure 5-17, but does not show the CUSTOMER and WAREHOUSE entities. Analysts often use this technique to simplify a DFD and reduce unnecessary clutter. Because the missing symbols appear on the parent DFD, you can refer to that diagram to identify the source or destination of the data flows.

BALANCING EXAMPLES   Balancing ensures that the input and output data flows of the parent DFD are maintained on the child DFD. For example, Figure 5-19 shows two DFDs: The order system diagram 0 is shown at the top of the figure, and the exploded diagram 3 DFD is shown at the bottom.

The two DFDs are balanced, because the child diagram at the bottom has the same input and output flows as the parent process 3 shown at the top. To verify the balancing, notice that the parent process 3, APPLY PAYMENT, has one incoming data flow from an external entity, and three outgoing data flows to external entities. Now examine the child DFD, which is diagram 3. Now, ignore the internal data flows and count the data flows to and from external entities. You will see that the three processes maintain the same one incoming and three outgoing data flows as the parent process.

Order System Diagram 0 DFD

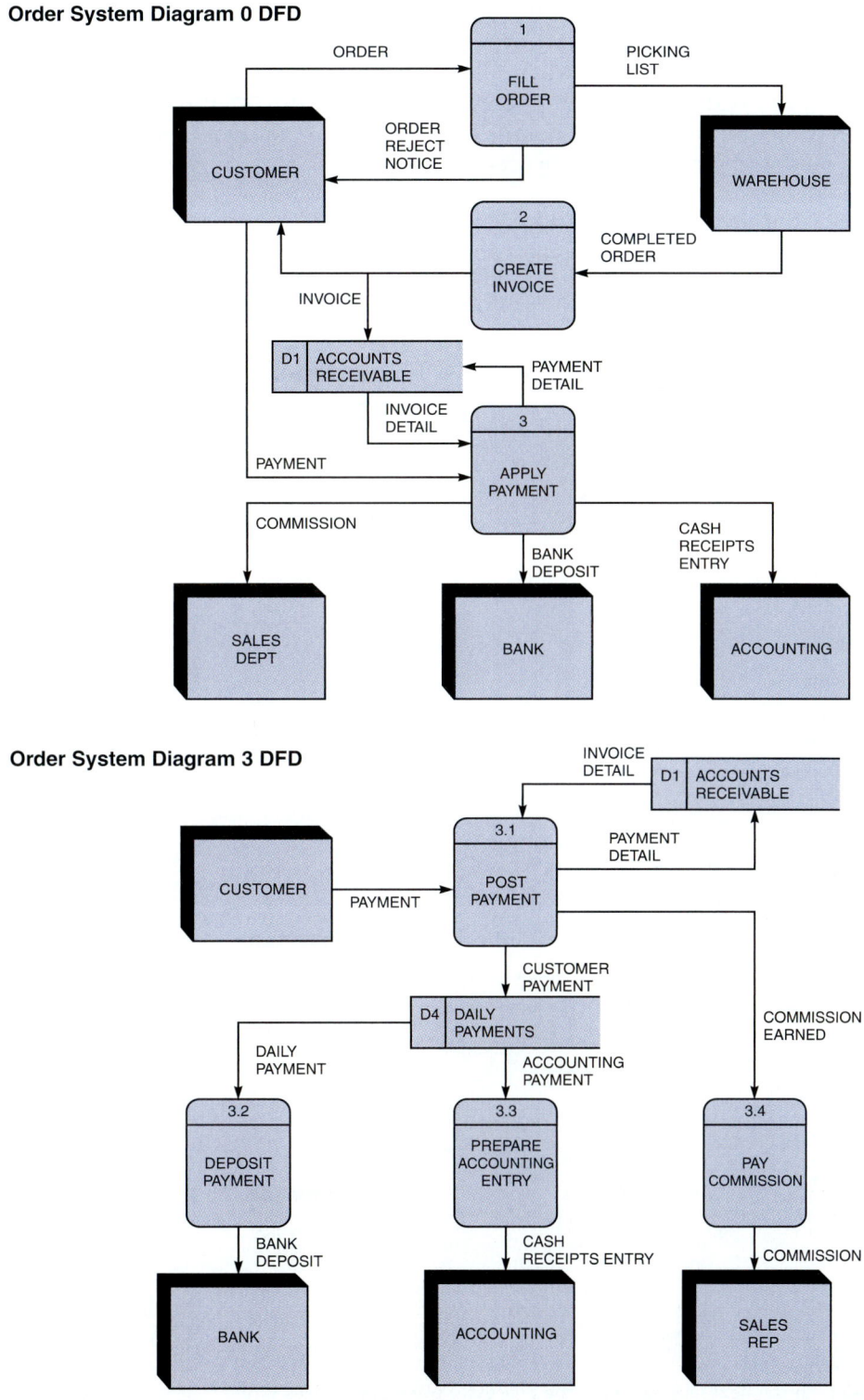

Order System Diagram 3 DFD

FIGURE 5-19   The order system diagram 0 is shown at the top of the figure, and exploded diagram 3 DFD (for the APPLY PAYMENT process) is shown at the bottom. The two DFDs are balanced, because the child diagram at the bottom has the same input and output flows as the parent process 3 shown at the top.

Another example of balancing is shown in Figures 5-20 and 5-21 on the next page. The DFDs in these figures were created using Visible Analyst, a popular CASE tool.

Figure 5-20 shows a sample context diagram. The process 0 symbol has two input flows and two output flows. Notice that process 0 can be considered as a black box, with no internal detail shown. In Figure 5-21, process 0 (the parent DFD) is exploded into the next level of detail. Now three processes, two data stores, and four internal data flows are visible. Notice that the details of process 0 are shown inside a dashed line, just as if you could see inside the process.

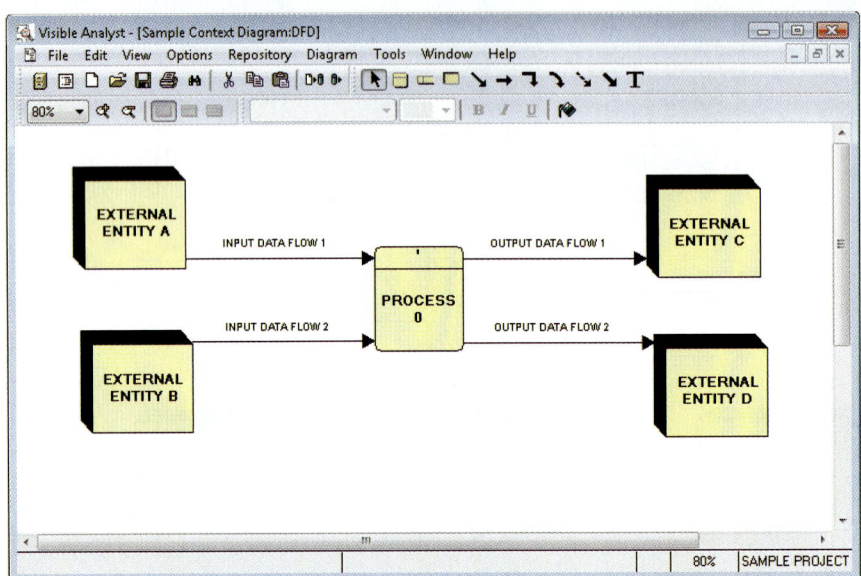

FIGURE 5-20 Example of a parent DFD diagram, showing process 0 as a black box.

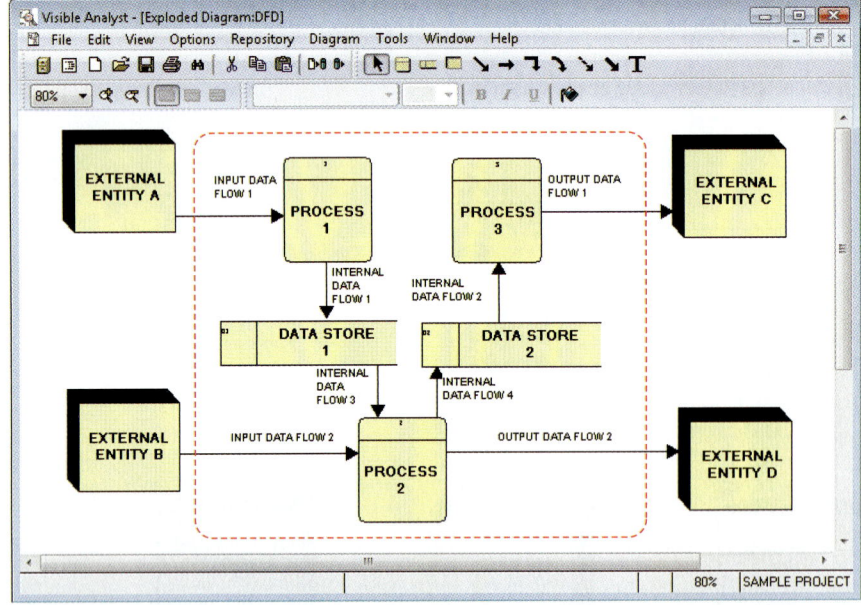

FIGURE 5-21 In the next level of detail, the process 0 black box reveals three processes, two data stores, and four internal data flows — all of which are shown inside the dashed line.

The DFDs in Figures 5-20 and 5-21 are balanced, because the four data flows into and out of process 0 are maintained on the child DFD. The DFDs also are leveled, because each internal process is numbered to show that it is a child of the parent process.

CASE IN POINT 5.1: BIG TEN UNIVERSITY

You are the IT director at Big Ten University. As part of a training program, you decide to draw a DFD that includes some obvious mistakes to see whether your newly hired junior analysts can find them. You came up with the diagram 0 DFD shown in Figure 5-22. Based on the rules explained in this chapter, how many problems should the analysts find?

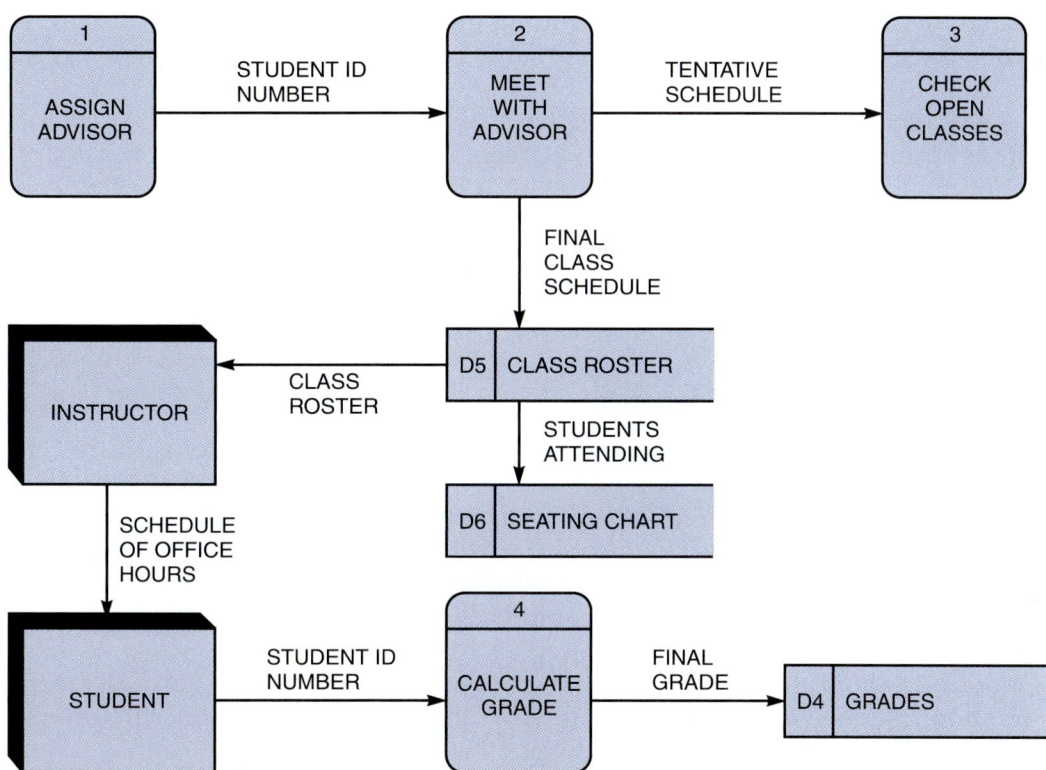

FIGURE 5-22   What are the mistakes in this diagram 0 DFD?

DATA DICTIONARY

A set of DFDs produces a logical model of the system, but the details within those DFDs are documented separately in a data dictionary, which is the second component of structured analysis.

A **data dictionary**, or **data repository**, is a central storehouse of information about the system's data. An analyst uses the data dictionary to collect, document, and organize specific facts about the system, including the contents of data flows, data stores, entities, and processes. The data dictionary also defines and describes all data elements and meaningful combinations of data elements. A **data element**, also called a **data item** or **field**, is the smallest piece of data that has meaning within an information system. Examples of data elements are student grade, salary, Social Security number, account balance, and company name. Data elements are combined into **records**, also called **data structures**. A record is a meaningful combination of related data elements that is included in a data flow or retained in a data store. For example, an auto parts store inventory record might include part number, description, supplier code, minimum and maximum stock levels, cost, and list price.

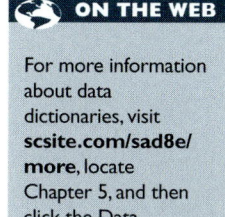

Significant relationships exist among the items in a data dictionary. For example, data stores and data flows are based on data structures, which in turn are composed of data elements. Data flows are connected to data stores, entities, and processes. Accurately documenting these relationships is essential so the data dictionary is consistent with the DFDs. You can use CASE software to help you document the design.

Using CASE Tools for Documentation

TOOLKIT TIME

The CASE tools in Part 2 of the Systems Analyst's Toolkit can help you document business functions and processes. To learn more about these tools, turn to Part 2 of the four-part Toolkit that follows Chapter 12.

The more complex the system, the more difficult it is to maintain full and accurate documentation. Fortunately, modern CASE tools simplify the task. For example, in the Visible Analyst CASE tool, documentation automatically flows from the modeling diagrams into the central repository, along with information entered by the user. This section contains several examples of Visible Analyst screens that show the data repository and its contents.

A CASE repository ensures data consistency, which is especially important where multiple systems require the same data. In a large company, for example, the sales, accounting, and shipping systems all might use a data element called CUSTOMER NUMBER. Once the CUSTOMER NUMBER element has been defined in the repository, it can be accessed by other processes, data flows, and data stores. The result is that all systems across the enterprise can share data that is up to date and consistent. You will learn more about CASE tools in Part 2 of the Systems Analyst's Toolkit.

Documenting the Data Elements

You must document every data element in the data dictionary. Some analysts like to record their notes on online or manual forms. Others prefer to enter the information directly into a CASE tool. Several of the DFDs and data dictionary entries that appear in this chapter were created using a popular CASE tool called Visible Analyst. Although other CASE tools might use other terms or display the information differently, the objective is the same: to provide clear, comprehensive information about the data and processes that make up the system.

Figure 5-23 shows how the analyst used an online documentation form to record information for the SOCIAL SECURITY NUMBER data element. Notice that the figure caption identifies eight specific characteristics for this data element.

1. Online or manual documentation entries often indicate which system is involved. This is not necessary with a CASE tool because all information is stored in one file that is named for the system.
2. The data element has a standard label that provides consistency throughout the data dictionary.
3. The data element can have an alternative name, or alias.
4. This entry indicates that the data element consists of nine numeric characters.
5. Depending on the data element, strict limits might be placed on acceptable values.
6. The data comes from the employee's job application.
7. This entry indicates that only the payroll department has authority to update or change this data.
8. This entry indicates the individual or department responsible for entering and changing data.

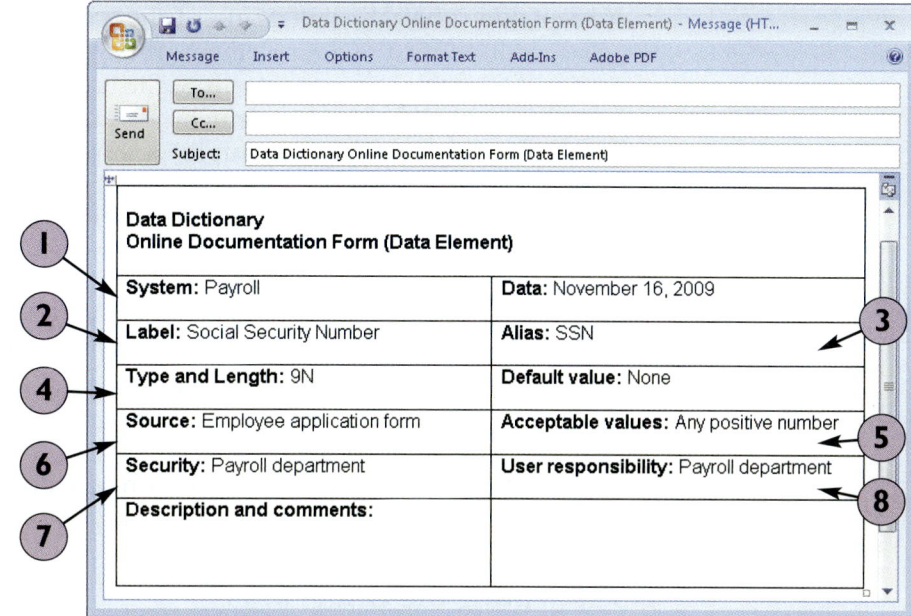

FIGURE 5-23 Using an online documentation form, the analyst has recorded information for a data element named SOCIAL SECURITY NUMBER. Later, the analyst will create a data dictionary entry using a CASE tool.

Figure 5-24 shows a sample screen that illustrates how the SOCIAL SECURITY NUMBER data element might be recorded in the Visible Analyst data dictionary.

Regardless of the terminology or method, the following attributes usually are recorded and described in the data dictionary:

Data element name or label. The data element's standard name, which should be meaningful to users.

Alias. Any name(s) other than the standard data element name; this alternate name is called an **alias**. For example, if you have a data element named CURRENT BALANCE, various users might refer to it by alternate names such as OUTSTANDING BALANCE, CUSTOMER BALANCE, RECEIVABLE BALANCE, or AMOUNT OWED.

Type and length. **Type** refers to whether the data element contains numeric, alphabetic, or character values. **Length** is the maximum number of characters for an alphabetic or character data element or the maximum number of digits and number of decimal positions for a numeric data element. In addition to text and numeric data, sounds and images also can be stored in digital form. In some systems, these binary data objects are managed and processed just as traditional data elements are. For example, an employee record might include a digitized photo image of the person.

Default value. The value for the data element if a value otherwise is not entered for it. For example, all new customers might have a default value of $500 for the CREDIT LIMIT data element.

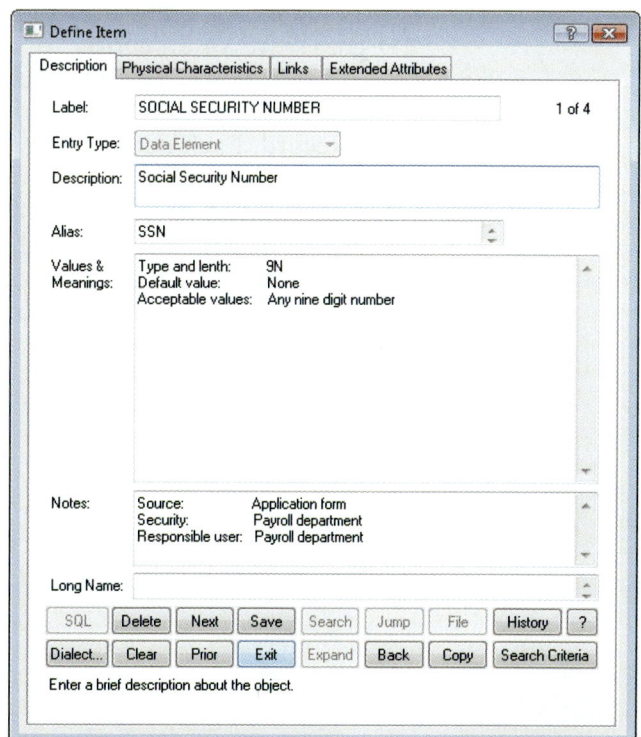

FIGURE 5-24 A Visible Analyst screen describes the data element named SOCIAL SECURITY NUMBER. Notice that many of the items were entered from the online form shown in Figure 5-23 on page 217.

Acceptable values. Specification of the data element's **domain,** which is the set of values permitted for the data element; these values either can be specifically listed or referenced in a table, or can be selected from a specified range of values. You also would indicate if a value for the data element is optional. Some data elements have additional **validity rules.** For example, an employee's salary must be within the range defined for the employee's job classification.

Source. The specification for the origination point for the data element's values. The source could be a specific form, a department or outside organization, another information system, or the result of a calculation.

Security. Identification for the individual or department that has access or update privileges for each data element. For example, only a credit manager has the authority to change a credit limit, while sales reps are authorized to access data in a read-only mode.

Responsible user(s). Identification of the user(s) responsible for entering and changing values for the data element.

Description and comments. This part of the documentation allows you to enter additional notes.

Documenting the Data Flows

In addition to documenting each data element, you must document all data flows in the data dictionary. Figure 5-25 shows a definition for a data flow named COMMISSION. The information on the manual form at the top was entered into the CASE tool data dictionary at the bottom of Figure 5-25.

Although terms can vary, the typical attributes are as follows:

Data flow name or label. The data flow name as it appears on the DFDs.

Description. Describes the data flow and its purpose.

Alternate name(s). Aliases for the DFD data flow name(s).

Origin. The DFD beginning, or source, for the data flow; the origin can be a process, a data store, or an entity.

Destination. The DFD ending point(s) for the data flow; the destination can be a process, a data store, or an entity.

Record. Each data flow represents a group of related data elements called a record or data structure. In most data dictionaries, records are defined separately from the data flows and data stores. When records are defined, more than one data flow or data store can use the same record, if necessary.

Volume and frequency. Describes the expected number of occurrences for the data flow per unit of time. For example, if a company has 300 employees, a TIME CARD data flow would involve 300 transactions and records each week, as employees submit their work hour data.

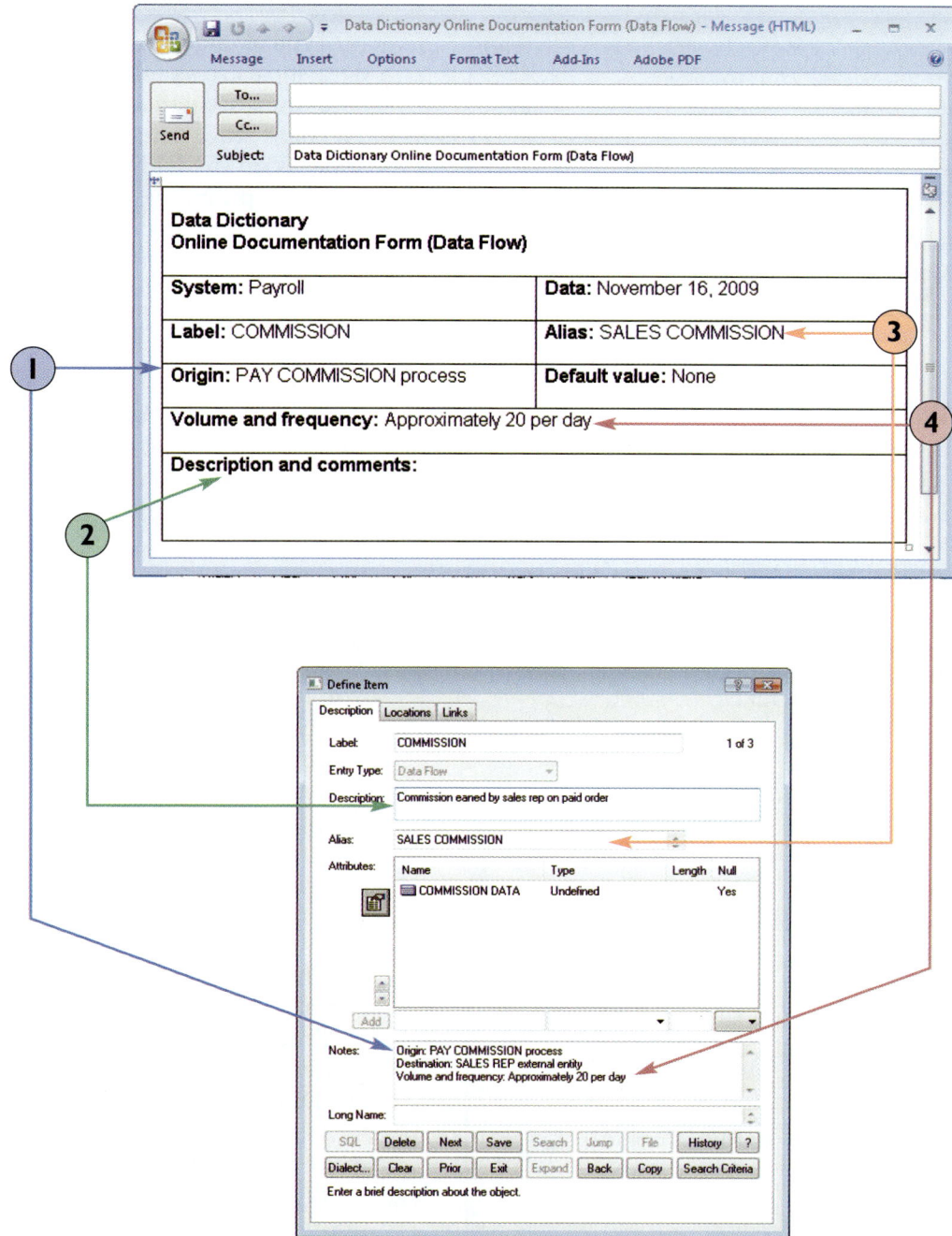

FIGURE 5-25 In the upper screen, an analyst has entered four items of information in an online documentation form. The lower screen shows the same four items entered into a Visible Analyst data dictionary form.

Documenting the Data Stores

You must document every DFD data store in the data dictionary. Figure 5-26 on the next page shows the definition of a data store named IN STOCK.

1. This data store has an alternative name, or alias.
2. For consistency, data flow names are standardized throughout the data dictionary.
3. It is important to document these estimates, because they will affect design decisions in subsequent SDLC phases.

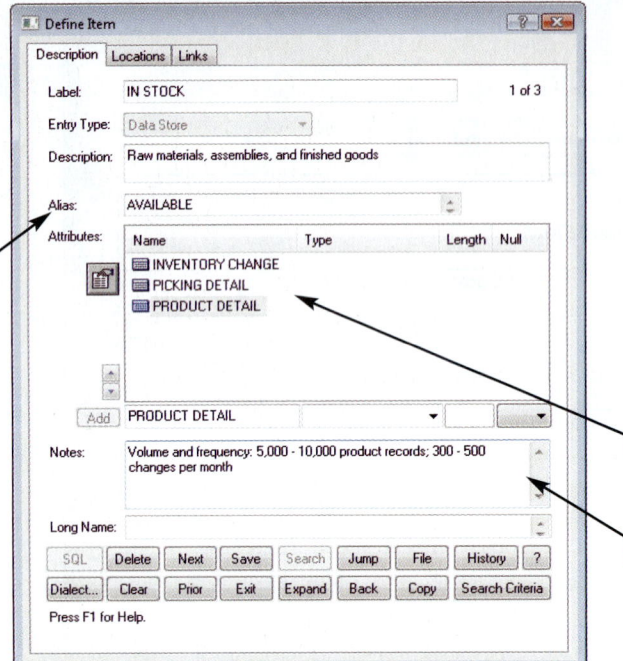

FIGURE 5-26 Visible Analyst screen that documents a data store named IN STOCK.

1. The process number identifies this process. Any subprocesses are numbered 1.1, 1.2, 1.3, and so on.
2. These data flows will be described specifically elsewhere in the data dictionary.

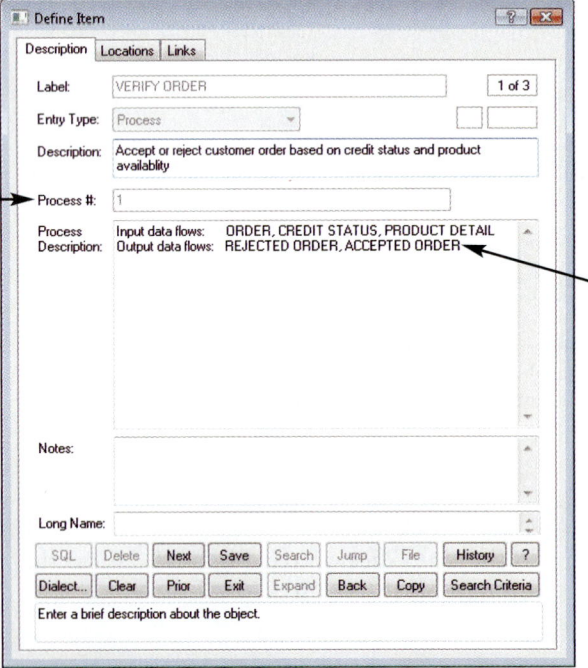

FIGURE 5-27 Visible Analyst screen that describes a process named VERIFY ORDER.

Typical characteristics of a data store are as follows:

Data store name or label. The data store name as it appears on the DFDs.

Description. Describes the data store and its purpose.

Alternate name(s). Aliases for the DFD data store name.

Attributes. Standard DFD names that enter or leave the data store.

Volume and frequency. Describes the estimated number of records in the data store and how frequently they are updated.

Documenting the Processes

You must document every process, as shown in Figure 5-27. Your documentation includes a description of the process's characteristics and, for functional primitives, a process description, which is a model that documents the processing steps and business logic.

The following are typical characteristics of a process:

Process name or label. The process name as it appears on the DFDs.

Description. A brief statement of the process's purpose.

Process number. A reference number that identifies the process and indicates relationships among various levels in the system.

Process description. This section includes the input and output data flows. For functional primitives, the process description also documents the processing steps and business logic. You will learn how to write process descriptions in the next section.

Documenting the Entities

By documenting all entities, the data dictionary can describe all external entities that interact with the system. Figure 5-28 shows a definition for an external entity named WAREHOUSE.

Typical characteristics of an entity include the following:

Entity name. The entity name as it appears on the DFDs.

Description. Describe the entity and its purpose.

Alternate name(s). Any aliases for the entity name.

Input data flows. The standard DFD names for the input data flows to the entity.

Output data flows. The standard DFD names for the data flows leaving the entity.

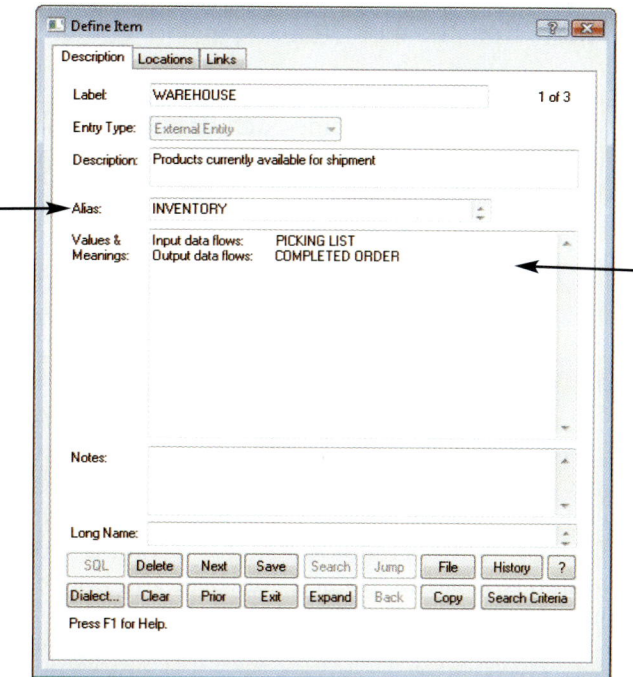

1. The external entity also can have an alternative name, or alias, if properly documented.
2. For consistency, these data flow names are standardized throughout the data dictionary.

FIGURE 5-28 Visible Analyst screen that documents an external entity named WAREHOUSE.

Documenting the Records

A record is a data structure that contains a set of related data elements that are stored and processed together. Data flows and data stores consist of records that you must document in the data dictionary. You define characteristics of each record, as shown in Figure 5-29.

Typical characteristics of a record include the following:

Record or data structure name. The record name as it appears in the related data flow and data store entries in the data dictionary.

Definition or description. A brief definition of the record.

Alternate name(s). Any aliases for the record name.

Attributes. A list of all the data elements included in the record. The data element names must match exactly what you entered in the data dictionary.

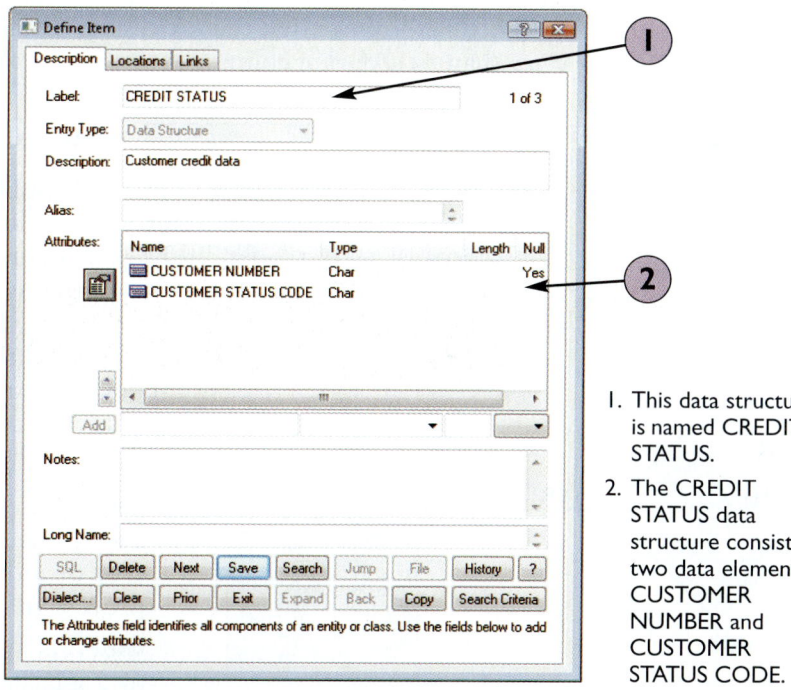

1. This data structure is named CREDIT STATUS.
2. The CREDIT STATUS data structure consists of two data elements: CUSTOMER NUMBER and CUSTOMER STATUS CODE.

FIGURE 5-29 Visible Analyst screen that documents a record, or data structure named CREDIT STATUS.

Data Dictionary Reports

The data dictionary serves as a central storehouse of documentation for an information system. A data dictionary is created when the system is developed, and is updated constantly as the system is implemented, operated, and maintained. In addition to describing each data element, data flow, data store, record, entity, and process, the data dictionary documents the relationships among these components. You can obtain many valuable reports from a data dictionary, including the following:

- An alphabetized list of all data elements by name
- A report describing each data element and indicating the user or department that is responsible for data entry, updating, or deletion
- A report of all data flows and data stores that use a particular data element
- Detailed reports showing all characteristics of data elements, records, data flows, processes, or any other selected item stored in the data dictionary

PROCESS DESCRIPTION TOOLS

A **process description** documents the details of a functional primitive, and represents a specific set of processing steps and business logic. Using a set of process description tools, you create a model that is accurate, complete, and concise. Typical process description tools include structured English, decision tables, and decision trees. When you analyze a functional primitive, you break the processing steps down into smaller units in a process called modular design.

It should be noted that this chapter deals with structured analysis, but the process description tools also can be used in object-oriented development, which is described in Chapter 6. You learned in Chapter 1 that O-O analysis combines data and the processes that act on the data into things called objects, that similar objects can be grouped together into classes, and that O-O processes are called methods. Although O-O programmers use different terminology, they create the same kind of modular coding structures, except that the processes, or methods, are stored inside the objects, rather than as separate components.

FIGURE 5-30 Sequence structure.

Modular Design

Modular design is based on combinations of three **logical structures**, sometimes called **control structures**, which serve as building blocks for the process. Each logical structure must have a single entry and exit point. The three structures are called sequence, selection, and iteration. A rectangle represents a step or process, a diamond shape represents a condition or decision, and the logic follows the lines in the direction indicated by the arrows.

1. **Sequence.** The completion of steps in sequential order, one after another, as shown in Figure 5-30. One or more of the steps might represent a subprocess that contains additional logical structures.

2. **Selection.** The completion of one of two or more process steps based on the results of a test or condition. In the example shown in Figure 5-31, the system tests the input, and if the hours are greater than 40, it performs the CALCULATE OVERTIME PAY process.

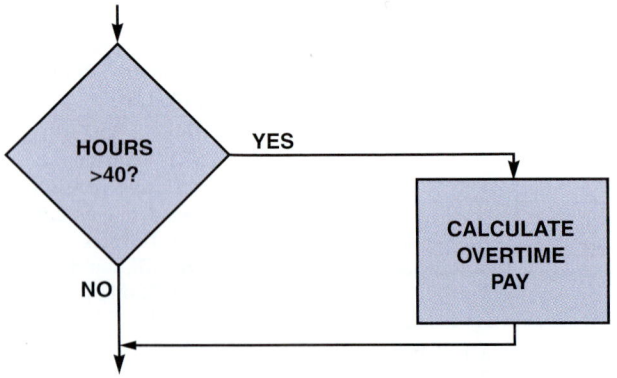

FIGURE 5-31 Selection structure.

3. **Iteration.** The completion of a process step that is repeated until a specific condition changes, as shown in Figure 5-32. An example of iteration is a process that continues to print paychecks until it reaches the end of the payroll file. Iteration also is called **looping**.

Sequence, selection, and iteration structures can be combined in various ways to describe processing logic.

Structured English

Structured English is a subset of standard English that describes logical processes clearly and accurately. When you use structured English, you must conform to the following rules:

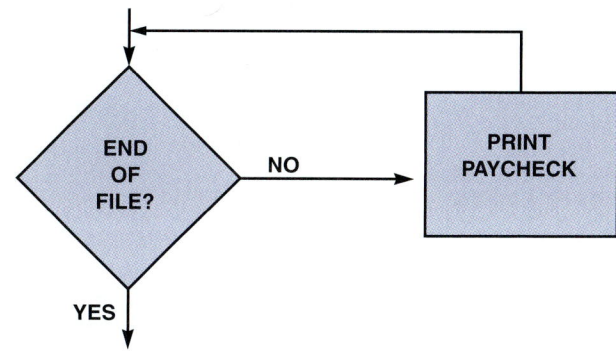

FIGURE 5-32 Iteration structure.

- Use only the three building blocks of sequence, selection, and iteration.

- Use indentation for readability.

- Use a limited vocabulary, including standard terms used in the data dictionary and specific words that describe the processing rules.

An example of structured English appears in Figure 5-33, which shows the VERIFY ORDER process that was illustrated earlier. In Figure 5-33, structured English was added to describe the processing logic.

Structured English might look familiar to programming students because it resembles **pseudocode**, which is used in program design. Although the techniques are similar, the primary purpose of structured English is to describe the underlying business logic, while programmers, who are concerned with coding, mainly use pseudocode as a shorthand notation for the actual code.

Figure 5-34 on the next page shows another example of structured English. After you study the sales promotion policy, notice that the structured English version describes the processing logic that the system must apply. Following these structured English rules ensures that your process descriptions are understandable to users who must confirm that the process is correct, as well as to other analysts and programmers who must design the information system from your descriptions.

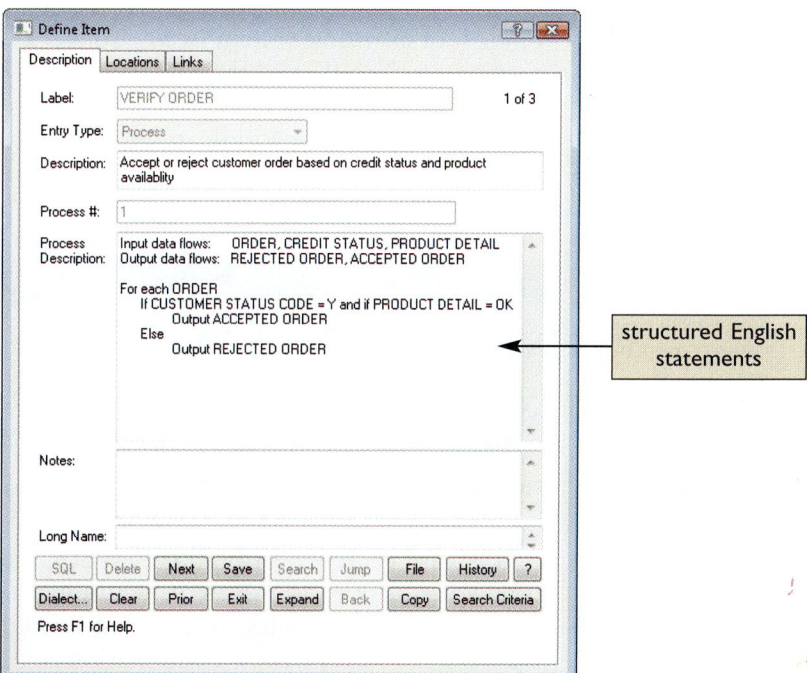

FIGURE 5-33 The VERIFY ORDER process description includes logical rules and a structured English version of the policy. Notice the alignment and indentation of the logic statements.

ON THE WEB

For more information about structured English, visit **scsite.com/ sad8e/more**, locate Chapter 5, and then click the Structured English link.

SAMPLE OF A SALES PROMOTION POLICY

- Preferred customers who order more than $1,000 are entitled to a 5% discount, and an additional 5% discount if they used our charge card.

- Preferred customers who do not order more than $1,000 receive a $25 bonus coupon.

- All other customers receive a $5 bonus coupon.

STRUCTURED ENGLISH VERSION OF THE SALES PROMOTION POLICY

```
IF customer is a preferred customer, and
        IF customer orders more than $1,000 then
                Apply a 5% discount, and
                IF customer uses our charge card, then
                        Apply an additional 5% discount
        ELSE
                Award a $25 bonus coupon
ELSE
        Award a $5 bonus coupon
```

FIGURE 5-34 Sample of a policy with logical rules, and a structured English version of the policy. Notice the alignment and indentation of the logic statements.

Decision Tables

A **decision table** shows a logical structure, with all possible combinations of conditions and resulting actions. Analysts often use decision tables, in addition to structured English, to describe a logical process and ensure that they have not overlooked any logical possibility.

A simple example of a decision table based on the VERIFY ORDER process is shown in Figure 5-35. When documenting a process, it is important to consider every possible outcome to ensure that you have overlooked nothing. From the structured English description shown in Figure 5-33, we know that an accepted order requires that credit status is OK and the product is in stock. Otherwise, the order is rejected. The decision table shown in Figure 5-35 shows all the possibilities. To create the decision table, follow the steps indicated in the figure.

1. Place the name of the process in a heading at the top left.
2. Enter the conditions under the heading, with one condition per line, to represent the customer status and availability of products.
3. Enter all potential combinations of Y/N (for yes and no) for the conditions. Each column represents a numbered possibility called a rule.
4. Place an X in the action entries area for each rule to indicate whether to accept or reject the order.

VERIFY ORDER Process

	1	2	3	4
Credit status is OK	Y	Y	N	N
Product is in stock	Y	N	Y	N
Accept order	X			
Reject order		X	X	X

FIGURE 5-35 Example of a simple decision table showing the processing logic of the VERIFY ORDER process.

Because each condition has two possible values, the number of rules doubles each time you add a condition. For example, one condition creates only two rules, two conditions create four rules, three conditions create eight rules, and so on. As shown in Figure 5-36, four possible combinations exist, but only one rule — rule 1 — permits an accepted order output.

The first table in Figure 5-36 shows eight rules. Because some rules are duplicates, however, the table can be simplified. To reduce the number of rules, you must look closely at each combination of conditions and actions. If you have rules with three conditions, only one or two of them may control the outcome, and the other conditions do not matter. You can indicate that with dashes (-) as shown in the second table in Figure 5-36. Then you can combine and renumber the rules, as shown in the final table.

In the example, rules 1 and 2 can be combined because credit status is OK, and the waiver is not needed. Rules 3, 4, 7, and 8 also can be combined because the product is not in stock, and credit status does not matter. The result is that instead of eight possibilities, only four logical rules are created that control the VERIFY ORDER process.

In addition to multiple conditions, decision tables can have more than two possible outcomes. An example is presented in the SALES PROMOTION POLICY decision table shown in Figure 5-37 on the next page. The decision table shown here is based on the sales promotion policy described in Figure 5-34. Here three conditions exist: Was the customer a preferred customer, did the customer order more than $1,000, and did the customer use our charge card? Based on these three conditions, four possible actions can occur, as shown in the table.

Decision tables often are the best way to describe a complex set of conditions. Many analysts use decision tables because they are easy to construct and understand, and programmers find it easy to work from a decision table when developing code.

ON THE WEB

For more information about decision tables, visit **scsite.com/ sad8e/more**, locate Chapter 5, and then click the Decision Tables link.

VERIFY ORDER *Process with Credit Waiver (Initial version)*

	1	2	3	4	5	6	7	8
Credit status is OK	Y	Y	Y	Y	N	N	N	N
Product is in stock	Y	Y	N	N	Y	Y	N	N
Waiver from credit manager	Y	N	Y	N	Y	N	Y	N
Accept order	X	X			X			
Reject order			X	X		X	X	X

VERIFY ORDER *Process with Credit Waiver (With rules marked for combination)*

	1	2	3	4	5	6	7	8
Credit status is OK	Y	Y	-	-	N	N	-	-
Product is in stock	Y	Y	N	N	Y	Y	N	N
Waiver from credit manager	-	-	-	-	Y	N	-	-
Accept order	X	X			X			
Reject order			X	X		X	X	X

1. Because the product is not in stock, the other conditions do not matter.
2. Because the other conditions are met, the waiver does not matter.

VERIFY ORDER *Process with Credit Waiver (After rule combination and simplification)*

	1 (COMBINES PREVIOUS 1, 2)	2 (PREVIOUS 5)	3 (PREVIOUS 6)	4 (COMBINES PREVIOUS 3, 4, 7, 8)
Credit status is OK	Y	N	N	-
Product is in stock	Y	Y	Y	N
Waiver from credit manager	-	Y	N	-
Accept order	X	X		
Reject order			X	X

FIGURE 5-36 This example is more complex, because the credit manager can waive the credit status requirement in certain cases. To ensure that all possibilities are covered, notice that the first condition provides an equal number of Ys and Ns, the second condition alternates Y and N pairs, and the third condition alternates single Ys and Ns.

Sales Promotion Policy (Initial version)

				1	2	3	4	5	6	7	8
Preferred customer				Y	Y	Y	Y	N	N	N	N
Ordered more than $1,000				Y	Y	N	N	Y	Y	N	N
Used our charge card				Y	N	Y	N	Y	N	Y	N
5% discount				X	X						
Additional 5% discount				X							
$25 bonus coupon								X	X		
$5 bonus coupon										X	X	X	X

FIGURE 5-37 Sample decision table based on the sales promotion policy described in Figure 5-34 on page 224. This is the initial version of the table, before simplification.

CASE IN POINT 5.2: ROCK SOLID OUTFITTERS (PART 1)

Leah Jones is the IT manager at Rock Solid Outfitters, a medium-sized supplier of outdoor climbing and camping gear. Steve Allen, the marketing director, has asked Leah to develop a special Web-based promotion. As Steve described it to Leah, Rock Solid will provide free shipping for any customer who either completes an online survey form or signs up for the Rock Solid online newsletter. Additionally, if a customer completes the survey *and* signs up for the newsletter, Rock Solid will provide a $10 merchandise credit for orders over $100. Leah has asked you to develop a decision table that will reflect the promotional rules that a programmer will use. She wants you to show all possibilities, then to simplify the results to eliminate any combinations that would be unrealistic or redundant.

Decision Trees

A **decision tree** is a graphical representation of the conditions, actions, and rules found in a decision table. Decision trees show the logic structure in a horizontal form that resembles a tree with the roots at the left and the branches to the right. Like flowcharts, decision trees are useful ways to present the system to management. Decision trees and decision tables provide the same results, but in different forms. In many situations, a graphic is the most effective means of communication, as shown in Figure 5-38.

Figure 5-39 shows the same SALES PROMOTION POLICY conditions and actions shown in Figure 5-37. A decision tree is read from left to right, with the conditions along the various branches and the actions at the far right. Because the example has two conditions with four resulting sets of actions, the example has four terminating branches at the right side of the tree.

FIGURE 5-38 Analysts and managers often use graphical representations to show the process under consideration.

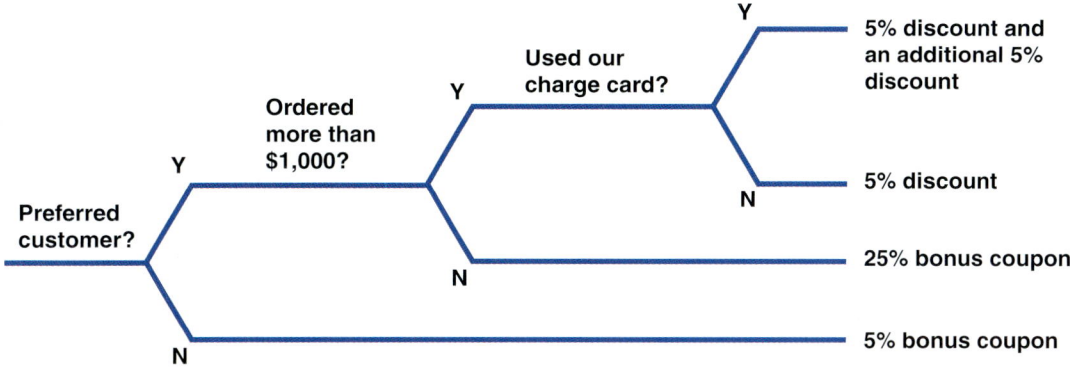

FIGURE 5-39 Sample decision tree. Like a decision table, a decision tree illustrates the action to be taken based on certain conditions, but presents it graphically. This decision tree is based on the sales promotion policy described in Figures 5-34 and 5-37 on pages 224 and 226.

Whether to use a decision table or a decision tree often is a matter of personal preference. A decision table might be a better way to handle complex combinations of conditions. On the other hand, a decision tree is an effective way to describe a relatively simple process.

CASE IN POINT 5.3:   ROCK SOLID OUTFITTERS (PART 2)

Leah Jones, the IT manager at Rock Solid Outfitters, thinks you did a good job on the decision table task she assigned to you. Now she wants you to use the same data to develop a decision tree that will show all the possibilities for the Web-based promotion described in Part 1 of the case. She also wants you to discuss the pros and cons of decisions tables versus decision trees.

LOGICAL VERSUS PHYSICAL MODELS

While structured analysis tools are used to develop a logical model for a new information system, such tools also can be used to develop physical models of an information system. A physical model shows how the system's requirements are implemented. During the systems design phase, you create a physical model of the new information system that follows from the logical model and involves operational tasks and techniques.

Sequence of Models

What is the relationship between logical and physical models? Think back to the beginning of the systems analysis phase, when you were trying to understand the existing system. Rather than starting with a logical model, you first studied the physical operations of the existing system to understand how the current tasks were carried out. Many systems analysts create a physical model of the current system and then develop a logical model of the current system before tackling a logical model of the new system. Performing that extra step allows them to understand the current system better.

Four-Model Approach

Many analysts follow a **four-model approach,** which means that they develop a physical model of the current system, a logical model of the current system, a logical model of the new system, and a physical model of the new system. The major benefit of the four-model approach is that it gives you a clear picture of current system functions before you make any modifications or improvements. That is important because mistakes made early in systems development will affect later SDLC phases and can result in unhappy users and additional costs. Taking additional steps to avoid these potentially costly mistakes can prove to be well worth the effort. Another advantage is that the requirements of a new information system often are quite similar to those of the current information system, especially where the proposal is based on new computer technology rather than a large number of new requirements. Adapting the current system logical model to the new system logical model in these cases is a straightforward process.

The only disadvantage of the four-model approach is the added time and cost needed to develop a logical and physical model of the current system. Most projects have very tight schedules that might not allow time to create the current system models. Additionally, users and managers want to see progress on the new system — they are much less concerned about documenting the current system. As a systems analyst, you must stress the importance of careful documentation and resist the pressure to hurry the development process at the risk of creating serious problems later.

CASE IN POINT 5.4: TIP TOP STAFFING

Tip Top Staffing supplies employees to hundreds of IT firms that require specialized skills for specific projects. Systems analysts Lisa Nuevo and Bill Goodman are working on the logical model of Tip Top's billing and records system, using DFDs, a data dictionary, and process descriptions. At some point while working on the logical model of the system, Lisa felt that some improvements should be made in the data forms that Tip Top uses to obtain information about job applicants. Was the subject of improving the forms a physical implementation issue? Is Lisa going off on a tangent by considering *how* something will be done, instead of sticking to *what* will be done?

A QUESTION OF ETHICS

This is your first week in your new job at Safety Zone, a leading producer of IT modeling software. Your prior experience with a smaller competitor gave you an edge in landing the job, and you are excited about joining a larger company in the same field.

So far, all is going well and you are getting used to the new routine. However, you are concerned about one issue. In your initial meeting with the IT manager, she seemed very interested in the details of your prior position, and some of her questions made you a little uncomfortable. She did not actually ask you to reveal any proprietary information, but she made it clear that Safety Zone likes to know as much as possible about its competitors.

Thinking about it some more, you try to draw a line between information that is OK to discuss, and topics such as software specifics or strategy that should be considered private. This is the first time you have ever been in a situation like this. How will you handle it?

CHAPTER SUMMARY

During data and process modeling, a systems analyst develops graphical models to show how the system transforms data into useful information. The end product of data and process modeling is a logical model that will support business operations and meet user needs. Data and process modeling involves three main tools: data flow diagrams, a data dictionary, and process descriptions.

Data flow diagrams (DFDs) graphically show the movement and transformation of data in the information system. DFDs use four symbols: The process symbol transforms data; the data flow symbol shows data movement; the data store symbol shows data at rest; and the external entity symbol represents someone or something connected to the information system. Various rules and techniques are used to name, number, arrange, and annotate the set of DFDs to make them consistent and understandable.

A set of DFDs is like a pyramid with the context diagram at the top. The context diagram represents the information system's scope and its external connections but not its internal workings. Diagram 0 displays the information system's major processes, data stores, and data flows and is the exploded version of the context diagram's process symbol, which represents the entire information system. Lower-level DFDs show additional detail of the information system through the leveling technique of numbering and partitioning. Leveling continues until you reach the functional primitive processes, which are not decomposed further and are documented with process descriptions. All diagrams must be balanced to ensure their consistency and accuracy.

The data dictionary is the central documentation tool for structured analysis. All data elements, data flows, data stores, processes, entities, and records are documented in the data dictionary. Consolidating documentation in one location allows you to verify the information system's accuracy and consistency more easily and generate a variety of useful reports.

Each functional primitive process is documented using structured English, decision tables, and decision trees. Structured English uses a subset of standard English that defines each process with combinations of the basic building blocks of sequence, selection, and iteration. You also can document the logic by using decision tables or decision trees.

Structured analysis tools can be used to develop a logical model during one systems analysis phase, and a physical model during the systems design phase. Many analysts use a four-model approach, which involves a physical model of the current system, a logical model of the current system, a logical model of the new system, and a physical model of the new system.

Key Terms and Phrases

alias *217*
balancing *210*
black box *199*
black hole *200*
business logic *198*
business rules *198*
child diagram *209*
context diagram *205*
control structures *222*
data dictionary *215*
data element *215*
data flow *199*
data flow diagram (DFD) *198*
data item *215*
data repository *215*
data store *201*
data structures *215*
decision table *224*
decision tree *226*
decomposing *210*
diagram 0 *207*
diverging data flow *209*
domain *218*
entity *203*
exploding *210*
field *215*
four-model approach *228*
functional primitive *209*

Gane and Sarson *198*
gray hole *201*
iteration *223*
length *217*
leveling *210*
logical model *196*
logical structures *222*
looping *223*
modular design *222*
parent diagram *209*
partitioning *210*
physical model *196*
process *198*
process 0 *205*
process description *222*
pseudocode *223*
records *215*
selection *222*
sequence *222*
sink *203*
source *203*
spontaneous generation *200*
structured English *223*
terminators *203*
type *217*
validity rules *218*
Yourdon *198*

Learn It Online

Instructions: To complete the Learn It Online exercises, start your browser, click the Address bar, and then enter the Web address **scsite.com/sad8e/learn**. When the Systems Analysis and Design Learn It Online page is displayed, follow the instructions in the exercises below. Each exercise has instructions for saving your results, either for your own records or for submission to your instructor.

1 Chapter Reinforcement

TF, MC, and SA

Below SAD Chapter 5, click one of the Chapter Reinforcement links for Multiple Choice, True/False, or Short Answer. Answer each question and submit to your instructor.

2 Flash Cards

Below SAD Chapter 5, click the Flash Cards link and read the instructions. Type 20 (or a number specified by your instructor) in the Number of playing cards text box, type your name in the Enter your Name text box, and then click the Flip Card button. When the flash card is displayed, read the question and then click the ANSWER box arrow to select an answer. Flip through the Flash Cards. If your score is 15 (75%) correct or greater, click Print on the File menu to print your results. If your score is less than 15 (75%) correct, then redo this exercise by clicking the Replay button.

3 Practice Test

Below SAD Chapter 5, click the Practice Test link. Answer each question, enter your first and last name at the bottom of the page, and then click the Grade Test button. When the graded practice test is displayed on your screen, click Print on the File menu to print a hard copy. Continue to take practice tests until you score 80% or better.

4 Who Wants To Be a Computer Genius?

Below SAD Chapter 5, click the Computer Genius link. Read the instructions, enter your first and last name at the bottom of the page, and then click the Play button. When your score is displayed, click the PRINT RESULTS link to print a hard copy.

5 Wheel of Terms

Below SAD Chapter 5, click the Wheel of Terms link. Read the instructions, and then enter your first and last name and your school name. Click the PLAY button. When your score is displayed on the screen, right-click the score and then click Print on the shortcut menu to print a hard copy.

6 Crossword Puzzle Challenge

Below SAD Chapter 5, click the Crossword Puzzle Challenge link. Read the instructions, and then click the Continue button. Work the crossword puzzle. When you are finished, click the Submit button. When the crossword puzzle is redisplayed, submit it to your instructor.

Case-Sim: SCR Associates

Background

SCR Associates is a consulting firm that offers IT solutions and training. SCR needs an information system to manage operations at its new training center. The new system will be called TIMS (Training Information Management System). As a newly hired systems analyst, you will report to Jesse Baker, systems group man-

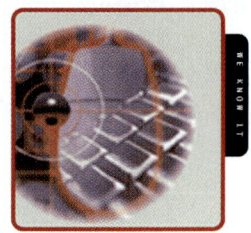

ager. You will work on various tasks and practice the skills you learned in this chapter.

Using the Case

The SCR Associates case study is a Web-based simulation that allows you to practice your skills in a real-world environment. The case study transports you to SCR's company intranet, where you can complete 12 work sessions, each aligning with a chapter. As you work on the case, you will receive e-mail and voice mail messages, obtain information from SCR's online libraries, and perform various tasks. The first time you enter the SCR Case, you should go to the starting page at **scsite.com/sad8e/scr** for detailed instructions.

Preview: Session 5

You recently completed requirements modeling tasks for the new Training Information Management System (TIMS). Now you are ready to begin data and process modeling, which will produce a logical model of the new system. You will create DFDs, develop a data dictionary, and use decision tables and trees.

To start a work session, you log on to the SCR intranet. When you enter your name and password, an opening screen displays links to the work sessions, and you select Session 5. At this point, you check your e-mail and voice mail messages carefully. Then you begin working on your task list, which includes the following items:

Tasks: Data and Process Modeling

1. *Jesse wants to see a context diagram and a diagram 0 DFD for TIMS.*

2. *Need to review the JAD session summary again! Try to identify four main TIMS functions and draw a lower-level DFD for each process.*

3. *Prepare a reply to Jesse's message about CASE tools. Search the Internet to find two more alternatives.*

4. *Prepare a decision table and a decision tree that show the logical rules described in Jesse's message about fees and discounts.*

FIGURE 5-40   Task list: Session 5.

Chapter Exercises

Review Questions

1. Describe data and process modeling, and name the main data and process modeling techniques.
2. Describe the Gane and Sarson symbols used for processes, data flows, data stores, and entities. Give four examples of typical names for processes, data flows, data stores, and entities.
3. What is the relationship between a context diagram and diagram 0, and which symbol is not used in a context diagram?
4. What is meant by an exploded DFD?
5. Describe a data dictionary and give examples of how and when it is used.
6. Explain the DFD leveling technique.
7. What is a balanced DFD?
8. Describe the steps in creating a decision table.
9. Discuss the pros and cons of decision tables versus decision trees.
10. What is structured English?

Discussion Topics

1. Suppose you were assigned to develop a logical model of the registration system at a school or college. Would you be better off using a top-down approach, or would a bottom-up strategy be better? What would influence your decision?
2. Some systems analysts find it better to start with a decision table, then construct a decision tree. Others believe it is easier to do it in the reverse order. Which do you prefer? Why?
3. A systems analyst attended a weeklong workshop on structured analysis. When she returned to her job, she told her boss that structured analysis was not worth the time to learn and use on the job. Her view was that it was too academic and had too much new terminology to be useful in a practical setting. Do you agree or disagree? Defend your position.
4. This chapter describes a black box concept that allows more detail to be shown as a process is exploded. Can the concept be applied in business management generally, or is it limited to information systems design? Provide reasons and examples with your answer.

Projects

1. Draw a context diagram and a diagram 0 DFD that represent the registration system at your school or an imaginary school.
2. On the Internet, locate at least three firms that offer CASE tools. Write e-mail messages to the companies to find out whether they offer demonstration copies or student versions of their products.
3. Suppose that you want to demonstrate a decision table to someone who has never seen one. Think of an example, with two or three conditions, from everyday life. Draw a decision table that captures all possible outcomes.
4. The data flow symbols shown on page 199 were designed by Ed Yourdon, a well-known IT author, lecturer, and consultant. Many IT professionals consider him to be among the most influential men and women in the software field. Learn more about Mr. Yourdon by visiting his Web site at www.yourdon.com, and write a brief review of his accomplishments.

Apply Your Knowledge

The Apply Your Knowledge section contains four mini-cases. Each case describes a situation, explains your role in the case, and asks you to respond to questions. You can answer the questions by applying knowledge you learned in the chapter.

1 Digital Consulting

Situation:

You are a senior systems analyst at Digital Consulting, a growing IT consulting firm. You are leading the development team for a major client. You need to explain structured analysis to your two newly hired junior analysts (Sara and Mike) before meeting with the client tomorrow afternoon.

1. Describe the rules for creating a context diagram.
2. Make a basic list of dos and don'ts when developing DFDs.
3. Explain the importance of leveling and balancing.
4. Ask Sara and Mike to review the order system context diagram on page 206, and compare it with the order system diagram 0 DFD on page 210. Then ask them to answer the following questions: (a) How many external entities are shown in each diagram? (b) In each diagram, how many data flows connect to the external entities? (c) How many subprocesses are identified in the diagram 0 DFD? (d) Could the data store have been shown in the context diagram? Why or why not?

2 Precision Tools

Situation:

Precision Tools sells a line of high-quality woodworking tools. When customers place orders on the company's Web site, the system checks to see if the items are in stock, issues a status message to the customer, and generates a shipping order to the warehouse, which fills the order. When the order is shipped, the customer is billed. The system also produces various reports.

1. Draw a context diagram for the order system.
2. Draw a diagram 0 DFD for the order system.
3. Name four attributes that you can use to define a process in the order system.
4. Name four attributes that you can use to define an entity in the order system.

3 Claremont School

Situation:

The Claremont School course catalog reads as follows: "To enroll in CIS 288, which is an advanced course, a student must complete two prerequisites — CIS 110 and CIS 286. A student who completes either one of these prerequisites and obtains the instructor's permission, however, will be allowed to take CIS 288."

1. Create a decision table that describes the Claremont School course catalog regarding eligibility for CIS 288. Show all possible rules.
2. Simplify the table you just created. Describe the results.
3. Draw a simplified decision tree to represent the Claremont School catalog. Describe the results.
4. Why might you use a decision tree rather than a decision table?

4 City Bus Lines

Situation:

City Bus Lines is developing an information system that will monitor passenger traffic, peak travel hours, and equipment requirements. The IT manager wants you to document a process called BALANCE that determines whether extra buses currently are needed on a particular route. The BALANCE process automatically assigns additional buses to that route, but *only* if all other routes are operating on schedule. In any case, a supervisor can override the automatic BALANCE process if he or she so desires.

1. Create a decision table that describes the bus transfer process.
2. Draw a decision tree that describes the bus transfer process.
3. Name four attributes that you can use to define a data flow in the bus information system.
4. Name four attributes that you can use to define a data store in the bus information system.

Case Studies

Case studies allow you to practice specific skills learned in the chapter. Each chapter contains several case studies that continue throughout the textbook, and a chapter capstone case.

NEW CENTURY HEALTH CLINIC

New Century Health Clinic offers preventive medicine and traditional medical care. In your role as an IT consultant, you will help New Century develop a new information system.

Background

You began the systems analysis phase at New Century Health Clinic by completing a series of interviews, reviewing existing reports, and observing office operations. (Your instructor may provide you with a sample set of interview summaries.)

As you learned, the doctors, nurses, and physical therapists provide services and perform various medical procedures. All procedures are coded according to Current Procedure Terminology, which is published by the American Medical Association. The procedure codes consist of five numeric digits and a two-digit suffix, and are used for all billing and insurance claims.

From your fact-finding, you determined that seven reports are required at the clinic. The first report is the daily appointment list for each provider. The list shows all scheduled appointment times, patient names, and services to be performed, including the procedure code and description. A second daily report is the call list, which shows the patients who are to be reminded of their next day's appointments. The call list includes the patient name, telephone number, appointment time, and provider name. The third report is the weekly provider report that lists each of the providers and the weekly charges generated, plus a month-to-date (MTD) and a year-to-date (YTD) summary.

The fourth report is the statement — a preprinted form that is produced monthly and mailed in a window envelope. Statement header information includes the statement date, head of household name and address, the previous month's balance, the total household charges MTD, the total payments MTD, and the current balance. The bottom section of the statement lists all activity for the month in date order. For each service performed, a line shows the patient's name, the service date, the procedure code and description, and the charge. The statement also shows the date and amount of all payments and insurance claims. When an insurance payment is received, the source and amount are noted on the form. If the claim is denied or only partially paid, a code is used to explain the reason. A running balance appears at the far right of each activity line.

The associates also require two insurance reports: the weekly Insurance Company Report and the monthly Claim Status Summary. In addition to these six reports, the office staff would like to have mailing labels and computer-generated postcards for sending reminders to patients when it is time to schedule their next appointment. Reminders usually are mailed twice monthly. Now you are ready to organize the facts you gathered and prepare a system requirements document that represents a logical model of the proposed system. Your tools will include DFDs, a data dictionary, and process descriptions.

Assignments

1. Prepare a context diagram for New Century's information system.
2. Prepare a diagram 0 DFD for New Century. Be sure to show numbered processes for handling appointment processing, payment and insurance processing, report processing, and records maintenance. Also, prepare lower-level DFDs for each numbered process.

3. Prepare a list of data stores and data flows needed for the system. Under each data store, list the data elements required.

4. Prepare a data dictionary entry and process description for one of the system's functional primitives.

PERSONAL TRAINER, INC.

Personal Trainer, Inc., owns and operates fitness centers in a dozen Midwestern cities. The centers have done well, and the company is planning an international expansion by opening a new "supercenter" in the Toronto area. Personal Trainer's president, Cassia Umi, hired an IT consultant, Susan Park, to help develop an information system for the new facility. During the project, Susan will work closely with Gray Lewis, who will manage the new operation.

Background

Susan Park has completed a preliminary investigation and performed the fact-finding tasks that were described in Chapters 2 and 4. Now, she will use the results to develop a logical model of the proposed information system.

Assignments

Before you perform the following tasks, you should review the information provided in Chapters 2 and 4 of the case.

1. Prepare a context diagram for the new system.
2. Prepare a diagram 0 DFD for the new system.
3. Write a brief memo that explains the importance of leveling a set of DFDs.
4. Write a brief memo that explains the importance of balancing a set of DFDs.

CHAPTER CAPSTONE CASE: SoftWear, Limited

SoftWear, Limited (SWL), is a continuing case study that illustrates the knowledge and skills described in each chapter. In this case study, the student acts as a member of the SWL systems development team and performs various tasks.

Background

Rick Williams, a systems analyst, and Carla Moore, a programmer/analyst, continued their work on the SWL payroll system project. After completing detailed interviews and other fact-finding activities, Rick and Carla now understand how the current system operates and the new requirements desired by users. They are ready to organize and document their findings by preparing a logical model of the payroll system.

Data Flow Diagrams

After they completed the preliminary investigation, Rick and Carla felt that they knew more about the system entities and how they interacted.

The two analysts knew that the payroll department issues paychecks based on timesheet data submitted by department heads, and that each employee receives a W-2 form at the end of the year. They also knew that the human resources department prepares employee status changes, and the payroll department enters the pay data. The diagram also noted the output of state and federal government reports and internal reports to SWL's finance and payroll departments. The credit union and the SWL stock transfer department reports and fund transfers also were included.

Using this information, Rick and Carla prepared a sketch of a context diagram and scheduled a meeting for the next day with Amy Calico, director of the payroll department, to discuss the diagram. At the meeting, Amy made several comments:

- The human resources department would be setting up additional ESIP deduction choices for employees under a new 401(k) plan. Human resources also would receive ESIP reports from the payroll system.

- The payroll department enters timesheet data received from department heads, who do not interact directly with the system. Rather than showing the department head entity symbol on the context diagram, the input data flow from the payroll department should be expanded and called PAY DETAIL.

- State and federal reporting requirements differ, so they should be treated as two separate entities. Also, periodic changes in government tax rates should be shown as inputs to the payroll system.

- All accounting reports, except for an overall financial summary, should be distributed to the accounting department instead of to the finance department. The accounting department also should receive a copy of the payroll report.

- The bank returns cleared payroll checks to the payroll department once a month. Amy reminded the analysts that the payroll system handles the reconciliation of payroll checks.

After discussing Amy's comments, Rick and Carla prepared the final version of the payroll system context diagram shown in Figure 5-41.

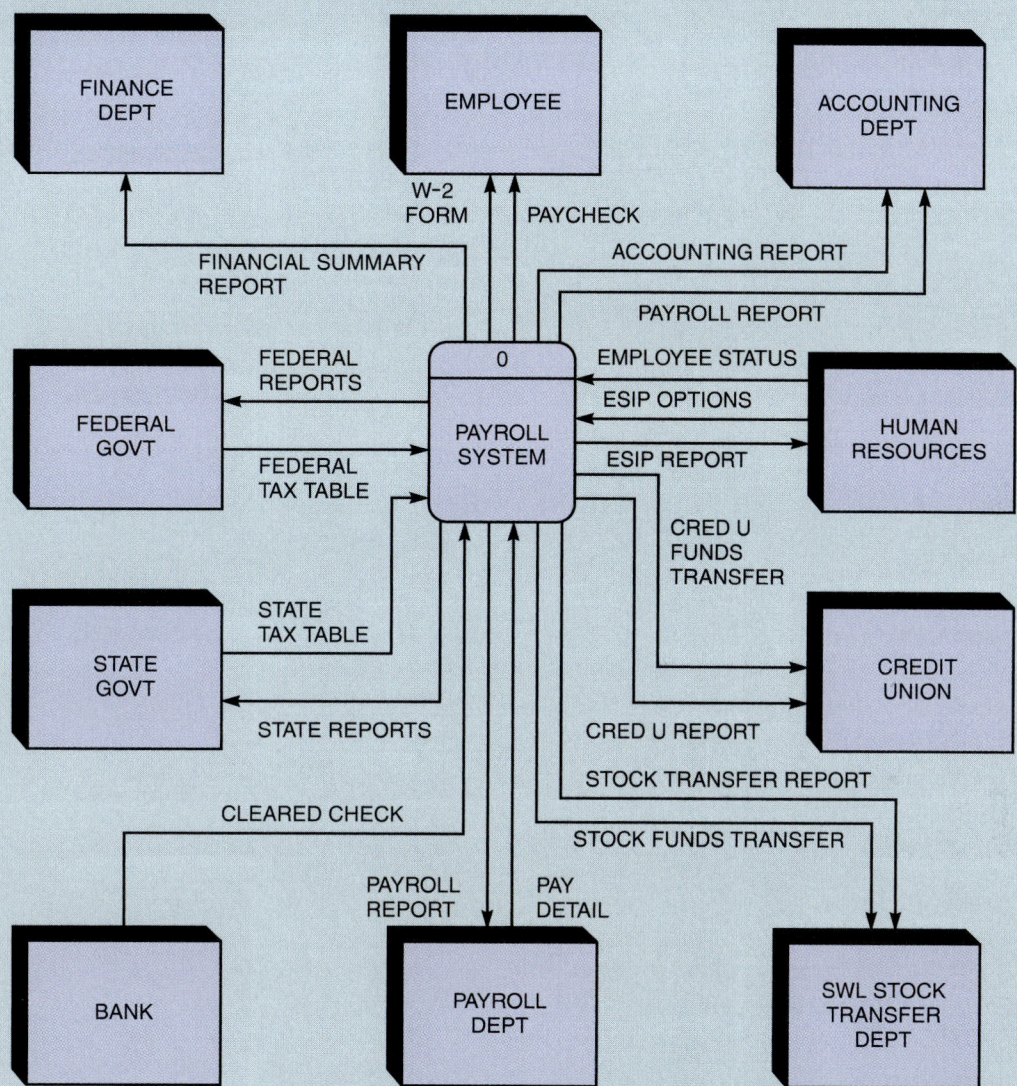

FIGURE 5-41 Final context diagram for SoftWear, Limited's payroll system.

CHAPTER CAPSTONE CASE: SoftWear, Limited (continued)   SWL

While their conversation with Amy Calico still was fresh in her mind, Carla proposed that they construct the diagram 0 DFD. After going through several draft versions, they completed the diagram 0 shown in Figure 5-42. They identified four processes: the check reconciliation subsystem, the pay employee subsystem, the payroll accounting subsystem, and a subsystem that would handle all voluntary deductions, which they called the ESIP deduction subsystem.

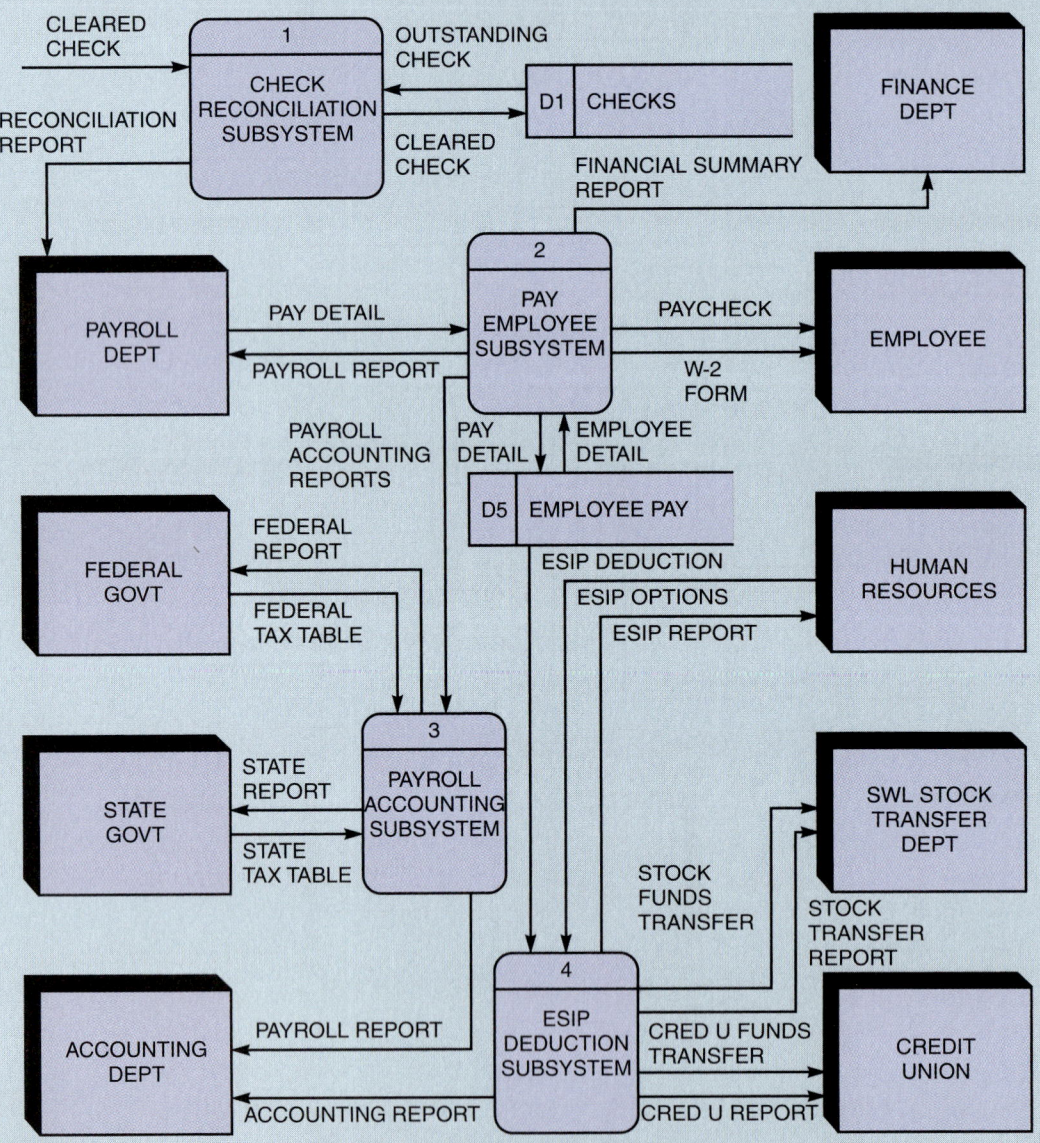

FIGURE 5-42   Diagram 0 DFD for SoftWear, Limited's payroll system.

CHAPTER CAPSTONE CASE: SoftWear, Limited (continued)   SWL

Over the next few days, Rick concentrated on partitioning the pay employee subsystem and the ESIP subsystem, while Carla developed the lower-level diagrams for the other two subsystems.

At that point, Rick considered the problem of applying certain deductions on a monthly cycle, even though the deductions were made weekly. To provide flexibility, he decided to use two separate processes, as shown in Figure 5-43. When he finished, his diagram 4 DFD contained the two processes EXTRACT DEDUCTION and APPLY DEDUCTION, as well as a local data store, UNAPPLIED DEDUCTIONS. Several local data flows also were included. The first process, EXTRACT DEDUCTION, would deduct the proper amount in each pay period. The deductions would be held in the temporary data store

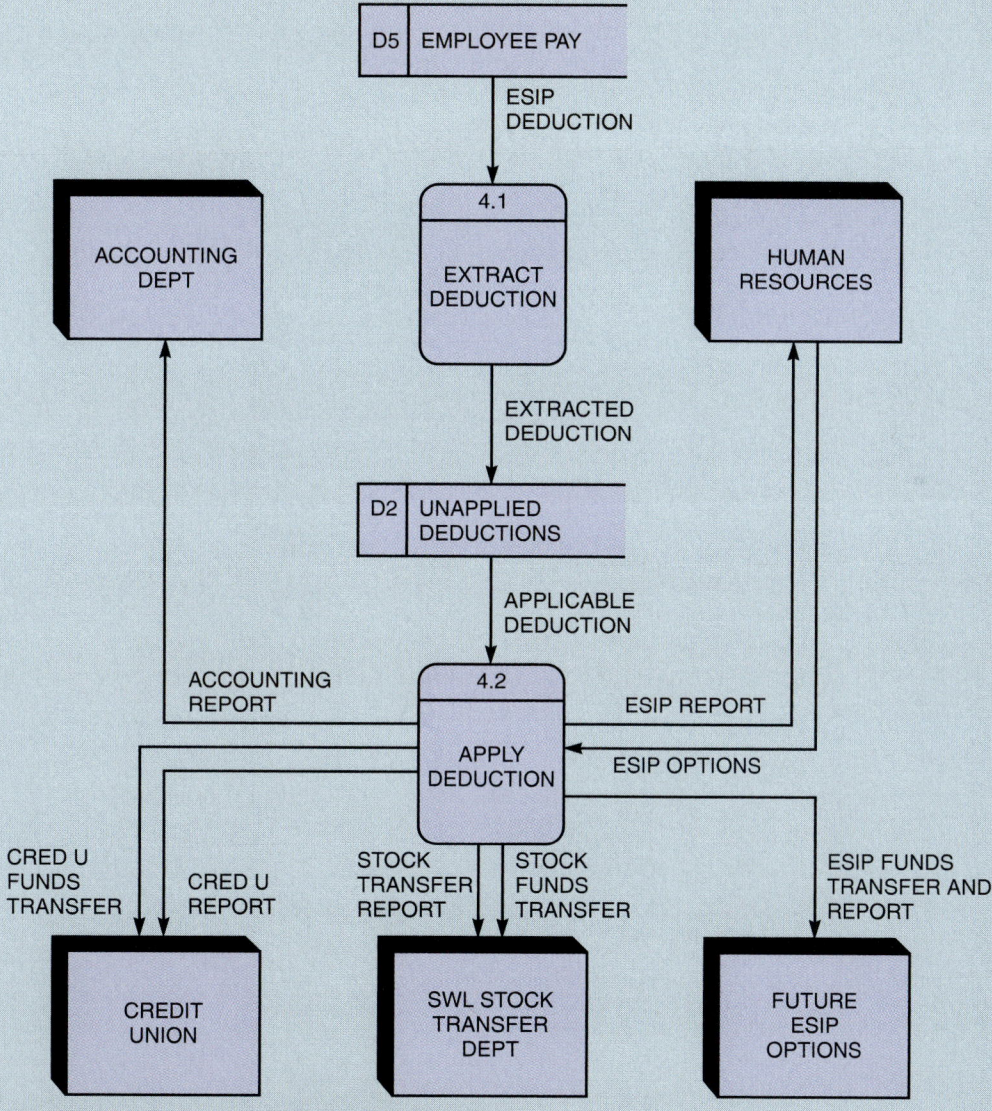

FIGURE 5-43 Diagram 4 DFD for SoftWear, Limited's payroll system shows the detail of process 4, the ESIP DEDUCTION SUBSYSTEM.

CHAPTER CAPSTONE CASE: SoftWear, Limited (continued)   SWL

and then applied in the APPLY DEDUCTION process on a weekly or monthly basis, depending on the type of deduction. Rick decided that those processes were functional primitives and he did not need to partition them further. That task completed the logical model of the new SWL payroll system.

Rick and Carla also considered the physical design of the ESIP deduction subsystem that would be completed later. They knew that it would be necessary to add some new forms and to redesign others. They saw that the human resources department would need a new form for enrollments or deduction changes for the credit union, SWL stock purchase plan, or any new ESIP choices that might be offered in the future. The payroll department then could use the form as its official notification. To provide for future expansion and add flexibility, the human resources department also would need a form to notify payroll of any new type of deduction, with a deduction code, the name of the deduction, and the payroll cycle involved. Rick anticipated that the new system would eliminate problems with improper deductions, while adding flexibility and reducing maintenance costs.

Data Dictionary and Process Descriptions

As they constructed the DFDs for the payroll system, Rick and Carla also developed the data dictionary entries with supporting process descriptions.

Using the Visible Analyst CASE tool, Rick documented the PROCESS 4 ESIP deduction subsystem shown in Figure 5-44. Then he defined the data flow called ESIP REPORT that originates in the APPLY DEDUCTION process and connects to the HUMAN RESOURCES entity, as shown in Figure 5-45. He also documented the ESIP OPTIONS data store shown in Figure 5-46, which consists of eight data elements, six of which are visible in the figure. Rick and Carla then spent the next two days documenting the rest of the data flows and entities for the ESIP deduction subsystem, along with the data elements and data structures.

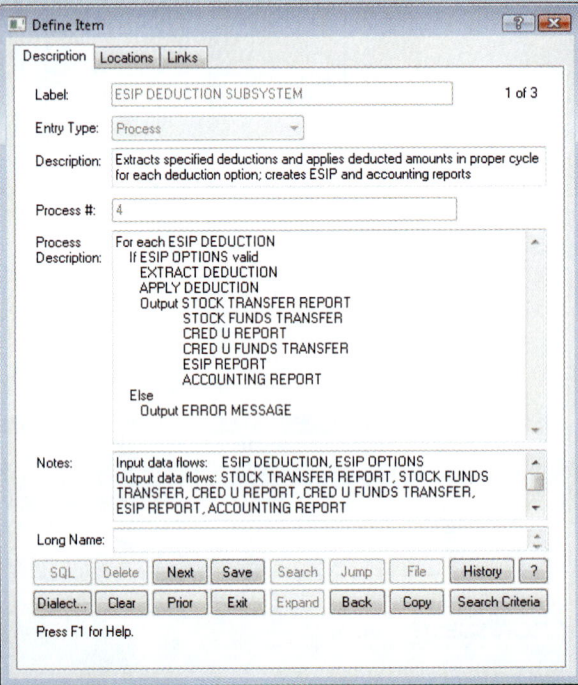

FIGURE 5-44 Data dictionary definition for process 4, the ESIP DEDUCTION SUBSYSTEM.

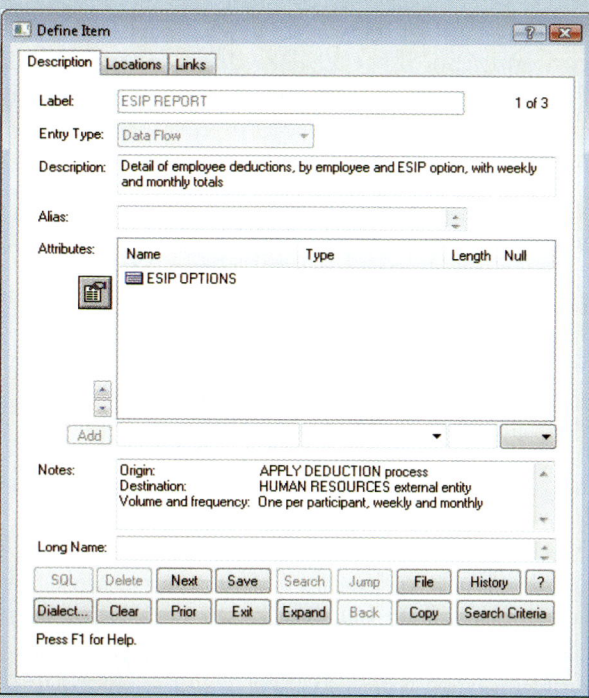

FIGURE 5-45 Data dictionary definition for the ESIP REPORT data flow.

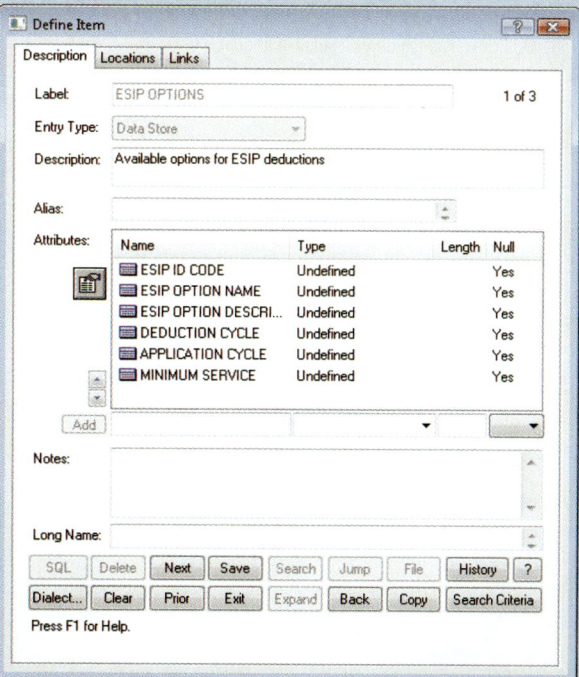

FIGURE 5-46 Data dictionary definition for the ESIP OPTIONS data store.

CHAPTER CAPSTONE CASE: SoftWear, Limited (continued)   SWL

After completing the documentation of the ESIP deduction subsystem, Carla and Rick met with Amy to review the logical model for the subsystem. After a thorough discussion of all proposed changes and processing, Amy approved the model.

Rick and Carla continued their analysis and documentation of the payroll system over the next several days. As they completed a model of a portion of the information system, they would meet with the appropriate users at SWL to review the model, obtain user input, make necessary adjustments to the model, and obtain the users' approval. After Rick and Carla finished the complete payroll information system logical model, they turned their attention to completing the rest of the system requirements document.

SWL Team Tasks

Suppose that you are working with Rick and Carla when a new systems request comes in. SWL's vice president of marketing, Amy Neal, wants to change the catalog mailing program and provide a reward for customers who use the Internet.

Amy's plan specifies that customers will remain on SWL's mailing list if they either requested a catalog, ordered from SWL in the last two years, or signed the guest register on SWL's new Web site. To encourage Internet visitors, customers who register on the Web site also will receive a special discount certificate.

To document the requirements, Rick wants you to design a decision table. Initially, it appears to have eight rules, but you notice that some of those rules are duplicates, or might not be realistic combinations.

1. Design the decision table with all possibilities.
2. Simplify the table by combining rules where appropriate.
3. Draw a decision tree that reflects Amy Neal's policy.
4. Create a set of structured English statements that accurately describes the policy.

Manage the SWL Project

You have been asked to manage SWL's new information system project. One of your most important activities will be to identify project tasks and determine when they will be performed. Before you begin, you should review the SWL case in this chapter. Then list and analyze the tasks, as follows:

LIST THE TASKS Start by listing and numbering at least 10 tasks that the SWL team needs to perform to fulfill the objectives of this chapter. Your list can include SWL Team Tasks and any other tasks that are described in this chapter. For example, Task 3 might be to Identify the system entities, and Task 6 might be to Draw a context diagram.

ANALYZE THE TASKS Now study the tasks to determine the order in which they should be performed. First identify all concurrent tasks, which are not dependent on other tasks. In the example shown in Figure 5-47, Tasks 1, 2, 3, 4, and 5 are concurrent tasks, and could begin at the same time if resources were available.

Other tasks are called dependent tasks, because they cannot be performed until one or more earlier tasks have been completed. For each dependent task, you must identify specific tasks that need to be completed before this task can begin. For example, you would want to identify the system entities before you could draw a context diagram, so Task 6 cannot begin until Task 3 is completed, as Figure 5-47 shows.

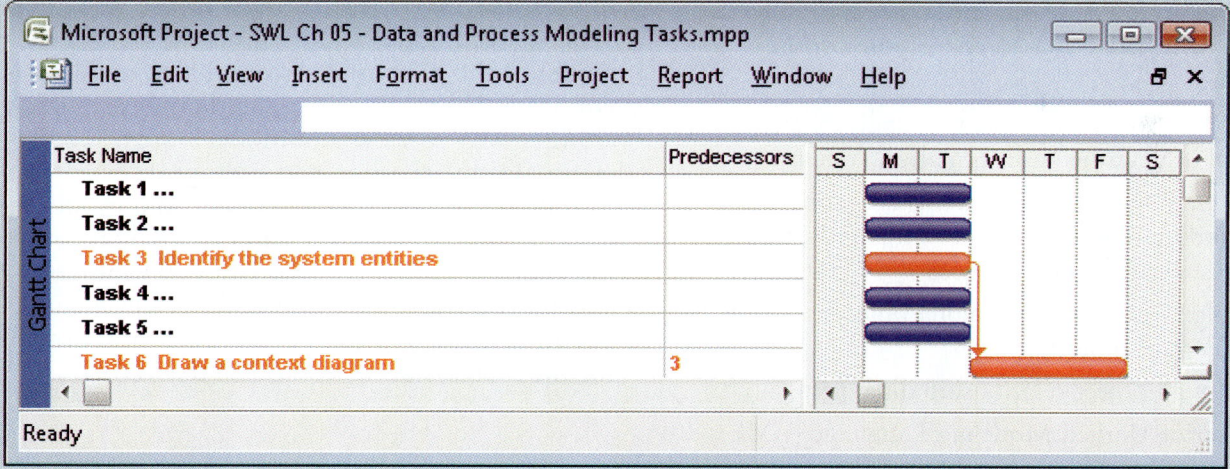

FIGURE 5-47 Tasks 1, 2, 3, 4, and 5 are concurrent tasks that could be performed at the same time. Task 6 is a dependent task that cannot be performed until Task 3 has been completed.

Chapter 3 describes project management tools, techniques, and software. To learn more, you can visit the Features section on your Student Study Tool CD-ROM, or the project management resources library at **scsite.com/sad8e/project**. On the Web, Microsoft offers demo versions, training, and tips for using Project 2007. You also can visit the OpenWorkbench.org site to learn more about this free, open-source software.

CHAPTER 6 Object Modeling

Chapter 6 is the third of four chapters in the systems analysis phase of the SDLC. This chapter discusses object modeling techniques that analysts use to create a logical model. In addition to structured analysis, object-oriented analysis is another way to represent and design an information system.

INTRODUCTION

OBJECTIVES

When you finish this chapter, you will be able to:

- Explain how object-oriented analysis can be used to describe an information system

- Define object modeling terms and concepts, including objects, attributes, methods, messages, classes, and instances

- Explain relationships among objects and the concept of inheritance

- Draw an object relationship diagram

- Describe Unified Modeling Language (UML) tools and techniques, including use cases, use case diagrams, class diagrams, sequence diagrams, state transition diagrams, and activity diagrams

- Explain the advantages of using CASE tools in developing the object model

- Explain how to organize an object model

In Chapter 5, you learned how to use structured analysis techniques to develop a data and process model of the proposed system. Now, in Chapter 6, you learn about object-oriented analysis, which is another way to view and model system requirements. In this chapter, you use object-oriented techniques to document, analyze, and model the information system. In Chapter 7, which concludes the systems analysis phase, you will evaluate alternatives, develop the system requirements document, learn about prototyping, and prepare for the systems design phase of the SDLC.

CHAPTER INTRODUCTION CASE: Mountain View College Bookstore

Background: Wendy Lee, manager of college services at Mountain View College, wants a new information system that will improve efficiency and customer service at the three college bookstores.

In this part of the case, Florence Fullerton (systems analyst) and Harry Boston (student intern) are talking about object-oriented concepts, tools, and techniques.

Participants:	Florence and Harry
Location:	Mountain View College Cafeteria, Tuesday afternoon, November 2, 2009
Project status:	Florence and Harry have completed data and processing modeling, and are discussing object-oriented techniques that they can use to develop an object model of the new system.
Discussion topics:	Object-oriented concepts, tools, and techniques

Florence: Hi, Harry. I want to chat with you about object-oriented analysis before we finish the systems analysis phase. Would this be a good time to talk?

Harry: *Sure. I know that object-oriented analysis is another way of viewing the system, but I don't know much about it.*

Florence: Well, object-oriented analysis describes an information system by identifying things called objects. An object represents a real person, place, event, or transaction. For example, in the bookstore, when a student purchases a textbook, the student is an object, the textbook is an object, and the purchase transaction itself is an object.

Harry: *That sounds a little like the entities we identify in structured analysis.*

Florence: Yes, but there's a major difference! In structured analysis we treat data and the processes that affect the data separately. Objects, on the other hand, contain the data *and* the processes, called methods, that can add, modify, or change the data. To make it even more interesting, one object can send a message to another object to request some action or response. For example, a driver object adjusts the cruise control, which sends one or more messages to the car object telling it to maintain a steady speed.

Harry: *I get it. So objects can be people, things, or events?*

Florence: Yes. To show how the system works, we use a special modeling language called the UML. We even show human actors as stick figures that interact with business functions called use cases.

Harry: *I'd like to give it a try.*

Florence: No problem. Although we'll still use structured analysis, it will be interesting to model the system in object-oriented terms. Let's get started on our task list:

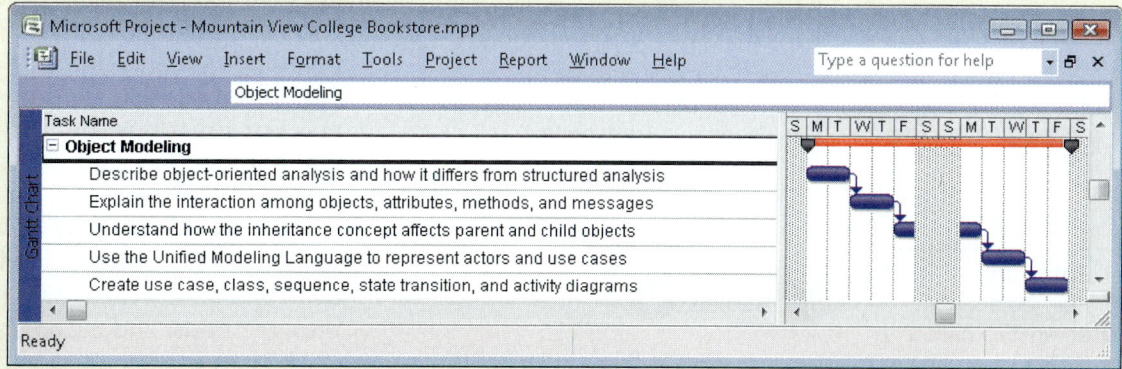

FIGURE 6-1 Typical object modeling task list.

OVERVIEW OF OBJECT-ORIENTED ANALYSIS

As you learned in Chapter 1, the most popular systems development options are structured analysis, object-oriented analysis (O-O), and agile methods, also called adaptive methods. The table in Figure 1-23 on page 18 shows the three alternatives and describes some pros and cons for each approach. As the table indicates, O-O methodology is popular because it integrates easily with object-oriented programming languages such as Java, Smalltalk, C++, and Perl. Programmers also like O-O code because it is modular, reusable, and easy to maintain.

Object-oriented (O-O) analysis describes an information system by identifying things called objects. An **object** represents a real person, place, event, or transaction. For example, when a patient makes an appointment to see a doctor, the patient is an object, the doctor is an object, and the appointment itself is an object.

Object-oriented analysis is a popular approach that sees a system from the viewpoint of the objects themselves as they function and interact. The end product of object-oriented analysis is an **object model**, which represents the information system in terms of objects and object-oriented concepts.

Object-Oriented Terms and Concepts

In Chapter 4, you learned that the **Unified Modeling Language (UML)** is a widely used method of visualizing and documenting an information system. In this chapter, you use the UML to develop object models. Your first step is to understand basic object-oriented terms, including objects, attributes, methods, messages, classes, and instances. In this chapter, you will learn how systems analysts use those terms to describe an information system.

An object represents a person, place, event, or transaction that is significant to the information system. In Chapter 5, you created DFDs that treated data and processes separately. An object, however, includes data *and* the processes that affect that data. For example, a customer object has a name, an address, an account number, and a current balance. Customer objects also can perform specific tasks, such as placing an order, paying a bill, and changing their address.

An object has certain **attributes**, which are characteristics that describe the object. For example, if you own a car, it has attributes such as make, model, and color. An object also has **methods**, which are tasks or functions that the object performs when it receives a **message**, or command, to do so. For example, your car performs a method called OPERATE WIPERS when you send a message by moving the proper control. Figure 6-2 shows examples of attributes, methods, and messages for a car object.

A **class** is a group of similar objects. For example, Chevy Cobalts belong to a class called CAR. An **instance** is a specific member of a class. *Your* Chevy Cobalt, therefore, is an instance of the CAR class. At an auto dealership, like the one shown in

ON THE WEB

For more information about object-oriented analysis, visit **scsite.com/ sad8e/more**, locate Chapter 6, and then click the Object-Oriented Analysis link.

ON THE WEB

For more information about the Unified Modeling Language, visit **scsite.com/ sad8e/more**, locate Chapter 6, and then click the Unified Modeling Language link.

Examples of Interaction Between Objects

Messages
The driver object sends messages to the car object, such as Clean the windshield or Slow down

Attributes
The car object has characteristics called attributes, such as make, model, and color

Methods
Operate wipers
Apply brakes

FIGURE 6-2 Objects have attributes, can send and receive messages, and perform actions called methods.

Figure 6-3, you might observe many instances of the CAR class, the TRUCK class, the MINIVAN class, and the SPORT UTILITY VEHICLE class, among others. Although the term "object" usually refers to a particular instance, systems analysts sometimes use the term to refer to a class of objects. Usually the meaning is understood from the context and the way the term is used.

FIGURE 6-3 At an auto dealership, you can observe the CAR class, the TRUCK class, the MINIVAN class, and the SPORT UTILITY VEHICLE class.

Objects

Consider how the UML describes a family with parents and children. The UML represents an object as a rectangle with the object name at the top, followed by the object's attributes and methods.

Figure 6-4 shows a PARENT object with certain attributes such as name, age, sex, and hair color. If there are two parents, then there are two instances of the PARENT object. The PARENT object can perform methods, such as reading a bedtime story, driving the car pool van, or preparing a school lunch. When a PARENT object receives a message, it performs an action, or method. For example, the message GOOD NIGHT from a child might tell the PARENT object to read a bedtime story, while the message DRIVE from another parent signals that it is the PARENT object's turn to drive in the car pool.

Continuing with the family example, the CHILD object in Figure 6-5 possesses the same attributes as the PARENT object and an additional attribute that shows the number of siblings. A CHILD object performs certain methods, such as picking up toys, eating

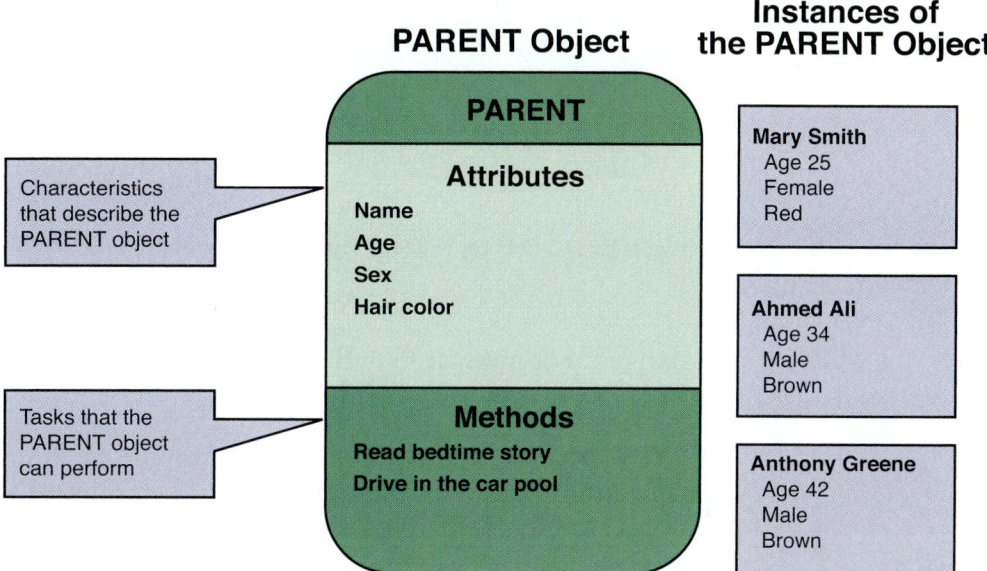

FIGURE 6-4 The PARENT object includes four attributes and two methods. Mary Smith, Ahmed Ali, and Anthony Greene are instances of the PARENT object.

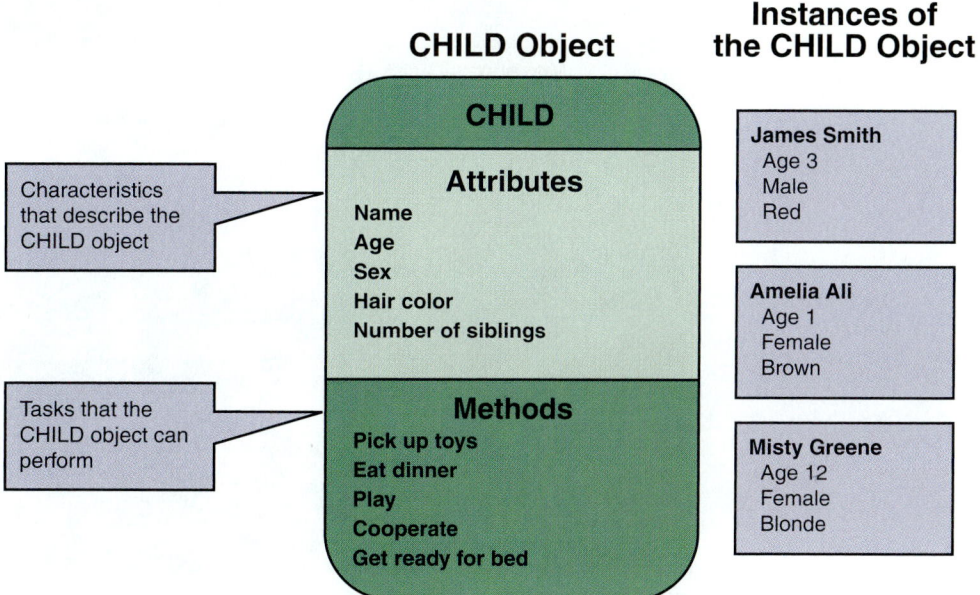

FIGURE 6-5 The CHILD object includes five attributes and five methods. James Smith, Amelia Ali, and Misty Greene are instances of the CHILD object.

DOG Object

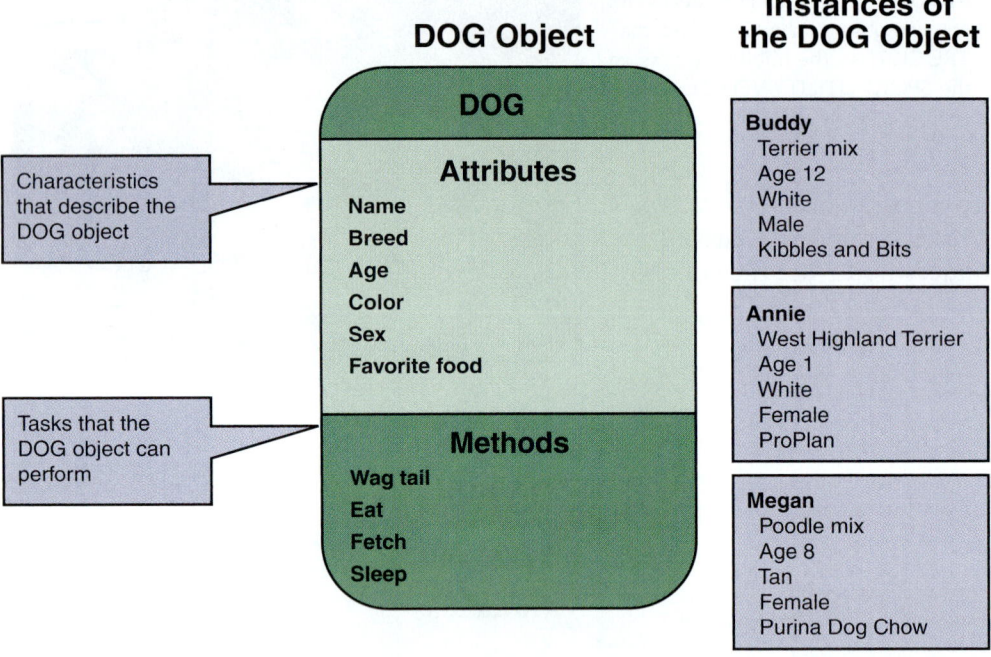

Instances of the DOG Object

Buddy
Terrier mix
Age 12
White
Male
Kibbles and Bits

Annie
West Highland Terrier
Age 1
White
Female
ProPlan

Megan
Poodle mix
Age 8
Tan
Female
Purina Dog Chow

FIGURE 6-6 The DOG object includes six attributes and four methods. Buddy, Annie, and Megan are instances of the DOG object.

dinner, playing, cooperating, and getting ready for bed. To signal the CHILD object to perform those tasks, a parent can send certain messages that the CHILD object will understand. For example, the DINNER'S READY message tells a CHILD object to come to the table, while the SHARE WITH YOUR BROTHER/SISTER message tells a CHILD object to cooperate with other CHILD objects.

The family also might have a DOG object, as shown in Figure 6-6. That object can have attributes such as name, breed, age, color, sex, and favorite food. The DOG object can perform methods such as wagging its tail, eating, fetching, and sleeping. The message GOOD DOG, when directed to the DOG object, signals it to wag its tail. Similarly, the DINNER'S READY message signals the DOG object to run to its food bowl.

Now consider an example of a fitness center, as shown in Figure 6-7, and the objects that interact with the fitness center's enrollment system. A typical fitness center might have students, instructors, fitness-class schedules, and a registration process.

FIGURE 6-7 A typical fitness center might have students, instructors, fitness-class schedules, and a registration process.

STUDENT and INSTRUCTOR objects are shown in Figure 6-8. Each STUDENT object has the following attributes: student number, name, address, telephone, date of birth, fitness record, and status. In addition, a STUDENT can add a fitness-class; drop a fitness-class; change an address, telephone, or status; and update his or her fitness record.

The INSTRUCTOR object in Figure 6-8 has the following attributes: instructor number, name, address, telephone, fitness-classes taught, availability, private lesson fee, and status. An INSTRUCTOR object can teach a fitness-class, and change his or her availability, address, telephone, private lesson fee, or status.

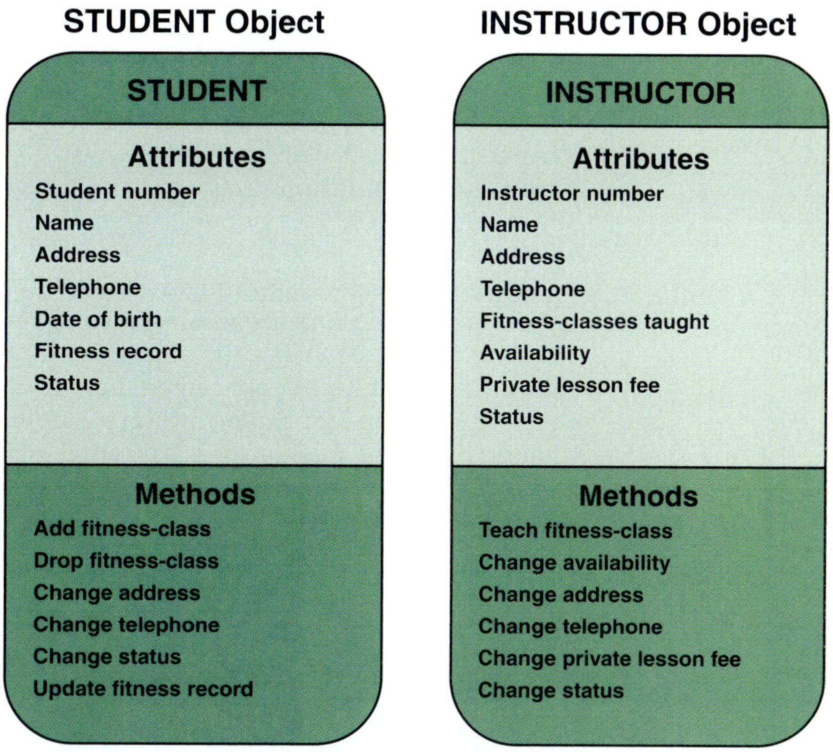

The FITNESS-CLASS SCHEDULE object shown in Figure 6-9 includes data about fitness classes, including fitness-class number, date, time, type, location, instructor number, and maximum enrollment. The FITNESS-CLASS SCHEDULE object includes the methods that can add or delete a fitness class, or change a fitness-class date, time, instructor, location, or enrollment.

The REGISTRATION RECORD object shown in Figure 6-10 on the next page includes the student number, fitness-class number, registration date, fee, and status. The REGISTRATION RECORD object includes methods to add a REGISTRATION instance when a student enrolls, or drop a REGISTRATION instance if the fitness class is canceled or for nonpayment. Notice that if a student registers for three fitness classes, the result is three instances of the REGISTRATION RECORD object. The REGISTRATION RECORD object also includes a method of notifying students and instructors of information.

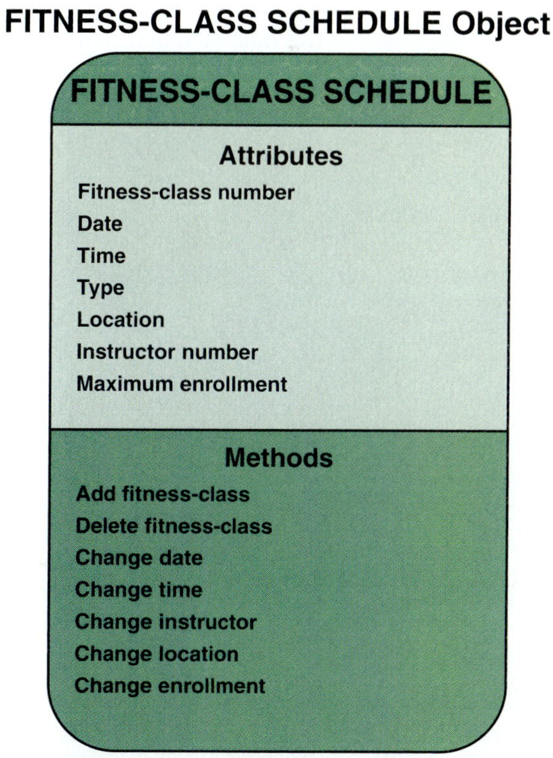

Attributes

If objects are similar to nouns, attributes are similar to adjectives that describe the characteristics of an object. How many attributes are needed? The answer depends on the business requirements of the information system and its users. Even a relatively simple object, such as an inventory item, might have a part number, description, supplier, quantity on hand, minimum stock level, maximum stock level, reorder time, and so on. Some objects might have a few attributes; others might have dozens.

STUDENT Object

INSTRUCTOR Object

STUDENT	INSTRUCTOR
Attributes	**Attributes**
Student number	Instructor number
Name	Name
Address	Address
Telephone	Telephone
Date of birth	Fitness-classes taught
Fitness record	Availability
Status	Private lesson fee
	Status
Methods	**Methods**
Add fitness-class	Teach fitness-class
Drop fitness-class	Change availability
Change address	Change address
Change telephone	Change telephone
Change status	Change private lesson fee
Update fitness record	Change status

FIGURE 6-8 The STUDENT object includes seven attributes and six methods. The INSTRUCTOR object includes eight attributes and six methods.

FITNESS-CLASS SCHEDULE Object

FITNESS-CLASS SCHEDULE
Attributes
Fitness-class number
Date
Time
Type
Location
Instructor number
Maximum enrollment
Methods
Add fitness-class
Delete fitness-class
Change date
Change time
Change instructor
Change location
Change enrollment

FIGURE 6-9 The FITNESS-CLASS SCHEDULE object includes seven attributes and seven methods.

Systems analysts define an object's attributes during the systems design process. In an object-oriented system, objects can inherit, or acquire, certain attributes from other objects. When you learn about relationships between objects and classes, you will understand how that occurs.

Objects can have a specific attribute called a **state**. The state of an object is an adjective that describes the object's current status. For example, depending on the state, a student can be a future student, a current student, or a past student. Similarly, a bank account can be active, inactive, closed, or frozen.

Methods

A method defines specific tasks that an object can perform. Just as objects are similar to nouns and attributes are similar to adjectives, methods resemble verbs that describe *what* and *how* an object does something.

Consider a server who prepares fries in a fast-food restaurant, as shown in Figure 6-11. A systems analyst might describe the operation as a method called MORE FRIES, as shown in Figure 6-12. The MORE FRIES method includes the steps required to heat the oil, fill the fry basket with frozen potato strips, lower it into the hot oil, check for readiness, remove the basket when ready and drain the oil, pour the fries into a warming tray, and add salt.

Figure 6-13 shows another example of a method. At the fitness center, an ADD STUDENT method adds a new instance of the STUDENT class. Notice that nine steps are required to add the new instance and record the necessary data.

REGISTRATION RECORD Object

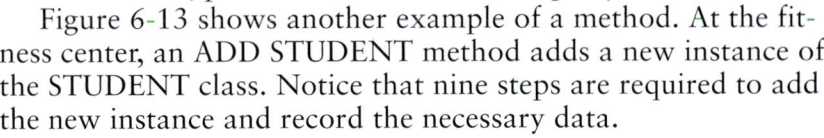

REGISTRATION RECORD

Attributes

Student number

Fitness-class number

Registration date

Fee

Status

Methods

Add student

Drop student

Notify instructor of add

Notify instructor of drop

Notify all of fitness-class cancellations

FIGURE 6-10 The REGISTRATION object includes five attributes and five methods.

FIGURE 6-11 In a fast-food restaurant, preparing more fries is a common task.

Method: MORE FRIES	Steps:
	1. Heat oil
	2. Fill fry basket with frozen potato strips
	3. Lower basket into hot oil
	4. Check for readiness
	5. When ready raise basket and let drain
	6. Pour fries into warming tray
	7. Add salt

FIGURE 6-12 The MORE FRIES method requires the server to perform seven specific steps.

Messages

A message is a command that tells an object to perform a certain method. For example, the message ADD STUDENT directs the STUDENT class to add a STUDENT instance. The STUDENT class understands that it should add the student number, name, and other data about that student, as shown in Figure 6-14. Similarly, a message named DELETE STUDENT tells the STUDENT class to delete a STUDENT instance.

The same message to two different objects can produce different results. The concept that a message gives different meanings to different objects is called **polymorphism**. For example, in Figure 6-15, the message GOOD NIGHT signals the PARENT object to read a bedtime story, but the same message to the DOG object tells the dog to sleep. The GOOD NIGHT message to the CHILD object signals it to get ready for bed.

You can view an object as a **black box**, because a message to the object triggers changes within the object without specifying how the changes must be carried out. A gas pump is an example of a black box. When you select the economy grade at a pump, you do not need to think about how the pump determines the correct price and selects the right fuel, as long as it does so properly.

The black box concept is an example of **encapsulation**, which means that all data and methods are self-contained. A black box does not want or need outside interference. By limiting access to internal processes, an object prevents its internal code from being altered by another object or process. Encapsulation allows objects to be used as modular components anywhere in the system, because objects send and receive messages but do not alter the internal methods of other objects.

Object-oriented designs typically are implemented with object-oriented programming languages. A major advantage of O-O designs is that systems analysts can save time and avoid errors by using modular objects, and programmers can translate the designs into code, working with reusable program modules that have been tested and verified.

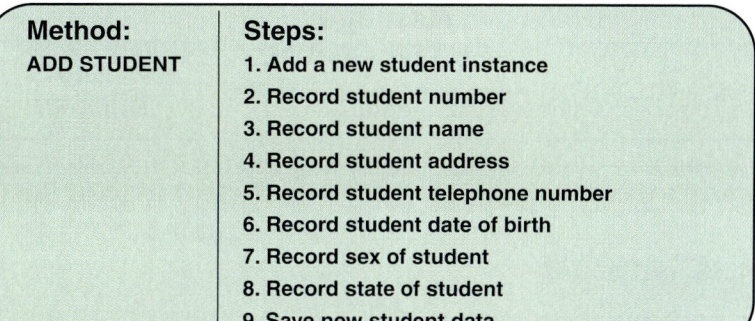

Method:	Steps:
ADD STUDENT	1. Add a new student instance
	2. Record student number
	3. Record student name
	4. Record student address
	5. Record student telephone number
	6. Record student date of birth
	7. Record sex of student
	8. Record state of student
	9. Save new student data

FIGURE 6-13 In the fitness center example, the ADD STUDENT method requires the STUDENT object to perform nine specific steps.

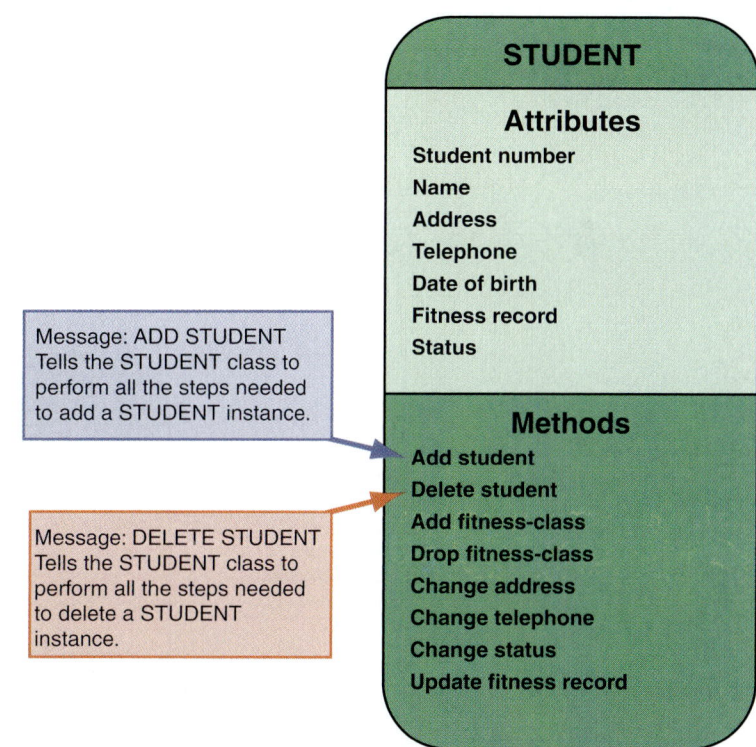

FIGURE 6-14 The message ADD STUDENT signals the STUDENT class to perform the ADD STUDENT method. The message DELETE STUDENT signals the STUDENT class to perform the DELETE STUDENT method.

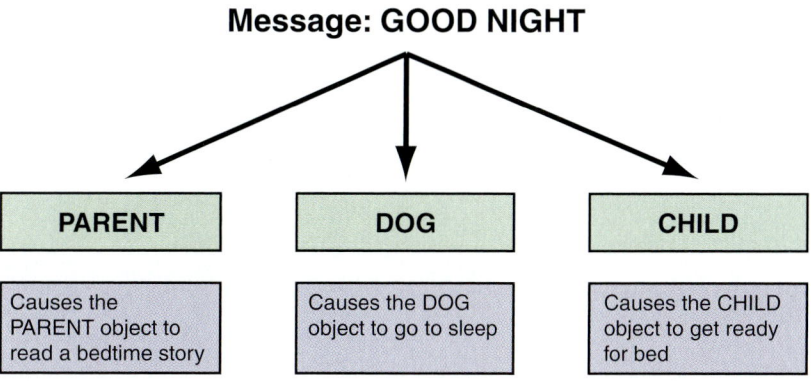

FIGURE 6-15 In an example of polymorphism, the message GOOD NIGHT produces different results, depending on which object receives it.

FIGURE 6-16 In a school information system, an INSTRUCTOR object sends an ENTER GRADE message to an instance of the STUDENT RECORD class.

ON THE WEB

For more information about polymorphism, visit **scsite.com/ sad8e/more**, locate Chapter 6, and then click the Polymorphism link.

For example, in Figure 6-16, an INSTRUCTOR object sends an ENTER GRADE message to an instance of the STUDENT RECORD class. Notice that the INSTRUCTOR object and STUDENT RECORD class could be reused, with minor modifications, in other school information systems where many of the attributes and methods would be similar.

Classes

An object belongs to a group or category called a class. All objects within a class share common attributes and methods, so a class is like a blueprint, or template for all the objects within the class. Objects within a class can be grouped into **subclasses**, which are more specific categories within a class. For example, TRUCK objects represent a subclass within the VEHICLE class, along with other subclasses called CAR, MINIVAN, and SCHOOL BUS, as shown in Figure 6-17. Notice that all four subclasses share common traits of the VEHICLE class, such as make, model, year, weight, and color. Each subclass also can possess traits that are uncommon, such as a load limit for the TRUCK or an emergency exit location for the SCHOOL BUS.

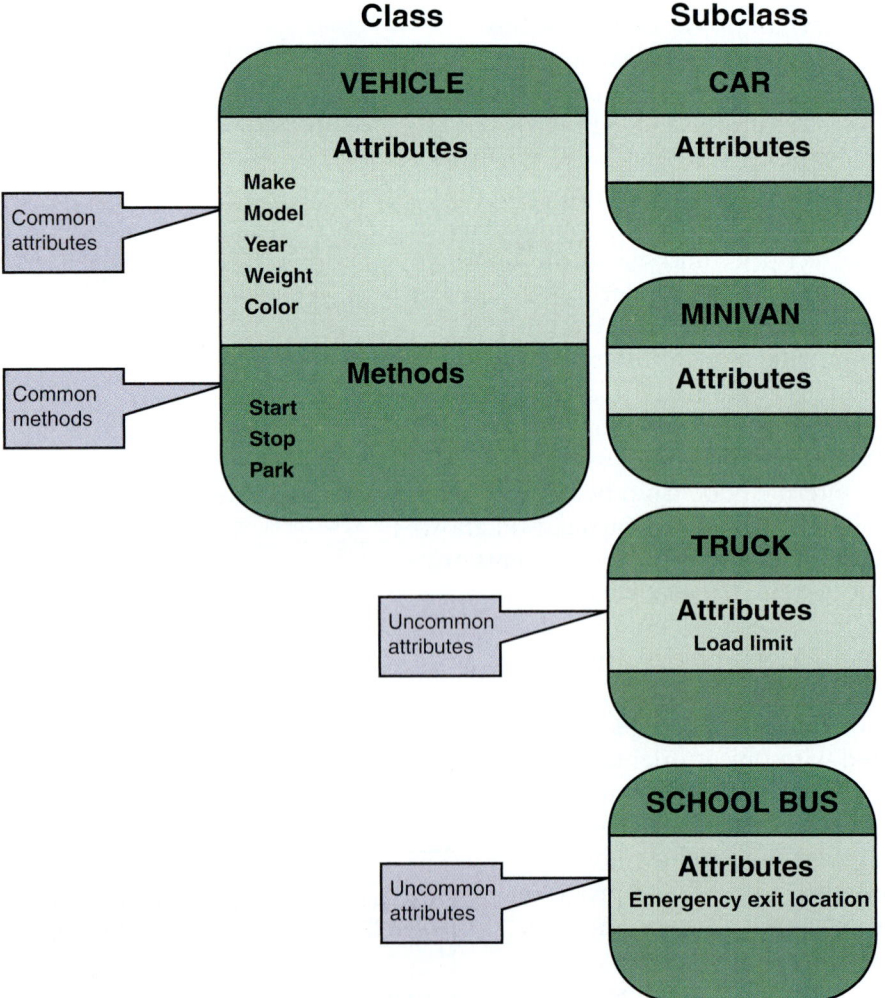

FIGURE 6-17 The VEHICLE class includes common attributes and methods. CAR, TRUCK, MINIVAN, and SCHOOL BUS are instances of the VEHICLE class.

In the fitness center example shown in Figure 6-18, INSTRUCTOR objects represent a subclass within the EMPLOYEE class. The EMPLOYEE class also can contain MANAGER and OFFICE STAFF subclasses, because a manager and staff members are employees. All INSTRUCTOR, MANAGER, and OFFICE STAFF objects contain similar information (such as employee name, title, and pay rate) and perform similar tasks (such as getting hired and changing an address or telephone number).

A class can belong to a more general category called a **superclass**. For example, a NOVEL class belongs to a superclass called BOOK, because all novels are books. The NOVEL class can have subclasses called HARDCOVER and PAPERBACK. Similarly, as shown in Figure 6-19, the EMPLOYEE class belongs to the PERSON superclass, because every employee is a person, and the INSTRUCTOR class is a subclass of EMPLOYEE.

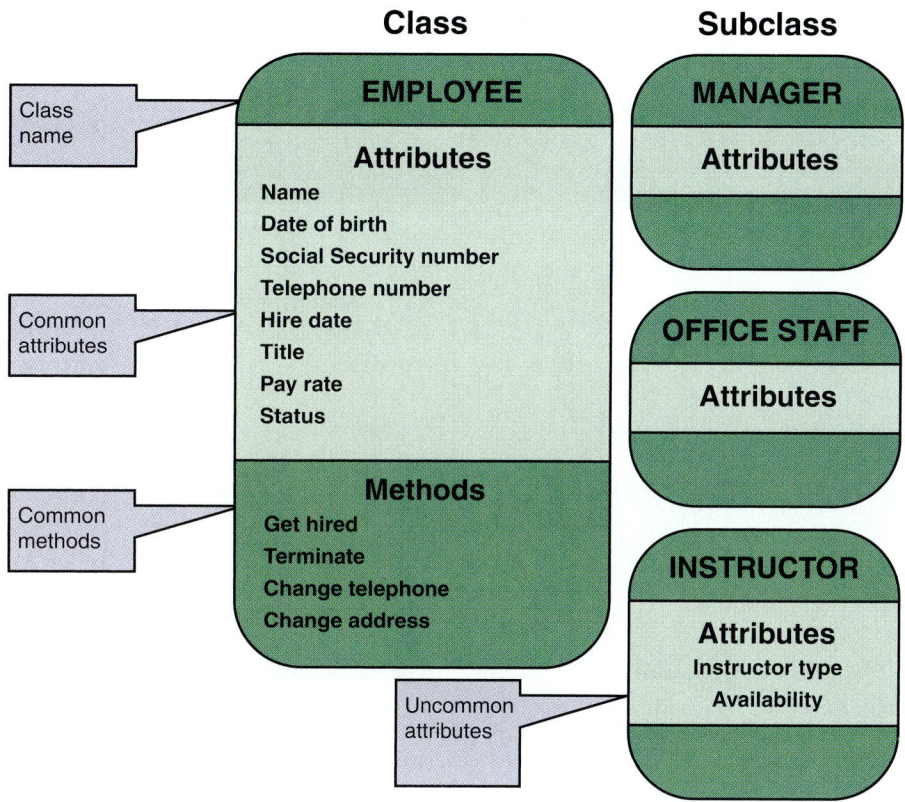

FIGURE 6-18 The fitness center EMPLOYEE class includes common attributes and methods. INSTRUCTOR, MANAGER, and OFFICE STAFF are subclasses within the EMPLOYEE class.

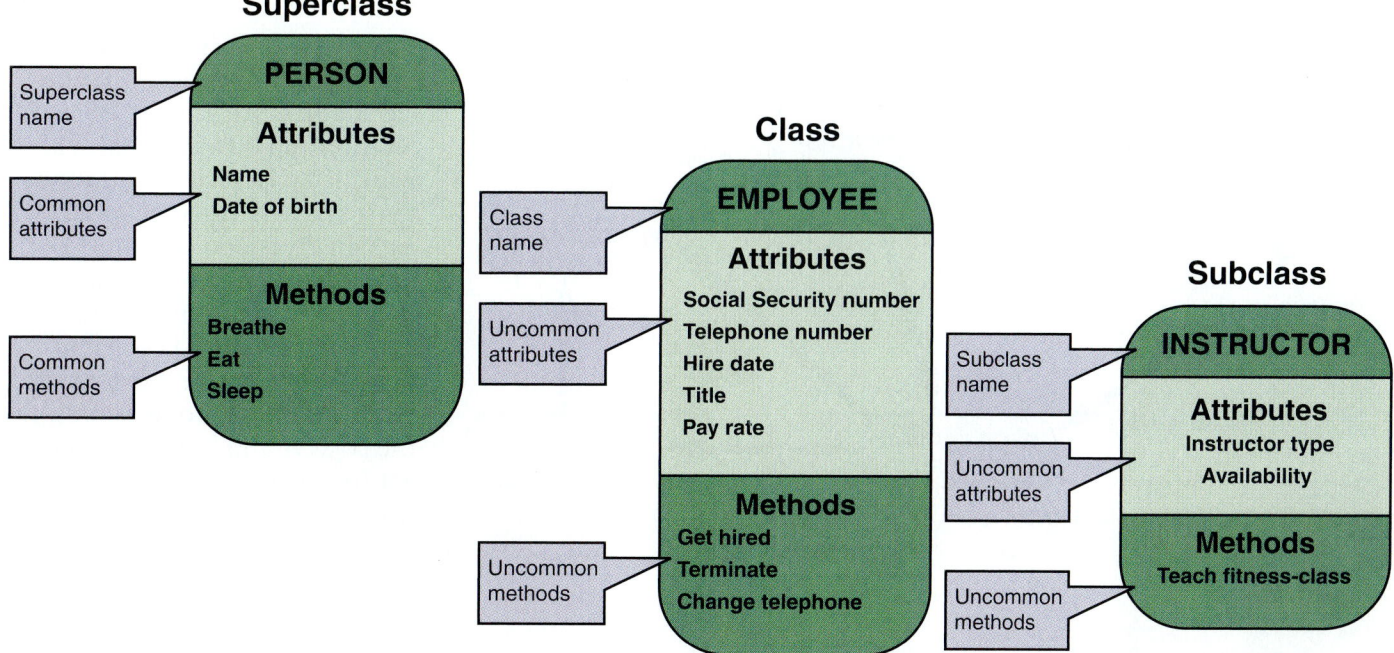

FIGURE 6-19 At the fitness center, the PERSON superclass includes common attributes and methods. EMPLOYEE is a class within the PERSON superclass. INSTRUCTOR is a subclass within the EMPLOYEE class.

Inheritance

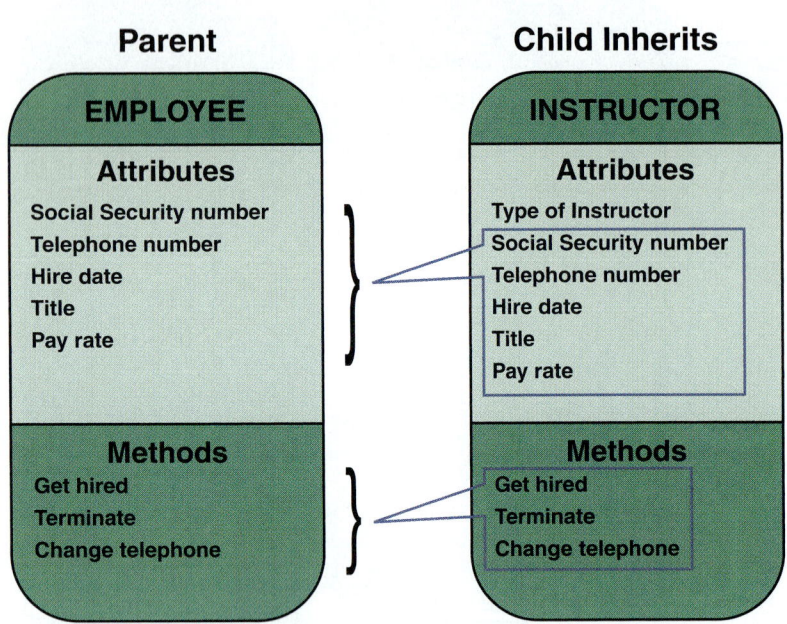

FIGURE 6-20 An inheritance relationship exists between the INSTRUCTOR and EMPLOYEE objects. The INSTRUCTOR (child) object inherits characteristics from the EMPLOYEE (parent) class and can have additional attributes of its own.

RELATIONSHIPS AMONG OBJECTS AND CLASSES

Relationships enable objects to communicate and interact as they perform business functions and transactions required by the system. Relationships describe what objects need to know about each other, how objects respond to changes in other objects, and the effects of membership in classes, superclasses, and subclasses. Some relationships are stronger than others (just as a relationship between family members is stronger than one between casual acquaintances). The strongest relationship is called inheritance. **Inheritance** enables an object, called a **child**, to derive one or more of its attributes from another object, called a **parent**. In the example in Figure 6-20, the INSTRUCTOR object (child) inherits many traits from the EMPLOYEE object (parent), including SOCIAL SECURITY NUMBER, TELEPHONE NUMBER, and HIRE DATE. The INSTRUCTOR object also can possess additional attributes, such as TYPE OF INSTRUCTOR. Because all employees share certain attributes, those attributes are assumed through inheritance and do not need to be repeated in the INSTRUCTOR object.

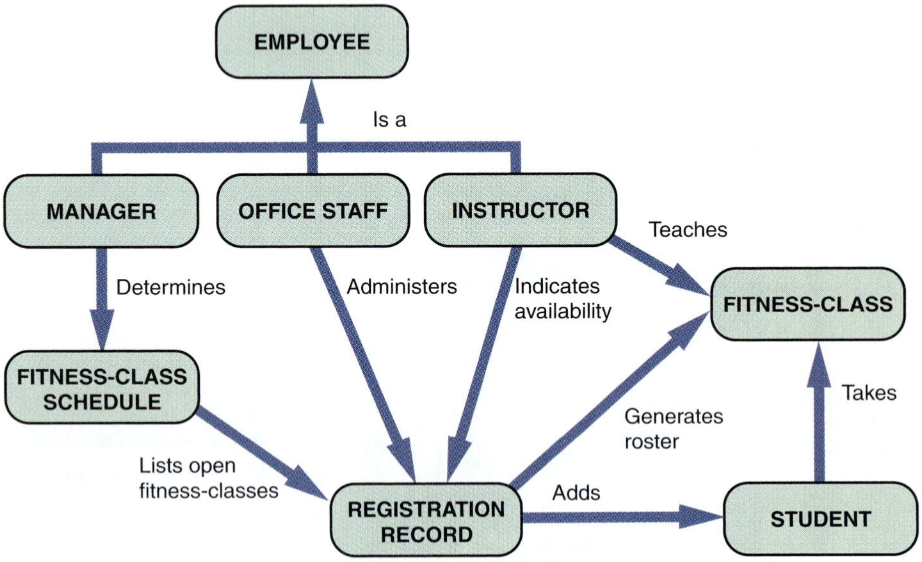

Object Relationship Diagram

After you identify the objects, classes, and relationships, you are ready to prepare an object relationship diagram that will provide an overview of the system. You will use that model as a guide as you continue to develop additional diagrams and documentation. Figure 6-21 shows an object relationship diagram for a fitness center. Notice that the model shows the objects and how they interact to perform business functions and transactions.

FIGURE 6-21 Object relationship diagram for the fitness center.

OBJECT MODELING WITH THE UNIFIED MODELING LANGUAGE

ON THE WEB

For more information about use case modeling, visit **scsite.com/sad8e/ more**, locate Chapter 6, and then click the Use Case Modeling link.

Just as structured analysis uses DFDs to model data and processes, systems analysts use the Unified Modeling Language (UML) to describe object-oriented systems.

In Chapter 4, you learned that the UML is a popular technique for documenting and modeling a system. The UML uses a set of symbols to represent graphically the various components and relationships within a system. Although the UML can be used for business process modeling and requirements modeling, it mainly is used to support object-oriented system analysis and to develop object models.

Use Case Modeling

A **use case** represents the steps in a specific business function or process. An external entity, called an **actor**, initiates a use case by requesting the system to perform a function or process. For example, in a medical office system, a PATIENT (actor) can MAKE APPOINTMENT (use case), as shown in Figure 6-22.

Notice that the UML symbol for a use case is an oval with a label that describes the action or event. The actor is shown as a stick figure, with a label that identifies the actor's role. The line from the actor to the use case is called an association, because it links a particular actor to a use case. Figure 6-23 shows use case examples of a passenger making an airline reservation, a customer placing an order, and a bus dispatcher changing a student's pickup address.

Use cases also can interact with other use cases. When the outcome of one use case is incorporated by another use case, we say that the second case *uses* the first case. The UML indicates the relationship with a hollow-headed arrow that *points at* the use case being used. Figure 6-24 on the next page shows an example where a student adds a fitness class and PRODUCE FITNESS-CLASS ROSTER *uses* the results of ADD FITNESS-CLASS to generate a new fitness-class roster. Similarly, if an instructor changes his or her availability, UPDATE INSTRUCTOR INFORMATION *uses* the CHANGE AVAILABILITY use case to update the instructor's information.

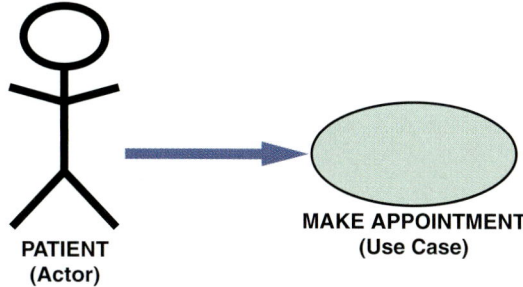

PATIENT (Actor) **MAKE APPOINTMENT (Use Case)**

FIGURE 6-22 In a medical office system, a PATIENT (actor) can MAKE APPOINTMENT (use case).

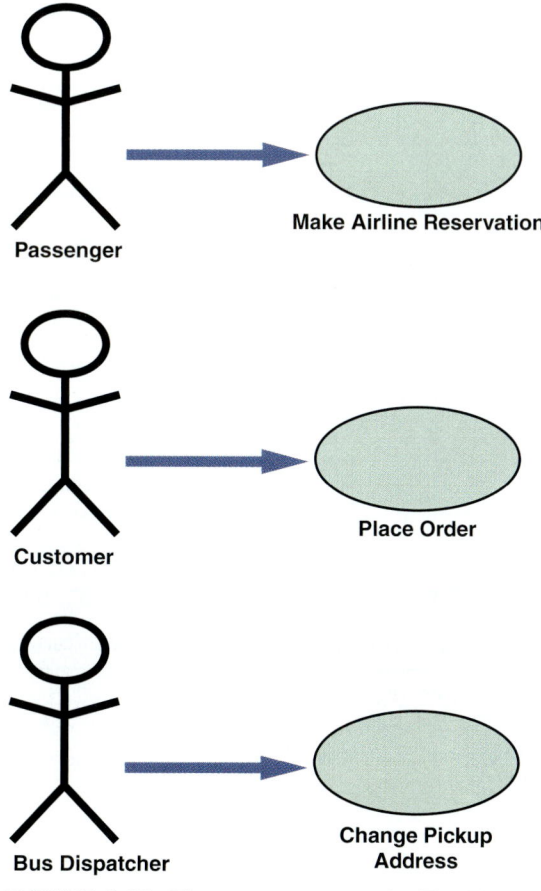

Passenger **Make Airline Reservation**

Customer **Place Order**

Bus Dispatcher **Change Pickup Address**

FIGURE 6-23 Three use case examples. The UML symbol for a use case is an oval. The actor is shown as a stick figure.

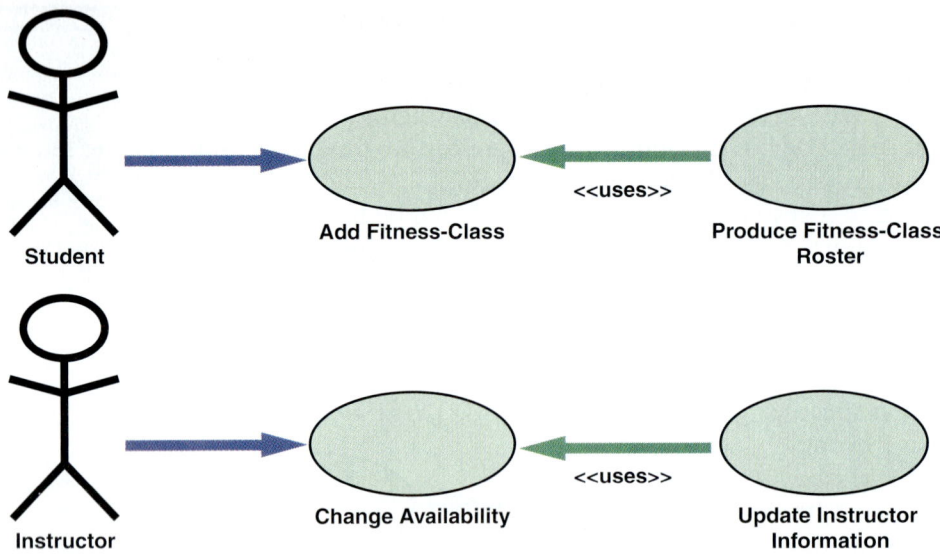

FIGURE 6-24   When a student adds a class, PRODUCE FITNESS-CLASS ROSTER uses the results of ADD CLASS to generate a new class roster. When an instructor changes his or her availability, UPDATE INSTRUCTOR INFORMATION uses the CHANGE AVAILABILITY use case to update the instructor's information.

To create use cases, you start by reviewing the information that you gathered during the requirements modeling phase. Your objective is to identify the actors and the functions or transactions they initiate. For each use case, you also develop a **use case description** in the form of a table. A use case description documents the name of the use case, the actor, a description of the use case, a step-by-step list of the tasks and actions required for successful completion, a description of alternative courses of action, preconditions, postconditions, and assumptions. Figure 6-25 shows an example of the ADD NEW STUDENT use case.

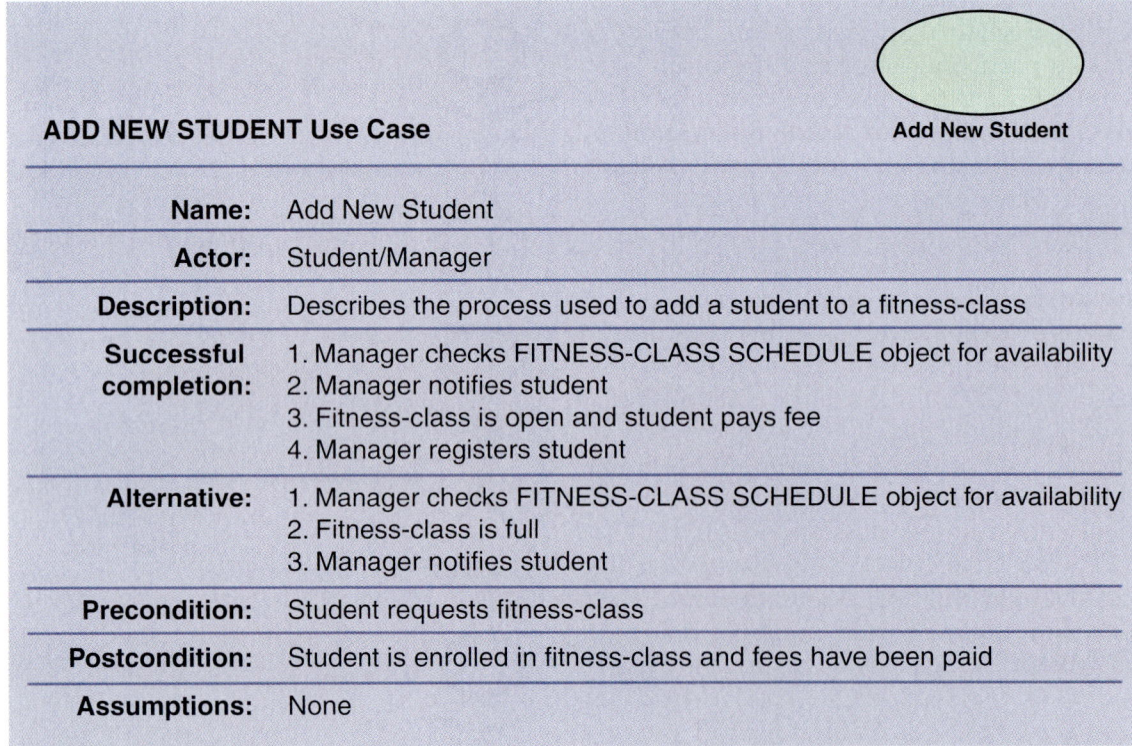

ADD NEW STUDENT Use Case		Add New Student
Name:	Add New Student	
Actor:	Student/Manager	
Description:	Describes the process used to add a student to a fitness-class	
Successful completion:	1. Manager checks FITNESS-CLASS SCHEDULE object for availability 2. Manager notifies student 3. Fitness-class is open and student pays fee 4. Manager registers student	
Alternative:	1. Manager checks FITNESS-CLASS SCHEDULE object for availability 2. Fitness-class is full 3. Manager notifies student	
Precondition:	Student requests fitness-class	
Postcondition:	Student is enrolled in fitness-class and fees have been paid	
Assumptions:	None	

FIGURE 6-25   The ADD NEW STUDENT use case description documents the process used to add a current student into an existing class.

When you identify use cases, try to group all the related transactions into a single use case. For example, when a hotel customer reserves a room, the reservation system blocks a room, updates the occupancy forecast, and sends the customer a confirmation. Those events are all part of a single use case called RESERVE ROOM, and the specific actions are step-by-step tasks within the use case.

CASE IN POINT 6.1: HILLTOP MOTORS

You have been hired by Hilltop Motors as a consultant to help the company plan a new information system. Hilltop is an old-line dealership, and the prior owner was slow to change. A new management team has taken over, and they are eager to develop a first-class system. Right now, you are reviewing the service department, which is going though a major expansion. You decide to create a model of the service department in the form of a use case diagram. The main actors in the service operation are customers, service writers who prepare work orders and invoices, and mechanics who perform the work. You are meeting with the management team tomorrow morning. Create an initial draft of the diagram to present to them at that time.

Use Case Diagrams

A **use case diagram** is a visual summary of several related use cases within a system or subsystem. Consider a typical auto service department, as shown in Figure 6-26. The service department involves customers, service writers who prepare work orders and invoices, and mechanics who perform the work. Figure 6-27 on the next page shows a possible use case diagram for the auto service department.

ON THE WEB

For more information about use case diagrams, visit **scsite.com/sad8e/ more**, locate Chapter 6, and then click the Use Case Diagrams link.

FIGURE 6-26 A typical auto service department might involve customers, service writers who prepare work orders, and mechanics who perform the work.

Use Case Diagram: Auto Service Department

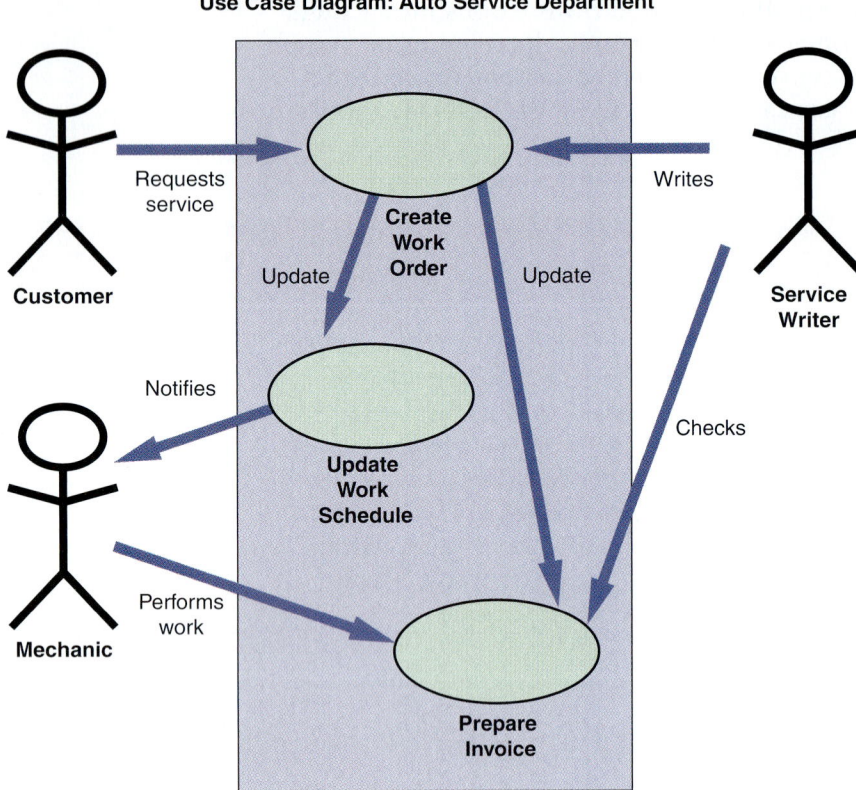

FIGURE 6-27 A use case diagram to handle work at an auto service department.

Use Case Diagram: Create Bus Route

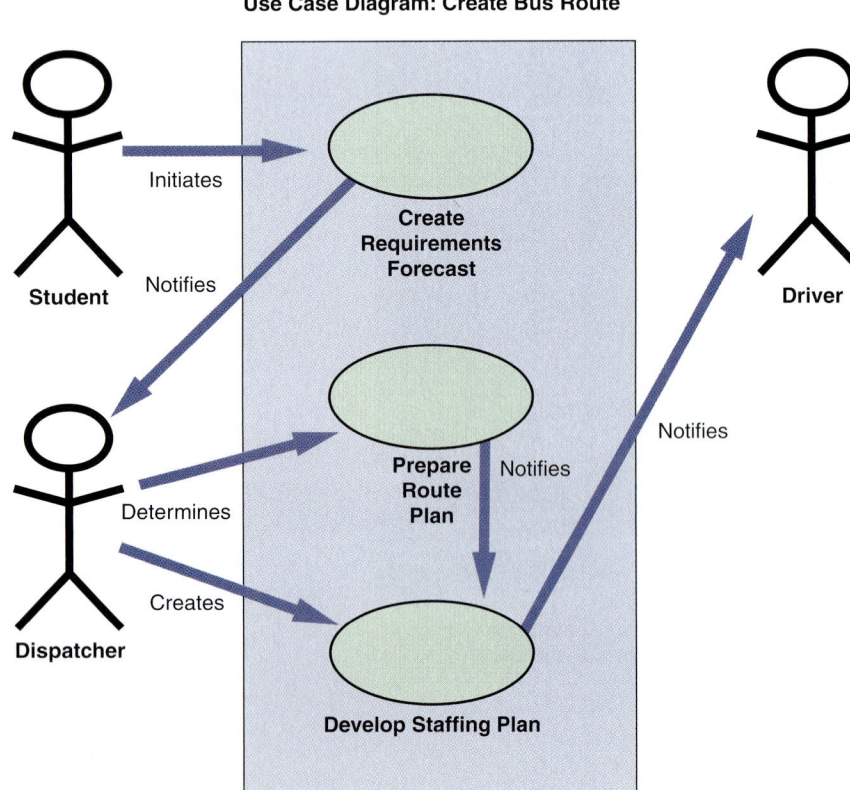

FIGURE 6-28 A use case diagram to create a school bus route.

When you create a use case diagram, the first step is to identify the system boundary, which is represented by a rectangle. The **system boundary** shows what is included in the system (inside the rectangle) and what is not included in the system (outside the rectangle). After you identify the system boundary, you place the use cases on the diagram, add the actors, and show the relationships. Figure 6-28 shows a use case diagram for a school bus system that creates a new bus route.

Class Diagrams

A **class diagram** represents a detailed view of a single use case, shows the classes that participate in the use case, and documents the relationship among the classes. Like a DFD, a class diagram is a logical model, which evolves into a physical model and finally becomes a functioning information system. In structured analysis, entities, data stores, and processes are transformed into data structures and program code. Similarly, class diagrams evolve into code modules, data objects, and other system components.

In a class diagram, each class appears as a rectangle, with the class name at the top, followed by the class's attributes and methods. Lines show relationships between classes and have labels identifying the action that relates the two classes. When you construct the diagram, the first step is to review the use case and identify the classes that participate in the underlying business transaction.

The class diagram also includes a concept called **cardinality**, which describes how instances of one class relate to instances of another class. For example, an employee might have earned no vacation days or one vacation day or many vacation days. Similarly, an employee might have no spouse or one spouse. Figure 6-29 shows various UML notations and cardinality examples. Notice that in Figure 6-29, the first column shows a UML notation symbol that identifies the relationship shown in the second column. The third column provides a typical example of the relationship, which is described in the last column. In the first row of the figure, the UML notation *0..** identifies a *zero or many* relation. The example is that an employee can have no payroll deductions or many deductions.

UML Notation	Nature of the Relationship	Example		Description
0..*	Zero or many	Employee — Payroll Deduction 1 0..*		An employee can have no payroll deductions or many deductions.
0..1	Zero or one	Employee — Spouse 1 0..1		An employee can have no spouse or one spouse.
1	One and only one	Office Manager — Sales Office 1 1		An office manager manages one and only one office.
1..*	One or many	Order — Item Ordered 1 1..*		One order can include one or many items ordered.

FIGURE 6-29 Examples of UML notations that indicate the nature of the relationship between instances of one class and instances of another class.

You will learn more about cardinality in Chapter 9, which discusses data design.

Figure 6-30 shows a class diagram for a sales order use case. Notice that the sales office has one sales manager who can have anywhere from zero to many sales reps. Each sales rep can have anywhere from zero to many customers, but each customer has only one sales rep.

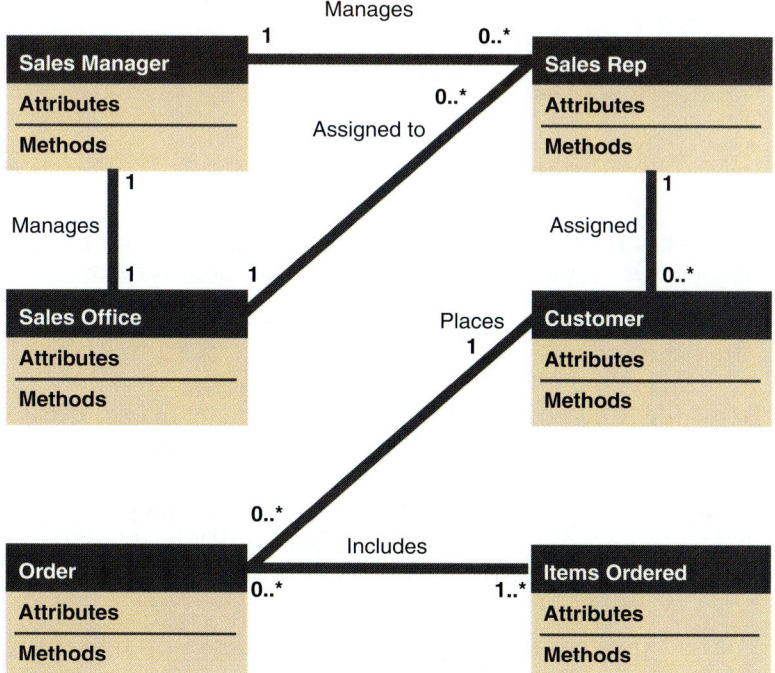

FIGURE 6-30 Class diagram for a sales order use case (attributes and methods omitted for clarity).

CASE IN POINT 6.2: TRAIN THE TRAINER, INC.

Train the Trainer develops seminars and workshops for corporate training managers, who in turn train their employees. Your job at Train the Trainer is to put together the actual training materials. Right now, you are up against a deadline. The new object modeling seminar has a chapter on cardinality, and the client wants you to come up with at least three more examples for each of the four cardinality categories listed in Figure 6-29 on the previous page. The four categories are *zero or many*, *zero or one*, *one and only one*, and *one or many*. Even though you are under pressure, you are determined to use examples that are realistic and familiar to the students. What examples will you submit?

Sequence Diagrams

A **sequence diagram** is a dynamic model of a use case, showing the interaction among classes during a specified time period. A sequence diagram graphically documents the use case by showing the classes, the messages, and the timing of the messages. Sequence diagrams include symbols that represent classes, lifelines, messages, and focuses. These symbols are shown in Figure 6-31.

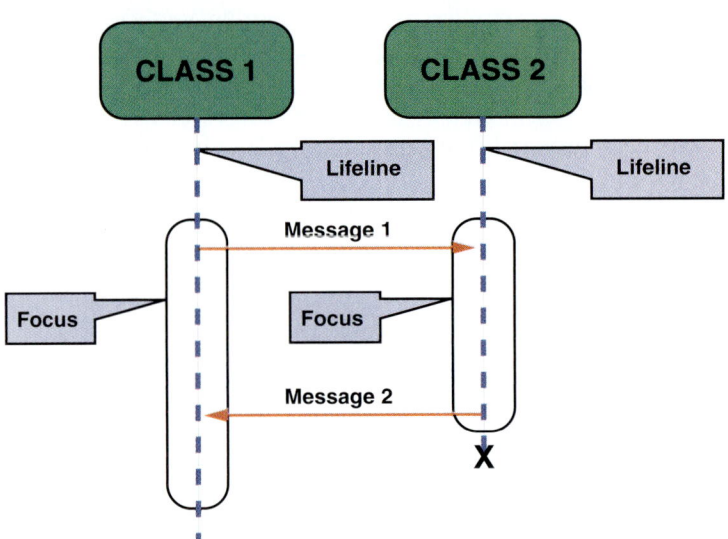

FIGURE 6-31 A sequence diagram with two classes. Notice the X that indicates the end of the CLASS 2 lifeline. Also notice that each message is represented by a line with a label that describes the message, and that each class has a focus that shows the period when messages are sent or received.

CLASSES A class is identified by a rectangle with the name inside. Classes that send or receive messages are shown at the top of the sequence diagram.

LIFELINES A lifeline is identified by a dashed line. The **lifeline** represents the time during which the object above it is able to interact with the other objects in the use case. An *X* marks the end of the lifeline.

MESSAGES A message is identified by a line showing direction that runs between two objects. The label shows the name of the message and can include additional information about the contents.

FOCUSES   A focus is identified by a narrow vertical shape that covers the lifeline. The **focus** indicates when an object sends or receives a message.

The fitness center example shown in Figure 6-32 shows a sequence diagram for the ADD NEW STUDENT use case. Notice that the vertical position of each message indicates the timing of the message.

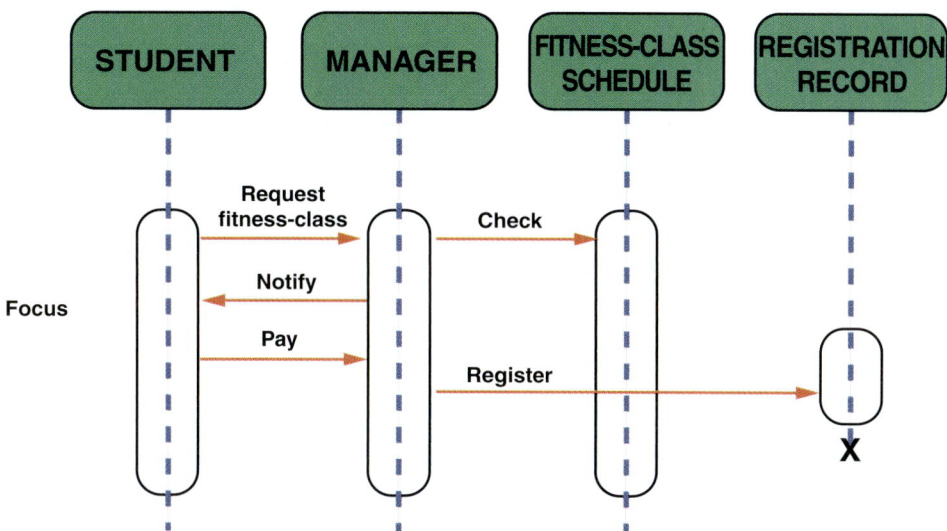

FIGURE 6-32   The sequence diagram for the ADD NEW STUDENT use case. The use case description for ADD NEW STUDENT is shown in Figure 6-25 on page 258.

State Transition Diagrams

Earlier in this chapter you learned that state refers to an object's current status. A **state transition** diagram shows how an object changes from one state to another, depending on events that affect the object. All possible states must be documented in the state transition diagram, as shown in Figure 6-33. A bank account, for example, could be opened as a NEW account, change to an ACTIVE or EXISTING account, and eventually become a CLOSED or FORMER account. Another possible state for a bank account could be FROZEN, if the account's assets are legally attached.

In a state transition diagram, the states appear as rounded rectangles with the state names inside. The small circle to the left is the initial state, or the point where the object first interacts with the system. Reading from left to right, the lines show direction and describe the action or event that causes a transition from one state to another. The circle at the right with a hollow border is the final state.

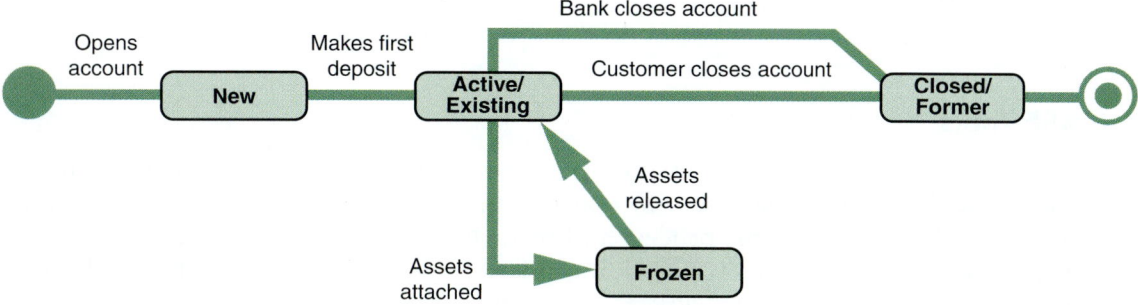

FIGURE 6-33   An example of a state transition diagram for a bank account.

Activity Diagrams

An **activity diagram** resembles a horizontal flowchart that shows the actions and events as they occur. Activity diagrams show the order in which the actions take place and identify the outcomes. Figure 6-34 shows an activity diagram for a cash withdrawal at an ATM machine. Notice that the customer initiates the activity by inserting an ATM card and requesting cash. Activity diagrams also can display multiple use cases in the form of a grid, where classes are shown as vertical bars and actions appear as horizontal arrows.

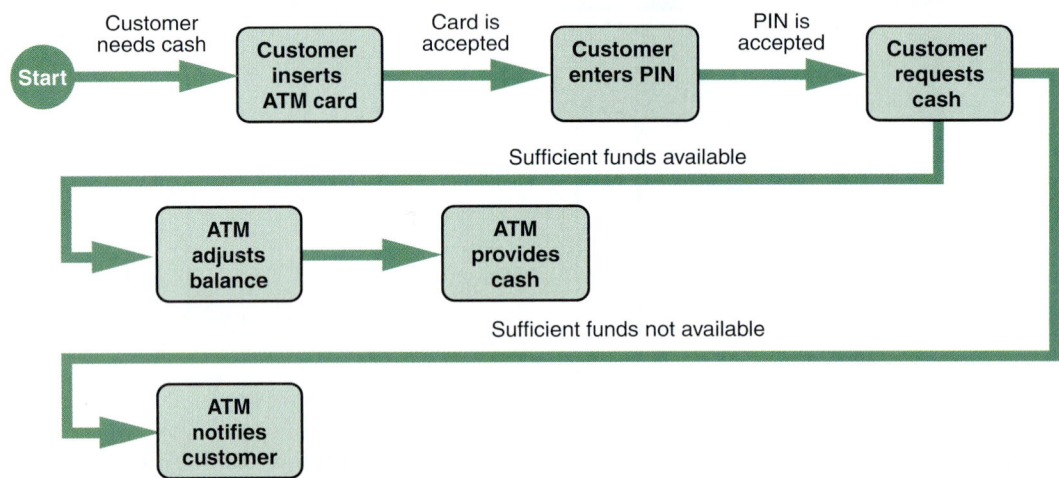

FIGURE 6-34 An activity diagram shows the actions and events involved in withdrawing cash from an ATM machine.

Sequence diagrams, state transition diagrams, and activity diagrams are dynamic modeling tools that can help a systems analyst understand how objects behave and interact with the system.

CASE IN POINT 6.3: TravelBiz

Jack Forester and Lisa Turner are systems analysts in the IT department of TravelBiz, a nationwide travel agency that specializes in business travel. TravelBiz has decided to expand into the vacation travel market by launching a new business division called TravelFun. The IT director assigned Jack and Lisa to create a flexible, efficient information system for the new division. Jack wants to use traditional analysis and modeling techniques for the project. Lisa, on the other hand, wants to use an object-oriented methodology. Which approach would you suggest and why?

⚒ TOOLKIT TIME

The CASE tools in Part 2 of the Systems Analyst's Toolkit can help you develop and maintain complex information systems. To learn more about these tools, turn to Part 2 of the four-part Toolkit that follows Chapter 12.

CASE Tools

Object modeling requires many types of diagrams to represent the proposed system. Creating the diagrams by hand is time consuming and tedious, so systems analysts rely on CASE tools to speed up the process and provide an overall framework for documenting the system components. In addition, CASE tools ensure consistency and provide common links so that once objects are described and used in one part of the design, they can be reused multiple times without further effort.

ORGANIZING THE OBJECT MODEL

In this chapter, you learned how to use object-oriented tools and techniques to build a logical model of the information system. Now you are ready to organize your diagrams and documentation so the object model is easily read and understood. If you used a CASE tool to develop the design, much of this work will be performed automatically by the CASE software.

There are many ways to proceed with the task of organizing the object model, and experience will be your best teacher. After you identify the system's objects, classes, and relationships, you should develop an object relationship diagram that provides an overview of the system. If you do not use a CASE-generated model, each diagram or object definition should be supported by clear, relevant documentation that can be accessed easily by anyone who reviews the object model. For example, you should organize your use cases and use case diagrams so they can be linked to the appropriate class, state transition, sequence, and activity diagrams. Your diagrams and documentation are the foundation for the system's design, so accuracy is important. Remember that it is much easier to repair a diagram now than to change the software later.

CASE IN POINT 6.4: CYBER ASSOCIATES

One of your responsibilities at Cyber Associates, an IT consulting firm, is to assign new systems analysts to various tasks and projects. Some of the senior people believe that inexperienced analysts should start with object-oriented techniques, which are easier to learn and apply. Others think that an analyst should learn structured analysis first, and then proceed to object-oriented skills. What is your viewpoint, and why?

A QUESTION OF ETHICS

Last month, your company launched a peer review process for IT projects. At the end of each project, team members rate the performance of the overall team, and his or her co-workers individually. The stated goal was to obtain honest, peer-based feedback. Unfortunately, like many good ideas, there was a downside. Although the input is anonymous, the results are submitted to the entire team. Some members, including you, are uncomfortable with the new process because it could encourage cliques and actually undermine a team-based culture. Others see it as an opportunity for honest input.

One team member, who is a close friend of yours, is not very popular with her teammates. To make matters worse, she recently had some personal problems that affected her work, and she is worried that her ratings will be quite negative. She has not specifically asked you about your feedback, but you know she is hoping for a favorable review from you. Even though her work was not great, you don't want to see her get hurt by a process that you yourself are not comfortable with.

Is this a question of ethics versus friendship? Would it be wrong to tilt the scales in her favor just a bit?

CHAPTER SUMMARY

This chapter introduces object modeling, which is a popular technique that describes a system in terms of objects. Objects represent real people, places, events, and transactions. Unlike structured analysis, which treats data and processes separately, objects include data and processes that can affect the data. During the implementation process, systems analysts and programmers transform objects into program code modules that can be optimized, tested, and reused as often as necessary.

Object-oriented terms include classes, attributes, instances, messages, and methods. Classes include objects that have similar attributes, or characteristics. Individual members of a class are called object instances. Objects within a class can be grouped into subclasses, which are more specific categories within the class. A class also can belong to a more general category called a superclass.

Objects can send messages, or commands, that require other objects to perform certain methods, or tasks. The concept that a message gives different meanings to different objects is called polymorphism. An object resembles a black box, with encapsulated, or self-contained, data and methods. The strongest relationship between objects is inheritance.

After you identify the objects, classes, and relationships, you prepare an object relationship diagram that shows the objects and how they interact to perform business functions and transactions.

The Unified Modeling Language (UML) is a widely used method of visualizing and documenting an information system. UML techniques include use cases, use case diagrams, class diagrams, sequence diagrams, state transition diagrams, and activity diagrams.

A use case describes a business situation initiated by an actor, who interacts with the information system. Each use case represents a specific transaction, or scenario. A use case diagram is a visual summary of related use cases within a system or subsystem. A class diagram represents a detailed view of a single use case, showing the classes that participate in the underlying business transaction, and the relationship among class instances, which is called cardinality. A sequence diagram is a dynamic model of a use case, showing the interaction among classes during a specified time period. Sequence diagrams include lifelines, messages, and focuses. A state transition diagram shows how an object changes from one state to another, depending on events that affect the object. An activity diagram resembles a horizontal flowchart that shows actions and events as they occur in a system.

CASE tools provide an overall framework for system documentation. CASE tools can speed up the development process, ensure consistency, and provide common links that enable objects to be reused.

At the end of the object modeling process, you organize your use cases and use case diagrams and create class, sequence, state transition, and activity diagrams.

Key Terms and Phrases

activity diagram *264*

actor *257*

attributes *248*

black box *253*

cardinality *261*

child *256*

class *248*

class diagram *260*

encapsulation *253*

focus *263*

inheritance *256*

instance *248*

lifeline *262*

message *248*

methods *248*

object *248*

object model *248*

object-oriented (O-O) analysis *248*

parent *256*

polymorphism *253*

relationships *256*

sequence diagram *262*

state *252*

state transition *263*

subclass *254*

superclass *255*

system boundary *260*

Unified Modeling Language (UML) *248*

use case *257*

use case description *258*

use case diagram *259*

Learn It Online

Instructions: To complete the Learn It Online exercises, start your browser, click the Address bar, and then enter the Web address **scsite.com/sad8e/learn**. When the Systems Analysis and Design Learn It Online page is displayed, follow the instructions in the exercises below. Each exercise has instructions for saving your results, either for your own records or for submission to your instructor.

1 Chapter Reinforcement

TF, MC, and SA

Below SAD Chapter 6, click one of the Chapter Reinforcement links for Multiple Choice, True/False, or Short Answer. Answer each question and submit to your instructor.

2 Flash Cards

Below SAD Chapter 6, click the Flash Cards link and read the instructions. Type 20 (or a number specified by your instructor) in the Number of Playing Cards text box, type your name in the Enter your Name text box, and then click the Flip Card button. When the flash card is displayed, read the question and then click the ANSWER box arrow to select an answer. Flip through the Flash Cards. If your score is 15 (75%) correct or greater, click Print on the File menu to print your results. If your score is less than 15 (75%) correct, then redo this exercise by clicking the Replay button.

3 Practice Test

Below SAD Chapter 6, click the Practice Test link. Answer each question, enter your first and last name at the bottom of the page, and then click the Grade Test button. When the graded practice test is displayed on your screen, click Print on the File menu to print a hard copy. Continue to take practice tests until you score 80% or better.

4 Who Wants To Be a Computer Genius?

Below SAD Chapter 6, click the Computer Genius link. Read the instructions, enter your first and last name at the bottom of the page, and then click the Play button. When your score is displayed, click the PRINT RESULTS link to print a hard copy.

5 Wheel of Terms

Below SAD Chapter 6, click the Wheel of Terms link. Read the instructions, and then enter your first and last name and your school name. Click the PLAY button. When your score is displayed on the screen, right-click the score and then click Print on the shortcut menu to print a hard copy.

6 Crossword Puzzle Challenge

Below SAD Chapter 6, click the Crossword Puzzle Challenge link. Read the instructions, and then click the Continue button. Work the crossword puzzle. When you are finished, click the Submit button. When the crossword puzzle is redisplayed, submit it to your instructor.

Case-Sim: SCR Associates

Background

SCR Associates is a consulting firm that offers IT solutions and training. SCR needs an information system to manage operations at its new training center. The new system will be called TIMS (Training Information Management System). As a newly hired systems analyst, you will report to Jesse Baker, systems group manager.
You will work on various tasks and practice the skills you learned in this chapter.

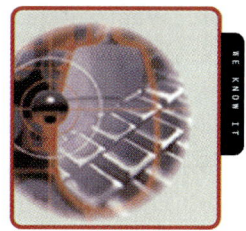

Using the Case

The SCR Associates case study is a Web-based simulation that allows you to practice your skills in a real-world environment. The case study transports you to SCR's company intranet, where you can complete 12 work sessions, each aligning with a chapter. As you work on the case, you will receive e-mail and voice mail messages, obtain information from SCR's online libraries, and perform various tasks. The first time you enter the SCR case, you should go to the starting page at **scsite.com/sad8e/scr** for detailed instructions.

Preview: Session 6

In the last session, you used data and process modeling techniques to develop a logical model of the new system. Now you will apply your object modeling skills and create various diagrams and documentation for the new TIMS system. You will review background material and develop an object-oriented model.

To start a work session, you log on to the SCR intranet. When you enter your name and password, an opening screen displays links to the work sessions, and you select Session 6. At this point, you check your e-mail and voice mail messages carefully. You then begin working on your task list, which includes the following items:

Tasks: Object Modeling

1. I need to review Jesse's e-mail message regarding object modeling and the JAD session summary. Then she wants me to identify possible use cases and actors, and create a use case diagram for the TIMS system.

2. She also wants me to select one of the use cases and create a class diagram.

3. I will need a sequence diagram for the selected use case.

4. Jesse asked for a state transition diagram that describes typical student states and how they change based on certain actions and events.

FIGURE 6-35 Task list: Session 6.

Chapter Exercises

Review Questions

1. What is object-oriented analysis, and what are some advantages of using this technique?
2. Define an object, and give an example.
3. Define an attribute, and give an example.
4. Define a method, and give an example.
5. Define encapsulation, and explain the benefits it provides.
6. Define polymorphism, and give an example.
7. Define a class, subclass, and superclass, and give examples.
8. Define an actor, and give an example.
9. Define a use case and a use case diagram, and give examples.
10. Define the term *black box*, and explain why it is an important concept in object-oriented analysis.

Discussion Topics

1. The chapter mentioned that systems analysts and programmers transform objects into program code modules that can be optimized, tested, and reused. Modular design is a very popular design concept in many industries. What other examples of modular design can you suggest?
2. You are an IT consultant, and you are asked to create a new system for a small real estate brokerage firm. Your only experience is with traditional data and process modeling techniques. This time, you decide to try an object-oriented approach. How will you begin? How are the tasks different from traditional structured analysis?
3. You are creating a system for a bowling alley to manage information about its leagues. During the modeling process, you create a state transition diagram for an object called LEAGUE BOWLERS. What are the possible states of a league bowler, and what happens to a bowler who quits the league and rejoins the following season?
4. A debate is raging at the IT consulting firm where you work. Some staff members believe that it is harder for experienced analysts to learn object-modeling techniques, because the analysts are accustomed to thinking about data and processes as separate entities. Others believe that solid analytical skills are easily transferable and do not see a problem in crossing over to the newer approach. What do you think, and why?

Projects

1. Search the Internet for information about the history and development of UML.
2. Contact the IT staff at your school or at a local business to learn whether the organization uses object-oriented programming languages. If so, determine what languages and versions are used and why they were selected.
3. Search the Internet for information about groups and organizations that support and discuss object-oriented methods and issues.
4. Search the Internet for information about CASE tools that provide UML support.

Apply Your Knowledge

The Apply Your Knowledge section contains four mini-cases. Each case describes a situation, explains your role in the case, and asks you to respond to questions. You can answer the questions by applying knowledge you learned in the chapter.

1 Hertford Post Office

Situation:

Hertford has a typical small town post office that sells stamps, rents post office boxes, and delivers mail to postal customers.

1. Identify possible actors and use cases involved in the post office functions.
2. Create a use case diagram for the post office operation.
3. Select one of the use cases and create a class diagram.
4. Create a sequence diagram for the use case you selected.

2 New Branch School District

Situation:

The New Branch School District operates a fleet of 40 buses that serve approximately 1,000 students in grades K–12. The bus operation involves 30 regular routes, plus special routes for activities, athletic events, and summer sessions. The district employs 12 full-time drivers and 25 to 30 part-time drivers. A dispatcher coordinates the staffing and routes and relays messages to drivers regarding students and parents who call about pickup and drop-off arrangements.

1. Identify possible actors and use cases involved in school bus operations.
2. Create a use case diagram for the school bus system.
3. Create a sequence diagram for the use case you selected.
4. Create a state transition diagram that describes typical student states and how they change based on specific actions and events.

3 **Pleasant Creek Community College Registration System**

Situation:

Pleasant Creek Community College has a typical school registration process. Student support services include faculty advisors and tutors. The administration has asked you, as IT manager, to develop an object-oriented model for a new registration system.

1. List possible objects in the new registration system, including their attributes and methods.
2. Identify possible use cases and actors.
3. Create a use case diagram that shows how students register.
4. Create a state transition diagram that describes typical student states and how they change based on specific actions and events.

4 **Student Bookstore at Pleasant Creek Community College**

Situation:

The bookstore staff at Pleasant Creek Community College works hard to satisfy students, instructors, and the school's business office. Instructors specify textbooks for particular courses, and the bookstore orders the books and sells them to students. The bookstore wants you to develop an object-oriented model for a new bookstore information management system.

1. List possible objects in the bookstore operation, including their attributes and methods.
2. Identify possible use cases and actors.
3. Select one of the use cases that you identified in step 2 and create a sequence diagram.
4. Create an object relationship diagram that provides an overview of the system, including how textbooks are selected by instructors, approved by a department head, and sold to students by the bookstore.

Case Studies

Case studies allow you to practice specific skills learned in the chapter. Each chapter contains several case studies that continue throughout the textbook, and a chapter capstone case.

NEW CENTURY HEALTH CLINIC

New Century Health Clinic offers preventive medicine and traditional medical care. In your role as an IT consultant, you will help New Century develop a new information system.

Background

You began the systems analysis phase at New Century Health Clinic by completing a series of interviews, reviewing existing reports, and observing office operations. Then, in Chapter 5, you acquired more information and developed a set of DFDs, process descriptions, and a data dictionary.

Now you decide to practice the object modeling skills you learned in this chapter. Before you begin, go back to Chapter 5 and review the New Century background material and fact-finding results. Also, your instructor may provide you with a complete set of interview summaries that you can use to perform your assignments. Then complete the following tasks.

Assignments

1. Identify possible use cases and actors, and create a use case diagram for the New Century Health Clinic system.
2. Select one of the use cases and create a class diagram.
3. Create a sequence diagram for the use case that you selected.
4. Create a state transition diagram that describes typical patient states and how they change based on specific actions and events.

PERSONAL TRAINER, INC.

Personal Trainer, Inc., owns and operates fitness centers in a dozen Midwestern cities. The centers have done well, and the company is planning an international expansion by opening a new "supercenter" in the Toronto area. Personal Trainer's president, Cassia Umi, hired an IT consultant, Susan Park, to help develop an information system for the new facility. During the project, Susan will work closely with Gray Lewis, who will manage the new operation.

Background

Working as an IT consultant for Personal Trainer, Susan Park used data and process modeling tools to create a logical model of the proposed information system. Now she wants to build an object-oriented view of the system using O-O tools and techniques.

Assignments

Before you perform the following tasks, you should review the information and background in Chapters 1 and 2, and the fact-finding summary of the case provided in Chapter 4.

1. Identify possible use cases and actors, and create a use case diagram for the Personal Trainer information system.
2. Select one of the use cases and create a class diagram.
3. Create an object relationship diagram for the system.
4. Create a state transition diagram that describes typical member states and how they change based on specific actions and events.

CHAPTER CAPSTONE CASE: SoftWear, Limited

SoftWear, Limited (SWL), is a continuing case study that illustrates the knowledge and skills described in each chapter. In this case study, the student acts as a member of the SWL systems development team and performs various tasks.

Background

Rick Williams, a systems analyst, and Carla Moore, a programmer/analyst, completed a set of DFDs representing a data and process model of the SWL payroll system project. Rick had recently attended a workshop on object modeling techniques, and suggested that he and Carla should experiment with object-oriented analysis. After he explained the concepts and techniques to Carla, she agreed that it was a good opportunity to gain some experience, and they decided to give it a try.

Rick and Carla began by reviewing the data they had collected earlier, during requirements modeling. They studied the DFDs and the data dictionary, and they identified the people, events, and transactions that they would represent as classes. They identified employees, human resources transactions, time sheet entries, payroll actions, and stock transfers. They defined attributes and methods for each of those classes, as shown in Figure 6-36. When they were finished, they reviewed the results. They noticed that the structured DFDs did not show a department head as an entity. Rick remembered that department heads submitted time sheets to the payroll department, and the payroll clerks actually entered the data into the system. Because they were looking at the system in a different way, they decided to include department heads as a subclass of the EMPLOYEE class.

EMPLOYEE	HR TRANSACTION	TIME SHEET ENTRY
Employee number	Employee number	Employee number
Employee name	Employee name	Week ending date
Address	State	Hours worked
Telephone number		
Date of birth		
Sex		
Title, rate of pay		
Deductions		
State		
Add new	Add new	Add new
Change name	Change state	Correct hours
Change address	Notify	Generate report
Change telephone		
Change deductions		
Change state		

PAYROLL ACTION	STOCK TRANSFER	DEPT HEAD
Employee number	Employee number	Employee number
Hours worked	Stock holdings	Employee name
Overtime hours	Stock contribution	Address
Rate of pay		Telephone number
Overtime rate		Date of birth
Deductions		Sex
Contributions		Title, rate of pay
Federal tax withheld		Deductions
State tax withheld		State
Local tax withheld		
Generate checks	Purchase stock	Add new
Change deductions	Sell stock	Change name
Change contributions	Change contribution	Change address
Change federal rate	Generate report	Change telephone
Change local rate		Change deductions
Change state rate		Change state
Change rate of pay		Manages work
Generate W-2		Submits time sheets
Notify		
Calculate		

FIGURE 6-36 SWL payroll system classes.

Use Cases

The next step was for Rick and Carla to define the use cases. They decided to list all situations that involved the EMPLOYEE object. They realized that a new employee gets hired, an employee gets promoted, an employee gets a raise, an employee gets fired, an employee retires, an employee changes his or her name, and an employee requests a contributions change.

CHAPTER CAPSTONE CASE: SoftWear, Limited (continued) **SWL**

They also decided to create use cases for the PAYROLL ACTION object. The scenarios included these: Change an employee's deductions, change an employee's contributions, change the federal tax rate, change the state tax rate, change the local tax rate, calculate weekly gross pay, calculate weekly taxes, calculate weekly contributions, generate weekly paychecks, and notify the stock transfer department of change in contributions.

Next, they identified the actors involved in each use case. Rick suggested that EMPLOYEE and DEPARTMENT HEAD were needed, and Carla added the PAYROLL CLERK. After they defined the use cases and the actors, they started to create a description for each use case. They created a table for each use case showing the use case name, actors, description, successful completion, alternatives, preconditions, postconditions, and assumptions.

Creating use case descriptions was hard work, and they found that they had to return frequently to their documentation and fact-finding results. The first batch of use case descriptions involved the EMPLOYEE actor, as shown in Figure 6-37. The use cases were GET HIRED and RECEIVE RAISE. Then they created use case descriptions for RECEIVE PROMOTION and TERMINATE, which are shown in Figure 6-38 on the next page. Carla thought that the descriptions had given them plenty of practice, so they agreed to go on to the next step.

Get Hired Use Case

Name:	Get Hired
Actor:	Employee
Description:	Describes the process used to add a new hire
Successful completion:	1. Employee gets hired 2. Employee fills out new hire package 3. Human resources department enters employee data and completes the HR transaction
Alternative:	None
Precondition:	Employee has been approved for hire
Postcondition:	Employee has been enrolled in benefits programs and is in active employee status
Assumptions:	Employee accepts position and begins service

Receive Raise Use Case

Name:	Receive Raise
Actor:	Employee
Description:	Describes the change to an employee's pay rate
Successful completion:	1. Employee gets a raise 2. Human resources department changes employee data to the employee object and the human resources records
Alternative:	None
Precondition:	Employee has been approved for a raise
Postcondition:	Employee's pay rate is changed on all records
Assumptions:	None

FIGURE 6-37 Use case descriptions for GET HIRED and RECEIVE RAISE use cases.

CHAPTER CAPSTONE CASE: SoftWear, Limited (continued)

Receive Promotion Use Case

Name:	Receive Promotion
Actor:	Employee
Description:	Describes change to employee title
Successful completion:	1. Employee gets promoted 2. Human resources department changes employee data and completes HR transaction
Alternative:	None
Precondition:	Employee has been approved for promotion
Postcondition:	Employee title is changed
Assumptions:	Employee accepts position

Terminate Use Case

Name:	Terminate
Actor:	Employee
Description:	Describes termination process
Successful completion:	1. Employee terminates 2. Human resources department changes employee data and completes HR transaction 3. Employee issued final paycheck and benefits
Alternative:	None
Precondition:	Termination approved for employee
Postcondition:	Employees status is changed on all records
Assumptions:	None

FIGURE 6-38 Use case descriptions for RECEIVE PROMOTION and TERMINATE use cases.

Now they were ready to develop a use case diagram that would provide a visual summary of related use cases. Rick suggested that they limit the use cases to three per diagram and Carla agreed, adding that they wanted the final package to be easy to read. They decided to document an employee contribution change, and they created the use case diagram shown in Figure 6-39. Their diagram included three related use cases inside a system boundary: REQUEST CONTRIBUTION CHANGE, NOTIFY PAYROLL, and NOTIFY STOCK TRANSFER. The use case diagram depicts the EMPLOYEE actor initiating a change in the contribution amount.

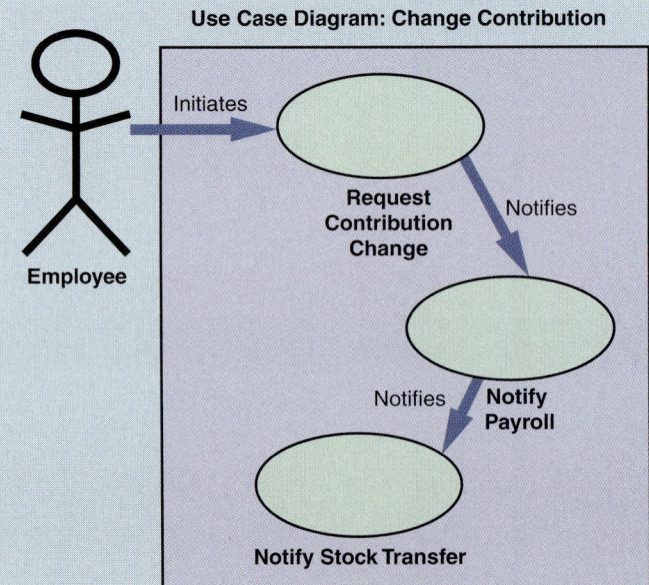

FIGURE 6-39 Use case diagram for the CHANGE CONTRIBUTION scenario.

Next, they decided to create a use case diagram to describe how the payroll is generated. Their use case diagram is shown in Figure 6-40. The diagram includes three use cases: CREATE TIMESHEET, CALCULATE PAYROLL, and GENERATE PAYCHECK. In their diagram, the DEPARTMENT HEAD actor creates a new instance of the TIMESHEET ENTRY object, which notifies the CALCULATE PAYROLL use case, which is initiated by the PAYROLL CLERK. The GENERATE PAYCHECK use case then issues a paycheck to the EMPLOYEE actor.

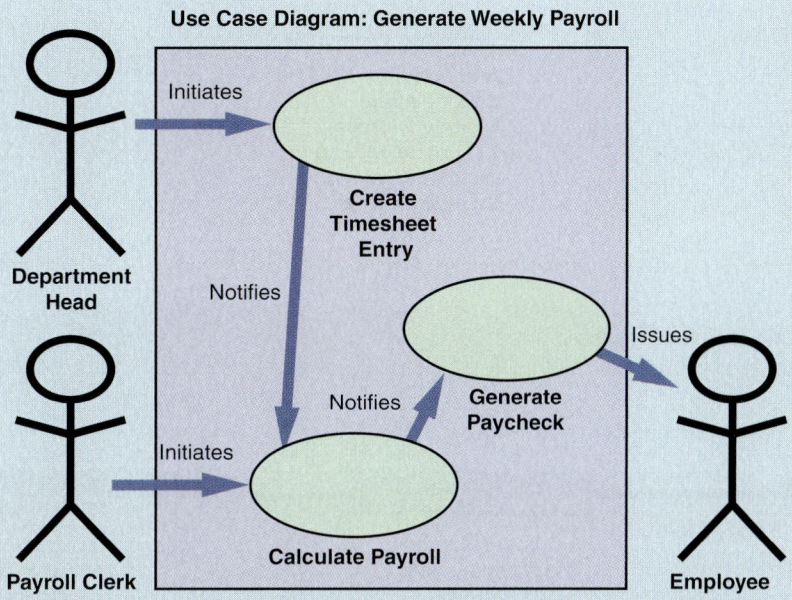

FIGURE 6-40 Use case diagram for the GENERATE WEEKLY PAYROLL scenario.

Class Diagrams

The use cases and use case diagrams helped Rick and Carla understand how the various objects and classes interacted. At this point, Carla suggested that they construct class diagrams for each use case. The class diagrams document the classes and relationships involved in an individual use case. They agreed to start with the GENERATE PAYROLL use case, because it was quite complex and would test their new skills. When they finished, they saw that the class diagram included five classes and several different types of relationships. Their class diagram is shown in Figure 6-41 on the next page.

CHAPTER CAPSTONE CASE: SoftWear, Limited (continued)   **SWL**

HR TRANSACTION ————— Maintained for ————— **EMPLOYEE**

HR TRANSACTION
Employee number
Employee name
State
Add new
Change state
Notify

1   1

EMPLOYEE
Employee number
Employee name
Address
Telephone number
Date of birth
Sex
Title, rate of pay
Deductions
State
Add new
Change name
Change address
Change telephone
Change deductions
Change state
0..*

0..*

Manages

1

0..*

Notifies   0..*

Based on   Submits

PAYROLL ACTION
Employee number
Hours worked
Overtime hours
Rate of pay
Overtime rate
Deductions
Contributions
Federal tax withheld
State tax withheld
Local tax withheld
Generate checks
Change deductions
Change contributions
Change federal rate
Change local rate
Change state rate
Change rate of pay
Generate W-2
Notify
Calculate

1..*   1

TIMESHEET ENTRY
Employee number
Week ending date
Hours worked
Add new
Correct hours
Generate report

1..*   1

DEPT HEAD
Employee number
Employee name
Address
Telephone number
Date of birth
Sex
Title, rate of pay
Deductions
State
Add new
Change name
Change address
Change telephone
Change deductions
Change state
Manages work
Submits timesheets

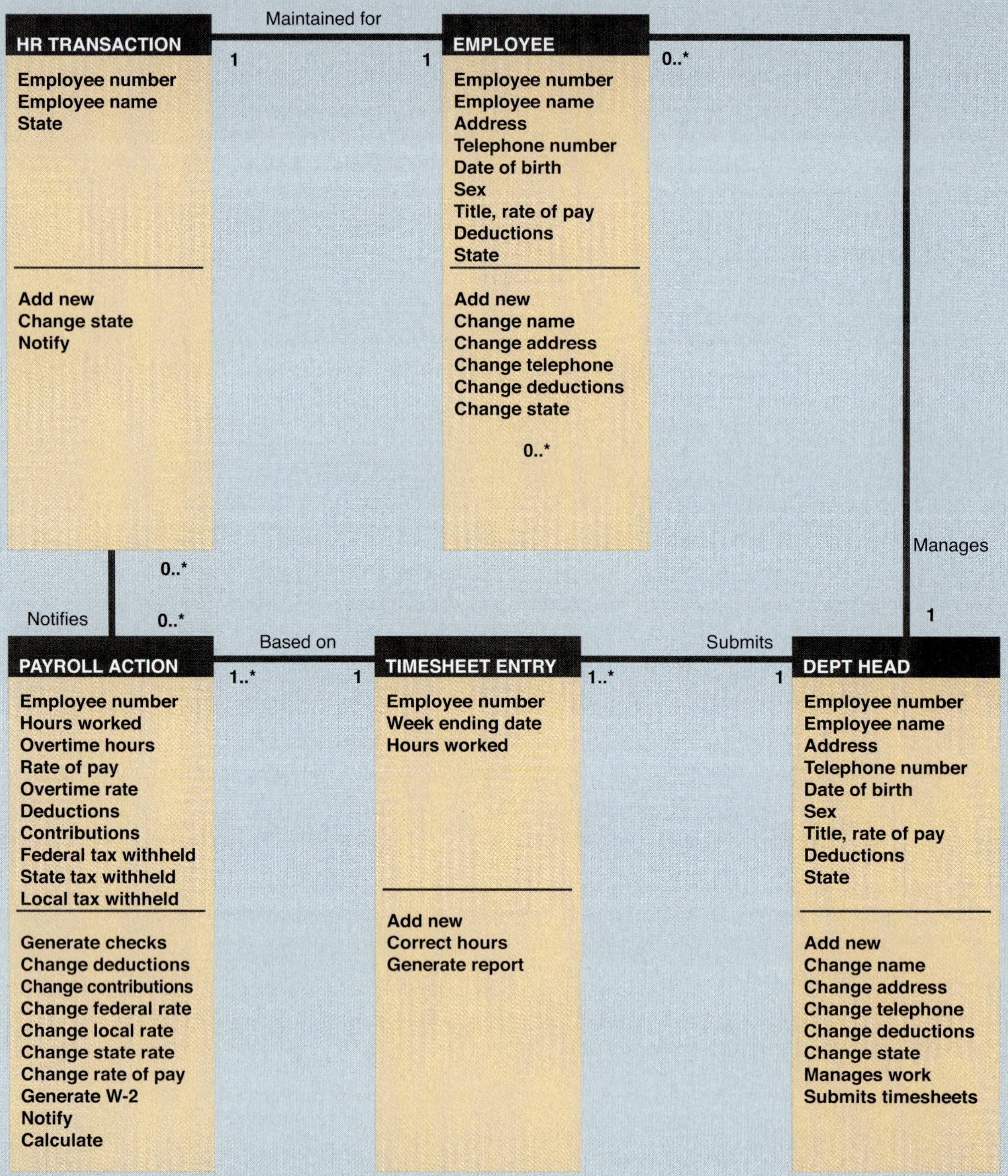

FIGURE 6-41   The GENERATE WEEKLY PAYROLL class diagram includes five classes and various types of relationships among the classes.

CHAPTER CAPSTONE CASE: SoftWear, Limited (continued)

SWL

Sequence Diagrams

Next, the pair decided to create a sequence diagram. Carla was eager to see how a sequence diagram would help them visualize the time frame in which events occur. They created a diagram for the CHANGE CONTRIBUTIONS method in the EMPLOYEE object. The sequence diagram in Figure 6-42 shows the steps that occur when an employee changes benefits contributions. Notice that the diagram includes the messages sent and the lifeline of the objects. Rick and Carla were satisfied that they could create sequence diagrams easily, and they decided to move on to the state transition diagram.

State Transition Diagram

Rick explained that a state transition diagram shows how an object's state, or status, changes as a result of various actions or events. The state transition diagram they created in Figure 6-43 shows the status of an employee from the time the employee is hired to the time he or she quits, is fired, or retires. Notice that the employee is a PROSPECTIVE employee until all physicals are passed and all paperwork is processed, and then he or she becomes a CURRENT employee. Once employment ends for any reason, the individual becomes a PAST employee. At this point, even if the employee returns to the company later on he or she will come in as a new instance of the EMPLOYEE object. Rick and Carla were surprised at how easy that was, and they decided to try an activity diagram.

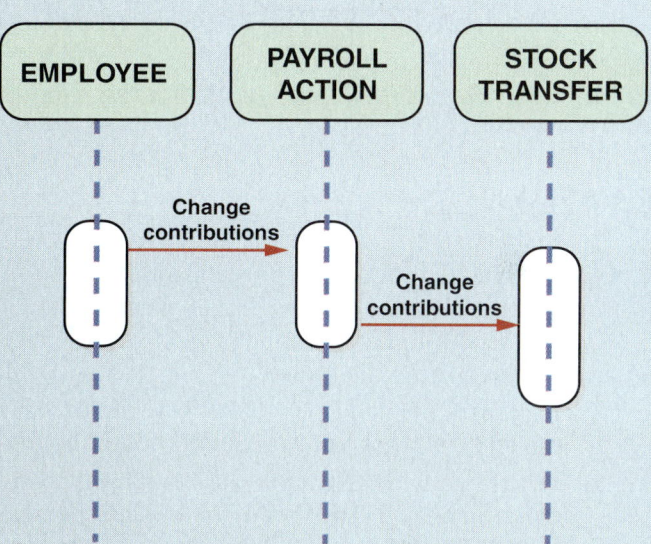

FIGURE 6-42 Sequence diagram for the CHANGE CONTRIBUTIONS scenario.

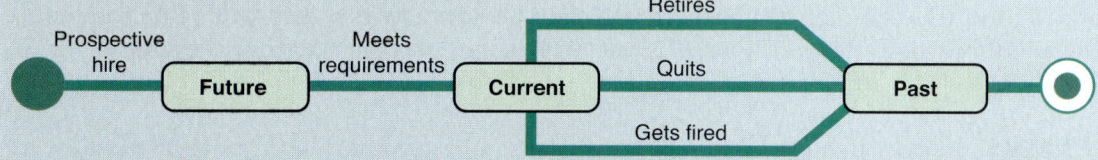

FIGURE 6-43 State transition diagram shows changes in employee status caused by actions and events.

Activity Diagram

Rick suggested that they create an activity diagram showing some of the situations they had explored in detail. Their diagram showed the interaction between objects during certain scenarios and enabled them to visualize system activity, as shown in Figure 6-44 on the next page. They agreed that the technique gave them additional object modeling experience that would be valuable in the future. At that point, they packaged all the diagrams in a folder and saved the overall object model for future reference.

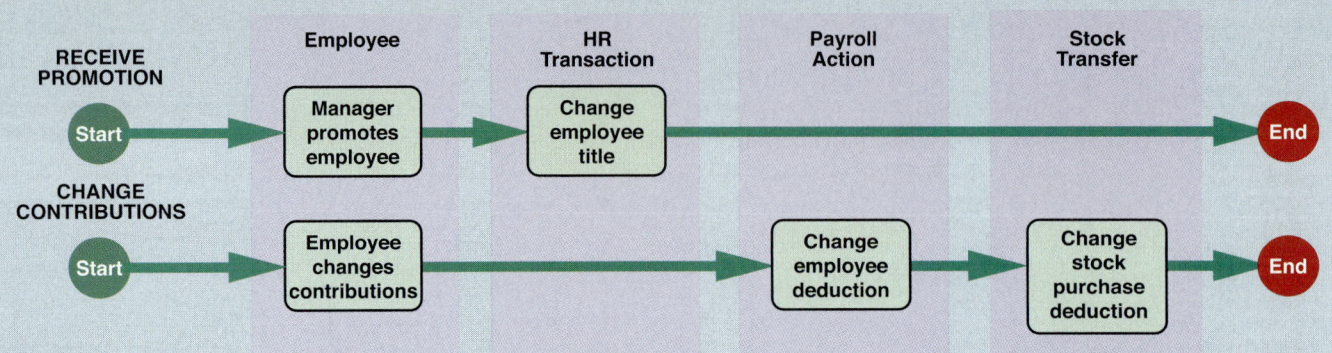

FIGURE 6-44 Activity diagram shows the RECEIVE PROMOTION scenario and the CHANGE CONTRI-BUTIONS scenario.

Team Tasks

1. Rick is interested in your views on the future of object-oriented analysis and design. He is scheduled to make a presentation on the topic next week at a meeting of IT professionals. He asked you to do some research, using the Internet and industry publications, and send him an e-mail message describing the current use of object-oriented analysis and trends for the future.

2. As a team member, you know how important it can be to have a well-organized object model. The team has asked you to handle this task. How will you go about it?

3. When you worked on the class diagrams, you had to understand and apply the concept of cardinality. How would you explain this concept to a new team member?

4. List all the different types of diagrams you used to create the object model, with a brief explanation of each diagram.

Manage the SWL Project

You have been asked to manage SWL's new information system project. One of your most important activities will be to identify project tasks and determine when they will be performed. Before you begin, you should review the SWL case in this chapter. Then list and analyze the tasks, as follows:

LIST THE TASKS Start by listing and numbering at least 10 tasks that the SWL team needs to perform to fulfill the objectives of this chapter. Your list can include SWL Team Tasks and any other tasks that are described in this chapter. For example, Task 3 might be to Identify the actors, and Task 6 might be to Draw a use case diagram.

ANALYZE THE TASKS Now study the tasks to determine the order in which they should be performed. First identify all concurrent tasks, which are not dependent on other tasks. In the example shown in Figure 6-45, Tasks 1, 2, 3, 4, and 5 are concurrent tasks, and could begin at the same time if resources were available.

Other tasks are called dependent tasks, because they cannot be performed until one or more earlier tasks have been completed. For each dependent task, you must identify specific tasks that need to be completed before this task can begin. For example, you would want to identify the actors before you could draw a use case diagram, so Task 6 cannot begin until Task 3 is completed, as Figure 6-45 shows.

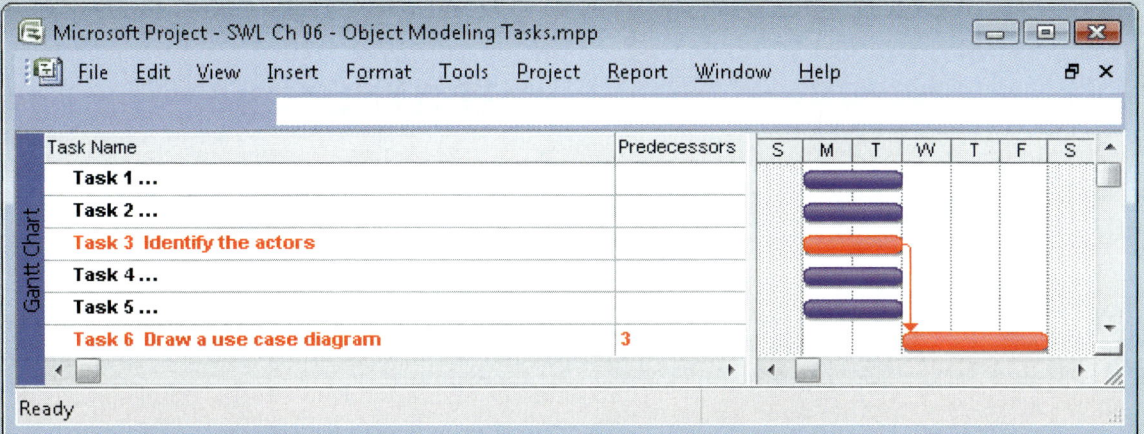

FIGURE 6-45 Tasks 1, 2, 3, 4, and 5 are concurrent tasks that could be performed at the same time. Task 6 is a dependent task that cannot be performed until Task 3 has been completed.

Chapter 3 describes project management tools, techniques, and software. To learn more, you can visit the Features section on your Student Study Tool CD-ROM, or the project management resources library at **scsite.com/sad8e/project**. On the Web, Microsoft offers demo versions, training, and tips for using Project 2007. You also can visit the OpenWorkbench.org site to learn more about this free, open-source software.

CHAPTER 7 Development Strategies

Chapter 7 is the final chapter in the systems analysis phase of the SDLC. This chapter describes software trends, acquisition and development strategies, traditional versus Web-based development, outsourcing versus in-house development, the system requirements document, prototyping, and preparing for the transition to the next SDLC phase — systems design.

INTRODUCTION

OBJECTIVES

When you finish this chapter, you will be able to:

- Describe the concept of Software as a Service
- Define Web 2.0 and cloud computing
- Explain software acquisition alternatives, including traditional and Web-based software development strategies
- Describe software outsourcing options, including offshore outsourcing and the role of service providers
- Explain advantages and disadvantages of in-house software development
- Explain cost-benefit analysis and financial analysis tools
- Explain the differences between a request for proposal (RFP) and a request for quotation (RFQ)
- Describe the system requirements document
- Explain the transition from systems analysis to systems design, and the importance of prototyping
- Discuss guidelines for systems design
- Describe software development trends

The main objective of the systems analysis phase is to build a logical model of the new information system. In Chapters 4, 5, and 6, you learned about requirements modeling, data and process modeling, and object modeling. Chapter 7 describes the remaining activities in the systems analysis phase, which include evaluation of alternative solutions, preparation of the system requirements document, and presentation of the system requirements document to management. The chapter also describes the transition to systems design, prototyping, and systems design guidelines. The chapter concludes with a discussion of trends in software development.

CHAPTER INTRODUCTION CASE: Mountain View College Bookstore

Background: Wendy Lee, manager of college services at Mountain View College, wants a new information system that will improve efficiency and customer service at the three college bookstores.

In this part of the case, Florence Fullerton (systems analyst) and Harry Boston (student intern) are talking about development strategies for the new system.

Participants:	Florence and Harry
Location:	Florence's office, Monday morning, November 16, 2009
Project status:	Florence and Harry developed a logical model that includes data flow diagrams, a data dictionary, and process descriptions. They also created an object model. Now they are ready to discuss development strategies for the new bookstore system.
Discussion topics:	Web-based versus traditional development, cost-benefit analysis, steps in purchasing a software package, transition to systems design, and systems design guidelines

Florence:	Good morning, Harry. Are you ready for the next step?
Harry:	*Sure. Now that we have a logical model of the bookstore system, what comes next?*
Florence:	We're at a transition point between the logical design, which describes what the new system will do, and the physical design phase, which describes how it will be done, including the user interface and physical components. Before we start the physical design, we have to study various systems development options and make a recommendation to Wendy.
Harry:	*What are the options?*
Florence:	Well, some large organizations use Web-based systems hosted by outside vendors who supply and maintain the software. In a sense, the customer rents the application. I checked with our IT director, and she feels we're not ready for that approach. She wants us to implement a system on the college LAN, and then migrate to a Web-based system later. That brings us to the next set of questions.
Harry:	*Such as?*
Florence:	We need to consider our role in the development process. We can build the system ourselves, which is called in-house development. Or we can purchase a software package, which might need some degree of modification to meet our needs. Or we could consider outsourcing options, including hiring an IT consultant to help with development tasks. Either way, we need to do a cost-benefit study.
Harry:	*What about the transition from logical to physical design that you mentioned?*
Florence:	The idea is to take our logical design, which is similar to an architect's proposal, and translate it into a physical design, which is more like a working blueprint. If we decide to develop the system in-house, we'll probably build a prototype, or working model of the system. If we decide to purchase a package, we'll follow a series of steps that will help us select the best product. We'll also talk about systems design guidelines.
Harry:	*When you mention the idea of a blueprint, it sounds like we're getting ready to pick up our tools and go to work.*
Florence:	We sure are. Here's a task list to get us started:

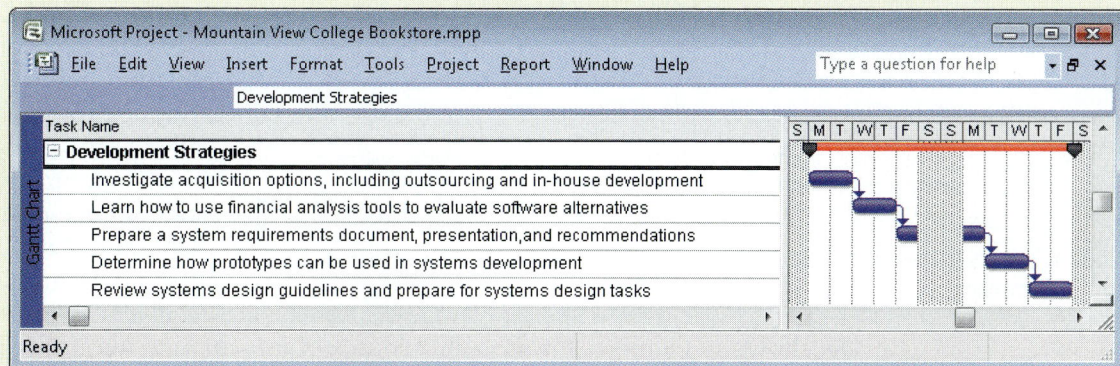

FIGURE 7-1   Typical development strategies task list.

DEVELOPMENT STRATEGIES OVERVIEW

Just a few years ago, a typical company either developed software itself, purchased a software package (which might need some modification), or hired consultants or outside resources to perform the work. Today, that company has many more choices, including application service providers, Web-hosted software options, and firms that offer a variety of enterprise-wide software solutions.

Selecting the best development path is an important decision that requires companies to consider three key topics: the impact of the Internet, software outsourcing options, and in-house software development alternatives. These topics are reviewed in the following sections.

THE IMPACT OF THE INTERNET

The Internet has triggered enormous changes in business methods and operations, and software acquisition is no exception. This section examines a trend that views Software as a Service, the changing marketplace for software, and how Web-based development compares to traditional methods. The section concludes with a description of Internet-related trends, including Web 2.0 and cloud computing.

Software as a Service

In the traditional model, software vendors develop and sell application packages to customers. Typically, customers purchase licenses that give them the right to use the software under the terms of the license agreement. Although this model still accounts for most software acquisition, a new model, called **Software as a Service (SaaS)**, is changing the picture dramatically.

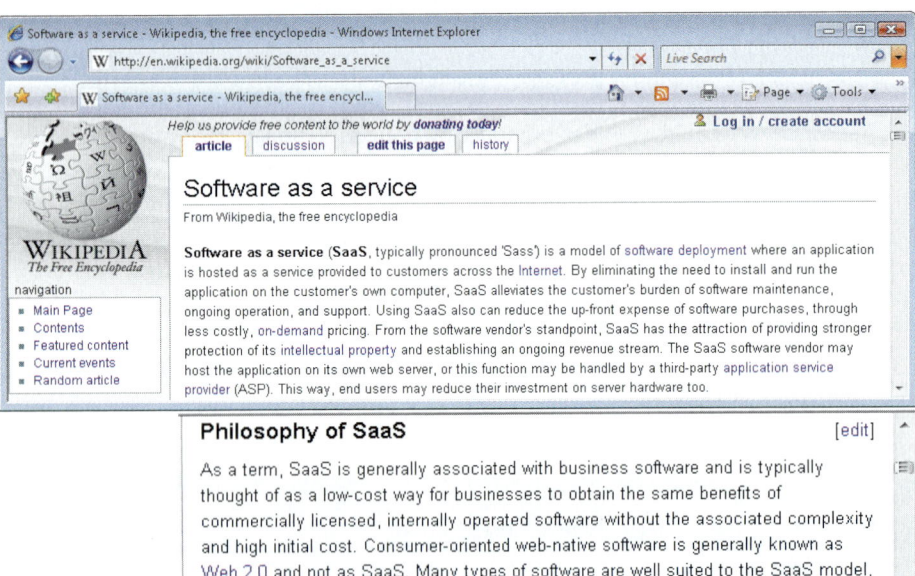

As the Wikipedia article shown in Figure 7-2 points out, SaaS is a model of software deployment where an application is hosted as a service provided to customers over the Internet. The article also notes that SaaS reduces the customer's need for software maintenance, operation, and support.

In a highly competitive marketplace, major vendors constantly strive to deliver new and better solutions. For example, as shown in Figure 7-3, Microsoft claims that its SaaS platform offers the best solution and the most business value. Microsoft also mentions a larger concept called Software-plus-Services, which combines three elements: SaaS, service-oriented development, and Web 2.0, which is discussed later in this section.

FIGURE 7-2 SaaS is a software deployment strategy where an application is hosted as a service to customers.

The Web Host Industry Review shown in Figure 7-4 is an online source of information about SaaS products, trends, and events. In a published report, the Review quoted a Gartner, Inc. prediction that

25% of all new business software will be deployed as a service by 2011, while the value of the SaaS industry will grow to $40 billion.

Traditional vs. Web-Based Systems Development

As a systems analyst, you must consider whether development will take place in a Web-centric framework, or in a traditional environment. This section provides an overview of some of the similarities and differences.

In an Internet-based system, the Web becomes an integral part of the application, rather than just a communication channel, and systems analysts need new application development tools and solutions to handle the new systems. In Chapter 1, you learned that two major Web-based development environments are IBM's **WebSphere** and Microsoft's **.NET**, shown in Figure 7-5 on the next page. IBM describes WebSphere as a set of products specifically designed to support e-business applications across multiple computing platforms. Microsoft regards .NET as a component of the Windows operating system that provides a platform-independent software environment.

Although there is a major trend toward Web-based architecture, many firms rely on traditional systems, either because they are legacy applications that are not easily replaced, or because they do not require a Web component to satisfy user needs. If you need to choose, you should consider some key differences between traditional and Web-based system development. Building the application in a Web-based environment can offer greater benefits, and sometimes greater risks, compared to a traditional environment. The following sections list some characteristics of traditional versus Web-based development.

TRADITIONAL DEVELOPMENT In a traditional systems development environment:

- Systems design is influenced by compatibility issues, including existing hardware and software platforms and legacy system requirements.

- Systems are designed to run on local and wide-area company networks.

- Systems often utilize Internet links and resources, but Web-based features are treated as enhancements rather than core elements of the design.

- Development typically follows one of three main paths: in-house development, purchase of a software package with possible modification, or use of outside consultants.

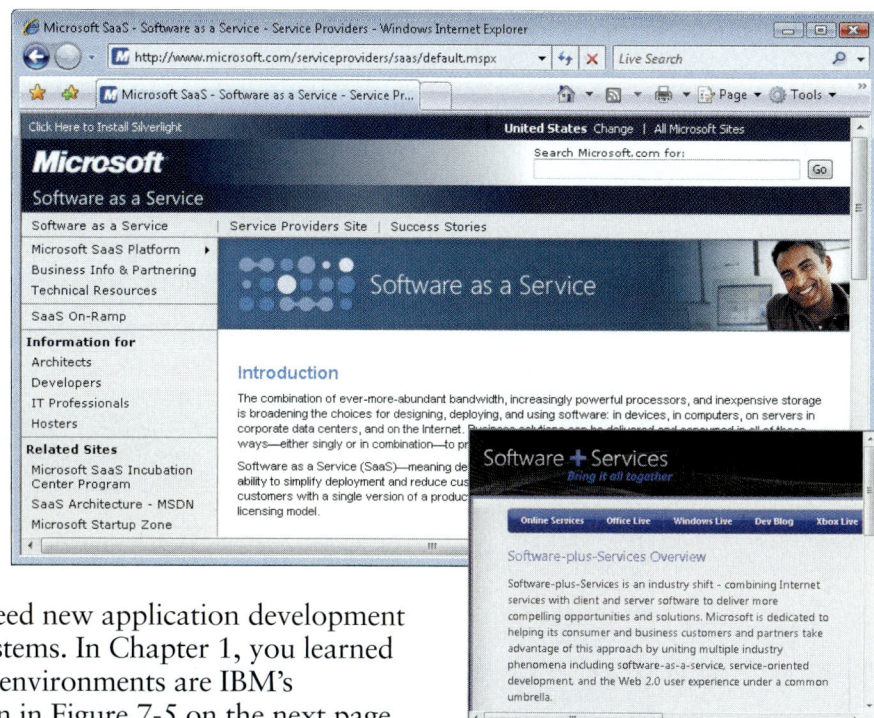

FIGURE 7-3 Microsoft's SaaS platform promises better value and a new concept called Software-plus-Services.

FIGURE 7-4 The Web Host Industry Review (WHIR) is a clearinghouse for SaaS information.

- Scalability can be affected by telecommunications limitations and local network constraints.

- Many applications require substantial desktop computing power and resources.

- Security issues usually are less complex than with Web-based systems, because the system operates on a private telecommunication network, rather than the Internet.

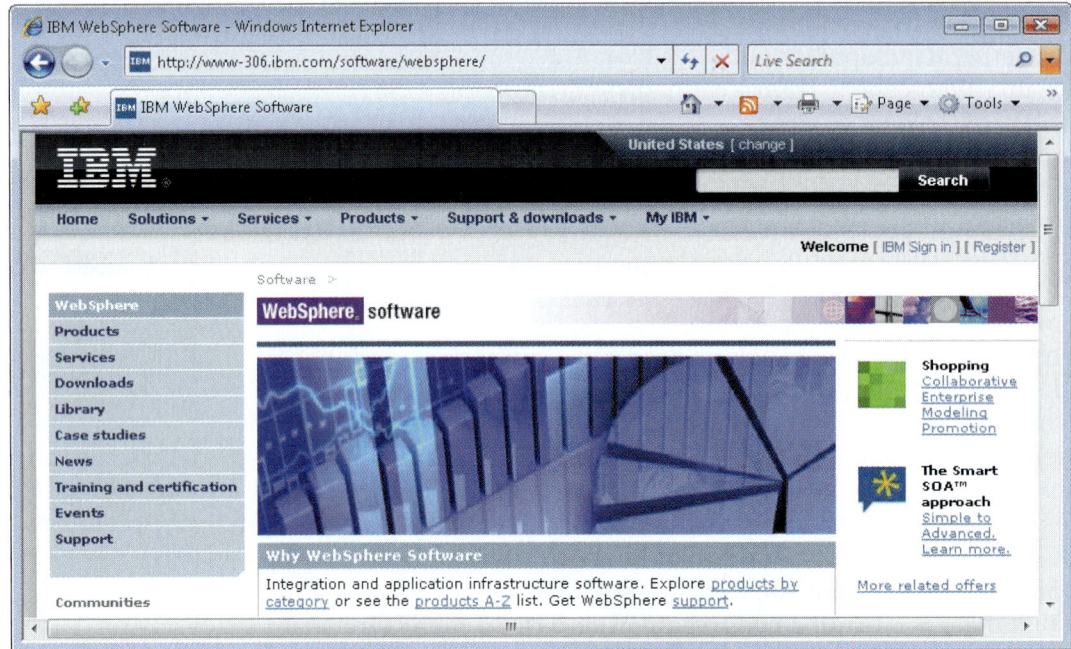

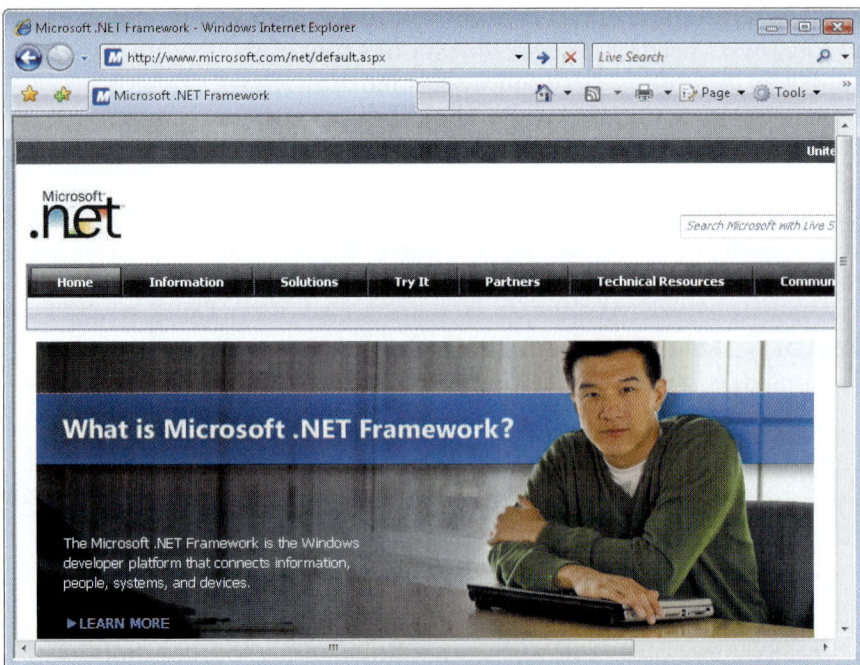

FIGURE 7-5 IBM's WebSphere and Microsoft's .NET are Web-based software development environments.

WEB-BASED DEVELOPMENT In a Web-based systems development environment:

- Systems are developed and delivered in an Internet-based framework such as .NET or WebSphere.

- Internet-based development treats the Web *as* the platform, rather than just a communication channel.

- Web-based systems are easily scalable, and can run on multiple hardware environments.

- Large firms tend to deploy Web-based systems as enterprise-wide software solutions for applications such as customer relationship management, order processing, and materials management.

- Web-based software treats the software application as a service that is less dependent on desktop computing power and resources.

- When companies acquire Web-based software as a *service* rather than a *product* they purchase, they can limit in-house involvement to a minimum and have the vendor install, configure, and maintain the system by paying agreed-upon fees.

- Web-based software usually requires additional layers, called **middleware**, to communicate with existing software and legacy systems.

ON THE WEB

For more information about Web 2.0, visit **scsite.com/ sad8e/more**, locate Chapter 7, and then click the Web 2.0 link.

ON THE WEB

For more information about cloud computing, visit **scsite.com/sad8e/ more**, locate Chapter 7, and then click the Cloud Computing link.

Looking to the Future: Web 2.0 and Cloud Computing

In the dynamic world of IT, no area is more fluid than Internet technology. Two examples of evolving trends are Web 2.0 and cloud computing. Systems analysts should be aware of these concepts and consider them as they plan large-scale systems. Web 2.0 and cloud computing are discussed in more detail in Chapter 10, System Architecture.

Many IT professionals use the term **Web 2.0** to describe a second generation of the World Wide Web that will enable people to collaborate, interact, and share information much more dynamically. This new environment is based on continuously available user applications rather than static HTML Web pages, without limitations regarding the number of users or how they will be able to access, modify, and exchange data. The Web 2.0 platform will enhance interactive experiences, including wikis and blogs, and social-networking applications such as MySpace and Facebook.

The term **cloud computing** refers to the cloud symbol that indicates a network, or the Internet. Some industry leaders predict that cloud computing will offer an overall online software and data environment supported by supercomputer technology. If so, cloud computing would be an ultimate form of SaaS, delivering services and data to users who would need only an Internet connection and a browser. According to the *Business Week* article shown in Figure 7-6, cloud computing could bring enormous computing power to business and personal Internet users.

FIGURE 7-6 Many IT observers believe that cloud computing is a powerful trend.

OUTSOURCING

Outsourcing is the transfer of information systems development, operation, or maintenance to an outside firm that provides these services, for a fee, on a temporary or long-term basis. Outsourcing can refer to relatively minor programming tasks, the rental of software from a service provider, the outsourcing of a basic business process (often called **business process outsourcing, or BPO**), or the handling of a company's entire IT function. Numerous firms and organizations offer information about outsourcing topics and issues. For example, the Outsourcing Center, shown in Figure 7-7, provides free research, case studies, database directories, market intelligence, and updates on trends and best practices in outsourcing as a strategic business solution.

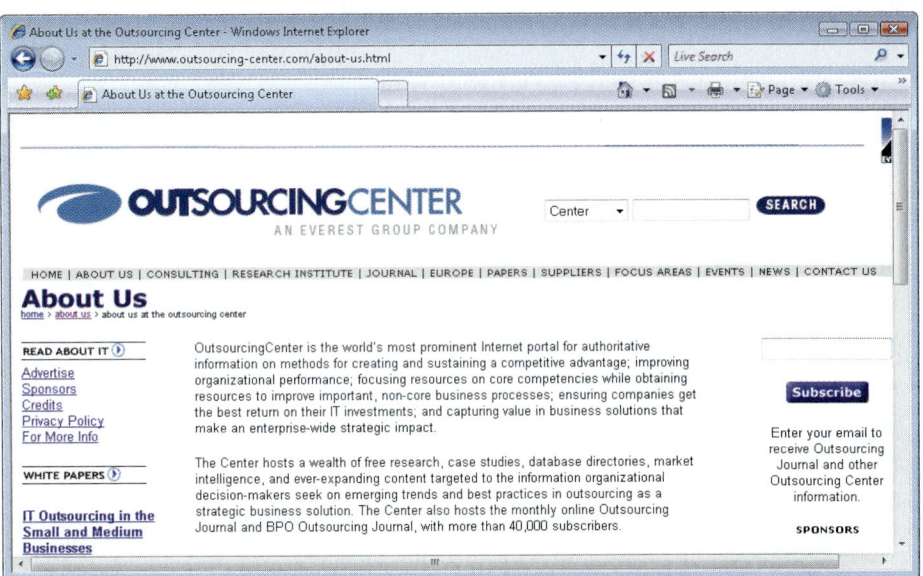

FIGURE 7-7 The Outsourcing Center is dedicated to providing information about outsource trends and practices.

The Growth of Outsourcing

Traditionally, firms outsourced IT tasks as a way of controlling costs and dealing with rapid technological change. While those reasons still are valid, outsourcing has become part of an overall IT strategy for many organizations. The outsourcing trend also has affected software vendors, who have adjusted their marketing accordingly. For example, Oracle Corporation offers a service called Oracle On Demand, which provides E-business applications, as shown in Figure 7-8. Oracle also cites data that shows that businesses spend up to 80% of their IT budgets maintaining existing software and systems, which forces IT managers "... to spend time managing tedious upgrades instead of revenue-generating IT projects."

A firm that offers outsourcing solutions is called a **service provider**. Some service providers concentrate on specific software applications; others offer business services such as order processing and customer billing. Still others offer enterprise-wide software solutions that integrate and manage functions such as accounting, manufacturing, and inventory control.

Two popular outsourcing options involve application service providers and firms that offer Internet business services. These terms are explained in the following sections.

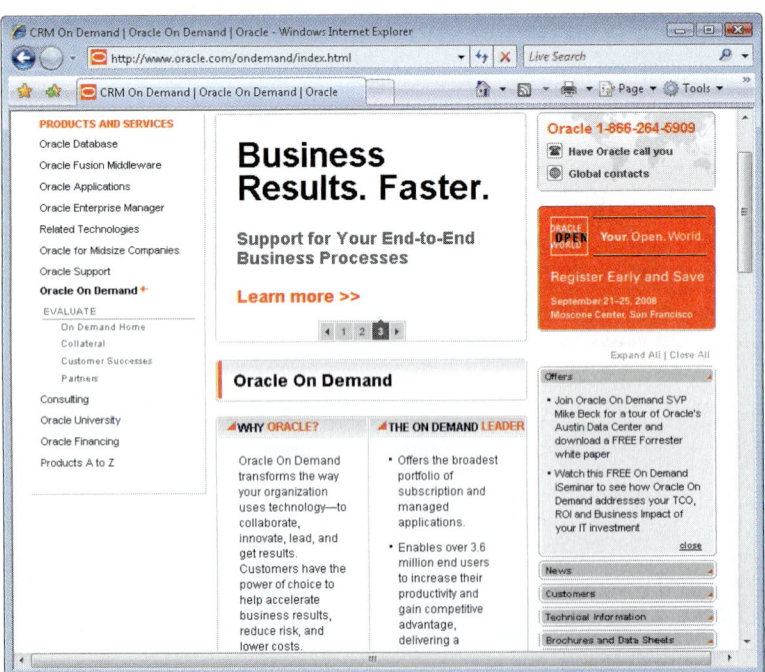

FIGURE 7-8 Oracle Corporation offers a fixed-fee outsourcing plan called Oracle On Demand.

APPLICATION SERVICE PROVIDERS   An **application service provider (ASP)** is a firm that delivers a software application, or access to an application, by charging a usage or subscription fee. An ASP provides more than a license to use the software; it *rents* an operational package to the customer. ASPs typically provide commercially available software such as databases and accounting packages. If a company uses an ASP to supply a data management package, for example, the company does not have to design, develop, implement, or maintain the package. ASPs represent a rapidly growing trend, using the Internet as the primary delivery channel.

ON THE WEB

For more information about application service providers, visit **scsite.com/sad8e/more**, locate Chapter 7, and then click the Application Service Providers link.

INTERNET BUSINESS SERVICES   Some firms offer **Internet business services (IBS)**, which provide powerful Web-based support for transactions such as order processing, billing, and customer relationship management. Another term for IBS is **managed hosting**, because system operations are managed by the outside firm, or host.

An IBS solution is attractive to customers because it offers online data center support, mainframe computing power for mission-critical functions, and universal access via the Internet. Many firms, such as Rackspace, compete in the managed hosting market, as shown in Figure 7-9.

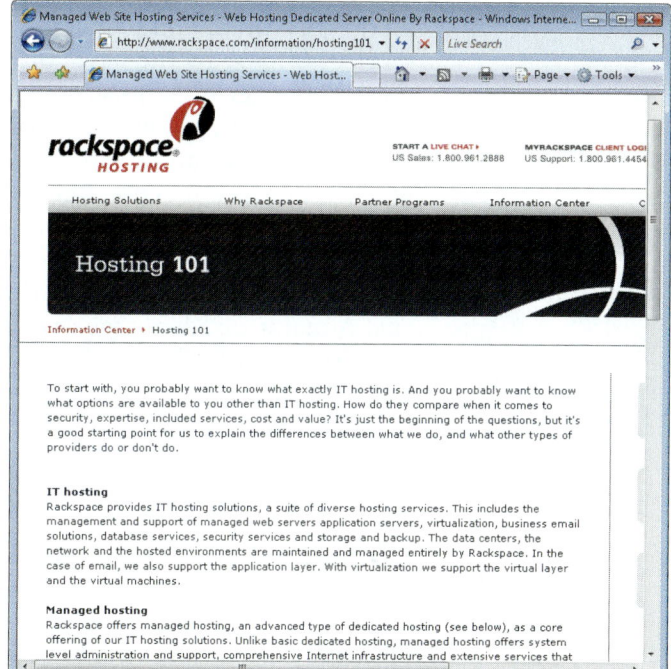

Outsourcing Fees

Firms that offer Software as a Service, rather than a product, have developed fee structures that are based on how the application is used by customers during a specific time period. Several models exist, including fixed fee, subscription, and usage or transaction. A **fixed fee model** uses a set fee based on a specified level of service and user support. An example of a fixed fee model is Oracle's On Demand service. A **subscription model** has a variable fee based on the number of users or workstations that have access to the application. Finally, a **usage model** or **transaction model** charges a variable fee based on the volume of transactions or operations performed by the application.

FIGURE 7-9   Rackspace Corporation offers multi-platform managed hosting and IBS services.

When a company considers outsourcing, it should estimate usage characteristics to determine which fee structure would be most desirable, and then attempt to negotiate a service provider contract based on that model.

Outsourcing Issues and Concerns

When a company decides to outsource IT functions, it takes an important step that can affect the firm's resources, operations, and profitability. Mission-critical IT systems should be outsourced only if the result is a cost-attractive, reliable, business solution that fits the company's long-term business strategy and involves an acceptable level of risk. Moving IT work overseas raises even more issues, including potential concerns about control, culture, communication, and security.

In addition to long-term strategic consequences, outsourcing also can affect day-to-day company operations and can raise some concerns. For example, a company must turn over sensitive data to an external service provider and trust the provider to maintain security, confidentiality, and quality. Also, before outsourcing, a company must review carefully issues relating to insurance, potential liability, licensing and information ownership, warranties, and disaster recovery.

Most important, a company considering outsourcing must realize that the solution can be only as good as the outsourcing firm that provides the service. A dynamic economy can give rise to business failures and uncertainty about the future. In this climate, it is especially important to review the history and financial condition of an outsourcing firm before making a commitment.

Mergers and acquisitions also can affect outsourcing clients. For example, after their merger, Compaq and Hewlett-Packard restructured and streamlined the products and services offered by the new company. Even with large, financially healthy firms such as these, a merger or acquisition can have some impact on clients and customers. If stability is important, an outsourcing client should consider these issues.

Outsourcing can be especially attractive to a company whose volume fluctuates widely, such as a defense contractor. In other situations, a company might decide to outsource application development tasks to an IT consulting firm if the company lacks the time or expertise to handle the work on its own. Outsourcing relieves a company of the responsibility of adding IT staff in busy times and downsizing when the workload lightens. A major disadvantage of outsourcing is that it raises employee concerns about job security. Talented IT people usually prefer positions where the firm is committed to in-house IT development — if they do not feel secure, they might decide to work directly for the service provider.

ON THE WEB

For more information about outsourcing, visit **scsite.com/ sad8e/more**, locate Chapter 7, and then click the Outsourcing link.

Offshore Outsourcing

Offshore outsourcing, or **global outsourcing**, refers to the practice of shifting IT development, support, and operations to other countries. In a trend similar to the outflow of manufacturing jobs over a several-decade period, many firms are sending IT work overseas at an increasing rate.

For example, Dartmouth professor Matthew Slaughter has noted that IT work will move offshore even faster than manufacturing, because it is easier to ship work across networks and telephone lines and put consultants on airplanes than it is to ship bulky raw materials, build factories, and deal with tariffs and transportation issues. Several years ago, the IT consulting firm Gartner, Inc., accurately forecast the steady growth of offshore outsourcing, and predicted that outsourcing would evolve from labor-intensive maintenance and support to higher-level systems development and software design.

In addition to exporting IT jobs, many large multinational firms, including Microsoft and IBM, have opened technical centers in India and other countries. Some observers believe that India might gain as many as 2 million IT jobs in the next decade.

The main reason for offshore outsourcing is the same as domestic outsourcing: lower bottom-line costs. Offshore outsourcing, however, involves some unique risks and concerns. For example, workers, customers, and shareholders in some companies have protested this trend, and have raised public awareness of possible economic impact. Even more important, offshore outsourcing involves unique concerns regarding project control, security issues, disparate cultures, and effective communication with critical functions that might be located halfway around the globe.

CASE IN POINT 7.1: TURNKEY SERVICES

Turnkey Services is an application service provider that offers payroll and tax preparation services for hundreds of businesses in the Midwest. The firm is considering a major expansion into accounting and financial services, and is looking into the possibility of supporting this move by hiring IT subcontractors in several foreign countries. Peter Belmont, Turnkey's president, has asked you to help him reach a decision. Specifically, he wants you to cite the pros and cons of offshore outsourcing. He expects you to perform Internet research on this topic, and he wants you to present your views at a meeting of Turnkey managers next week.

IN-HOUSE SOFTWARE DEVELOPMENT OPTIONS

In addition to numerous outsourcing options, a company can choose to develop its own systems, or purchase, possibly customize, and implement a software package. These development alternatives are shown in Figure 7-10. Although many factors influence this decision, the most important consideration is the total cost of ownership (TCO), which was explained in Chapter 4. In addition to these options, companies also develop user applications designed around commercial software packages, such as Microsoft Office, to improve user productivity and efficiency.

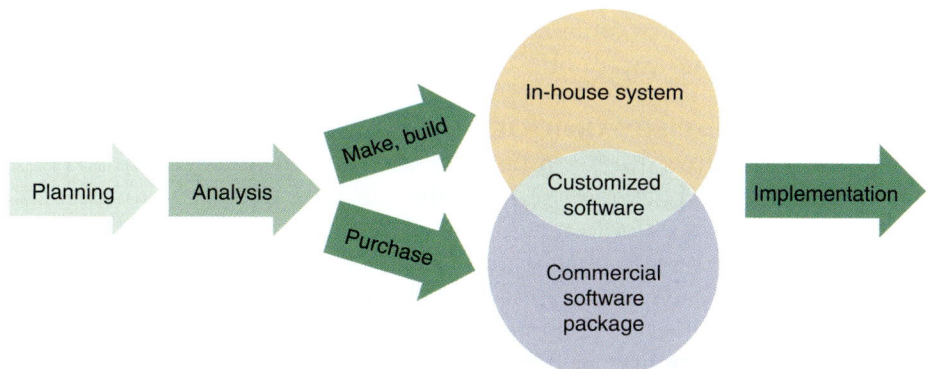

FIGURE 7-10 Instead of outsourcing, a company can choose to develop a system in-house, or purchase and possibly customize a commercial package.

Make or Buy Decision

The choice between developing versus purchasing software often is called a **make or buy**, or **build or buy** decision. The company's IT department makes, builds, and develops **in-house software**. A **software package** is obtained from a vendor or application service provider.

    The package might be a standard commercial program or a customized package designed specifically for the purchaser. Companies that develop software for sale are called **software vendors**. A firm that enhances a commercial package by adding custom features and configuring it for a particular industry is called a **value-added reseller (VAR)**.

    Software packages are available for every type of business activity. A software package that can be used by many different types of organizations is called a **horizontal application**. An accounting package is a good example of a horizontal application because it can be utilized by many different businesses, or separate divisions that exist in large, diversified companies.

ON THE WEB

For more information about value-added resellers, visit **scsite.com/ sad8e/more**, locate Chapter 7, and then click the Value-Added Resellers link.

FIGURE 7-11 Hotel chains require vertical applications to support reservation systems and information needs that are unique to the hotel industry.

In contrast, a software package developed to handle information requirements for a specific type of business is called a **vertical application**. For example, organizations with special system requirements include colleges, banks, hospitals, insurance companies, construction companies, real estate firms, and airlines. A hotel chain might require a vertical application for its guest reservation system, as shown in Figure 7-11, and use horizontal applications for basic business needs, such as payroll processing and accounts payable.

Of the in-house software acquisition options — developing a system, buying a software package, or customizing a software package — each has advantages, disadvantages, and cost considerations, as shown in Figure 7-12. These software acquisition options are described in detail in the following sections.

Developing Software In-House

With an enormous variety of software packages available to handle horizontal and vertical business operations, why would a firm choose to develop its own software? Typically, companies choose in-house development to satisfy unique business requirements, to minimize changes in business procedures and policies, to meet constraints of existing systems and existing technology, and to develop internal resources and capabilities.

SATISFY UNIQUE BUSINESS REQUIREMENTS Companies often decide to develop software in-house because no commercially available software package can meet their unique business requirements. A college, for example, needs a course scheduling system based on curriculum requirements, student demand, classroom space, and available instructors. A package delivery company needs a system to identify the best combination of routes and loading patterns for the company's fleet of delivery trucks. If existing software packages cannot handle those requirements, then in-house developed software might be the only choice.

REASONS FOR IN-HOUSE DEVELOPMENT	REASONS FOR PURCHASING A SOFTWARE PACKAGE
Satisfy unique business requirements	Lower costs
Minimize changes in business procedures and policies	Requires less time to implement
Meet constraints of existing systems	Proven reliability and performance benchmarks
Meet constraints of existing technology	Requires less technical development staff
Develop internal resources and capabilities	Future upgrades provided by the vendor
Satisfy unique security requirements	Obtain input from other companies

FIGURE 7-12 Companies consider various factors when comparing in-house development with the purchase of a software package.

MINIMIZE CHANGES IN BUSINESS PROCEDURES AND POLICIES A company also might choose to develop its own software if available packages will require changes in current business operations or processes. Installing a new software package almost always requires some degree of change in how a company does business; however, if the installation of a purchased package will be too disruptive, the organization might decide to develop its own software instead.

MEET CONSTRAINTS OF EXISTING SYSTEMS Any new software installed must work with existing systems. For example, if a new budgeting system must interface with an existing accounting system, finding a software package that works correctly with the existing accounting system might prove difficult. If so, a company could develop its own software to ensure that the new system will interface with the old system.

MEET CONSTRAINTS OF EXISTING TECHNOLOGY Another reason to develop software in-house is that the new system must work with existing hardware and legacy systems. That could require a custom design not commercially available. Some companies have older microcomputer workstations that cannot handle graphics-intensive software or high-speed Internet access. In that situation, the company either must upgrade the environment or must develop in-house software that can operate within the constraints of the existing hardware. As a systems analyst, you addressed the issue of technical feasibility during the preliminary investigation. Now, in the systems analysis phase, you must examine the advantages and disadvantages of in-house software development to decide whether it is justifiable.

DEVELOP INTERNAL RESOURCES AND CAPABILITIES By designing a system in-house, companies can develop and train an IT staff that understands the organization's business functions and information support needs. Many firms feel that in-house IT resources and capabilities provide a competitive advantage because an in-house team can respond quickly when business problems or opportunities arise. For example, if a company lacks internal resources, it must depend on an outside firm for vital business support. Also, outsourcing options might be attractive, but a series of short-term solutions would not necessarily translate into lower TCO over the long term. Top managers often feel more comfortable with an internal IT team to provide overall guidance and long-term stability.

Purchasing a Software Package

If a company decides not to outsource, a commercially available software package might be an attractive alternative to developing its own software. Advantages of purchasing a software package over developing software in-house include lower costs, less time to implement a system, proven reliability and performance benchmarks, less technical development staff, future upgrades that are provided by the vendor, and the ability to obtain input from other companies who already have implemented the software.

LOWER COSTS Because many companies use software packages, software vendors spread the development costs over many customers. Compared with software developed in-house, a software package almost always is less expensive, particularly in terms of initial investment.

REQUIRES LESS TIME TO IMPLEMENT When you purchase a package, it already has been designed, programmed, tested, and documented. The in-house time normally spent on those tasks, therefore, is eliminated. Of course, you still must install the software and integrate it into your systems environment, which can take a significant amount of time.

PROVEN RELIABILITY AND PERFORMANCE BENCHMARKS If the package has been on the market for any length of time, any major problems probably have been detected already and corrected by the vendor. If the product is popular, it almost certainly has been rated and evaluated by independent reviewers.

REQUIRES LESS TECHNICAL DEVELOPMENT STAFF Companies that use commercial software packages often are able to reduce the number of programmers and systems analysts on the IT staff. Using commercial software also means that the IT staff can concentrate on systems whose requirements cannot be satisfied by software packages.

FUTURE UPGRADES PROVIDED BY THE VENDOR Software vendors regularly upgrade software packages by adding improvements and enhancements to create a new version or release. A new release of a software package, for example, can include drivers to support a new laser printer or a new type of data storage technology. In many cases, the vendor receives input and suggestions from current users when planning future upgrades.

INPUT FROM OTHER COMPANIES Using a commercial software package means that you can contact users in other companies to obtain their input and impressions. You might be able to try the package or make a site visit to observe the system in operation before making a final decision.

Customizing a Software Package

If the standard version of a software product does not satisfy a company's requirements, the firm can consider adapting the package to meet its needs. Three ways to customize a software package are:

1. You can purchase a basic package that vendors will customize to suit your needs. Many vendors offer basic packages in a standard version with add-on components that are configured individually. A vendor offers options when the standard application will not satisfy all customers. A human resources information system is a typical example, because each company handles employee compensation and benefits differently. If you need assistance in making a determination, firms such as Ideas International offer services to help you select and configure a system, as shown in Figure 7-13.

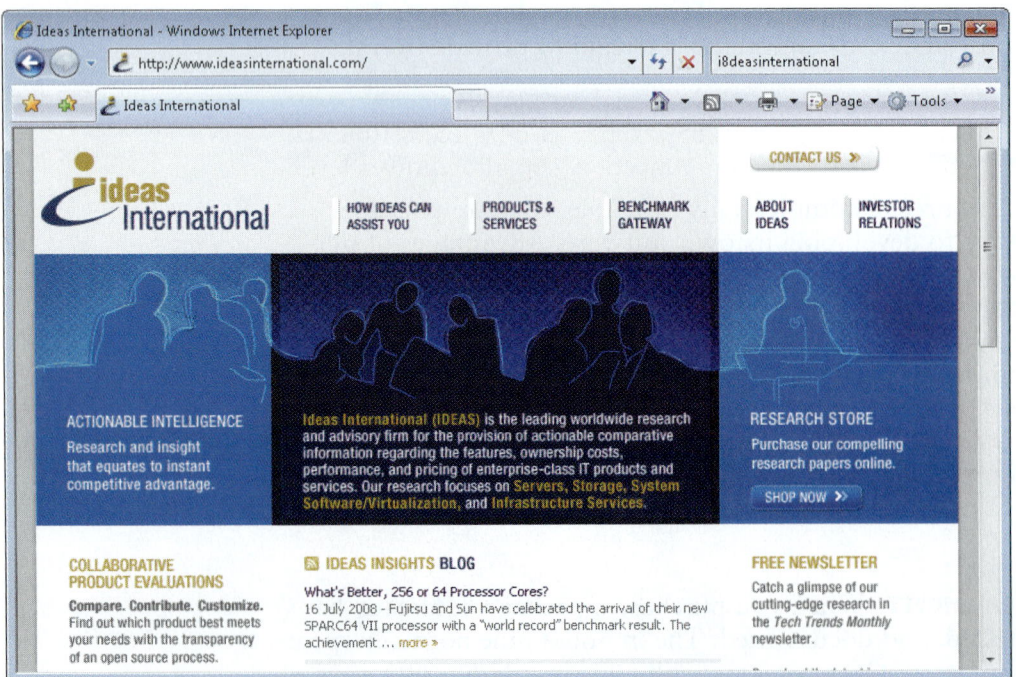

FIGURE 7-13 Firms such as Ideas International offer services to help customers select and configure a system.

2. You can negotiate directly with the software vendor to make enhancements to meet your needs by paying for the changes.

3. You can purchase the package and make your own modifications, if this is permissible under the terms of the software license. A disadvantage of this approach is that systems analysts and programmers might be unfamiliar with the software and will need time to learn the package and make the modifications correctly.

Additionally, some advantages of purchasing a standard package can be lost if the product must be customized. If the vendor does the customizing, the modified package probably will cost more and take longer to obtain. Another issue is future support: Although vendors regularly upgrade their standard software packages, they might not upgrade a customized version. In addition, if the modifications are done by the company purchasing the software, when a new release of the package becomes available, the company might have to modify the new version.

Creating User Applications

Business requirements sometimes can be fulfilled by a user application, rather than a formal information system or commercial package. User applications are examples of user productivity systems, which were discussed in Chapter 1.

A **user application** utilizes standard business software, such as Microsoft Word or Microsoft Excel, which has been configured in a specific manner to enhance user productivity. For example, to help a sales rep respond rapidly to customer price requests, an IT support person can set up a form letter with links to a spreadsheet that calculates incentives and discounts. In addition to configuring the software, the IT staff can create a **user interface**, which includes screens, commands, controls, and features that enable users to interact more effectively with the application. User interface design is described in Chapter 8.

In some situations, user applications offer a simple, low-cost solution. Most IT departments have a backlog of projects, and IT solutions for individuals or small groups do not always receive a high priority. At the same time, application software is more powerful, flexible, and user-friendly than ever. Companies such as Microsoft and Corel offer software suites and integrated applications that can exchange data with programs that include tutorials, wizards, and Help features to guide less experienced users who know what they need to do but do not know how to make it happen.

Many companies empower lower-level employees by providing more access to data and more powerful data management tools. The main objective is to allow lower-level employees more access to the data they require to perform their jobs, with no intervention from the IT department. This can be accomplished by creating effective user interfaces for company-wide applications such as accounting, inventory, and sales systems. Another technique is to customize standard productivity software, such as Microsoft Word or Microsoft Excel, to create user applications. In either case, empowerment makes the IT department more productive because it can spend less time responding to the daily concerns and data needs of users and more time on high-impact systems development projects that support strategic business goals.

Empowerment reduces costs and makes good business sense, but companies that adopt this approach must provide the technical support that empowered users require. In most large and medium-sized companies, a **help desk**, or **information center** (**IC**), within the IT department is responsible for providing user support. The IC staff offers services such as hotline assistance, training, and guidance to users who need technical help.

Once they learn an application, many users can perform tasks that once required a programmer. Some user applications have powerful **screen generators** and **report generators** that allow users to design their own data entry forms and reports. For example, as shown in Figure 7-14 on the next page, Microsoft Access includes a Form

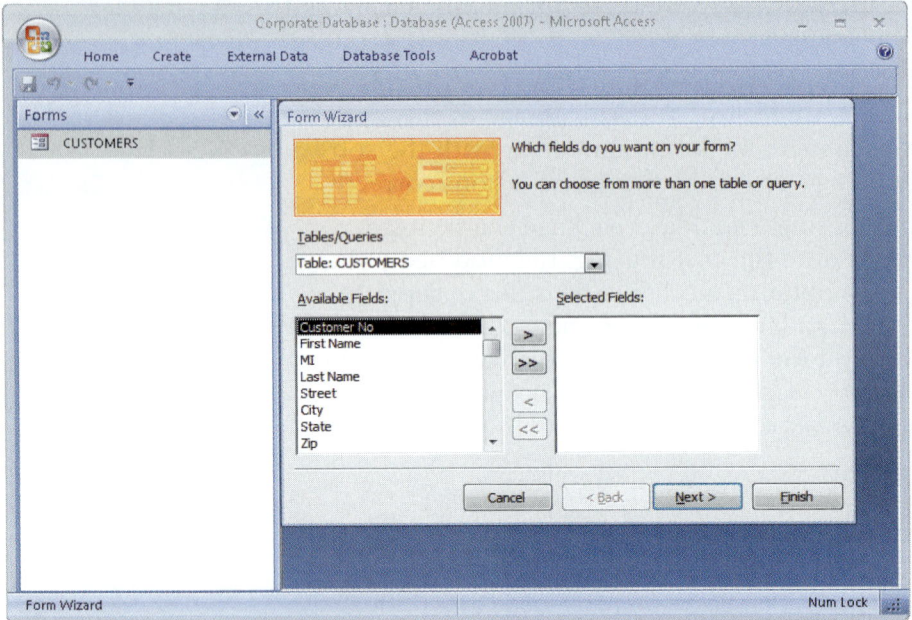

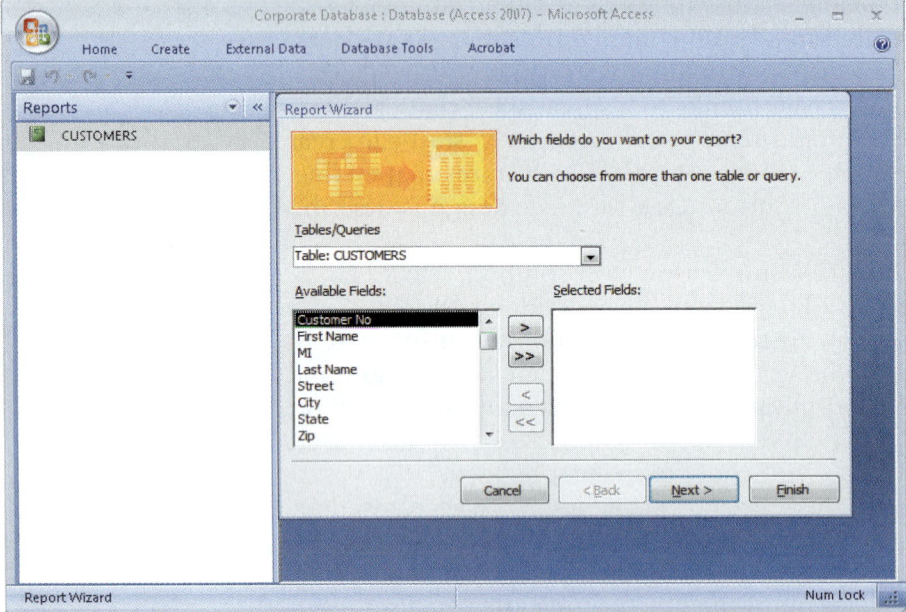

FIGURE 7-14 Microsoft Access includes Form Wizard and Report Wizard tools that ask a series of questions, and then create the form or report.

Wizard and a Report Wizard, which are menu-driven tools that can create screen forms and reports. These design tools allow users to design specific input and output views that meet their operational needs — with little or no assistance required from the IT staff.

Users typically require spreadsheets, database management programs, and other software packages to meet their information needs. If user applications access corporate data, you must provide appropriate controls to ensure data security and integrity. For example, some files should be hidden totally from view; others should have **read-only properties** so users can view, but not change, the data. For security reasons, companies usually restrict user applications to PC-based systems within a user's department.

ROLE OF THE SYSTEMS ANALYST

At this point in the systems development process, the company must decide whether to use an outsourcing option, develop software in-house, acquire a software package, develop user applications, or select some combination of these solutions. The decision will affect the remaining SDLC phases and your involvement as a systems analyst. The decision to develop software in-house, for example, will require more participation from the systems analyst than outsourcing or choosing a commercial package. Management usually makes a determination after receiving written recommendations from the IT staff and a formal presentation, which is described later in this chapter.

Even a single system can use a mix of software alternatives. For example, a company might purchase a standard software package to process its payroll, and then develop its own software to handle the interface between the payroll package and the company's in-house manufacturing cost analysis system.

The evaluation and selection of alternatives is not a simple process. The objective is to obtain the product with the lowest total cost of ownership, but actual cost and performance can be difficult to forecast. With a large number of choices, how do you select the best alternative?

When selecting hardware and software, systems analysts often work as an **evaluation and selection team**. A team approach ensures that critical factors are not overlooked and that a sound choice is made. The evaluation and selection team also must include users, who will participate in the selection process and feel a sense of ownership in the new system.

The primary objective of the evaluation and selection team is to eliminate system alternatives that will not meet requirements, rank the alternatives that are feasible, and present the viable alternatives to management for a final decision. The process begins with a careful study of the costs and benefits of each alternative, as explained in the following section.

ANALYZING COST AND BENEFITS

In Chapter 2, you learned that economic feasibility is one of the four feasibility measurements that are made during the preliminary investigation of a systems request. Now, at the end of the systems analysis phase of the SDLC, you must apply financial analysis tools and techniques to evaluate development strategies and decide how the project will move forward. Part 3 of the Systems Analyst's Toolkit describes three popular tools, which are payback analysis, return on investment (ROI), and net present value (NPV). These tools, and others, can be used to determine total cost of ownership (TCO), which was described in Chapter 4. At this stage, you will identify specific systems development strategies and choose a course of action. For example, a company might find that its total cost of ownership will be higher if it develops a system in-house, compared with outsourcing the project or using an ASP.

An accurate forecast of TCO is critical, because nearly 80% of total costs occur *after* the purchase of the hardware and software, according to Gartner, Inc. An IT department can develop its own TCO estimates, or use TCO calculation tools offered by vendors. For example, as shown in Figure 7-15 on the next page, HP and Oracle offer an online TCO calculator that includes a questionnaire and a graphical display of results.

Financial Analysis Tools

Part 3 of the Systems Analyst's Toolkit explains how to use three main cost analysis tools: payback analysis, return on investment (ROI), and net present value (NPV). **Payback analysis** determines how long it takes an information system to pay for itself through reduced costs and increased benefits. **Return on investment (ROI)** is a percentage rate that compares the total net benefits (the return) received from a project to the total costs (the investment) of the project. The **net present value (NPV)** of a project is the total value of the benefits minus the total value of the costs, with both costs and benefits adjusted to reflect the point in time at which they occur.

ON THE WEB

For more information about financial analysis tools, visit **scsite.com/ sad8e/more**, locate Chapter 7, and then click the Financial Analysis Tools link.

TOOLKIT TIME

The Financial Analysis tools in Part 3 of the Systems Analyst's Toolkit can help you analyze project costs, benefits, and economic feasibility. To learn more about these tools, turn to Part 3 of the four-part Toolkit that follows Chapter 12.

CASE IN POINT 7.2: STERLING ASSOCIATES

Joan Sterling is CEO and principal stockholder of Sterling Associates, which specializes in advising clients on IT projects and information systems development. Joan is creating a brochure for prospective new clients. She wants you to develop a section that describes payback analysis, ROI, and NPV in simple terms, and mentions the pros and cons of each financial analysis tool. She suggested that you start by reviewing the material in Part 3 of the Systems Analyst's Toolkit.

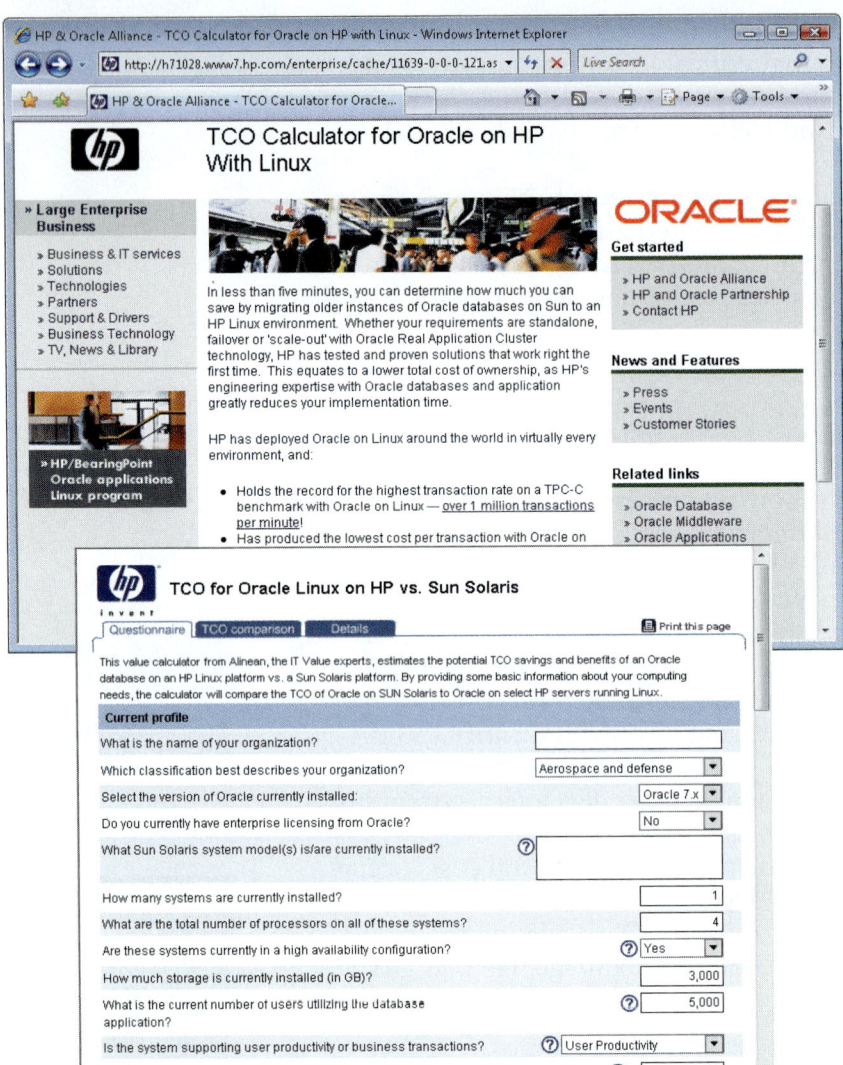

Cost-Benefit Analysis Checklist

Companies use all three financial analysis tools to evaluate various development strategies. The best way to apply the tools is to develop a cost-benefit checklist with the following steps:

- List each development strategy being considered.

- Identify all costs and benefits for each alternative. Be sure to indicate when costs will be incurred and benefits realized.

- Consider future growth and the need for scalability.

- Include support costs for hardware and software.

- Analyze various software licensing options, including fixed fees and formulas based on the number of users or transactions.

- Apply the financial analysis tools to each alternative.

- Study the results and prepare a report to management.

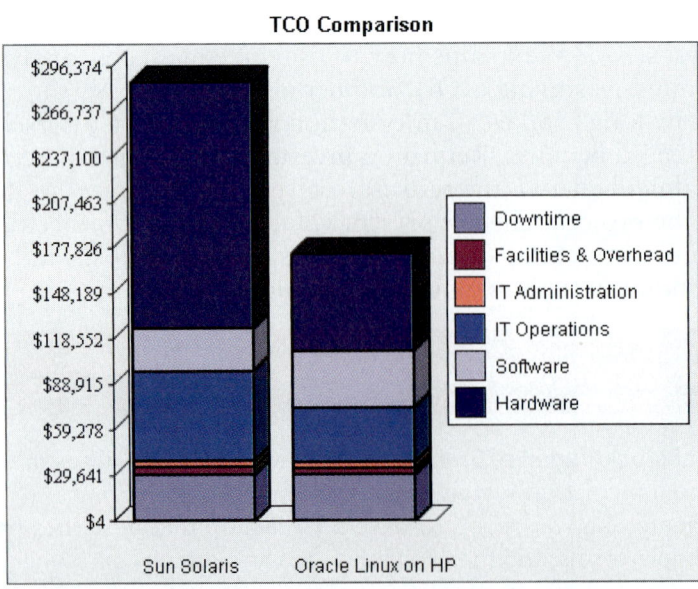

FIGURE 7-15 HP and Oracle offer an online TCO calculator.

THE SOFTWARE ACQUISITION PROCESS

Although each situation is different, the following section describes a typical example of the issues and tasks involved in software acquisition.

Step 1: Evaluate the Information System Requirements

Based on your analysis of the system requirements, you must identify the system's key features; consider network and Web-related issues; estimate volume and future growth; specify any hardware, software, or personnel constraints; and prepare a request for proposal or quotation.

IDENTIFY KEY FEATURES Whether you are considering in-house development or outsourcing options, you must develop a clear, detailed list of features that can serve as an overall specification for the system. Using the data you gathered during fact-finding, which was discussed in Chapter 4, you must list all system requirements and critical features. This information will be included in the system requirements document, which is the end product of the SDLC systems analysis phase.

CONSIDER NETWORK AND WEB-RELATED ISSUES As you evaluate the system requirements, you must consider network and Web-related issues. You must decide whether the system will run on a network, the Internet, or a company intranet, and build these requirements into the design. Also, you must determine whether the system will exchange data with vendor or customer systems, and ensure that the system will be compatible.

ESTIMATE VOLUME AND FUTURE GROWTH You need to know the current volume of transactions and forecast future growth. Figure 7-16 shows volume estimates for an order processing system. In addition to current levels, the figure displays two forecasts, one based on the existing order processing procedures and another that assumes a new Web site is operational.

Online Order Processing System Estimated Activity During Next 12-Month Period

	CURRENT LEVEL	FUTURE GROWTH BASED ON EXISTING ORDER PROCESSING PROCEDURES	FUTURE GROWTH ASSUMING NEW WEB SITE IS OPERATIONAL
Customers	36,500	40,150	63,875
Daily Orders	1,435	1,579	2,811
Daily Order Lines	7,715	7,893	12,556
Sales Reps	29	32	12
Order Processing Support Staff	2	4	3
Products	600	650	900

FIGURE 7-16 Volume estimate for an order processing system showing current activity levels and two forecasts: one based on the existing order processing procedures and another that assumes a new Web site is operational.

A comparison of the two forecasts shows that the Web site will generate more new customers, process almost 80% more orders, and substantially reduce the need for sales reps and support staff. If you are considering in-house development, you must make sure that your software and hardware can handle future transaction volumes and data storage requirements. Conversely, if you are considering outsourcing, volume and usage data is essential to analyze ASP fee structures and develop cost estimates for outsourcing options.

SPECIFY HARDWARE, SOFTWARE, OR PERSONNEL CONSTRAINTS You must determine whether existing hardware, software, or personnel issues will affect the acquisition decision. For example, if the firm has a large number of legacy systems or if an ERP strategy has been adopted, these factors will have an impact on the decision. Also, you must investigate the company's policy regarding outsourcing IT functions, and whether outsourcing is part of a long-term strategy. With regard to personnel issues, you must define in-house staffing requirements to develop, acquire, implement, and maintain the system — and determine whether the company is willing to commit to those staffing levels versus an outsourcing option.

PREPARE A REQUEST FOR PROPOSAL OR QUOTATION To obtain the information you need to make a decision, you should prepare a request for proposal or a request for quotation. The two documents are similar but used in different situations, based on whether or not you have selected a specific software product.

A **request for proposal (RFP)** is a document that describes your company, lists the IT services or products you need, and specifies the features you require. An RFP helps ensure that your organization's business needs will be met. An RFP also spells out the service and support levels you require. Based on the RFP, vendors can decide if they have a product that will meet your needs. RFPs vary in size and complexity, just like the systems they describe. An RFP for a large system can contain dozens of pages with unique requirements and features. You can use an RFP to designate some features as essential and others as desirable. An RFP also requests specific pricing and payment terms.

Figure 7-17 shows an example of a ready-made RFP template offered by Infotivity Technologies. Notice that the vendor can choose from a range of responses, and also add comments.

Figure 7-18 shows a software template developed by RFP Evaluation Centers. This organization provides free examples of RFP templates, sample letters, and related literature.

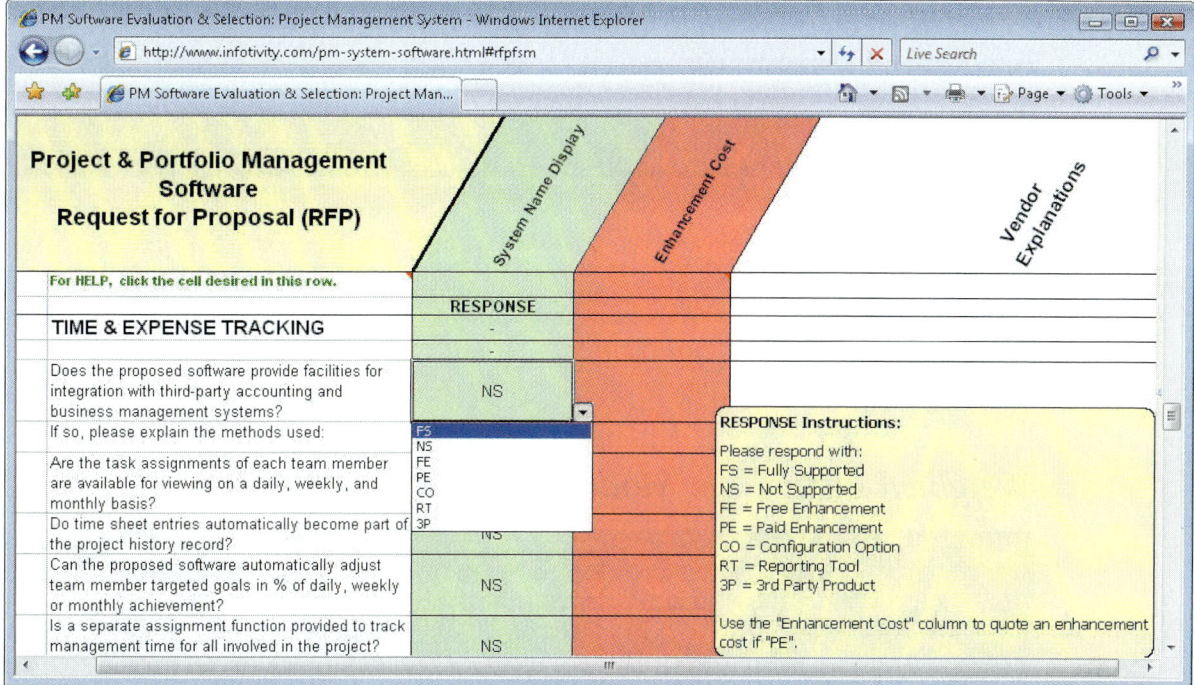

FIGURE 7-17 Infotivity Technologies offers a ready-made RFP template that allows a wide range of reponses and comments.

When you evaluate several responses to an RFP, you might find it helpful to use an evaluation model. An **evaluation model** is a technique that uses a common yardstick to measure and compare vendor ratings.

Figure 7-19 on the next page shows two evaluation models for a network project. The evaluation model at the top of the figure simply lists the key elements and each vendor's score. The model at the bottom of the figure adds a weight factor. In this example, each element receives a rating based on its relative importance. Although the initial scores are the same in both models, notice that vendor A has the highest point total in the top example, but vendor C emerges as the best in the weighted model.

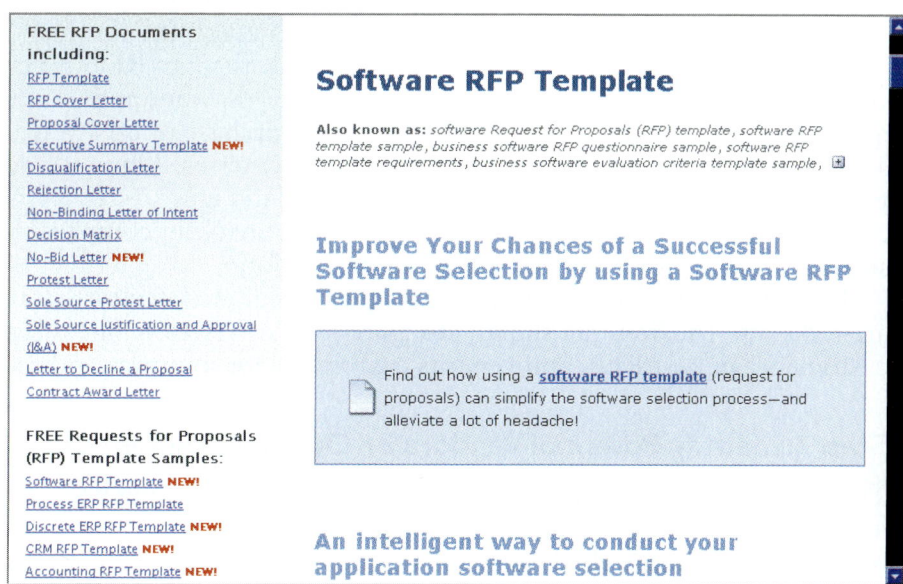

FIGURE 7-18 The RFP Evaluation Centers site offers free examples of RFP templates, sample letters, and related literature.

Unweighted Evaluation Model for a Network Project

Instructions: Rate each vendor on a scale from 1(low) to 10 (high), then add vendor scores to calculate total points.			
	VENDOR A	**VENDOR B**	**VENDOR C**
Price	6	5	9
Completion Date	2	5	8
Layout/Design	8	8	5
References	10	6	3
TOTAL POINTS	26	24	25

Weighted Evaluation Model for a Network Project

Instructions: Rate each vendor on a scale from 1(low) to 10 (high), then multiply the vendor's score by the weight factor. Add vendor scores to calculate total points.			
	WEIGHT FACTOR	**VENDOR A**	**VENDOR B**	**VENDOR C**
Price	25	6 * 25 = 150	5 * 25 = 125	9 * 25 = 225
Completion Date	25	2 * 25 = 50	5 * 25 = 125	8 * 25 = 200
Layout/Design	35	8 * 35 = 280	8 * 35 = 280	5 * 35 = 175
References	15	10 * 15 = 150	6 * 15 = 90	3 * 15 = 45
TOTAL POINTS	100	630	620	645

FIGURE 7-19 The three vendors have the same initial ratings, but the two evaluation models produce different results. In the unweighted model at the top of the figure, vendor A has the highest total points. However, after applying weight factors, vendor C is the winner, as shown in the model at the bottom of the figure.

Evaluation models can be used throughout the SDLC, and you will find them a valuable tool. You can use a spreadsheet program to build an evaluation model, experiment with different weighting factors, and graph the results.

A **request for quotation (RFQ)** is more specific than an RFP. When you use an RFQ, you already know the specific product or service you want and you need to obtain price quotations or bids. RFQs can involve outright purchase or a variety of leasing options and can include maintenance or technical support terms. Many vendors provide convenient RFQ forms on their Web sites, as shown in Figure 7-20. RFPs and RFQs have the same objective: to obtain vendor replies that are clear, comparable, and responsive so you can make a well-informed selection decision.

In today's fast-paced IT marketplace, traditional methods for obtaining RFPs often are too slow. The Web site shown in Figure 7-21 offers an online meeting place where customers can post RFPs and vendors can reply with solutions and bids.

Step 2: Identify Potential Vendors or Outsourcing Options

The next step is to identify potential vendors or outsourcing providers. The Internet is a primary marketplace for all IT products and services, and you can find descriptive information on the Web about all major products and acquisition alternatives.

If you need to locate vertical applications for specific industries, you can research industry trade journals or Web sites to find reviews for industry-specific software. Industry trade groups often can direct you to companies that offer specific software solutions.

Another approach is to work with a consulting firm. Many IT consultants offer specialized services that help companies select software packages. A major advantage of using a consultant is that you can tap into broad experience that is difficult for any one company to acquire. Consultants can be located by contacting professional organizations or industry sources, or simply by searching the Internet. Using a consultant involves additional expense but can prevent even more costly mistakes.

Another valuable resource is the Internet bulletin board system that contains thousands of forums, called **newsgroups**, that cover every imaginable topic. Newsgroups are excellent sources of information and good places to exchange ideas with other analysts and IT professionals. You can search the Web for newsgroups that interest you, or you can visit the sites of specific companies, such as Microsoft, that provide a valuable source of information for IT professionals, including blogs, technical chats, newsgroups, Webcasts, and other resources, as shown in Figure 7-22 on the next page.

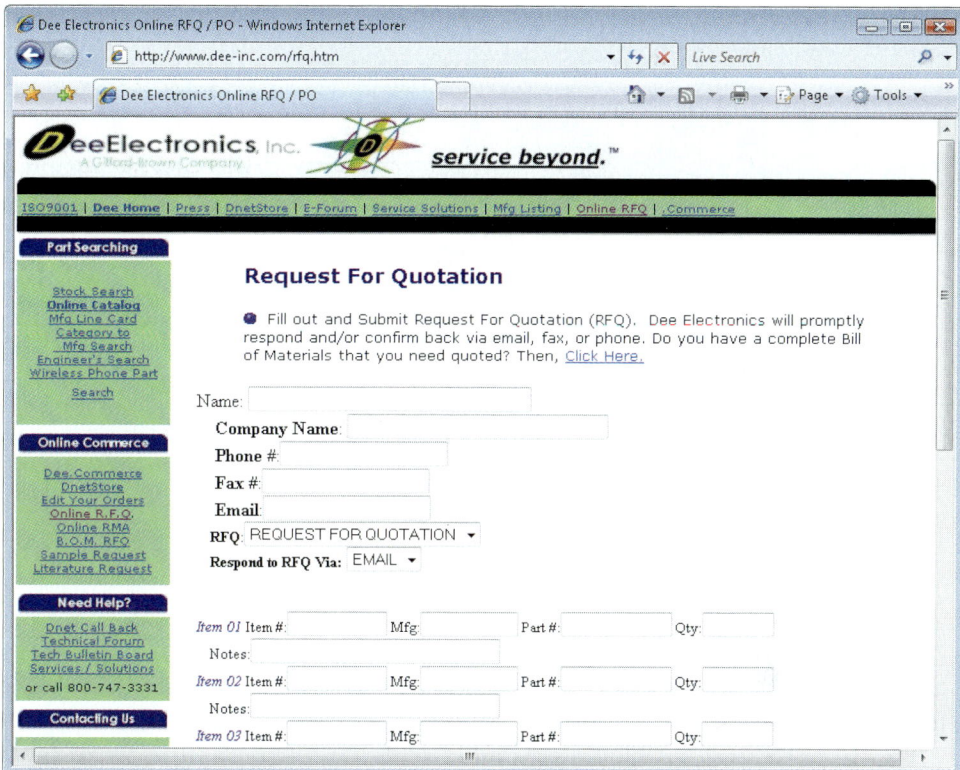

FIGURE 7-20 Many vendors provide convenient RFQ forms on their Web sites, as shown in this example.

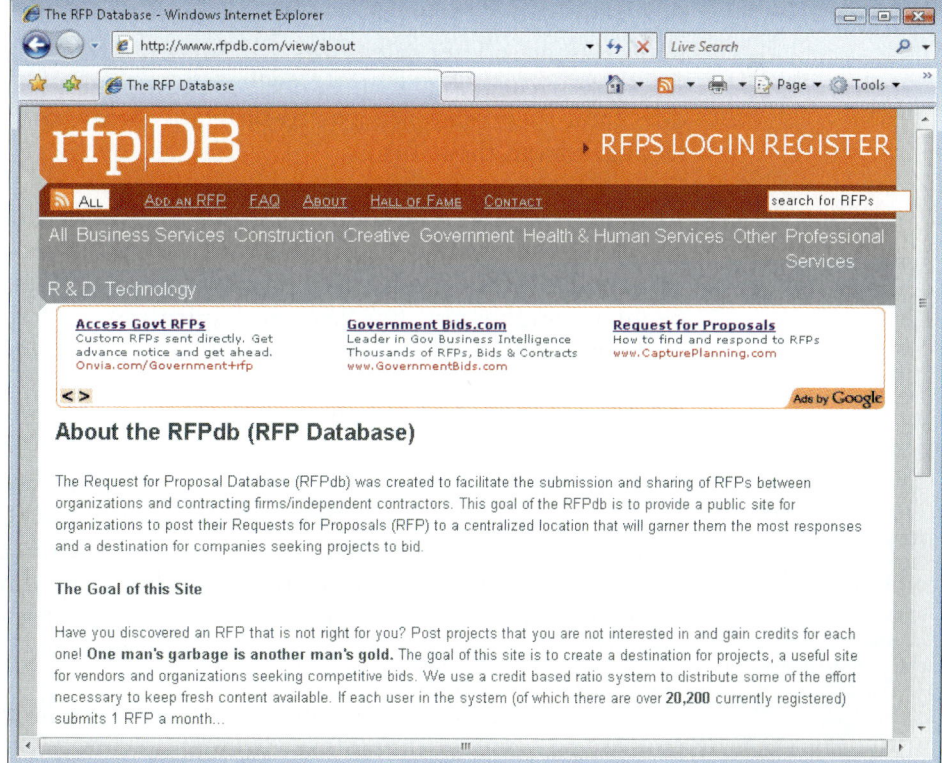

FIGURE 7-21 The rfpDB site offers an online meeting place where customers post RFPs and vendors can respond.

FIGURE 7-22 Microsoft Communities is an excellent resource for IT professionals.

Step 3: Evaluate the Alternatives

After identifying the alternatives, you must select the one that best fits the company's needs. You should obtain information about the options from as many sources as possible, including vendor presentations and literature, product documentation, trade publications, and companies that perform software testing and evaluation. To learn more about particular software packages, search the Internet using keywords that describe the application. Web sites maintained by consultants and software publishers often include product references and links to vendors. As part of the evaluation process, you should try to obtain information from existing users, test the application, and benchmark the package.

EXISTING USERS You can contact existing users to obtain feedback and learn about their experiences. For large-scale software packages, ASPs and vendors typically supply user references. User references are important because you need to know whether the software package has worked well for companies like yours. Be aware that some vendors limit their reference lists to satisfied clients, so you can expect mostly positive feedback from those firms.

APPLICATION TESTING If a software package is one of the options, find out if it is possible for users in your organization to try the product. For horizontal applications or a small system, using a demonstration copy to enter a few sample transactions could be an acceptable test. For vertical applications or large systems, a team of IT staff and users might need several days or weeks to perform tests.

BENCHMARKING To determine whether a package can handle a certain transaction volume efficiently, you can perform a benchmark test. A **benchmark** measures the time a package takes to process a certain number of transactions. For example, a benchmark test can measure the time needed to post 1,000 sales transactions.

    If you use benchmarks, remember that a benchmark test is conducted in a controlled environment, which might not resemble the actual day-to-day situation at your company. Although benchmarking cannot predict your specific results, benchmark testing is a good way to measure relative performance of two or more competing products in a standard environment.

    Many IT publications publish regular reviews of individual packages, including benchmark tests, and often have annual surveys covering various categories of software. Some of the publications shown in Figure 7-23 also offer online versions and additional Web-based features, search capability, and IT links.

ON THE WEB

For more information about benchmark tests, visit **scsite.com/ sad8e/more**, locate Chapter 7, and then click the Benchmark Tests link.

You also can obtain information from independent firms that benchmark various software packages and sell comparative analyses of the results, as shown in Figure 7-24 on the next page. The Transaction Processing Performance Council (TPC) is an example of a non-profit organization that publishes standards and reports for its members and the general public, while InfoSizing is an IT consulting firm that offers analysis of performance benchmarks.

Finally, you should match each package against the RFP features and rank the choices. If some features are more important than others, give them a higher weight using an evaluation model similar to the one shown in Figure 7-19 on page 302.

FIGURE 7-23 Many IT publications provide specialized information, reviews of individual software packages, and benchmark tests. These reviews are important to the systems analyst because they can provide an unbiased opinion of the package.

Step 4: Perform Cost-Benefit Analysis

Review the suggestions in this chapter and in Part 3 of the Systems Analyst's Toolkit, and develop a spreadsheet to identify and calculate TCO for each option you are considering. Be sure to include all costs, using the volume forecasts you prepared. If you are considering outsourcing options, carefully study the alternative fee structure models described earlier. If possible, prepare charts to show the results graphically, and build in what-if capability so you can gauge the impact if one or more variables change.

If you are considering a software package, be sure to consider acquisition options. When you purchase software, what you are buying is a **software license** that gives you the right to use the software under certain terms and conditions. For example, the license could allow you to use the software only on a single computer, a specified number of computers, a network, or an entire site, depending on the terms of the agreement. Other license restrictions could prohibit you from making the software available to others or modifying the program. For desktop applications, software license terms and conditions usually cannot be modified. For large-scale systems, license agreement terms often can be negotiated.

Also consider user support issues, which can account for a significant part of TCO. If you select an outsourcing alternative, the arrangement probably will include certain technical support and maintenance. If you choose in-house development, you must consider the cost of providing these services on your own. If you purchase a software package, consider a supplemental **maintenance agreement**, which offers additional support and assistance from the vendor. The agreement might provide full support for a period of time or list specific charges for particular services. Some software packages provide free technical support for a period of time. Afterward, support is offered with a charge per occurrence, or per minute or hour of technical support time. Some software vendors contact registered owners whenever a new release is available and usually offer the new release at a reduced price.

TOOLKIT TIME

The Financial Analysis tools in Part 3 of the Systems Analyst's Toolkit can help you analyze project costs, benefits, and economic feasibility. To learn more about these tools, turn to Part 3 of the four-part Toolkit that follows Chapter 12.

Step 5: Prepare a Recommendation

You should prepare a recommendation that evaluates and describes the alternatives, together with the costs, benefits, advantages, and disadvantages of each option. At this point, you may be required to submit a formal system requirements document and deliver a presentation. You should review the suggestions for presenting written

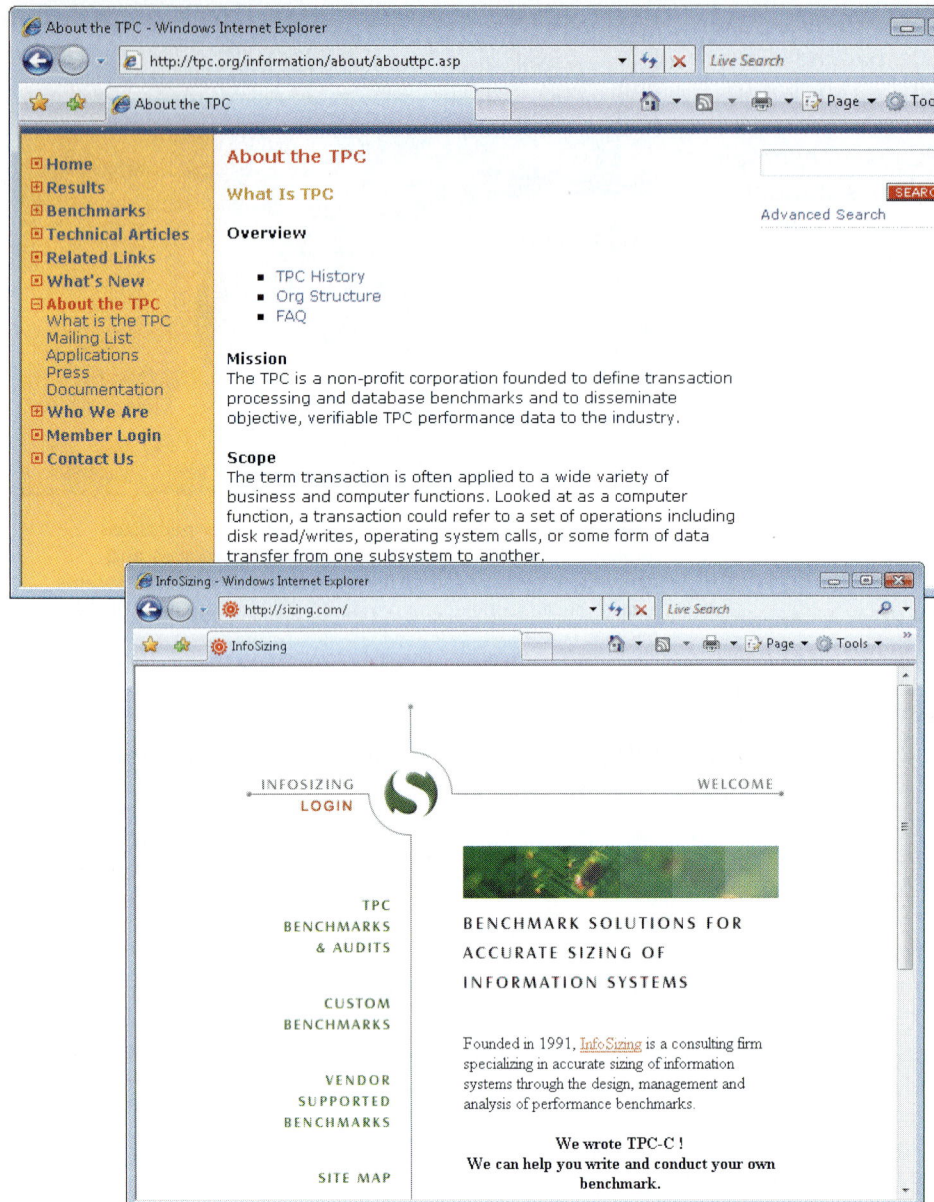

FIGURE 7-24 The Transaction Processing Performance Council is a non-profit organization that publishes standards and reports for its members and the general public, while InfoSizing is an IT consulting firm that offers analysis of performance benchmarks.

proposals and oral presentations in Part 1 of the Systems Analyst's Toolkit. Additional suggestions about preparing the system requirements document and the management presentation are contained in the following section.

Step 6: Implement the Solution

Implementation tasks will depend on the solution selected. In-house options will require more time and effort than outsourcing alternatives. For large systems or network installations, the process can require considerable time and effort. Your installation strategy should be planned well in advance, especially if any disruption of normal business operations is expected. If the software package is customized, then the task will be more complex and difficult.

Before the new software becomes operational, you must complete all implementation steps, including loading, configuring, and testing the software; training users; and converting data files to the new system's format. Chapter 11 will discuss implementation strategies and techniques in more detail.

CASE IN POINT 7.3: DOUG'S SPORTING GOODS

Doug's Sporting Goods sells hiking and camping supplies. The company has grown considerably in the last two years. Doug Sawyer, the company's founder and president, wants to develop a customer order entry system and hired your IT consulting firm to advise him about software alternatives. Doug is leaning toward in-house development because he does not want to depend on outside vendors and suppliers for technical support and upgrades. Doug also says that he is not interested in selling on the Web, but that could change in the future.

Doug wants to meet with you tomorrow to make a decision. What will you say to Doug at the meeting?

COMPLETION OF SYSTEMS ANALYSIS TASKS

To complete the systems analysis phase, you must prepare the system requirements document and your presentation to management.

System Requirements Document

The **system requirements document**, or **software requirements specification**, contains the requirements for the new system, describes the alternatives that were considered, and makes a specific recommendation to management. This important document is the starting point for measuring the performance, accuracy, and completeness of the finished system before entering the systems design phase.

The system requirements document is like a contract that identifies what the system developers must deliver to users. Recall that system requirements are identified during the fact-finding process, and a system requirements checklist is created at that time. Various examples of system requirements are listed on pages 151–153 in Chapter 4. You should write the system requirements document in language that users can understand so they can offer input, suggest improvements, and approve the final version.

Because the system requirements document can be lengthy, you should format and organize it so it is easy to read and use. The system requirements document should include a cover page and a detailed table of contents. You also can add an index and a glossary of terms to make the document easier to use. The content of the system requirements document will depend on the company and the complexity of the system.

Presentation to Management

The presentation to management at the end of the systems analysis phase is one of the most critical milestones in the systems development process. At this point, managers make key decisions that affect the future development of the system.

Prior to the management presentation, you might give two other presentations: one to the principal individuals in the IT department to keep them posted, and another presentation to users to answer their questions and invite feedback. The system requirements document is the basis for all three presentations, and you should distribute the document (or a summary) in advance so the recipients can review it.

When preparing your presentation, you should review the suggestions in Part 1 of the Systems Analyst's Toolkit, which will help you design and deliver a successful presentation. If you plan a slide presentation, you should review the Toolkit guidelines for effective presentations. In addition to the techniques found in the Toolkit, also keep the following suggestions in mind:

- Begin your presentation with a brief overview of the purpose and primary objectives of the system project, the objectives of this presentation, and what decisions need to be made.

- Summarize the primary viable alternatives. For each alternative, describe the costs, advantages, and disadvantages.

- Explain why the evaluation and selection team chose the recommended alternative.

- Allow time for discussion and for questions and answers.

- Obtain a final decision from management or agree on a timetable for the next step in the process.

The object of the management presentation is to obtain approval for the development of the system and to gain management's full support, including necessary financial resources. Management probably will choose one of five alternatives: develop an in-house system, modify a current system, purchase or customize a software package, perform

TOOLKIT TIME

The Communication Tools in Part 1 of the Systems Analyst's Toolkit can help you develop better documents, reports, and presentations. To learn more about these tools, turn to Part 1 of the four-part Toolkit that follows Chapter 12.

additional systems analysis work, or stop all further work. Depending on their decision, your next task as a systems analyst will be one of the following:

1. Implement an outsourcing alternative. If outsourcing is selected, you will work with representatives of the service provider to achieve a smooth transition to the new environment.

2. Develop an in-house system. Begin the systems design phase for the new system, which is described in Chapters 8, 9, and 10.

3. Purchase or customize a software package. Negotiate the purchase terms with the software vendor for management approval. Then, if the package will be used without modification, you can begin planning the systems implementation phase. If you must make modifications to the package, your next step is to start the systems design phase. If the vendor will make the modifications, then your next step is to start planning the testing and documentation of the modifications as part of the systems implementation phase, which is described in Chapter 11.

4. Perform additional systems analysis work. Management might want you to investigate certain alternatives further, explore alternatives not examined, develop a prototype, reduce the project scope because of cost constraints, or expand the project scope based on new developments. If necessary, you will perform the additional work and schedule a follow-up presentation.

5. Stop all further work. The decision might be based on your recommendation, a shift in priorities or costs, or for other reasons. Whatever the reason, if that is management's decision, then you have no additional tasks for the project other than to file all your research in a logical location so it can be retrieved if the project is reopened in the future.

After the presentation and management decision, you will begin a transition to the systems design phase of the SDLC. If you are developing an in-house system or modifying a package, you will build a model of the proposed system and start designing the system's output, input, files, and data structures. The following sections describe several tools and techniques that can assist you in that process, including prototyping, CASE tools, and alternative graphical tools.

THE TRANSITION TO SYSTEMS DESIGN

If management decides to develop the system in-house, then the transition to the systems design phase begins. In a smaller company, you might be assigned full responsibility for the design tasks. In a large IT group, you might work as a member of the design team even though you had not participated in the earlier SDLC phases. The following sections discuss preparation for systems design and the relationship between analysis and design tasks.

The chapter concludes with a list of systems design guidelines and a description of prototyping methods and tools.

Preparing for Systems Design Tasks

When systems design begins, it is essential to have an accurate and understandable system requirements document. Your document contains the design for the new system and is the starting point for the systems design phase. Errors, omissions, ambiguities, and other problems will affect the quality of the finished product. As you proceed to the design phase, you must be certain that you performed a thorough and accurate systems analysis and communicated the results in your system requirements document.

The Relationship Between Logical and Physical Design

You develop the logical design of an information system during the systems analysis phase of the SDLC. The **logical design** defines the functions and features of the system and the relationships among its components. The logical design includes the output that must be produced by the system, the input needed by the system, and the processes that must be performed by the system without regard to how tasks will be accomplished *physically*.

As previously discussed, a logical design defines *what* must take place, not *how* it is to be accomplished. Logical designs do not address the actual methods of implementation. The logical design for a customer records system, for example, describes the data that must be entered for each customer, specifies that records must be displayed in customer number order, and explains what information to produce for a customer status report. Specifications for the actual input, or entry, of data, the sorting method, the physical process of creating the report, and the exact format of the report are not part of the logical design.

In contrast to the logical design, the **physical design** of an information system is a plan for the actual implementation of the system. You develop the physical design during the systems design phase of the SDLC. The physical design is built on the system's logical design and describes a specific implementation, much like a working blueprint describes the actual construction of a building.

In a typical system, the physical design describes the actual processes of entering, verifying, and storing data; the physical layout of data files; the sorting procedures; the exact format of reports; and so on. Whereas logical design is concerned with *what* the system must accomplish, physical design is concerned with *how* the system will meet those requirements.

Because logical and physical designs are related so closely, good systems design is impossible without careful, accurate systems analysis. In fact, the design phase typically cannot begin until the analysis work is complete. Although some overlap is possible, it usually is better to complete the analysis phase before moving on to systems design; however, each situation is different. For example, you might return to fact-finding if you discover that you overlooked an important issue, if users have significant new needs, or if legal or governmental requirements change.

SYSTEMS DESIGN GUIDELINES

The systems analyst must understand the logical design of the system before beginning the physical design of any one component. The first step is to review the system requirements document, which is especially important if you did not work on the previous phase or if a substantial amount of time has passed since the analysis phase was completed. After you review the system requirements, you are ready to start the actual design process. What should you do first, and why?

The best place to begin is with data design, which defines the physical data structures, elements, and relationships. When data design is complete, you move to the user interface, which affects the interaction between the user and the system. As you develop the user interface, you will work on specific input and output design tasks. Then you will work on the architecture that will translate the design into code modules.

STEP	ACTIVITY	DESCRIPTION
I	**Review system requirements.**	Become familiar with the logical design.
2	**Design the system.**	
	• User interface	Design an overall user interface, including screens, commands, controls, and features that enable users to interact with an application.
	• Input processes	Determine how data will be input to the system and design necessary source documents.
	• Input and output	Design the physical layout for each input and output formats and reports screen and printed report.
	• Data	Determine how data will be organized, stored, maintained, updated, accessed, and used.
	• System architecture	Determine processing strategies and methods, client/server interaction, network configuration, and Internet/intranet interface issues.
3	**Present the system design.**	Create the systems design specification document, in which you describe the proposed system design, the anticipated benefits of the system, and the estimated development and implementation costs.

FIGURE 7-25 The systems design phase of the SDLC consists of three main steps.

Because the components of a system are interdependent, the design phase is not a series of clearly defined steps. Although you might start with one component, it is not unusual to work on several components at the same time. For example, making a decision to change a report format might require other changes in data design or input screens.

The final step in systems design is to prepare a systems design specification and present your results to management. Those tasks and other systems design activities are listed and described in Figure 7-25.

Systems Design Objectives

The goal of **systems design** is to build a system that is effective, reliable, and maintainable.

A system is effective if it satisfies the defined requirements and constraints. The system also must be accepted by users who use it to support the organization's business objectives.

A system is reliable if it adequately handles errors, such as input errors, processing errors, hardware failures, or human mistakes. Ideally, *all* errors can be prevented. Unfortunately, no system is completely foolproof, whether it is a payroll system, a telephone switching system, an Internet access system, or a space shuttle navigation system. A more realistic approach to building a reliable system is to plan for errors, detect them as early as possible, allow for their correction, and prevent them from damaging the system itself.

A system is maintainable if it is well designed, flexible, and developed with future modifications in mind. No matter how well a system is designed and implemented, at some point it will need to be modified. Modifications will be necessary to correct problems, to adapt to changing user requirements, to enhance the system, or to take advantage of changing technology. Your systems design must be capable of handling future modifications or the system soon will be outdated and fail to meet requirements.

As shown in Figure 7-26, design considerations involve users, data, and architecture.

SYSTEMS DESIGN CONSIDERATIONS

User Considerations

- Consider points where users interact with the system
- Anticipate future user, system, and organizational needs

Data Considerations

- Enter data where and when it occurs
- Verify data where it is input
- Use automated data entry methods whenever possible
- Control access for data entry
- Report every instance of entry and change of data
- Enter data only once

Architecture Considerations

- Use a modular design
- Design independent modules that perform a single function

FIGURE 7-26 Good design results in systems that are effective, reliable, and maintainable. Design considerations involve users, data, and architecture.

USER CONSIDERATIONS Of the many issues you must consider during systems design, your most important goal is to make the system acceptable to users, or user-friendly. Throughout the design process, the essential factor to consider is how decisions will affect users. Always remember that you are designing the system for the users, and keep these basic points in mind:

Carefully consider any point where users receive output from, or provide input to, the system. Above all, the user interface must be easy to learn and use. Input processes should be well documented, easy to follow, intuitive, and forgiving of errors. Output should be attractive and easy to understand, with an appropriate level of detail.

Anticipate future needs of the users, the system, and the organization. Suppose that an employee master file contains a one-character field to indicate each employee's category. The field currently has two valid values: *F* indicates a full-time employee, and *P* indicates a part-time employee. Depending on the field value, either FULL-TIME or PART-TIME will print as the value on various reports. While those two values could be programmed, or **hard-coded**, into the report programs, designing a separate table with category codes and captions is a better choice. The hard-coded solution is straightforward, but if the organization adds another value, such as *X* for FLEXTIME, a programmer would have to change all the report programs. If a separate table for codes and captions is used, it can be changed easily without requiring modifications to the reports.

Provide flexibility. Suppose that a user wants a screen display of all customer balances that exceed $5,000 in an accounts receivable system. How should you design that feature? The program could be coded to check customer balances against a fixed value of 5000, which is a simple solution for both the programmer and the user because no extra keystrokes are required to produce the display. That approach, however, is inflexible. For instance, if a user later needs a list of customers whose balances exceed $7,500 rather than $5,000, the existing system cannot provide the information. To accommodate the request, the programmer would have to change and retest the program and rewrite new documentation.

On the other hand, you could design the program to produce a report of all customers whose balance exceeds a specific amount entered by the user. For example, if a user wants to display customers with balances of more than $7,500, he or she can enter the number 7500 in a parameter query. A **parameter** is a value that the user enters whenever the query is run, which provides flexibility, enables users to access information easily, and costs less.

A good systems design can combine both approaches. For example, you could design the program to accept a variable amount entered by the user but start with a default value of 5,000. A **default** is a value that the system displays automatically. Users can press the ENTER key to accept the default value, or enter another value.

Often the best design strategy is to come up with several alternatives, so users can decide what will work best for them. Again, always remember to design the system with the users in mind.

CASE IN POINT 7.4: DOWNTOWN!

Downtown! is a rapidly growing Web-based retailer with about 100 management and technical support employees at its headquarters office in Florida. Mary Estrada, the firm's IT manager, is planning a new information system that will give users better access to sales and marketing data and trends. She has a concern, however. She knows that users often request reports but use only a small portion of the data. In many offices she sees inboxes filled with printed reports gathering dust. Mary asked for your opinion: What if new system users could design most of their own reports without assistance from the IT staff, by using a powerful, user-friendly report writer program? Do you think they would request as many reports or the same types of reports? What are the pros and cons of giving users total control over output?

DATA CONSIDERATIONS Data entry and storage considerations are important parts of the systems design. Here are some guidelines to follow:

Data should be entered into the system where and when it occurs because delays cause data errors. For example, employees in the receiving department should enter data about incoming shipments when the shipments arrive, and sales clerks should enter data about new orders when they take the orders.

Data should be verified when it is entered, to catch errors immediately. The input design should specify a data type such as alphabetic, numeric, or alphanumeric and a range of acceptable values for each data entry item. If an incorrect data value is entered, the system should recognize and flag it immediately. The system also should allow corrections at any time. Some errors, for example, are most easily corrected right at entry while the original source documents are at hand or the customer is on the telephone. Other errors may need further investigation, so users must be able to correct errors at a later time.

Automated methods of data entry should be used whenever possible. Receiving department employees in many companies, for example, use scanners to capture data about merchandise received. Automated data entry methods, such as the RFID scanner shown in Figure 7-27, reduce input errors and improve employee productivity.

Access for data entry should be controlled and all entries or changes to critical data values should be reported. Dollar fields and many volume fields are considered critical data fields. Examples of critical volumes include the number of checks processed, the number of medical prescriptions dispensed, or the number of insurance premium payments received. Reports that trace the entry of and changes to critical data values are called **audit trails** and are essential in every system.

Every instance of entry and change to data should be logged. For example, the system should record when a customer's credit limit was established, and by whom, which is necessary to construct the history of a transaction.

Data should be entered into a system only once. If input data for a payroll system also is needed for a human resources system, you should design a program interface between the systems so data can be transferred automatically.

Data duplication should be avoided. In an inventory file, for example, the suppliers' addresses should not be stored with every part record. Otherwise, the address of a vendor who supplies 100 different parts is repeated 100 times in the data file. Additionally, if the vendor's address changes, all 100 parts records must be updated. Data duplication also can produce inconsistencies. If those 100 stored addresses for the vendor are not identical, how would a user know which version is correct? In Chapter 9, you will learn about data design and a technique called normalization, which is a set of rules that can help you identify and avoid data design problems when you create a database.

ARCHITECTURE CONSIDERATIONS In addition to the issues affecting users and data, you should consider the following suggestions:

Use a modular design. In a modular design, you create individual processing components, called modules, which connect to a higher-level program or process. In a traditional, structured design, each module represents a specific process or subprocess shown on a DFD and documented in a process description. If you are using an object-oriented design, as described in Chapter 6, object classes are represented by code modules. You will learn more about modular design in Chapter 10, which describes system architecture.

Design modules that perform a single function are easier to understand, implement, and maintain. Independent modules also provide greater flexibility because they can be developed and tested individually, and then combined at a later point in the development process. Modular design is helpful especially when developing large-scale systems, because separate teams of analysts and programmers can work on different areas of the project and then integrate the modules to create a finished system.

FIGURE 7-27 Automated data entry methods, such as the RFID scanner shown above, reduce input errors and improve employee productivity.

Design Trade-Offs

You will find that design goals often conflict with each other. In the systems design phase, you constantly must analyze alternatives and weigh trade-offs. To make a system easier to use, for example, programming requirements might be more complex. Making a system more flexible might increase maintenance requirements. Meeting one user's requirements could make it harder to satisfy another user's needs.

Most design trade-off decisions that you will face come down to the basic conflict of quality versus cost. Although every project has budget and financial constraints, you should avoid decisions that achieve short-term savings but might mean higher costs later. For example, if you try to reduce implementation costs by cutting back on system testing or user training, you can create higher operational costs in the future. If necessary, you should document and explain the situations carefully to management and discuss the possible risks. Each trade-off must be considered individually, and the final result must be acceptable to users, the systems staff, and company management.

FIGURE 7-28 Wind tunnel testing is a typical example of prototyping.

PROTOTYPING

Prototyping produces an early, rapidly constructed working version of the proposed information system, called a **prototype**. Prototyping, which involves a repetitive sequence of analysis, design, modeling, and testing, is a common technique that can be used to design anything from a new home to a computer network. For example, engineers use a prototype to evaluate an aircraft design before production begins, as shown in the wind tunnel testing in Figure 7-28.

User input and feedback is essential at every stage of the systems development process. Prototyping allows users to examine a model that accurately represents system outputs, inputs, interfaces, and processes. Users can "test-drive" the model in a risk-free environment and either approve it or request changes. In some situations, the prototype evolves into the final version of the information system; in other cases, the prototype is intended only to validate user requirements and is discarded afterward.

Perhaps the most intense form of prototyping occurs when agile methods are used. As you learned in Chapter 1, agile methods build a system by creating a series of prototypes and constantly adjusting them to user requirements. As the agile process continues, developers revise, extend, and merge earlier versions into the final product. An agile approach emphasizes continuous feedback, and each incremental step is affected by what was learned in the prior steps.

Prototyping Methods

Systems analysts use two different prototyping methods: system prototyping and design prototyping. **System prototyping** produces a full-featured, working model of the information system. As Figure 7-29 shows, a system prototype is ready for the implementation phase of the SDLC.

FIGURE 7-29 The end product of system prototyping is a working model of the information system, ready for implementation.

While agile methods represent the latest approach to system prototyping, rapid application development (RAD), which is described in Chapter 4, remains a popular strategy. Using RAD methods, a team of users, managers, and IT staff members works together to develop a model of the information system that evolves into the completed system. The RAD team defines, analyzes, designs, and tests prototypes using a highly interactive process, which is shown in Figure 4-5 on page 144.

Systems analysts also use prototyping to verify user requirements, after which the prototype is discarded and implementation continues, as shown in Figure 7-30. The approach is called **design prototyping**, or **throwaway prototyping**. In this case, the prototyping objectives are more limited, but no less important. The end product of design prototyping is a user-approved model that documents and benchmarks the features of the finished system.

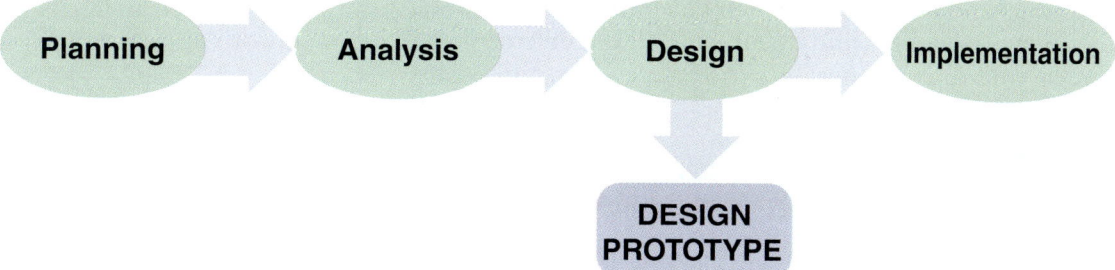

FIGURE 7-30 The end product of design prototyping is a user-approved model that documents and benchmarks the features of the finished system.

Design prototyping makes it possible to capture user input and approval while continuing to develop the system within the framework of the SDLC. Systems analysts typically use design prototyping as they construct outputs, inputs, and user interfaces, as discussed in Chapter 8.

Whenever possible, you should allow users to experiment with a prototype and provide feedback on how well it meets their needs. This approach can increase development costs, but the expense will be offset by lower costs during subsequent SDLC phases. Prototyping offers many benefits, including the following:

- Users and systems developers can avoid misunderstandings.

- System developers can create accurate specifications for the finished system based on the prototype.

- Managers can evaluate a working model more effectively than a paper specification.

- Systems analysts can use a prototype to develop testing and training procedures before the finished system is available.

- Prototyping reduces the risk and potential financial exposure that occur when a finished system fails to support business needs.

Although most systems analysts believe that the advantages of prototyping far outweigh any disadvantages, you should consider the following potential problems:

- The rapid pace of development can create quality problems, which are not discovered until the finished system is operational.

- Other system requirements, such as reliability and maintainability, cannot be tested adequately using a prototype.

- In very complex systems, the prototype becomes unwieldy and difficult to manage.

Prototyping Tools

Systems analysts can use powerful tools to develop prototypes. Most prototyping is done using CASE tools, application generators, report generators, screen generators, and fourth-generation languages (4GLs). In a **fourth-generation language (4GL)**, the commands tend to resemble natural statements that people use. For example, a 4GL statement might be PRINT ALL PRODUCTS WHERE CODE = IN STOCK AND STATUS = OK.

In combination, the tools provide a framework for rapid, efficient software development, called a **fourth-generation environment.**

Part 2 of the Systems Analyst's Toolkit describes CASE tools in more detail and explains how systems analysts can use them to speed the development process, reduce costs, and avoid design errors. In a fourth-generation environment, the development tools are highly interactive. For example, systems analysts use CASE tools to create a series of

ON THE WEB

For more information about software development trends, visit **scsite.com/sad8e/more**, locate Chapter 7, and then click the Software Development Trends link.

diagrams and definitions, which generate a data dictionary automatically. The data dictionary organizes and documents all data elements and interacts with application, screen, and report generators to produce a system prototype.

Limitations of Prototypes

The final version of the system typically demands higher-level performance than the prototype can provide. A prototype is a functioning system, but it is less efficient than a fully developed system. Because it is a model, rather than a completed system, the prototype will have slower processing speeds and response times. The prototype also might lack security requirements, exception and error-handling procedures, and other required functions. Despite those limitations, systems developers can upgrade the prototype into the final information system by adding the necessary capability. Otherwise, the prototype is discarded and the remaining SDLC phases are completed.

Even when it does not evolve into the finished system, a prototype helps to ensure that the final product will meet all requirements. Satisfying system requirements is the ultimate goal of systems development, and prototyping is an extremely valuable tool during the process.

SOFTWARE DEVELOPMENT TRENDS

TechTarget.com publishes news and information of interest to IT professionals, and offers articles, white papers, Webcasts, downloads, and a searchable library. In a recent review of software development trends by Jennette Mullaney and Michelle Davidson, shown in Figure 7-31, outsourcing and agile methods headed the list of hot topics to watch.

As the article points out, outsourcing volume will grow rapidly, but detailed specifications and quality requirements will be essential to success. At the same time, agile development is gaining considerable momentum as developers recognize the need for close collaboration between developers, users, and customers.

A review of current online topics being discussed in the IT community also includes the following:

- Software quality will be more important than ever, and intense modeling will be part of the quality assurance process. As you learned in Chapters 4, 5, and 6, an accurate model can help ensure that user requirements will be satisfied. Software testing also will receive more emphasis in the future. In today's post-9/11 environment, many firms will need to do more testing to ensure that their systems are not vulnerable to physical attack or electronic intrusion. Software testing is covered in Chapter 11, Managing Systems Implementation, and security issues are discussed in more detail in Chapter 12, Managing Systems Support and Security.

- Project management will be a major focus of IT managers. With increased pressure for quality software that meets budget, schedule, and quality requirements, project managers will be key players. In this environment, there will be even more emphasis on project management training and credentials.

FIGURE 7-31 The TechTarget.com site offers news and information from IT professionals.

TOOLKIT TIME

The Internet Resource tools in Part 4 of the Systems Analyst's Toolkit can help you in using the Internet to stay abreast of current IT trends and to build your systems analysis skills. To learn more about these tools, turn to Part 4 of the four-part Toolkit that follows Chapter 12.

- Service-oriented architecture (SOA) will become an important factor in future development. **Service-oriented architecture (SOA)** is an architectural style whose goal is to achieve loose coupling among interacting software objects that provide services. **Loose coupling** means that objects can interact, but are essentially independent. A common example is a DVD and a DVD player— if you want to watch your DVD, you put it into a DVD player and watch your video, because the player provides a DVD playing service. But loose coupling allows you to you to replace your DVD player with another, or to play your videos on more than one player. The concept of loose coupling is discussed in more detail in Chapter 10, System Architecture.

- Growth in open-source software such as Linux has increased demand for powerful open-source development tools, while traditional development languages such as C and C++ are becoming less popular. There is a growing open-source community that supports and promotes vendor-neutral open-source development.

- Developers will use more **Web services**, which are modular applications such as currency converters or language translators. Most Web services are based on a combination of HTML and XML. You learned in Chapter 1 that HTML is a platform-independent language that controls the way information is presented on a browser, and that Extensible Markup Language (XML) provides a common data description language that allows easy Web-based communication between different types of hardware and software.

- Programmers will continue to use dynamic languages such as Java, Python, Perl, Ruby, and Visual Basic, among others, and new languages will evolve.

As a systems analyst, you will be affected by rapidly changing technology, and you will want to know about IT trends and developments. Part 4 of the Systems Analyst's Toolkit contains tips and techniques that you can use to access Web-based information and use it to help build your skills and success.

A QUESTION OF ETHICS

Sally works as a junior analyst for a medium-sized IT consulting firm. Her manager, Bob, has asked her to draft a response to an RFP from a large company that is seeking IT consulting services in connection with a new accounting system.

As Sally worked on the RFP, she noticed a specific question about her firm's recent experience on this type of system. To the best of her knowledge, the firm has only worked on one other accounting project in the last three years. When Bob saw Sally's draft response, he was upset about the way she answered the question. "You don't have to be quite that candid," he said. "Even though we only had one *formal* project, we do have several people who worked on accounting systems before they came here."

"Yes," Sally replied, "But that isn't what the question is asking." As he left her office, Bob's final comment was, "If we want that job, we'll have to come up with a better answer." Thinking about it, Sally isn't comfortable with anything but a straight answer. Is this an ethical question? What are Sally's options?

CHAPTER SUMMARY

This chapter describes system development strategies, the preparation and presentation of the system requirements document, and the transition to the systems design phase of the SDLC.

An important trend that views Software as a Service (SaaS), rather than a product, has created new software acquisition options. Systems analysts must consider Web-based development environments such as .NET and WebSphere, and various outsourcing options, including application service providers and Internet business services. Application service providers (ASPs) charge subscription fees for providing application software packages. Internet business services (IBSs) offer powerful Web-based servers, software hosting, and IT support services to customers.

Traditional systems must function in various hardware and software environments, be compatible with legacy systems, and operate within the constraints of company networks and desktop computing capability. Such systems utilize Internet links and resources as enhancements. In contrast, Internet-based systems treat the Web *as* the platform, rather than just a communication channel. Many large companies use Web-based systems to handle enterprise-wide applications. Compared to traditional systems, Web-based systems are more scalable, less dependent on specific hardware and software, and more adaptable to outsourcing the operation and support of a software application.

The next Web generation will be called Web 2.0, and will greatly enhance information-sharing, user collaboration, and social-networking applications such as MySpace and Facebook. Another trend, called cloud computing because of the commonly used cloud symbol for the Internet, describes an overall online software and data environment, powered by supercomputer technology, that will be an ultimate form of Software as a Service.

If a company chooses to handle its own software development needs, it can create in-house systems, or purchase (and possibly customize) commercially available software packages from a software vendor or value-added reseller (VAR).

Compared with developing an in-house system, an existing commercial software package can be an attractive alternative, because a package generally costs less, takes less time to implement, has a proven track record, and is upgraded frequently. In-house development or customizing a software package might be the best choice when a standard software package cannot meet specific business requirements or constraints. In addition to customizing software packages, companies can create user applications based on standard software that has been specially configured to enhance user productivity.

The systems analyst's role in the software development process depends on the specific development strategy. In-house development requires much more involvement than outsourcing or choosing a commercial package.

The most important factor in choosing a development strategy is total cost of ownership (TCO). Financial analysis tools include payback analysis, which determines how long it takes for a system to pay for itself through reduced costs and increased benefits; return on investment (ROI), which compares a project's total return with its total costs; and net present value (NPV), which analyzes the value of a project by adjusting costs and benefits to reflect the time that they occur.

The process of acquiring software involves a series of steps: evaluate the system requirements, consider network and Web-related issues, identify potential software vendors or outsourcing options, evaluate the alternatives, perform cost-benefit analysis, prepare a recommendation, and implement the solution. During software acquisition, a company can use a request for proposal (RFP) or a request for quotation (RFQ). An RFP invites vendors to respond to a list of system requirements and features; an RFQ seeks bids for a specific product or service.

The system requirements document is the deliverable, or end product, of the systems analysis phase. The document details all system requirements and constraints, recommends the best solution, and provides cost and time estimates for future development work. The system requirements document is the basis for the management presentation. At this point, the firm might decide to develop an in-house system, modify the current system, purchase or customize a software package, perform additional systems analysis work, or stop all further work.

As you prepared for the transition from the systems analysis to systems activities, you learned that a prototype is a working model of the proposed system that you can use to verify the system requirements with users or as a basis for the new system.

You learned that a set of interactive tools, called a fourth-generation environment, can help you construct the prototype. A fourth-generation environment includes screen generators, report writers, application or code generators, and fourth-generation languages, all of which interact with a data dictionary developed with CASE tools. You also reviewed a set of systems design guidelines and suggestions, including user considerations, data considerations, and processing considerations. Finally, you learned about trends in software development, including outsourcing, agile development, and various other topics.

Key Terms and Phrases

application service provider (ASP) *289*

audit trail *312*

benchmark *304*

build or buy *291*

business process outsourcing (BPO) *288*

cloud computing *287*

default *312*

design prototyping *314*

evaluation and selection team *297*

evaluation model *301*

fixed fee model *289*

fourth-generation environment *315*

fourth-generation language (4GL) *315*

global outsourcing *290*

hard-coded *311*

help desk *295*

horizontal application *291*

in-house software *291*

information center (IC) *295*

Internet business services (IBS) *289*

logical design *309*

loose coupling *317*

maintenance agreement *305*

make or buy *291*

managed hosting *289*

middleware *287*

.NET *285*

net present value (NPV) *297*

newsgroup *303*

offshore outsourcing *290*

outsourcing *288*

parameter *312*

payback analysis *297*

physical design *309*

prototype *314*

prototyping *314*

read-only properties *296*

report generator *295*

request for proposal (RFP) *300*

request for quotation (RFQ) *302*

return on investment (ROI) *297*

screen generator *295*

service-oriented architecture (SOA) *317*

service provider *288*

Software as a Service (SaaS) *284*

software license *305*

software package *291*

software requirements specification *307*

software vendor *291*

subscription model *289*

system prototyping *314*

system requirements document *307*

systems design *310*

throwaway prototyping *314*

transaction model *289*

usage model *289*

user application *295*

user interface *295*

value-added reseller (VAR) *291*

vertical application *292*

Web 2.0 *287*

WebSphere *285*

Web services *317*

Learn It Online

Instructions: To complete the Learn It Online exercises, start your browser, click the Address bar, and then enter **scsite.com/sad8e/learn**. When the Systems Analysis and Design Learn It Online page is displayed, follow the instructions in the exercises below. Each exercise has instructions for saving your results, either for your own records or for submission to your instructor.

1 Chapter Reinforcement

TF, MC, and SA

Below SAD Chapter 7, click one of the Chapter Reinforcement links for Multiple Choice, True/False, or Short Answer. Answer each question and submit to your instructor.

2 Flash Cards

Below SAD Chapter 7, click the Flash Cards link and read the instructions. Type 20 (or a number specified by your instructor) in the Number of playing cards text box, type your name in the Enter your Name text box, and then click the Flip Card button. When the flash card is displayed, read the question and then click the ANSWER box arrow to select an answer. Flip through the Flash Cards. If your score is 15 (75%) correct or greater, click Print on the File menu to print your results. If your score is less than 15 (75%) correct, then redo this exercise by clicking the Replay button.

3 Practice Test

Below SAD Chapter 7, click the Practice Test link. Answer each question, enter your first and last name at the bottom of the page, and then click the Grade Test button. When the graded practice test is displayed on your screen, click Print on the File menu to print a hard copy. Continue to take practice tests until you score 80% or better.

4 Who Wants To Be a Computer Genius?

Below SAD Chapter 7, click the Computer Genius link. Read the instructions, enter your first and last name at the bottom of the page, and then click the Play button. When your score is displayed, click the PRINT RESULTS link to print a hard copy.

5 Wheel of Terms

Below SAD Chapter 7, click the Wheel of Terms link. Read the instructions, and then enter your first and last name and your school name. Click the PLAY button. When your score is displayed on the screen, right-click the score and then click Print on the shortcut menu to print a hard copy.

6 Crossword Puzzle Challenge

Below SAD Chapter 7, click the Crossword Puzzle Challenge link. Read the instructions, and then click the Continue button. Work the crossword puzzle. When you are finished, click the Submit button. When the crossword puzzle is redisplayed, submit it to your instructor.

Case-Sim: SCR Associates

Background

SCR Associates is a consulting firm that offers IT solutions and training. SCR needs an information system to manage operations at its new training center. The new system will be called TIMS (Training Information Management System). As a newly hired systems analyst, you will

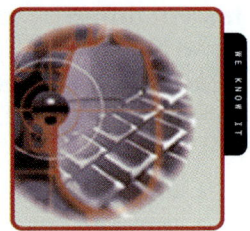

report to Jesse Baker, systems group manager. You will work on various tasks and practice the skills you learned in this chapter.

Using the Case

The SCR Associates case study is a Web-based simulation that allows you to practice your skills in a real-world environment. The case study transports you to SCR's company intranet, where you can complete 12 work sessions, each aligning with a chapter. As you work on the case, you will receive e-mail and voice mail messages, obtain information from SCR's online libraries, and perform various tasks. The first time you enter the SCR Case, you should go to the starting page at **scsite.com/sad8e/scr** for detailed instructions.

Preview: Session 7

As you consider various development strategies for the TIMS system, you receive specific directions from your supervisor, Jesse Baker. She wants you to determine whether vertical software packages exist, and she wants you to explore outsourcing options for the new system. She also expects you to conduct a cost-benefit analysis of developing TIMS in-house, and she wants your input on outsourcing and prototyping.

To start a work session, you log on to the SCR intranet. When you enter your name and password, an opening screen displays links to the work sessions, and you select Session 7. At this point, you check your e-mail and voice mail messages carefully. Then you begin working on your task list, which includes the following items:

Tasks: Development Strategies

1. *Determine whether vertical software packages exist for training operations management. Search the Internet and draft a message describing the results.*

2. *Investigate the possibility of outsourcing the TIMS system. List the options, together with advantages and disadvantages of each.*

3. *Follow Jesse's e-mail instructions about calculating payback, ROI, and NPV for the TIMS system.*

4. *Jesse wants my thoughts on how we can use prototyping for TIMS. She also wants me to prepare a system requirements document and a management presentation.*

FIGURE 7-32   Task list: Session 7.

Chapter Exercises

Review Questions

1. Describe the trend that views software as a service rather than a product. What effect has this trend had on software acquisition options?
2. Explain the difference between horizontal and vertical application software.
3. What is the most common reason for a company to choose to develop its own information system? Give two other reasons why a company might choose the in-house approach.
4. What is an RFP, and how does it differ from an RFQ?
5. What is the purpose of a benchmark test?
6. Explain software licenses and maintenance agreements.
7. What decisions might management reach at the end of the systems analysis phase, and what would be the next step in each case?
8. What is a prototype, and how do systems developers use prototyping?
9. What is a fourth-generation environment?
10. Explain the relationship between logical and physical design.

Discussion Topics

1. As more companies outsource systems development, will there be less need for in-house systems analysts? Why or why not?
2. Suppose you tried to explain the concept of throwaway prototyping to a manager, and she responded by asking, "So, is throwaway prototyping a waste of time and money?" How would you reply?
3. Select a specific type of vertical application software to investigate. Visit local computer stores and use the Internet to determine what software packages are available. Describe the features of those packages.
4. Select a specific type of horizontal application software to investigate. Visit local computer stores and use the Internet to determine what software packages are available. Describe the features of those packages.

Projects

1. The text mentions several firms and organizations that offer IT benchmarking. Locate another benchmarking firm on the Internet, and describe its services.
2. Turn to Part 3 of the Systems Analyst's Toolkit and review the concept of net present value (NPV). Determine the NPV for the following: An information system will cost $95,000 to implement over a one-year period and will produce no savings during that year. When the system goes online, the company will save $30,000 during the first year of operation. For the next four years, the savings will be $20,000 per year. Assuming a 12% discount rate, what is the NPV of the system?
3. Visit the IT department at your school or at a local company and determine whether the information systems were developed in-house or purchased as software packages. If packages were acquired, determine what customizing was done, if any. Write a brief memo describing the results of your visit.
4. To create user applications as described in this chapter, systems analysts often use macros. Microsoft defines a macro as "a series of commands and instructions that you group together as a single command to accomplish a task automatically." Learn more about macros by using the Help feature in Microsoft Word, and suggest three tasks that might be performed by macros.

Apply Your Knowledge

The Apply Your Knowledge section contains four mini-cases. Each case describes a situation, explains your role in the case, and asks you to respond to questions. You can answer the questions by applying knowledge you learned in the chapter.

1 Top Sail Realty

Situation:

Top Sail Realty is one of the largest time-sharing and rental brokers for vacation cottages along the North Carolina coast. After 10 successful years of matching up owners and renters, Top Sail decided to acquire a computerized reservation and booking system. Top Sail's owner read an article about software packages, and she asked you, as an IT consultant, for your advice.

1. Should Top Sail implement a Web-based system? Why or why not?
2. What software acquisition options are available to Top Sail?
3. Do you consider the reservations system to be a horizontal or a vertical application? Give reasons for your answer.
4. When you evaluate software packages, what steps will you follow?

2 One Way Movers, Inc.

Situation:

As IT manager at One Way, you scheduled a management presentation next week. You prepared and distributed a system requirements document, and you anticipate some intense questioning at the meeting.

1. When planning your presentation, what are some techniques you will use?
2. Based on the suggestions in the Systems Analyst's Toolkit, what visual aids can you use during your presentation?
3. In deciding on your proposal, what options does management have?
4. If management decides to purchase or customize a software package, what steps will you take?

3 Tangible Investments Corporation

Situation:

Tangible Investments Corporation needs a new customer billing system. As project leader, you decided to create a prototype that users can evaluate before the final design is implemented. You plan to use a traditional structured analysis methodology. To prepare for your meeting with top management tomorrow, you need to review the following topics.

1. Explain the main purpose of prototyping.
2. Explain why a prototype might or might not evolve into the final version of the system.
3. Describe the tools typically used in developing prototypes.
4. List three advantages and three disadvantages of prototyping.

4 IT Flash Magazine

Situation:

You are a staff writer at IT Flash Magazine, a popular online newsletter aimed at IT professionals. Your editor has asked you to prepare a special report for next week's edition. Specifically, she wants you to research the subject of software outsourcing, and other significant trends that might affect software development in the future. If possible, she wants you to cite specific sources for your information, including IT employment statistics and employment forecasts from the Bureau of Labor Statistics.

1. Search for information about software outsourcing generally, using the search techniques described in Part 4 of the Systems Analyst's Toolkit.
2. Visit the Bureau of Labor Statistics site at bls.gov and search for information about employment trends affecting systems analysts, computer programmers, and software engineers.
3. Does the Bureau of Labor Statistics offer any comments or insights into the subject of outsourcing generally? What conclusions does it reach?
4. In your report, comment on whether the offshore outsourcing of IT jobs is just another step in the progression that began with manufacturing jobs, or represents a whole new trend. Be sure to cite Web research sources and your own reasons.

Case Studies

Case studies allow you to practice specific skills learned in the chapter. Each chapter contains several case studies that continue throughout the textbook, and a chapter capstone case.

NEW CENTURY HEALTH CLINIC

New Century Health Clinic offers preventive medicine and traditional medical care. In your role as an IT consultant, you will help New Century develop a new information system.

Background

Based on your earlier recommendations, New Century decided to continue the systems development process for a new information system that would improve operations, decrease costs, and provide better service to patients.

Now, at the end of the systems analysis phase, you are ready to prepare a system requirements document and give a presentation to the New Century associates. Many of the proposed system's advantages were described during the fact-finding process. Those include smoother operation, better efficiency, and more user-friendly procedures for patients and New Century staff.

You also must examine tangible costs and benefits to determine the economic feasibility of several alternatives. If New Century decides to go ahead with the development process, the main options are to develop the system in-house or purchase a vertical package and configure it to meet New Century's needs. You have studied those choices and put together some preliminary figures.

You know that New Century's current workload requires three hours of office staff overtime per week at a base rate of $8.50 per hour. In addition, based on current projections, New Century will need to add another full-time clerical position in about six months. Neither the overtime nor the additional job will be needed if New Century implements the new system. The current manual system also causes an average of three errors per day, and each error takes about 20 minutes to correct. The new system should eliminate those errors.

Based on your research, you estimate by working full-time you could complete the project in about 12 weeks. Your consulting rate, which New Century agreed to, is $30 per hour. If you design the new system as a database application, you can expect to spend about $2,500 for a networked commercial package. After the system is operational and the staff is trained, New Century should be able to handle routine maintenance tasks without your assistance.

As an alternative to in-house development, a vertical software package is available for about $9,000. The vendor offers a lease-purchase package of $3,000 down, followed by two annual installments of $3,000 each. If New Century buys the package, it would take you about four weeks to install, configure, and test it, working full-time. The vendor provides free support during the first year of operation, but then New Century must sign a technical support agreement at an annual cost of $500. Although the package contains many of the features that New Century wants, most of the reports are pre-designed and it would be difficult to modify their layouts.

No matter which approach is selected, New Century probably will need you to provide about 10 hours of initial training and support each week for the first three months of operation. After the new system is operational, it will need routine maintenance, file backups, and updating. These tasks will require about four hours per week and can be performed by a clinic staff member. In both cases, the necessary hardware and network installation will cost about $5,000.

In your view, the useful life of the system will be about five years, including the year in which the system becomes operational.

Assignments

You scheduled a presentation to New Century in one week, and you must submit a system requirements document during the presentation. Prepare both the written documentation and the presentation. (To give a successful presentation, you will need to learn the skills described in Part 1 of the Systems Analyst's Toolkit.) Your oral and written presentation must include the following tasks:

1. Provide an overview of the proposed system, including costs and benefits, with an explanation of the various cost-and-benefit types and categories.
2. Develop an economic feasibility analysis, using payback analysis, ROI, and present value (assume a discount rate of 10%).
3. Prepare a context diagram and diagram 0 for the new system.
4. Provide a brief explanation of the various alternatives that should be investigated if development continues, including in-house development and any other possible strategies.

You may wish to include other material to help your audience understand the new system and make a decision on the next step.

Presentation Rules

The following presentation rules should be considered:

- Use suitable visual aids.
- Use presentation software, if possible.
- Distribute handouts before, during, or after the presentation.
- Follow the guidelines in Part 1 of the Systems Analyst's Toolkit.
- Keep your presentation to 30 minutes, including 5 minutes for questions.

Rules for the System Requirements Document

Consider the following rules while preparing the system requirements document:

- Follow the guidelines in Part 1 of the Systems Analyst's Toolkit.
- Include charts, graphs, or other helpful visual information in the document.
- Spell check and carefully proofread the entire document.

PERSONAL TRAINER, INC.

Personal Trainer, Inc., owns and operates fitness centers in a dozen Midwestern cities. The centers have done well, and the company is planning an international expansion by opening a new "supercenter" in the Toronto area. Personal Trainer's president, Cassia Umi, hired an IT consultant, Susan Park, to help develop an information system for the new facility. During the project, Susan will work closely with Gray Lewis, who will manage the new operation.

Background

During data and process modeling, Susan Park developed a logical model of the proposed system. She drew an entity-relationship diagram and constructed a set of leveled and balanced DFDs. Now Susan is ready to consider various development strategies for the new system. She will investigate traditional and Web-based approaches and weigh the advantages and disadvantages of in-house development versus other alternatives. As she moves ahead to the systems design phase, she will review design guidelines, consider the use of prototypes, and analyze the possible use of codes.

Before You Begin ...

Review the facts presented in the Personal Trainer case study in Chapters 2, 4, and 5. Use that information to complete the following tasks.

Assignments

1. Should the new system be designed as a Web-based system? Why or why not? What are some specific issues and options that Susan should consider in making a decision?

2. Assume that Cassia Umi, Personal Trainer's president, has asked Susan to prepare a system requirements document and deliver a presentation to the management team. What should be the main elements of the system requirements document? Also, based on the suggestions in Part 1 of the Systems Analyst's Toolkit, what visual aids should Susan use during her presentation?

3. Should Susan use a prototype during systems design? What options does she have, and how would you advise her?

4. Susan wants to prepare a presentation that will calculate the total cost of ownership for the system. What financial analysis tools are available to her, and what are the advantages (and possible disadvantages) of each tool?

CUTTING EDGE

Chapter 3 discusses software change control, which is the process of managing and controlling changes after management approves the system requirements document. You should review the material about this topic now, before completing the Cutting Edge case study.

Background

Cutting Edge is a company that develops and sells accounting software tailored to specific businesses and industries. Product manager Michelle Kellogg is leading a development team working on a specialized package for building contractors. The systems analysis phase was completed on March 6. The package is scheduled for release on August 1.

On April 1, Michelle received a request from Cutting Edge's CEO for a new feature to be added to the package. She asked Michelle to analyze the change and determine its impact on the project.

Michelle reviewed the results at a meeting on April 15. Adding the new feature to the package now increases the development cost by $28,000 and delays the release date by one month. Michelle also evaluated a second alternative: Develop the package without the requested change and add the feature in a follow-up release, which requires two additional months to complete and costs $66,000.

Assignments

1. Because the project is in the systems design phase and no programming has been started, why would incorporating the requested change add one month and $28,000 to the project?

2. If the change were delayed until after the initial release of the package, why would it require two additional months and $66,000 more to implement a new version with the desired feature?

3. What are the advantages and disadvantages of each alternative?

4. Suppose you are Michelle and you must recommend action on the change request. What factors would you consider, and what would you recommend?

CHAPTER CAPSTONE CASE: SoftWear, Limited

SoftWear, Limited (SWL), is a continuing case study that illustrates the knowledge and skills described in each chapter. In this case study, the student acts as a member of the SWL systems development team and performs various tasks.

Background

Systems analyst Rick Williams and programmer/analyst Carla Moore continued to work on a logical model of the payroll system. Meanwhile, the information systems department recently purchased and installed Visible Analyst, a CASE toolkit that supports logical and physical modeling. Rick and Carla traveled to Massachusetts to attend a one-week workshop to learn how to use the package.

After returning from their trip, Rick and Carla decided to create the logical model for the payroll system with Visible Analyst. They felt that the time spent now would pay off in later phases of the project. Rick and Carla used the manual DFDs they created in Chapter 5 to create computerized DFDs using Visible Analyst. Now all related items for the new system are stored in the CASE tool.

Over the next month, Rick and Carla looked at various alternatives and spent their time evaluating the potential solutions. They determined that the best solution was to purchase a payroll package, but the ESIP processing was so unique that none of the available software packages would handle SWL's specific requirements. They concluded that SWL should purchase a payroll package and develop the ESIP system in-house. Jane Rossman and Ann Hon agreed with their recommendation.

The systems analysts completed work on the logical model, alternative evaluations, and cost and time estimates and then prepared the system requirements document for the payroll system. The document was printed and distributed and a management presentation was scheduled at the end of the following week.

At this point, the IT team members were confident that they had done a good job. They had worked closely with SWL users throughout the development process and received user approval on important portions of the document as it was being prepared. They developed visual aids, rehearsed the presentation, and then tried to anticipate questions that management might ask.

Carla gave the management presentation. She recommended that SWL purchase a payroll package sold by Pacific Software Solutions and that ESIP processing be developed in-house to interface with the payroll package.

During the presentation, Carla and Rick answered questions on several points, including the economic analysis they had done. Michael Jeremy, vice president of finance, was especially interested in the method they used to calculate payback analysis, return on investment, and net present value for the new system.

Robert Lansing, SWL's president, arrived for the last part of the presentation. When the presentation ended, he asked the top managers how they felt about the project, and they indicated support for the proposal made by the IT department. The next step was to negotiate a contract with Pacific Software Solutions and for Rick and Carla to begin systems design for the ESIP processing component.

SWL Team Tasks

1. Although the presentation was successful, Rick and Carla ask you to create a checklist of presentation dos and don'ts that would be helpful for IT staff people who deliver presentations.
2. Rick and Carla also want you to review the DFDs that they prepared to see if you have any suggestions for improvement. If you have access to a copier, make a copy of the DFDs shown in Chapter 5 and then write your notes directly on the diagrams.

CHAPTER CAPSTONE CASE: SoftWear, Limited (continued)

3. Michael Jeremy, vice president of finance, was interested in the financial analysis tools that Rick and Carla used in the presentation. Rick has asked you to write a memo to Mr. Jeremy explaining each tool, with a specific description of how it is used, and what results can be obtained. Before you do this, you should review the material in Part 3 of the Systems Analyst's Toolkit.

4. Although SWL decided to develop the ESIP system in-house, Ann Hon, director of information technology, has requested a report on the trend toward outsourcing software development. Perform Internet research to get up-to-date information about this topic, and prepare a memo for Ms. Hon. Be sure to cite your sources of information.

Manage the SWL Project

You have been asked to manage SWL's new information system project. One of your most important activities will be to identify project tasks and determine when they will be performed. Before you begin, you should review the SWL case in this chapter. Then list and analyze the tasks, as follows:

LIST THE TASKS Start by listing and numbering at least ten tasks that the SWL team needs to perform to fulfill the objectives of this chapter. Your list can include SWL Team Tasks and any other tasks that are described in this chapter. For example, Task 3 might be to Evaluate system requirements, and Task 6 might be to Prepare an RFP.

ANALYZE THE TASKS Now study the tasks to determine the order in which they should be performed. First identify all concurrent tasks, which are not dependent on other tasks. In the example shown in Figure 7-33, Tasks 1, 2, 3, 4, and 5 are concurrent tasks, and could begin at the same time if resources were available.

Other tasks are called dependent tasks, because they cannot be performed until one or more earlier tasks have been completed. For each dependent task, you must identify specific tasks that need to be completed before this task can begin. For example, you would want to evaluate system requirements before you could prepare an RFP, so Task 6 cannot begin until Task 3 is completed, as Figure 7-33 shows.

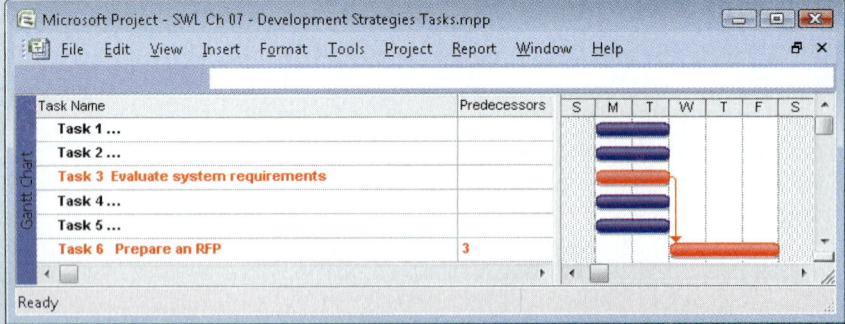

FIGURE 7-33 Tasks 1, 2, 3, 4, and 5 are concurrent tasks that could be performed at the same time. Task 6 is a dependent task that cannot be performed until Task 3 has been completed.

Chapter 3 describes project management tools, techniques, and software. To learn more, you can visit the Features section on your Student Study Tool CD-ROM, or the project management resources library at **scsite.com/sad8e/project**. On the Web, Microsoft offers demo versions, training, and tips for using Project 2007. You also can visit the OpenWorkbench.org site to learn more about this free, open-source software.

PHASE 3 SYSTEMS DESIGN

DELIVERABLE
System Design Specification

TOOLKIT SUPPORT
Primary tools: Communications and CASE tools
Other tools as required

As the Dilbert cartoon suggests, you should understand a problem before you decide that a database is the solution. You will learn more about systems design topics, including database design, in the systems design phase.

Systems design is the third of five phases in the systems development life cycle. In the previous phase, systems analysis, you developed a logical model of the new system. Now you will work on a physical design that will meet the specifications described in the system requirements document. Your tasks will include output and user interface design, data design, and system architecture. The deliverable for this phase is the system design specification.

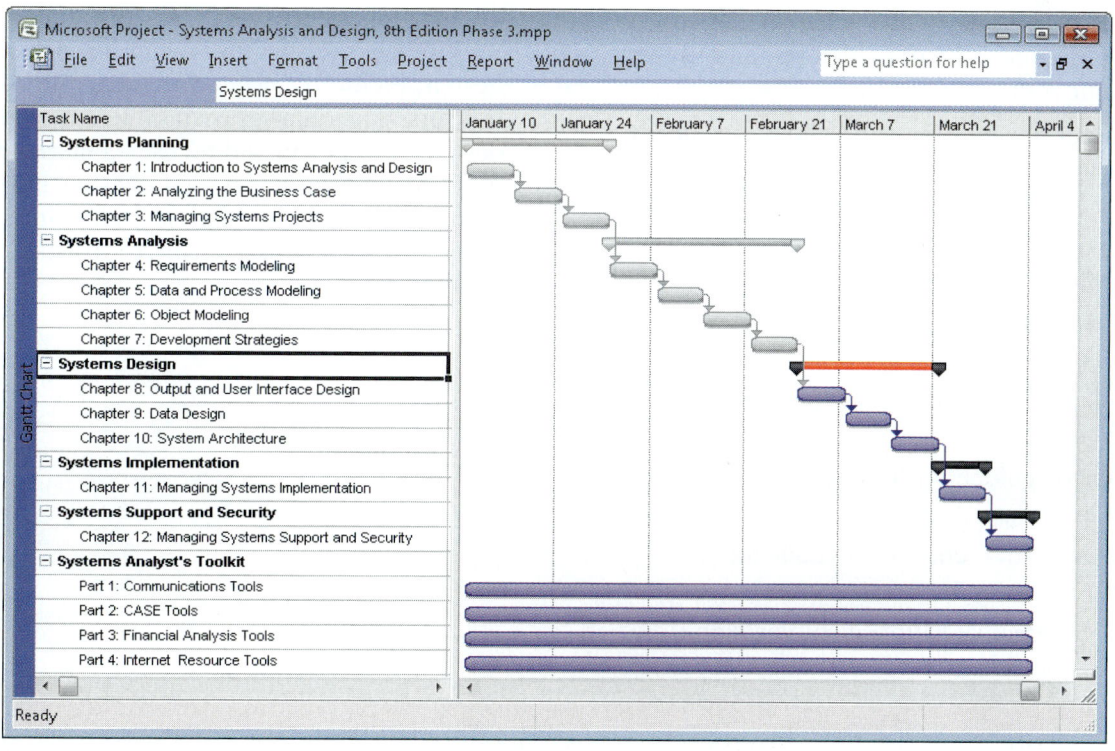

CHAPTER 8 Output and User Interface Design

Chapter 8 is the first of three chapters in the systems design phase of the SDLC. This chapter explains how to design the desired system output and how to construct an effective user interface that includes suitable input screens and procedures. The chapter stresses the importance of user feedback and involvement in all design decisions.

INTRODUCTION

OBJECTIVES

When you finish this chapter, you will be able to:

- Discuss output design issues and various types of output

- Design various types of reports, and suggest output controls and security

- Explain the concept of user interface design and human-computer interaction, including the basic principles of user-centered design

- List specific guidelines for user interface design

- Describe user interface techniques, including screen elements and controls

- Explain input design concepts, techniques, and methods

- Describe guidelines for data entry screen design

- Use validation checks for reducing input errors

- Design effective source documents and input controls

Output and user interface design is the first task in the systems design phase of the SDLC. Output design focuses on user needs for screen and printed forms of output, while user interface design stresses user interaction with the computer, including input design and procedures.

This chapter begins with a discussion of output design, including printed reports and other system outputs. The chapter then covers user interface design concepts, including user-centered design principles and guidelines. The chapter concludes with a discussion of input design, including input methods, volume, screen design, error controls, and source document design.

CHAPTER INTRODUCTION CASE: Mountain View College Bookstore

Background: Wendy Lee, manager of college services at Mountain View College, wants a new information system that will improve efficiency and customer service at the three college bookstores.

In this part of the case, Florence Fullerton (systems analyst) and Harry Boston (student intern) are talking about output and user interface design issues.

Participants:	Florence and Harry
Location:	Mountain View College Cafeteria, Monday afternoon, November 30, 2009.
Project status:	Florence and Harry have examined development strategies for the new bookstore system. After performing cost-benefit analysis, they recommended in-house development of the new bookstore system. Now they are ready to begin the systems design phase by working on output and user interface design for the new system.
Discussion topics:	Output design issues, and user interface concepts and principles

Florence: Hi, Harry. Ready to start work on output and user interface design?

Harry: *Sure. I guess we start with output, because that's the most important issue to users.*

Florence: Right. We'll deal with on-screen and printed output. The other big issue with users is how they interact with the system. The user interface involves human-computer interaction. As you know, we've come a long way from character-based screens. We'll create a graphical interface to make it easy for users to learn and work with the new system. The object is to design everything from a user's point of view.

Harry: *How do we do that?*

Florence: Well, many sources of information about effective design concepts and principles are available. IBM's user bill of rights is a good example. We'll study those, and then ask our own users for their input and suggestions.

Harry: *What about input and data entry?*

Florence: Good question. You've heard the old saying, "garbage in, garbage out." User interface principles apply to user input generally, but repetitive data entry deserves special attention. We need to create screen forms that are logical and easy to understand, as well as data entry validation checks. We also need to review any source documents that will be filled in manually.

Harry: *Anything else?*

Florence: Yes. The bookstore system probably will have some confidential data regarding budgets and markup policies, so we'll have to consider output control and security. If you're ready, here's a task list to get us started:

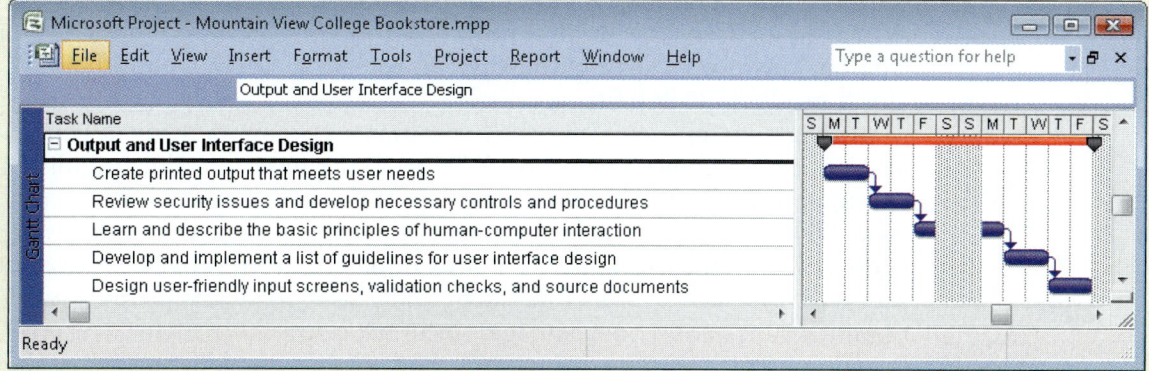

FIGURE 8-1   Typical output and user interface design tasks.

OUTPUT DESIGN

Before designing output, ask yourself several questions:

- What is the purpose of the output?
- Who wants the information, why is it needed, and how will it be used?
- What specific information will be included?
- Will the output be printed, viewed on-screen, or both? What type of device will the output go to?
- When will the information be provided, and how often must it be updated?
- Do security or confidentiality issues exist?

The design process should not begin until you have answered those questions. Some of the information probably was gathered during the systems analysis phase. To complete your understanding, you should meet with users to find out exactly what kind of output is needed. You can use prototypes and mock-ups to obtain feedback throughout the design process. Your answers will affect your output design strategies, as you will see in the next section.

Types of Output

Although business information systems still provide most output as screen displays and printed matter, technology is having an enormous impact on how people communicate and obtain information. This trend is especially important to firms that use information technology to lower their costs, improve employee productivity, and communicate effectively with their customers.

In addition to screen output and printed matter, output can take many forms. The system requirements document probably identified user output needs. Now, in the systems design phase, you will create the actual forms, reports, documents, and other types of output. During this process, you must consider the format and how it will be delivered, stored, and retrieved. The following sections explain various output types and the technologies that are available to systems developers.

INTERNET-BASED INFORMATION DELIVERY Millions of firms use the Internet to reach new customers and markets around the world. To support the explosive growth in e-commerce, Web designers must provide user-friendly screen interfaces that display output and accept input from customers. For example, a business can link its inventory system to its Web site so the output from the inventory system is displayed as an online catalog. Customers visiting the site can review the items, obtain current prices, and check product availability.

Another example of Web-based output is a system that provides customized responses to product or technical questions. When a user enters a product inquiry or requests technical support, the system responds with appropriate information from an on-site knowledge base. Web-based delivery allows users to download a universe of files and documents to support their information needs. For example, the Web provides consumers with instant access to brochures, product manuals, and parts lists; while prospective home buyers can obtain instant quotes on mortgages, insurance, and other financial services.

To reach prospective customers and investors, companies also use a live or prerecorded **Webcast**, which is an audio or video media file distributed over the Internet. Radio and TV stations also use this technique to broadcast program material to their audiences.

E-MAIL E-mail is an essential means of internal and external business communication. Employees send and receive e-mail on local or wide area networks, including the Internet. Companies send new product information to customers via e-mail, and financial services

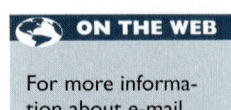

ON THE WEB

For more information about e-mail, visit **scsite.com/sad8e/more**, locate Chapter 8, and then click the E-Mail link.

companies use e-mail messages to confirm online stock trades. Employees use e-mail to exchange documents, data, and schedules and to share business-related information they need to perform their jobs. In many firms, e-mail has virtually replaced traditional memos and printed correspondence.

BLOGS   Web-based logs, called **blogs**, are another form of Web-based output. Because blogs are journals written from a particular point of view, they not only deliver facts to Web readers, but also provide opinions. Blogs are useful for posting news, reviewing current events, and promoting products.

INSTANT MESSAGING   This popular form of communication is another way for individuals and companies to communicate effectively over the Internet. Although some users feel that it can be a distraction, others like the constant flow of communication, especially as a team member in a collaborative situation.

WIRELESS DEVICES   Messages and data can be transmitted to a wide array of mobile devices, including PDAs, handheld computers, smart cell phones, and similar wireless products that combine portable computing power, multimedia capability, and Internet access.

DIGITAL AUDIO, IMAGES, AND VIDEO   Sounds, images, and video clips can be captured, stored in digital format, and transmitted as output to users who can reproduce the content.

Audio output can be attached to an e-mail message or inserted as an audio clip in a Microsoft Word document, as shown in Figure 8-2 on the next page. In addition, many firms use automated systems to handle voice transactions and provide information to customers. For example, using a telephone keypad, a customer can confirm an airline seat assignment, check a credit card balance, or determine the current price of a mutual fund.

If a picture is worth a thousand words, then digital images and video clips certainly are high-value output types that offer a whole new dimension. For example, an insurance adjuster with a digital camera phone can take a picture, submit the image via a wireless device, and receive immediate authorization to pay a claim on the spot. If images are a valuable form of output, video clips are even better in some situations. For example, video clips provide online virtual tours that allow realtors to show off the best features of homes they are marketing. The user can zoom in or out, and rotate the image in any direction.

PODCASTS   A **podcast** is a specially formatted digital audio file that can be downloaded by Internet users from a variety of content providers. Many firms use podcasts as sales and marketing tools, and to communicate with their own employees. Using software such as iTunes, you can receive a podcast, launch the file on your computer, and store it on your portable player. Podcasts can include images, sounds, and video.

AUTOMATED FACSIMILE SYSTEMS   An **automated facsimile** or **faxback** system allows a customer to request a fax using e-mail, via the company Web site, or by telephone. The response is transmitted in a matter of seconds back to the user's fax machine. Although most users prefer to download documents from the Web, many organizations, including the U.S. Department of Transportation, still offer an automated faxback service as another way to provide immediate response 24 hours a day.

COMPUTER OUTPUT TO MICROFILM (COM)   Computer output to microfilm (COM) is often used by large firms to scan and store images of original documents to provide high-quality records management and archiving. COM systems are especially important for legal reasons, or where it is necessary to display a signature, date stamp, or other visual features of a document.

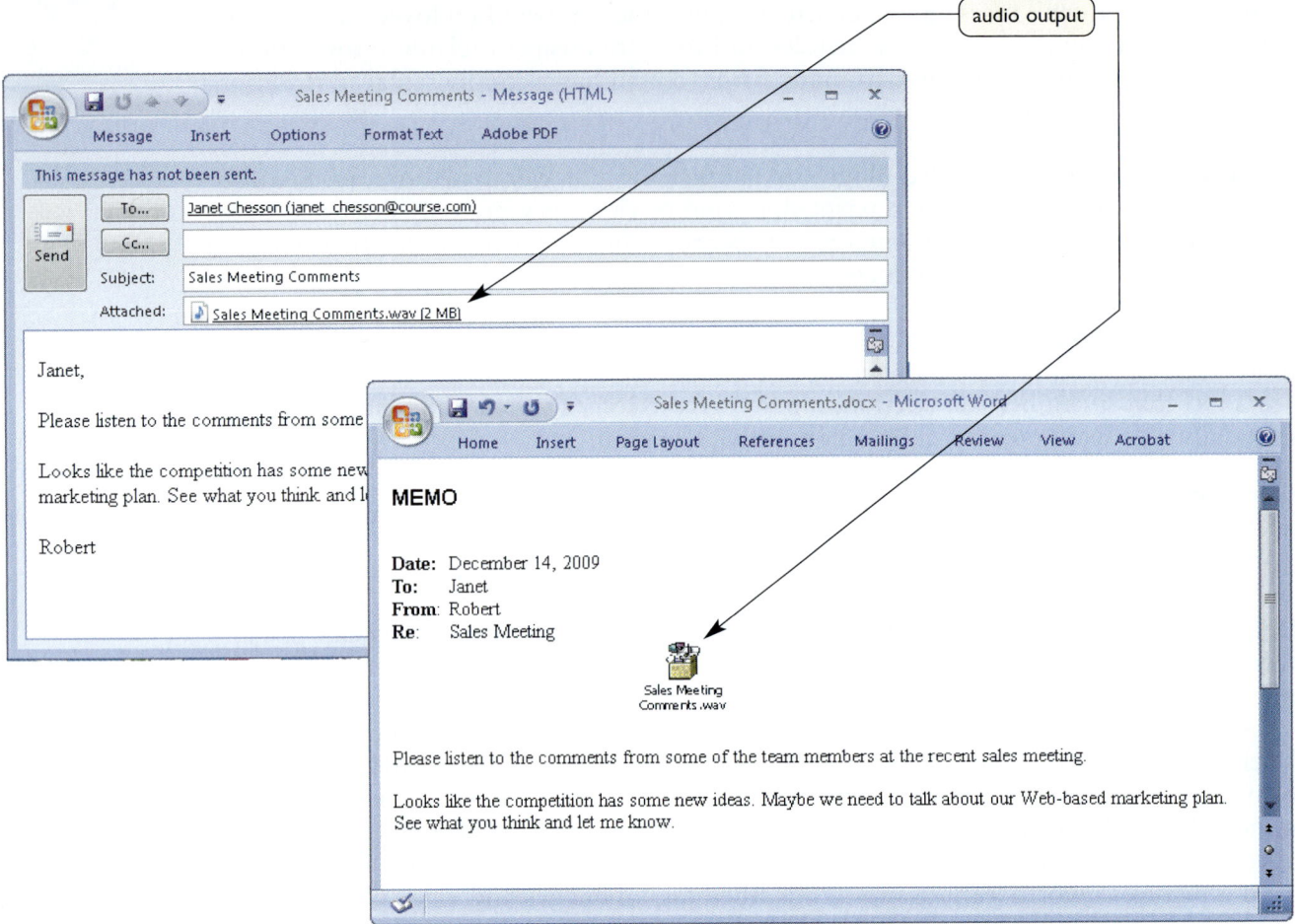

FIGURE 8-2 Audio output can be attached to an e-mail message or inserted as an audio clip in a Microsoft Word document. Either way, the recipient can double-click the link or icon and a media player opens the file. In addition to audio output, the same technology can be used with video clips.

COMPUTER OUTPUT TO DIGITAL MEDIA This process is used when many paper documents must be scanned, stored in digital format, and retrieved quickly. For example, if an insurance company stores thousands of paper application forms, special software can treat the documents as data and extract information from a particular column or area on the form. Digital storage media can include magnetic tape, CDs, DVDs, and high-density laser disks.

SPECIALIZED FORMS OF OUTPUT An incredibly diverse marketplace requires many forms of specialized output. Consider the following examples:

- Retail point-of-sale terminals that handle computer-based credit card transactions, print receipts, and update inventory records
- Automatic teller machines (ATMs) that can process bank transactions and print deposit and withdrawal slips
- Special-purpose printers that can produce labels, employee ID cards, driver's licenses, gasoline pump receipts, and, in some states, lottery tickets
- Plotters that can produce high-quality images such as blueprints, maps, and electronic circuit diagrams

- Digitized information that can be embedded in employee identification cards
- Programmable devices such as MP3 players and DVD players that can display digital output

In today's interconnected world, output from one system often becomes input into another system. For example, within a company, production data from the manufacturing system becomes input to the inventory system. The same company might transmit employee W-2 tax data to the IRS system electronically. A company employee might use tax preparation software to file a tax return online, receive a refund deposited directly into his or her bank account, and see the deposit reflected on the bank's information system.

Although digital technology has opened new horizons in business communications, printed material still is a common type of output, and specific considerations apply to it. For those reasons, printed and screen output are discussed in a separate section, which follows.

PRINTED AND SCREEN OUTPUT

Although many organizations strive to reduce the flow of paper and printed reports, few firms have been able to eliminate printed output totally. Because they are portable, printed reports are convenient, and even necessary in some situations. Many users find it handy to view screen output, then print the information they need for a discussion or business meeting. Printed output also is used in **turnaround documents**, which are output documents that are later entered back into the same or another information system. In some areas, your telephone or utility bill, for example, might be a turnaround document printed by the company's billing system. When you return the required portion of the bill with your check, the bill is scanned into the company's accounts receivable system to record the payment accurately.

Overview of Report Design

Designers use a variety of styles, fonts, and images to produce reports that are attractive and user friendly. Whether printed or viewed on-screen, reports must be easy to read and well organized. Rightly or wrongly, some managers judge an entire project by the quality of the reports they receive.

Database programs such as Microsoft Access include a variety of report design tools, including a Report Wizard, which is a menu-driven feature that designers can use to create reports quickly and easily. Microsoft Access also provides a comprehensive guide to designing reports, as shown in Figure 8-3.

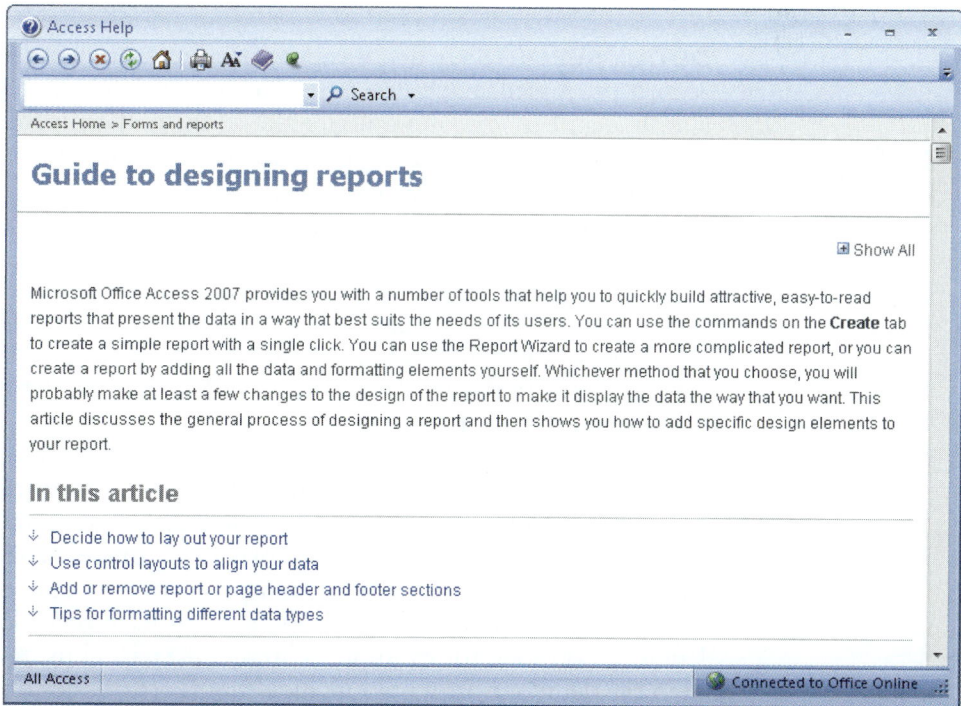

FIGURE 8-3 The Microsoft Access Guide offers valuable tips and suggestions for designing reports.

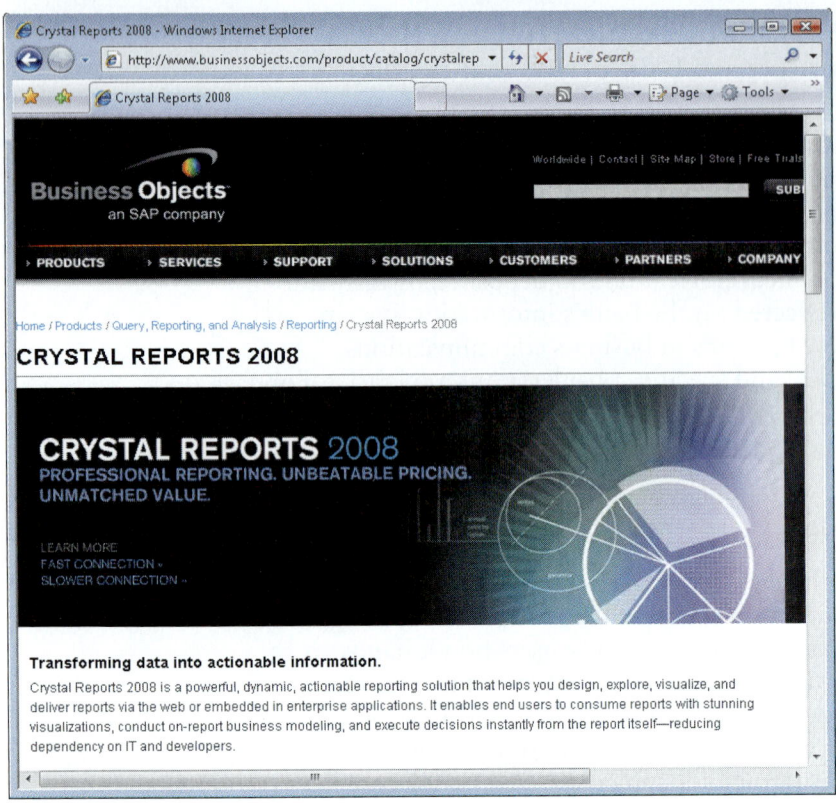

FIGURE 8-4 Crystal Reports is a popular, powerful, report design package.

In addition to built-in design tools, popular software packages such as Crystal Reports offer powerful features that help designers deal with professional-level design issues across the enterprise, as shown in Figure 8-4.

Although the vast majority of reports are designed graphically, some systems still produce one or more **character-based reports** that use a character set with fixed spacing. Printing character-based reports on high-speed impact printers is a fast, inexpensive method for producing large-scale reports, such as payroll or inventory reports, or registration rosters at a school. This is especially true if multiple copies are required.

Types of Reports

To be useful, a report must include the information that a user needs. From a user's point of view, a report with too little information is of no value. Too much information, however, can make a report confusing and difficult to understand.

ON THE WEB

For more information about printed output, visit **scsite.com/ sad8e/more**, locate Chapter 8, and then click the Printed Output link.

When designing reports, the essential goal is to match the report to the user's specific information needs. Depending on their job functions, users might need one or more of the reports described in the following sections.

DETAIL REPORTS A **detail report** produces one or more lines of output for each record processed. Each line of output printed is called a **detail line**. Figure 8-5 shows a simple detail report of employee hours for a chain of retail stores. Notice that one detail line prints for each employee. All the fields in the record do not have to be printed, nor do the fields have to be printed in the sequence in which they appear in the record. An employee paycheck that has multiple output lines for a single record is another example of a detail report.

A well-designed detail report should provide totals for numeric fields. Notice that the report shown in Figure 8-5 lacks subtotals and grand totals for regular hours, overtime hours, and total hours. Figure 8-6 shows the same report with subtotals and grand totals added. In the example, the STORE NUMBER field is called a **control field** because it controls the output.

When the value of a control field changes, a control break occurs. A **control break** usually causes specific actions, such as printing subtotals for a group of records. That type of detail report is called a **control break report**. To produce a control break report, the records must be arranged, or sorted, in control field order. The sorting can be done by the report program itself, or in a previous procedure.

Because it contains one or more lines for each record, a detail report can be quite lengthy. Consider, for example, a large auto parts business. If the firm stocks 3,000 parts, then the detail report would include 3,000 detail lines on approximately 50 printed pages. A user who wants to locate any part in short supply has to examine 3,000 detail lines to find the critical items. A much better alternative is to produce an exception report.

EMPLOYEE HOURS
WEEK ENDING DATE: 6/26/09 PAGE 1

STORE NUMBER	EMPLOYEE NAME	POSITION	REGULAR HOURS	OVERTIME HOURS	TOTAL HOURS
8	Andres, Marguerite	Clerk	20.0	0.0	20.0
8	Bogema, Michelle	Clerk	12.5	0.0	12.5
8	Davenport, Kim	Asst Mgr	40.0	5.0	45.0
8	Lemka, Susan	Clerk	32.7	0.0	32.7
8	Ramirez, Rudy	Manager	40.0	8.5	48.5
8	Ullery, Ruth	Clerk	20.0	0.0	20.0
17	De Martini, Jennifer	Clerk	40.0	8.4	48.4
17	Haff, Lisa	Manager	40.0	0.0	40.0
17	Rittenbery, Sandra	Clerk	40.0	11.0	51.0
17	Wyer, Elizabeth	Clerk	20.0	0.0	20.0
17	Zeigler, Cecille	Clerk	32.0	0.0	32.0

detail lines

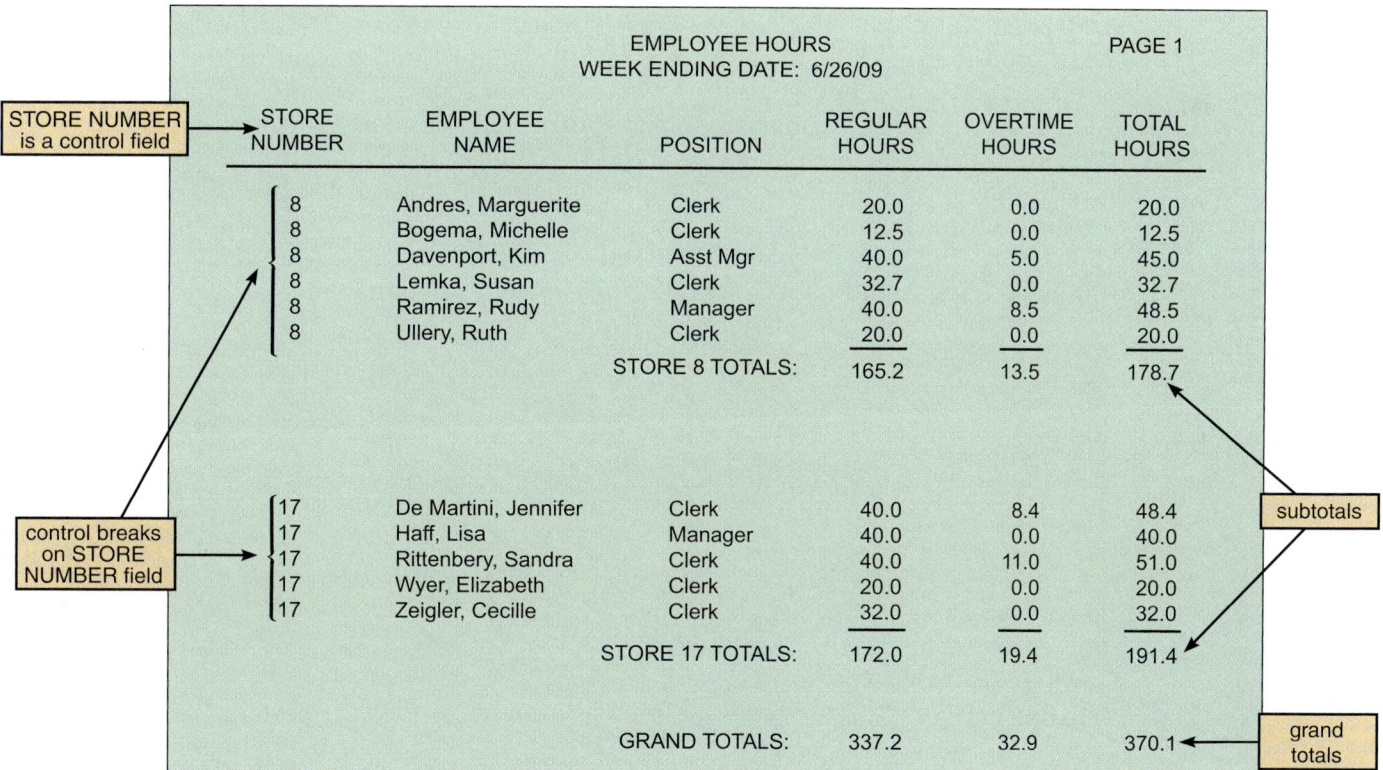

FIGURE 8-5 A detail report with one printed line per employee.

EMPLOYEE HOURS
WEEK ENDING DATE: 6/26/09 PAGE 1

STORE NUMBER	EMPLOYEE NAME	POSITION	REGULAR HOURS	OVERTIME HOURS	TOTAL HOURS
8	Andres, Marguerite	Clerk	20.0	0.0	20.0
8	Bogema, Michelle	Clerk	12.5	0.0	12.5
8	Davenport, Kim	Asst Mgr	40.0	5.0	45.0
8	Lemka, Susan	Clerk	32.7	0.0	32.7
8	Ramirez, Rudy	Manager	40.0	8.5	48.5
8	Ullery, Ruth	Clerk	20.0	0.0	20.0
		STORE 8 TOTALS:	165.2	13.5	178.7
17	De Martini, Jennifer	Clerk	40.0	8.4	48.4
17	Haff, Lisa	Manager	40.0	0.0	40.0
17	Rittenbery, Sandra	Clerk	40.0	11.0	51.0
17	Wyer, Elizabeth	Clerk	20.0	0.0	20.0
17	Zeigler, Cecille	Clerk	32.0	0.0	32.0
		STORE 17 TOTALS:	172.0	19.4	191.4
		GRAND TOTALS:	337.2	32.9	370.1

STORE NUMBER is a control field

control breaks on STORE NUMBER field

subtotals

grand totals

FIGURE 8-6 This detail report contains the same data as Figure 8-5, but provides much more information. Control breaks are used to separate the data for each store, with subtotals and grand totals for numeric fields.

EXCEPTION REPORTS An **exception report** displays only those records that meet a specific condition or conditions. Exception reports are useful when the user wants information only on records that might require action, but does not need to know the details. For example, a credit manager might use an exception report to identify only those customers with past due accounts, or a customer service manager might want a report on all packages that were not delivered within a specified time period. Figure 8-7 on the next page shows an exception report that includes information only for those employees who worked overtime, instead of listing information for all employees.

STORE NUMBER	POSITION	EMPLOYEE NAME	OVERTIME HOURS
		OVERTIME REPORT	PAGE 1
		WEEK ENDING DATE: 6/26/09	
8	Asst Mgr	Davenport, Kim	5.0
	Manager	Ramirez, Rudy	8.5
		STORE 8 TOTALS:	13.5
17	Clerk	De Martini, Jennifer	8.4
	Clerk	Rittenbery, Sandra	11.0
		STORE 17 TOTALS:	19.4
		GRAND TOTAL:	32.9

FIGURE 8-7 An exception report that shows information *only* for employees who worked overtime.

SUMMARY REPORTS Upper-level managers often want to see total figures and do not need supporting details. A sales manager, for example, might want to know total sales for each sales representative, but not want a detail report listing every sale made by them. In that case, a **summary report** is appropriate. Similarly, a personnel manager might need to know the total regular and overtime hours worked by employees in each store but might not be interested in the number of hours worked by each employee. For the personnel manager, a summary report such as the one shown in Figure 8-8 would be useful. Generally, reports used by individuals at higher levels in the organization include less detail than reports used by lower-level employees.

STORE NUMBER	REGULAR HOURS	OVERTIME HOURS	TOTAL HOURS
EMPLOYEE HOURS SUMMARY			PAGE 1
WEEK ENDING DATE: 6/26/09			
8	181.2	13.5	194.7
17	172.0	19.4	191.4
TOTALS:	337.2	32.9	370.1

FIGURE 8-8 A summary report displays totals without showing details.

User Involvement in Report Design

Printed reports are an important way of delivering information to users, so recipients should approve all report designs in advance. To avoid problems, you should submit each design for approval as you complete it, rather than waiting until you finish all report designs.

When designing a report, you should prepare a sample report, which is called a **mock-up**, or prototype, for users to review. The sample should include typical field values and contain enough records to show all the design features. Depending on the type of printed output, you can use a word processor or a report generator to create mock-up reports. After a report design is approved, you should document the design by creating **a report analysis form**, which contains information about the fields, data types and lengths, report frequency and distribution, and other comments.

Report Design Principles

Printed reports must be attractive, professional, and easy to read. Good report design, like any other aspect of the user interface, requires effort and attention to detail. To produce a well-designed report, the analyst must consider several topics, including report headers and footers, page headers and footers, column headings and alignment, column spacing, field order, and grouping of detail lines. Figure 8-9 shows an example of those design features.

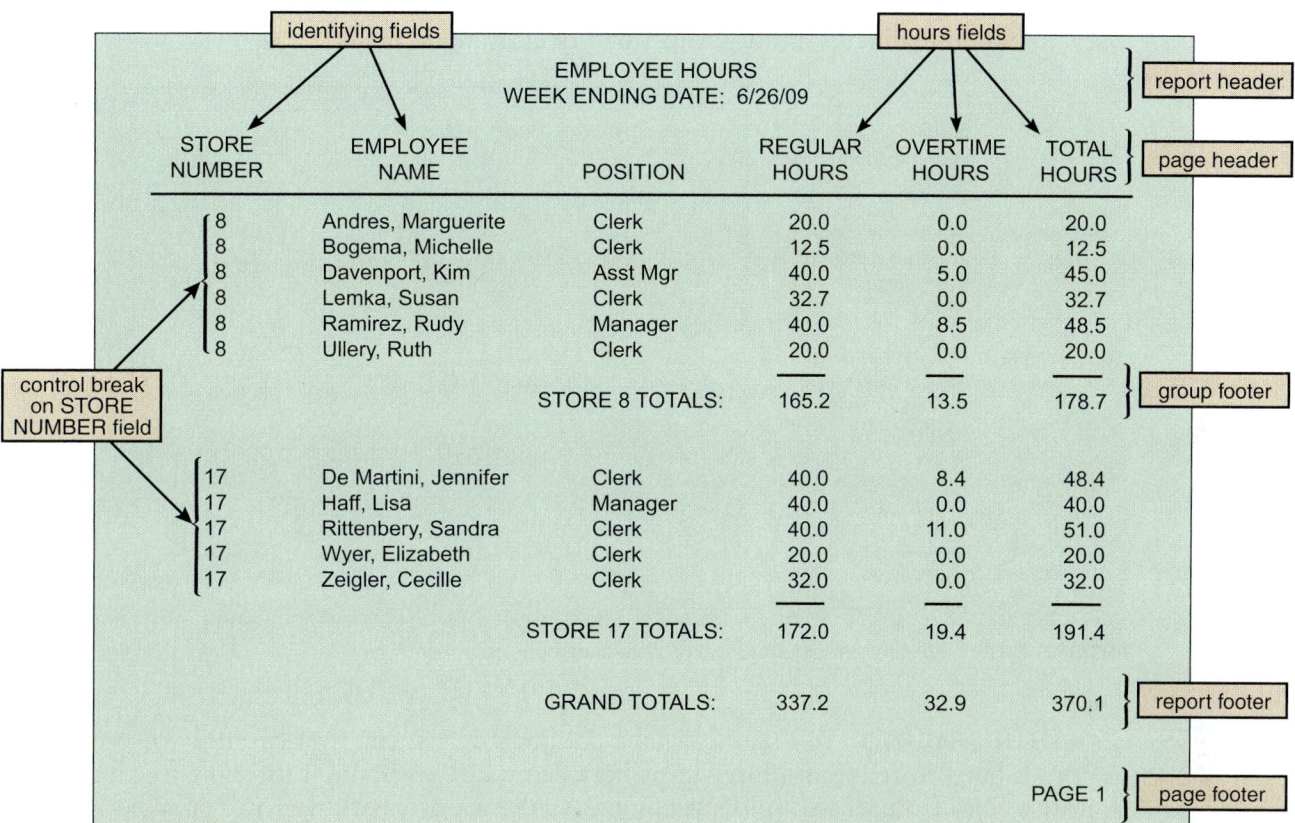

FIGURE 8-9 The Employee Hours report is a detail report with control breaks, subtotals, and grand totals. Notice that a report header identifies the report, a page header contains column headings, a group footer contains subtotals for each store, a report footer contains grand totals, and a page footer identifies the page number.

REPORT HEADERS AND FOOTERS Every report should have a report header and a report footer. The **report header**, which appears at the beginning of the report, identifies the report, and contains the report title, date, and other necessary information. The **report footer**, which appears at the end of the report, can include grand totals for numeric fields and other end-of-report information, as shown in Figure 8-9.

PAGE HEADERS AND FOOTERS Every page should include a **page header**, which appears at the top of the page and includes the column headings that identify the data. The headings should be short but descriptive. Avoid abbreviations unless you know that users will understand them clearly. Either a page header or a **page footer**, which appears at the bottom of the page, is used to display the report title and the page number.

COLUMN HEADING ALIGNMENT Figure 8-10 shows several column heading alignment options. In Example 1, the left-justified column headings do not work well with numeric fields because the amount 1.25 would print past the right edge of the AMOUNT heading. In Example 2, the right-justified headings cause a problem with alphanumeric fields, because none of the characters in a short name would print under any part of the NAME heading. Centering headings over *maximum* field widths, as shown in Example 3, is not ideal when many of the actual values are shorter than the maximum width. Many designers prefer Example 4, where headings are left-justified over alphanumeric fields and right-justified over numeric fields.

Example 1:	Column headings are left-justified over maximum field widths.	NAME XXXXXXXXXXXXXXXXXXXXXXXX	NUMBER ZZZ9	AMOUNT ZZZ,ZZ9.99
Example 2:	Column headings are right-justified over maximum field widths.	NAME XXXXXXXXXXXXXXXXXXXXXXXX	NUMBER ZZZ9	AMOUNT ZZZ,ZZ9.99
Example 3:	Column headings are centered over maximum field widths.	NAME XXXXXXXXXXXXXXXXXXXXXXXX	NUMBER ZZZ9	AMOUNT ZZZ,ZZ9.99
Example 4:	Column headings are left-justified over alphanumeric fields and right-justified over numeric fields.	NAME XXXXXXXXXXXXXXXXXXXXXXXX	NUMBER ZZZ9	AMOUNT ZZZ,ZZ9.99

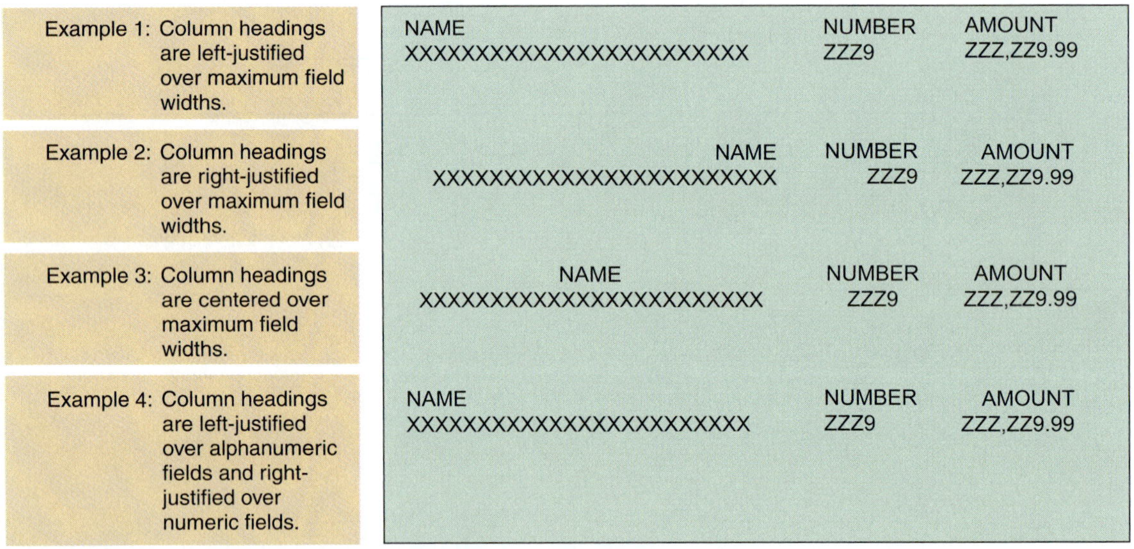

FIGURE 8-10 Four different column heading alignment options.

COLUMN SPACING You should space columns of information carefully. A crowded report is hard to read, and large gaps between columns make it difficult for the eye to follow a line. Columns should stretch across the report, with uniform spacing and suitable margins at top, bottom, right, and left. Some report designers use landscape orientation when working with a large number of columns; others prefer to break the information into more than one report. In some cases, a smaller point size will solve the problem, but the report must remain readable and acceptable to users.

FIELD ORDER Fields should be displayed and grouped in a logical order. The report shown in Figure 8-9 on the previous page, for example, shows the detail lines printed in alphabetical order within store number, so the store number is in the left column, followed by the employee name. The employee position relates to the employee's name, so the items are adjacent. The three hours fields also are placed together.

GROUPING DETAIL LINES Often, it is meaningful to arrange detail lines in groups, based on a control field. For example, using the department number as a control field, individual employees can be grouped by department. You can print a **group header** above the first detail line and a **group footer** after the last detail line in a group, as shown in Figure 8-9.

Database programs such as Microsoft Access make it easy to create groups and subgroups based on particular fields. You also can have the report calculate and display totals, averages, record counts, and other data for any group or subgroup. For example, in a large company, you might want to see total sales and number of sales broken down by product within each of the 50 states. The screen shown in Figure 8-11 is an Access Help screen that refers to a step-by-step process for creating multilevel grouping.

FIGURE 8-11   Microsoft Access includes an easy-to-use tool for grouping data.

CONSISTENT DESIGN Good report design includes standards to ensure that reports are uniform and consistent. When a system produces multiple reports, each report should share common design elements. For example, the date and page numbers should print in the same place on each report page. Abbreviations used in reports also should be consistent. For example, when indicating a numeric value, it is confusing for one report to use #, another *NO*, and a third *NUM*. Items in a report also should be consistent. If one report displays the inventory location as a shelf number column followed by a bin number column, that same layout should be used on all inventory location reports.

Report Design Example

Revisit the Employee Hours report shown in Figure 8-9 on page 341. Although the report follows many of the design guidelines discussed, you still could improve it. Too much detail is on the page, forcing users to search for the information they need. Can you see any material that you could eliminate?

If most employees do not work overtime, then overtime hours should stand out. You can do that by not printing 0.0 when overtime hours are zero. Repeating the store number for each employee also is unnecessary, because the employees are grouped by store number. Another way to avoid repeating the store number is to use a group header to identify each store, and eliminate the STORE NUMBER field altogether. Finally, most of the employees in a store are clerks. The manager and assistant manager titles would stand out better if the word Clerk were not printed for all clerical employees. Those changes have been made in Figure 8-12, and the EMPLOYEE NAME and POSITION columns were exchanged to avoid a large gap between names and the REGULAR HOURS column.

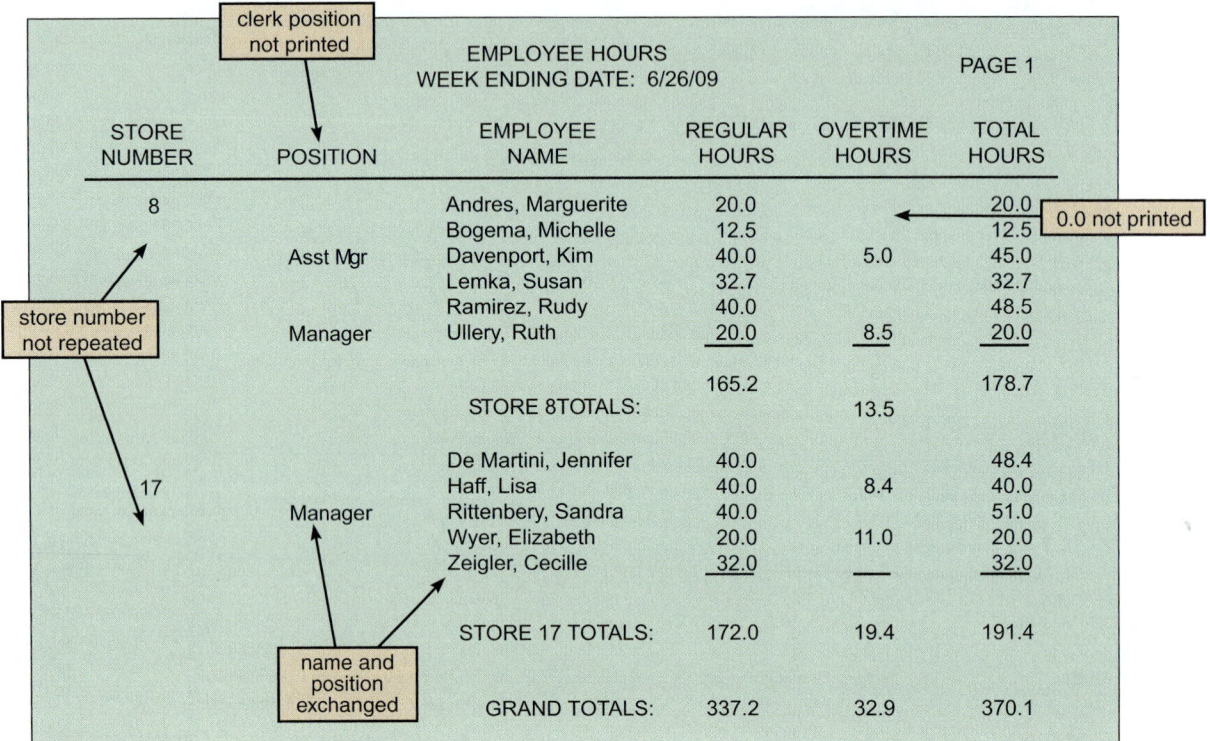

FIGURE 8-12 An improved version of the Employee Hours report shown in Figure 8-9 on page 341.

CASE IN POINT 8.1: LAZY EDDIE

Lynn Jennings is the IT manager at Lazy Eddie, a chain that specializes in beanbag chairs and recliners. She asked Jan Lauten, a senior systems analyst, to review the large number of printed reports that are distributed to Lazy Eddie's 35 store managers. "Jan, I just can't believe that our people really read all of those reports," Lynn said. "We constantly add new reports, and we never seem to eliminate the old ones. Sometimes I think all we're doing is keeping the paper companies in business!" Jan replied, "I agree, but what can we do? The managers say they want the reports, but I always see them stacked on top of file cabinets. I've never seen anyone read a report."

"I have an idea," Lynn said. "I want you to come up with a procedure that requires users to review and justify their information needs to see if they really use the reports we send them. You could design a form that asks if the information still is required, and why. Try to get users to decide if a report is worth the cost of producing it. Do you think you can do it?"

"Sure I can," Jan replied. When Jan returned to her office, she wondered where to begin. What advice would you give to Jan?

Output Control and Security

Output must be accurate, complete, current, and secure. Companies use various **output control** methods to maintain output integrity and security. For example, every report should include an appropriate title, report number or code, printing date, and time period covered. Reports should have pages that are numbered consecutively, identified as *Page nn of nn*, and the end of the report should be labeled clearly. Control totals and record counts should be reconciled against input totals and counts. Reports should be selected at random for a thorough check of correctness and completeness. All processing errors or interruptions must be logged so they can be analyzed.

Output security protects privacy rights and shields the organization's proprietary data from theft or unauthorized access. To ensure output security, you must perform several important tasks. First, limit the number of printed copies and use a tracking procedure to account for each copy. When printed output is distributed from a central location, you should use specific procedures to ensure that the output is delivered to authorized recipients only. That is especially true when reports contain sensitive information, such as payroll data. All sensitive reports should be stored in secure areas. All pages of confidential reports should be labeled appropriately.

As shown in Figure 8-13, it is important to shred sensitive reports, out-of-date reports, and output from aborted print runs. Blank check forms must be stored in a secure location and be inventoried regularly to verify that no forms are missing. If signature stamps are used, they must be stored in a secure location away from the forms storage location.

In most organizations, the IT department is responsible for output control and security measures. Systems analysts must be concerned with security issues as they design, implement, and support information systems. Whenever possible, security should be designed into the system by using passwords, shielding sensitive data, and controlling user access. Physical security always will be necessary, especially in the case of printed output that is tangible and can be viewed and handled easily.

Enterprise-wide data access creates a whole new set of security and control issues. In the past, many firms installed diskless workstations, as shown in the photo in Figure 8-14. A **diskless workstation** is a network terminal that supports a full-featured user interface, but limits the printing or copying of data, except to certain network resources that can be monitored and controlled. This concept worked well with terminals that had limited hardware and software features.

However, over time, the number of removable media devices has expanded greatly, along with a wide variety of physical interfaces such as USB, FireWire, and PCMCIA, and many wireless interfaces, such as Wi-Fi and Bluetooth. A popular security solution is the use of a network-based application, often called a **port protector**, that controls access to and from workstation interfaces. The SafeGuard® PortProtector is shown in Figure 8-14.

FIGURE 8-13 To maintain output security, it is important to shred sensitive material.

ON THE WEB

For more information about output control and security, visit **scsite.com/sad8e/ more**, locate Chapter 8, and then click the Output Control and Security link.

FIGURE 8-14 A diskless workstation can limit the printing or copying of data. Port protector software can enhance data security by controlling the functionality of terminal ports and interfaces.

For more information about user interface design, visit **scsite.com/ sad8e/more**, locate Chapter 8, and then click the User Interface Design link.

USER INTERFACE DESIGN

Although output design involves a separate set of physical design issues, it is an integral part of a larger concept called a user interface (UI). A **user interface (UI)** describes how users interact with a computer system, and consists of all the hardware, software, screens, menus, functions, output, and features that affect two-way communications between the user and the computer.

Figure 8-15 suggests an interesting viewpoint that interface designers should keep in mind: Industry leader IBM believes that the best interfaces are the ones that users do not even notice — they make sense because they do what users expect them to do.

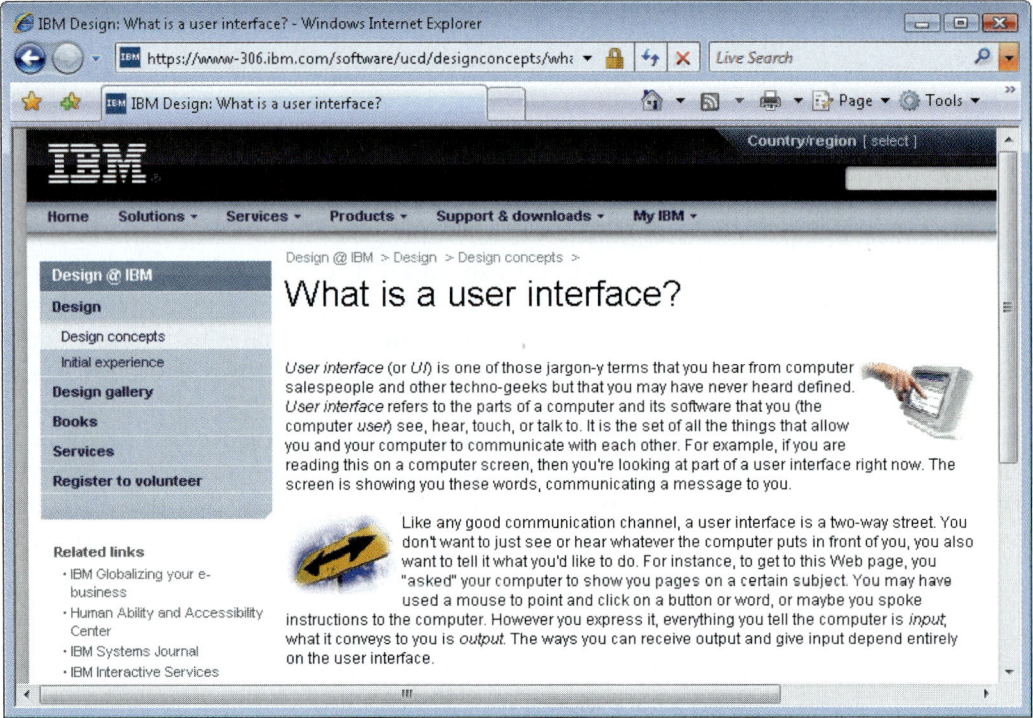

FIGURE 8-15   IBM devotes a great deal of effort to create user interfaces that are simple and easy to learn.

Evolution of the User Interface

When developing older systems, analysts typically designed all the printed and screen output first, then worked on the inputs necessary to produce the results. Often, the user interface mainly consisted of **process-control** screens that allowed the user to send commands to the system. That approach worked well with traditional systems that simply transformed input data into structured output.

As information management evolved from centralized data processing to dynamic, enterprise-wide systems, the primary focus also shifted — from the IT department to the users themselves. The IT group became a supplier of information technology, rather than a supplier of information. Today, the main focus is on users within and outside the company, how they communicate with the information system, and how the system supports the firm's business operations. Figure 8-16 compares a traditional, processing-centered information system with a modern, user-centered system. Notice that the IT department, which was the main interface for user information requests, has become a system facilitator that maintains and supports the system for its users.

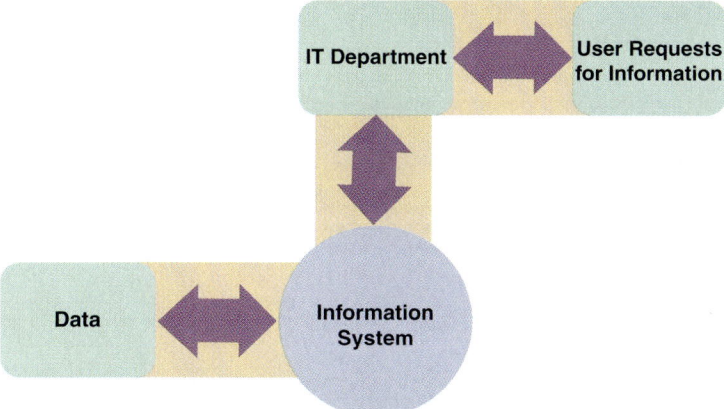

Traditional, Processing-Centered Information System Model

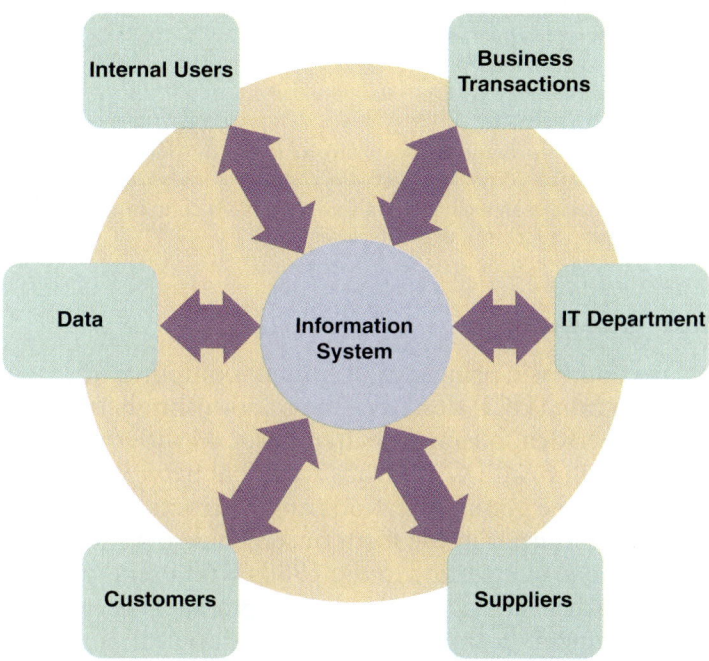

Modern, User-Centered Information System Model

FIGURE 8-16 Compare the traditional, processing-centered system at the top of the figure with the modern, user-centered information system at the bottom. Notice the change in the role of the IT department.

In a **user-centered** system, the distinction blurs between input, output, and the interface itself. Most users work with a varied mix of input, screen output, and data queries as they perform their day-to-day job functions. Because all those tasks require interaction with the computer system, the user interface is a vital element in the systems design phase. Ergosoft laboratories is one of many firms that offer consulting services and software solutions to help companies develop successful user interfaces, as shown in Figure 8-17 on the next page.

User interface design requires an understanding of human-computer interaction and user-centered design principles, which are discussed in the next section. Input and output design topics are covered later in this chapter.

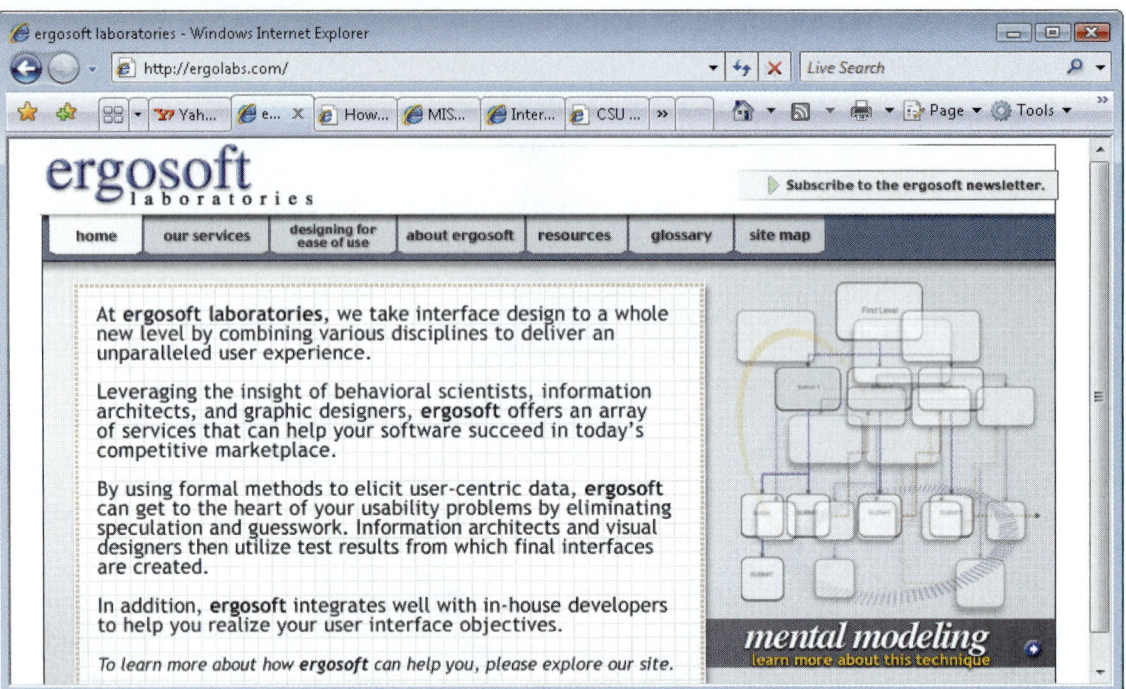

FIGURE 8-17 Ergosoft laboratories is an example of a firm that offers consulting services and software solutions to help companies develop successful user interfaces.

Human-Computer Interaction

A user interface is based on basic principles of human-computer interaction. **Human-computer interaction (HCI)** describes the relationship between computers and people who use them to perform business-related tasks, as shown in Figure 8-18. HCI concepts apply to everything from PC desktops to global networks. In its broadest sense, a user interface includes all the communications and instructions necessary to enter input to the system and to obtain output in the form of screen displays or printed reports.

The human-computer interface started in the 1980s with users typing complex commands in green text on a black screen. Then came the **graphical user interface (GUI)**, which was a huge improvement, because it used icons, graphical objects, and pointing devices. Today, designers strive to translate user behavior, needs, and desires into an interface that users don't really notice. As IBM points out in Figure 8-15 on page 346, the best user interfaces are *"almost transparent — you can see right though the interface to your own work."*

As a systems analyst, you will design user interfaces for in-house developed software and customize interfaces for various commercial packages and user productivity applications. Your main objective is to create a user-friendly design that is easy to learn and use.

Industry leaders Microsoft and IBM both devote considerable resources to user interface research. Figure 8-19 describes

FIGURE 8-18 A user interface is based on basic principles of human-computer interaction, which describes the relationship between computers and people who use them to perform business-related tasks.

Microsoft's Redmond labs, where engineers observe volunteers who participate in software usability studies.

At its Almaden Research Center, IBM conducts usability testing and studies human-computer interaction, as shown in Figure 8-20. According to IBM, its User Sciences & Experience Research (USER) lab focuses on improving ease of use and exploring new ways of using computers.

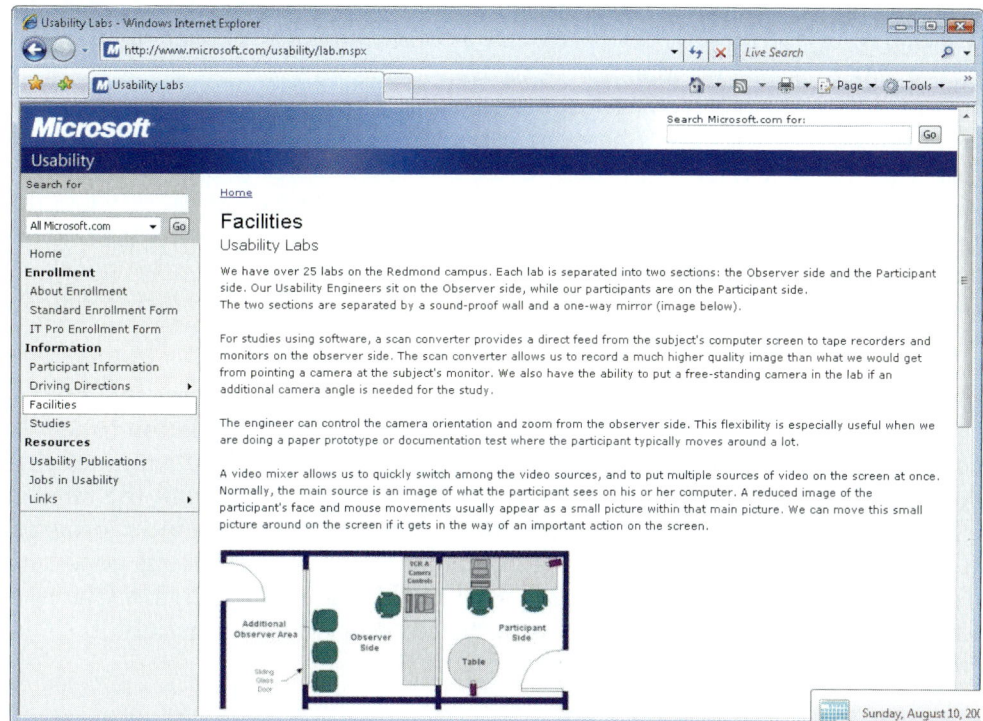

FIGURE 8-19   At its Redmond labs, Microsoft engineers observe volunteers who participate in software usability studies.

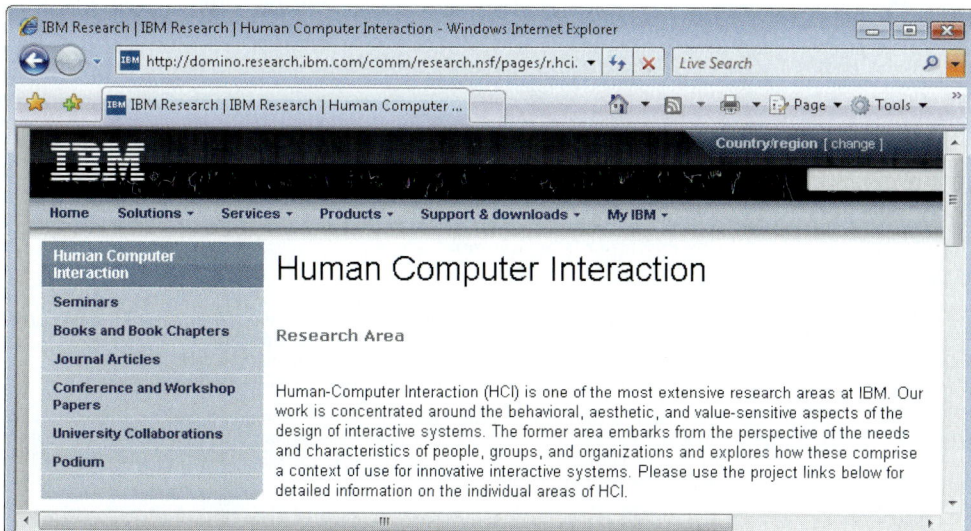

FIGURE 8-20   IBM claims that HCI is one of the most extensive research areas at the company.

CASE IN POINT 8.2: CASUAL OBSERVER SOFTWARE

Casual Observer Software's main product is a program that monitors and analyzes user keystrokes and mouse clicks to learn more about the way employees use their computer systems. The problem is that some users feel this is an unwarranted intrusion into their privacy, and they prefer not to be observed. Some even fear that the data would be used for other reasons, including performance appraisal. You are a consultant who has been hired by a client firm that is trying to decide whether or not to use this software.

Before you advise the client, go back and review the Microsoft usability lab shown in Figure 8-19 on the previous page, where the users being studied in the Redmond labs were willing participants. Then, refer to Chapter 4, Requirements Modeling, page 161, and consider the Hawthorne Effect, which suggests that employees might behave differently when they know they are being observed. Finally, think about the ethical issues that might be involved in this situation. What will you advise your client, and why?

For more information about human-computer interaction, visit **scsite.com/ sad8e/more**, locate Chapter 8, and then click the Human-Computer Interaction link.

IBM believes that the user interface evolution will lead to computers that truly are *consumer products* that are simple and natural for the general population to use. This will occur, in IBM's view, because computers will function in a friendlier, more predictable way — much like a telephone or video player. Most important, the interface will be based on the perspective of a user rather than a computer engineer, programmer, or systems analyst. To understand the magnitude of this shift in thinking, consider the powerful statement shown in Figure 8-21, where IBM usability expert Dr. Clare-Marie Karat states that "in this new computer age, the customer is not only right, the customer has rights." These rights are listed in Figure 8-22.

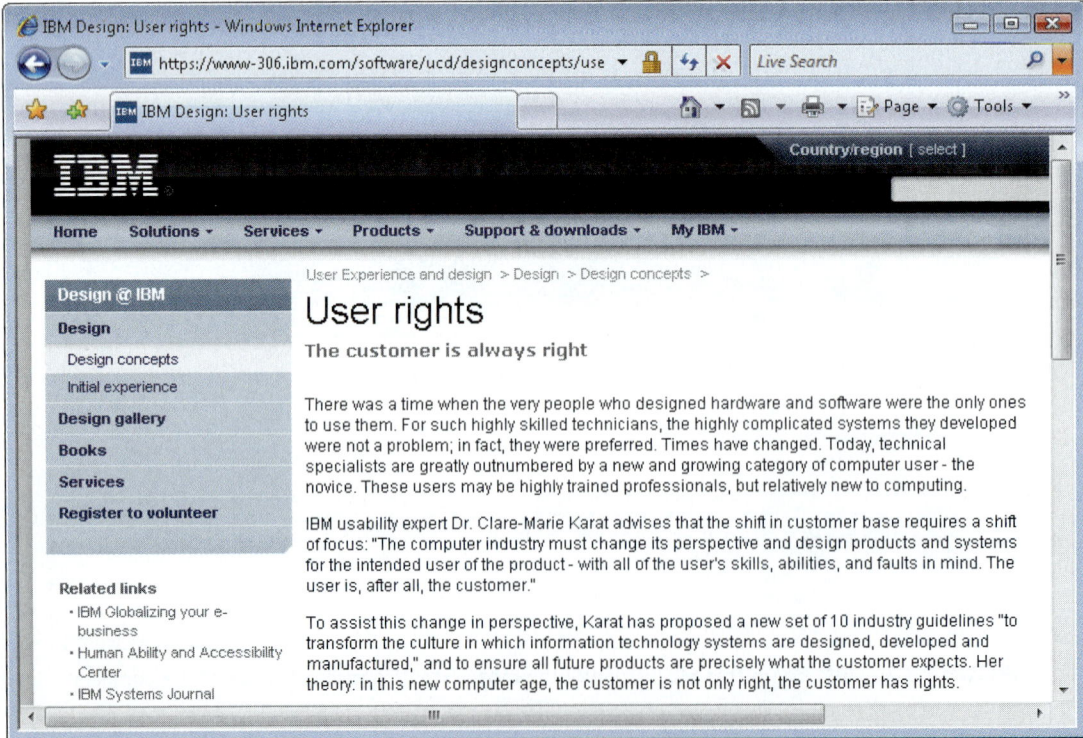

FIGURE 8-21 IBM strongly believes that computers must be user-friendly, and that users have specific rights.

User Rights

1. **Perspective:** The user always is right. If there is a problem with the use of the system, the system is the problem, not the user.
2. **Installation:** The user has the right to install and uninstall software and hardware systems easily without negative consequences.
3. **Compliance:** The user has the right to a system that performs exactly as promised.
4. **Instruction:** The user has the right to easy-to-use instructions (user guides, online or contextual help, and error messages) for understanding and utilizing a system to achieve desired goals and recover efficiently and gracefully from problem situations.
5. **Control:** The user has the right to be in control of the system and to be able to get the system to respond to a request for attention.
6. **Feedback:** The user has the right to a systen that provides clear, understandable, and accurate information regarding the task it is performing and the progress toward completion.
7. **Dependencies:** The user has the right to be informed clearly about all systems requirements for successfully using software or hardware.
8. **Scope:** The user has the right to know the limits of the system's capabilities.
9. **Assistance:** The user has the right to communicate with the technology provider and receive a thoughtful and helpful response when raising concerns.
10. **Usability:** The user should be the master of software and hardware technology, not vice versa. Products should be natural and intuitive to use.

Source: http://www-306.ibm.com/software/ucd/designconcepts/userrights.html

FIGURE 8-22 User rights suggested by IBM's Dr. Clare-Marie Karat.

Basic Principles of User-Centered Design

Although IT professionals have different views about interface design, most would agree that good design depends on eight basic principles, which are described in the following sections.

UNDERSTAND THE UNDERLYING BUSINESS FUNCTIONS The interface designer must understand the underlying business functions and how the system supports individual, departmental, and enterprise goals. The overall objective is to design an interface that helps users to perform their jobs. A good starting point might be to analyze a functional decomposition diagram (FDD). As you learned in Chapter 4, an FDD is a graphical representation of business functions that starts with major functions, and then breaks them down into several levels of detail. An FDD can provide a checklist of user tasks that you must include in the interface design.

MAXIMIZE GRAPHICAL EFFECTIVENESS Studies show that people learn better visually. The immense popularity of Microsoft Windows is largely the result of a graphical user interface that is easy to learn and use. Now that GUIs have become universal in application packages, users expect in-house software also to have GUIs. A well-designed GUI can help users learn a new system rapidly, and work with the system effectively. Also, in a GUI environment, a user can display and work with multiple windows on a single screen and transfer data between programs. Because GUIs are used for data entry as well as for process control, they must follow the guidelines for data entry screen design discussed later in this chapter.

PROFILE THE SYSTEM'S USERS A systems analyst should understand user experience, knowledge, and skill levels. If a wide range of capability exists, the interface should be flexible enough to accommodate novices as well as experienced users.

THINK LIKE A USER To develop a user-centered interface, the designer must learn to think like a user and see the system through a user's eyes. The interface should use terms and metaphors that are familiar to users. Users are likely to have real-world experience with many other machines and devices that provide feedback, such as automobiles, ATM machines, and microwave ovens. Based on that experience, users will expect useful, understandable feedback from a computer system.

USE PROTOTYPING   From a user's viewpoint, the interface is the most critical part of the system design because it is where he or she interacts with the system — perhaps for many hours each day. It is essential to construct models and prototypes for user approval. An interface designer should obtain as much feedback as possible, as early as possible. You can present initial screen designs to users in the form of a **storyboard**, which is a sketch that shows the general screen layout and design. The storyboard can be created with software, or drawn freehand. Users must test all aspects of the interface design and provide feedback to the designers. User input can be obtained in interviews, via questionnaires, and by observation. Interface designers also can obtain data, called **usability metrics**, by using software that can record and measure user interaction with the system.

DESIGN A COMPREHENSIVE INTERFACE   The user interface should include all tasks, commands, and communications between users and the information system. The screen in Figure 8-23 shows the main options for a student registration system. Each screen option leads to another screen, with more options. The objective is to offer a reasonable number of choices that a user easily can comprehend. Too many options on one screen can confuse a user — but too few options increase the number of submenu levels and complicate the navigation process. Often, an effective strategy is to present the most common choice as a default, but allow the user to select other options.

CONTINUE THE FEEDBACK PROCESS   Even after the system is operational, it is important to monitor system usage and solicit user suggestions. You can determine if system features are being used as intended by observing and surveying users. Sometimes, full-scale operations highlight problems that were not apparent when the prototype was tested. Based on user feedback, Help screens might need revision and design changes to allow the system to reach its full potential.

TOOLKIT TIME

The Communication Tools in Part 1 of the Systems Analyst's Toolkit can help you communicate effectively with users. To learn more about these tools, turn to Part 1 of the four-part Toolkit that follows Chapter 12.

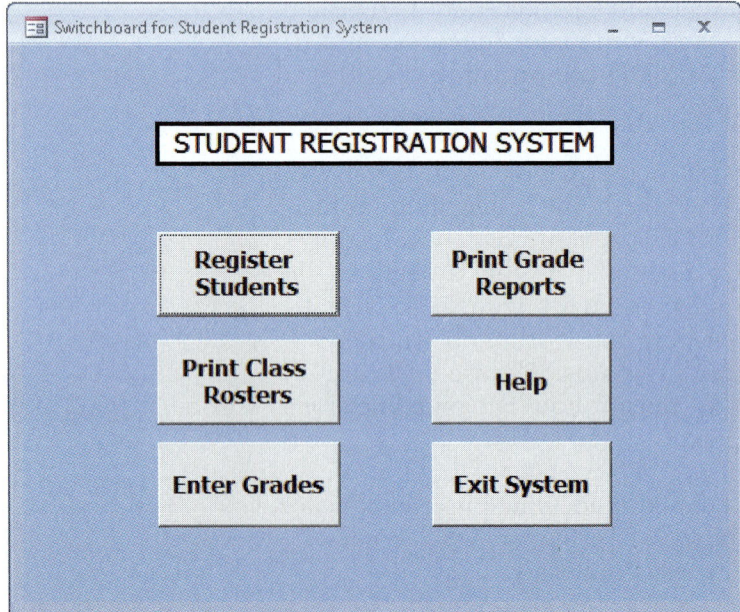

FIGURE 8-23   The screen displays the main options for a school's student registration system.

DOCUMENT THE INTERFACE DESIGN   You should document all screen designs for later use by programmers. If you are using a CASE tool or screen generator, number the screen designs and save them in a hierarchy similar to a menu tree. User-approved sketches and storyboards also can be used to document the user interface.

By applying basic user-centered design principles, a systems analyst can plan, design, and deliver a successful user interface.

Guidelines for User Interface Design

It is important to design a user interface that is easy to use, attractive, and efficient. When you create a user interface, you should follow eight basic guidelines. These guidelines also apply to data entry screen design, which is discussed later in this chapter.

1. Focus on basic objectives.
2. Build an interface that is easy to learn and use.
3. Provide features that promote efficiency.
4. Make it easy for users to obtain help or correct errors.
5. Minimize input data problems.
6. Provide feedback to users.
7. Create an attractive layout and design.
8. Use familiar terms and images.

Good user interface design is based on a combination of ergonomics, aesthetics, and interface technology. **Ergonomics** describes how people work, learn, and interact with computers; **aesthetics** focuses on how an interface can be made attractive and easy to use; and **interface technology** provides the operational structure required to carry out the design objectives. As shown in Figure 8-24, Cognetics Corporation offers user interface design services. Cognetics stresses that an interface must be effective, efficient, engaging, error tolerant, and easy to learn.

The following sections provide examples of the basic user interface design guidelines. As mentioned earlier, many of the specific points also apply to data entry screen design, which is discussed on page 363.

FOCUS ON BASIC OBJECTIVES

- Facilitate the system design objectives, rather than calling attention to the interface.
- Create a design that is easy to learn and remember.
- Design the interface to improve user efficiency and productivity.
- Write commands, actions, and system responses that are consistent and predictable.
- Minimize data entry problems.
- Allow users to correct errors easily.
- Create a logical and attractive layout.

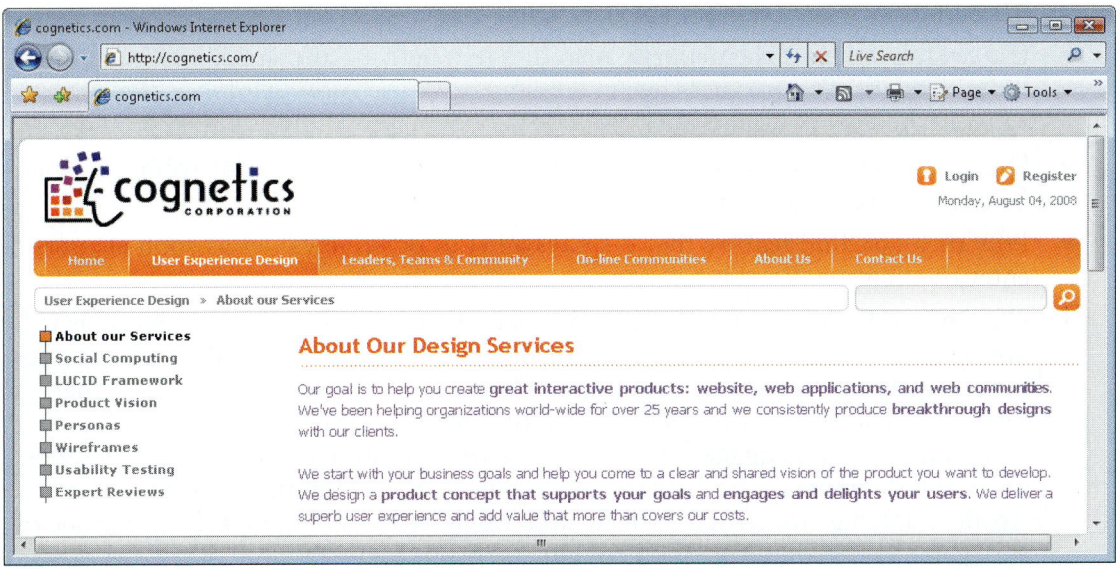

FIGURE 8-24 Cognetics Corporation believes that an interface must be effective, efficient, engaging, error tolerant, and easy to learn.

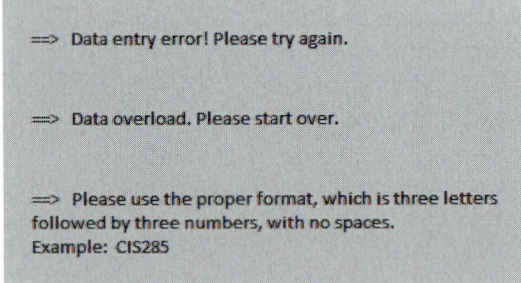

==> Data entry error! Please try again.

==> Data overload. Please start over.

==> Please use the proper format, which is three letters followed by three numbers, with no spaces.
Example: CIS285

FIGURE 8-25 In the example at the top, only one of the six icons, the *Home* image, is familiar and predictable. In the bottom screen, the first two messages are not very helpful, but the third message would be easy for a user to understand.

BUILD AN INTERFACE THAT IS EASY TO LEARN AND USE

- Label clearly all controls, buttons, and icons.

- Select only those images that users can understand easily, if you use images to identify icons or controls. Also, provide on-screen instructions that are logical, concise, and clear. For example, the top screen in Figure 8-25 shows six control buttons, but only the home image has an obvious meaning. In the bottom screen, notice the difference in the messages: The first two provide little or no information; the third one is easy to understand.

  - Show all commands in a list of menu items, but dim any commands that are not currently available.

  - Make it easy to navigate or return to any level in the menu structure.

PROVIDE FEATURES THAT PROMOTE EFFICIENCY

- Organize tasks, commands, and functions in groups that resemble actual business operations. You should group functions and submenu items in a multilevel menu hierarchy, or tree, that is logical and reflects how users typically perform the tasks. Figure 8-26 shows an example of a menu hierarchy for an order tracking system.

- Create alphabetical menu lists or place the selections used frequently at the top of the menu list. No universally accepted approach to menu item placement exists. The best strategy is to design a prototype and obtain feedback from users. Some applications even allow menus to show recently used commands first. Some users like that feature, but others might find it distracting. The best approach is to offer a choice, and let users decide.

- Provide shortcuts so experienced users can avoid multiple menu levels. You can create shortcuts using hot keys that allow a user to press the ALT key + the underlined letter of a command.

- Use default values if the majority of values in a field are the same. For example, if 90% of the firm's customers live in Albuquerque, use *Albuquerque* as the default value in the City field.

- Use a duplicate value function that enables users to insert the value from the same field in the previous record.

- Provide a fast-find feature that displays a list of possible values as soon as users enter the first few letters.

- Use a natural language feature that allows users to type commands or requests in normal English phrases. For example, Microsoft Office products allow users to request Help by typing a question into a dialog box. The software then uses natural language technology to retrieve a list of topics that match the request. Most users like natural language features because they do not have to memorize a series of complex commands and syntax. According to the American Association for Artificial Intelligence (AAAI), whose Web site is shown in Figure 8-27, the value of being able to communicate with computers in everyday "natural" language cannot be overstated. Natural language technology is used in speech recognition systems, text-to-speech synthesizers, automated voice response systems, Web search engines, text editors, and language instruction materials.

MAKE IT EASY FOR USERS TO OBTAIN HELP OR CORRECT ERRORS

- Ensure that Help is always available. Help screens should provide information about menu choices, procedures, shortcuts, and errors.

- Provide user-selected Help and context-sensitive Help. **User-selected** Help displays information when the user requests it. By making appropriate choices through the menus and submenus, the user eventually reaches a screen with the desired information. Figure 8-28 on the next page shows the main Help screen for the student registration system. **Context-sensitive** Help offers assistance for the task in progress. Figure 8-29 shows a Help dialog box that is displayed if a user requests Help while entering data into the ADVISOR ASSIGNED field. Clicking the Close button returns the user to the current task.

- Provide a direct route for users to return to the point from where Help was requested. Title every Help screen to identify the topic, and keep Help text simple and concise. Insert blank lines between paragraphs to make Help easier to read, and provide examples where appropriate.

- Include contact information, such as a telephone extension or e-mail address if a department or Help desk is responsible for assisting users.

- Require user confirmation before data deletion (*Are you sure?*) and provide a method of recovering data that is deleted inadvertently. Build in safeguards that prevent critical data from being changed or erased.

- Provide an Undo key or a menu choice that allows the user to eradicate the results of the most recent command or action.

- When a user-entered command contains an error, highlight the erroneous part and allow the user to make the correction without retyping the entire command.

- Use hypertext links to assist users as they navigate through Help topics.

Customer Order Tracking System

FIGURE 8-26 Tasks, commands, and functions should be organized in logical groups, such as this menu hierarchy for a customer order tracking system.

Main Menu
- Customers
  - Add a New Customer
  - Update Customer Data
  - Delete a Customer
- Orders
  - Enter a New Order
  - Modify Order Data
  - Cancel an Order
- Products
  - Enter a New Product
  - Update Product Data
  - Delete a Product

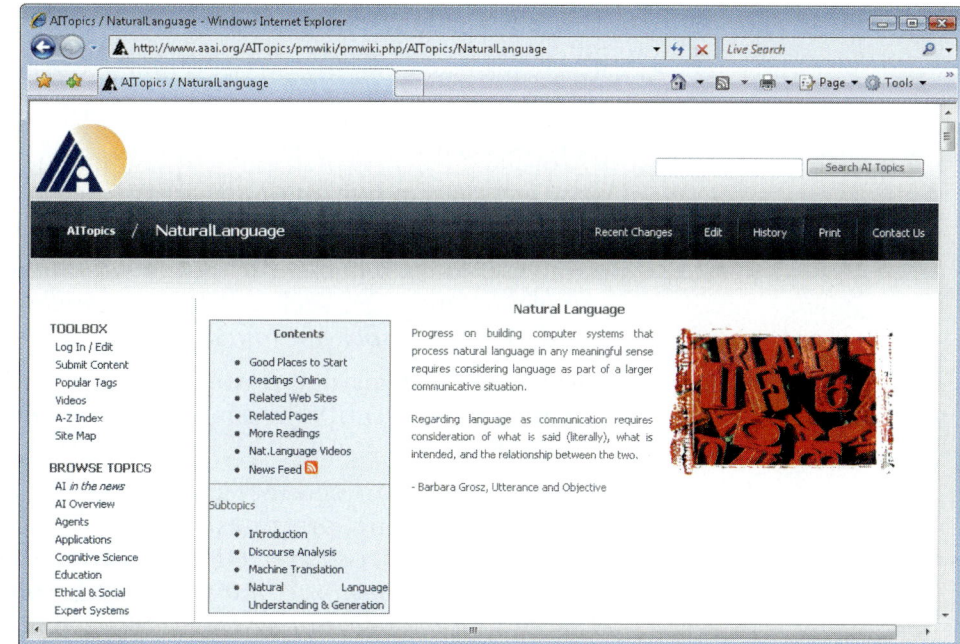

FIGURE 8-27 The American Association for Artificial Intelligence (AAAI) sees natural language processing as a key element of the communication process between users and their computers.

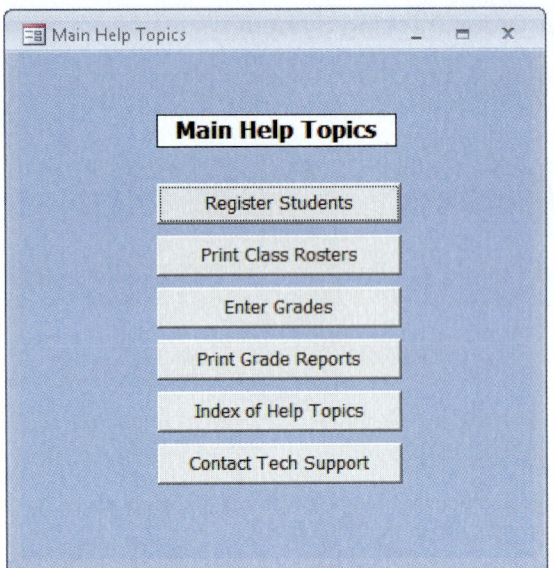

FIGURE 8-28 The main Help screen for a student registration system.

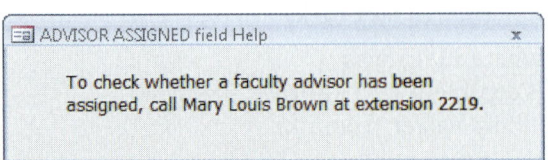

FIGURE 8-29 A context-sensitive dialog box displays if a user requests help while entering data into the ADVISOR ASSIGNED field. Clicking the Close button returns the user to the task.

MINIMIZE INPUT DATA PROBLEMS

- Provide data validation checks. More information on data validation techniques is provided in the section on input design later in this chapter.

- Display event-driven messages and reminders. Just as context-sensitive Help is important to users, it is desirable to display an appropriate message when it is time for the user to perform a certain task. For example, when exiting the system, a message might ask users if they want a printed report of the data entered during the recent session.

- Establish a list of predefined values that users can click to select. Predefined values prevent spelling errors, avoid inappropriate data in a field, and make the user's job easier — the input screen displays a list of acceptable values and the user simply points and clicks the choice.

- Build in rules that enforce data integrity. For example, if the user tries to enter an order for a new customer, the customer must be added before the system will accept the order data.

- Use **input masks**, which are templates or patterns that make it easier for users to enter data. Microsoft Access 2007 provides standard input masks for fields such as dates, telephone numbers, ZIP codes, and Social Security numbers. In addition, you can create custom input masks for any type of data, as shown in Figure 8-30. Notice that a mask can manipulate the input data and apply a specific format. For example, if a user enters text in lowercase letters, the input mask >L<???????????? will automatically capitalize the first letter.

PROVIDE FEEDBACK TO USERS

- Display messages at a logical place on the screen, and be consistent.

- Alert users to lengthy processing times or delays. Give users an on-screen progress report, especially if the delay is lengthy.

- Allow messages to remain on the screen long enough for users to read them. In some cases, the screen should display messages until the user takes some action.

- Let the user know whether the task or operation was successful or not. For example, use messages such as *Update completed*, *All transactions have been posted*, or *ID Number not found*.

- Provide a text explanation if you use an icon or image on a control button. This helps the user to identify the control button when moving the mouse pointer over the icon or image.

- Use messages that are specific, understandable, and professional. Avoid messages that are cute, cryptic, or vague, such as: *ERROR — You have entered an unacceptable value*, or *Error DE4-16*. Better examples are: *Enter a number between 1 and 5*; *Customer number must be numeric. Please re-enter a numeric value*; or *Call the Accounting Department, Ext. 239 for assistance*.

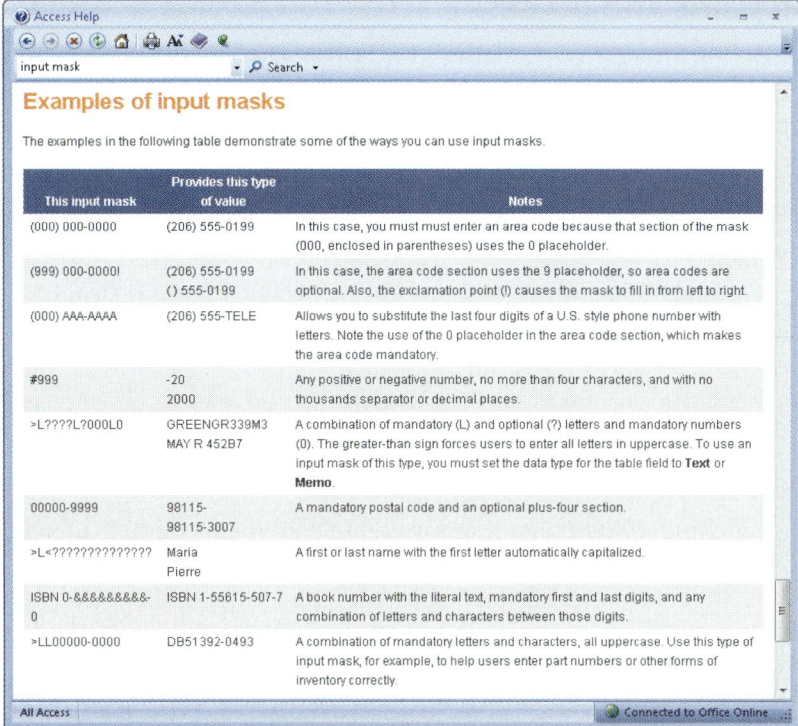

FIGURE 8-30 Using Microsoft Access 2007, you can create custom input masks as shown in these examples.

CREATE AN ATTRACTIVE LAYOUT AND DESIGN

- Use appropriate colors to highlight different areas of the screen; avoid gaudy and bright colors.

- Use special effects sparingly. For example, animation and sound might be effective in some situations, but too many special effects can be distracting and annoying to a user, especially if he or she must view them repeatedly.

- Use hyperlinks that allow users to jump to related topics.

- Group related objects and information. Visualize the screen the way a user will see it, and simulate the tasks that the user will perform.

- Screen density is important. Keep screen displays uncluttered, with enough white space to create an attractive, readable design.

- Display titles, messages, and instructions in a consistent manner and in the same general locations on all screens.

- Use consistent terminology. For example, do not use the terms *delete*, *cancel*, and *erase* to indicate the same action. Similarly, the same sound always should signal the same event.

- Ensure that commands always will have the same effect. For example, if the *BACK* control button returns a user to the prior screen, the *BACK* command always should perform that function throughout the application.

- Ensure that similar mouse actions will produce the same results throughout the application. The results of pointing, clicking, and double-clicking should be consistent and predictable.

- When the user enters data that completely fills the field, do not move automatically to the next field. Instead, require the user to confirm the entry by pressing the ENTER key or TAB key at the end of every fill-in field.

USE FAMILIAR TERMS AND IMAGES

- Remember that users are accustomed to a pattern of *red = stop*, *yellow = caution*, and *green = go*. Stick to that pattern and use it when appropriate to reinforce on-screen instructions.

- Provide a keystroke alternative for each menu command, with easy-to-remember letters, such as File, Exit, and Help.

- Use familiar commands if possible, such as Cut, Copy, and Paste.

- Provide a Windows look and feel in your interface design if users are familiar with Windows-based applications.

- Avoid complex terms and technical jargon; instead, select terms that come from everyday business processes and the vocabulary of a typical user.

User Interface Controls

The designer can include many control features, such as menu bars, toolbars, dialog boxes, text boxes, toggle buttons, list boxes, scroll bars, drop-down list boxes, option buttons, check boxes, command buttons, spin bars, and calendar controls, among others. Figure 8-31 shows a data entry screen for the student registration system. The screen design uses several features that are described in the following section.

The **menu bar** at the top of the screen displays the main menu options. Some software packages allow you to create customized menu bars and toolbars. You can add a shortcut feature that lets a user select a menu command either by clicking the desired choice or by pressing the ALT key + the underlined letter. Some forms also use a **toolbar** that contains icons or buttons that represent shortcuts for executing common commands.

A **command button** initiates an action such as printing a form or requesting Help. For example, when a user clicks the Find Student command button in Figure 8-31, a dialog box opens with instructions, as shown in Figure 8-32.

Other design features include dialog boxes, text boxes, toggle buttons, list boxes, scroll bars, drop-down list boxes, option or radio buttons, check boxes, and calendar controls. These features are described as follows:

- A **dialog box** allows a user to enter information about a task that the system will perform.

- A **text box** can display messages or provide a place for a user to enter data.

- A **toggle button** is used to represent on or off status — clicking the toggle button switches to the other status.

- A **list box** displays a list of choices that the user can select. If the list does not fit in the box, a **scroll bar** allows the user to move through the available choices.

- A **drop-down list box** displays the current selection; when the user clicks the arrow, a list of the available choices displays.

- An **option button**, or **radio button**, represents one choice in a set of options. The user can select only one option at a time, and selected options show a black dot.

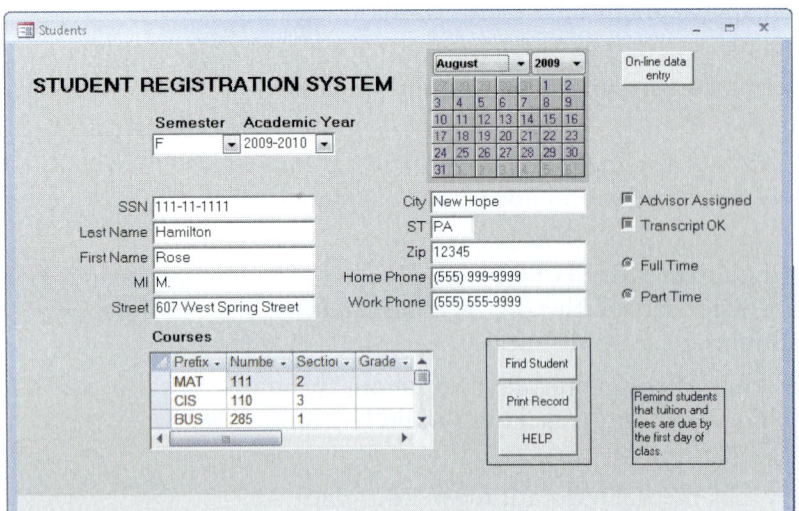

FIGURE 8-31 A data entry screen for the student registration system. This screen uses several design features that are described in the text.

- A **check box** is used to select one or more choices from a group. Selected options are represented by a checkmark or an X.

- A **calendar control** allows the user to select a date that the system will use as a field value.

Screen design requires a sense of aesthetics as well as technical skills. You should design screens that are attractive, easy to use, and workable. You also should obtain user feedback early and often as the design process continues.

The opening screen is especially important because it introduces the application and allows users to view the main options. When designing an opening screen, you can use a main form that functions as a switchboard. A **switchboard** uses command buttons that enable users to navigate the system and select from groups of related tasks. Figure 8-33 shows the switchboard and a data entry screen for a project management system. Notice the drop-down list box that allows users to enter a status code simply by clicking a selection.

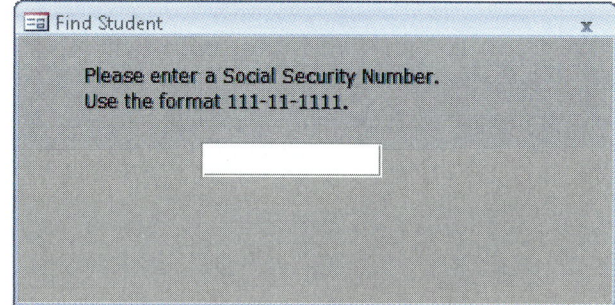

FIGURE 8-32 When a user clicks the Find Student command button, a dialog box is displayed with instructions.

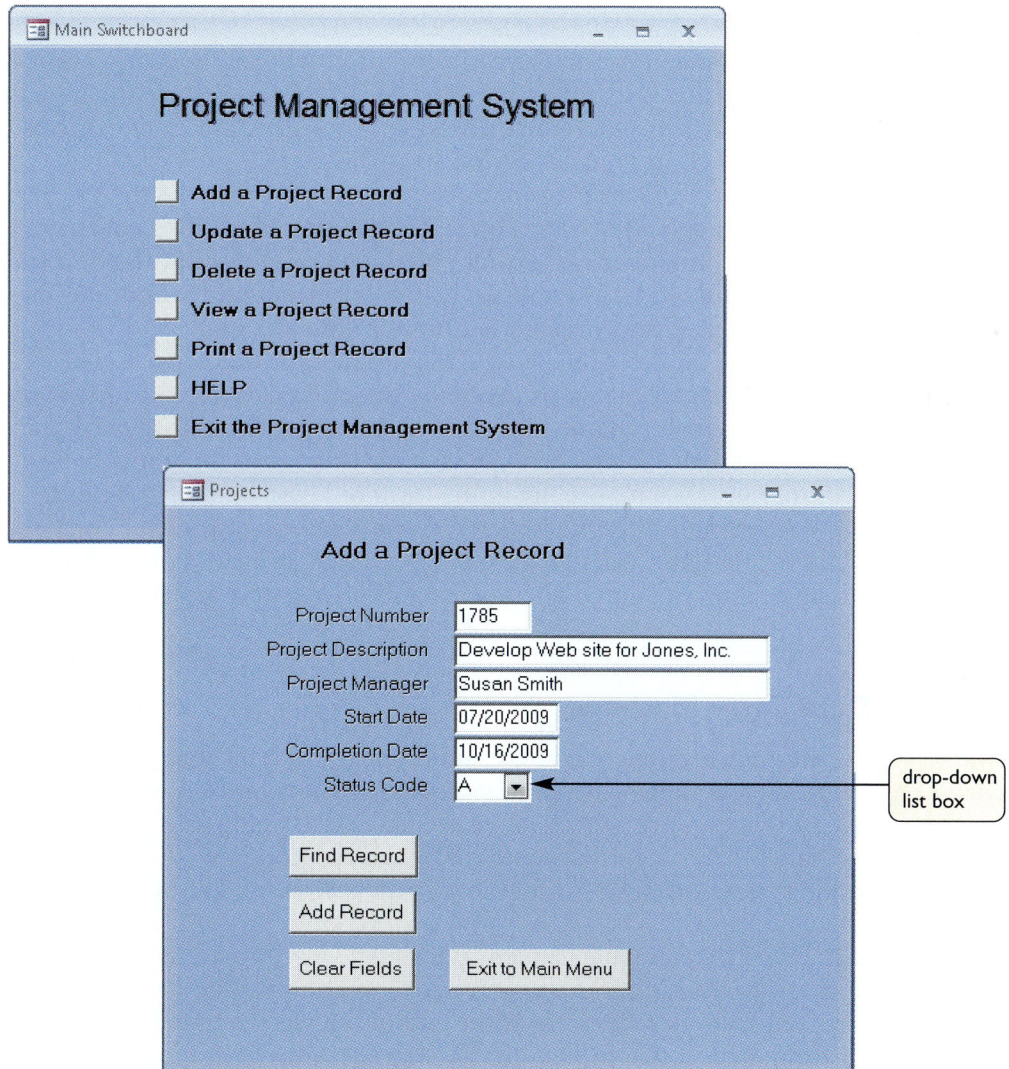

FIGURE 8-33 An example of a switchboard and data entry screen for a project management system.

INPUT DESIGN

ON THE WEB

For more information about input devices, visit **scsite.com/ sad8e/more**, locate Chapter 8, and then click the Input Devices link.

Input technology has changed dramatically in recent years. Today, many input devices and techniques are available, as shown in Figure 8-34. Businesses use the new technology to speed up the input process, reduce costs, and capture data in new forms, such as the digital signature shown in Figure 8-35 on page 362.

No matter how data enters an information system, the quality of the output is only as good as the quality of the input. The concept, sometimes known as **garbage in, garbage out** (GIGO), is familiar to IT professionals, who know that the best time to avoid problems is when the data is entered. The main objective of input design is to ensure the quality, accuracy, and timeliness of input data.

During input design, you determine how data will be captured and entered into the system. **Data capture** uses an automated or manually operated device to identify source data and convert it into computer-readable form. Examples of data capture devices include credit card scanners and bar code readers. **Data entry** is the process of manually entering data into the information system, usually in the form of keystrokes or mouse clicks.

Input design has six main objectives: Select a suitable input and data entry method, reduce input volume, design attractive data entry screens, use validation checks to reduce input errors, design required source documents and forms, and develop effective input controls.

Input and Data Entry Methods

ON THE WEB

For more information about data entry, visit **scsite.com/sad8e/ more**, locate Chapter 8, and then click the Data Entry link.

Input processes should be efficient, timely, and logical. Systems analysts apply business process engineering techniques when studying transactions and business operations to determine how and when data should enter the system. Usually, the first decision is whether to use batch or online input methods. Each method has advantages and disadvantages, and the systems analyst must consider the following factors.

BATCH INPUT Using **batch input**, data entry usually is performed on a specified time schedule, such as daily, weekly, monthly, or longer. For example, batch input occurs when a payroll department collects time cards at the end of the week and enters the data as a **batch**. Another example is a school that enters all grades for the academic term in a batch.

ONLINE INPUT Although batch input is used in specific situations, most business activity requires **online data entry**. The online method offers major advantages, including the

Input Device	Description
Biological feedback device	Device that creates a digital image of biological data such as fingerprints, retina patterns, or facial characteristics.
Brain-computer interface	Translates neural brain signals into digital commands. Possible applications include access for disabled persons, virtual reality, and gaming.
Data collection device	Fixed or portable devices that can read data on-site. Examples include warehouse inventory control points and terminals used by package delivery drivers, some of which can capture and store a signature digitally.
Digital camera	Device that records photographs in digital form rather than using traditional film; the resulting data file can be stored, displayed, or manipulated by the computer.
Electronic whiteboard	Electronic version of a standard whiteboard that can capture and store text or graphics that are displayed on the board, and function as a touch screen.
Graphic input device	Includes light pens, digitizers, and graphics tablets that allow drawings to be translated into digital form that can be processed by a computer.
Handheld computer stylus	Device that allows users to form characters on the screen of a handheld computer. Handwriting recognition software can translate the characters into digital input.
Input devices for physically challenged users	Examples include on-screen keyboard displays that allow users to point to and click a character, head-mounted pointers, foot pedals, and breath-sensitive controls.
Internet workstation	Enables the user to provide input to Web-based intranet or Internet recipients; can be integrated with information system output or personal computer applications.
Keyboard	Most common input device.
MICR (magnetic ink character recognition)	Technology used primarily in the banking industry to read magnetic ink characters printed on checks.
Microphone	Converts sound waves into digital information that can be stored, transmitted, attached to e-mail and documents, and manipulated in other ways. Users can enter data and issue commands using spoken words.
Motion sensors	Can estimate an object's three-dimensional position, speed, and axis of movement and transmit the data to a computer. Used for virtual reality programs and gaming applications such as the Nintendo Wii.
Mouse	Wired or wireless pointing device with control buttons and a wheel that allows the user to scroll the viewing screen. A trackball is a stationary version of a mouse. Instead of moving the device itself, the user rotates the ball.
Pointing stick	Pressure-sensitive pointing device located between the keys on a notebook computer. A pointing stick resembles a button or pencil eraser.
Radio Frequency Identification (RFID)	Uses high-frequency radio waves to track physical objects. RFID can be used to track items in a wide range of information systems, including supply chain management, inventory control, and point-of-sale. RFID also can be used to monitor the movement of people or objects in a security management system.
Scanner/optical recognition	Devices that read printed bar codes, characters, or images. Optical character recognition (OCR) can convert scanned material to digital information.
Telephone	Technology that allows users to press telephone buttons or speak selected words to choose options in a system, such as electronic funds transfer, shop-at-home purchases, or registration for college courses.
Terminal	Device that might be dumb (screen and keyboard only) or intelligent (screen, keyboard, independent processing).
Touch screen or pad	Sensors that allow users to select options by touching specific locations on the screen. Touch pads also are used as pointing devices on notebook computers. An iPod is a popular example of a device that uses touch screen input.
Video input	Video camera input, in digital form, that can be stored and replayed later.

FIGURE 8-34 Input devices can be very traditional, or based on the latest technology.

FIGURE 8-35 When a customer's signature is stored in digital form, it becomes input to the information system.

immediate validation and availability of data. A popular online input method is **source data automation**, which combines online data entry and automated data capture using input devices such as **RFID tags** or **magnetic data strips**. Source data automation is fast and accurate, and minimizes human involvement in the translation process.

Many large companies use a combination of source data automation and a powerful communication network to manage global operations instantly. Some common examples of source data automation are:

- Businesses that use point-of-sale (POS) terminals equipped with bar code scanners and magnetic swipe scanners to input credit card data.

- Automatic teller machines (ATMs) that read data strips on bank cards.

- Factory employees who use magnetic ID cards to clock on and off specific jobs so the company can track production costs accurately.

- Hospitals that imprint bar codes on patient identification bracelets and use portable scanners when gathering data on patient treatment and medication.

- Retail stores that use portable bar code scanners to log new shipments and update inventory data.

- Libraries that use handheld scanners to read optical strips on books.

TRADEOFFS Although online input offers many advantages, it does have some disadvantages. For example, unless source data automation is used, manual data entry is slower and more expensive than batch input because it is performed at the time the transaction occurs and often done when computer demand is at its highest.

The decision to use batch or online input depends on business requirements. For example, hotel reservations must be entered and processed immediately, but hotels can enter their monthly performance figures in a batch. In fact, some input occurs naturally in batches. A cable TV provider, for example, receives customer payments in batches when the mail arrives.

Input Volume

To reduce input volume, you must reduce the number of data items required for each transaction. Data capture and data entry require time and effort, so when you reduce input volume, you avoid unnecessary labor costs, get the data into the system more quickly, and decrease the number of errors. The following guidelines will help reduce input volume:

1. Input necessary data only. Do not input a data item unless it is needed by the system. A completed order form, for example, might contain the name of the clerk who took the order. If that data is not needed by the system, the user should not enter it.

2. Do not input data that the user can retrieve from system files or calculate from other data. In the order system example shown in Figure 8-36, the system generates an order number and logs the current date and time. Then the user enters a customer ID. If the entry is valid, the system displays the customer name so the user can verify it. The user then enters the item and quantity. Note that the description, price, extended price, total price, sales tax, and grand total are retrieved automatically or calculated by the system.

3. Do not input constant data. If orders are in batches with the same date, then a user should enter the order date only once for the first order in the batch. If orders are entered online, then the user can retrieve the order date automatically using the current system date.

4. Use codes. Codes are shorter than the data they represent, and coded input can reduce data entry time. For example, in the order system example in Figure 8-36, the Customer ID and Item fields are codes. You will learn more about various types of codes in Chapter 9, Data Design.

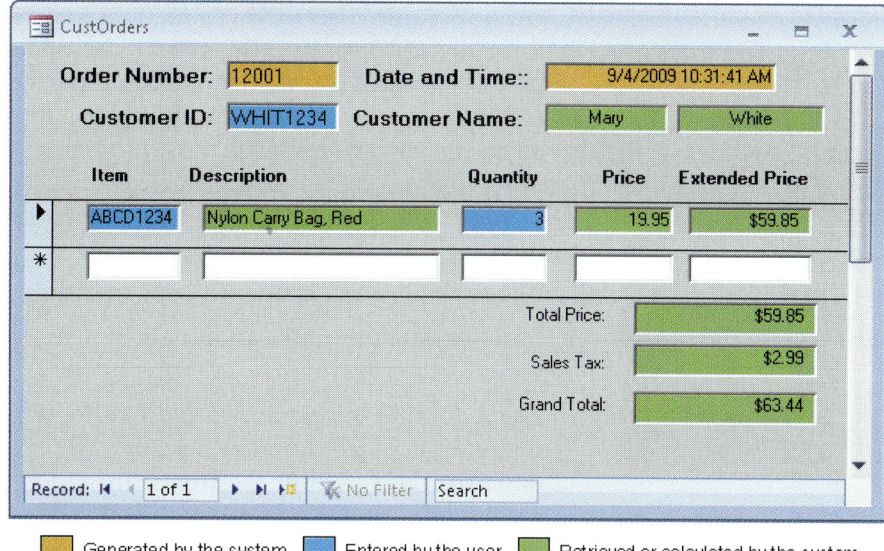

FIGURE 8-36 In this data screen for customer orders, the system generates an order number and logs the current date and time. The user enters a customer ID. If the entry is valid, the system displays the customer name so the user can verify it. The user then enters the item and quantity. Note that the description, price, extended price, total price, sales tax, and grand total are retrieved automatically or calculated by the system.

Designing Data Entry Screens

Some users work with many features of the user interface; others spend most of their time entering data. This section discusses interface guidelines and concepts that primarily relate to repetitive data entry. Notice that many of the guidelines are based on general principles of interface design discussed in this chapter.

The most effective method of online data entry is **form filling**, in which a blank form that duplicates or resembles the source document is completed on the screen. The user enters the data and then moves to the next field. The following guidelines will help you design data entry screens that are easy to learn and use.

1. Restrict user access to screen locations where data is entered. For example, when the screen in Figure 8-36 appears, the system should position the insertion point in the first data entry location. After the operator enters a Customer ID, the insertion point should move automatically to the entry location for the next field (Item). A user should be able to position the insertion point only in places where data is entered on the form.

2. Provide a descriptive caption for every field, and show the user where to enter the data and the required or maximum field size. Typically, white boxes show the location and length of each field. Other methods used to indicate field locations are video highlighting, underscores, special symbols, or a combination of these features.

3. Display a sample format if a user must enter values in a field in a specific format. For example, provide an on-screen instruction to let users know that the date format is MMDDYY, and provide an example if the user must enter **separators**, such as slashes. It is better to use an input mask, so users simply can enter 112709 to represent November 27, 2009.

4. Require an ending keystroke for every field. Pressing the ENTER key or the TAB key should signify the end of a field entry. Avoid a design that moves automatically to the next item when the field is full. The latter approach requires an ending keystroke *only* when the data entered is less than the maximum field length. It is confusing to use two different data entry procedures.

5. Do not require users to type leading zeroes for numeric fields. For example, if a three-digit project number is 045, the operator should be able to type 45 instead of 045 before pressing the ENTER key. An exception to that rule might occur when entering a date, where a leading zero is needed to identify single-digit months or days, such as 06-04-2010.

6. Do not require users to type trailing zeroes for numbers that include decimals. For example, when a user types a value of 98, the system should interpret the value as 98.00 if the field has been formatted to include numbers with two decimal places. The decimal point is needed *only* to indicate nonzero decimal places, such as 98.76.

7. Display default values so operators can press the ENTER key to accept the suggested value. If the default value is not appropriate, the operator can change it.

8. Use a default value when a field value will be constant for successive records or throughout the data entry session. For example, if records are input in order by date, the date used in the first transaction should be used as the default date until a new date is entered, at which time the new date becomes the default value.

9. Display a list of acceptable values for fields, and provide meaningful error messages if the user enters an unacceptable value. An even better method, which was described in the user interface design section, is to provide a drop-down list box containing acceptable values that allows the user to select a value by clicking.

10. Provide a way to leave the data entry screen at any time without entering the current record. This feature is available in the screen shown in Figure 8-37, which is an enhanced version of the data entry screen shown in Figure 8-36 on the previous page. Notice that the new version has command buttons that provide flexibility and allow the user to perform various functions. For example, clicking the Cancel Order Without Entering button cancels the current order and moves the insertion point back to the beginning of the form.

11. Provide users with an opportunity to confirm the accuracy of input data before entering it by displaying a message such as, *Add this record? (Y/N)*. A positive response *(Y)* adds the record, clears the entry fields, and positions the insertion point in the first field so the user can input another record. If the response is negative *(N)*, the current record is not added and the user can correct the errors.

12. Provide a means for users to move among fields on the form in a standard order or in any order they choose. For example, when a user opens the form shown in Figure 8-37, the insertion point automatically will be in the first field. After the user fills in each field and confirms the entry, the insertion point moves to the next field, in a predetermined order. In a graphical user interface (GUI), the user can override the standard field order and select field locations using the mouse or arrow keys.

13. Design the screen form layout to match the layout of the source document. If the source document fields start at the top of the form and run down in a column, the input screen should use the same design.

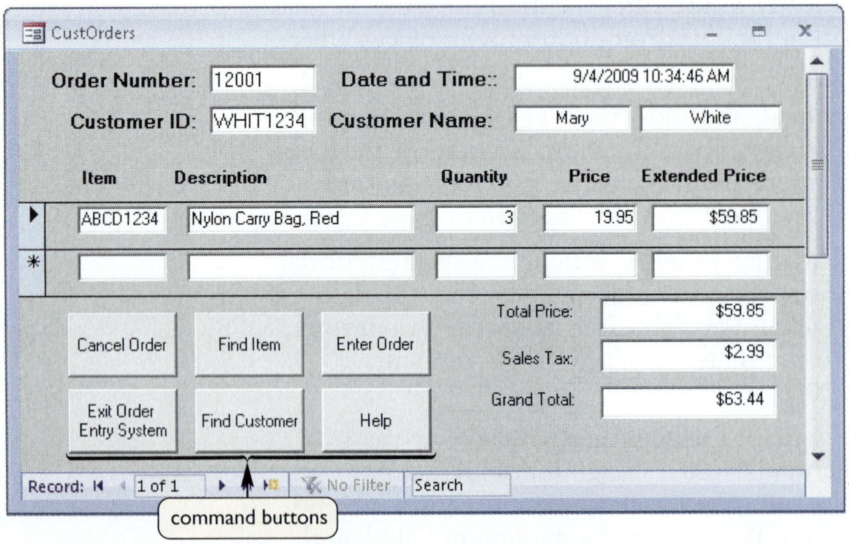

FIGURE 8-37 This is an enhanced version of the data entry screen shown in Figure 8-36. The new version has command buttons that allow the user to perform various functions.

14. Allow users to add, change, delete, and view records. Figure 8-37 shows a screen that can be used for entering orders, finding items, and finding customers. After the operator enters a customer identification code, the order form displays current values for all appropriate fields. Then the operator can view the data, make changes, enter the order, or cancel without ordering. Messages such as: *Apply these changes? (Y/N)* or *Delete this record? (Y/N)* should require users to confirm the actions. Highlighting the letter *N* as a default response will avoid problems if the user presses the ENTER key by mistake.

15. Provide a method to allow users to search for specific information, as shown in Figure 8-37.

Input Errors

Reducing the number of input errors improves data quality. One way to reduce input errors is to eliminate unnecessary data entry, as discussed earlier in the chapter. For example, a user cannot misspell a customer name if it is not entered, or is entered automatically based on the user entering the customer ID. Similarly, an outdated item price cannot be used if the item price is retrieved from a master file instead of being entered manually.

The best defense against incorrect data is to identify and correct errors before they enter the system by using data validation checks. A **data validation check** improves input quality by testing the data and rejecting any entry that fails to meet specified conditions. You can design at least eight types of data validation checks into the input process.

1. A **sequence check** is used when the data must be in some predetermined sequence. If the user must enter work orders in numerical sequence, for example, then an out-of-sequence order number indicates an error, or if the user must enter transactions chronologically, then a transaction with an out-of-sequence date indicates an error.

2. An **existence check** is used for mandatory data items. For example, if an employee record requires a Social Security number, an existence check would not allow the user to save the record until he or she enters a suitable value in the Social Security number field.

3. A **data type check** tests to ensure that a data item fits the required data type. For example, a numeric field must have only numbers or numeric symbols, and an alphabetic field can contain only the characters A through Z (or a through z).

4. A **range check** tests data items to verify that they fall between a specified minimum and maximum value. The daily hours worked by an employee, for example, must fall within the range of 0 to 24. When the validation check involves a minimum or a maximum value, but not both, it is called a **limit check**. Checking that a payment amount is greater than zero, but not specifying a maximum value, is an example of a limit check.

5. A **reasonableness check** identifies values that are questionable, but not necessarily wrong. For example, input payment values of $.05 and $5,000,000.00 both pass a simple limit check for a payment value greater than zero, and yet both values could be errors. Similarly, a daily hours worked value of 24 passes a 0 to 24 range check; however, the value seems unusual, and the system should verify it using a reasonableness check.

6. A **validity check** is used for data items that must have certain values. For example, if an inventory system has 20 valid item classes, then any input item that does not match one of the valid classes will fail the check. Verifying that a customer number on an order matches a customer number in the customer file is another type of validity check. Because the value entered must refer to another value, that type of check also is called **referential integrity**, which is explained in

Chapter 9, Data Design. Another validity check might verify that a new customer number does *not* match a number already stored in the customer master file.

7. A **combination check** is performed on two or more fields to ensure that they are consistent or reasonable when considered together. Even though all the fields involved in a combination check might pass their individual validation checks, the combination of the field values might be inconsistent or unreasonable. For example, if an order input for 30 units of a particular item has an input discount rate applicable only for purchases of 100 or more units, then the combination is invalid; either the input order quantity or the input discount rate is incorrect.

8. **Batch controls** are totals used to verify batch input. Batch controls might check data items such as record counts and numeric field totals. For example, before entering a batch of orders, a user might calculate the total number of orders and the sum of all the order quantities. When the batch of orders is entered, the order system also calculates the same two totals. If the system totals do not match the input totals, then a data entry error has occurred. Unlike the other validation checks, batch controls do not identify specific errors. For example, if the sum of all the order quantities does not match the batch control total, you know only that one or more orders in that batch were entered incorrectly or not input. The batch control totals often are called **hash totals**, because they are not meaningful numbers themselves, but are useful for comparison purposes.

Source Documents

A **source document** is a form used to request and collect input data, trigger or authorize an input action, and provide a record of the original transaction. During the input design stage, you develop source documents that are easy to complete and inexpensive.

Consider a time when you struggled to complete a poorly designed form. You might have encountered insufficient space, confusing instructions, or poor organization, all symptoms of incorrect **form layout**. Good form layout makes the form easy to complete and provides enough space, both vertically and horizontally, for users to enter the data. A form should indicate data entry positions clearly using blank lines or boxes and descriptive captions. Figure 8-38 shows several techniques for using line and boxed captions in source

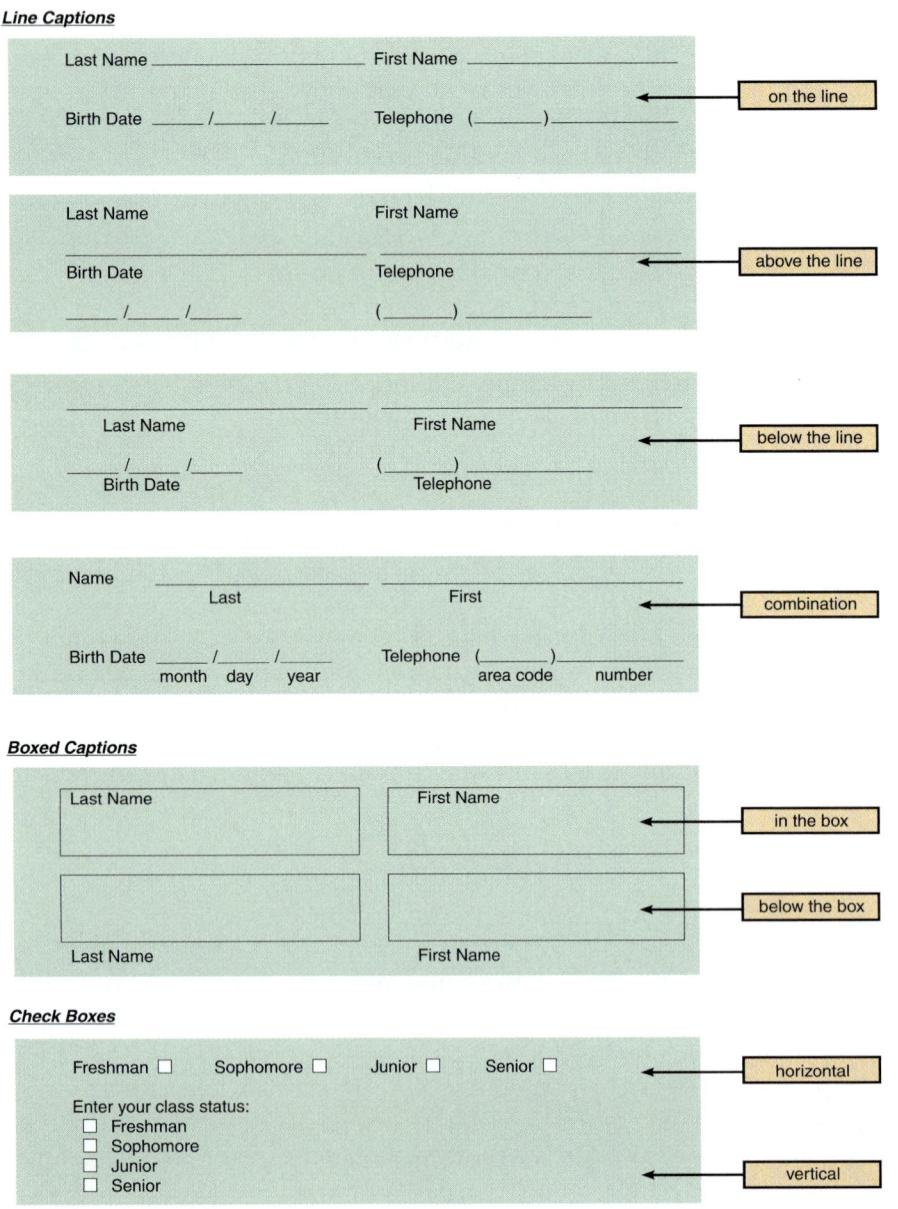

FIGURE 8-38 Examples of caption techniques for source documents.

documents, and an example of check boxes, which are effective when a user must select choices from a list.

The placement of information on a form also is important. Source documents typically include most of the zones shown in Figure 8-39. The **heading zone** usually contains the company name or logo and the title and number of the form. The **control zone** contains codes, identification information, numbers, and dates that are used for storing completed forms. The **instruction zone** contains instructions for completing the form. The main part of the form, called the **body zone**, usually takes up at least half of the space on the form and contains captions and areas for entering variable data. If totals are included on the form, they appear in the **totals zone**. Finally, the **authorization zone** contains any required signatures.

Information should flow on a form from left to right and top to bottom to match the way users read documents naturally. That layout makes the form easy to use for the individual who completes the form, and for users who enter data into the system using the completed form. You can review samples of source document design that appear in the SWL case study on pages 379–385.

The same user-friendly design principles also apply to printed forms such as invoices and monthly statements, except that heading information usually is preprinted. You should make column headings short but descriptive, avoid nonstandard abbreviations, and use reasonable spacing between columns for better readability.

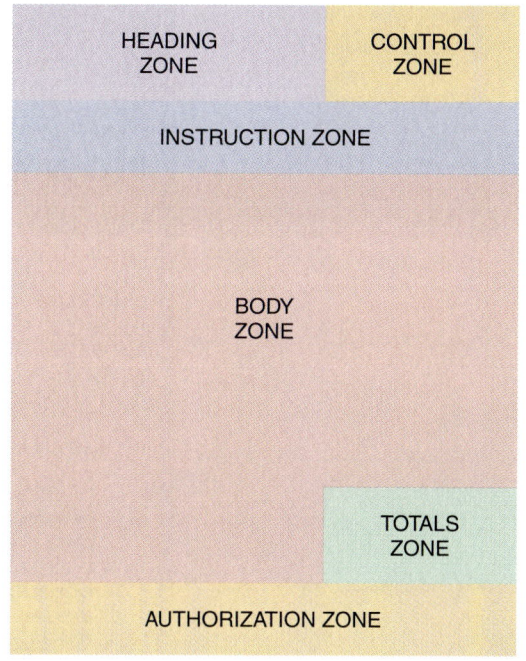

FIGURE 8-39 Source document zones.

The order and placement of printed fields should be logical, and totals should be identified clearly. When designing a preprinted form, you should contact the form's vendor for advice on paper sizes, type styles and sizes, paper and ink colors, field placement, and other important form details. Your goal is to design a form that is attractive, readable, and effective.

Layout and design also are important on Web-based forms, and you can find many resources that will help you design efficient, user-friendly forms. For example, Figure 8-40 describes a book by Luke Wroblewski, a well-known author and consultant. His Web site offers valuable suggestions, guidelines, and examples.

A major challenge of Web-based form design is that most people read and interact differently with on-screen information compared to paper forms. In the view of Dr. Jakob Nielsen, a pioneer in Web usability design, users simply do not read on the Web, as shown in Figure 8-41 on the next page. Dr. Nielsen believes that users *scan* a page, picking out individual words and sentences. As a result, Web designers must use **scannable text** to capture and hold a user's attention. On his site, Dr. Nielson offers several suggestions for creating scannable text. Also notice that Dr. Nielsen employs various usability metrics to measure user responses, comprehension, and recall.

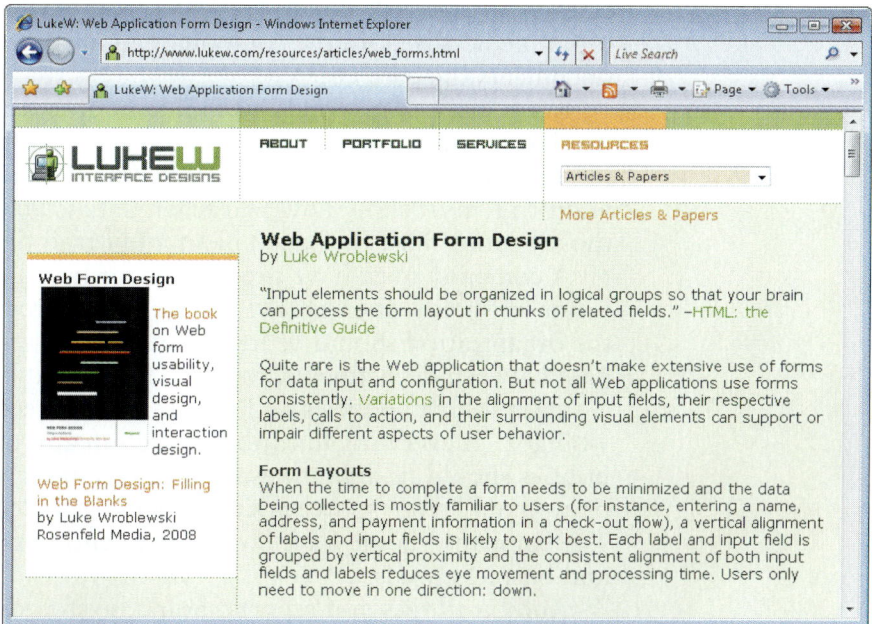

FIGURE 8-40 Luke Wroblewski's Web site is a good source of information about form design.

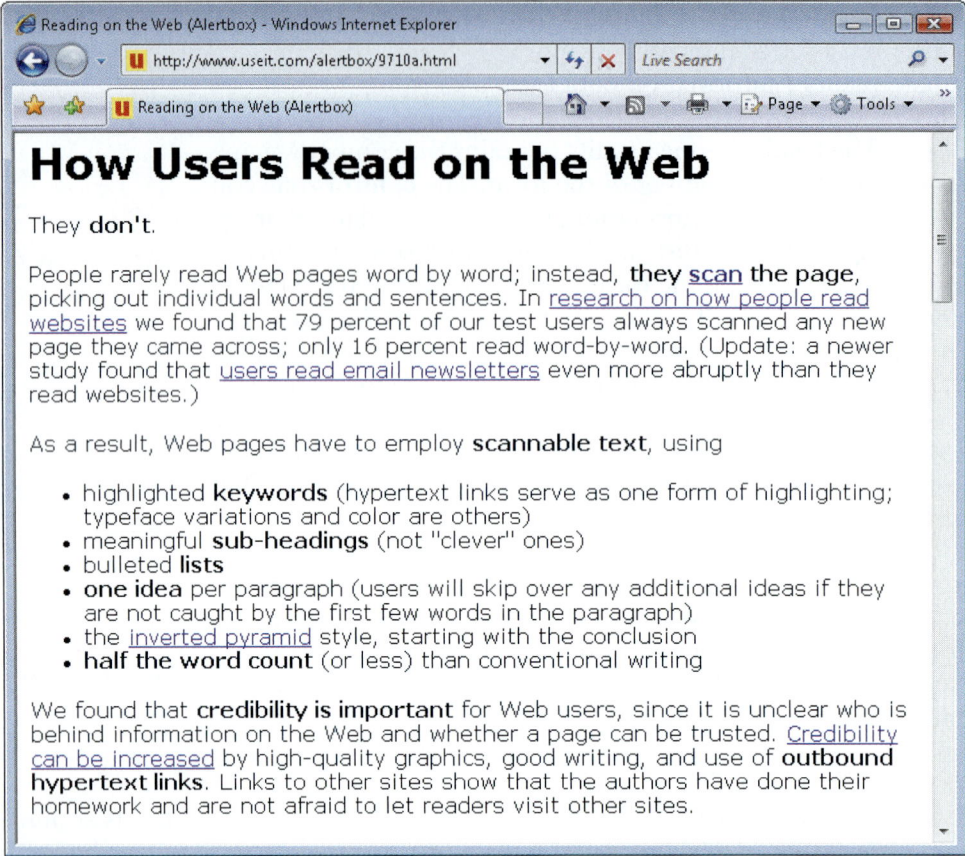

FIGURE 8-41   Dr. Jakob Nielsen believes that users scan Web material rather than reading it. He suggests that Web designers must use scannable text and employ usability metrics to measure the results.

Input Control

Input control includes the necessary measures to ensure that input data is correct, complete, and secure. You must focus on input control during every phase of input design, starting with source documents that promote data accuracy and quality. When a batch input method is used, the computer can produce an input log file that identifies and documents the data entered.

Every piece of information should be traceable back to the input data that produced it. That means that you must provide an **audit trail** that records the source of each data item and when it entered the system. In addition to recording the original source, an audit trail must show how and when data is accessed or changed, and by whom. All those actions must be logged in an audit trail file and monitored carefully.

A company must have procedures for handling source documents to ensure that data is not lost before it enters the system. All source documents that originate from outside the organization should be logged when they are received. Whenever source documents pass between departments, the transfer should be recorded.

Data security policies and procedures protect data from loss or damage, which is a vital goal in every organization. If the safeguards are not 100% effective, data recovery utilities should be able to restore lost or damaged data. Once data is entered, the company should store source documents in a safe location for some specified length of time. The company should have a **records retention policy** that meets all legal requirements and business needs.

Audit trail files and reports should be stored and saved. Then, if a data file is damaged, you can use the information to reconstruct the lost data. Data security also involves protecting data from unauthorized access. System sign-on procedures should prevent unauthorized individuals from entering the system, and users should change their passwords regularly. Having several levels of access also is advisable. For example, a data entry person might be allowed to *view* a credit limit, but not *change* it. Sensitive data can be **encrypted**, or coded, in a process called **encryption**, so only users with decoding software can read it.

CASE IN POINT 8.4: BOOLEAN TOYS

When should a systems analyst decide a design issue, and when should users be allowed to select what works best for them? The field of ergonomics is concerned with improving the work environment and studying how users interact with their environment.

Suppose you are a systems analyst studying the order processing system at Boolean Toys, a fast-growing developer of software for preschool children. You know that many data entry users have complained about the input screens. Some users would prefer to rearrange the order of the fields; others would like to change the background color on their screens; still others want shortcuts that would allow them to avoid a series of introductory screens.

What if Boolean's users could customize their own data entry screens without assistance from the IT staff by using a menu-driven utility program? What would be the pros and cons of such an approach?

A QUESTION OF ETHICS

Jacob thought that he did a good job of designing the company's tech support Web page, but Emily, his supervisor, isn't so sure. She is concerned that Jacob's design is very similar to a page used by the company's major competitor, and she asked him whether he had used any HTML code from that site in his design. Although Jacob didn't copy any of the code, he did examine it in his Web browser to see how they handled some design issues.

Emily asked Jacob to investigate Web page copyright issues, and report back to her. In his research, he learned that outright copying would be a copyright violation, but merely viewing other sites to get design ideas would be permissible. What is not so clear is the gray area in the middle. Jacob asked you, as a friend, for your opinion on this question: Even if no actual copying is involved, are there ethical constraints on how far you should go in using the creative work of others? How would you answer Jacob?

CHAPTER SUMMARY

The purpose of systems design is to create a physical model of the system that satisfies the logical design requirements that were defined during the systems analysis phase. The chapter began with a discussion of output design issues and a description of various types of output, such as Web-based information delivery, audio output, instant messaging, podcasts, e-mail, and other specialized forms of output.

You learned about various types of printed reports, including detail, exception, and summary reports. The chapter also described the features and sections of reports, including control fields, control breaks, report headers and footers, page headers and footers, and group headers and footers. Examples of report design were presented, along with suggestions about designing character-based reports. Also, you learned about output control and the various measures you can take to achieve adequate output control to ensure that information is correct, complete, and secure.

You learned about human-computer interaction concepts and graphical user interfaces (GUIs), which use graphical objects and techniques that allow users to communicate with the system. User-centered design principles require an analyst to understand the business functions, maximize graphical effectiveness, profile the system's users, think like a user, use prototyping, design a comprehensive interface, continue the feedback process, and document the interface design.

The chapter listed several interface design guidelines, which suggested that you focus on basic objectives, make the interface easy to learn and use, provide features that promote efficiency, make it easy for users to obtain Help and correct errors, minimize input data problems, provide feedback to users, create an attractive layout and design, and use familiar terms and images.

You learned that an interface can include various controls, including menu bars, toolbars, drop-down list boxes, dialog boxes, toggle buttons, list boxes, option buttons, check boxes, command buttons, and spin bars. Controls are placed on a main switchboard, which is like a graphical version of a main menu.

During input design, you learned about batch and online input methods. Input design includes selecting appropriate input media and methods, developing efficient input procedures, reducing input volume, and avoiding input errors. In carrying out those tasks, the systems analyst must consider three key procedures: data capture, data entry, and input methods. Data capture involves identifying and recording source data. Data entry involves converting source data into a computer-readable form and entering it into the system. Many input methods are available, including optical and voice recognition systems, biological feedback devices, motion sensors, and a variety of graphical input devices.

An effective way to reduce input errors is to reduce input volume. You also can reduce errors by using well-designed data entry screens and by using data validation checks that verify data sequence, existence, range and limit, reasonableness, validity, combination, and batch controls.

You learned about source document design and the various zones in a source document, including the heading zone, the control zone, the instruction zone, the body zone, the totals zone, and the authorization zone. Finally, you learned about input control, including audit trails, data security, privacy issues, and records retention policies.

Key Terms and Phrases

aesthetics *353*

audit trail *368*

authorization zone *367*

automated facsimile *335*

batch *360*

batch control *366*

batch input *360*

blog *335*

body zone *367*

calendar control *359*

character-based report *338*

check box *359*

combination check *366*

command button *358*

computer output to microfilm (COM) *335*

context-sensitive *355*

control break *338*

control break report *338*

control field *338*

control zone *367*

data capture *360*

data entry *360*

data security *368*

data type check *365*

data validation check *365*

detail line *338*

detail report *338*

dialog box *358*

diskless workstation *345*

drop-down list box *358*

encrypted *369*

encryption *369*

ergonomics *353*

exception report *339*

existence check *365*

faxback *335*

form filling *363*

form layout *366*

garbage in, garbage out (GIGO) *360*

graphical user interface (GUI) *348*

group footer *342*

group header *342*

hash totals *366*

heading zone *367*

human-computer interaction (HCI) *348*

input control *368*

input masks *356*

instruction zone *367*

interface technology *353*

limit check *365*

list box *358*

magnetic data strip *362*

menu bar *358*

mock-up *340*

online data entry *360*

option button *358*

output control *345*

output security *345*

page footer *341*

page header *341*

podcast *335*

port protector *345*

process-control *346*

radio button *358*

range check *365*

reasonableness check *365*

records retention policy *368*

referential integrity *365*

report analysis form *340*

report footer *341*

report header *341*

RFID tag *362*

scannable text *367*

scroll bar *358*

separator *363*

sequence check *365*

source data automation *362*

source document *366*

storyboard *352*

summary report *340*

switchboard *359*

text box *358*

toggle button *358*

toolbar *358*

totals zone *367*

turnaround document *337*

usability metrics *352*

user-centered *347*

user interface (UI) *346*

user-selected *355*

validity check *365*

Webcast *334*

Learn It Online

Instructions: To complete the Learn It Online exercises, start your browser, click the Address bar, and then enter the Web address **scsite.com/sad8e/learn**. When the Systems Analysis and Design Learn It Online page is displayed, follow the instructions in the exercises below. Each exercise has instructions for saving your results, either for your own records or for submission to your instructor.

1   Chapter Reinforcement

TF, MC, and SA

Below SAD Chapter 8, click one of the Chapter Reinforcement links for Multiple Choice, True/False, or Short Answer. Answer each question and submit to your instructor.

2   Flash Cards

Below SAD Chapter 8, click the Flash Cards link and read the instructions. Type 20 (or a number specified by your instructor) in the Number of playing cards text box, type your name in the Enter your Name text box, and then click the Flip Card button. When the flash card is displayed, read the question and then click the ANSWER box arrow to select an answer. Flip through the Flash Cards. If your score is 15 (75%) correct or greater, click Print on the File menu to print your results. If your score is less than 15 (75%) correct, then redo this exercise by clicking the Replay button.

3   Practice Test

Below SAD Chapter 8, click the Practice Test link. Answer each question, enter your first and last name at the bottom of the page, and then click the Grade Test button. When the graded practice test is displayed on your screen, click Print on the File menu to print a hard copy. Continue to take practice tests until you score 80% or better.

4   Who Wants To Be a Computer Genius?

Below SAD Chapter 8, click the Computer Genius link. Read the instructions, enter your first and last name at the bottom of the page, and then click the Play button. When your score is displayed, click the PRINT RESULTS link to print a hard copy.

5   Wheel of Terms

Below SAD Chapter 8, click the Wheel of Terms link. Read the instructions, and then enter your first and last name and your school name. Click the PLAY button. When your score is displayed on the screen, right-click the score and then click Print on the shortcut menu to print a hard copy.

6   Crossword Puzzle Challenge

Below SAD Chapter 8, click the Crossword Puzzle Challenge link. Read the instructions, and then click the Continue button. Work the crossword puzzle. When you are finished, click the Submit button. When the crossword puzzle is redisplayed, submit it to your instructor.

Case-Sim: SCR Associates

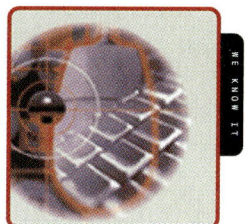

Background

SCR Associates is a consulting firm that offers IT solutions and training. SCR needs an information system to manage operations at its new training center. The new system will be called TIMS (Training Information Management System). As a newly hired systems analyst, you will report to Jesse Baker, systems group manager. You will work on various tasks and practice the skills you learned in this chapter.

Using the Case

The SCR Associates case study is a Web-based simulation that allows you to practice your skills in a real-world environment. The case study transports you to SCR's company intranet, where you can complete 12 work sessions, each aligning with a chapter. As you work on the case, you will receive e-mail and voice mail messages, Obtain information from SCR's online libraries, and perform various tasks. The first time you enter the SCR case, you should go to the starting page at **scsite.com/sad8e/scr** for detailed instructions.

Preview: Session 8

Now that the overall data design is complete, Jesse Baker wants you to work on output and user interface design. You will consider user needs, and apply principles of human-computer interaction to build a user-centered interface that is easy to learn and use. You also will consider data validation checks, source documents, forms, and reports.

To start a work session, you log on to the SCR intranet. When you enter your name and password, an opening screen displays links to the work sessions, and you select Session 8. At this point, you check your e-mail and voice mail messages carefully. Then you begin working on your task list, which includes the following items:

Tasks: Output and User Interface Design

1. *Create a detail report that will display all SCR courses in alphabetical order, with the course name and the instructor name in a group header; the Social Security number, name, and telephone number of each current student in the detail section; and the student count in a group footer.*

2. *Create a switchboard design with control buttons that lead to students, instructors, courses, course schedules, and course rosters. Allow a user to add, update, or delete records in each area. Jesse wants to see storyboards that show the proposed screens.*

3. *Suggest data validation checks for data entry screens.*

4. *Create a source document for an SCR mail-in registration form. Also need a design for a Web-based course registration form.*

FIGURE 8-42 Task list: Session 8.

Chapter Exercises

Review Questions

1. List and describe various types of output, including technology-based forms of information delivery.
2. Define detail reports, exception reports, and summary reports. Explain the concept of a control field and how it is used to produce a control-break report.
3. Explain the concept of human-computer interaction (HCI).
4. Describe eight principles for a user-centered interface design.
5. Explain each of the data validation checks mentioned in this chapter.
6. Describe six types of user interface controls, and provide an example of how you could use each type in a data entry screen.
7. Explain the concept of a GUI and a switchboard. How does a GUI design differ from a character-based screen design?
8. Explain batch and online input methods. Define source data automation and provide an example.
9. What are five principles of source document design?
10. Provide four guidelines for reducing input volume.

Discussion Topics

1. Some systems analysts maintain that source documents are unnecessary. They say that all input can be entered directly into the system, without wasting time in an intermediate step. Do you agree? Can you think of any situations where source documents are essential?
2. Some systems analysts argue, "Give users what they ask for. If they want lots of reports and reams of data, then that is what you should provide. Otherwise, they will feel that you are trying to tell them how to do their jobs." Others say, "Systems analysts should let users know what information can be obtained from the system. If you listen to users, you'll never get anywhere, because they really don't know what they want and don't understand information systems." What do you think of these arguments?
3. Suppose your network support company employs 75 technicians who travel constantly and work at customer sites. Your task is to design an information system that provides technical data and information to the field team. What types of output and information delivery would you suggest for the system?
4. A user interface can be quite restrictive. For example, the interface design might not allow a user to exit to a Windows desktop or to log on to the Internet. Should a user interface include such restrictions? Why or why not?

Projects

1. Visit the administrative office at your school or a local company. Ask to see examples of output documents, such as computer-printed invoices, form letters, or class rosters. Analyze the design and appearance of each document, and try to identify at least one possible improvement for each.
2. Search the Web to find an example of an attractive user interface. Document your research and discuss it with your class.
3. Examine various application software packages to find examples of good (or bad) user interface design. Document your research and discuss it with your class.
4. Search your own files or review other sources to find good (or bad) examples of source document design. Document your research and discuss it with your class.

Apply Your Knowledge

The Apply Your Knowledge section contains four mini-cases. Each case describes a situation, explains your role in the case, and asks you to respond to questions. You can answer the questions by applying knowledge you learned in the chapter.

1 North Shore Boat Sales

Situation:

North Shore Boat Sales sells new and used boats and operates a Web-based boat brokerage business in Toronto. The company has grown, and North Shore needs a new information system to manage the inventory, the brokerage operation, and information about prospective buyers and sellers. Dan Robeson, the owner, asked you to design samples of computer screens and reports that the new system might produce.

1. Design a switchboard that includes the main information management functions that North Shore might require. Create a storyboard with a design layout that allows customers to perform the following functions: Obtain information about new boats, obtain information about used boats, send an e-mail to North Shore, learn more about the company, or review links to other marine-related sites.
2. Prospective buyers might want to search for boats by type, size, price range, or manufacturer. Develop a screen design that would permit those choices.
3. Suggest reports that might be useful to North Shore's management.
4. Suggest the general layout for a Web-based source document that prospective sellers could use to describe their boats. The information should include boat type (sail or power), manufacturer, year, length, type of engine, hull color, and asking price.

2 Terrier News

Situation:

Terrier News is a monthly newsletter devoted to various breeds of terriers and topics of interest to terrier owners and breeders. Annie West, the editor and publisher, asked you to help her design a system to enter and manage the hundreds of classified ads that Terrier News publishes. Some ads are for dogs wanted; some are for dogs for sale; and some offer products and services.

1. Design a suitable source document for ads that are telephoned or mailed in.
2. Explain user-centered design principles in a brief memo to Annie.
3. Suggest at least four user interface design guidelines that could be used for the new system.
4. Suggest several types of controls that might be used on the switchboard you plan to design. Explain why you chose each control, and create a storyboard that shows the switchboard layout.

3 **Sky-High Internet Services**

Situation:

Sky-High Internet Services is a leading Internet service provider in a metropolitan area. The new customer billing system has caused an increase in complaints. Tammy Jones, the office manager, asked you to investigate the situation. After interviewing data entry operators and observing the online data input process, you are fairly certain that most errors occur when data is entered.

1. Write a brief memo to Tammy explaining the importance of data validation during the input process.
2. Suggest at least three specific data validation checks that might help reduce input errors.
3. Would a batch input system offer any advantages? Write a brief memo to Tammy stating your views.
4. Suppose that Sky-High is predicting 25% annual growth, on a current base of 90,000 customers. If the growth pattern holds, how many customers will Sky-High have in three years? If it takes about 12 minutes to enter a new customer into the system, how many additional data entry operators will be needed to handle the growth next year? Assume that an operator works about 2,000 hours per year. Also assume a 30% annual attrition rate for existing customers.

4 **Castle Point Antique Auction**

Situation:

Castle Point Antique Auction operates a successful Web site that offers an auction forum for buyers and sellers of fine antiques. Monica Creighton, the owner, asked you to help her design some new documents and reports.

1. Suggest the general layout for a Web-based source document that prospective bidders would submit. The information should include user ID, password, name, address, telephone, e-mail address, item number, bid offered, and method of payment (money order, check, American Express, MasterCard, or Visa).
2. Suggest the general layout for a Web-based source document that prospective sellers could use to describe their antiques. The information should include the user ID, password, item, dimensions, origin, condition, and asking price.
3. Write a brief memo to Monica explaining the difference between detail reports, exception reports, and summary reports. Suggest at least one example of each type of report that she might want to consider.
4. Suggest several types of data validation checks that could be used when input data is entered.

Case Studies

Case studies allow you to practice specific skills learned in the chapter. Each chapter contains several case studies that continue throughout the textbook, and a chapter capstone case.

NEW CENTURY HEALTH CLINIC

The associates at New Century Health Clinic approved your recommendations for a new computer system. Your next step is to develop a design for the new system, including output and user interface issues.

Background

To complete the output and user interface design for the new information system at New Century, you should review the DFDs and object-oriented diagrams you prepared previously, and the rest of the documentation from the systems analysis phase. Perform the following tasks.

Assignments

1. Dr. Jones has asked you to create a monthly Claim Status Summary report. He wants you to include the insurance company number, the patient number and name, the procedure date, the procedure code and description, the fee, the date the claim was filed, the amount of the claim, the amount of reimbursement, and the amount remaining unpaid. He wants you to group the data by insurance company number, with subtotals by company and grand totals for each numeric field. When you design the report, make sure to include a mock-up report and a report analysis form.

2. Design the daily appointment list and a monthly statement to make it readable and visually attractive. Include a mock-up report and a report analysis form for each report.

3. Determine the data required for a new patient. Design an input source document that will be used to capture the data and a data entry screen to input the information.

4. What data validation checks would the clinic need for the new patient data entry screen? Write a brief memo with your recommendations.

PERSONAL TRAINER, INC.

Personal Trainer, Inc., owns and operates fitness centers in a dozen Midwestern cities. The centers have done well, and the company is planning an international expansion by opening a new "supercenter" in the Toronto area. Personal Trainer's president, Cassia Umi, hired an IT consultant, Susan Park, to help develop an information system for the new facility. During the project, Susan will work closely with Gray Lewis, who will manage the new operation.

Background

Following the decision to use an in-house team to develop a design prototype, Susan began to work on the physical design for Personal Trainer's new information system. At this stage, she is ready to begin working with Gray on the output and user interface design. Together, Susan and Gray will seek to develop a user-centered design that is easy to learn and use. Personal Trainer users will include managers, fitness instructors, support staff, and members themselves.

Assignments

1. Create a detail report that will display all Personal Trainer courses in alphabetical order, with the course name and the instructor name in a group header; the Social Security number, name, and telephone number of each current student in the detail section; and the student count in a group footer.

2. Create a switchboard design with control buttons that lead to members, fitness instructors, activities and services, schedules, and fitness class rosters. Allow a user to add, update, or delete records in each area.

3. Suggest context-sensitive and specific Help for the switchboard and lower-level menus and forms. Prepare storyboards that show the proposed screens. Also suggest at least six types of data validation checks for data entry screens.

4. Design a mail-in source document that members can use to register for fitness classes. Also design a Web-based registration form.

VIDEO SUPERSTORE

Video Superstore has hired you to design two online data entry screens. Based on what you know about the operation of a video rental store, complete the following assignments.

Assignments

1. Design a weekly operations summary report that will include overall data on rentals, new customers, late charges, and anything else you think a store manager might want to review. Be sure to include numeric activity and dollar totals.

2. Design a data entry screen for entering new members.

3. Design a video rental input screen. In addition to the video data, the video rental form must include the following fields: Member Number, Name, and Date.

4. Suggest at least three data validation checks that might help reduce input errors for the video rental system.

CHAPTER CAPSTONE CASE: SoftWear, Limited

SoftWear, Limited (SWL), is a continuing case study that illustrates the knowledge and skills described in each chapter. In this case study, the student acts as a member of the SWL systems development team and performs various tasks.

Background

SoftWear, Limited, decided to use the payroll package developed by Pacific Software Solutions and customize it by adding its own ESIP system to handle SWL's Employee Savings and Investment Plan.

Because most of the payroll requirements would be handled by Pacific's payroll package, the IT team decided that Carla Moore would work on the new ESIP modules, and Rick Williams would concentrate on the rest of the payroll system. A new systems analyst, Becky Evans, was assigned to help Rick with the payroll package.

Pacific Software Solutions offered free training for new customers, so Rick and Becky attended a three-day train-the-trainer workshop at Pacific's site in Los Angeles. When they returned, they began developing a one-day training session for SWL users, including people from the accounting, payroll, and human resources departments. The initial training would include the features and processing functions of the new payroll package that was scheduled to arrive the following week.

Carla's first task was to work on the ESIP outputs. She had to design several reports: the ESIP Deduction Register, the ESIP Payment Summary, and the checks that SWL sends to the credit union and the stock purchase plan. Carla also needed to develop an ESIP Accounting Summary as a data file to be loaded into the accounting system.

Carla learned that standard SWL company checks were used to make the payments to the credit union and stock purchase department. In addition, the output entry to the accounting system had been specified by the accounting department in a standard format for all entries into that system.

To prepare the new ESIP Deduction Register, Carla reviewed the documentation from the systems analysis phase to make sure that she understood the logical design and the available data fields.

Carla decided to use a monthly format because accounting would apply some deductions on a monthly cycle. She started her design with a standard SWL report heading. Then she added a control heading and footing for each employee's Social Security number. The total of employee deductions would be compared with the transferred ESIP funds to make sure that they matched. After preparing a rough layout, Carla used a report generator (Microsoft Access) to create the design shown in Figure 8-43.

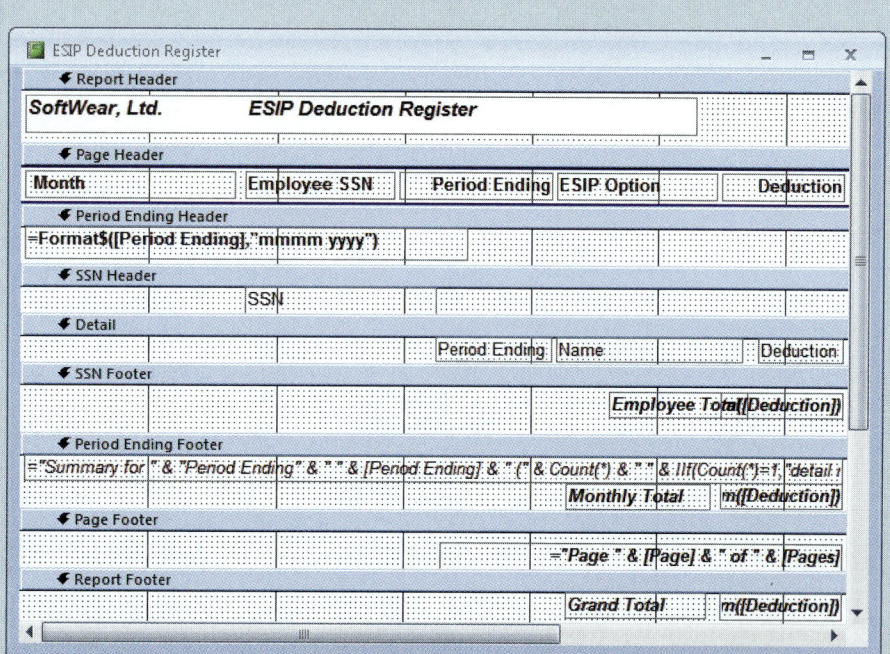

FIGURE 8-43 The layout for the ESIP Deduction Register that Carla created. Notice that SSN is a control break field. An SSN header introduces the group of detail records, and an SSN footer includes subtotals for the group.

CHAPTER CAPSTONE CASE: SoftWear, Limited (continued)

Then she prepared the mock-up report shown in Figure 8-44, using test data for the month of June 2009.

During the systems analysis phase, Carla learned that the accounting department required a control report to show the ESIP deduction amounts that were not yet applied. The control report is used to verify the amount of the checks or fund transfers SWL makes to the credit union and stock purchase plan. The accounting department needs the control

SoftWear, Ltd.	**ESIP Deduction Register**			
Month	**Employee SSN**	**Period Ending**	**ESIP Option**	**Deduction**
June 2009				
	111-11-1111			
		6/5/2009	Stock Purchase Plan	$15.00
		6/5/2009	Credit Union	$20.00
		6/12/2009	Credit Union	$25.00
		6/12/2009	Stock Purchase Plan	$15.00
		6/19/2009	Stock Purchase Plan	$15.00
		6/19/2009	Credit Union	$20.00
		6/26/2009	Stock Purchase Plan	$15.00
		6/26/2009	Credit Union	$20.00
			Employee Total	**$145.00**
	222-22-2222			
		6/5/2009	Stock Purchase Plan	$10.00
		6/5/2009	Credit Union	$30.00
		6/12/2009	Stock Purchase Plan	$10.00
		6/12/2009	Credit Union	$30.00
		6/19/2009	Stock Purchase Plan	$10.00
		6/19/2009	Credit Union	$30.00
		6/26/2009	Credit Union	$30.00
		6/26/2009	Stock Purchase Plan	$10.00
			Employee Total	**$160.00**
	333-33-3333			
		6/5/2009	ESIP: Option 1	$9.99
		6/12/2009	ESIP: Option 1	$9.99
		6/19/2009	ESIP: Option 1	$9.99
		6/26/2009	ESIP: Option 1	$9.99
			Employee Total	**$39.96**
Summary for Period Ending 6/26/2009 (20 detail records)				
			Monthly Total	**$344.96**
			Grand Total	**$344.96**

FIGURE 8-44 Mock-up for the ESIP Deduction Register.

Chapter Capstone Case: SoftWear, Limited (continued)

report to verify the accounting system outputs and balance the ESIP deduction totals against the payroll system's Payroll Register report.

Figure 8-45 shows a mock-up of the ESIP Payment Summary report. Carla met with Buddy Goodson, director of accounting, to review the design. Buddy was pleased with the report and felt it would be acceptable to the company's outside auditors. Buddy met with the auditing firm later that week and secured the team's approval.

As Carla turned her attention to the user interface for the ESIP system, she realized that she would need to develop two new source documents: an ESIP Option Form and an ESIP Deduction Authorization Form. She also planned to design a main switchboard and all necessary screen forms.

SoftWear, Ltd.* *ESIP Payment Summary Report

Last Deduction Period in Month: 6/26/2009

ESIP Code	Name	*Deduction*
CREDUN	Credit Union	*$205.00*
ESIP01	ESIP: Option 1	*$39.96*
SWLSTK	Stock Purchase Plan	*$100.00*
	Grand Total:	***$344.96***

FIGURE 8-45 Mock-up of the ESIP Payment Summary Report.

Carla started by designing the ESIP Option Form that could be used for adding new ESIP options and modifying existing ones when authorized by the vice president of human resources. She included an ESIP ID code, an ESIP option name, and other information as shown in Figure 8-46. For each field, she established a range of acceptable values.

SWL

Employee Saving and Investment Plan
ESIP Option Form

ESIP ID ____________

INSTRUCTIONS: Recommendations must be reviewed by SWL's legal counsel before submission.

ESIP Option Name: ____________
Description: ____________
Deduction Cycle: ____________
Application Cycle: ____________
Minimum Service: ____________
Minimum Deduction Amount: $____________
Maximum Deduction Percentage: %____________

Approval: ____________ Date: ____________
(Vice President - Human Resources)

FIGURE 8-46 The ESIP Option Form for adding new ESIP options and modifying existing ones.

CHAPTER CAPSTONE CASE: SoftWear, Limited (continued) SWL

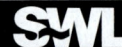

Next, Carla worked on a data entry screen based on the ESIP Option Form. Using SWL's existing screen design standards, she quickly developed the screen mock-up shown in Figure 8-47. Her design allowed users to add, delete, save, clear, or find a record by clicking the appropriate command button.

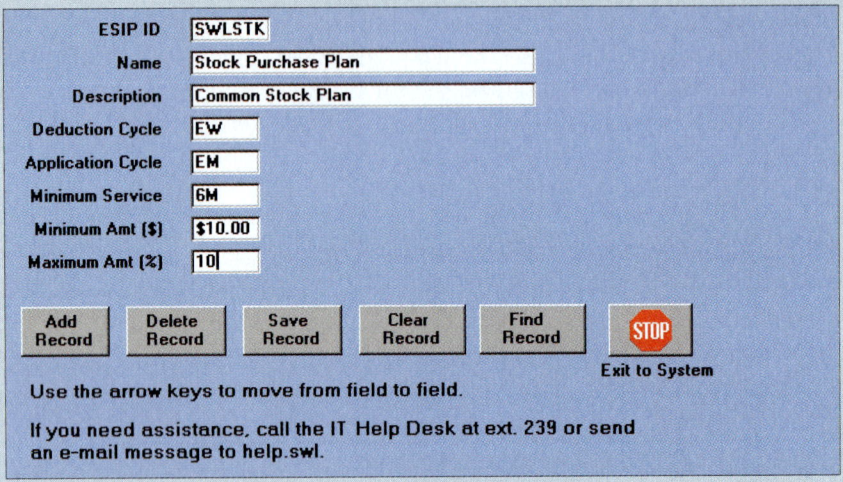

FIGURE 8-47 Carla's ESIP data entry screen form design.

The ESIP Deduction Authorization Form was more complicated than the ESIP Option Form because it required more data and signatures. It took several hours for Carla to design the form shown in Figure 8-48. She divided the form into three sections: The employee completes the information in the top section; human resources completes the middle section; and payroll representatives complete the bottom section.

Carla designed a data entry screen based on the ESIP Deduction Authorization Form shown in Figure 8-48 and made the screen consistent with the other ESIP screen designs. Now a user could add, delete, save, clear, or find a record by clicking the appropriate command buttons. Carla also provided instructions to the operator for exiting from the system.

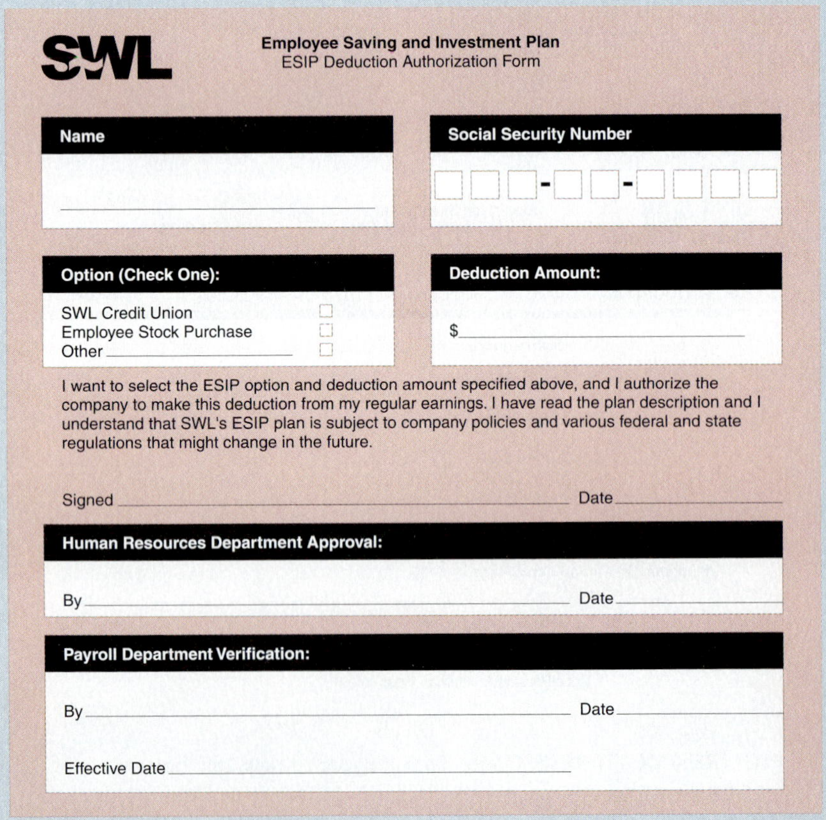

SWL

Employee Saving and Investment Plan
ESIP Deduction Authorization Form

Name

Social Security Number

☐☐☐-☐☐-☐☐☐☐

Option (Check One):

SWL Credit Union ☐
Employee Stock Purchase ☐
Other ________ ☐

Deduction Amount:

$________

I want to select the ESIP option and deduction amount specified above, and I authorize the company to make this deduction from my regular earnings. I have read the plan description and I understand that SWL's ESIP plan is subject to company policies and various federal and state regulations that might change in the future.

Signed ________ Date ________

Human Resources Department Approval:

By ________ Date ________

Payroll Department Verification:

By ________ Date ________

Effective Date ________

FIGURE 8-48 The ESIP Deduction Authorization Form.

CHAPTER CAPSTONE CASE: SoftWear, Limited (continued)   SWL

Carla decided to create a series of mock-ups to show users how the new ESIP deduction screen would work. Figure 8-49 shows the screen after the user has entered an employee's Social Security number. In Figure 8-50, the system has retrieved the employee's name, Sean R. Fitzpatrick, so the user can verify it against the source document.

The users approved the new design, with one suggestion — the system date should be added automatically as the entry date. Carla made the change and then designed the switchboard shown in Figure 8-51, based on comments that users made during the design process. Now that she had a working model, Carla went back to the users to show them the complete package.

SWL Team Tasks

1. Review the mock-up report shown in Figure 8-44 on page 380. When Carla showed this report design to Mike Feiner, director of human resources, he said that he wanted to see the data grouped by the type of ESIP deduction with the appropriate subtotals. Carla wants you to modify the report design to satisfy his request. You can use Microsoft Access, a report generator, or simply construct a sample layout using any word processing or drawing program. Be sure to show the placement and grouping of all fields.

2. Carla Moore also wants employees to have an online information request form that they can use to learn more about ESIP options and request up-to-date balances for their ESIP accounts.

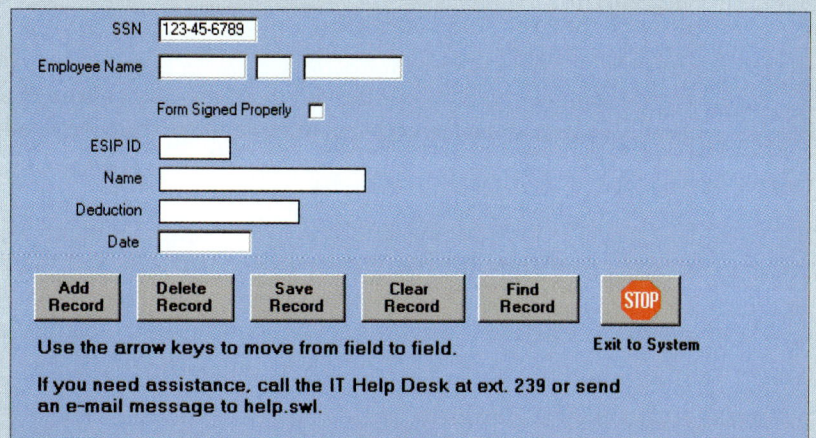

FIGURE 8-49   Data entry screen for ESIP deductions. Here, the user has entered an employee Social Security number and is about to press the ENTER key. Notice that the form has several command buttons, including a STOP button that exits the program.

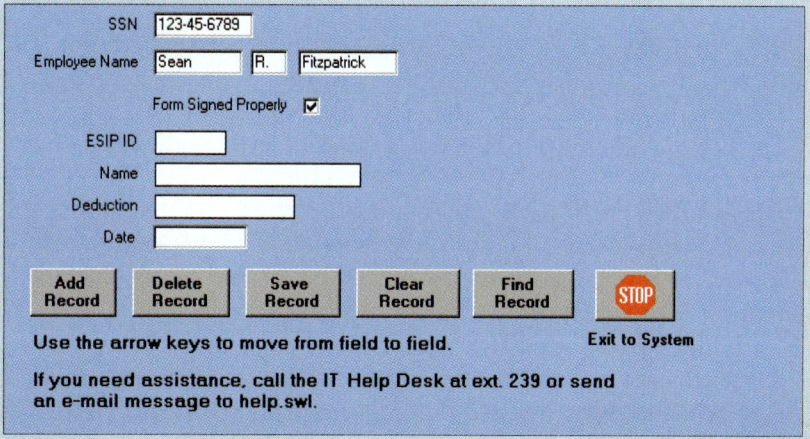

FIGURE 8-50   After the user presses the ENTER key, the system retrieves the employee name and displays it so the user can verify it. Notice that the user must check a box to verify that the form has been signed properly.

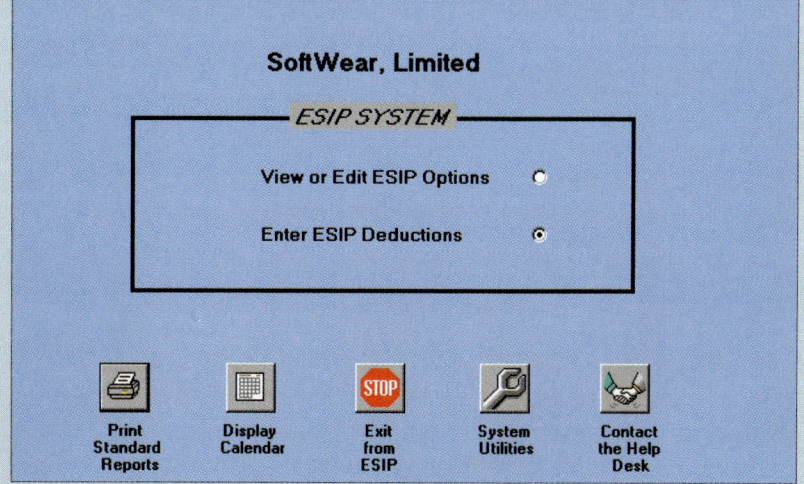

FIGURE 8-51   The ESIP switchboard includes option buttons and command buttons to select various processing choices.

CHAPTER CAPSTONE CASE: SoftWear, Limited (continued)

Follow the guidelines and suggestions in this chapter, and design an online screen form for Carla.

3. In addition to being available online, Carla wants the information request form to be available as a paper source document, which can be used by employees who do not have easy access to the online form. Follow the guidelines and suggestions in this chapter, and design a paper source document for Carla.

4. Carla wants an update on usability, how users read on the Web. Review the material in this chapter and visit the Web site shown in Figure 8-41 on page 368, and any others that provide more information on this topic. Summarize the results of your research in a memo to Carla.

Manage the SWL Project

You have been asked to manage SWL's new information system project. One of your most important activities will be to identify project tasks and determine when they will be performed. Before you begin, you should review the SWL case in this chapter. Then list and analyze the tasks, as follows:

LIST THE TASKS Start by listing and numbering at least 10 tasks that the SWL team needs to perform to fulfill the objectives of this chapter. Your list can include SWL Team Tasks and any other tasks that are described in this chapter. For example, Task 3 might be to Find out what output is needed, and Task 6 might be to Design an output screen.

ANALYZE THE TASKS Now study the tasks to determine the order in which they should be performed. First identify all concurrent tasks, which are not dependent on other tasks. In the example shown in Figure 8-52, Tasks 1, 2, 3, 4, and 5 are concurrent tasks, and could begin at the same time if resources were available.

Other tasks are called dependent tasks, because they cannot be performed until one or more earlier tasks have been completed. For each dependent task, you must identify specific tasks that need to be completed before this task can begin. For example, you would want to find out what output is needed before you could design an output screen, so Task 6 cannot begin until Task 3 is completed, as Figure 8-52 shows.

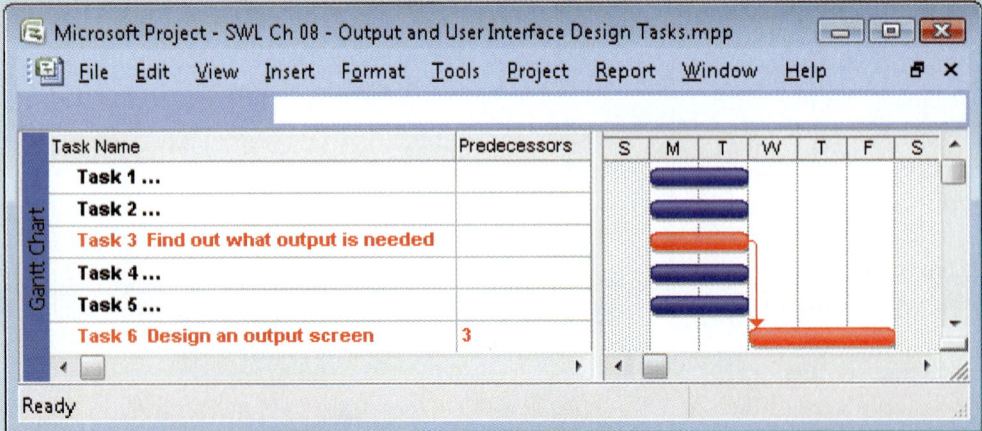

FIGURE 8-52 Tasks 1, 2, 3, 4, and 5 are concurrent tasks that could be performed at the same time. Task 6 is a dependent task that cannot be performed until Task 3 has been completed.

CHAPTER CAPSTONE CASE: SoftWear, Limited (continued) SWL

Chapter 3 describes project management tools, techniques, and software. To learn more, you can visit the Features section on your Student Study Tool CD-ROM, or the project management resources library at **scsite.com/sad8e/project.** On the Web, Microsoft offers demo versions, training, and tips for using Project 2007. You also can visit the OpenWorkbench.org site to learn more about this free, open-source software.

CHAPTER 9 Data Design

Chapter 9 is the second of three chapters in the systems design phase of the SDLC. In this chapter, you will focus on data design skills that are necessary to construct the physical model of the information system.

INTRODUCTION

OBJECTIVES

When you finish this chapter, you will be able to:

- Explain data design concepts and structures
- Describe file processing systems
- Explain database systems and define the components of a database management system (DBMS)
- Describe Web-based data design
- Explain data design terminology, including entities, fields, common fields, records, files, tables, and key fields
- Describe data relationships, draw an entity-relationship diagram, define cardinality, and use cardinality notation
- Explain the concept of normalization
- Explain the importance of codes and describe various coding schemes
- Describe relational and object-oriented database models
- Explain data warehousing and data mining
- Differentiate between logical and physical storage and records
- Explain data control measures

In the systems analysis phase, you created data flow diagrams and object models to create a logical design for the information system. In this part of the systems design phase, you will develop a physical plan for data organization, storage, and retrieval.

This chapter begins with a review of data design concepts and terminology, then discusses file-based systems and database systems, including Web-based databases. You will learn how to create entity-relationship diagrams that show the relationships among data elements, and you will learn how to use normalization concepts to build an effective data design. You also will learn about codes that can be used to represent data items. The chapter concludes with a discussion of data storage and access, including strategic tools such as data warehousing and data mining, physical design issues, logical and physical records, data storage formats, and data control.

CHAPTER INTRODUCTION CASE: Mountain View College Bookstore

Background: Wendy Lee, manager of college services at Mountain View College, wants a new information system that will improve efficiency and customer service at the three college bookstores.

In this part of the case, Florence Fullerton (systems analyst) and Harry Boston (student intern) are talking about data design issues.

Participants:	Florence and Harry
Location:	Mountain View College Cafeteria, Monday morning, December 7, 2009
Project status:	Florence and Harry have completed their output and user interface design tasks and are ready to continue the systems design phase by working on data design for the new system.
Discussion topics:	Data design terms and concepts, cardinality, relational databases, normalization, Web-based design, codes, and physical design issues

Florence:	Good morning, Harry. Now that we have a logical model of the bookstore information system, we're ready for the next step. We have to select an overall data design strategy. I think we should start by looking at a relational database, rather than a file processing design.
Harry:	*What are the pros and cons?*
Florence:	Well, in some situations file processing systems are better, especially when you have to process large numbers of records in a sequence. But in our case, I think a relational database would be more powerful and flexible.
Harry:	*I know what a database is, but what do you mean by the term relational?*
Florence:	In a relational database, all the entities — the individual people, places, events, and transactions — are stored in separate locations called tables, which are related or linked together. That means you have to enter an item of data only once, and you can access all the data items just as if they were all stored in a single location.
Harry:	*That makes sense. How do we decide what data goes where?*
Florence:	We'll start by creating an entity-relationship diagram. Then we'll develop a set of table designs that follow a set of rules called normalization.
Harry:	*I've heard that term before. Aren't there several different levels of normalization, called normal forms?*
Florence:	Yes, and we want our data to be in what's called third normal form, which is what most business-related systems use. We also will consider using various codes to represent data items.
Harry:	*I know that we decided to build the system to run on the college network and then migrate to a Web-based system in the future. But shouldn't we use a design that will make it easy to migrate to the Web?*
Florence:	Yes, and a relational database will be the most flexible approach.
Harry:	*Sounds good. Any other issues?*
Florence:	Well, we need to consider some physical design issues, too. Here's a task list to get us started:

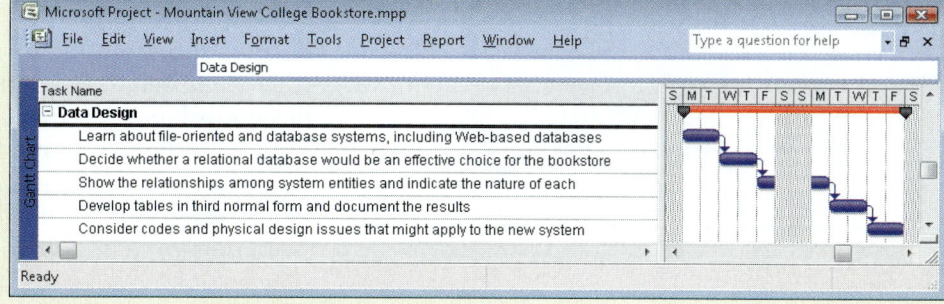

FIGURE 9-1 Typical data design task list.

DATA DESIGN CONCEPTS

Before constructing an information system, a systems analyst must understand basic data design concepts, including data structures and the characteristics of file processing and database systems, including Web-based database design.

Data Structures

A **data structure** is a framework for organizing and storing data in an information system. Data structures consist of files or tables that are linked in various ways. Each **file** or **table** contains data about people, places, things, or events that interact with the information system. For example, one file or table might contain data about customers, and other files or tables might store data about products, orders, suppliers, or employees. Depending on how the system's files and tables are organized and linked, an information system is called either a file processing system or a database management system.

A **file processing system**, also called **a file-oriented system**, stores and manages data in one or more separate files. Examples of two file processing systems are shown in Figure 9-2. The figure shows an auto repair shop with a Job Records system that uses a JOB data file, and an Employee Records system that uses a MECHANIC data file. In the example shown, the Job Records system uses a file that contains all the data necessary to answer inquiries and generate reports about work performed at the shop. Similarly, the Employee Records system uses a file that stores all the data necessary to answer inquiries and generate reports about shop employees. A major disadvantage of file-oriented systems is that the same data is stored in more than one place. For example, notice that three items of information (Mechanic No, Name, and Pay Rate) are stored in both data files.

Now consider Figure 9-3, which shows how the same auto repair shop might use a database system instead of a file-oriented system. A **database system** consists of linked tables that form one overall data structure. Compared to file processing, a database system offers much greater flexibility and efficiency. For example, in the file processing designs shown in Figure 9-2, the two files are unrelated — but in the database design in Figure 9-3, the tables are linked by the Mechanic No field. This link allows information to be accessed from either table as if the two tables were one large table, making it unnecessary to store duplicate information in both tables.

File processing systems still exist to handle specific applications, but the vast

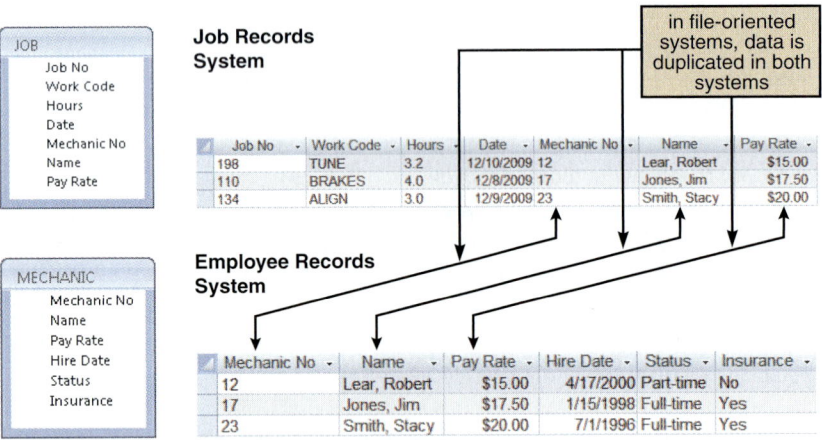

FIGURE 9-2 Example of an auto repair shop that uses two separate file-oriented systems: a Job Records system (with a JOB data file) and an Employee Records system (with a MECHANIC data file). Notice that three items of information must be duplicated in both data files.

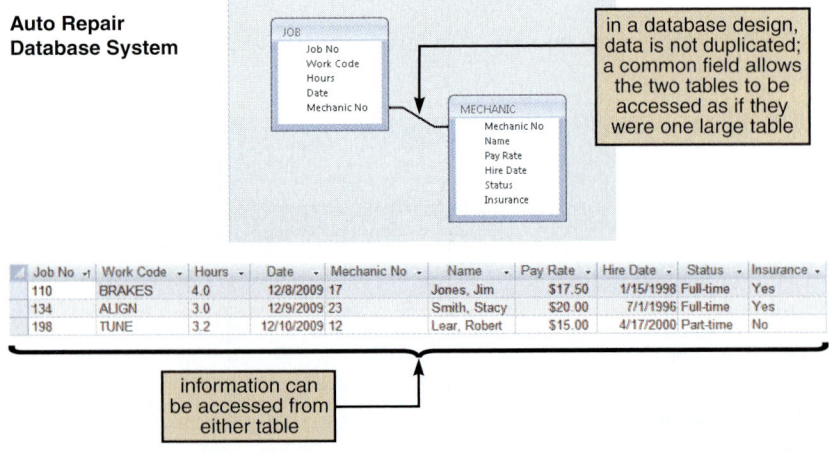

FIGURE 9-3 Example of a database design for the auto repair shop that links two data tables and avoids duplication. Notice that the Mechanic No field provides a link between the two tables, and information can be accessed from either table.

majority of information systems today are designed as databases. The pros and cons of file processing compared with database systems are discussed in the following section.

Overview of File Processing

Although file processing is an older approach, you should understand how these systems were designed, constructed, and maintained. Some companies still use file processing to handle large volumes of structured data on a regular basis. Many older legacy systems utilized file processing designs because that approach was well suited to mainframe hardware and batch input. Although much less common today, file processing can be efficient and cost-effective in certain situations. For example, consider a credit card company that posts thousands of daily transactions from a TRANSACTION file to customer balances stored in a CUSTOMER file, as shown in Figure 9-4. For that relatively simple process, file processing is highly effective.

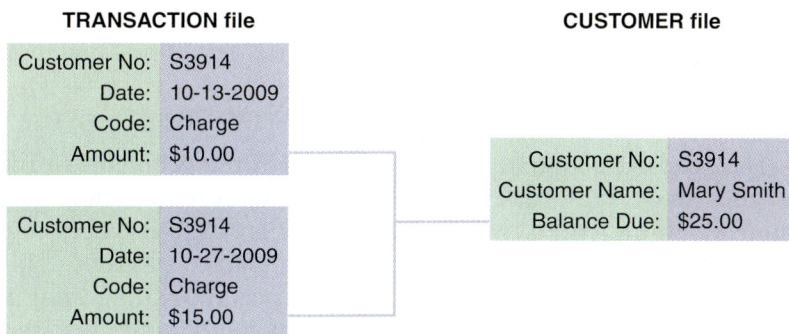

FIGURE 9-4   A credit card company might use a file processing system to post daily sales transactions from a TRANSACTION file to the CUSTOMER file.

In a typical file processing environment, a company might have three departments, each with its own information system and data files. Three potential problems exist in a file processing environment. The first problem is **data redundancy**, which means that data common to two or more information systems is stored in several places. Data redundancy requires more storage space, and maintaining and updating data in several locations is expensive.

Second, **data integrity** problems can occur if updates are not applied in every file. Changing the data in only one of the systems will cause inconsistent data and result in incorrect information in the second system.

The third problem is the rigid data structure of a typical file processing environment. Businesses must make decisions based on company-wide data, and managers often require information from multiple business units and departments. In a file processing environment, that means retrieving information from independent, file-based systems, which is slow and inefficient.

A file-oriented information system can contain various types of files, including master files, table files, transaction files, work files, security files, and history files.

- A **master file** stores relatively permanent data about an entity. For example, a PRODUCT master file contains one logical record for each product the company sells. The quantity field in each record might change daily, but other data, such as the product's code, name, and description, would not change.

- A **table file** contains reference data used by the information system. As with master files, table files are relatively static and are not updated by the information system. Examples of table files include tax tables and postage rate tables.

- A **transaction file** stores records that contain day-to-day business and operational data. A transaction file is an input file that updates a master file; after the update is completed, the transaction file has served its purpose. An example of a transaction file is a charges and payments file that updates a customer balance file.

- A **work file** is a temporary file created by an information system for a single task. Examples of work files include sorted files and report files that hold output reports until they are printed.

- A **security file** is created and saved for backup and recovery purposes. Examples of security files include audit trail files and backups of master, table, and transaction files. New security files must be created regularly to replace outdated files.

- A **history file** is a file created for archiving purposes. For example, students who have not registered for any course in the last two semesters might be deleted from the active student master file and added to an inactive student file, which is a history file that can be used for queries or reports.

These problems do not exist in a database system, which is explained in the following section.

The Evolution from File Systems to Database Systems

A properly designed database system offers a solution to the problems of file processing. A database provides an overall framework that avoids data redundancy and supports a real-time, dynamic environment, two potential problems of a file processing system.

In a file processing environment, data files are designed to fit individual business systems. In contrast, in a database environment, several systems can be built around a single database. Figure 9-5 shows a database environment with a database serving five separate information systems.

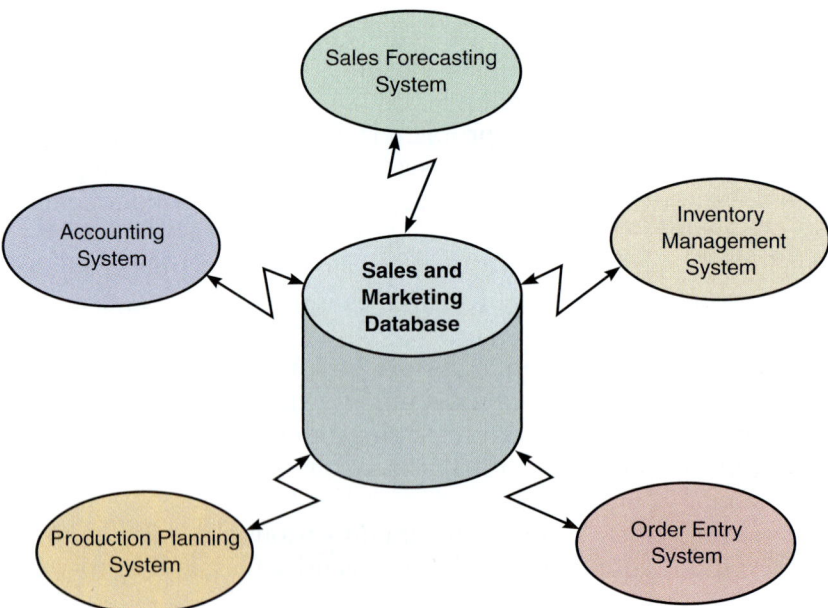

FIGURE 9-5 A typical database environment might consist of a database serving five separate business systems.

A **database management system (DBMS)** is a collection of tools, features, and interfaces that enables users to add, update, manage, access, and analyze the contents of a set of data. From a user's point of view, the main advantage of a DBMS is that it offers timely, interactive, and flexible data access. Specific DBMS advantages include the following:

- **Scalability**, which means that a system can be expanded, modified, or downsized easily to meet the rapidly changing needs of a business enterprise. For example, if a company decides to add data about secondary suppliers of material it uses, a new table can be added to the relational database and linked with a common field.

- Better support for client/server systems. In a **client/server** system, processing is distributed throughout the organization. Client/server systems require the power and flexibility of a database design. You will learn more about client/server systems in Chapter 10.

- Economy of scale. Database design allows better utilization of hardware. If a company maintains an enterprise-wide database, processing is less expensive using a powerful mainframe server instead of using several smaller computers. The inherent efficiency of high-volume processing on larger computers is called **economy of scale**.

- Flexible data sharing. Data can be shared across the enterprise, allowing more users to access more data. A database can be highly flexible, allowing users to view the same information in different ways. Users are empowered because they have access to the information they need to do their jobs.

- Enterprise-wide application. Typically, a DBMS is managed by a person called a **database administrator (DBA)**, who assesses overall requirements and maintains the database for the benefit of the entire organization rather than a single department or user. Database systems can support enterprise-wide applications more effectively than file processing systems.

- Stronger standards. Effective database administration helps ensure that standards for data names, formats, and documentation are followed uniformly throughout the organization.

- Controlled redundancy. Redundancy means storing data in more than one place, which can result in inconsistency and data errors. Because the data is stored in a set of related tables, data items do not need to be duplicated in multiple locations. Even where some duplication is desirable for performance reasons, or disaster recovery, the database approach allows control of the redundancy.

- Better security. The DBA can define authorization procedures to ensure that only legitimate users can access the database and can allow different users to have different levels of access. Most DBMSs provide sophisticated security support.

- Increased programmer productivity. Programmers do not have to create the underlying file structure for a database. That allows them to concentrate on logical design and, therefore, a new database application can be developed more quickly than in a file-oriented system.

- Data independence. Systems that interact with a DBMS are relatively independent of how the physical data is maintained. That design provides the DBA flexibility to alter data structures without modifying information systems that use the data.

Although the trend is toward enterprise-wide database design, many companies still use a combination of centralized DBMSs and smaller, department-level database systems. Why is this so? Most large businesses view data as a company-wide resource that must be accessible to users throughout the company. At the same time, other factors encourage a decentralized design, including network expense; a reluctance to move away from smaller, more flexible systems; and a realization that enterprise-wide DBMSs can be highly complex and expensive to maintain. The compromise, in many cases, is a client/server design, where processing is shared among several computers. Client/server systems are described in detail in Chapter 10. As with many design decisions, the best solution depends on the individual circumstances.

DBMS COMPONENTS

A DBMS provides an interface between a database and users who need to access the data. Although users are concerned primarily with an easy-to-use interface and support for their business requirements, a systems analyst must understand all of the components of a DBMS. In addition to interfaces for users, database administrators, and related systems, a DBMS also has a data manipulation language, a schema and sub-schemas, and a physical data repository, as shown in Figure 9-6.

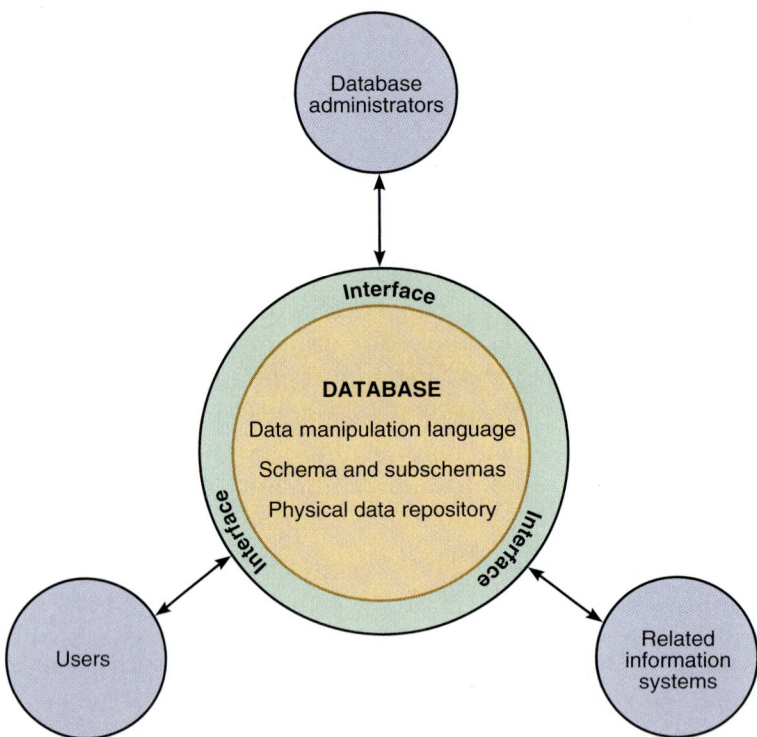

FIGURE 9-6 In addition to interfaces for users, database administrators, and related information systems, a DBMS also has a data manipulation language, a schema and subschemas, and a physical data repository.

Interfaces for Users, Database Administrators, and Related Systems

When users, database administrators, and related information systems request data and services, the DBMS processes the request, manipulates the data, and provides a response.

USERS Users typically work with predefined queries and switchboard commands, but also use query languages to access stored data. A **query language** allows a user to specify a task without specifying how the task will be accomplished. Some query languages use natural language commands that resemble ordinary English sentences. With a **query by example (QBE)** language, the user provides an example of the data requested. Many database programs also generate **SQL (Structured Query Language)**, which is a language that allows client workstations to communicate with servers and mainframe computers. Figure 9-7 on the next page shows a QBE request for all Redfire Metallic or Vapor Silver Metallic 2009 Ford Fusions with navigation. The QBE request generates the SQL commands shown at the bottom of Figure 9-7.

DATABASE ADMINISTRATORS A DBA is responsible for DBMS management and support. DBAs are concerned with data security and integrity, preventing unauthorized access, providing backup and recovery, audit trails, maintaining the database, and supporting user needs. Most DBMSs provide utility programs to assist the DBA in creating and updating data structures, collecting and reporting patterns of database usage, and detecting and reporting database irregularities.

RELATED INFORMATION SYSTEMS A DBMS can support several related information systems that provide input to, and require specific data from, the DBMS. Unlike a user interface, no human intervention is required for two-way communication between the DBMS and the related systems.

Data Manipulation Language

A **data manipulation language (DML)** controls database operations, including storing, retrieving, updating, and deleting data. Most commercial DBMSs, such as Oracle and IBM's DB/2, use a DML. Some database products, such as Microsoft Access, also provide an easy-to-use graphical environment that enables users to control operations with menu-driven commands.

Schema

The complete definition of a database, including descriptions of all fields, tables, and relationships, is called a **schema**. You also can define one or more subschemas. A **subschema** is a view of the database used by one or more systems or users. A subschema defines only those portions of the database that a particular system or user needs or is allowed to access. For example, to protect individual privacy, you might not want to allow a project management system to retrieve employee pay rates. In that case, the project management system subschema would not include the pay rate field. Database designers also use subschemas to restrict the level of access permitted. For example, specific users, systems, or locations might be permitted to create, retrieve, update, or delete data, depending on their needs and the company's security policies.

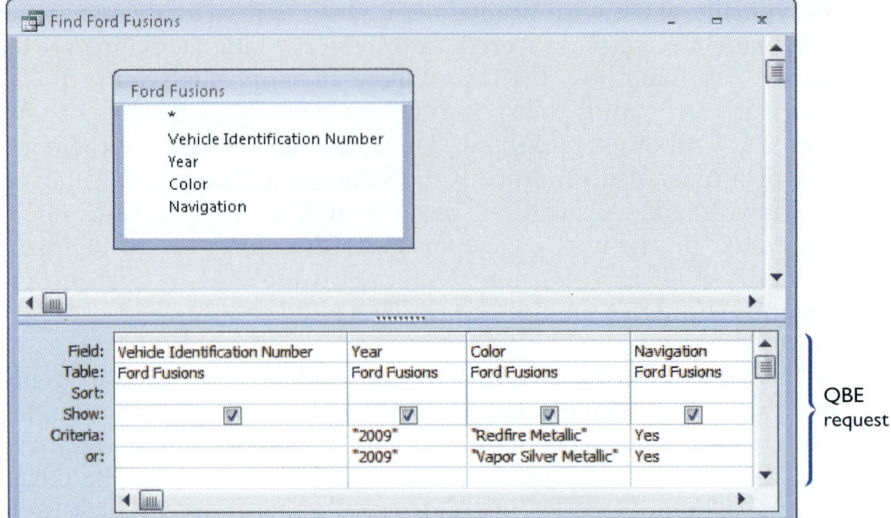

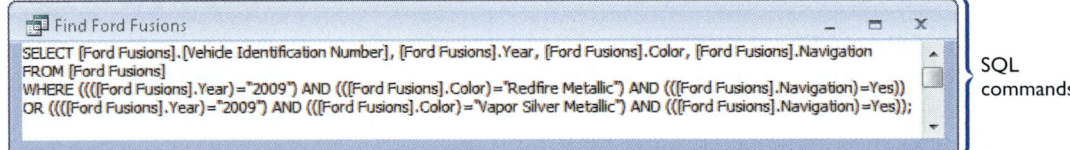

FIGURE 9-7   Using QBE, a user can request a list of all Redfire Metallic or Vapor Silver Metallic 2009 Ford Fusions with navigation.

Physical Data Repository

In Chapter 5, you learned about a data dictionary, which describes all data elements included in the logical design. At this stage of the systems development process, the data dictionary is transformed into a physical data repository, which also contains the schema and subschemas. The physical repository might be centralized, or it might be distributed at several locations. In addition, the stored data might be managed by a single DBMS, or several systems. To resolve potential database connectivity and access problems, companies use ODBC-compliant software that enables communication among various systems and DBMSs. **ODBC,** which stands for **open database connectivity,** is an industry-standard protocol that makes it possible for software from different vendors to interact and exchange data. ODBC uses SQL statements that the DBMS understands and can execute, similar to the ones shown in Figure 9-7. Another common standard is called **JDBC,** or **Java database connectivity**. JDBC enables Java applications to exchange data with any database that uses SQL statements and is JDBC-compliant.

You will learn more about physical design issues in Chapter 10, which discusses system architecture, and in Chapter 11, which discusses system implementation and data conversion.

WEB-BASED DATABASE DESIGN

The concept of Web-based systems was discussed in Chapter 7, Development Strategies. In this chapter, you will revisit this concept and examine the main components of a Web-based database system. The following sections discuss the characteristics of Web-based design, Internet terminology, connecting a database to the Web, and data security on the Web.

Characteristics of Web-Based Design

Figure 9-8 lists some major characteristics of Web-based database design. In a Web-based design, the Internet serves as the front end, or interface, for the database management system. Internet technology provides enormous power and flexibility because the system is not tied to any specific combination of hardware and software. Access to the database requires only a Web browser and an Internet connection. Web-based systems are popular because they offer ease of access, cost-effectiveness, and worldwide connectivity — all of which are vital to companies that must compete in a global economy.

Web-Based Database Design Characteristics

CHARACTERISTIC	EXPLANATION
Global access	The Internet enables worldwide access, using existing infrastructure and standard telecommunications protocols.
Ease of use	Web browsers provide a familiar interface that is user-friendly and easily learned.
Multiple platforms	Web-based design is not dependent on a specific combination of hardware or software. All that is required is a browser and an Internet connection.
Cost effectiveness	Initial investment is relatively low because the Internet serves as the communication network. Users require only a browser, and Web-based systems do not require powerful workstations. Flexibility is high because numerous outsourcing options exist for development, hosting, maintenance, and system support.
Security issues	Security is a universal issue, but Internet connectivity raises special concerns. These can be addressed with a combination of good design, software that can protect the system and detect intrusion, stringent rules for passwords and user identification, and vigilant users and managers.
Adaptability issues	The Internet offers many advantages in terms of access, connectivity, and flexibility. Migrating a traditional database design to the Web, however, can require design modification, additional software, and some added expense.

FIGURE 9-8   Web-based design characteristics include global access, ease of use, multiple platforms, cost effectiveness, security issues, and adaptability issues. In a Web-based design, the Internet serves as the front end, or interface, for the database management system. Access to the database requires only a Web browser and an Internet connection.

Internet Terminology

To understand Web-based data design, it is helpful to review some basic Internet terms and concepts. To access information on the Internet, a person uses a **Web browser,** which is an application that enables the user to navigate, or browse, the Internet and display Web pages on his or her local computer. A **Web page** is a text document written in **HTML (Hypertext Markup Language)**. HTML uses formatting codes called **tags,** which specify how the text and visual elements will be displayed in a Web browser. Web pages are stored on a **Web server,** which is a computer that receives requests and makes Web pages available to users. Together, the Web server and the Web pages are referred to as a **Web site.**

In addition to maintaining a Web site, many companies use intranets and extranets to support business operations and communications. An **intranet** is a private, company-owned

ON THE WEB

For more information about HTML, visit **scsite.com/ sad8e/ more**, locate Chapter 9, and then click the HTML link.

network to provide Web-based access to internal users. An **extranet** is an extension of a company intranet that allows access by external users, such as customers and suppliers. Extranets are typical examples of B2B (business-to-business) data sharing and EDI (electronic data interchange), which were discussed in Chapter 1, where you also learned about Extensible Markup Language (XML). XML is a flexible data description language that allows Web-based communication between different hardware and software environments. Because intranets and extranets use the same **protocols**, or data transmission standards, as the Internet, they are called **Web-centric**.

The Internet and company intranets/extranets are forms of client/server architecture. In a client/server design, tasks are divided between **clients**, which are workstations that users interact with, and **servers**, which are computers that supply data, processing, and services to the client workstations. Client/server architecture is discussed in more detail in Chapter 10, System Architecture.

Connecting a Database to the Web

To access data in a Web-based system, the database must be connected to the Internet or intranet. The database and the Internet speak two different languages, however. Databases are created and managed by using various languages and commands that have nothing to do with HTML, which is the language of the Web. The objective is to connect the database to the Web and enable data to be viewed and updated.

To bridge the gap, it is necessary to use **middleware**, which is software that integrates different applications and allows them to exchange data. Middleware can interpret client requests in HTML form and translate the requests into commands that the database can execute. When the database responds to the commands, middleware translates the results into HTML pages that can be displayed by the user's browser, as shown in Figure 9-9. Notice that the four steps in the process can take place using the Internet or a company intranet as the communications channel.

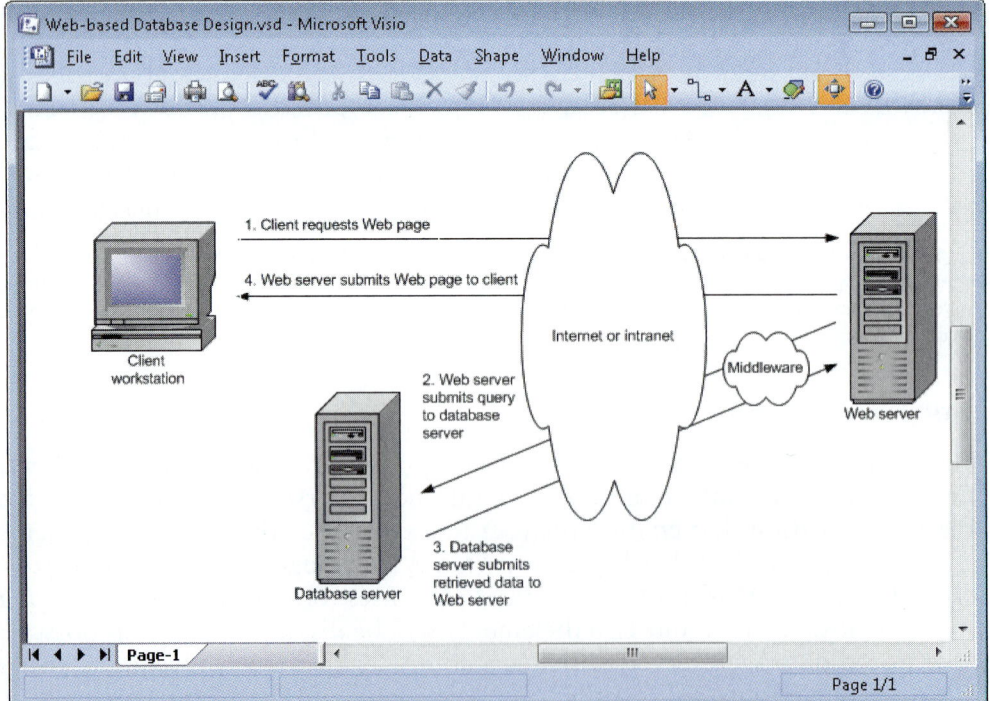

FIGURE 9-9 When a client workstation requests a Web page (1), the Web server uses middleware to generate a data query to the database server (2). The database server responds (3) and middleware translates the retrieved data into an HTML page that can be sent by the Web server and displayed by the user's browser (4).

A popular example of middleware is Adobe ColdFusion, which is shown in Figure 9-10. Middleware is discussed in more detail in Chapter 10.

Data Security

Web-based data must be secure, yet easily accessible to authorized users. To achieve this goal, well-designed systems provide security at three levels: the database itself, the Web server, and the telecommunication links that connect the components of the system.

Data security is discussed in this chapter and in Chapter 12, Managing System Support and Security.

DATA DESIGN TERMINOLOGY

Using the concepts discussed in the previous section, a systems analyst can select a design approach and begin to construct the system. The first step is to understand data design terminology.

FIGURE 9-10 Adobe ColdFusion 8 is a popular example of middleware.

Definitions

Figure 9-11 shows a CUSTOMER entity, which is represented by a table, or file, with eight fields and three records. Notice that the Customer ID field serves as the primary key. These terms are defined in the following sections.

ENTITY An **entity** is a person, place, thing, or event for which data is collected and maintained. For example, an online sales system may include entities named CUSTOMER, ORDER, PRODUCT, and SUPPLIER. When you prepared DFDs during the systems analysis phase, you identified various entities and data stores. Now you will consider the relationships among the entities.

CUSTOMER Table or File

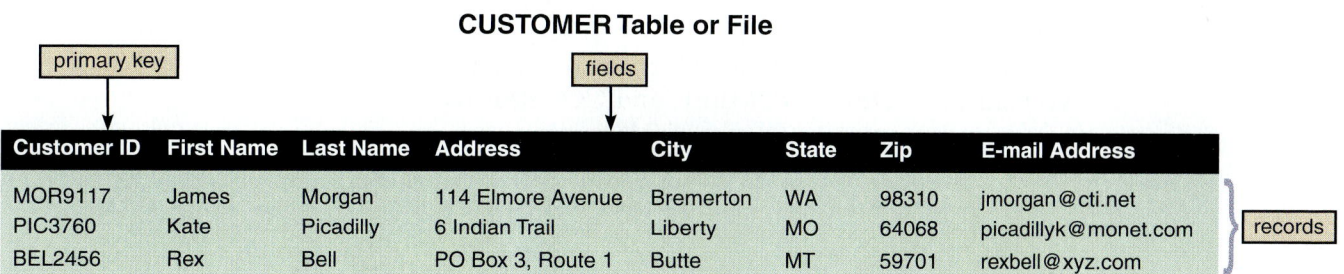

Customer ID	First Name	Last Name	Address	City	State	Zip	E-mail Address
MOR9117	James	Morgan	114 Elmore Avenue	Bremerton	WA	98310	jmorgan@cti.net
PIC3760	Kate	Picadilly	6 Indian Trail	Liberty	MO	64068	picadillyk@monet.com
BEL2456	Rex	Bell	PO Box 3, Route 1	Butte	MT	59701	rexbell@xyz.com

FIGURE 9-11 The CUSTOMER entity is represented by a table or file with eight fields, three records, and a primary key.

TABLE OR FILE   Data is organized into tables or files. A table, or file, contains a set of related records that store data about a specific entity. Tables and files are shown as two-dimensional structures that consist of vertical columns and horizontal rows. Each column represents a field, or characteristic of the entity, and each row represents a record, which is an individual instance, or occurrence of the entity. For example, the CUSTOMER table in Figure 9-11 on the previous page has eight columns, or fields, that describe the entity and three rows, or records, each representing a specific customer.

Although they can have different meanings in a specific context, the terms *table* and *file* often can be used interchangeably.

FIELD   A **field**, also called an **attribute**, is a single characteristic or fact about an entity. In the example shown in Figure 9-11, the CUSTOMER entity has eight fields that store the Customer ID, First Name, Last Name, Address, City, State, Zip, and E-mail Address.

A **common field** is an attribute that appears in more than one entity. Common fields can be used to link entities in various types of relationships.

RECORD   A **record**, also called a **tuple** (rhymes with couple), is a set of related fields that describes one instance, or occurrence of an entity, such as one customer, one order, or one product. A record might have one or dozens of fields, depending on what information is needed.

Key Fields

During the systems design phase, you use **key fields** to organize, access, and maintain data structures. The four types of keys are primary keys, candidate keys, foreign keys, and secondary keys.

PRIMARY KEY   A **primary key** is a field or combination of fields that uniquely and minimally identifies a particular member of an entity. For example, in a customer table the customer number is a unique primary key because no two customers can have the same customer number. That key also is minimal because it contains no information beyond what is needed to identify the customer. In Figure 9-11, Customer ID is an example of a primary key based on a single field.

A primary key also can be composed of two or more fields. For example, if a student registers for three courses, his or her student number will appear in three records in the registration system. If one of those courses has 20 students, 20 separate records will exist for that course number — one record for each student who registered.

In the registration file, neither the student number nor the course ID is unique, so neither field can be a primary key. To identify a specific student in a specific course, the primary key must be a combination of student number and course ID. In that case, the primary key is called a **combination key**. A combination key also can be called a **composite key**, a **concatenated key**, or a **multi-valued key**.

Figure 9-12 shows four different tables. The first three tables have single-field primary keys. Notice that in the fourth table, however, the primary key is a combination of two fields: STUDENT-NUMBER and COURSE-ID.

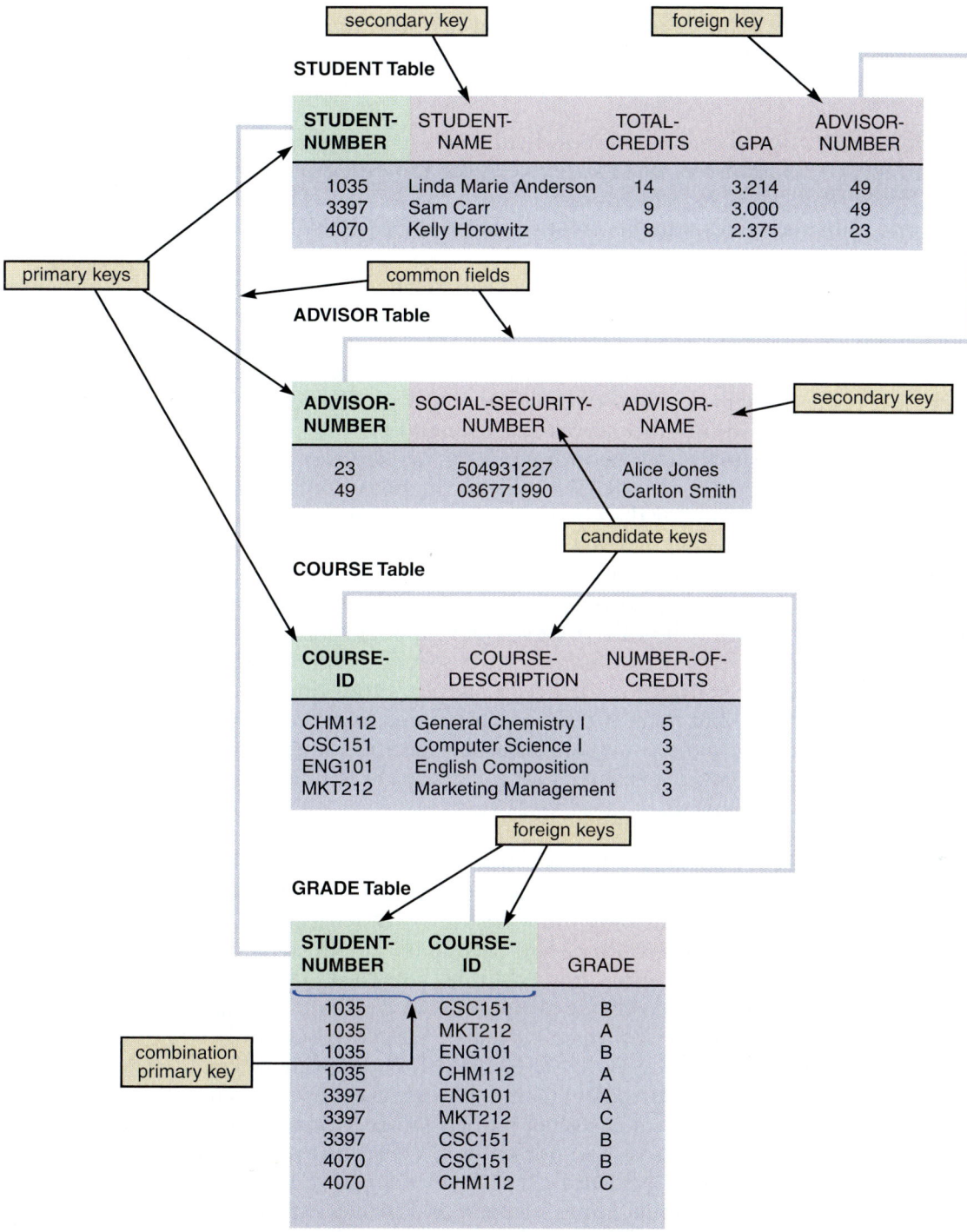

FIGURE 9-12   Examples of common fields, primary keys, candidate keys, foreign keys, and secondary keys.

CANDIDATE KEY   Sometimes you have a choice of fields or field combinations to use as the primary key. Any field that could serve as a primary key is called a **candidate key**. For example, if every employee has a unique employee number, then you could use either the employee number or the Social Security number as a primary key. Because you can designate only one field as a primary key, you should select the field that contains the least amount of data and is the easiest to use. Any field that is not a primary key or a candidate key is called a **nonkey field**.

The primary keys shown in Figure 9-12 on the previous page also are candidate keys. Two other candidate keys exist: the SOCIAL-SECURITY-NUMBER field in the ADVISOR table and the COURSE-DESCRIPTION field in the COURSE table.

FOREIGN KEY Recall that a common field exists in more than one table and can be used to form a relationship, or link, between the tables. For example, in Figure 9-12, the ADVISOR-NUMBER field appears in both the STUDENT table and the ADVISOR table and joins the tables together. Notice that ADVISOR-NUMBER is a primary key in the ADVISOR table, where it uniquely identifies each advisor, and is a foreign key in the STUDENT table. A **foreign key** is a field in one table that must match a primary key value in another table in order to establish the relationship between the two tables.

Unlike a primary key, a foreign key need not be unique. For example, Carlton Smith has advisor number 49. The value 49 must be a unique value in the ADVISOR table because it is the primary key, but 49 can appear any number of times in the STUDENT table, where the advisor number serves as a foreign key.

Figure 9-12 also shows how two foreign keys can serve as a composite primary key in another table. Consider the GRADE table at the bottom of the figure. The two fields that form the primary key for the GRADE table are both foreign keys: the STUDENT-NUMBER field, which must match a student number in the STUDENT table, and the COURSE-ID field, which must match one of the course IDs in the COURSE table.

How can these two foreign keys serve as a primary key in the GRADE table? When you study the table, you will notice that student numbers and course IDs can appear any number of times, but the *combination* of a specific student and a specific course occurs only once. For example, student 1035 appears four times and course CSC151 appears three times — but there is only *one* combined instance of student 1035 *and* course CSC151. Because the combination of the specific student (1035) and the specific course (CSC151) is unique, it ensures that the grade (B) will be assigned to the proper student in the proper course.

SECONDARY KEY A **secondary key** is a field or combination of fields that can be used to access or retrieve records. Secondary key values are not unique. For example, if you need to access records for only those customers in a specific ZIP code, you would use the ZIP code field as a secondary key. Secondary keys also can be used to sort or display records in a certain order. For example, you could use the GPA field in a STUDENT file to display records for all students in grade point order.

The need for a secondary key arises because a table can have only one primary key. In a CUSTOMER file, the CUSTOMER-NUMBER is the primary key, so it must be unique. You might know a customer's name, but not the customer's number. For example, you might want to access a customer named James Morgan in the table shown in Figure 9-11 on page 397, but you do not know his customer number. If you search the table using the CUSTOMER-NAME field as a secondary key, you can retrieve the records for all customers named James Morgan and then select the correct one.

In Figure 9-12, student name and advisor names are identified as secondary keys, but other fields also could be used. For example, to find all students who have a particular advisor, you could use the ADVISOR-NUMBER field in the STUDENT file as a secondary key.

Referential Integrity

Validity checks can help avoid data input errors. One type of validity check, called **referential integrity**, is a set of rules that avoids data inconsistency and quality problems. In a relational database, referential integrity means that a foreign key value

For more information about referential integrity, visit **scsite.com/sad8e/more**, locate Chapter 9, and then click the Referential Integrity link.

cannot be entered in one table unless it matches an existing primary key in another table. For example, referential integrity would prevent you from entering a customer order in an order table unless that customer already exists in the customer table. Without referential integrity, you might have an order called an **orphan**, because it had no related customer.

In the example shown in Figure 9-12 on page 399, referential integrity will not allow a user to enter an advisor number (foreign key value) in the STUDENT table unless a valid advisor number (primary key value) already exists in the ADVISOR table.

Referential integrity also can prevent the deletion of a record if the record has a primary key that matches foreign keys in another table. For example, suppose that an advisor resigns to accept a position at another school. You cannot delete the advisor from the ADVISOR table while records in the STUDENT file still refer to that advisor number. Otherwise, the STUDENT records would be orphans. To avoid the problem, students must be reassigned to other advisors by changing the value in the ADVISOR-NUMBER field; then the advisor record can be deleted.

When creating a relational database, you can build referential integrity into the design. Figure 9-13 shows a Microsoft Access screen that identifies a common field and allows the user to enforce referential integrity rules.

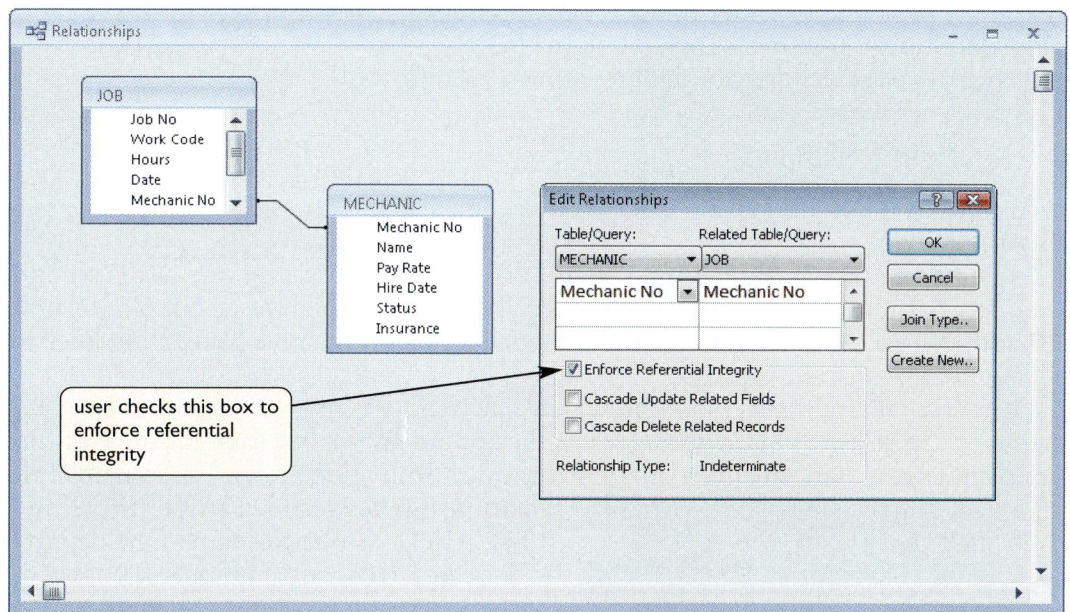

FIGURE 9-13 Microsoft Access allows a user to specify that referential integrity rules will be enforced in a relational database design.

ENTITY-RELATIONSHIP DIAGRAMS

Recall that an entity is a person, place, thing, or event for which data is collected and maintained. For example, entities might be customers, sales regions, products, or orders. An information system must recognize the relationships among entities. For example, a *customer* entity can have several instances of an *order* entity, and an *employee* entity can have one instance, or none, of a *spouse* entity.

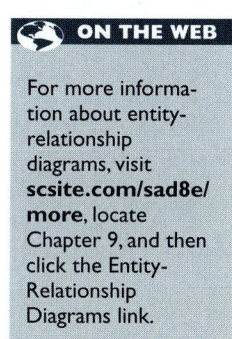

An **entity-relationship diagram (ERD)** is a model that shows the logical relationships and interaction among system entities. An ERD provides an overall view of the system and a blueprint for creating the physical data structures.

Drawing an ERD

The first step is to list the entities that you identified during the systems analysis phase and to consider the nature of the relationships that link them. At this stage, you can use a simplified method to show the relationships between entities.

Although there are different ways to draw ERDs, a popular method is to represent entities as rectangles and relationships as diamond shapes. The entity rectangles are labeled with singular nouns, and the relationship diamonds are labeled with verbs, usually in a top-to-bottom and left-to-right fashion. For example, in Figure 9-14, a doctor entity *treats* a patient entity. Unlike data flow diagrams, entity-relationship diagrams depict relationships, not data or information flows.

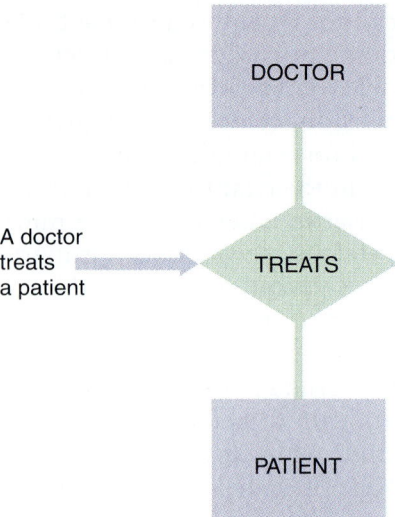

FIGURE 9-14 In an entity-relationship diagram, entities are labeled with singular nouns and relationships are labeled with verbs. The relationship is interpreted as a simple English sentence.

Types of Relationships

Three types of relationships can exist between entities: one-to-one, one-to-many, and many-to-many.

A **one-to-one relationship**, abbreviated **1:1**, exists when exactly one of the second entity occurs for each instance of the first entity. Figure 9-15 shows examples of several 1:1 relationships. A number 1 is placed alongside each of the two connecting lines to indicate the 1:1 relationship.

A **one-to-many relationship**, abbreviated **1:M**, exists when one occurrence of the first entity can relate to many instances of the second entity, but each instance of the second entity can associate with only one instance of the first entity. For example, the relationship between DEPARTMENT and EMPLOYEE is one-to-many: One department can have many employees, but each employee works in only one department at a time. Figure 9-16 shows several 1:M relationships. The line connecting the *many* entity is labeled with the letter M, and the number 1 labels the other connecting line. How many is *many*? The first 1:M relationship shown in Figure 9-16 shows the entities INDIVIDUAL and AUTOMOBILE. One individual might own five automobiles, or one, or none. Thus, *many* can mean any number, including zero.

A **many-to-many relationship**, abbreviated **M:N**, exists when one instance of the first entity can relate to many instances of the second entity, and one instance of the second entity can relate to many instances of the first entity. The relationship between STUDENT and CLASS, for example, is

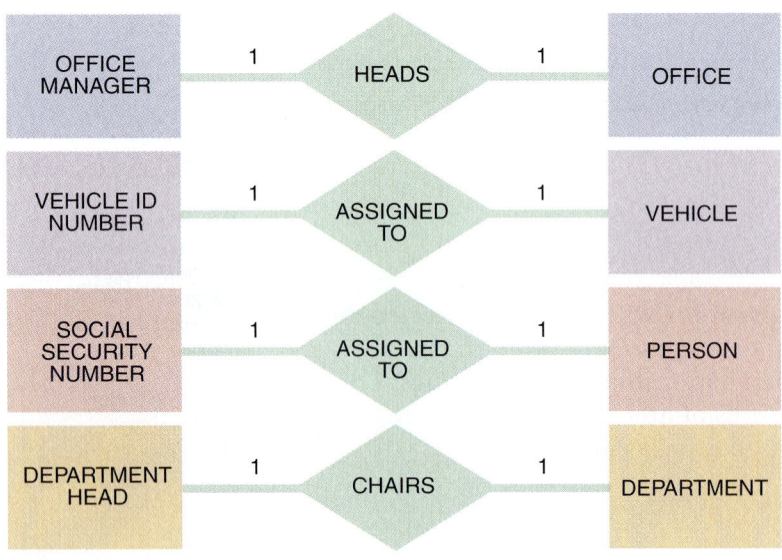

FIGURE 9-15 Examples of one-to-one (1:1) relationships.

many-to-many — one student can take many classes, and one class can have many students enrolled. Figure 9-17 shows several M:N entity-relationships. One of the connecting lines is labeled with the letter M, and the letter N labels the other connection.

Notice that an M:N relationship is different from 1:1 or 1:M relationships because the event or transaction that links the two entities is actually a third entity, called an **associative entity** that has its own characteristics. In the first example in Figure 9-17, the ENROLLS IN symbol represents a REGISTRATION entity that records each instance of a specific student enrolling in a specific course. Similarly, the RESERVES SEAT ON symbol represents a RESERVATION entity that records each instance of a specific passenger reserving a seat on a specific flight. In the third example, the LISTS symbol represents an ORDER-LINE entity that records each instance of a specific product listed in a specific customer order.

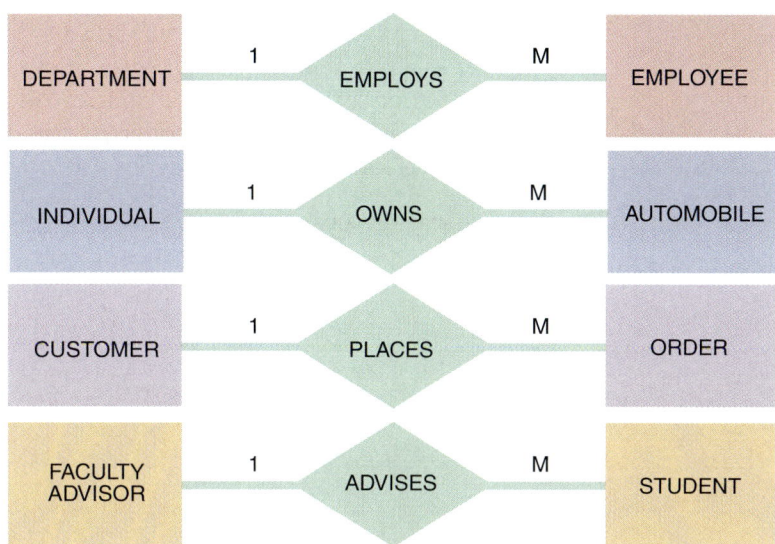

FIGURE 9-16 Examples of one-to-many (1:M) relationships.

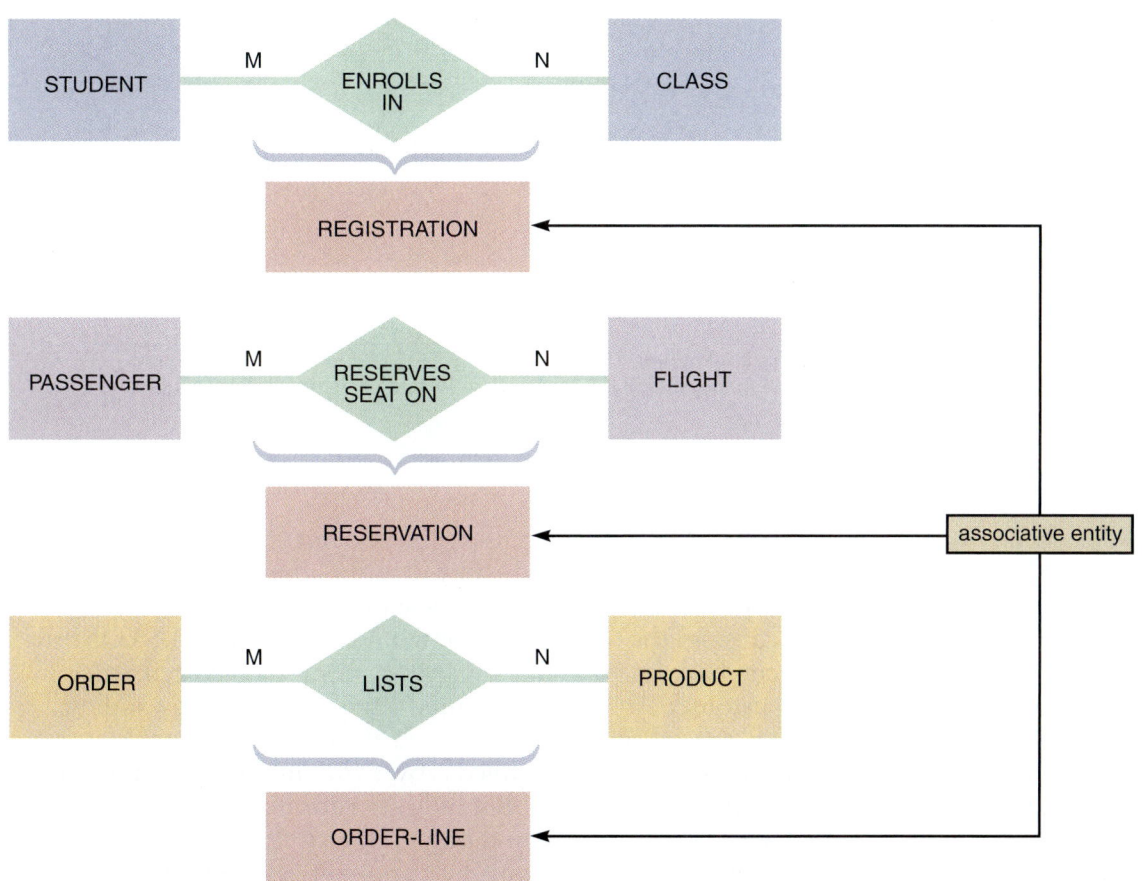

FIGURE 9-17 Examples of many-to-many (M:N) relationships. Notice that the event or transaction that links the two entities is an associative entity with its own set of attributes and characteristics.

Figure 9-18 shows an ERD for a sales system. Notice the various entities and relationships shown in the figure, including the associative entity named ORDER-LINE. The detailed nature of these relationships is called cardinality. As an analyst, you must understand cardinality in order to create a data design that accurately reflects all relationships among system entities.

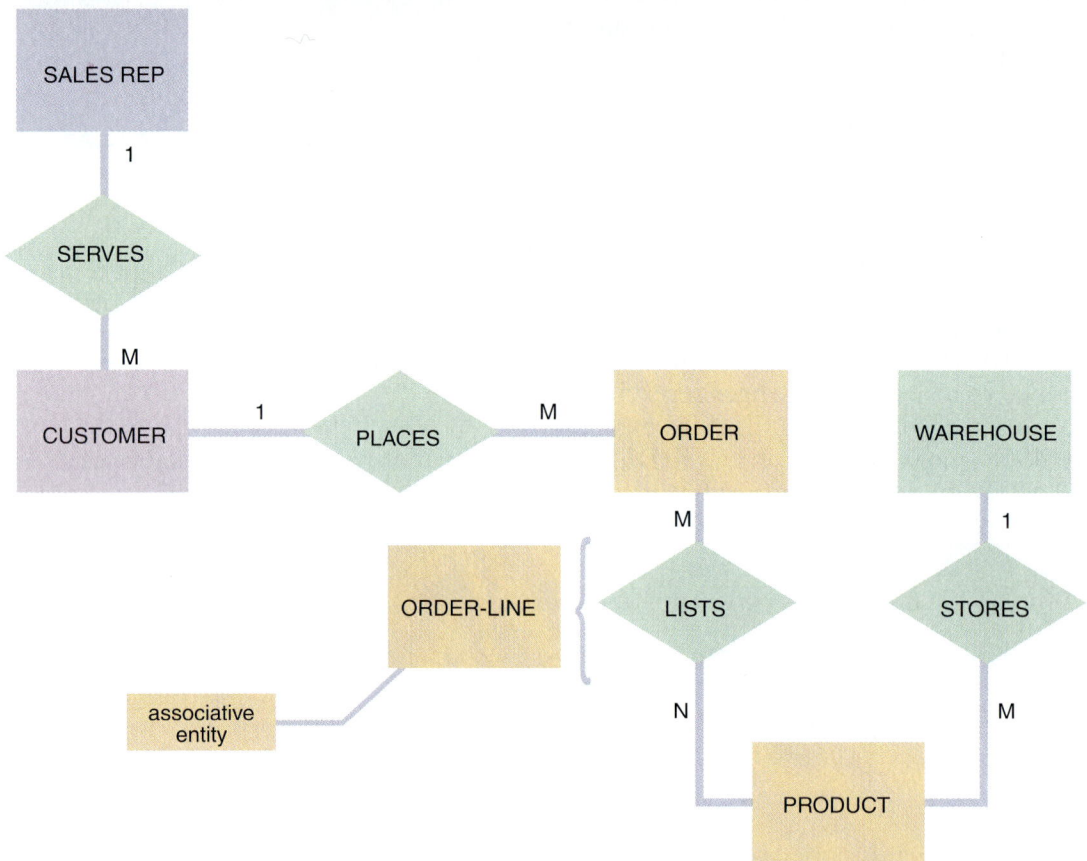

FIGURE 9-18   An entity-relationship diagram for SALES REP, CUSTOMER, ORDER, PRODUCT, and WAREHOUSE. Notice that the ORDER and PRODUCT entities are joined by an associative entity named ORDER-LINE.

Cardinality

For more information about cardinality, visit **scsite.com/sad8e/ more**, locate Chapter 9, and then click the Cardinality link.

After an analyst draws an initial ERD, he or she must define the relationships in more detail by using a technique called cardinality. **Cardinality** describes the numeric relationship between two entities and shows how instances of one entity relate to instances of another entity. For example, consider the relationship between two entities: CUSTOMER and ORDER. One customer can have one order, many orders, or none, but each order must have one and only one customer. An analyst can model this interaction by adding **cardinality notation**, which uses special symbols to represent the relationship.

A common method of cardinality notation is called **crow's foot notation** because of the shapes, which include circles, bars, and symbols, that indicate various possibilities. A single bar indicates one, a double bar indicates one and only one, a circle indicates zero, and a crow's foot indicates many. Figure 9-19 shows various cardinality symbols, their meanings, and the UML representations of the relationships. As you learned in Chapter 4, the **Unified Modeling Language (UML)** is a widely used method of visualizing and documenting software systems design.

In Figure 9-20, four examples of cardinality notation are shown. In the first example, one and only one CUSTOMER can place anywhere from zero to many of the ORDER entity. In the second example, one and only one ORDER can include one ITEM ORDERED or many. In the third example, one and only one EMPLOYEE can have one SPOUSE or none. In the fourth example, one EMPLOYEE, or many employees, or none, can be assigned to one PROJECT, or many projects, or none.

Most CASE products support the drawing of ERDs from entities in the data repository. Figure 9-21 on the next page shows part of a library system ERD drawn using the Visible Analyst CASE tool. Notice that crow's foot notation is used to show the nature of the relationships, which are described in both directions.

Now that you understand database elements and their relationships, you can start designing tables. The first step is the normalization of your table designs, which is described next.

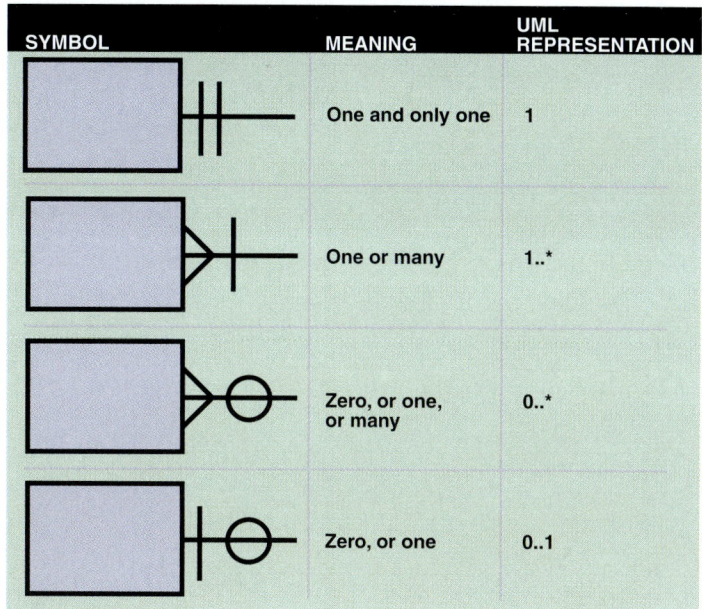

SYMBOL	MEANING	UML REPRESENTATION
	One and only one	1
	One or many	1..*
	Zero, or one, or many	0..*
	Zero, or one	0..1

FIGURE 9-19 Crow's foot notation is a common method of indicating cardinality. The four examples show how you can use various symbols to describe the relationships between entities.

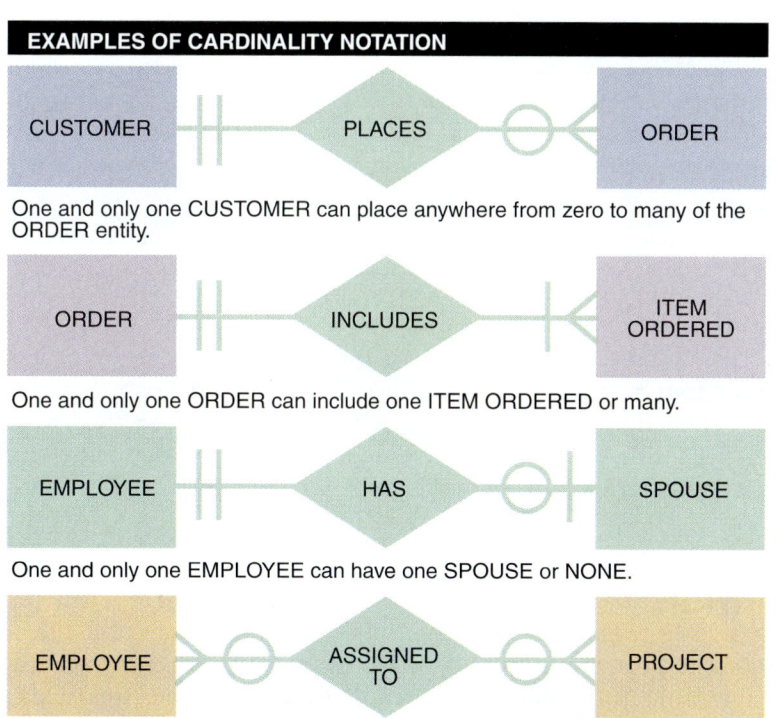

EXAMPLES OF CARDINALITY NOTATION

CUSTOMER — PLACES — ORDER

One and only one CUSTOMER can place anywhere from zero to many of the ORDER entity.

ORDER — INCLUDES — ITEM ORDERED

One and only one ORDER can include one ITEM ORDERED or many.

EMPLOYEE — HAS — SPOUSE

One and only one EMPLOYEE can have one SPOUSE or NONE.

EMPLOYEE — ASSIGNED TO — PROJECT

One EMPLOYEE, or many employees, or none, can be assigned to one PROJECT, or many projects, or none.

FIGURE 9-20 In the first example of cardinality notation, one and only one CUSTOMER can place anywhere from zero to many of the ORDER entity. In the second example, one and only one ORDER can include one ITEM ORDERED or many. In the third example, one and only one EMPLOYEE can have one SPOUSE or none. In the fourth example, one EMPLOYEE, or many employees, or none, can be assigned to one PROJECT, or many projects, or none.

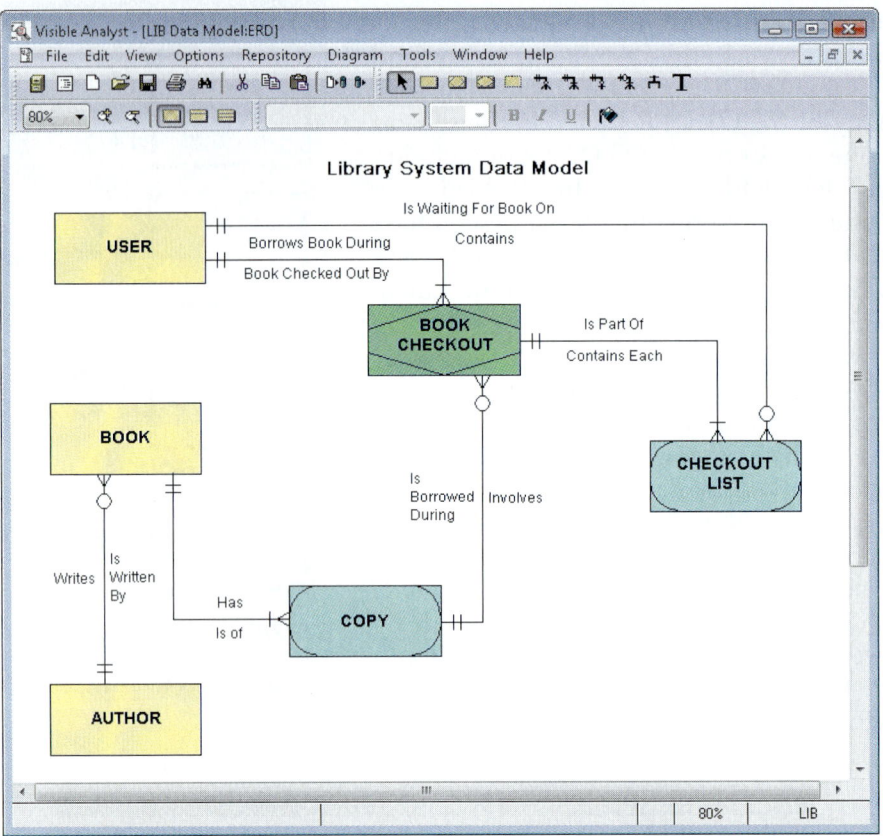

FIGURE 9-21 An ERD for a library system drawn with Visible Analyst. Notice that crow's foot notation has been used and relationships are described in both directions.

CASE IN POINT 9.1: TopText Publishing

TopText Publishing is a textbook publishing company with a headquarters location, a warehouse, and three sales offices that each have a sales manager and sales reps. TopText sells to schools, colleges, and individual customers. Many authors write more than one book for TopText, and some books are written by more than one author. TopText maintains an active list of more than 100 books, each identified by a universal code called an ISBN number. You have been asked to draw an ERD for the TopText information system, and to include cardinality notation.

NORMALIZATION

ON THE WEB

For more information about normalization, visit **scsite.com/ sad8e/more**, locate Chapter 9, and then click the Normalization link.

Normalization is the process of creating table designs by assigning specific fields or attributes to each table in the database. A **table design** specifies the fields and identifies the primary key in a particular table or file. Working with a set of initial table designs, you use normalization to develop an overall database design that is simple, flexible, and free of data redundancy. Normalization involves applying a set of rules that can help you identify and correct inherent problems and complexities in your table designs. The concept of normalization is based on the work of Edgar Codd, a British computer scientist who formulated the basic principles of relational database design.

The normalization process typically involves four stages: unnormalized design, first normal form, second normal form, and third normal form. The three normal forms constitute a progression in which third normal form represents the best design. Most business-related databases must be designed in third normal form.

Standard Notation Format

Designing tables is easier if you use a **standard notation format** to show a table's structure, fields, and primary key. The standard notation format in the following examples starts with the name of the table, followed by a parenthetical expression that contains the field names separated by commas. The primary key field(s) is underlined, like this:

NAME (<u>FIELD 1</u>, FIELD 2, FIELD 3)

Repeating Groups and Unnormalized Designs

During data design, you must be able to recognize a repeating group of fields. A **repeating group** is a set of one or more fields that can occur any number of times in a single record, with each occurrence having different values.

Repeating groups often occur in manual documents prepared by users. For example, consider a school registration form with the student's information at the top of the form, followed by a list of courses the student is taking. If you were to design a table based on this registration form, the courses would represent a repeating group of values for each student.

An example of a repeating group is shown in Figure 9-22. The first two records in the ORDER table contain multiple products, which represent a repeating group of fields. Notice that in addition to the order number and date, the records with multiple products contain repetitions of the product number, description, and number ordered. You can think of a repeating group as a set of child (subsidiary) records contained within the parent (main) record.

ORDER (UNNORMALIZED)

primary key (ORDER-NUM); primary key for repeating group (PRODUCT-NUM); repeating groups

RECORD#	<u>ORDER-NUM</u>	ORDER-DATE	PRODUCT-NUM	PRODUCT-DESC	NUM-ORDERED
1	40311	03112005	304	All-purpose gadget	7
			633	Assembly	1
			684	Super gizmo	4
2	40312	03112005	128	Steel widget	12
			304	All-purpose gadget	3
3	40313	03122005	304	All-purpose gadget	144

FIGURE 9-22 In the ORDER table design, records 1 and 2 have repeating groups because they contain several products. ORDER-NUM is the primary key for the ORDER table, and PRODUCT-NUM serves as a primary key for the repeating group. Because it contains a repeating group, the ORDER table design is unnormalized.

A table design that contains a repeating group is called **unnormalized**. The standard notation method for representing an unnormalized design is to enclose the repeating group of fields within a second set of parentheses. An example of an unnormalized table would look like this:

NAME (<u>FIELD 1</u>, FIELD 2, FIELD 3, (REPEATING FIELD 1, REPEATING FIELD 2))

Now review the unnormalized ORDER table design shown in Figure 9-22. Following the notation guidelines, you can describe the design as follows:

ORDER (<u>ORDER-NUM</u>, ORDER-DATE, (<u>PRODUCT-NUM</u>, PRODUCT-DESC, NUM-ORDERED))

The notation indicates that the ORDER table design contains five fields, which are listed within the outer parentheses. The ORDER-NUM field is underlined to show that it is the primary key. The PRODUCT-NUM, PRODUCT-DESC, and NUM-ORDERED fields are enclosed within an inner set of parentheses to indicate that they are fields within a repeating group. Notice that PRODUCT-NUM also is underlined because it acts as the primary key of the repeating group. If a customer orders three different products in one order, then the fields PRODUCT-NUM, PRODUCT-DESC, and NUM-ORDERED repeat three times, as shown in Figure 9-22 on the previous page.

First Normal Form

A table is in **first normal form (1NF)** if it does not contain a repeating group. To convert an unnormalized design to 1NF, you must expand the table's primary key to include the primary key of the repeating group.

For example, in the ORDER table shown in Figure 9-22, the repeating group consists of three fields: PRODUCT-NUM, PRODUCT-DESC, and NUM-ORDERED. Of the three fields, only PRODUCT-NUM can be a primary key because it uniquely identifies each instance of the repeating group. The product description cannot be a primary key because it might or might not be unique. For example, a company might sell a large number of parts with the same descriptive name, such as *washer*, relying on a coded part number to identify uniquely each washer size.

When you expand the primary key of ORDER table to include PRODUCT-NUM, you eliminate the repeating group and the ORDER table is now in 1NF, as shown:

> ORDER (<u>ORDER-NUM</u>, ORDER-DATE, <u>PRODUCT-NUM</u>, PRODUCT-DESC, NUM-ORDERED)

Figure 9-23 shows the ORDER table in 1NF. Notice that when you eliminate the repeating group, additional records emerge — one for each combination of a specific order and a specific product. The result is more records, but a greatly simplified design. In the new version, the repeating group for order number 40311 has become three separate records, and the repeating group for order number 40312 has become two separate records. Therefore, when a table is in 1NF, each record stores data about a single instance of a specific order and a specific product.

ORDER IN 1NF

combination — primary key

RECORD#	ORDER- NUM	ORDER- DATE	PRODUCT- NUM	PRODUCT- DESC	NUM- ORDERED
1	40311	03112007	304	All-purpose gadget	7
2	40311	03112007	633	Assembly	1
3	40311	03112007	684	Super gizmo	4
4	40312	03112007	128	Steel widget	12
5	40312	03112007	304	All-purpose gadget	3
6	40313	03122007	304	All-purpose gadget	144

repeating groups have been eliminated

FIGURE 9-23 The ORDER table as it appears in 1NF. The repeating groups have been eliminated. Notice that the repeating group for order 40311 has become three separate records, and the repeating group for order 40312 has become two separate records. The 1NF primary key is a combination of ORDER-NUM and PRODUCT-NUM, which uniquely identifies each record.

Also notice that the 1NF design shown in Figure 9-23 has a combination primary key. The primary key of the 1NF design cannot be the ORDER-NUM field alone, because the order number does not uniquely identify each product in a multiple-item order. Similarly, PRODUCT-NUM cannot be the primary key, because it appears more than once if several orders include the same product. Because each record must reflect a specific product in a specific order, you need *both* fields, ORDER-NUM and PRODUCT-NUM, to identify a single record uniquely. Therefore, the primary key is the *combination* of two fields: ORDER-NUM and PRODUCT-NUM.

Second Normal Form

To understand second normal form (2NF), you must understand the concept of functional dependence. For example, field X is **functionally dependent** on field Y if the value of field X depends on the value of field Y. For example, in Figure 9-23, the ORDER-DATE value is functionally dependent on the ORDER-NUM, because for a particular order number, there can be only one value for that order's date. In contrast, a product description is not dependent on the order number. For a particular order number, there might be several product descriptions — one for each item ordered.

A table design is in **second normal form (2NF)** if it is in 1NF *and* if all fields that are not part of the primary key are functionally dependent on the *entire* primary key. If any field in a 1NF table depends on only one of the fields in a combination primary key, then the table is not in 2NF.

Notice that if a 1NF design has a primary key that consists of only one field, the problem of partial dependence does not arise — because the entire primary key is a single field. Therefore, a 1NF table with a single-field primary key is automatically in 2NF.

Now reexamine the 1NF design for the ORDER table shown in Figure 9-23:

ORDER (<u>ORDER-NUM</u>, ORDER-DATE, <u>PRODUCT-NUM</u>, PRODUCT-DESC, NUM-ORDERED)

Recall that the primary key is the combination of the order number and the product number. The NUM-ORDERED field depends on the *entire* primary key, because NUM-ORDERED refers to a specific product number *and* a specific order number. In contrast, the ORDER-DATE field depends on the order number, which is only a part of the primary key. Similarly, the PRODUCT-DESC field depends on the product number, which also is only a part of the primary key. Because some fields are not dependent on the *entire* primary key, the design is not in 2NF.

A standard process exists for converting a table from 1NF to 2NF. The objective is to break the original table into two or more new tables and reassign the fields so that each nonkey field will depend on the entire primary key in its table. To accomplish this, you follow these steps:

1. First, create and name a separate table for each field in the existing primary key. For example, in Figure 9-23, the ORDER table's primary key has two fields, ORDER-NUM and PRODUCT-NUM, so you must create two tables. The ellipsis (...) indicates that fields will be assigned later. The result is:

   ORDER (<u>ORDER-NUM</u>,...)

   PRODUCT (<u>PRODUCT-NUM</u>,...)

2. Next, create a new table for each possible combination of the original primary key fields. In the Figure 9-23 example, you would create and name a new table with a combination primary key of ORDER-NUM and PRODUCT-NUM. This table describes individual lines in an order, so it is named ORDER-LINE, as shown:

   ORDER-LINE (<u>ORDER-NUM</u>, <u>PRODUCT-NUM</u>,...)

3. Finally, study the three tables and place each field with its appropriate primary key, which is the minimal key on which it functionally depends. When you finish placing all the fields, remove any table that did not have any additional fields

assigned to it. The remaining tables are the 2NF version of your original table. In the Figure 9-23 example, the three tables would be shown as:

ORDER (<u>ORDER-NUM</u>, ORDER-DATE)

PRODUCT (<u>PRODUCT-NUM</u>, PRODUCT-DESC)

ORDER-LINE (<u>ORDER-NUM</u>, <u>PRODUCT-NUM</u>, NUM-ORDERED)

Figure 9-24 shows the 2NF table designs. By following the steps, you have converted the original 1NF table into three 2NF tables.

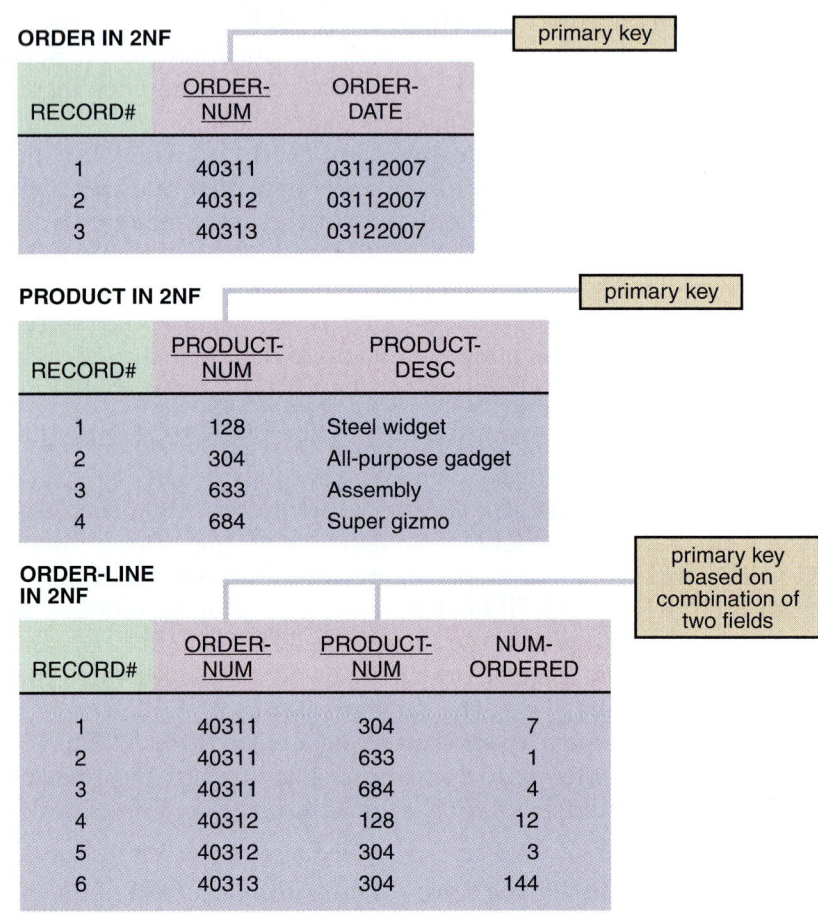

FIGURE 9-24 ORDER, PRODUCT, and ORDER-LINE tables in 2NF. All fields are functionally dependent on the primary key.

Why is it important to move from 1NF to 2NF? Four kinds of problems are found with 1NF designs that do not exist in 2NF:

- Consider the work necessary to change a particular product's description. Suppose 500 current orders exist for product number 304. Changing the product description involves modifying 500 records for product number 304. Updating all 500 records would be cumbersome and expensive.

- 1NF tables can contain inconsistent data. Because someone must enter the product description in each record, nothing prevents product number 304 from having different product descriptions in different records. In fact, if product number 304 appears in a large number of order records, some of the matching product descriptions might be inaccurate or improperly spelled. Even the presence or absence of a hyphen in the orders for *All-purpose gadget* would create consistency problems. If a data entry person must enter a term such as *IO1 Queue Controller* numerous times, it certainly is possible that some inconsistency will result.

- Adding a new product is a problem. Because the primary key must include an order number and a product number, you need values for both fields in order to add a record. What value do you use for the order number when you want to add a new product that has not been ordered by any customer? You could use a dummy order number, and then replace it with a real order number when the product is ordered to solve the problem, but that solution also creates difficulties.

- Deleting a product is a problem. If all the related records are deleted once an order is filled and paid for, what happens if you delete the only record that contains product number 633? The information about that product number and its description is lost.

Has the 2NF design eliminated all potential problems? To change a product description, now you can change just one PRODUCT record. Multiple, inconsistent values for the product description are impossible because the description appears in only one location. To add a new product, you simply create a new PRODUCT record, instead of creating a dummy order record. When you remove the last ORDER-LINE record for a particular product number, you do not lose that product number and its description because the PRODUCT record still exists. The four potential problems are eliminated, and the three 2NF designs are superior to both the original unnormalized table and the 1NF design.

Third Normal Form

A popular rule of thumb is that a design is in 3NF if every nonkey field depends on *the key, the whole key, and nothing but the key.* As you will see, a 3NF design avoids redundancy and data integrity problems that still can exist in 2NF designs.

Consider the following CUSTOMER table design, as shown in Figure 9-25:

CUSTOMER (<u>CUSTOMER-NUM</u>, CUSTOMER-NAME, ADDRESS, SALES-REP-NUM, SALES-REP-NAME)

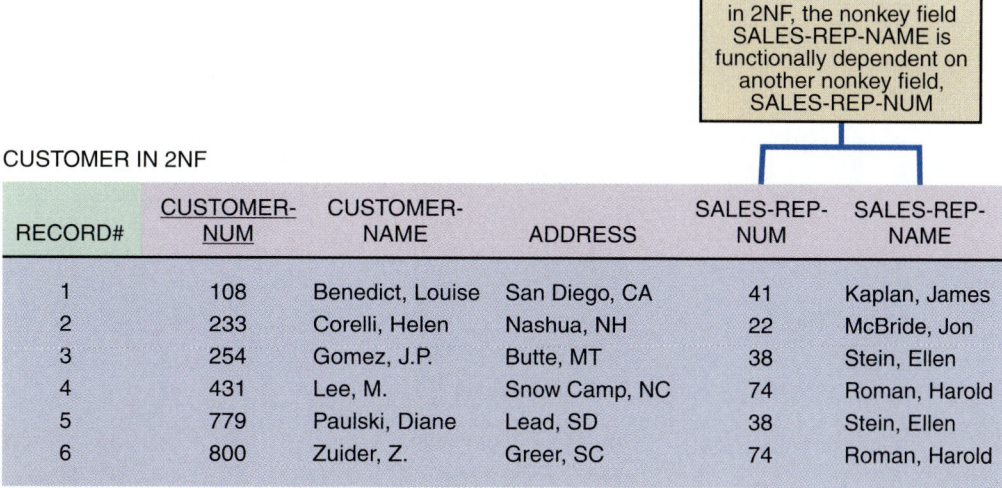

in 2NF, the nonkey field SALES-REP-NAME is functionally dependent on another nonkey field, SALES-REP-NUM

CUSTOMER IN 2NF

RECORD#	CUSTOMER- NUM	CUSTOMER- NAME	ADDRESS	SALES-REP- NUM	SALES-REP- NAME
1	108	Benedict, Louise	San Diego, CA	41	Kaplan, James
2	233	Corelli, Helen	Nashua, NH	22	McBride, Jon
3	254	Gomez, J.P.	Butte, MT	38	Stein, Ellen
4	431	Lee, M.	Snow Camp, NC	74	Roman, Harold
5	779	Paulski, Diane	Lead, SD	38	Stein, Ellen
6	800	Zuider, Z.	Greer, SC	74	Roman, Harold

FIGURE 9-25 2NF design for the CUSTOMER table.

The table is in 1NF because it has no repeating groups. The design also is in 2NF because the primary key is a single field. But the table still has four potential problems similar to the four 1NF problems described earlier. Changing the name of a sales rep still requires changing every record in which that sales rep name appears. Nothing about the design prohibits a particular sales rep from having different names in different records. In addition, because the sales rep name is included in the CUSTOMER table, you must create a dummy CUSTOMER record to add a new sales rep who has not yet been assigned any customers. Finally, if you delete all the records for customers of sales rep number 22, you will lose that sales rep's number and name.

Those potential problems are caused because the design is not in 3NF. A table design is in **third normal form (3NF)** if it is in 2NF and if no nonkey field is dependent on another nonkey field. Remember that a nonkey field is a field that is not a candidate key for the primary key. The CUSTOMER example in Figure 9-25 on the previous page is not in 3NF because one nonkey field, SALES-REP-NAME, depends on another nonkey field, SALES-REP-NUM.

To convert the table to 3NF, you must remove all fields from the 2NF table that depend on another nonkey field and place them in a new table that uses the nonkey field as a primary key. In the CUSTOMER example, the SALES-REP-NAME field depends on another field, SALES-REP-NUM, which is not part of the primary key. Therefore, to reach 3NF, you must remove SALES-REP-NAME and place it into a new table that uses SALES-REP-NUM as the primary key. As shown in Figure 9-26, the third normal form produces two separate tables:

CUSTOMER (<u>CUSTOMER-NUM</u>, CUSTOMER-NAME, ADDRESS, SALES-REP-NUM)

SALES-REP (<u>SALES-REP-NUM</u>, SALES-REP-NAME)

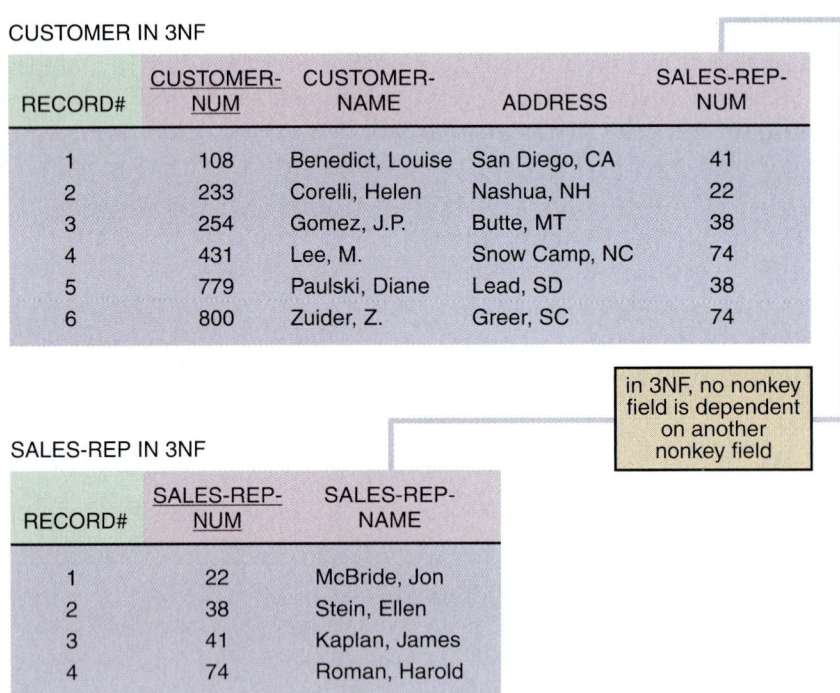

FIGURE 9-26 When the CUSTOMER table is transformed from 2NF to 3NF, the result is two tables: CUSTOMER and SALES-REP.

A Normalization Example

To show the normalization process, consider the familiar situation in Figure 9-27, which depicts several entities in a school advising system: ADVISOR, COURSE, and STUDENT. The relationships among the three entities are shown in the ERD in Figure 9-28. The following sections discuss normalization rules for these three entities.

Before you start the normalization process, you notice that the STUDENT table contains fields that relate to the ADVISOR and COURSE entities, so you decide to begin with the initial design for the STUDENT table, which is shown in Figure 9-29 on the next page. Notice that the table design includes the student number, student name, total credits taken, grade point average (GPA), advisor number, advisor name, and, for every course the student has taken, the course number, course description, number of credits, and grade received.

The STUDENT table in Figure 9-29 is unnormalized, because it has a repeating group. The STUDENT table design can be written as:

STUDENT (<u>STUDENT-NUMBER</u>, STUDENT-NAME, TOTAL-CREDITS, GPA,
    ADVISOR-NUMBER, ADVISOR-NAME,
(<u>COURSE-NUMBER</u>, COURSE-
    DESC, NUM-CREDITS, GRADE))

To convert the STUDENT record to 1NF, you must expand the primary key to include the key of the repeating group, producing:

FIGURE 9-27 A faculty advisor, who represents an entity, can advise many students, each of whom can register for one or many courses.

STUDENT (<u>STUDENT-NUMBER</u>, STUDENT-NAME, TOTAL-CREDITS, GPA,
    ADVISOR-NUMBER, ADVISOR-NAME, <u>COURSE-NUMBER</u>, COURSE-DESC,
    NUM-CREDITS, GRADE)

Figure 9-30 on the next page shows the 1NF version of the sample STUDENT data. Do any of the fields in the 1NF STUDENT record depend on only a portion of the primary key? The student name, total credits, GPA, advisor number, and advisor name all relate only to the student number and have no relationship to the course number. The course description depends on the course number, but not on the student number. Only the GRADE field depends on the entire primary key.

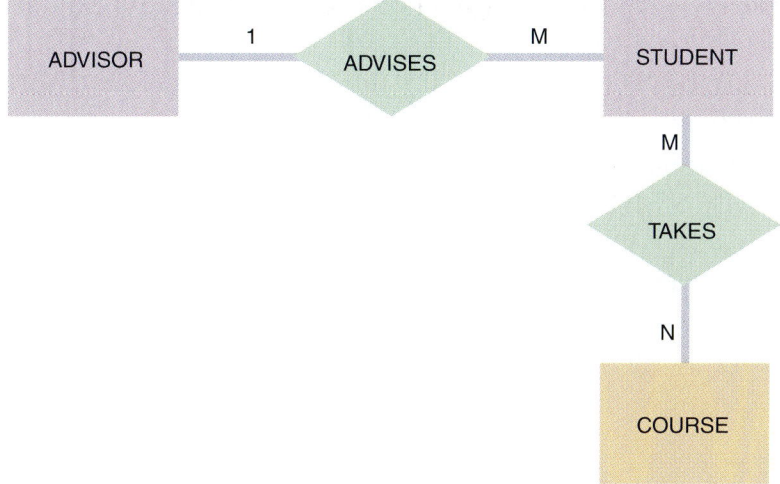

FIGURE 9-28 An initial entity-relationship diagram for ADVISOR, STUDENT, and COURSE.

STUDENT

STUDENT-NUMBER	STUDENT-NAME	TOTAL-CREDITS	GPA	ADVISOR-NUMBER	ADVISOR-NAME	COURSE-NUMBER	COURSE-DESC	NUM-CREDITS	GRADE
1035	Linda	47	3.647	49	Smith	CSC151	Computer Science I	4	B
						MKT212	Marketing Management	3	A
						ENG101	English Composition	3	B
						CHM112	General Chemistry I	4	A
						BUS105	Introduction to Business	2	A
3397	Sam	29	3.000	49	Smith	ENG101	English Composition	3	A
						MKT212	Marketing Management	3	C
						CSC151	Computer Science I	4	B
4070	Kelly	14	2.214	23	Jones	CSC151	Computer Science I	4	B
						CHM112	General Chemistry I	4	C
						ENG101	English Composition	3	C
						BUS105	Introduction to Business	2	C

repeating groups

FIGURE 9-29 The STUDENT table is unnormalized because it contains a repeating group that represents the courses each student has taken.

STUDENT

STUDENT-NUMBER	STUDENT-NAME	TOTAL-CREDITS	GPA	ADVISOR-NUMBER	ADVISOR-NAME	COURSE-NUMBER	COURSE-DESC	NUM-CREDITS	GRADE
1035	Linda	47	3.647	49	Smith	CSC151	Computer Science I	4	B
1035	Linda	47	3.647	49	Smith	MKT212	Marketing Management	3	A
1035	Linda	47	3.647	49	Smith	ENG101	English Composition	3	B
1035	Linda	47	3.647	49	Smith	CHM112	General Chemistry I	4	A
1035	Linda	47	3.647	49	Smith	BUS105	Introduction to Business	2	A
3397	Sam	29	3.000	49	Smith	ENG101	English Composition	3	A
3397	Sam	29	3.000	49	Smith	MKT212	Marketing Management	3	C
3397	Sam	29	3.000	49	Smith	CSC151	Computer Science I	4	B
4070	Kelly	14	2.214	23	Jones	CSC151	Computer Science I	4	B
4070	Kelly	14	2.214	23	Jones	CHM112	General Chemistry I	4	C
4070	Kelly	14	2.214	23	Jones	ENG101	English Composition	3	C
4070	Kelly	14	2.214	23	Jones	BUS105	Introduction to Business	2	C

FIGURE 9-30 The STUDENT table in 1NF. Notice that the primary key has been expanded to include STUDENT-NUMBER and COURSE-NUMBER. Also, the repeating group has been eliminated.

Following the 1NF – 2NF conversion process described earlier, you would create a new table for each field and combination of fields in the primary key, and place the other fields with their appropriate key. The result is:

STUDENT (<u>STUDENT-NUMBER</u>, STUDENT-NAME, TOTAL-CREDITS, GPA, ADVISOR-NUMBER, ADVISOR-NAME)

COURSE (<u>COURSE-NUMBER</u>, COURSE-DESC, NUM-CREDITS)

GRADE (<u>STUDENT-NUMBER</u>, <u>COURSE-NUMBER</u>, GRADE)

You now have converted the original 1NF STUDENT table to three tables, all in 2NF. In each table, every nonkey field depends on the entire primary key.

Figure 9-31 on the next page shows the 2NF STUDENT, COURSE, and GRADE designs and sample data. Are all three tables in 3NF? The COURSE and GRADE are in 3NF. STUDENT is not in 3NF, however, because the ADVISOR-NAME field depends on the ADVISOR-NUMBER field, which is not part of the STUDENT primary key. To convert STUDENT to 3NF, you remove the ADVISOR-NAME field from the STUDENT table and place it into a table with ADVISOR-NUMBER as the primary key.

Figure 9-32 on the next page shows the 3NF versions of the sample data for STUDENT, ADVISOR, COURSE, and GRADE. The final 3NF design is:

STUDENT (<u>STUDENT-NUMBER</u>, STUDENT NAME, TOTAL-CREDITS, GPA, ADVISOR-NUMBER)

ADVISOR (<u>ADVISOR-NUMBER</u>, ADVISOR-NAME)

COURSE (<u>COURSE-NUMBER</u>, COURSE-DESC, NUM-CREDITS)

GRADE (<u>STUDENT-NUMBER</u>, <u>COURSE-NUMBER</u>, GRADE)

Figure 9-33 on page 417 shows the complete ERD after normalization. Now there are four entities: STUDENT, ADVISOR, COURSE, and GRADE, which is an associative entity. If you go back to Figure 9-28 on page 413, which was drawn before you identified GRADE as an entity, you can see that the M:N relationship between STUDENT and COURSE has been converted into two 1:M relationships: one relationship between STUDENT and GRADE and the other relationship between COURSE and GRADE.

To create 3NF designs, you must understand the nature of first, second, and third normal forms. In your work as a systems analyst, you will encounter designs that are much more complex than the examples in this chapter. You also should know that normal forms beyond 3NF exist, but they rarely are used in business-oriented systems.

CASE IN POINT 9.2: CYBERTOYS

You handle administrative support for CyberToys, a small chain that sells computer hardware and software and specializes in personal service. The company has four stores located at malls and is planning more. Each store has a manager, a technician, and between one and four sales reps.

Bruce and Marcia Berns, the owners, want to create a personnel records database, and they asked you to review a table that Marcia designed. She suggested fields for store number, location, store telephone, manager name, and manager home telephone. She also wants fields for technician name and technician home telephone and fields for up to four sales rep names and sales rep home telephones.

Draw Marcia's suggested design and analyze it using the normalization concepts you learned in the chapter. What do you think of Marcia's design and why? What would you propose?

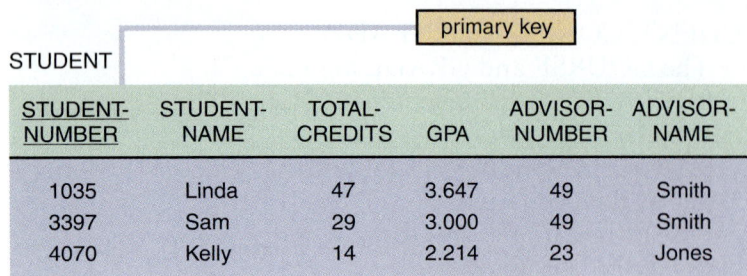

STUDENT

STUDENT-NUMBER	STUDENT-NAME	TOTAL-CREDITS	GPA	ADVISOR-NUMBER	ADVISOR-NAME
1035	Linda	47	3.647	49	Smith
3397	Sam	29	3.000	49	Smith
4070	Kelly	14	2.214	23	Jones

primary key

COURSE

COURSE-NUMBER	COURSE-DESC	NUM-CREDITS
BUS105	Introduction to Business	2
CHM112	General Chemistry I	4
CSC151	Computer Science I	4
ENG101	English Composition	3
MKT212	Marketing Management	3

primary key based on combination of two fields

GRADE

STUDENT-NUMBER	COURSE-NUMBER	GRADE
1035	CSC151	B
1035	MKT212	A
1035	ENG101	B
1035	CHM112	A
1035	BUS105	A
3397	ENG101	A
3397	MKT212	C
3397	CSC151	B
4070	CSC151	B
4070	CHM112	C
4070	ENG101	C
4070	BUS105	C

FIGURE 9-31 STUDENT, COURSE, and GRADE tables in 2NF. Notice that all fields are functionally dependent on the entire primary key of their respective tables.

STUDENT

STUDENT-NUMBER	STUDENT-NAME	TOTAL-CREDITS	GPA	ADVISOR-NUMBER
1035	Linda	47	3.647	49
3397	Sam	29	3.000	49
4070	Kelly	14	2.214	23

in 3NF, no nonkey field is dependent on another nonkey field

ADVISOR

ADVISOR-NUMBER	ADVISOR-NAME
23	Jones
49	Smith

COURSE

COURSE-NUMBER	COURSE-DESC	NUM-CREDITS
BUS105	Introduction to Business	2
CHM112	General Chemistry I	4
CSC151	Computer Science I	4
ENG101	English Composition	3
MKT212	Marketing Management	3

GRADE

STUDENT-NUMBER	COURSE-NUMBER	GRADE
1035	CSC151	B
1035	MKT212	A
1035	ENG101	B
1035	CHM112	A
1035	BUS105	A
3397	ENG101	A
3397	MKT212	C
3397	CSC151	B
4070	CSC151	B
4070	CHM112	C
4070	ENG101	C
4070	BUS105	C

FIGURE 9-32 STUDENT, ADVISOR, COURSE, and GRADE tables in 3NF. When the STUDENT table is transformed from 2NF to 3NF, the result is two tables: STUDENT and ADVISOR.

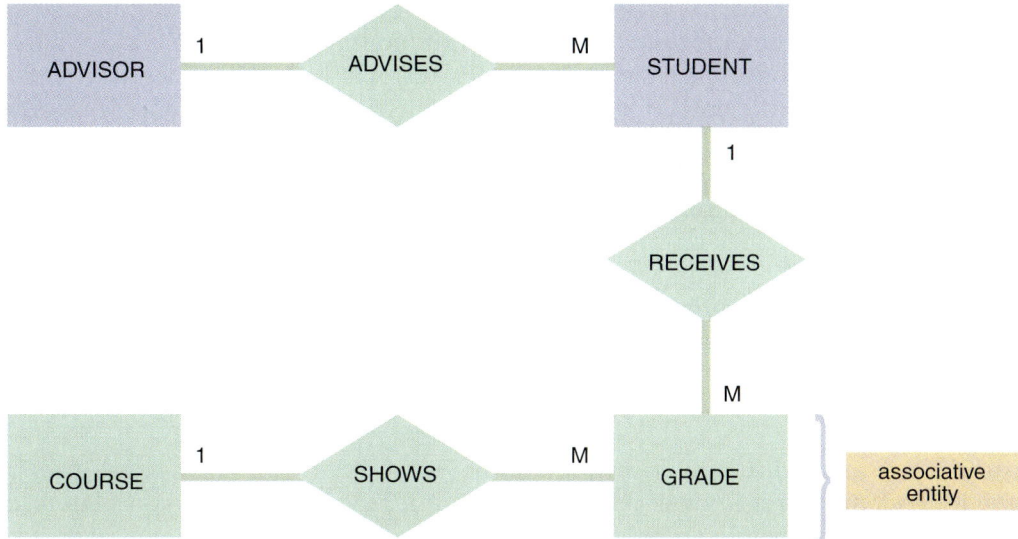

FIGURE 9-33 The entity-relationship diagram for STUDENT, ADVISOR, and COURSE after normalization. The GRADE entity was identified during the normalization process. GRADE is an associative entity that links the STUDENT and COURSE tables.

USING CODES DURING DATA DESIGN

A **code** is a set of letters or numbers that represents a data item. Codes can be used to simplify output, input, and data formats. During the data design process, you review existing codes and develop new ones that will be used to store and access data efficiently.

Overview of Codes

Because codes often are used to represent data, you encounter them constantly in your everyday life. Student numbers, for example, are unique codes to identify students in a school registration system. Three students with the name John Turner might be enrolled at your school, but only one is student number 268960.

Your ZIP code is another common example. A ZIP code contains multiple items of information compressed into nine digits. The first digit identifies one of ten geographical areas of the United States. The combination of the next three digits identifies a major city or major distribution point. The fifth digit identifies an individual post office, an area within a city, or a specific delivery unit. The last four digits identify a post office box or a specific street address.

For example, consider the ZIP code 27906-2624 shown in Figure 9-34. The first digit, 2, indicates a broad geographical area in the eastern United States. The digits, 790, indicate Elizabeth City, North Carolina. The fifth digit, 6, identifies the post office that services the College of the Albemarle. The last four digits, 2624, identify the post office box for the college.

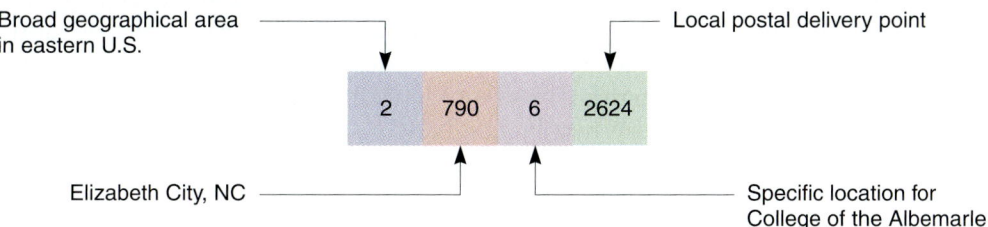

FIGURE 9-34 A ZIP code is an example of a significant digit code that uses subgroups to provide information.

As you can imagine, codes serve many useful purposes. Because codes often are shorter than the data they represent, they save storage space and costs, reduce data transmission time, and decrease data entry time. For example, ZIP codes are used to classify and sort mail efficiently. Codes also can be used to reveal or conceal information. The last two digits of a seven-digit part number, for example, can represent the supplier number. The coded wholesale price on a retail price tag is known to salespeople but generally not to customers.

Finally, codes can reduce data input errors in situations when the coded data is easier to remember and enter than the original source data, when only certain valid codes are allowed, and when something within the code itself can provide immediate verification that the entry is correct.

Types of Codes

Companies use many different coding methods. Because information system users must work with coded data, the codes should be easy to learn and apply. If you plan to create new codes or change existing ones, you first should obtain comments and feedback from users. The following section describes eight common coding methods.

1. **Sequence codes** are numbers or letters assigned in a specific order. Sequence codes contain no additional information other than an indication of order of entry into the system. For example, a human resource system issues consecutive employee numbers to identify employees. Because the codes are assigned in the order in which employees are hired, you can use the code to see that employee number 584 was hired after employee number 433. The code, however, does not indicate the starting date of either person's employment.

2. **Block sequence codes** use blocks of numbers for different classifications. College course numbers usually are assigned using a block sequence code. 100-level courses, such as Chemistry 110 and Mathematics 125, are freshman-level courses, whereas course numbers in the 200s indicate sophomore-level courses. Within a particular block, the sequence of numbers can have some additional meaning, such as when English 151 is the prerequisite for English 152.

3. **Alphabetic codes** use alphabet letters to distinguish one item from another based on a category, an abbreviation, or an easy-to-remember value, called a mnemonic code. Many classification codes fit more than one of the following definitions:

   a. **Category codes** identify a group of related items. For example, a local department store uses a two-character category code to identify the department in which a product is sold: GN for gardening supplies, HW for hardware, and EL for electronics.

   b. **Abbreviation codes** are alphabetic abbreviations. For example, standard state codes include NY for New York, ME for Maine, and MN for Minnesota. Some abbreviation codes are called **mnemonic codes** because they use a specific combination of letters that are easy to remember. Many three-character airport codes such as those pictured in Figure 9-35 are mnemonic codes: BOS represents Boston, SEA represents Seattle, and ANC represents Anchorage. Some airport codes are not mnemonic, such as ORD, which designates Chicago O'Hare International Airport, or HPN, which identifies the White Plains, New York Airport.

4. **Significant digit codes** distinguish items by using a series of subgroups of digits. ZIP codes, for example, are significant digit codes. Other such codes include inventory location codes that consist of a two-digit warehouse code, followed by a one-digit floor number code, a two-digit section code, a one-digit aisle number, and a two-digit bin number code. Figure 9-36 illustrates the inventory location code 11205327. What looks like a large eight-digit number is actually five separate numbers, each of which has significance.

5. **Derivation codes** combine data from different item attributes, or characteristics, to build the code. Most magazine subscription codes are derivation codes. One popular magazine's subscriber code consists of the subscriber's five-digit ZIP code, followed by the first, third, and fourth letters of the subscriber's last name, the last two digits of the subscriber's house number, and the first, third, and fourth letters of the subscriber's street name. The magazine's subscriber code for one particular subscriber is shown in Figure 9-37.

6. **Cipher codes** use a keyword to encode a number. A retail store, for example, might use a 10-letter word, such as CAMPGROUND, to code wholesale prices, where the letter C represents 1, A represents 2, and so on. Thus, the code, GRAND, indicates that the store paid $562.90 for the item.

7. **Action codes** indicate what action is to be taken with an associated item. For example, a student records program might prompt a user to enter or click an action code such as *D* (to display a record), *A* (to add a record), and *X* (to exit the program).

FIGURE 9-35 Airline baggage tags include three-letter codes that identify airports, and machine-readable bar codes that contain information about the passenger, the flight number, and other pertinent information.

Section code · Warehouse location code · Bin number

| 11 | 2 | 05 | 3 | 27 |

Floor number · Aisle number

FIGURE 9-36 Sample of a code that uses significant digits to pinpoint the location of an inventory item.

Developing a Code

Devising a code with too many features makes it difficult to remember, decipher, and verify. Keep the following suggestions in mind when developing a code:

1. Keep codes concise. Do not create codes that are longer than necessary. For example, if you need a code to identify each of 250 customers, you will not need a six-digit code.

2. Allow for expansion. A coding scheme must allow for reasonable growth in the number of assigned codes. If the company currently has eight warehouses, you should not use a one-digit code for the warehouse number. If three more warehouses are

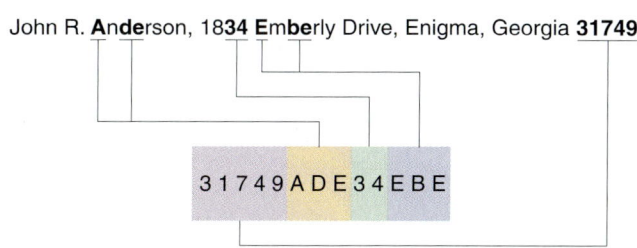

John R. **Ande**rson, 1**834 Em**berly Drive, Enigma, Georgia **31749**

3 1 7 4 9 A D E 3 4 E B E

FIGURE 9-37 A magazine subscriber code is derived from various parts of the name and address.

added, the code must be increased to two digits or changed to a character code in order to identify each location. The rule also applies to using a single letter as a character code; you might need more than 26 codes in the future.

3. Keep codes stable. Changes in codes can cause consistency problems and require data updates. During the changeover period, you will have to change all the stored occurrences of a particular code and all documents containing the old code, as users switch to the new code. Usually, both the old and new codes are used for an interim period, and special procedures are required to handle the two codes. For example, when area codes change, you can use either area code for a certain time period.

4. Make codes unique. Codes used for identification purposes must be unique to have meaning. If the code HW can indicate hardware or houseware, the code is not very useful.

5. Use sortable codes. If products with three-digit codes in the 100s or the 300s are of one type, while products with codes in the 200s are a different type, a simple sort will not group all the products of one type together. In addition, be careful that single-digit character codes will sort properly with double-digit codes — in some cases you must add a leading zero (01, 02, 03, and so on) to ensure that codes sort correctly.

6. Avoid confusing codes. Do not code some part numbers with two letters, a hyphen, and one digit, and others with one letter, a hyphen, and two digits. Avoid allowing both letters and numbers to occupy the same positions within a code because some of those are easily confused. It is easy to confuse the number zero (0) and the uppercase letter O, or the number one (1) with the lowercase letter L (l) or uppercase letter I. For example, the five-character code 5Z081 easily can be misread as 5ZO8I, or 52081, or even totally incorrectly as S2OBI.

7. Make codes meaningful. Codes must be easy to remember, useful for users, convenient to use, and easy to encode and interpret. Using SW as a code for the southwest sales region, for example, has far more meaning than the code 14. Using ENG as the code for the English department is easier to interpret and remember than either XVA or 132.

8. Use a code for a single purpose. Do not use a single code to classify two or more unrelated attributes. For example, if you use a single code to identify the combination of an employee's department *and* the employee's insurance plan type, users will have difficulty identifying all the subscribers of a particular plan, or all the workers in a particular department, or both. A separate code for each separate characteristic makes much more sense.

9. Keep codes consistent. For example, if the payroll system already is using two-digit codes for departments, do not create a new, different coding scheme for the personnel system. If the two systems already are using different coding schemes, you should try to convince the users to adopt a consistent coding scheme.

CASE IN POINT 9.3: DotCom Tools

DotCom Tools operates a small business that specializes in hard-to-find woodworking tools. The firm advertises in various woodworking magazines, and currently accepts mail and telephone orders. DotCom is planning a Web site that will be the firm's primary sales channel. The site will feature an online catalog, powerful search capabilities, and links to woodworking information and resources.

DotCom has asked you, an IT consultant, whether a set of codes would be advantageous and if so, what codes you would suggest. Provide at least two choices for a customer code and at least two choices for a product code. Be sure to describe your choices and provide some specific examples. Also include an explanation of why you selected these particular codes and what advantages they might offer.

STEPS IN DATABASE DESIGN

After normalizing your table designs and considering the use of codes, you are ready to create the database. The following steps can be used to create database and file designs. To highlight the steps, consider another familiar situation shown in Figure 9-38, which involves an information system for a video rental store.

1. *Create an initial ERD.* Start by reviewing DFDs and class diagrams to identify system entities. In addition, consider any data stores shown on DFDs to determine whether they might represent entities. Next, create a draft of the ERD. Carefully analyze each relationship to determine if it is 1:1, 1:M, or M:N. Figure 9-39 shows the initial ERD for the entities MEMBER and VIDEO in the video rental system.

2. *Assign all data elements to entities.* Verify that every data element in the data dictionary is associated logically with an entity. For the video rental system, the initial table designs with all data elements are listed under the ERD in Figure 9-39.

3. *Create 3NF designs for all tables, taking care to identify all primary, secondary, and foreign keys.* Generate the final ERD that will include new entities identified during normalization. Figure 9-40 on the next page shows the final ERD and the normalized designs. Notice that a new associative entity, RENTAL, was identified during normalization and the M:N relationship became two 1:M relationships.

FIGURE 9-38 A video rental involves several entities, including members and videos. In the store's information system, each entity is represented by a table that contains various fields.

```
MEMBER (MEMBER-NUMBER, NAME, ADDRESS, CITY, STATE, ZIP, HOME-TELEPHONE,
   WORK-TELEPHONE, CREDIT-CARD-CODE, CREDIT-CARD-NUMBER, (VIDEO-ID, TITLE,
   DATE-RENTED, DATE-RETURNED))
VIDEO (VIDEO-ID, TITLE)
```

FIGURE 9-39 The initial entity-relationship diagram and table designs for the video rental system. Notice that the MEMBER table contains a repeating group.

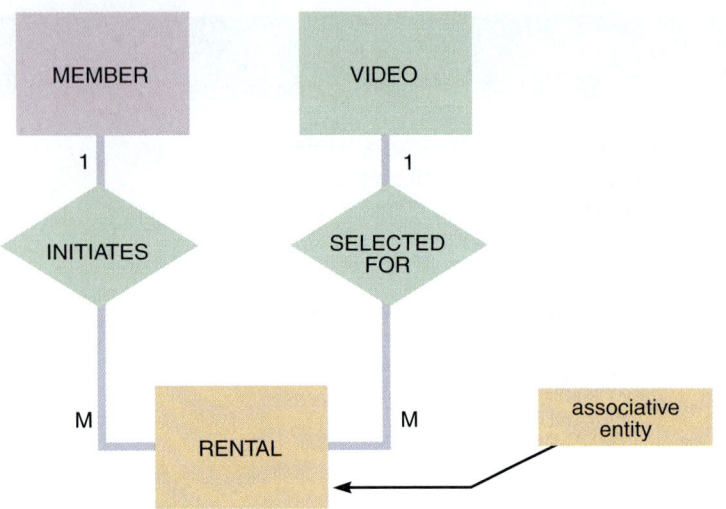

MEMBER (<u>MEMBER-NUMBER</u>, NAME, ADDRESS, CITY, STATE, ZIP, HOME-TELEPHONE,
   WORK-TELEPHONE, CREDIT-CARD-CODE, CREDIT-CARD-NUMBER)
VIDEO (<u>VIDEO-ID</u>, TITLE)
RENTAL (<u>MEMBER-NUMBER</u>, <u>VIDEO-ID</u>, DATE-RENTED, DATE-RETURNED)

FIGURE 9-40 The final entity-relationship diagram and normalized table designs for the video rental system. RENTAL has been identified as an associative entity that links the MEMBER and VIDEO tables.

4. *Verify all data dictionary entries.* Make sure that the data dictionary entries for all data stores, records, and data elements are documented completely and correctly. Also be sure that all codes that were developed or identified during the data design process are documented in the data dictionary.

After creating your final ERD and normalized table designs, you can transform them into a database. Your next step will be to consider the design and characteristics of various database models.

DATABASE MODELS

The two most popular database models are relational and object-oriented. Other models exist, but they are less common and generally are found on older, mainframe-based systems.

Relational databases can run on many platforms, including personal computers, and are well suited to client/server computing because they are so powerful and flexible. Object-oriented databases are modular, cost-effective, and a logical extension of the object-oriented analysis process.

Relational Databases

Earlier in this chapter, you learned that a relational database uses common fields, which are attributes that appear in more than one table, to establish relationships between the tables and form an overall data structure. This type of design, called a **relational model**, was introduced during the 1970s and became popular because it was flexible and powerful. Three decades later, the relational design still is the predominant model.

Figure 9-41 represents a relational database design for a company that performs on-site computer service. Figure 9-42 shows the tables, primary keys, and common fields in the relational database.

The design in Figure 9-42 uses many of the concepts described earlier in this chapter and demonstrates the power of a relational model. Separate tables exist for CUSTOMER, TECHNICIAN, SERVICE-CALL, SERVICE-PARTS-DETAIL, SERVICE-LABOR-DETAIL, PARTS, and LABOR-CODE tables. Notice that all the tables use a single field as a primary key, except the SERVICE-LABOR-DETAIL and

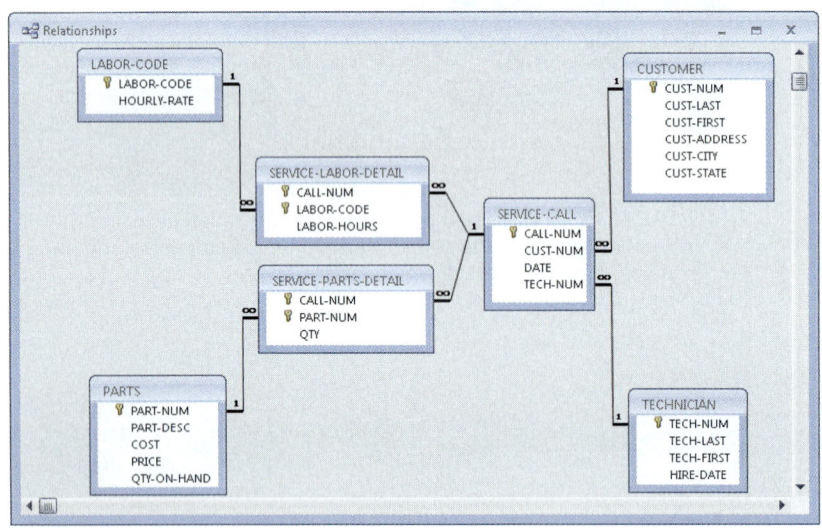

FIGURE 9-41 A relational database design for a computer service company uses common fields to link the tables and form an overall data structure. Notice the one-to-many notation symbols, and the primary keys, which are shown in bold.

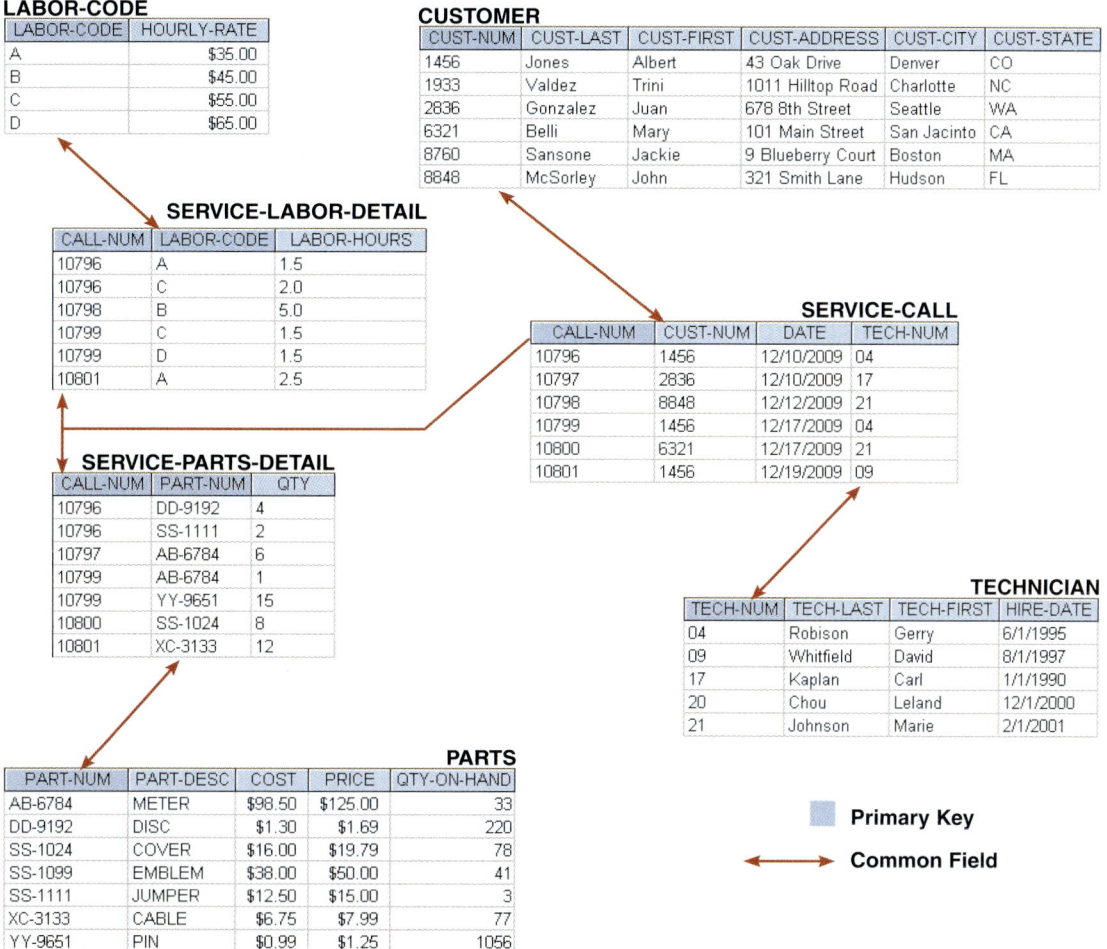

FIGURE 9-42 The tables, primary keys, and common fields for the database shown in Figure 9-41. The design is in 3NF. All nonkey fields are functionally dependent on a primary key, the whole key, and nothing but the key.

SERVICE-PARTS-DETAIL tables, in which the primary key requires a combination of two fields to identify each record uniquely.

Because all the tables are linked, a user can request data that meets specific conditions. To visualize how a relational DBMS handles complex queries, refer to Figure 9-42 and consider the following three examples. Suppose a user wants to see:

1. All customers who received service after 12/15/2009

2. All service calls on which technician Marie Johnson put in more than four hours of labor

3. The number and description of all parts sold to customers who live in Washington

In the first example, the DBMS looks in the SERVICE-CALL table and finds three records with a date later than 12/15/2009. Then, using CUSTOMER-NUM, which is a foreign key in the SERVICE-CALL table, the DBMS seeks matching primary key values in the CUSTOMER table to identify customers Albert Jones and Mary Belli.

In the second example, the DBMS locates Marie Johnson in the TECHNICIAN table and extracts her technician number, which is a primary key. Using her TECH-NUM value of 21, the DBMS identifies Marie's two service calls by seeking matching values of

21 in the TECH-NUM field in the SERVICE-CALL table. Next, the DBMS uses the CALL-NUM values of 10798 and 10800 to check for matching records with more than four hours in the SERVICE-LABOR-DETAIL table. In the example, only service call 10798 meets all the requirements.

In the third example, the DBMS first locates all records with the value *WA* in the CUST-STATE field in the CUSTOMER table. The process identifies customer 2836, Juan Gonzalez. Using the value 2836, which is a primary key value in the CUSTOMER table, the DBMS seeks matching values in the CUST-NUM field in the SERVICE-CALL table. After locating service call 10797, the DBMS seeks matching values in the PART-NUM field of the SERVICE-PARTS-DETAIL table. Then using the value AB-6784, the DBMS seeks the matching primary key value in the PARTS table and obtains the description from the PART-DESC field. The results indicate that six meters with part number AB-6784 were sold to Washington customers. Because it uses a relational model, the DBMS in Figure 9-42 on the previous page is very powerful and flexible.

New entities and attributes can be added at any time without restructuring the entire database. If the company wants to add new parts, new service codes, new customers, or new technicians, it can do so without affecting existing data or relationships. Updating information also is simple in a relational design. For example, suppose customer Trini Valdez changes her address from 1011 Hilltop Road to 23 Down Lane. Even though she is listed on 18 separate service calls, only one change is needed in the CUSTOMER table. Because a common field links the CUSTOMER and SERVICE-CALL tables, it is not necessary to make a change in 18 individual records.

Object-Oriented Databases

Ten years ago, virtually all information systems were designed and implemented as relational databases or file processing systems. As object-oriented analysis became popular during the late 1990s, some analysts began to describe systems in terms of objects. Chapter 6 describes object-oriented analysis and design concepts that will add to your overall understanding of data design.

Today, many systems developers are using **object-oriented database (OODB)** design as a natural extension of the object-oriented analysis process. Some IT professionals believe that OODBs will have a major impact and might eventually replace the relational approach. Most object-oriented developers support the design standards established by the **Object Management Group (OMG™)**, as shown in Figure 9-43. OMG™ is a nonprofit industry group that produces and maintains standards and specifications.

FIGURE 9-43 Many object-oriented developers support OODB design standards promulgated by the Object Management Group (OMG).

Each object in an OODB has a unique **object identifier**, which is similar to a primary key in a relational database. The identifier allows the object to interact with other objects and form relationships, as shown in Figure 9-44. Programmers use object-oriented languages, such as Java or Python, to describe and implement object-to-object relationships that resemble the relationships found in a traditional ERD.

DATA STORAGE AND ACCESS

Data storage and access involve strategic business tools, such as data warehousing and data mining software, as well as logical and physical storage issues, selection of data storage formats, and special considerations regarding storage of date fields.

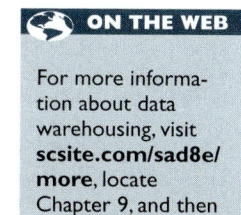

FIGURE 9-44 Objects have object identifiers that enable them to communicate and interact.

Strategic Tools for Data Storage and Access

Companies use data warehousing and data mining as strategic tools to help manage the huge quantities of data they need for business operations and decisions. A large number of software vendors compete for business in this fast-growing IT sector.

DATA WAREHOUSING Large firms maintain many databases, which might or might not be linked together into an overall structure. To provide rapid access to this information, companies use software packages that organize and store data in special configurations called data warehouses. A **data warehouse** is an integrated collection of data that can include seemingly unrelated information, no matter where it is stored in the company. Because it can link various information systems and databases, a data warehouse provides an enterprise-wide view to support management analysis and decision making.

A data warehouse allows users to specify certain **dimensions**, or characteristics. By selecting values for each characteristic, a user can obtain multidimensional information from the stored data. For example, in a typical company, most data is generated by transaction-based systems, such as order processing systems, inventory systems, and payroll systems. If a user wants to identify the customer on sales order 34071, he or she can retrieve the data easily from the order processing system by entering an order number.

ON THE WEB

For more information about data warehousing, visit **scsite.com/sad8e/more**, locate Chapter 9, and then click the Data Warehousing link.

On the other hand, suppose that a user wants to see May 2009 sales results for Sally Brown, the sales rep assigned to Jo-Mar Industries. The data is stored in two different systems: the sales information system and the human resources information system, as shown in Figure 9-45. Without a data warehouse, it would be difficult for a user to extract data that spans several information systems and time frames. Rather than accessing separate systems, a data warehouse stores transaction data in a format that allows users to retrieve and analyze the data easily.

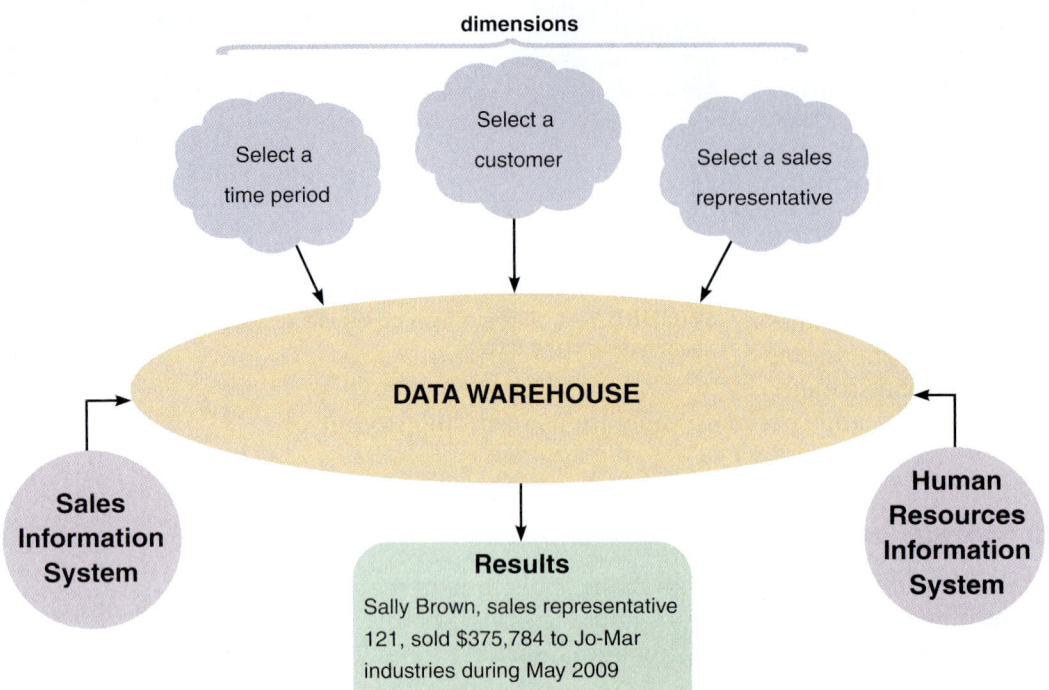

FIGURE 9-45 A data warehouse stores data from several systems. By selecting data dimensions, a user can retrieve specific information without having to know how or where the data is stored.

While a data warehouse typically spans the entire enterprise, many firms prefer to use a **data mart**, which is designed to serve the needs of a specific department, such as sales, marketing, or finance. Each data mart includes only the data that users in that department require to perform their jobs. There are pros and cons to both approaches, and the best solution usually depends on the specific situation.

One important advantage of a data warehouse is that it provides easy, flexible access for users to obtain the information they need. For example, as shown in Figure 9-46, the University of Southern California (USC) maintains a Web-based data warehouse that allows users to access raw data, create custom formats, select fields to display, and determine how the data should be sorted.

DATA MINING Data mining software looks for meaningful patterns and relationships among data. For example, data mining software could help a consumer products firm identify potential customers based on their prior purchases. Many firms offer data mining services and research, as shown in Figure 9-47.

The enormous growth in e-commerce has focused attention on data mining as a tool to analyze Web visitor behavior and traffic trends. In an article in *New Architect*, a Web-based magazine, Dan R. Greening noted that Web hosts typically possess a lot of information about visitors, but most of it is of little value. His article mentions that smart marketers and business analysts are using data mining techniques, which he describes as "machine learning algorithms that find buried patterns in databases, and report or act on those findings."

The author stresses privacy concerns, and suggests that Web hosts use a clear privacy statement for visitors. He concludes by saying that "The great advantage of Web marketing is that you can measure visitor interactions more effectively than in brick-and-mortar stores

ON THE WEB

For more information about data mining, visit **scsite.com/ sad8e/more**, locate Chapter 9, and then click the data mining link.

or direct mail. Data mining works best when you have clear, measurable goals." Some of the goals he suggests are:

- Increase average pages viewed per session

- Increase number of referred customers

- Reduce **clicks to close**, which means average page views to accomplish a purchase or obtain desired information

- Increase checkouts per visit

- Increase average profit per checkout

This type of data gathering is sometimes called **clickstream storage**. Armed with this information, a skillful Web designer could build a profile of typical new customers, returning customers, and customers who browse but do not buy. Although this information would be very valuable to the retailer, clickstream storage could raise serious legal and privacy issues if an unscrupulous firm sought to link a customer's Web behavior to a specific name or e-mail address, and then sell or otherwise misuse the information.

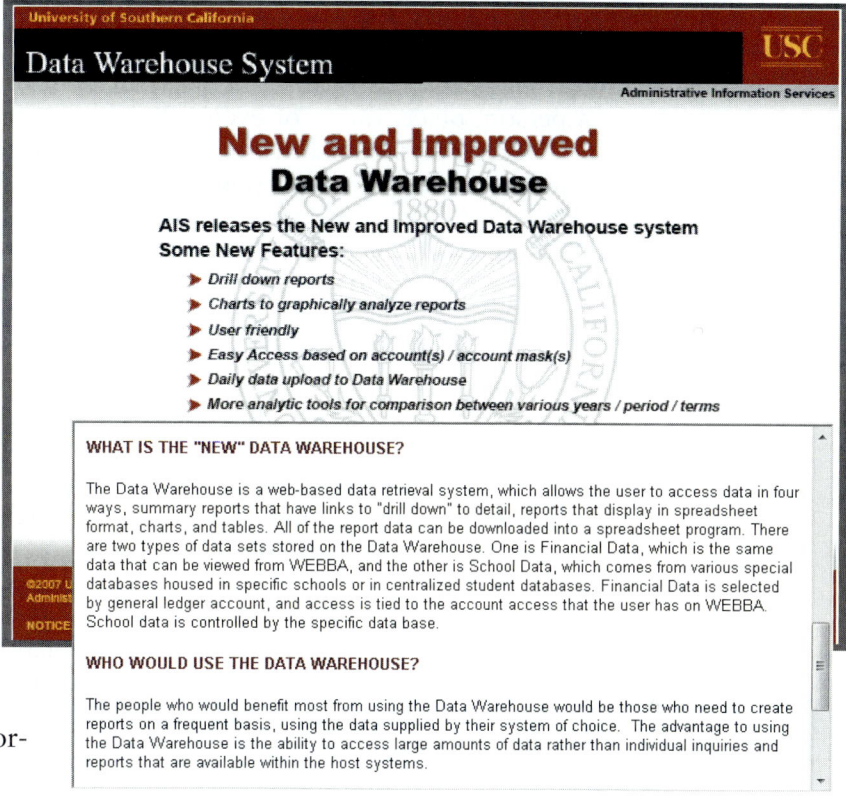

FIGURE 9-46 USC's Web-based data warehouse allows users to access and manipulate raw data to suit their needs.

Because it can detect patterns and trends in large amounts of data, data mining is a valuable tool for managers. The popular Wikipedia site contains an interesting example of data mining. According to the Wikipedia article, a hypothetical chain of supermarkets performed a detailed analysis of purchases, and found that beer and diapers were often purchased together. Without attempting to explain this correlation, the obvious tactic for a retailer would be to display these items in the same area of the store. Wikipedia states that this data mining technique is often called **market basket analysis**.

Logical and Physical Storage

It is important to understand the difference between logical storage and physical storage. **Logical storage** refers to data that a user can view, understand, and access, regardless of how or where that information actually is organized or stored. In contrast, **physical storage** is strictly hardware-related, because it involves the process of reading and writing binary data to physical media such as a hard drive, CD-ROM, or network-based storage device. For example, portions of a document might be stored in different physical locations on a hard drive, but the user sees the document as a single logical entity on the computer screen.

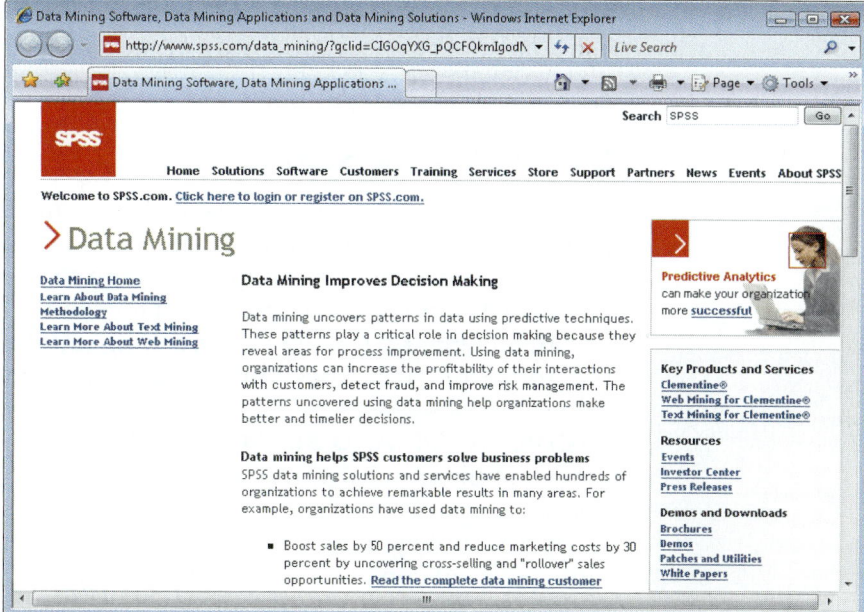

FIGURE 9-47 SPSS is an example of a firm that offers data mining software and solutions.

LOGICAL STORAGE Logical storage consists of alphabetic and numeric **characters**, such as the letter *A* or the number *9*. As you learned earlier in this chapter, a set of related characters forms a field, which describes a single characteristic, or attribute, of a person, place, thing, or event. A field also is called a **data element** or a **data item**.

When designing fields, you should provide space for the largest values that can be anticipated, without allocating unnecessarily large storage capacities that will not be used. For example, suppose you are designing a customer order entry system for a firm with 800 customers. It would be a mistake to limit the customer number field to three, or even four characters. Instead, you might consider a five-character field with leading zeros that could store customer numbers from *00001* to *99999*.

You also might consider a mix of alphabetic and numeric characters, which many people find easier to view and use. Alphabetic characters expand the storage capacity, because there are 26 possible values for each character position. Most airlines now use six alphabetic characters as a record locator, which has over 300 million possible values.

A **logical record** is a set of field values that describes a single person, place, thing, or event. For example, a logical customer record contains specific field values for a single customer, including the customer number, name, address, telephone number, credit limit, and so on. Application programs see a logical record as a group of related fields, regardless of how or where the data is stored physically.

The term *record* usually refers to a logical record. Whenever an application program issues a read or write command, the operating system supplies one logical record to the program or accepts one logical record from the program. The physical data might be stored on one or more servers, in the same building or thousands of miles away, but all the application program sees is the logical record — the physical storage location is irrelevant.

PHYSICAL STORAGE Physical storage involves a **physical record**, or **block**, which is the smallest data unit that can be handled by the operating system. The system reads or writes one physical record at a time. When the system *reads* a physical record, it loads the data from storage into a **buffer**, which is a segment of computer memory.

The physical storage location can be local, remote, or Web-based. Similarly, when the system *writes* a physical record, all data in the memory buffer is saved physically to a storage location. A physical record can contain more than one logical record, depending on the **blocking factor**. For example, a blocking factor of two means that each physical record will consist of two logical records. Some database programs, such as Microsoft Access, automatically write a physical record each time that a logical record is created or updated.

Data Coding and Storage

Computers represent data as **bits**, or **binary digits**, that have only two possible values: 1 (which indicates an electrical signal) and 0 (which indicates the absence of a signal). A computer understands a group of bits as a digital code that can be transmitted, received, and stored. Computers use various data coding and storage schemes, such as EBCDIC, ASCII, and binary. A more recent coding standard called Unicode also is popular. Also, the storage of dates raises some design issues that must be considered.

EBCDIC, ASCII, AND BINARY EBCDIC (pronounced EB-see-dik), which stands for Extended Binary Coded Decimal Interchange Code, is a coding method used on mainframe computers and high-capacity servers. **ASCII** (pronounced ASK-ee), which stands for American Standard Code for Information Interchange, is a coding method used on most personal computers. EBCDIC and ASCII both require eight bits, or one **byte**, for each character. For example, the name Ann requires three bytes of storage, the number *12,345* requires five bytes of storage, and the number *1,234,567,890* requires ten bytes of storage.

Compared with character-based formats, a **binary storage format** offers a more efficient storage method because it represents numbers as actual binary values, rather than as coded numeric digits. For example, an **integer format** uses only 16 bits, or two bytes, to represent the number *12,345* in binary form. A **long integer format** uses 32 bits, or four bytes, to represent the number *1,234,567,890* in binary form.

UNICODE Unicode is a more recent coding standard that uses two bytes per character, rather than one. This expanded scheme enables Unicode to represent more than 65,000 unique, multilingual characters. Why is this important? Consider the challenge of running a multinational information system, or developing a program that will be sold in Asia, Europe, and North America. Because it supports virtually all languages, Unicode has become a global standard.

Traditionally, domestic software firms developed a product in English, then translated the program into one or more languages. This process was expensive, slow, and error-prone. In contrast, Unicode creates translatable content right from the start. Today, most popular operating systems support Unicode, and the Unicode Consortium maintains standards and support, as shown in Figure 9-48. Although Unicode has many advantages, it also has some disadvantages. In fact, because of its size and certain typographical issues, Microsoft recommends that you use its Arial Unicode font " ... only when you can't use multiple fonts tuned for different writing systems."

STORING DATES What is the best way to store dates? The answer depends on how the dates will be displayed and whether they will be used in calculations.

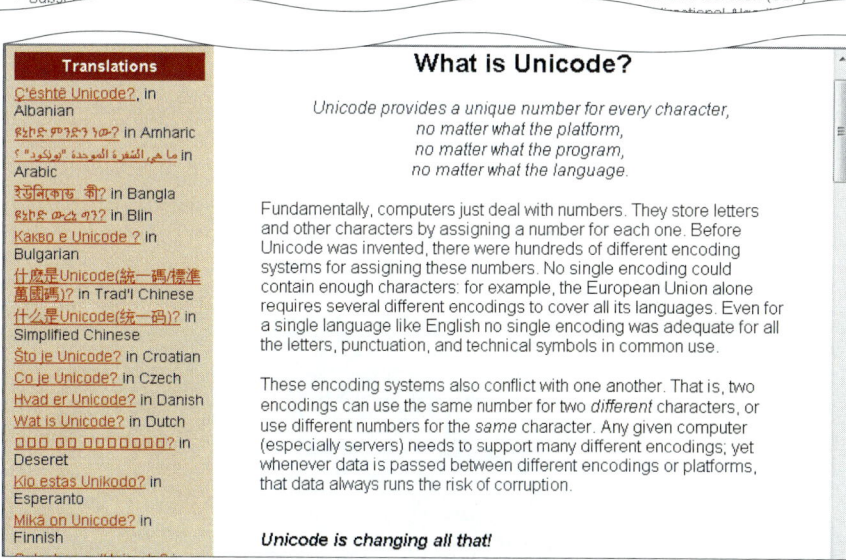

FIGURE 9-48 Unicode is an international coding format that represents characters as integers, using 16 bits per character. The Unicode Consortium maintains standards and support for Unicode.

At the beginning of the 21st century, many firms that used only two digits to represent the year were faced with a major problem called the **Y2K issue**. Based on that experience, most date formats now are based on the model established by the **International Organization for Standardization (ISO)**, which requires a format of four digits for the year, two for the month, and two for the day (YYYYMMDD). A date stored in that format can be sorted easily and used in comparisons. If a date in ISO form is larger than another date in the same form, then the first date is later. For example, 20100927 (September 27, 2010) is later than 20090713 (August 13, 2009).

But, what if dates must be used in calculations? For example, if a manufacturing order placed on June 23 takes three weeks to complete, when will the order be ready? If a payment due on August 13 is not paid until April 27 of the following year, exactly how late is the payment and how much interest is owed? In these situations, it is easier to use absolute dates.

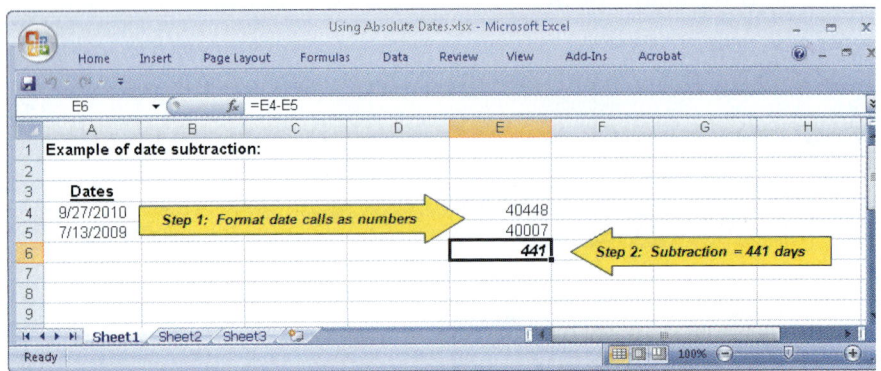

FIGURE 9-49 Microsoft Excel utilizes absolute dates that can be used in calculations, starting with January 1, 1900, which has a numeric value of 1. In the example shown, September 27, 2010 is displayed in number format as 40448, and July 13, 2009 is displayed as 40007. The difference between the dates is 441 days.

An **absolute date** is the total number of days from some specific base date. To calculate the number of days between two absolute dates, you subtract one date from the other. For example, if the base date is January 1, 1900, then September 27, 2010 has an absolute date value of 40448. Similarly, July 13, 2009 has an absolute date of 40007. If you subtract the earlier date value from the later one, the result is 441 days. You can use a spreadsheet to determine and display absolute dates easily, as shown in Figure 9-49.

DATA CONTROL

Just as it is important to secure the physical part of the system, as shown in Figure 9-50, file and database control must include all measures necessary to ensure that data storage is correct, complete, and secure. File and database control also is related to input and output techniques discussed earlier.

A well-designed DBMS must provide built-in control and security features, including subschemas, passwords, encryption, audit trail files, and backup and recovery procedures to maintain data. Your main responsibility is to ensure that the DBMS features are used properly.

FIGURE 9-50 System security involves the physical controls shown here and a range of software controls, including access codes, data encryption, passwords, and audit trails.

Earlier in this chapter, you learned that a subschema can be used to provide a limited view of the database to a specific user, or level of users. Limiting access to files and databases is the most common way of protecting stored data. Users must furnish a proper **user ID** and **password** to access a file or database. Different privileges, also called **permissions,** can be associated with different users, so some employees can be limited to read-only access, while other users might be allowed to update or delete data. For highly sensitive data, additional access codes can be established that restrict specific records or fields within records. Stored data also can be encrypted to prevent unauthorized access. **Encryption** is the process of converting readable data into unreadable characters to prevent unauthorized access to the data.

All system files and databases must be backed up regularly and a series of **backup** copies must be retained for a specified period of time. In the event of a file catastrophe, **recovery procedures** can be used to restore the file or database to its current state at the time of the last backup. **Audit log files,** which record details of all accesses and changes to the file or database, can be used to recover changes made since the last backup. You also can include **audit fields,** which are special fields within data records to provide additional control or security information. Typical audit fields include the date the record was created or modified, the name of the user who performed the action, and the number of times the record has been accessed.

CASE IN POINT 9.4: SOCCERMOM

SoccerMom Company sells a patented seat that spectators can take to youth soccer games. The seat folds so it is small enough to fit in the glove box of most vehicles. The company operates a factory in Kansas and also contracts its manufacturing projects to small firms in Canada and Mexico.

An unusual problem has occurred for this small multinational company: People are getting confused about dates in internal memos, purchase orders, and e-mail. Towson Hopkins handles all IT functions for SoccerMom. When he designed the company's database, he was not aware that the format for dates in Canada and Mexico was different from the format used in the United States. For example, in Canada and Mexico, the notation 7/1/09 indicates January 7, 2009, whereas in the United States the same notation indicates July 1, 2009. Although it seems like a small point, the date confusion has resulted in several order cancellations.

Towson has asked for your advice. You could suggest writing a simple program to convert the dates automatically or designing a switchboard command that would allow users to select a date format as data is entered. You realize, however, that SoccerMom might want to do business in other countries in the future. What would be the best course of action? Should SoccerMom adapt to the standard of each country, or should it maintain a single international format? What are the arguments for each option?

A QUESTION OF ETHICS

Olivia is the database manager at Tip Top Toys, a relatively small division of Worldwide Enterprises. Worldwide has nine other divisions, which include insurance, health care products, and financial planning services, to name a few.

Riccardo, corporate marketing director for Worldwide, has requested Tip Top's customer shopping data to target people who might be likely to purchase items or services from other Worldwide divisions. Olivia is not totally comfortable with this, and pointed out Tip Top's Web privacy policy, which states that "Tip Top Toys, a division of Worldwide Enterprises, will not share personal data with other companies without a customer's consent."

Riccardo replied that the statement only applies to outside companies – not other Worldwide divisions. He said he checked with the corporate legal department, and they agreed. Emily responded "Even if it is legally OK, it's not the *right* thing to do. Many people take our statement to mean that their data does not leave Tip Top. At the very least, we should give customers a choice, and share the data only with their consent."

Do you agree with Olivia? Why or why not?

CHAPTER SUMMARY

In this chapter, you continued your study of the systems design phase of the SDLC. You learned that files and tables contain data about people, places, things, or events that affect the information system. File processing systems, also called file-oriented systems, manage data stored in separate files, including master files, table files, transaction files, work files, security files, and history files.

A database consists of linked tables that form an overall data structure. A database management system (DBMS) is a collection of tools, features, and interfaces that enable users to add, update, manage, access, and analyze data in a database.

DBMS designs are more powerful and flexible than traditional file-oriented systems. A database environment offers scalability, support for organization-wide access, economy of scale, data sharing among user groups, balancing of conflicting user requirements, enforcement of standards, controlled redundancy, effective security, flexibility, better programmer productivity, and data independence. Large-scale databases are complex and require extensive security and backup/recovery features.

DBMS components include interfaces for users, database administrators, and related systems; a data manipulation language; a schema; and a physical data repository. Other data management techniques include data warehousing, which stores data in an easily accessible form for user access, and data mining, which looks for meaningful patterns and relationships among data. Data mining also includes clickstream storage, which records how users interact with a site, and market basket analysis, which can identify product relationships and consumer buying patterns.

In an information system, an entity is a person, place, thing, or event for which data is collected and maintained. A field, or attribute, is a single characteristic of an entity. A record, or tuple, is a set of related fields that describes one instance of an entity. Records are grouped into files (in a file-oriented system) and tables (in a database environment).

A primary key is the field or field combination that uniquely and minimally identifies a specific record; a candidate key is any field that could serve as a primary key. A foreign key is a field or field combination that must match the primary key of another file or table. A secondary key is a field or field combination used as the basis for sorting or retrieving records.

An entity-relationship diagram (ERD) is a graphic representation of all system entities and the relationships among them. The ERD is based on entities and data stores in DFDs prepared during the systems analysis phase. The three basic relationships represented in an ERD are one-to-one (1:1), one-to-many (1:M), and many-to-many (M:N). In a M:N relationship, the two entities are linked by an associative entity.

The relationship between two entities also is referred to as cardinality. A common form of cardinality notation is called crow's foot notation, which uses various symbols to describe the characteristics of the relationship.

Normalization is a process for avoiding problems in data design. A first normal form (1NF) record has no repeating groups. A record is in second normal form (2NF) if it is in 1NF and all nonkey fields depend on the entire primary key. A record is in third normal form (3NF) if it is in 2NF and if no field depends on a nonkey field.

Data design tasks include creating an initial ERD; assigning data elements to an entity; normalizing all table designs; and completing the data dictionary entries for files, records, and data elements. Files and database tables should be sized to estimate the amount of storage space they will require.

You learned that a code is a set of letters or numbers used to represent data in a system. By using codes, you can speed up data entry, reduce data storage space, and reduce transmission time. Codes also can be used to reveal or to conceal information. The main types of codes are sequence codes, block sequence codes, classification codes, alphabetic codes (including category codes, abbreviation codes, and mnemonic codes), significant digit codes, derivation codes, cipher codes, and action codes.

The most common database models are relational and object-oriented. The relational model is powerful and flexible, and provides the best support for client/server architecture. Object-oriented database (OODB) design is becoming more popular as a natural extension of the object-oriented analysis process. Each object in an OODB has a unique object identifier, which is similar to a primary key in a relational database and allows the object to interact with other objects and form relationships.

Logical storage is information seen through a user's eyes, regardless of how or where that information actually is organized or stored. Physical storage is hardware-related and involves reading and writing blocks of binary data to physical media. A logical record is a related set of field values that describes a single person, place, thing, or event. A physical record consists of one or more logical records, depending on the blocking factor. Data storage formats include EBCDIC, ASCII, binary, and Unicode. Dates can be stored in several formats, including ISO and absolute format.

File and database control measures include limiting access to the data, data encryption, backup/recovery procedures, audit-trail files, and internal audit fields.

Key Terms and Phrases

Learn It Online

Instructions: To complete the Learn It Online exercises, start your browser, click the Address bar, and then enter the Web address **scsite.com/sad8e/learn**. When the Systems Analysis and Design Learn It Online page is displayed, follow the instructions in the exercises below. Each exercise has instructions for saving your results, either for your own records or for submission to your instructor.

1 Chapter Reinforcement

TF, MC, and SA

Below SAD Chapter 9, click one of the Chapter Reinforcement links for Multiple Choice, True/False, or Short Answer. Answer each question and submit to your instructor.

2 Flash Cards

Below SAD Chapter 9, click the Flash Cards link and read the instructions. Type 20 (or a number specified by your instructor) in the Number of playing cards text box, type your name in the Enter your Name text box, and then click the Flip Card button. When the flash card is displayed, read the question and then click the ANSWER box arrow to select an answer. Flip through the Flash Cards. If your score is 15 (75%) correct or greater, click Print on the File menu to print your results. If your score is less than 15 (75%) correct, then redo this exercise by clicking the Replay button.

3 Practice Test

Below SAD Chapter 9, click the Practice Test link. Answer each question, enter your first and last name at the bottom of the page, and then click the Grade Test button. When the graded practice test is displayed on your screen, click Print on the File menu to print a hard copy. Continue to take practice tests until you score 80% or better.

4 Who Wants To Be a Computer Genius?

Below SAD Chapter 9, click the Computer Genius link. Read the instructions, enter your first and last name at the bottom of the page, and then click the Play button. When your score is displayed, click the PRINT RESULTS link to print a hard copy.

5 Wheel of Terms

Below SAD Chapter 9, click the Wheel of Terms link. Read the instructions, and then enter your first and last name and your school name. Click the PLAY button. When your score is displayed on the screen, right-click the score and then click Print on the shortcut menu to print a hard copy.

6 Crossword Puzzle Challenge

Below SAD Chapter 9, click the Crossword Puzzle Challenge link. Read the instructions, and then click the Continue button. Work the crossword puzzle. When you are finished, click the Submit button. When the crossword puzzle is redisplayed, submit it to your instructor.

Case-Sim: SCR Associates

Background

SCR Associates is a consulting firm that offers IT solutions and training. SCR needs an information system to manage operations at its new training center. The new system will be called TIMS (Training Information Management System). As a newly hired systems analyst, you will report to Jesse Baker, systems group manager. You will work on various tasks and practice the skills you learned in this chapter.

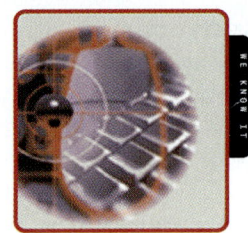

Using the Case

The SCR Associates case study is a Web-based simulation that allows you to practice your skills in a real-world environment. The case study transports you to SCR's company intranet, where you can complete 12 work sessions, each aligning with a chapter. As you work on the case, you will receive e-mail and voice mail messages, obtain information from SCR's online libraries, and perform various tasks. The first time you enter the SCR Case, you should go to the starting page at **scsite.com/sad8e/scr** for detailed instructions.

Preview: Session 9

Your supervisor, Jesse Baker, has asked you to begin working on data design tasks for the new information system, which will be implemented as a relational database. You will need to identify the entities, draw an ERD, design tables, and add sample data to each table.

To start a work session, you log on to the SCR intranet. When you enter your name and password, an opening screen displays links to the work sessions, and you select Session 9. At this point, you check your e-mail and voice mail messages carefully. Then you begin working on your task list, which includes the following items:

Tasks: Data Design

1. *List all the entities that interact with the TIMS system. Start by reviewing the data library, previous e-mail messages, DFDs, and other documentation.*

2. *Draw an ERD that shows cardinality relationships among the entities. Send the diagram to Jesse.*

3. *For each entity, Jesse wants to see table designs in 3NF. Use standard notation format to show the primary key and the other fields in each table.*

4. *Jesse wants to use sample data to populate fields for at least three records in each table. Better get started on this right away.*

FIGURE 9-51   Task list: Session 9.

Chapter Exercises

Review Questions

1. Explain the main differences between a file processing system and a database system.
2. What is a DBMS? Briefly describe the components of a DBMS.
3. Describe a primary key, candidate key, secondary key, foreign key, and common field.
4. What are entity-relationship diagrams and how are they used? What symbol is used to represent an entity in an ERD? What symbol is used for a relationship? What is cardinality, and what symbols do you use in the crow's foot notation method?
5. What are data warehousing and data mining? Are the terms related?
6. What is the criterion for a table design to be in first normal form? How do you convert an unnormalized design to 1NF?
7. What are the criteria for a table design to be in second normal form? How do you convert a 1NF design to 2NF?
8. What are the criteria for a table design to be in third normal form? How do you convert a 2NF design to 3NF?
9. Explain the difference between a logical record and a physical record.
10. How would a specific date, such as September 1, 2009, be represented as an absolute date?

Discussion Topics

1. Are there ethical issues to consider when planning a database? For example, should sensitive personal data (such as medical information) be stored in the same DBMS that manages employee salary and benefits data? Why or why not?
2. Suggest three typical business situations where referential integrity avoids data problems.
3. Consider an automobile dealership with three locations. Data fields exist for stock number, vehicle identification number, make, model, year, color, and invoice cost. Identify the possible candidate keys, the likely primary key, a probable foreign key, and potential secondary keys.
4. In the example shown in Figures 9-25 and 9-26 on pages 411 and 412, the 2NF customer table was converted to two 3NF tables. Verify that the four potential problems identified for 2NF tables were eliminated in the 3NF design.

Projects

1. Search the Internet to find information about data storage formats. Also do research on international date formats. Determine whether the date format used in the United States is the most common format.
2. Visit the IT department at your school or at a local business and determine whether the organization uses file processing systems, DBMSs, or both. Write a brief memo with your conclusions.
3. Use Microsoft Access or similar database software to create a DBMS for the imaginary company called TopText Publishing, which is described in Case In Point 9.1 on page 406. Add several sample records to each table and report to the class on your progress.
4. Visit the bookstore at your school or a bookstore in your area. Interview the manager or store employees to learn how the operation works and what entities are involved in bookstore operations. Remember that an entity is a person, place, thing, or event that affects the information system. Draw an ERD, including cardinality that describes the bookstore operations.

Apply Your Knowledge

The Apply Your Knowledge section contains four mini-cases. Each case describes a situation, explains your role in the case, and asks you to respond to questions. You can answer the questions by applying knowledge you learned in the chapter.

1 Pick and Shovel Construction Company

Situation:

Pick and Shovel Construction Company is a multistate building contractor specializing in medium-priced town homes. C. T. Scott, the owner, is in your office for the third time today to see how the new relational database project is coming along. Unfortunately, someone mentioned to C. T. that the delay had something to do with achieving *normalization*.

"Why is all this normalization stuff so important?" he asks. "The old system worked OK most of the time, and now you are telling me that we need all these special rules. Why is this necessary?"

1. How should you respond to C. T.? Write him a brief memo with your views.
2. Assume that the Pick and Shovel's main entities are its customers, employees, projects, and equipment. A customer can hire the company for more than one project, and employees sometimes work on more than one project at a time. Equipment, however, is assigned only to one project. Draw an ERD showing those entities.
3. Add cardinality notation to your ERD.
4. Create 3NF table designs.

2 Puppy Palace

Situation:

Puppy Palace works with TV and movie producers who need dogs that can perform special tricks, such as headstands, somersaults, ladder climbs, and various dog-and-pony tricks. Puppy Palace has about 16 dogs and a list of 50 tricks from which to choose. Each dog can perform one or more tricks, and many tricks can be performed by more than one dog. When a dog learns a new trick, the trainer assigns a skill level. Some customers insist on using dogs that score a 10, which is the highest skill level. As an IT consultant, you have been asked to suggest 3NF table designs. You are fairly certain that a M:N relationship exists between dogs and tricks.

1. Draw an ERD for the Puppy Palace information system.
2. Indicate cardinality.
3. Identify all fields you plan to include in the dogs and tricks tables. For example, in the dogs table, you might want breed, size, age, name, and so on. In the tricks table, you might want the trick name and description. You will need to assign a primary key in each table. *Hint*: Before you begin, review some database design samples in this chapter. You might spot a similar situation that requires an associative entity that you can use as a pattern. In addition, remember that numeric values work well in primary key fields.
4. Create 3NF table designs.

3 Mayville Public Library

Situation:

Mayville is a rural village with a population of 900. Until now, Mayville was served by a bookmobile from a larger town. The Mayville Village Council has authorized funds for a small public library, and you have volunteered to set up an information system for the library. Assume that the library will have multiple copies of certain books.

1. Draw an ERD for the Mayville library system.
2. Indicate cardinality.
3. Identify all fields you plan to include in the tables.
4. Create 3NF table designs.

4 Western Wear Outfitters

Situation:

Western Wear is a mail-order firm that offers an extensive selection of casual clothing for men and women. Western Wear plans to launch a new Web site, and the company wants to develop a new set of product codes. Currently, 650 different products exist, with the possibility of adding more in the future. Many products come in various sizes, styles, and colors. The marketing manager asked you to develop an individualized product code that can identify a specific item and its characteristics. Your initial reaction is that it can be done, but the code might be fairly complex. Back in your office, you give the matter some thought.

1. Design a code scheme that will meet the marketing manager's stated requirements.
2. Write a brief memo to the marketing manager suggesting at least one alternative to the code she proposed, and state your reasons.
3. Suggest a code scheme that will identify each Western Wear customer.
4. Suggest a code scheme that will identify each specific order.

Case Studies

Case studies allow you to practice specific skills learned in the chapter. Each chapter contains several case studies that continue throughout the textbook, and a chapter capstone case.

NEW CENTURY HEALTH CLINIC

New Century Health Clinic offers preventive medicine and traditional medical care. In your role as an IT consultant, you will help New Century develop a new information system.

Background

After completing the user interface, input, and output design for the new information system at New Century, you will consider data design issues. Begin by studying the DFDs and object-oriented diagrams you prepared previously and the rest of the documentation from the systems analysis phase. Perform the following tasks:

Assignments

1. Create an initial entity-relationship diagram for the New Century Health Clinic system.
2. Normalize your table designs.
3. If you identified any new entities during normalization, create a final entity-relationship diagram for the system.
4. Write a memo for your documentation file that contains your recommendation about whether a file processing or a database environment should be used. Attach copies of your ERD(s) and normalized designs.

PERSONAL TRAINER, INC.

Personal Trainer, Inc., owns and operates fitness centers in a dozen Midwestern cities. The centers have done well, and the company is planning an international expansion by opening a new "supercenter" in the Toronto area. Personal Trainer's president, Cassia Umi, hired an IT consultant, Susan Park, to help develop an information system for the new facility. During the project, Susan will work closely with Gray Lewis, who will manage the new operation.

Background

After evaluating various development strategies, Susan prepared a system requirements document and submitted her recommendations to Cassia Umi, Personal Trainer's president. During her presentation, Susan discussed several development strategies, including in-house development and outsourcing. She did not feel that a commercial software package would meet Personal Trainer's needs.

Based on her research, Susan felt it would be premature to select a development strategy at this time. Instead, she recommended to Cassia that an in-house team should develop a design prototype, using a relational database as a model. Susan said that the prototype would have two main objectives: It would represent a user-approved model of the new system, and it would identify all systems entities and the relationships among them. Susan explained that it would be better to design the basic system first, and then address other issues, including Web enhancements and implementation options. She proposed a three-step plan: data design, user interface design, and application architecture. She explained that systems analysts refer to this as the systems design phase of a development project.

Cassia agreed with Susan's recommendation, and asked her to go forward with the plan.

Assignments

1. Review the Personal Trainer fact-finding summary in Chapter 4 and draw an ERD with cardinality notation. Assume that system entities include members, activities and services, and fitness instructors.

2. Design tables in 3NF. As you create the database, include various codes for at least three of the fields.

3. Use sample data to populate the fields for at least three records in each table.

4. Recommend a date format for the new system. Should Personal Trainer adopt a single international standard, or should the format be determined by the country in which the center is located? Write a message to Susan with your recommendation.

FASTFLIGHT AIRLINES

FastFlight Airlines is a small air carrier operating in three northeastern states. FastFlight is computerizing its passenger reservation system. The data items must include reservation number, flight number, flight date, origin, destination, departure time, arrival time, passenger name, and seat number. For example, flight number 303 leaves Augusta, Maine, daily at 9:23 a.m. and arrives in Nashua, New Hampshire, at 10:17 a.m. A typical reservation number might be AXQTBC, for passenger Lisa Lane, in seat 4A, on flight 303 on 11/12/2009.

Assignments

1. Create an ERD for the reservations system.

2. Create 3NF table designs for the system.

3. For each of the entities identified, design tables and identify the possible candidate keys, the primary key, a probable foreign key, and potential secondary keys.

4. Use sample data to populate the fields for three records.

CHAPTER CAPSTONE CASE: SoftWear, Limited

SoftWear, Limited (SWL), is a continuing case study that illustrates the knowledge and skills described in each chapter. In this case study, the student acts as a member of the SWL systems development team and performs various tasks.

Background

Work continued on the ESIP system. Rick said that the system would be a client/server design that could support SWL's current and future business requirements. He also said that Ann Hon wanted to use the ESIP system as a prototype for developing other SWL systems in the future. Ann said that the new design would have to be powerful, flexible, and scalable. With that in mind, the team decided that a DBMS strategy was the best solution for SWL's future information systems requirements.

ERDs and Normalization

Rick asked Tom and Becky to draw an entity-relationship diagram with normalized designs. Tom and Becky used the Visible Analyst, a CASE tool, to produce the diagram shown in Figure 9-52. Rick noticed that only two entities were shown: EMPLOYEE and DEDUCTION. Rick suggested that the ESIP-OPTION and HUMAN RESOURCES entities should be added. Tom and Becky agreed. The second version of their ERD is shown in Figure 9-53.

With the ERD completed, Tom turned to the design of the EMPLOYEE table. He suggested the following design:

EMPLOYEE (<u>SSN</u>, EMPLOYEE-NAME, HIRE-DATE)

"The table obviously is in 1NF, because it has no repeating groups," Tom said. "It's also in 2NF, because it has a single field as the primary key. And I'm sure it's in 3NF, because the

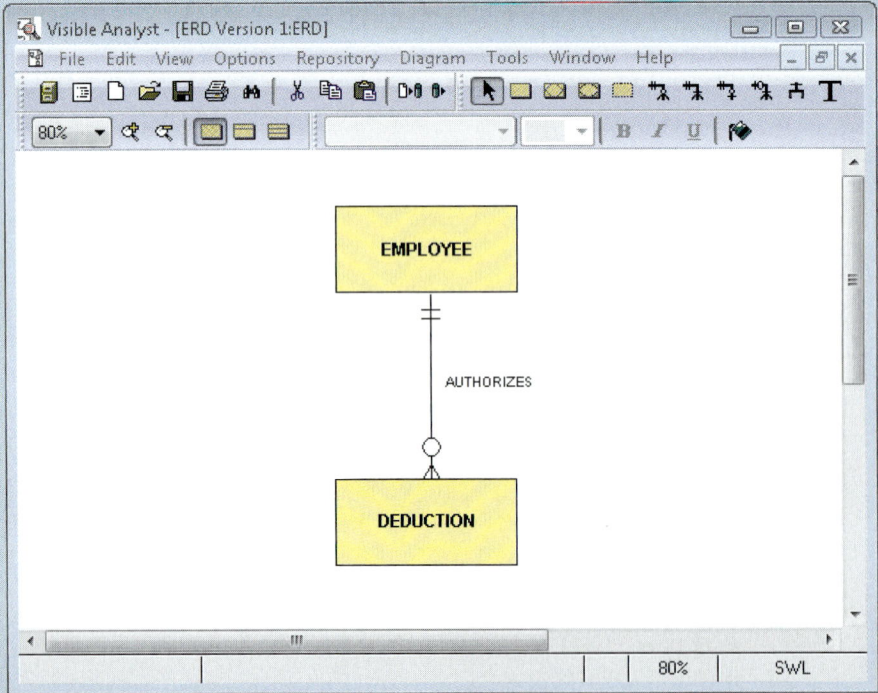

FIGURE 9-52 The initial ERD shows two entities: EMPLOYEE and DEDUCTION. Notice that crow's foot notation indicates that one and only one employee can authorize anywhere from zero to many deductions.

employee name and the hire date both depend on the Social Security number." Everyone agreed that this was the correct design. Tom and Becky turned their attention to designing the ESIP-OPTION table and later suggested the following design:

ESIP-OPTION (<u>OPTION-CODE</u>, OPTION-NAME, DESCRIPTION, DEDUCTION-CYCLE, APPLICATION-CYCLE, MIN-SERVICE, MIN-DEDUCTION, MAX-DEDUCTION)

The design appeared to meet the test for 3NF. Although seven fields existed in addition to the primary key, each field seemed to depend on the entire primary key. Finally, Becky proposed the following design for the DEDUCTION record:

DEDUCTION (<u>SSN</u>, <u>ESIP-OPTION</u>, <u>DATE</u>, EMPLOYEE-NAME, AMOUNT)

This time, Rick felt that the table was in 1NF, but not in 2NF because the EMPLOYEE-NAME field was dependent on only a part of the key rather than the entire key. Becky agreed with him and suggested that the EMPLOYEE-NAME field could be removed and accessed using the EMPLOYEE table. The SSN field could be used as a foreign key to match values in the EMPLOYEE table's primary key. To put the table into 2NF, she revised the DEDUCTION table design as follows:

DEDUCTION (<u>SSN</u>, <u>ESIP-OPTION</u>, <u>DATE</u>, AMOUNT)

With that change, everyone agreed that the table also was in 3NF because the AMOUNT field depended on the entire primary key. The next step was to work on a system design to interface with the payroll system and provide support for SWL's long-term information technology goals.

While Rick, Tom, and Becky were working on ERDs and normalization, Pacific Software delivered the payroll package that SWL ordered. Tom was assigned to work on installing and configuring the package and training users on the new payroll system.

Meanwhile, Rick felt that SWL should get more information about client/server design. With Ann Hon's approval, he contacted several IT consulting firms that advertised their client/server design expertise on the Internet. Rick and Becky met with three firms and recommended that SWL work with True Blue Systems, a consulting group with a local office in Raleigh, not far from SWL's headquarters. In addition to the design for the ESIP system, Rick suggested that the agenda should include a general discussion about a future SWL intranet with support for Web standards, and the possibility that employees could access their ESIP accounts from home via the Internet.

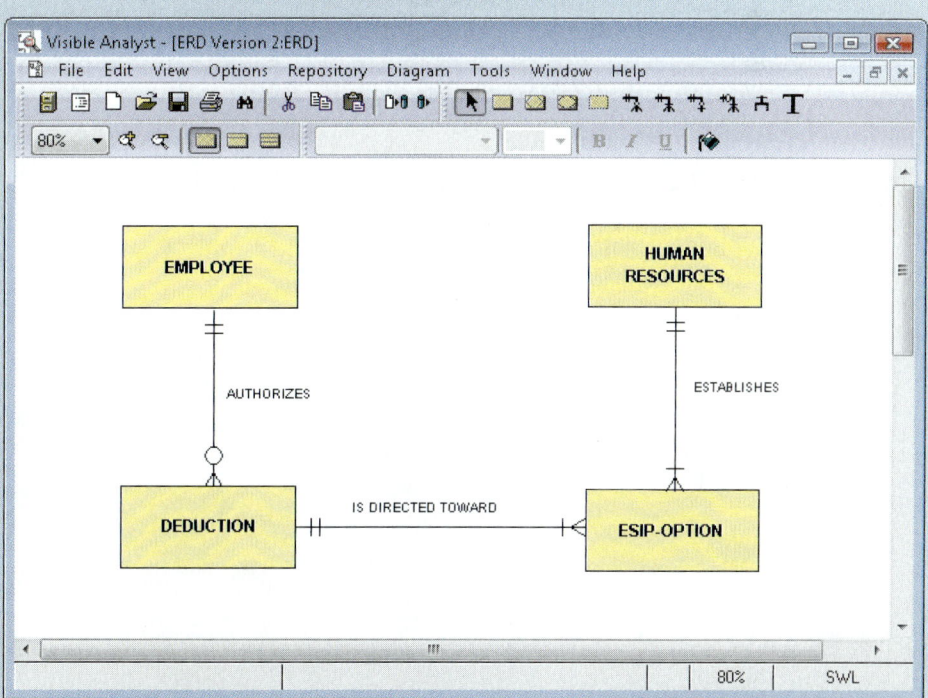

FIGURE 9-53 Second version of the ERD. Now, the DEDUCTION entity has relationships to two other entities: EMPLOYEE and ESIP-OPTION.

CHAPTER CAPSTONE CASE: SoftWear, Limited (continued)

SWL Team Tasks

1. Rick asked you to help him put together a brief progress update for Michael Jeremy and several other top managers. Specifically, Rick wants you to explain the concept of normalization without using a lot of technical jargon. Rick wants you to summarize the concept using plain English and simple examples.

2. At SWL, each employee is assigned to a specific department. Employees from several departments often are assigned to special project teams, however, when a new product is launched or for major marketing events. Carla wants to develop a project management system to track the projects, employees assigned, and accumulated project hours. She believes that employees and projects are in a M:N relationship. She showed you an initial design where all data is stored in a single table:

   PROJECT DATA (<u>PROJECT-NUMBER</u>, PROJECT-NAME, START-DATE,
        PROJECT-STATUS, (<u>EMPLOYEE-NUMBER</u>, EMPLOYEE-NAME, JOB-TITLE,
        DEPT-NUMBER, DEPT-NAME, PROJECT-HOURS))

   How would you describe Carla's design?

3. Carla wants you to create an ERD, including cardinality, for the project management system. She says that you probably will need to add an associative entity.

4. After you create the ERD in the previous step, design a table for each entity, in third normal form.

Manage the SWL Project

You have been asked to manage SWL's new information system project. One of your most important activities will be to identify project tasks and determine when they will be performed. Before you begin, you should review the SWL case in this chapter. Then list and analyze the tasks, as follows:

LIST THE TASKS Start by listing and numbering at least 10 tasks that the SWL team needs to perform to fulfill the objectives of this chapter. Your list can include SWL Team Tasks and any other tasks that are described in this chapter. For example, Task 3 might be to Identify all entities, and Task 6 might be to Create an initial ERD.

ANALYZE THE TASKS Now study the tasks to determine the order in which they should be performed. First identify all concurrent tasks, which are not dependent on other tasks. In the example shown in Figure 9-54, Tasks 1, 2, 3, 4, and 5 are concurrent tasks, and could begin at the same time if resources were available.

Other tasks are called dependent tasks, because they cannot be performed until one or more earlier tasks have been completed. For each dependent task, you must identify specific tasks that need to be completed before this task can begin. For example, you would want to identify all the entities before you could create an initial ERD, so Task 6 cannot begin until Task 3 is completed, as Figure 9-54 shows.

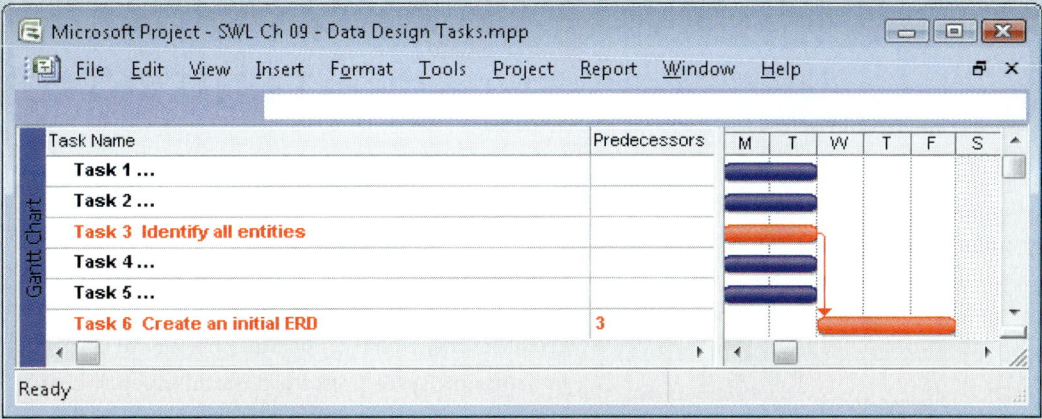

FIGURE 9-54 Tasks 1, 2, 3, 4, and 5 are concurrent tasks that could be performed at the same time. Task 6 is a dependent task that cannot be performed until Task 3 has been completed.

Chapter 3 describes project management tools, techniques, and software. To learn more, you can visit the Features section on your Student Study Tool CD-ROM, or the project management resources library at **scsite.com/sad8e/project**. On the Web, Microsoft offers demo versions, training, and tips for using Project 2007. You also can visit the OpenWorkbench.org site to learn more about this free, open-source software.

CHAPTER 10 System Architecture

Chapter 10 is the final chapter in the systems design phase of the SDLC. This chapter describes system architecture, which translates the logical design of an information system into a physical blueprint, or architecture. As you plan the system architecture, you will learn about servers, clients, processing methods, networks, and related issues.

INTRODUCTION

OBJECTIVES

When you finish this chapter, you will be able to:

- Provide a checklist of issues to consider when selecting a system architecture

- Describe servers, server-based processing, clients, and client-based processing

- Explain client/server architecture, including tiers, cost-benefit issues, and performance

- Compare in-house e-commerce development with packaged solutions

- Discuss the potential impact of cloud computing and Web 2.0

- Explain the difference between online and batch processing

- Define network topology, including hierarchical, bus, ring, and star models

- Explain network protocols and licensing issues

- Describe wireless networking, including wireless standards, topologies, and trends

- Describe the system design specification

At this point in the SDLC, your objective is to determine an overall architecture to implement the information system. You learned in Chapter 1 that an information system requires hardware, software, data, procedures, and people to accomplish a specific set of functions. An effective system combines those elements into an architecture, or design, that is flexible, cost-effective, technically sound, and able to support the information needs of the business. This chapter covers a wide range of topics that support the overall system design, just as a plan for a new home would include a foundation plan, building methods, wiring and plumbing diagrams, traffic flows, and costs.

System architecture translates the logical design of an information system into a physical structure that includes hardware, software, network support, processing methods, and security. The end product of the systems design phase is the system design specification. If this document is approved, the next step is systems implementation.

CHAPTER INTRODUCTION CASE: Mountain View College Bookstore

Background: Wendy Lee, manager of college services at Mountain View College, wants a new information system that will improve efficiency and customer service at the three college bookstores.

In this part of the case, Florence Fullerton (systems analyst) and Harry Boston (student intern) are talking about system architecture issues.

Participants:	Florence and Harry
Location:	Mountain View College cafeteria, Monday afternoon, January 11, 2010
Project status:	The team has completed data design tasks and user interface, output, and input design work. The last step in the systems design phase is to consider a system architecture for the bookstore system.
Discussion topics:	System architecture checklist, client/server architecture, processing methods, network issues, and system management tools

Florence:	Hi, Harry. Did you enjoy the holiday break?
Harry:	*I sure did. Now I'm ready to get back to work.*
Florence:	Good. As the last step in the systems design phase of the SDLC, we need to study the physical structure, or architecture, of the bookstore system. Our checklist includes enterprise resource planning, total cost of ownership, scalability, Web integration, legacy systems, processing methods, and security issues that could affect the system design.
Harry:	*Where do we start?*
Florence:	Well, the bookstore interfaces with many publishers and vendors, so we'll consider supply chain management, which is part of enterprise resource planning, or ERP for short.
Harry:	*What happens after we finish the checklist?*
Florence:	Then we'll define a client/server architecture. As I see it, the bookstore client workstations will share the processing with a server in the IT department. Also, we may need to look at middleware software to connect the new system with existing legacy systems, such as the college accounting system.
Harry:	*Anything else?*
Florence:	Yes. We need to select a network plan, or topology, so we'll know how to plan the physical cabling and connections — or possibly use wireless technology. When we're done, we'll submit a system design specification for approval.
Harry:	*Sounds good to me.*
Florence:	Good. Here's a task list to get us started:

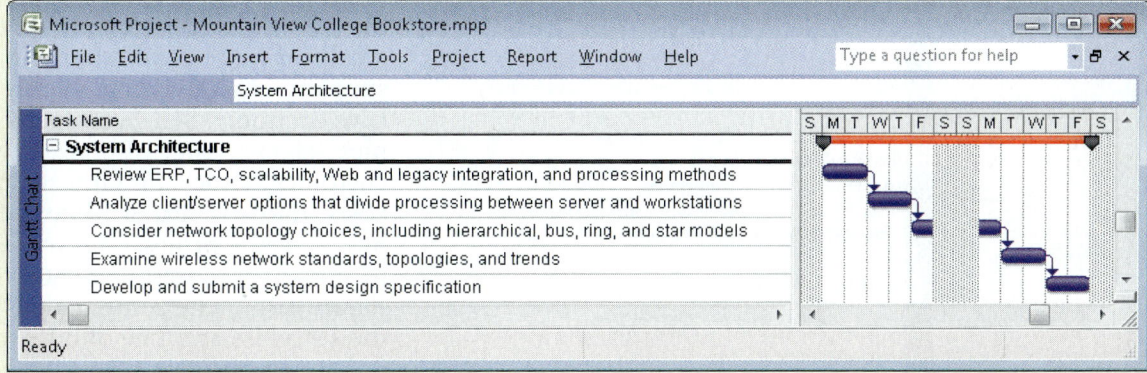

FIGURE 10-1 Typical system architecture tasks.

SYSTEM ARCHITECTURE CHECKLIST

Just as an architect begins a project with a list of the owner's requirements, a systems analyst must approach system architecture with an overall checklist. Before making a decision, the analyst must consider seven specific issues that will affect the architecture choice:

- Enterprise resource planning (ERP)
- Initial and total cost of ownership (TCO)
- Scalability
- Web integration
- Legacy system interface requirements
- Processing options
- Security issues

Enterprise Resource Planning

Many companies use **enterprise resource planning (ERP)** software, which was described in Chapter 1. The objective of ERP is to establish a company-wide strategy for using IT resources. ERP defines a specific architecture, including standards for data, processing, network, and user interface design. A main advantage of ERP is that it describes a specific hardware and software **environment**, also called a **platform**, that ensures connectivity and easy integration of future systems, including in-house software and commercial packages.

In the article in *CIO* magazine shown in Figure 10-2, Thomas Wailgum remarks that ERP isn't really about resources or planning — it is about the *enterprise*, and the notion that a single computer system can support various departments across the organization. Mr. Wailgum also notes that ERP doesn't replace departmental systems. Instead, it integrates them into an overall application, with a central database that allows departments to share data and communicate more effectively.

Many companies are extending internal ERP systems to their suppliers and customers, using a concept called **supply chain management (SCM)**. For example, in a totally integrated supply chain system, a customer order could cause a manufacturing system to schedule a work order, which in turn triggers a call for more parts from one or more suppliers. In a dynamic, highly competitive economy, SCM can help companies achieve faster response, better customer service, and lower operating costs. Industry leader Microsoft has introduced a product family called Microsoft Dynamics, as shown in Figure 10-3. The company describes the software as a line of integrated, adaptable business management solutions.

FIGURE 10-2 The article by Thomas Wailgum points out that the most important aspect of ERP is the focus on the overall enterprise.

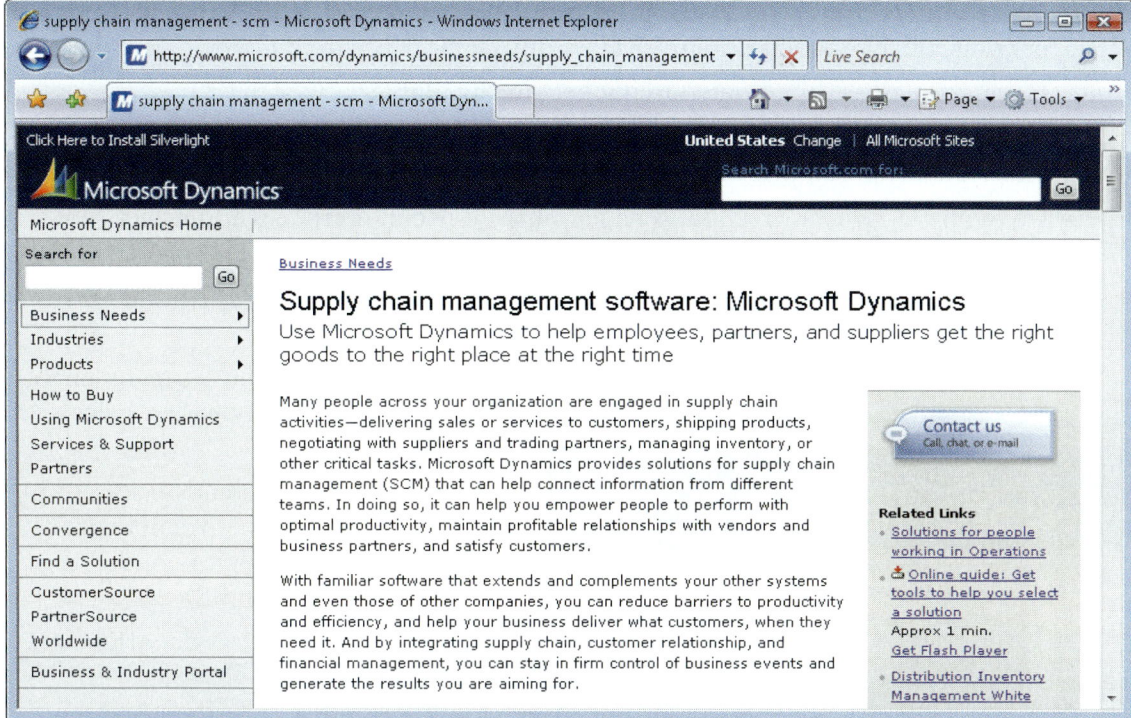

FIGURE 10-3   Microsoft Dynamics offers ERP and SCM solutions.

Most supply chain management systems depend on RFID technology for real-time input data. As you learned in Chapter 1, RFID allows companies to track incoming material, current production, and finished inventory by using small devices that respond to radio frequency signals.

CASE IN POINT 10.1:   ABC Systems

You are a systems analyst at ABC Systems, a fast-growing IT consulting firm that provides a wide range of services to companies that want to establish e-commerce operations. During the last 18 months, ABC acquired two smaller firms and set up a new division that specializes in supply chain management. Aligning ABC's internal systems was quite a challenge, and top management was not especially happy with the integration cost or the timetable. To avoid future problems, you have decided to suggest an ERP strategy, and you plan to present your views at the staff meeting tomorrow. ABC's management team is very informal and prefers a loose, flexible style of management. How will you persuade them that ERP is the way to go?

Initial Cost and TCO

You learned earlier about the importance of considering economic feasibility and TCO during systems planning and analysis. Now, during the final design stage, you make decisions that will have a major impact on the initial costs and TCO for the new system. At this point, you should review all previous cost estimates and ask the following questions:

- If in-house development was selected as the best alternative initially, is it still the best choice? Is the necessary technical expertise available, and does the original cost estimate appear realistic?

- If a specific package was chosen initially, is it still the best choice? Are newer versions or competitive products available? Have any changes occurred in pricing or support?

- Have any new types of outsourcing become available?

- Have any economic, governmental, or regulatory events occurred that could affect the proposed project?

- Have any significant technical developments occurred that could affect the proposed project?

- Have any major assumptions changed since the company made the build versus buy decision?

- Are there any merger or acquisition issues to consider, whereby the company might require compatibility with a specific environment?

- Have any new trends occurred in the marketplace? Are new products or technologies on the verge of being introduced?

- Have you updated the original TCO estimate? If so, are there any significant differences?

The answers to these questions might affect the initial cost and TCO for the proposed system. You should reanalyze system requirements and alternatives now, before proceeding to design the system architecture. To ensure that all costs are considered, you might want to review the direct and indirect cost concepts described by Gartner, Inc., in Figure 10-4. Gartner has been a leader in developing TCO methodology.

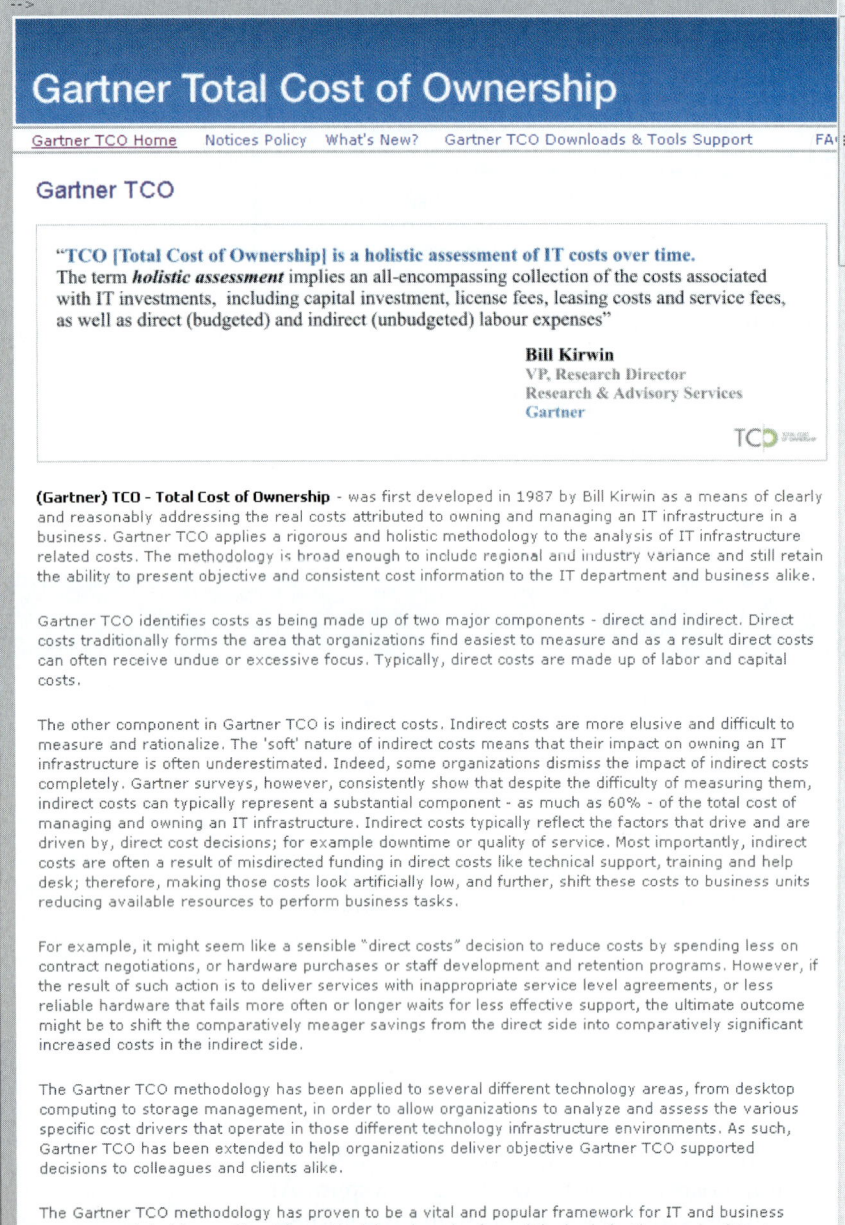

FIGURE 10-4 Gartner, Inc., emphasizes the importance of recognizing indirect as well as direct costs.

Scalability

Scalability, also called extensibility, refers to a system's ability to expand, change, or downsize easily to meet the changing needs of a business enterprise. Scalability is especially important in implementing systems that are volume-related, such as transaction processing systems. A scalable system is necessary to support a dynamic, growing business. For example, a scalable network could handle anywhere from a few dozen nodes to thousands of nodes; a scalable

DBMS could support the acquisition of a new sales division. When investing large amounts of money in a project, management is especially concerned about scalability issues that could affect the system's life expectancy.

ON THE WEB

For more information about TCO, visit **scsite.com/sad8e/ more**, locate Chapter 10, and then click the TCO link.

Web Integration

An information system includes **applications**, which are programs that handle the input, manage the processing logic, and provide the required output. The systems analyst must know if a new application will be part of an e-commerce strategy and the degree of integration with other Web-based components. As you learned earlier, a **Web-centric** architecture follows Internet design protocols and enables a company to integrate the new application into its e-commerce strategy. Even where e-commerce is not involved, a Web-centric application can run on the Internet or a company intranet or extranet. A Web-based application avoids many of the connectivity and compatibility problems that typically arise when different hardware environments are involved. In a Web-based environment, a firm's external business partners can use standard Web browsers to import and export data.

In Figure 10-5, notice that IBM uses the term **e-marketplaces** to describe Internet-based solutions that allow sellers and buyers to automate procurement processes and achieve substantial benefits.

Legacy System Interface Requirements

The new system might have to interface with one or more **legacy systems**, which are older systems that typically run on mainframe computers. When considering physical design, a systems analyst must determine how the new application will communicate with existing legacy systems. For example, a new marketing information system might need to report sales data to a server-based accounting system and obtain product cost data from a legacy manufacturing system.

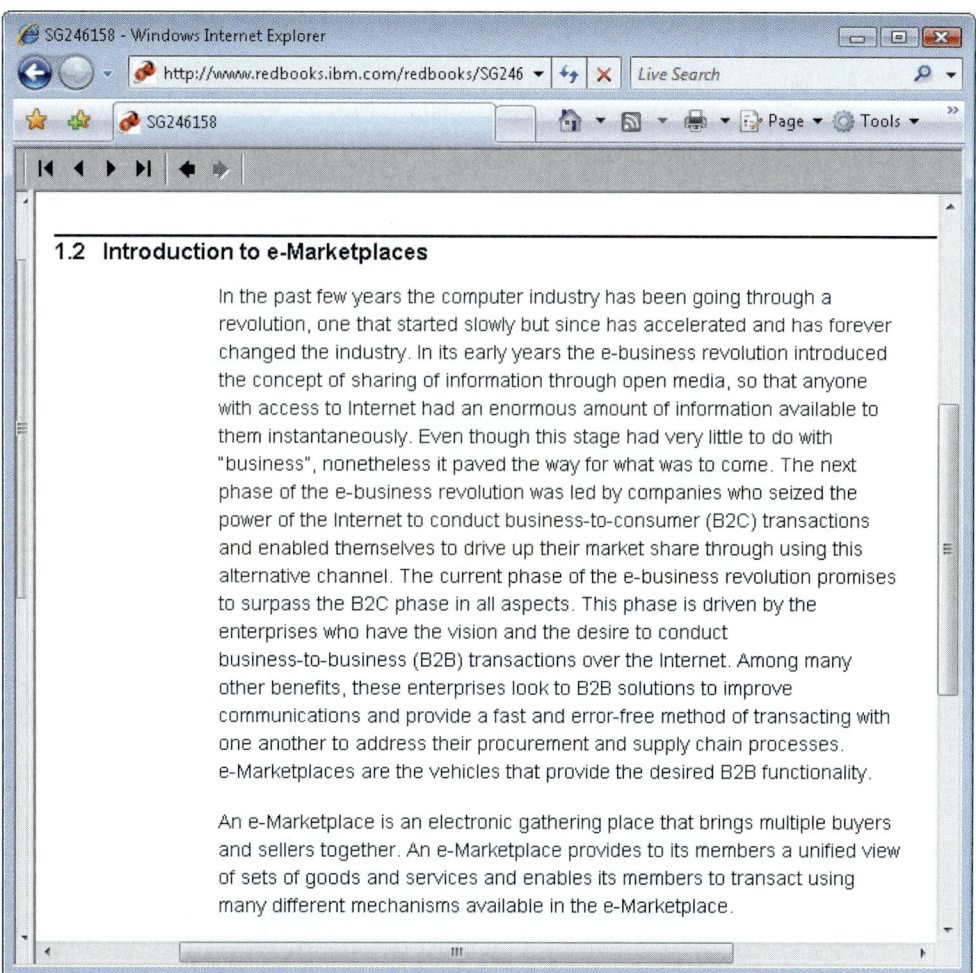

FIGURE 10-5   IBM uses the term e-marketplaces to describe Internet-based solutions that allow customers and suppliers to automate procurement processes and achieve substantial benefits.

ON THE WEB

For more information about legacy systems, visit **scsite.com/ sad8e/more**, locate Chapter 10, and then click the Legacy Systems link.

Interfacing a new system with a legacy system involves analysis of data formats and compatibility. In some cases, a company will need to convert legacy file data, which can be an expensive and time-consuming process. Finally, to select the best architecture, the analyst must know if the new application eventually will replace the legacy system.

Processing Options

In planning the architecture, designers also must consider how the system will process data — online or in batches. For example, a high-capacity transaction processing system, such as an order entry system, requires more network, processing, and data storage resources than a monthly billing system that handles data in batches. Also, if the system must operate online, 24 hours a day and seven days a week (24/7), provision must be made for backup and speedy recovery in the event of system failure.

The characteristics of online and batch processing methods are described later in this chapter, with examples of each type.

Security Issues

From the password protection shown in Figure 10-6 to complex intrusion detection systems, security threats and defenses are a major concern to a systems analyst. As the physical design is translated into specific hardware and software, the analyst must consider security issues and determine how the company will address them. Security is especially important when data or processing is performed at remote locations, rather than at a centralized facility. In mission-critical systems, security issues will have a major impact on system architecture and design.

Web-based systems introduce additional security concerns, as critical data must be protected in the Internet environment. Also, firms that use e-commerce applications must assure customers that their personal data is safe and secure. System security concepts and strategies are discussed in detail in Chapter 12, Managing Systems Support and Security.

FIGURE 10-6 User IDs and passwords are important elements of system security.

PLANNING THE ARCHITECTURE

Every information system involves three main functions: data storage and access methods, application programs to handle the processing logic, and an interface that allows users to interact with the system. Depending on the architecture, the three functions are performed on a server, on a client, or are divided between the server and the client. As you plan the system design, you must determine where the functions will be carried out and the advantages and disadvantages of each design approach. This section discusses server and client characteristics and how each design alternative handles system functions.

Servers

A **server** is a computer that supplies data, processing services, or other support to one or more computers, called **clients**. A system design where the server performs *all* the processing sometimes is described as **mainframe architecture**. Although the actual server does not have to be a mainframe, the term mainframe architecture typically describes a multiuser environment where the server is significantly more powerful than the clients. A systems analyst should know the history of mainframe architecture to understand the server's role in modern system design.

MAINFRAME HISTORY In the 1960s, mainframe architecture was the *only* system design available. In addition to centralized data processing, early systems performed all data input and output at a central location, often called a **data processing center**. Physical data was delivered or transmitted in some manner to the data processing center, where it was entered into the system. Users in the organization had no input or output capability, except for printed reports that were distributed by a corporate IT department.

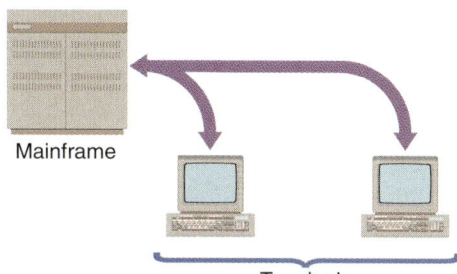

Mainframe

Terminals

FIGURE 10-7 In a centralized design, the remote user's keystrokes are transmitted to the mainframe, which responds by sending screen output back to the user's screen.

SERVER-BASED PROCESSING As network technology advanced and became affordable, companies installed terminals at remote locations, so that users could enter and access data from anywhere in the organization, regardless of where the centralized computer was located. A **terminal** included a keyboard and display screen to handle input and output, but lacked independent processing capability. In a centralized design, as shown in Figure 10-7, the remote user's keystrokes are transmitted from his or her terminal to the mainframe, which responds by sending screen output back to the user's screen.

A main advantage of server-based processing is that various types of terminals can communicate with the mainframe, and the design is not tied to a specific hardware platform. A disadvantage is that server-based processing typically uses character-based terminals that provide a limited interface for users. In a server-based system, all data storage, access, and application programs are located on the mainframe.

Today, mainframe architecture still is used in industries that require large amounts of data processing that can be done in batches at a central location. For example, a credit card company might run monthly statements in a batch, or a bank might use mainframe servers to update customer balances each night. In a blend of old and new technology, an Internet-based retail operation might use centralized data management at a customer service center to support and manage its online sales activity, as shown in Figure 10-8.

As server technology evolved, terminals also changed dramatically. Instead of simple input–output devices, a company might use a mix of PCs, handheld computers, and other specialized hardware that allows users to interact with a centralized server. In most companies, workstations that use powerful GUIs have replaced character-based terminals.

FIGURE 10-8 Internet-based retail operations such as Amazon.com use customer service centers to support online sales activity.

Clients

As PC technology exploded in the 1980s and 1990s, powerful microcomputers quickly appeared on corporate desktops. Users found that they could run their own word processing, spreadsheet, and database applications, without assistance from the IT group, in a mode called stand-alone computing. Before long, companies linked the stand-alone computers into networks that enabled the user clients to exchange data and perform local processing.

STAND-ALONE COMPUTING When an individual user works in **stand-alone** mode, the workstation performs all the functions of a server by storing, accessing, and processing data, as well as providing a user interface. Although stand-alone PCs improved employee productivity and allowed users to perform tasks that previously required IT department assistance, stand-alone computing was inefficient and expensive. Even worse, maintaining data on individual workstations raised major concerns about data security, integrity, and consistency. Without a central storage location, it was impossible to protect and back up valuable business data, and companies were exposed to enormous risks. In some cases, users who were frustrated by a lack of support and services from the IT department created and managed their own databases. In addition to security concerns, this led to data inconsistency and unreliability.

LOCAL AND WIDE AREA NETWORKS As technology became available, companies resolved the problems of stand-alone computing by joining clients into a **local area network (LAN)** that allows sharing of data and hardware resources, as shown in Figure 10-9. One or more LANs, in turn, can connect to a centralized server. Further advances in technology made it possible to create powerful networks that could use satellite links, high-speed fiber-optic lines, or the Internet to share data.

Printer   Scanner

LAN

Server   Client

Client   Client

FIGURE 10-9 A LAN allows sharing of data and hardware, such as printers and scanners.

ON THE WEB

For more information about local and wide area networks, visit **scsite.com/ sad8e/ more**, locate Chapter 10, and then click the Local and Wide Area Networks link.

A **wide area network (WAN)** spans long distances and can connect LANs that are continents apart, as shown in Figure 10-10. When a user accesses data on a LAN or WAN, the network is **transparent** because a user sees the data as if it were stored on his or her own workstation. Company-wide systems that connect one or more LANs or WANs are called **distributed systems**. The capabilities of a distributed system depend

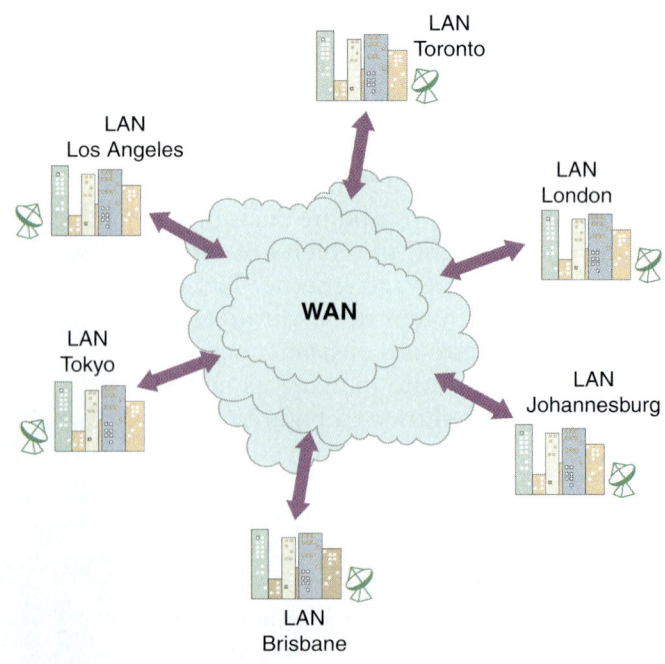

LAN
Toronto

LAN
Los Angeles

LAN
London

WAN

LAN
Tokyo

LAN
Johannesburg

LAN
Brisbane

FIGURE 10-10 A WAN can connect many LANs and link users who are continents apart.

on the power and capacity of the underlying data communication network. Compared to mainframe architecture, distributed systems increase concerns about data security and integrity because many individual clients require access to perform processing.

CLIENT-BASED PROCESSING In a typical LAN, clients share data stored on a local server that supports a group of users or a department. As LANs became popular, the most common LAN configuration was a file server design, as shown in Figure 10-11. In a **file server** design, also called a **file sharing architecture**, an individual LAN client has a copy of the application program installed locally, while the data is stored on a central file server. The client requests a copy of the data file and the server responds by transmitting the entire data file to the client. After performing the processing locally, the client returns the data file to the central file server where it is stored. File sharing designs are efficient only if the number of networked users is low and the transmitted data file sizes are relatively small. Because the entire data file is sent to each requesting client, a file server design requires significant network resources.

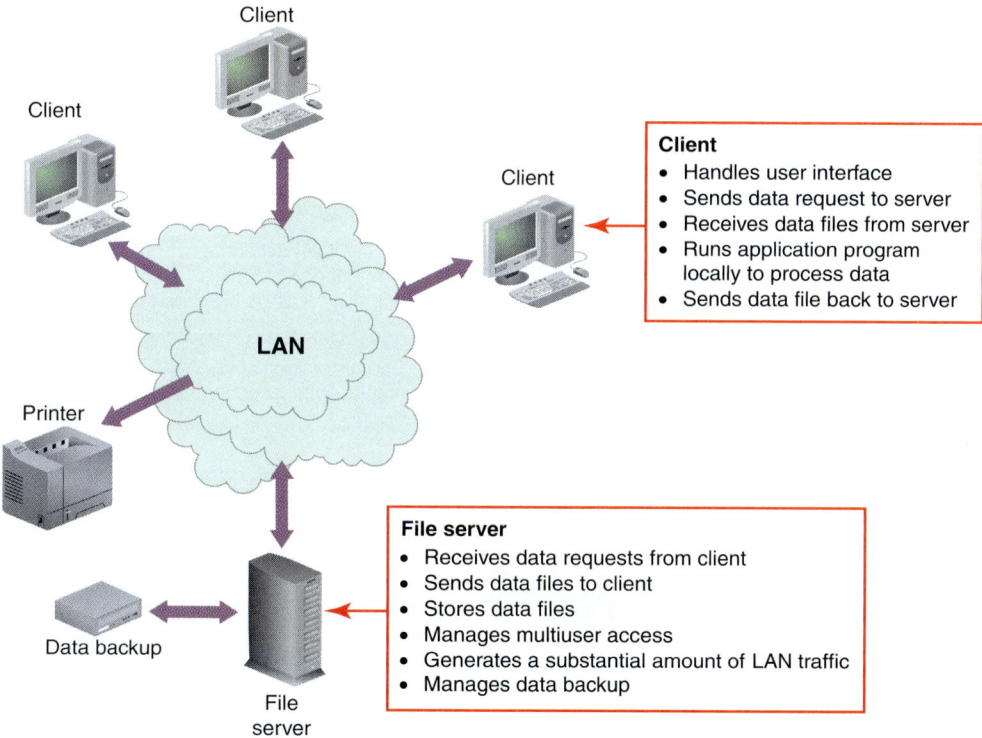

FIGURE 10-11 Example of a LAN file server design. The server stores and manages the data, while the clients run the application program and perform all the processing.

CLIENT/SERVER ARCHITECTURE

Today's interconnected world requires an information architecture that spans the entire enterprise. Whether you are dealing with a departmental network or a multinational corporation, as a systems analyst you will work with a distributed computing strategy called client/server architecture.

Overview

Although no standard definition exists, the term **client/server architecture** generally refers to systems that divide processing between one or more networked clients and a central server. In a typical client/server system, the client handles the entire user interface,

ON THE WEB

For more information about client/server architecture, visit **scsite.com/sad8e/ more**, locate Chapter 10, and then click the Client/Server Architecture link.

including data entry, data query, and screen presentation logic. The server stores the data and provides data access and database management functions. Application logic is divided in some manner between the server and the clients. In a client/server interaction, the client submits a request for information from the server, which carries out the operation and responds to the client. As shown in Figure 10-12, the data file is not transferred from the server to the client — only the request and the result are transmitted across the network. To fulfill a request from a client, the server might contact other servers for data or processing support, but that process is transparent to the client. The analogy can be made to a restaurant where the customer gives an order to a server, who relays the request to a cook, who actually prepares the meal.

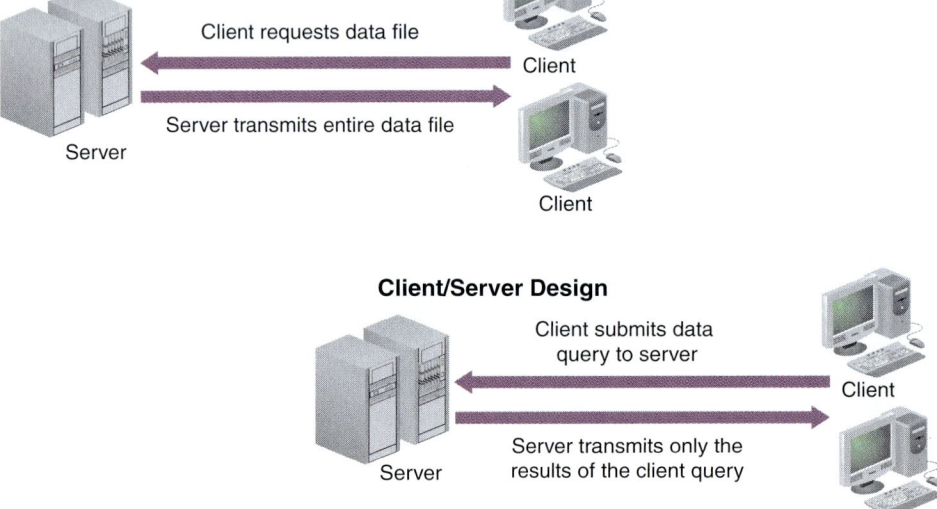

FIGURE 10-12 A file server design compared to a client/server design.

Comparison of Client/Server and Mainframe Systems		
Characteristics	**Client/Server**	**Mainframe**
Basic architecture	Very flexible	Very rigid
Application development	Flexible Fast Object-oriented	Highly structured Slow Traditional
User environment	PC-based GUI Empowers the user Improves productivity	Uses terminals Text interface Constrains the user Limited options
Security and control features	Decentralized Difficult to control	Centralized Easier to control
Processing options	Can be shared and configured in any form desired	Cannot be modified
Data storage options	Can be distributed to place data closer to users	All data is stored centrally
Hardware/software integration	Very flexible Multivendor model	Very rigid Single proprietary vendor

FIGURE 10-13 Comparison of the characteristics of client/server and mainframe systems.

Figure 10-13 lists some major differences between client/server and traditional mainframe systems. Many early client/server systems did not produce expected savings because few clear standards existed, and development costs often were higher than anticipated. Implementation was expensive because clients needed powerful hardware and software to handle shared processing tasks. In addition, many companies had an installed base of data, called **legacy data**, which was difficult to access and transport to a client/server environment.

As large-scale networks grew more powerful, client/server systems became more cost-effective. Many companies invested in client/server systems to achieve a unique combination of computing power, flexibility, and support for changing business operations. Today, client/server architecture is the dominant form of systems design, using Internet protocols and network models such as the ones described on pages 470–474. As businesses form new alliances with customers and suppliers, the client/server concept continues to expand to include clients and servers outside the organization.

Client/Server Design Styles

Client/server designs can take many forms, depending on the type of server and the relationship between the server and the clients. Figure 10-14 shows the client/server interaction for a database server, a transaction server, an object server, and a Web server. Notice that in each case, the processing is divided between the server and the clients. The nature of the communication depends on the type of server: A database server processes individual SQL commands, a transaction server handles a set of SQL commands, an object server exchanges object messages with clients, and a Web server sends and receives Internet-based communications.

Client/server design styles

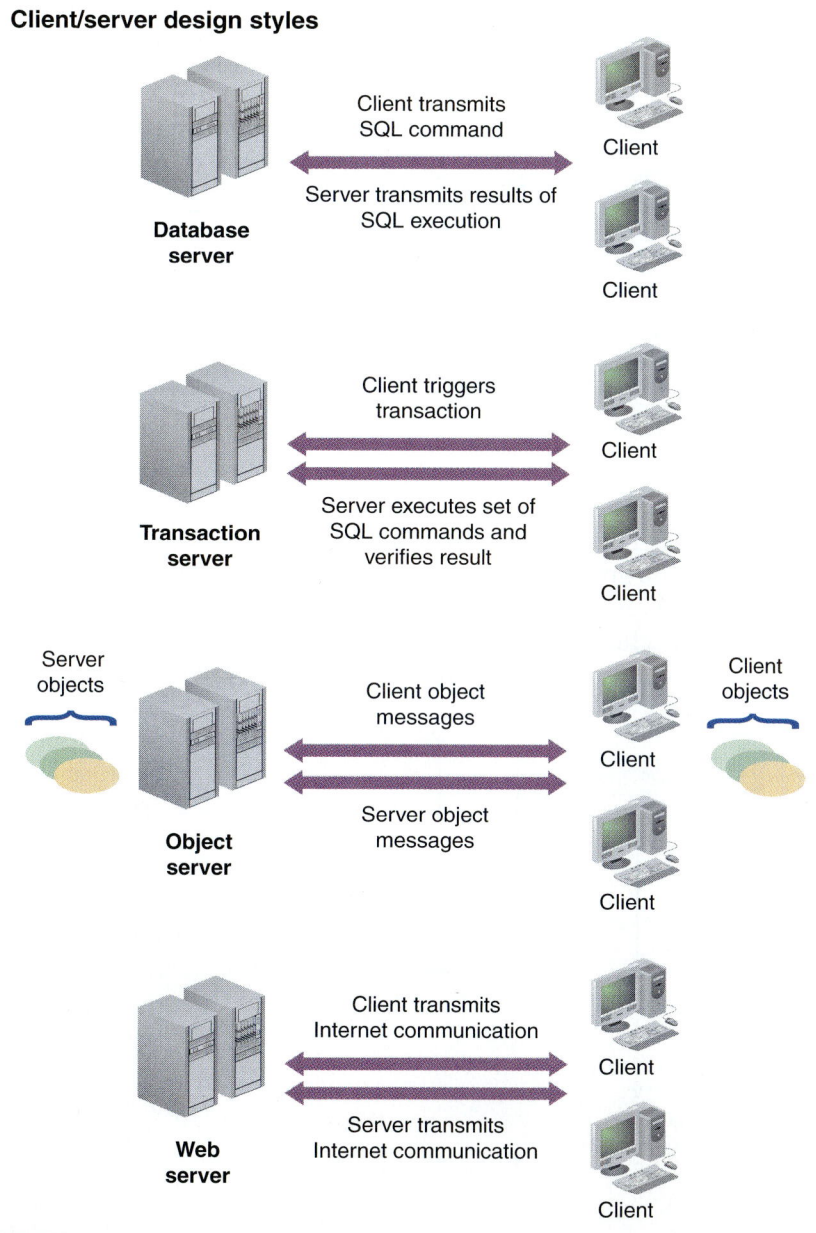

FIGURE 10-14 Client/server interaction for a database server, a transaction server, an object server, and a Web server.

Fat and Thin Clients

Client/server designs can be based on fat or thin clients. A **fat client**, also called a **thick client**, design locates all or most of the application processing logic at the client. A **thin client** design locates all or most of the processing logic at the server. What are the advantages and disadvantages of each design? Most IT experts agree that thin client designs provide better performance, because program code resides on the server, near the data. In contrast, a fat client handles more of the processing and must access and update the data more often. Compared with maintaining a central server, fat client TCO also is higher, because of initial hardware and software requirements and the ongoing expense of supporting and updating remote client computers. A fat client design, however, is simpler and less expensive to develop, because the architecture resembles traditional file server designs where all processing is performed at the client. Figure 10-15 compares the characteristics of fat and thin clients.

Characteristic	Fat Client	Thin Client
Network traffic	Higher, because the fat client must communicate more often with the server to access data and update processing results	Lower, because most interaction between code and data takes place at the server
Performance	Slower, because more network traffic is required	Faster, because less network traffic is required
Initial cost	Higher, because more powerful hardware is required	Lower, because workstation hardware requirements are not as stringent
Maintenance cost	Higher, because more program code resides on the client	Lower, because most program code resides on the central server
Ease of development	Easier, because systems resemble traditional file-server designs where all processing was performed at the client	More difficult, because developers must optimize the division of processing logic

FIGURE 10-15 Characteristics of fat and thin clients.

Client/Server Tiers

Early client/server designs were called two-tier designs. In a **two-tier** design, the user interface resides on the client, all data resides on the server, and the application logic can run either on the server or on the client, or be divided between the client and the server.

More recently, another form of client/server design, called a three-tier design, has become popular. In a **three-tier** design, the user interface runs on the client and the data is stored on the server, just as with a two-tier design. A three-tier design also has a middle layer between the client and server that processes the client requests and translates them into data access commands that can be understood and carried out by the server, as shown in Figure 10-16. You can think of the middle layer as an **application server**, because it provides the **application logic**, or **business logic**, required by the system. Three-tier designs also are called **n-tier** designs, to indicate that some designs use more than one intermediate layer.

The advantage of the application logic layer is that a three-tier design enhances overall performance by reducing the data server's workload. The

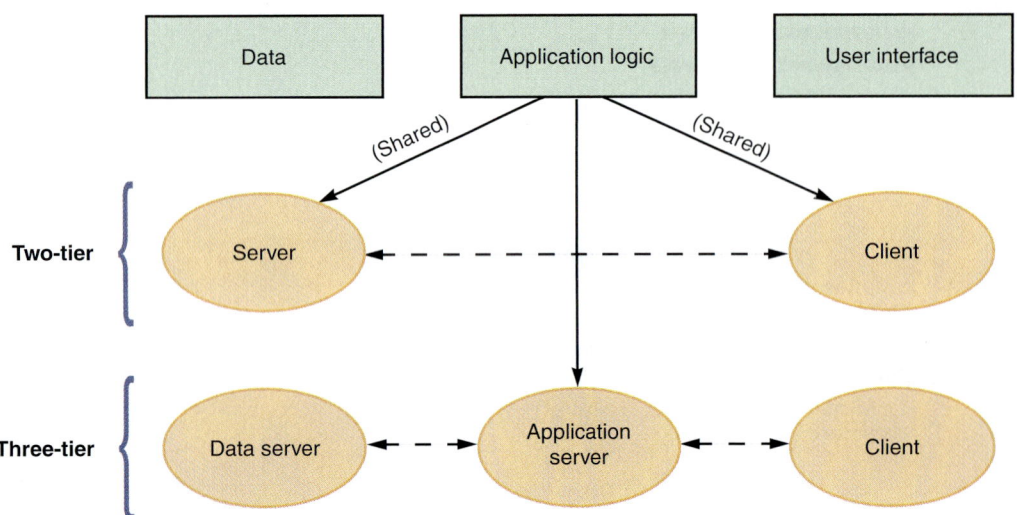

FIGURE 10-16 Characteristics of two-tier versus three-tier client/server design.

separate application logic layer also relieves clients of complex processing tasks. Because it can run on a mini-computer that is much more powerful than the typical client workstations, the middle layer is more efficient and cost-effective in large-scale systems. Figure 10-17 shows where the data, the application logic, and the user interface are located on various architectures. In a client/server system, the tiers communicate using software called middle-ware, which is described in the following section.

Architecture		Data	Application Logic	User Interface
Central data processing center	Server	X	X	X
	Client			
Central server with remote terminals	Server	X	X	
	Client			X
Stand-alone client	Server			
	Client	X	X	X
Two-tier client/server	Server	X	X	
	Client		X	X
Three-tier client/server	Data server	X		
	Application server		X	
	Client			X

FIGURE 10-17 The location of the data, the application logic, and the user interface depend on the type of architecture.

Middleware

In an n-tier system, special software called **middleware** enables the tiers to communicate and pass data back and forth. Some IT professionals refer to middleware as the glue that holds clients and servers together. The broader definition shown in Figure 10-18 on the next page states that middleware is software that mediates between an application program and a network.

Middleware provides a transparent interface that enables system designers to integrate dissimilar software and hardware. For example, middleware can link a departmental database to a Web server, which can be accessed by client computers via the Internet or a company intranet. Middleware also can integrate legacy systems and Web-based applications. For example, when a user enters a customer number on a Web-based inquiry form, middleware accesses a legacy accounting system and returns the results.

ON THE WEB

For more information about middleware, visit **scsite.com/ sad8e/more**, locate Chapter 10, and then click the Middleware link.

Cost-Benefit Issues

To support business requirements, information systems need to be scalable, powerful, and flexible. For most companies, client/server systems offer the best combination of features to meet those needs. Whether a business is expanding or downsizing, client/server systems enable the firm to scale the system in a rapidly changing environment. As the size of the business changes, it is easier to adjust the number of clients and the processing functions they perform than it is to alter the capability of a large-scale central server.

Client/server computing also allows companies to transfer applications from expensive mainframes to less-expensive client platforms. In addition, using common languages such as SQL, clients and servers can communicate across multiple platforms. That difference is important because many businesses have substantial investments in a variety of hardware and software environments.

Finally, compared to file server designs, client/server systems reduce network load and improve response times. For example, consider a user at a company headquarters who wants information about total sales figures. In a file server design, the system might need to transmit three separate sales transaction files from three regional offices in order to provide sales data that the client would process; in a client/server system, the server locates the data, performs the necessary processing, and responds immediately to the client's request. The data retrieval and processing functions are transparent to the client because they are done on the server, not the client.

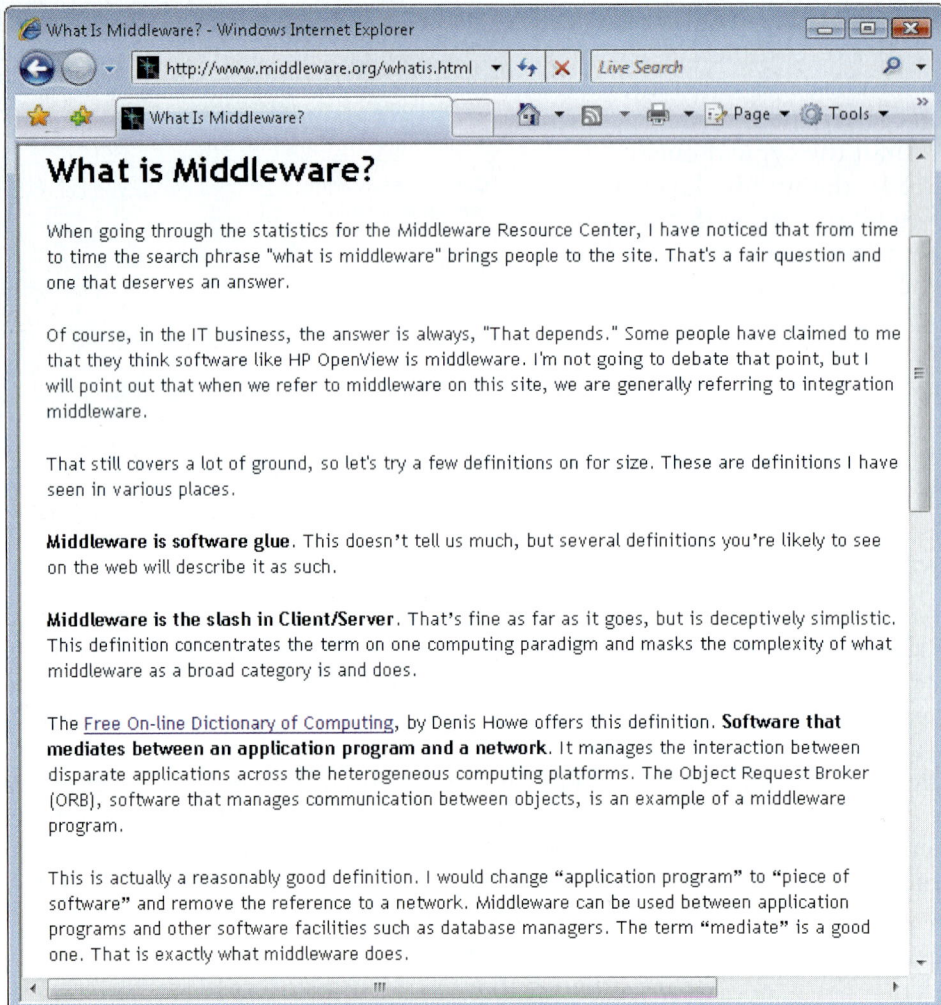

FIGURE 10-18 Middleware connects dissimilar applications and enables them to communicate and exchange data. Middleware also can integrate legacy systems and Web-based applications.

Client/Server Performance Issues

While it provides many important advantages over file-based systems, client/server architecture does involve performance issues that relate to the separation of server-based data and networked clients that must access the data.

Consider the difference between client/server design and a centralized environment, where a server-based program issues a command that is executed by the server's own CPU. Processing speed is enhanced because program instructions and data both travel on an internal system bus, which moves data more efficiently than an external network.

In contrast to the centralized system, a client/server design separates applications and data. Networked clients submit data requests to the server, which responds by sending data back to the clients. When the number of clients and the demand for services increases beyond a certain level, network capacity becomes a constraint, and system performance declines dramatically.

In the article shown in Figure 10-19, IBM states that the performance characteristics of a client/server system are not the same as a centralized processing environment. The article points out that as the communication between the clients and server increases, system performance degrades exponentially, as shown in the graph of average response time versus number of users. To deliver and maintain acceptable performance, system developers must anticipate the number of users, network traffic, server size and location, and design a client/server architecture that can support current and future business needs.

What is the answer to enhancing client/server performance? According to IBM, client/server systems must be designed so the client contacts the server only when necessary and makes as few trips as possible.

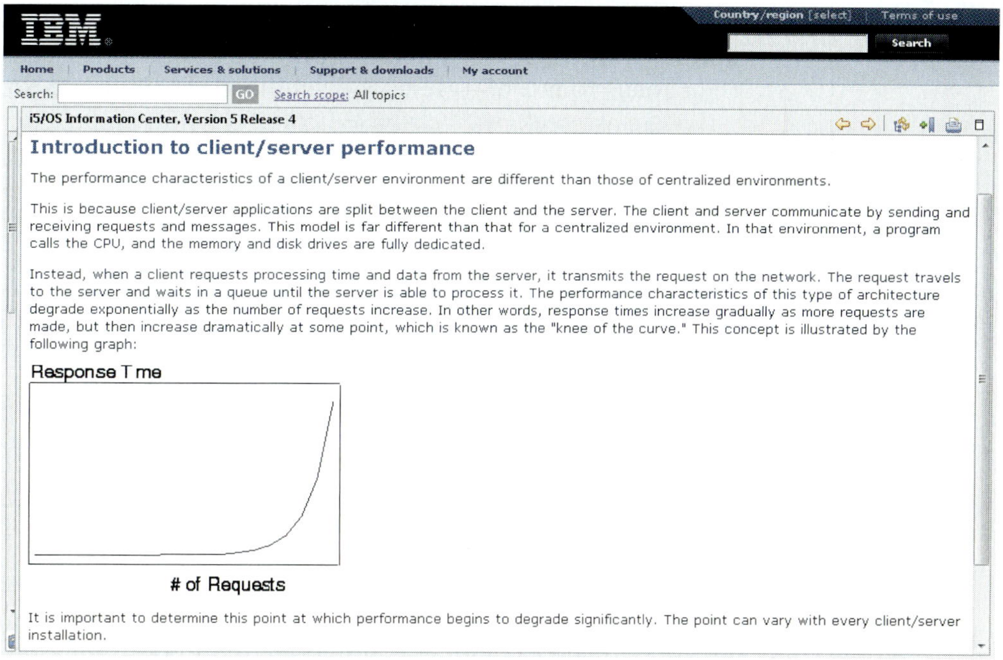

Another issue that affects client/server performance is data storage. Just as processing can be done at various places, data

FIGURE 10-19 According to IBM, client/server performance issues are far different than in a centralized environment. Response times increase gradually as more requests are made, but then increase dramatically at some point.

can be stored in more than one location using a **distributed database management system (DDBMS)**.

Using a DDBMS offers several advantages: Data stored closer to users can reduce network traffic; the system is scalable, so new data sites can be added without reworking the system design; and with data stored in various locations, the system is less likely to experience a catastrophic failure. A potential disadvantage of distributed data storage involves data security. It can be more difficult to maintain controls and standards when data is stored in various locations. In addition, the architecture of a DDBMS is more complex and difficult to manage. From a system design standpoint, the challenge is that companies often want it both ways — they want the control that comes with centralization *and* the flexibility associated with decentralization.

INTERNET-BASED ARCHITECTURE

The Internet has had an enormous impact on system architecture. The Internet has become more than a communication channel — many IT observers see it as a fundamentally different environment for system development.

Recall that in a traditional client/server system, the client handles the user interface, as shown in Figure 10-16 on page 458, and the server (or servers in a multi-tier system) handles the data and application logic. In a sense, part of the system runs on the client, part on the server. In contrast, in an Internet-based architecture, in addition to data and application logic, the entire user interface is provided by the Web server in the form of HTML coded documents that are interpreted and displayed by the client's browser. Shifting the responsibility for the interface from the client to the server simplifies the process of data transmission and results in lower hardware costs and complexities.

The trend toward Internet-based e-business is reshaping the IT landscape as more firms use the Web to build efficient, reliable, and cost-effective solutions. When planning new systems, analysts can use available and emerging technology to meet their company's business requirements.

The advantages of Internet-based architecture are changing fundamental ideas about how computer systems should be designed, and many IT experts are shifting their focus to a total online environment. At the same time, large numbers of individual users are seeking Web-based collaboration and social networking services to accomplish tasks that used to be done in person, over the phone, or by more traditional Internet channels. As you learned in Chapter 7, cloud computing and Web 2.0 are important concepts that reflect this online shift.

The following sections examine Web-based architecture, including in-house development, packaged solutions, e-business service providers, corporate portals, cloud computing, and Web 2.0. It is important to be aware of these trends, as they may predict where the IT industry is headed.

Developing E-Commerce Solutions In-House

In Chapter 7, you learned how to analyze advantages and disadvantages of in-house development versus purchasing a software package. The same basic principles apply to system design.

If you decide to proceed with an in-house solution, you must have an overall plan to help achieve your goals. How should you begin? Figure 10-20 offers guidelines for companies developing e-commerce strategies. An in-house solution usually requires a greater initial investment, but provides more flexibility for a company that must adapt quickly in a dynamic e-commerce environment. By working in-house, a company has more freedom to integrate with customers and suppliers and is less dependent on vendor-specific solutions.

Guidelines for In-house E-commerce Site Development
Analyze the company's business needs and develop a clear statement of your goals. Consider the experience of other companies with similar projects.
Obtain input from users who understand the business and technology issues involved in the project. Plan for future growth, but aim for ease of use.
Determine whether the IT staff has the necessary skills and experience to implement the project. Consider training, additional resources, and the use of consultants if necessary.
Consider integration requirements for existing legacy systems or enterprise resource planning. Select a physical infrastructure carefully, so it will support the application, now and later.
Develop the project in modular form so users can test and approve the functional elements as you go along.
Connect the application to existing in-house systems and verify interactivity.
Test every aspect of the site exhaustively. Consider a preliminary rollout to a pilot group to obtain feedback before a full launch.

FIGURE 10-20 Guidelines for companies developing e-commerce strategies.

For smaller companies, the decision about in-house Web development is even more critical, because this approach will require financial resources and management attention that many small companies might be unable or unwilling to commit. An in-house strategy, however, can provide valuable benefits, including the following:

• A unique Web site, with a look and feel consistent with the company's other marketing efforts

• Complete control over the organization of the site, the number of pages, and the size of the files

• A scalable structure to handle increases in sales and product offerings in the future

• More flexibility to modify and manage the site as the company changes

• The opportunity to integrate the firm's Web-based business systems with its other information systems, creating the potential for more savings and better customer service

Whether a firm uses an in-house or a packaged design, the decision about Web hosting is a separate issue. Although internal hosting has some advantages, such as greater control and security, the expense would be much greater, especially for a small- to medium-sized firm.

CASE IN POINT 10.2: SMALL POTATOES, INC.

Small Potatoes is a family-operated seed business that has grown rapidly. Small Potatoes specializes in supplying home gardeners with the finest seeds and gardening supplies. Until now, the firm has done all its business by placing ads in gardening and health magazines, and taking orders using a toll-free telephone number.

Now, the family has decided to establish a Web site and sell online, but there is some disagreement about the best way to proceed. Some say it would be better to develop the site on their own, and Betty Lou Jones, a recent computer science graduate, believes she can handle the task. Others, including Sam Jones, Betty's grandfather, feel it would be better to outsource the site and focus on the business itself. Suppose the family asked for your opinion. What would you say? What additional questions would you ask?

Packaged Solutions and E-Commerce Service Providers

If a small company is reluctant to take on the challenge and complexity of developing an Internet commerce site in-house, an alternative can be a packaged solution or an e-commerce service provider. This is true even for medium- to large-sized firms. Many vendors, including Microsoft and Intershop, offer turnkey systems for companies that want to get an e-business up and running quickly, as shown in Figure 10-21.

For large-scale systems that must integrate with existing applications, packaged solutions might be less attractive. Another alternative is to use an application service provider (ASP). As explained in Chapter 7, an ASP provides applications, or access to applications, by charging a usage or subscription fee. Today, many ASPs offer full-scale Internet business services for companies that decide to outsource those functions.

FIGURE 10-21 Microsoft and Intershop offer software solutions for companies that want to get an e-business up and running quickly.

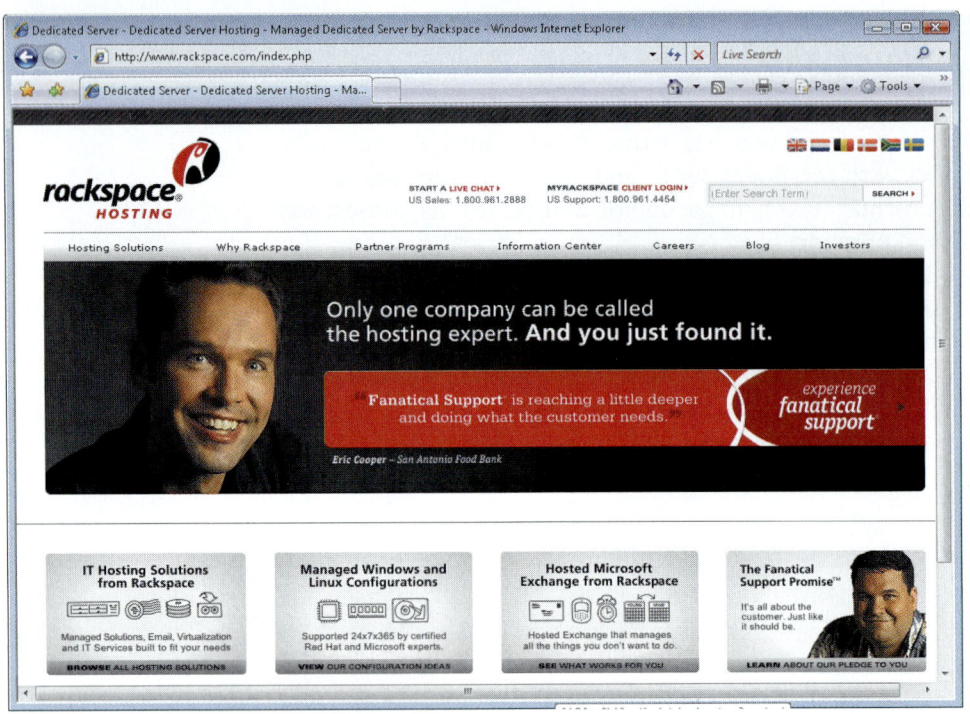

FIGURE 10-22 Rackspace is an example of a firm that offers Web hosting and maintenance solutions.

Another option is managed hosting, which also was discussed in Chapter 7. As shown in Figure 10-22, a solution provider such as Rackspace can host and maintain a corporate Web site. Rackspace states that its customers will "never have to implement, update, troubleshoot, patch, monitor, administer, backup data, or worry again."

A systems analyst confronts a bewildering array of products and strategies when implementing Internet-based systems. A good starting point might be to consider the experience of other companies in the same industry. Many firms, including Sybase, offer success stories and case studies of successful systems development, as shown in Figure 10-23. Although each situation is different, this type of research can provide valuable information about a vendor's products and services.

FIGURE 10-23 Success stories and case studies can provide valuable information about a vendor's products and services.

Corporate Portals

A **portal** is an entrance to a multifunction Web site. After entering a portal, a user can navigate to a destination using various tools and features provided by the portal designer. A corporate portal can provide access for customers, employees, suppliers, and the public. In a Web-based system, portal design provides an important link between the user and the system, and poor design can weaken system effectiveness and value. Figure 10-24 shows Oracle Portal, which the company describes as a unified and secure point of access to information and services. The main benefit of a portal, as Oracle points out, is that it provides an interface that empowers users, including employees, customers, partners, and suppliers, and enables self-service access to enterprise content and services.

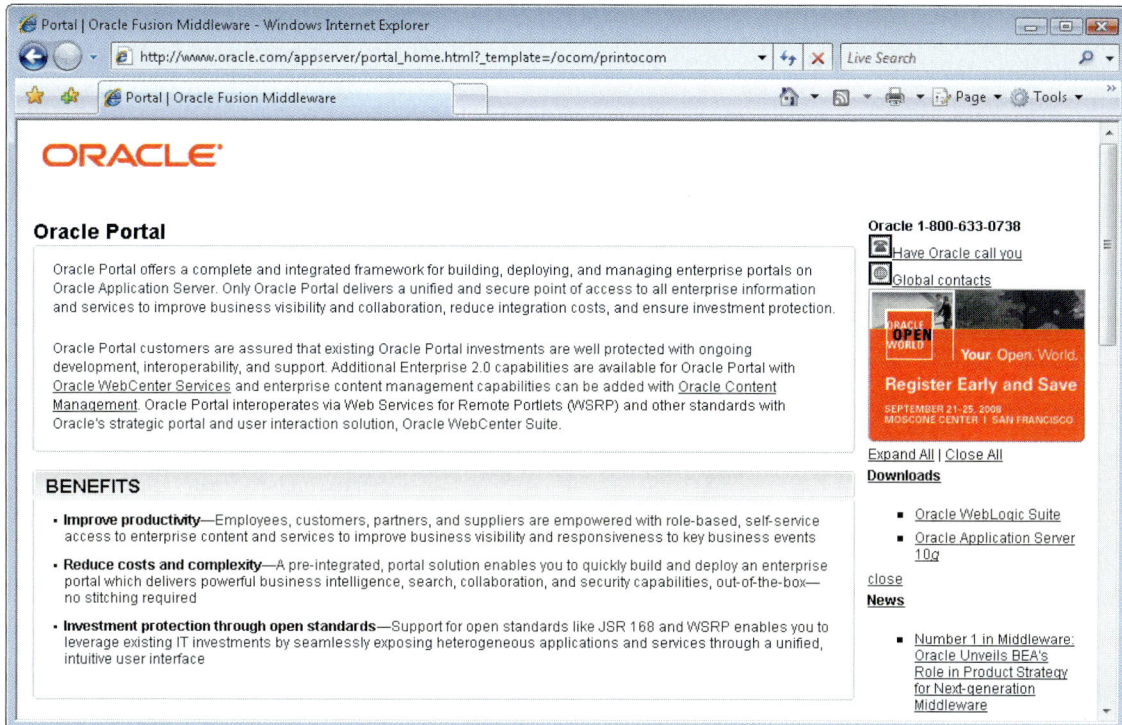

FIGURE 10-24   Oracle offers powerful enterprise portal software.

Cloud Computing

Cloud computing refers to the cloud symbol that often is used to represent the Internet. The cloud computing concept envisions a *cloud* of remote computers that provide a total online software and data environment that is hosted by third parties. A user's computer does not perform processing or computing tasks — the cloud does. This concept is in contrast to today's computing model, which is based on networks that strategically distribute processing and data across the enterprise. In a sense, the cloud of computers acts as one giant computer that performs tasks for users. As shown in Figure 10-25, a user logs into a local computer and is connected to the cloud, which performs the computing work. Instead of requiring specific hardware and software on the user's computer, cloud computing spreads the workload to powerful remote systems that are part of the cloud. The user appears to be working on a local system, but all computing is actually performed in the cloud. No updates or maintenance are required of the user, and there are no compatibility issues.

Cloud computing effectively eliminates compatibility issues, because the Internet itself is the platform. This architecture also provides **scaling on demand,** which matches resources to needs at any given time. For example, during peak loads, additional cloud servers might come on line automatically to support the workload.

User's notebook

FIGURE 10-25   In cloud computing, users connect to the Internet cloud to access personal content and services through an online software and data environment.

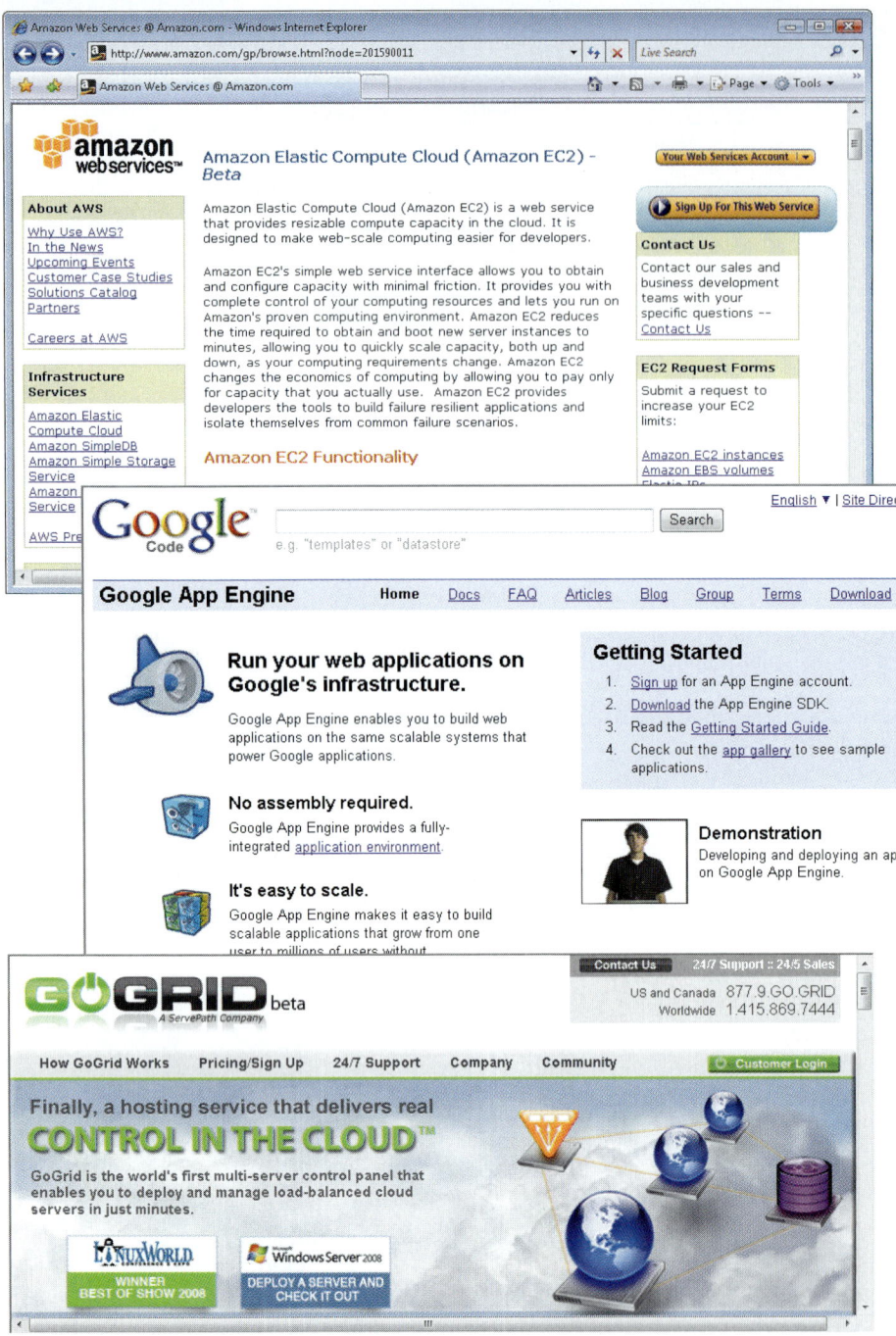

FIGURE 10-26   Three examples of current cloud computing platforms.

Cloud computing is an ideal platform for powerful Software as a Service (SaaS) applications. As you learned in Chapter 7, SaaS is a popular deployment method where software is not purchased but is paid for as a service, much like one pays for electricity or cable TV each month. In this architecture, updates and changes to services can be easily made by service providers without involving the users.

Even though cloud computing has many advantages, some concerns exist. First, cloud computing requires significantly more **bandwidth** (the amount of data that can be transferred in a fixed time period) than today's networks. Second, because cloud computing is Internet-based, if a user's Internet connection becomes unavailable, he or she will be unable to access any cloud-based services. In addition, there are security concerns associated with sending large amounts of data over the Internet, as well as concerns about storing it securely. Finally, there is the issue of control. Because a service provider hosts the resources and manages data storage and access, the provider has complete control of the system. Many firms are wary of handing over control of mission-critical data and systems to a third-party provider.

It remains to be seen whether cloud computing's advantages will outweigh its disadvantages. Technology advances continue to make cloud computing more feasible, desirable, and secure. As the IT industry moves toward Internet-based architectures, cloud computing's success will depend on how bandwidth, reliability, and security are addressed and how well cloud computing is received by users. As shown in Figure 10-26, examples of current cloud computing platforms include Amazon's Elastic Compute Cloud, Google's App Engine, and ServePath's GoGrid.

Web 2.0

The shift to Internet-based collaboration has been so powerful and compelling that it has been named **Web 2.0**. Web 2.0 is not a reference to a more technically advanced version of the current Web. Rather, Web 2.0 envisions a second generation of the Web that will enable people to collaborate, interact, and share information more dynamically.

Leading Web 2.0 author Tim O'Reilly has suggested that the strong interest in Web 2.0 is driven by the concept of the *Internet as a platform*. O'Reilly sees future Web 2.0 applications delivering software as a continuous service with no limitations on the number of users that can connect or how users can consume, modify, and exchange data. Social networking sites such as MySpace and Facebook are part of the Web 2.0 movement, as are wikis. A **wiki** is a Web-based repository of information that anyone can access, contribute to, or modify. In a sense, a wiki represents the collective knowledge of a group of people. One of the best-known wikis is Wikipedia.org, as shown in Figure 10-27.

One of the goals of Web 2.0 is to enhance creativity, interaction, and shared ideas. In this regard, the Web 2.0 concept resembles the agile development process and the open-source software movement. Web 2.0 communities and services are based on a body of data created by users. As users collaborate, new layers of information are added in an overall environment known as the **Internet operating system**. These layers can contain text, sound bytes, images, and video clips that are shared with the user community.

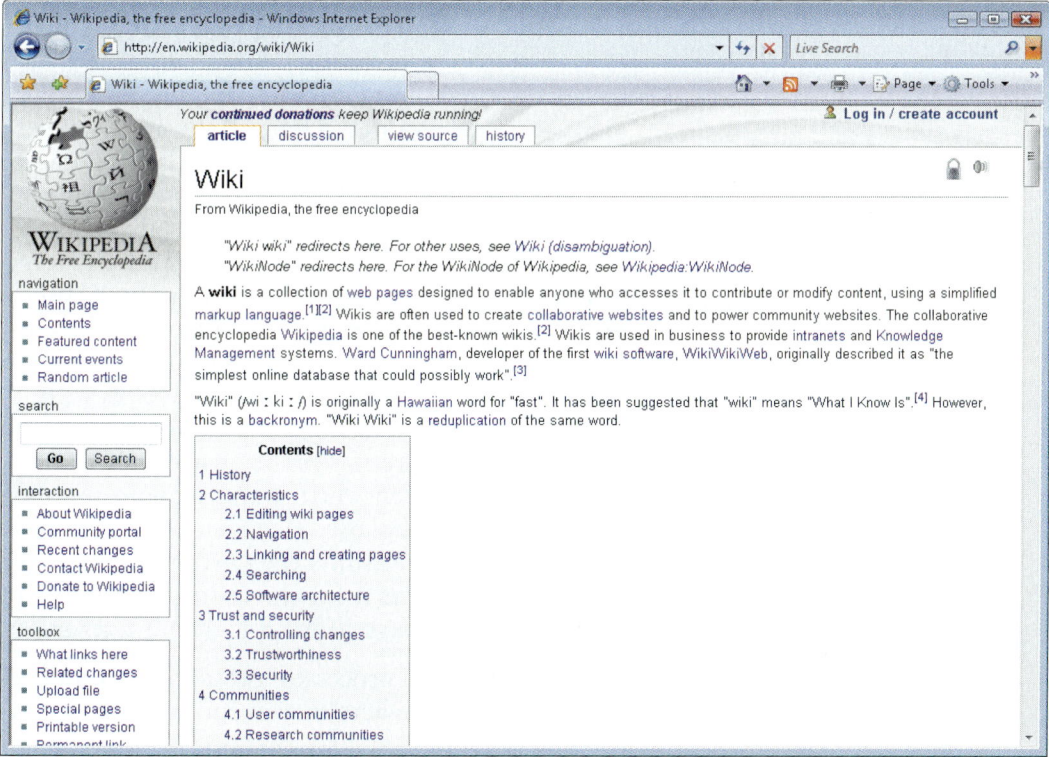

FIGURE 10-27 Wikipedia is an excellent example of a popular, well-known wiki.

PROCESSING METHODS

In selecting an architecture, the systems analyst must determine whether the system will be an online system, a batch processing system, or a combination of the two.

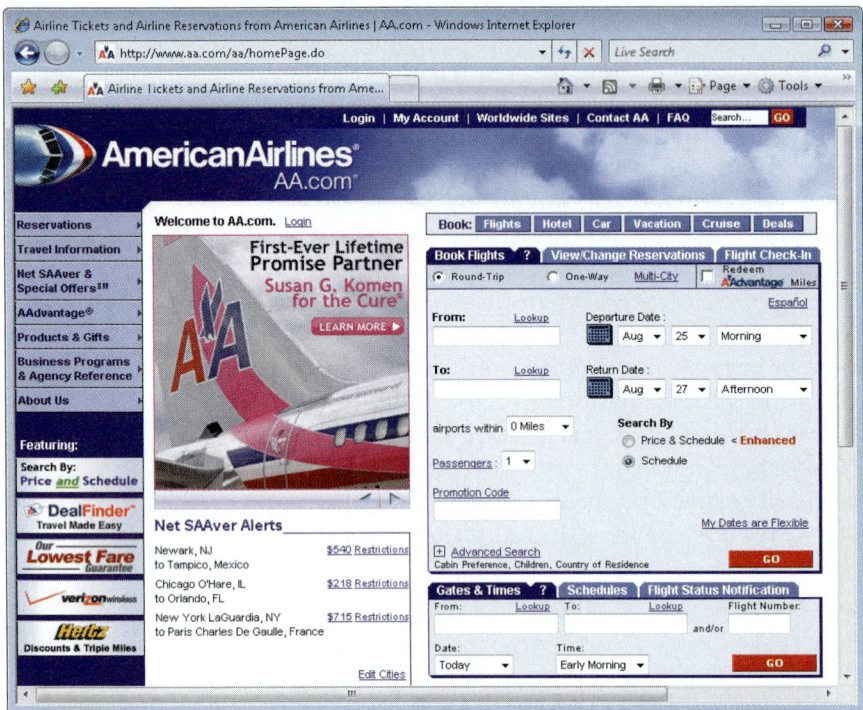

FIGURE 10-28 The American Airlines reservation system is an example of Web-based online processing.

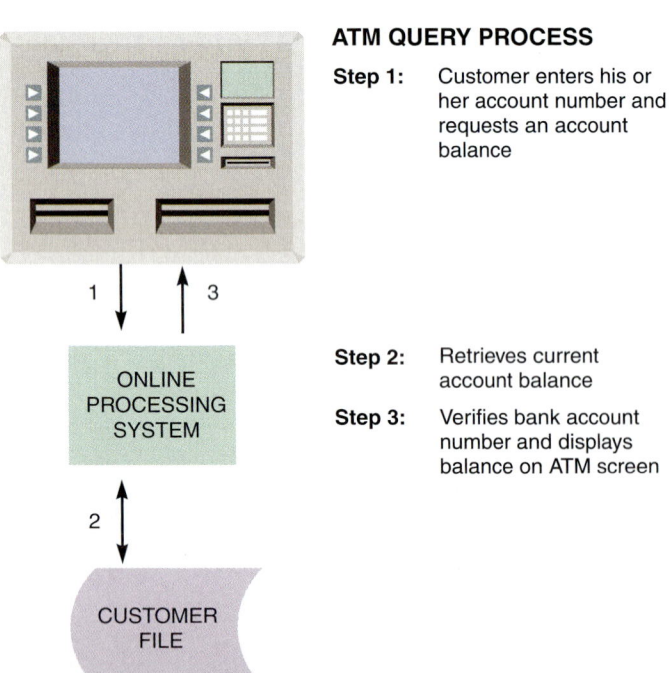

ATM QUERY PROCESS

Step 1: Customer enters his or her account number and requests an account balance

Step 2: Retrieves current account balance

Step 3: Verifies bank account number and displays balance on ATM screen

FIGURE 10-29 When a customer requests a balance, the ATM system verifies the account number, submits the query, retrieves the current balance, and displays the balance on the ATM screen.

Online Processing

Early computer systems relied mainly on batch processing, but the vast majority of systems today use online processing. An **online system** handles transactions when and where they occur and provides output directly to users. Because it is interactive, online processing avoids delays and allows a constant dialog between the user and the system.

An airline reservations system is a familiar example of online processing. When an online customer views the screen shown in Figure 10-28, he or she can enter the origin, destination, travel dates, and travel times. The system searches a database and responds by displaying available flights, times, and prices. The customer can make a reservation, enter a name, address, credit card information, and other required data and the system creates the reservation, assigns a seat, and updates the flight database immediately.

Online processing also can be used with file-oriented systems. Figure 10-29 shows what happens when a customer uses an ATM to inquire about an account balance. After the ATM verifies the customer's card and password, the customer enters the request (Step 1). Then, the system accesses the account master file using the account number as the primary key and retrieves the customer's record (Step 2). The system verifies the account number and displays the balance (Step 3). Data is retrieved and the system transmits the current balance to the ATM, which prints it for the customer. Online processing systems have four typical characteristics:

1. The system processes transactions completely when and where they occur.

2. Users interact directly with the information system.

3. Users can access data randomly.

4. The information system must be available whenever necessary to support business functions.

Batch Processing

In a **batch processing** system, data is collected and processed in groups, or batches. Although online processing is used for interactive business systems that require immediate data input and output, batch processing can handle other situations more efficiently. For example, batch processing typically is used for large amounts of data that must be processed on a routine schedule, such as paychecks or credit card transactions.

In batch processing, input transactions are grouped into a single file and processed together. For example, when a firm produces customer statements at the end of the month, a batch application might process many thousands of records in one run of the program. A batch processing system has several main characteristics: collect, group, and process transactions periodically; the IT operations group can run batch programs on a predetermined schedule, without user involvement, during regular business hours, at night, or on weekends; and batch programs require significantly fewer network resources than online systems.

CASE IN POINT 10.3: R/WAY TRUCKING COMPANY

You are the new IT manager at R/Way, a small but rapidly growing trucking company headquartered in Cleveland, Ohio. The company slogan is "Ship It R/Way — State-of-the-Art in Trucking and Customer Service." R/Way's information system currently consists of a file server and three workstations where freight clerks enter data, track shipments, and prepare freight bills. To perform their work, the clerks obtain data from the server and use database and spreadsheet programs stored on their PCs to process the data.

Unfortunately, your predecessor did not design a relational database. Instead, data is stored in several files, including one for shippers, one for customers, and one for shipments. The system worked well for several years, but cannot handle current volume or support online links for R/Way shippers and customers. The company president is willing to make changes, but he is reluctant to spend money on major IT improvements unless you can convince him that they are necessary.

What would you recommend and why?

Combined Online and Batch Processing

Even an online system can use batch processing to perform certain routine tasks. Online processing also can be used with file-oriented systems. Figure 10-30 shows a familiar **point-of-sale (POS)** terminal, and Figure 10-31 on the next page shows how a retail chain uses POS terminals to drive online and batch processing methods. Notice that the system uses online processing to handle data entry and inventory updates, while reports and accounting entries are performed in a batch.

The retail store system illustrates both online processing and batch processing of data. During business hours, the salesperson enters a sale on a POS terminal, which is part of an information system that handles daily sales transactions and maintains the online inventory file. When the salesperson enters the transaction, online processing occurs. The system performs calculations, updates the inventory file,

FIGURE 10-30 Retail point-of-sale terminals provide customer sales support and transaction processing capability.

POINT OF SALE (POS) PROCESSING

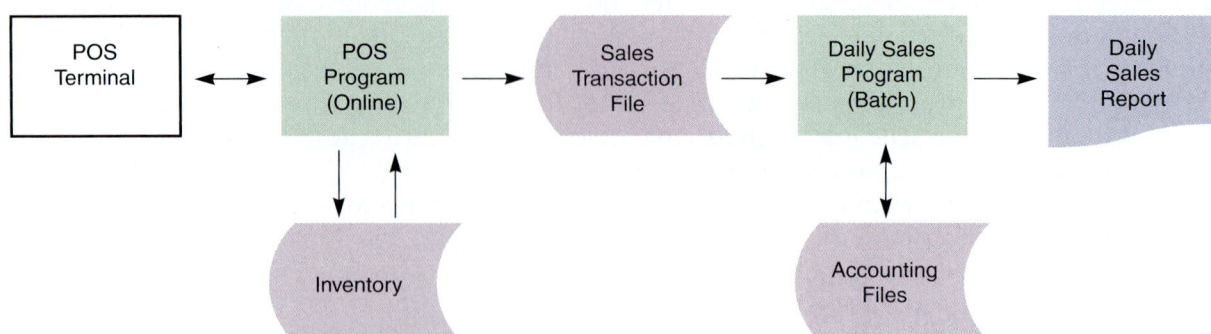

FIGURE 10-31 Many retailers use a combination of online and batch processing. When a salesperson enters the sale on the POS terminal, the online system retrieves data from the item file, updates the quantity in stock, and produces a sales transaction record. At the end of the day, a batch processing program produces a daily sales report and updates the accounting system.

and produces output on the POS terminal in the form of a screen display and a printed receipt. At the same time, each sales transaction creates input data for day-end batch processing.

When the store closes, the system uses the sales transactions to produce the daily sales report and related accounting entries using batch processing. Performing the processing online before all sales transactions are completed does not make sense. In that situation, a batch method provides better routine transaction processing, while an online approach supports point-of-sale processing, which must be done as it occurs.

In the retail store example, both online and batch processing are integral parts of the information system. Online processing offers an inherent advantage because data is entered and validated as it occurs, so the stored data is available sooner and always is up to date. Online processing is more expensive, however, and the effect of computer system downtime or slowdown while transactions are processed causes far more disruption than in batch processing. In addition, backup and recovery for online processing are more difficult. In many situations, batch processing is cost-effective, less vulnerable to system disruption, and less intrusive to normal operations. Many information systems will continue to use a combination of online and batch processing for some time to come.

NETWORK MODELS

A network allows the sharing of hardware, software, and data resources in order to reduce expenses and provide more capability to users. When planning a network design, you must consider network terms and concepts, including the OSI model, network modeling tools, network topology, network protocols, licensing issues, and wireless networks, which are covered in this section. Other important issues, such as network performance and security, are covered in Chapter 12, Managing Systems Support and Security.

The OSI Reference Model

Based on the discussion of system architecture earlier in this chapter, you already understand basic network terms such as client, server, LAN, WAN, file server design, client/server architecture, tiers, and middleware.

ON THE WEB

For more information about the OSI reference model, visit **scsite.com/ sad8e/more**, locate Chapter 10, and then click the OSI Reference Model link.

Before you study network topology, you should have a basic understanding of the **OSI (Open Systems Interconnection) model**, which describes how data actually moves from an application on one computer to an application on another networked computer. The OSI model consists of seven layers. Each layer performs a specific function, as shown in Figure 10-32.

LAYER NUMBER	NAME	DESCRIPTION
7	Application layer	Provides network services requested by a local workstation
6	Presentation layer	Ensures that data is uniformly structured and formatted for network transmission
5	Session layer	Defines control structures that manage the communications link between computers
4	Transport layer	Provides reliable data flow and error recovery
3	Network layer	Defines network addresses and determines how data packets are routed over the network
2	Data link layer	Defines specific methods of transmitting data over the physical layer, such as defining the start and end of a data frame
1	Physical layer	Contains physical components that carry data, such as cabling and connectors

FIGURE 10-32 In the OSI model, data proceeds down through the layers on the transmitting computer, then up through the layers on the receiving computer. Along the way, data may pass through one or more network routers that control the path from one network address to another.

It is important to understand that OSI is a logical model and is not tied to any specific physical environment or hardware. As a conceptual model, OSI offers a set of design standards that can promote interoperability among networks and products that are based on the OSI model.

Network Modeling Tools

As you translate the OSI logical model into a physical model of the networked system, you can use software tools, such as Microsoft Visio, which is a multipurpose drawing tool, to represent the physical structure and network components. Visio offers a wide variety of drawing types, styles, templates, and shapes. For example, Visio supplies templates for basic network designs, plus manufacturer-specific symbols for firms such as Cisco, IBM, Bay Systems, and Hewlett-Packard, among others.

Visio is an example of a CASE tool that can help you plan, analyze, design, and implement an information system. Visio can be used to create a simple network model, either by using drag-and-drop shapes displayed on the left of the screen, or by using provided wizards to walk through a step-by-step network design process. Figure 10-33 on the next page shows an example of a simple network model created using the drag-and-drop feature.

Network Topology

The way a network is configured is called the **network topology**. LAN and WAN networks typically are arranged in four patterns: hierarchical, bus, ring, and star. The concepts are the same regardless of the size of the network, but the physical implementation is different for a large-scale WAN that spans an entire business enterprise compared with a small LAN in a single department. The four topologies are described in the following sections.

TOOLKIT TIME

The CASE Tools in Part 2 of the Systems Analyst's Toolkit can help you document business functions and processes, develop graphical models, and provide an overall framework for information system development. To learn more about these tools, turn to Part 2 of the four-part Toolkit that follows Chapter 12.

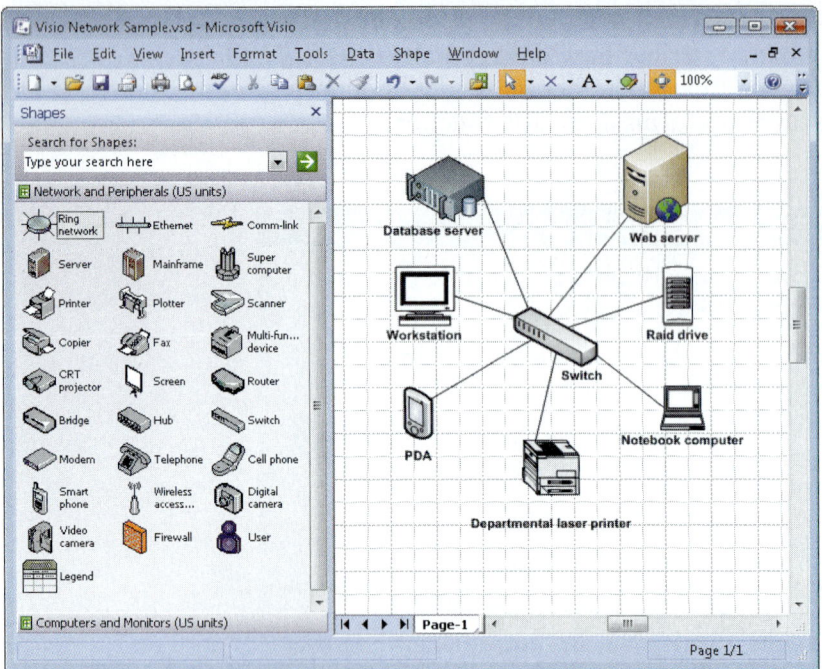

FIGURE 10-33 Microsoft Visio can be used to create a network drawing using drag-and-drop shapes displayed in the panel at the left of the screen.

HIERARCHICAL NETWORK In a **hierarchical network**, as shown in Figure 10-34, one or more powerful servers control the entire network. Departmental servers control lower levels of processing and network devices. An example of a hierarchical network might be a retail clothing chain, with a central computer that stores data about sales activity and inventory levels and local computers that handle store-level operations. The stores transmit data to the central computer, which analyzes sales trends, determines optimum stock levels, and coordinates a supply chain management system. In this situation, a hierarchical network might be used, because it mirrors the actual operational flow in the organization.

One disadvantage of a hierarchical network is that if a business adds additional processing levels, the network becomes more complex and expensive to operate and maintain. Hierarchical networks were often used in traditional mainframe-based systems, but are much less common today.

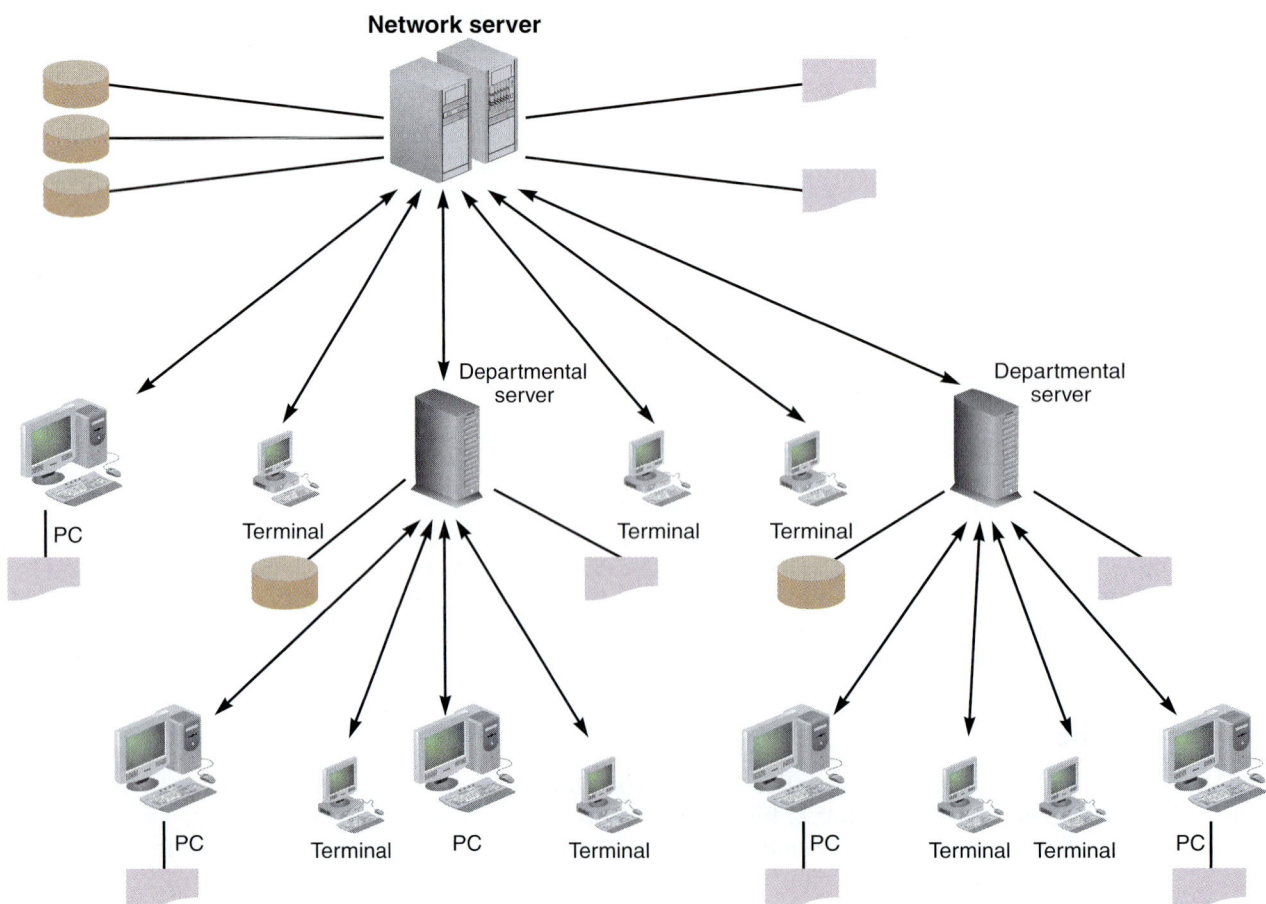

FIGURE 10-34 A hierarchical network with a single server that controls the network.

BUS NETWORK   In a **bus network**, as shown in Figure 10-35, a single communication path connects the central server, departmental servers, workstations, and peripheral devices. Information is transmitted in either direction between networked devices, and all messages travel over the same central bus. Bus networks require less cabling than other topologies, because only a single cable is used. Devices can also be attached or detached from the network at any point without disturbing the rest of the network. In addition, a failure in one workstation on the network does not necessarily affect other workstations on the network.

One major disadvantage of a bus network is that if the central bus becomes damaged or defective, the entire network shuts down. Another disadvantage is that overall performance declines as more users and devices are added, because all message traffic must flow along the central bus. This does not occur in the treelike structure of a hierarchical network or the hub-and-spoke design of a star network, where network paths are more isolated and independent.

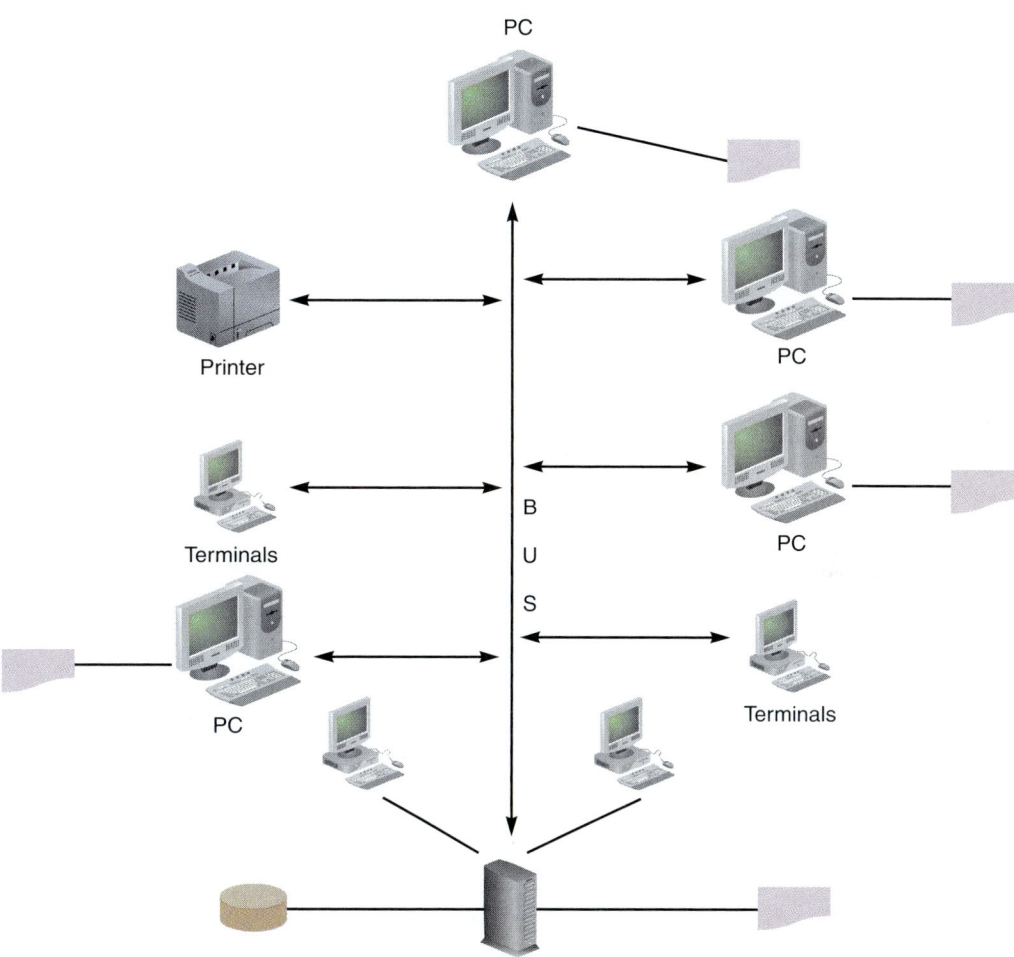

FIGURE 10-35   A bus network with all devices connected to a single communication path.

The bus network is one of the oldest LAN topologies, and is a simple way to connect multiple workstations. Before the proliferation of star networks, bus networks were very common. Today, the bus design is much less popular, but some firms have retained bus networks to avoid the expense of new wiring and hardware.

RING NETWORK   A **ring network**, as shown in Figure 10-36 on the next page, resembles a circle where the data flows in only one direction from one device to the next. In function, a ring network can be thought of as a bus network with the ends connected. One disadvantage of a ring network is that if a network device (such as a PC or a server) fails, the devices downstream from the failed device cannot communicate with the network. Although ring networks are less common than other topologies, they sometimes are used to tie local processing sites together. For example, workstations and servers in the accounting, sales, and shipping departments might perform local processing and then use a ring network to exchange data with other divisions within the company.

It is interesting to note that in a ring network implementation, the physical wiring can resemble a star pattern, using a central device called a **Multistation Access Unit (MAU)**. This unit internally wires the workstations into a logical ring, and manages the flow of data from one device to the next.

STAR NETWORK Because of its speed and versatility, the star network is by far the most popular LAN topology today. A **star network** has a central networking device called a **switch**, which manages the network and acts as a communications conduit for all network traffic. In the past, a device known as a **hub**

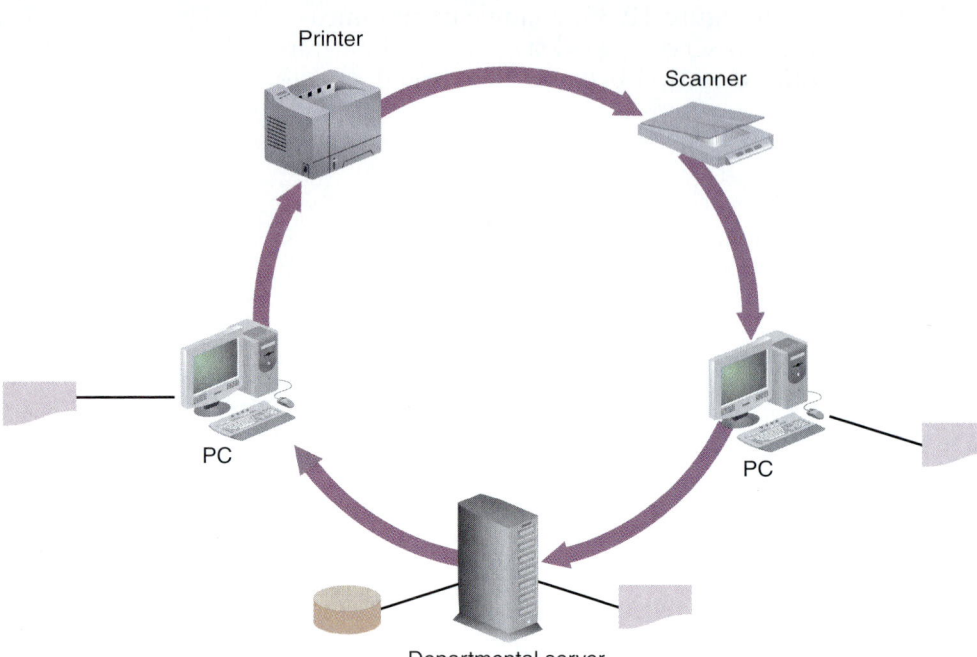

FIGURE 10-36 A ring network with a set of computers that send and receive data flowing in one direction.

was used to connect star networks, but a switch offers advanced technology and much better performance. A hub functions like a familiar multi-socket power strip, but with network devices such as servers, workstations, and printers plugged in rather than electrical appliances. The hub broadcasts network traffic, called **data frames**, to all connected devices. In contrast, a switch enhances network performance by sending traffic only to specific network devices that need to receive the data.

A star configuration, as shown in Figure 10-37, provides a high degree of network control, because all traffic flows into and out of the switch. An inherent disadvantage of the star design is that the entire network is dependent on the switch. However, in most large star networks, backup switches are available immediately in case of hardware failure.

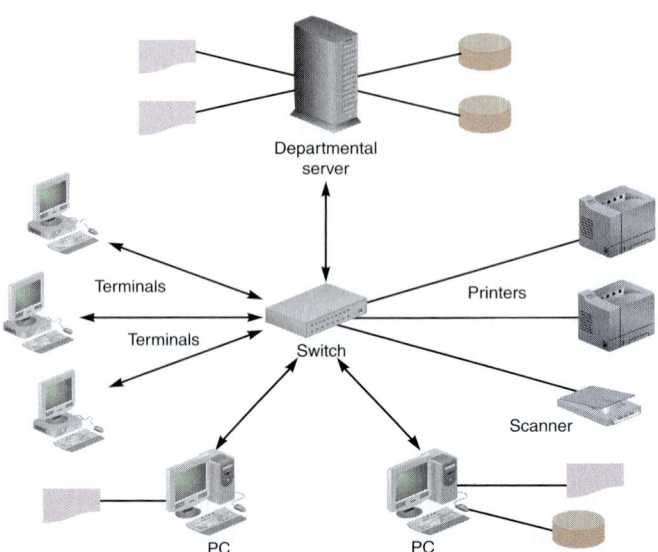

FIGURE 10-37 A typical star network with a switch, departmental server, and connected workstations.

Routers

Networks such as LANs or WANs can be interconnected using devices called routers. A **router** is a device that connects network segments, determines the most efficient data path, and guides the flow of data. Routers differ from switches in that they work at a higher OSI level (layer 3), dealing with IP packets, while switches handle data frames (layer 2). A typical router is shown in Figure 10-38.

Using a router, any network topology can connect to a larger, dissimilar network, such as the Internet. This connection is called a **gateway**. The example in Figure 10-39 shows a star topology, where the router links the network to the Internet. A device called a **proxy server** provides Internet connectivity for internal LAN users. The vast majority of business networks use routers to integrate the overall network architecture.

FIGURE 10-38 The Cisco VXR is an example of a powerful, rack-mountable router.

Network Protocols

In all cases, the network must use a **protocol**, which is a set of standards that govern network data transmission. A popular network protocol is **Transmission Control Protocol/Internet Protocol (TCP/IP)**. Originally developed by the U.S. Department of Defense to permit interconnection of military computers, today TCP/IP is the backbone of the Internet. Other older network protocols include NetBIOS, which was popular for LANs, and IPX, which is a protocol used by Novell Corporation for older NetWare products.

TCP/IP actually consists of many individual protocols that control the handling of files, mail, and Internet addresses, among others. A familiar example of a TCP/IP protocol is the **File Transfer Protocol (FTP)**, which provides a reliable means of copying files from one computer to another over a TCP/IP network, such as the Internet or an intranet.

ON THE WEB

For more information about network protocols, visit **scsite.com/sad8e/more**, locate Chapter 10, and then click the Network Protocols link.

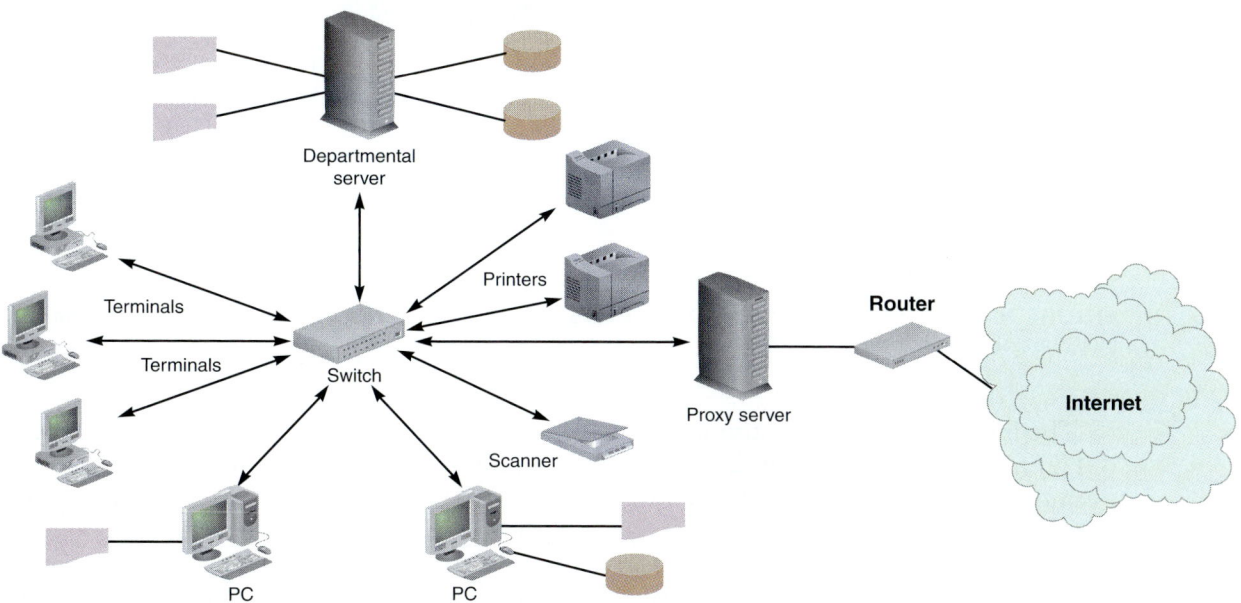

FIGURE 10-39 Routers can be used to connect LANs and WANs to other networks, such as the Internet.

Network Licensing Issues

When considering a network design, it is important to take into account software licensing restrictions. Various types of individual and site licenses are available from software vendors. Some vendors limit the number of users or the number of computers that can access the program simultaneously. You also must carefully investigate the capabilities of network software to ensure that it can handle the anticipated system traffic.

WIRELESS NETWORKS

Although a LAN provides enormous flexibility, a significant cost involves the initial expense of cabling and the inevitable wiring changes that occur in a dynamic organization. Many companies find wireless technology to be an attractive alternative. A **wireless local area network**, or **WLAN**, is relatively inexpensive to install and is well-suited to workgroups and users who are not anchored to a specific desk or location. Most notebook computers are equipped with built-in wireless capability, and it is relatively simple to add this feature to existing desktop computers and workstations in order to set up a wireless network.

Like their wired counterparts, wireless networks have certain standards and topologies, which are discussed in the following sections.

Wireless Network Standards

Wireless networks are based on various standards and protocols that still are evolving. The most popular of these is called IEEE **802.11**, which is a family of standards developed by the **Institute of Electrical and Electronics Engineers (IEEE)** for wireless LANs.

Current wireless networks are based on variations of the original 802.11 standard. Several versions, or **amendments**, were intended to improve bandwidth, range, and security. The table in Figure 10-40 contains a brief comparison of the IEEE 802.11 amendments. Note that maximum speed is measured in **Mbps (megabits per second)**.

STANDARD	INTRODUCED	MAXIMUM SPEED	APPROXIMATE RANGE	COMPATIBILITY
802.11b	1999	11 Mbps	100–300 feet	Early 802.11 version
802.11a	1999	54 Mbps	50–100 feet	Incompatible with 802.11b and 802.11g
802.11g	2003	54 Mbps	50–100 feet	Compatible with 802.11b
802.11n	2009–2010 (estimated)	200+ Mbps	150–300 feet	Compatible with all 802.11 standards

(Source: Wikipedia.org and IEEE.org)

FIGURE 10-40 This table shows various Wi-Fi standards and characteristics. Maximum speed is measured in Mbps (megabits per second).

The initial IEEE 802.11 standard, which was released in 1997, used a 2.4 gigahertz (GHz) frequency to transmit data, and offered only 2 Mbps of network bandwidth. Because of the low bandwidth, the original standard was not widely implemented and is virtually obsolete today. To increase bandwidth, the IEEE created two amendments, 802.11a and 802.11b, which offered considerable improvement in transmission speed. The original standard provided an early form of security protection, which was eventually superseded by newer technology. Wireless network security is discussed in detail in Chapter 12, Managing Systems Support and Security.

The **802.11a** version was based on a higher frequency of 5 GHz, and increased network bandwidth to a theoretical maximum of 54 Mbps. On one hand, this standard offered more bandwidth and less interference from devices such as microwaves and cordless phones. However, due to the higher cost of microchips needed for 802.11a devices, this standard never became as popular as its 802.11b counterpart. The **802.11b** standard used the original 2.4 GHz transmission frequency, but bandwidth only increased to a theoretical maximum speed of 11 Mbps. Still, this version met the needs of many users, and became popular. Eventually, the IEEE created a third version that offered the bandwidth gains of 802.11a but used the same frequency as 802.11b. Thus, in 2003, the **802.11g** standard was introduced. This version is used today by most WLANs.

Currently there is an emerging standard known as **802.11n**. Although it is still a draft specification, devices supporting the new standard are already on the market. The new technology, called **multiple input/multiple output (MIMO)**, is compatible with earlier 802.11 versions. MIMO uses multiple data streams and multiple antennas, and is expected to achieve speeds of 200+ Mbps while substantially increasing wireless range. The 802.11n amendment could greatly enhance wireless networking and bring about a large-scale expansion of WLANs as replacements for wired networks. However, some observers have commented that until the new standard is officially adopted by IEEE, many large corporations will delay implementation of 802.11n technology.

Wireless Network Topologies

Like wired networks, wireless networks also can be arranged in different topologies. The three major network topologies available for IEEE 802.11 WLANs are the Basic Service Set, the Extended Service Set, and the Independent Service Set.

The **Basic Service Set (BSS)**, also called the **infrastructure mode**, is shown is Figure 10-41. In this configuration, a central wireless device called an **access point** is used to serve all wireless clients. The access point is similar to a hub in the LAN star topology, except it provides network services to wireless clients instead of wired clients. Because access points use a single communications medium, the air, they broadcast all traffic to all clients, just as a hub would do in a wired network. Typically, the access point itself is connected to a wired network, so wireless clients can access the wired network.

The second wireless topology is the **Extended Service Set (ESS)**, as shown in Figure 10-42 on the next page. An Extended Service Set is made up of two or more Basic Service Set networks. Thus, using an ESS topology, wireless access can be expanded over a wide area. Each access point provides wireless services over a limited range. As a client moves away from one access point and closer to another, a process called **roaming** automatically allows the client to associate with the stronger access point, allowing for undisrupted service.

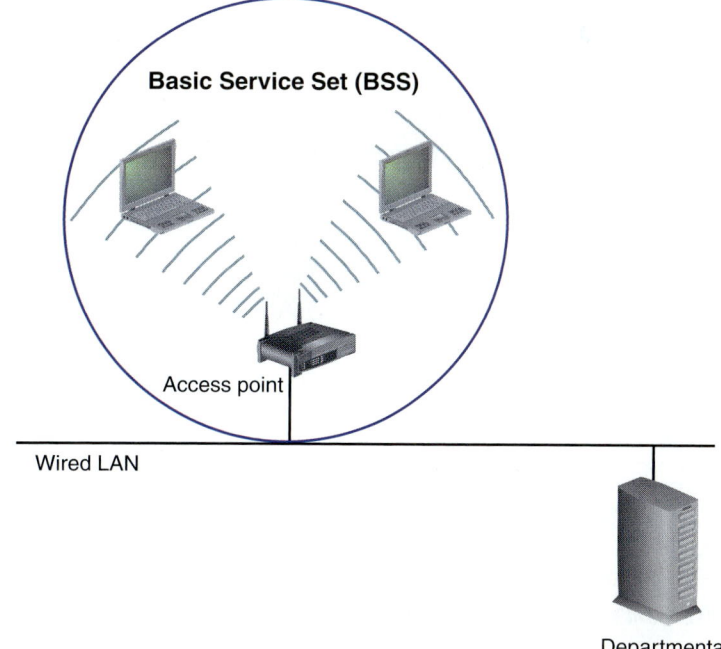

FIGURE 10-41 Basic Service Set (infrastructure mode).

Extended Service Set (ESS)

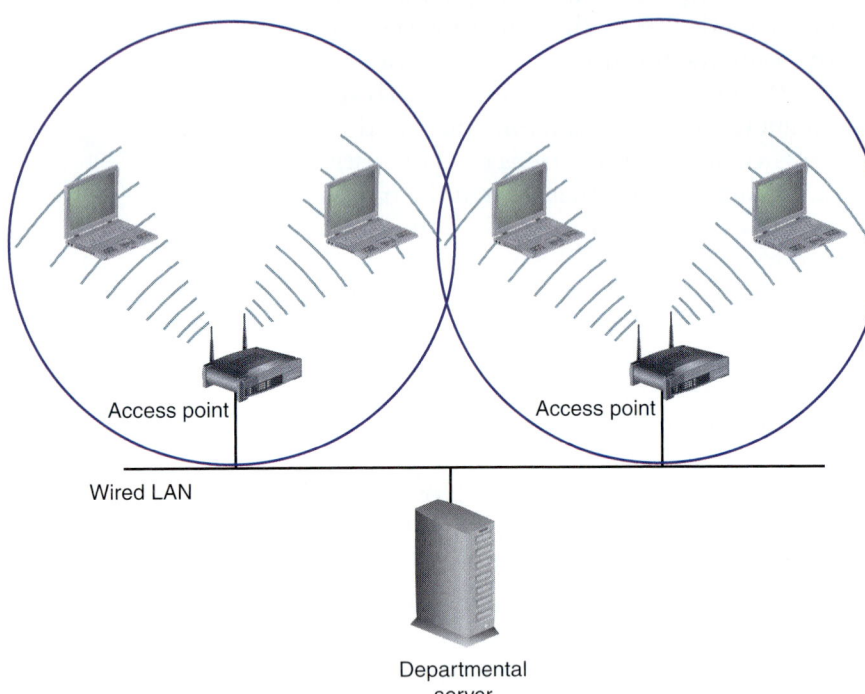

FIGURE 10-42 Extended Service Set.

Independent Service Set (ISS)

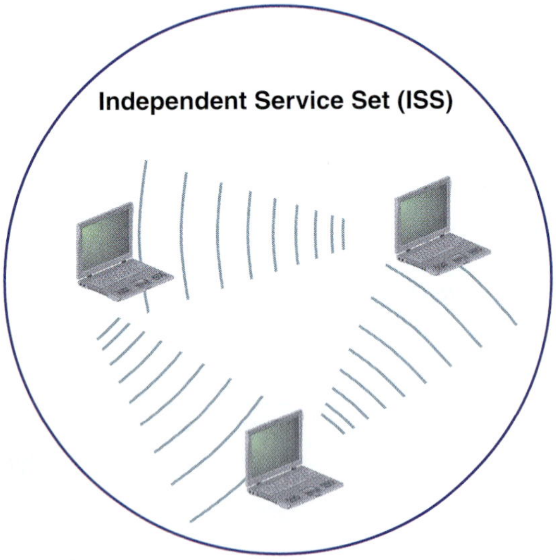

FIGURE 10-43 Independent Service Set (peer-to-peer mode).

The third wireless topology is the **Independent Service Set (ISS)**, as shown in Figure 10-43. In the ISS, also called **peer-to-peer mode**, no access point is used. Instead, wireless clients connect to each other directly. Most business WLANs use access points to provide wireless services, and do not utilize the Independent Service Set. However, ISS is well-suited to situations requiring quick data transfer among clients.

Wireless Trends

Wireless technology has brought explosive change to the IT industry, and will continue to affect businesses, individuals, and society. Even in the ever-changing world of IT, it would be difficult to find a more dynamic area than wireless technology.

With the growing popularity of 802.11, many firms offer networking products, services, and information. One of the most significant groups is the **Wi-Fi Alliance**, which maintains a Web site at *www.wi-fi.org*. According to the site, the Alliance is a nonprofit international association formed in 1999 to certify interoperability of wireless network products based on IEEE 802.11 specifications. Products that meet the requirements are certified as **Wi-Fi (wireless fidelity)** compatible. Currently the Wi-Fi Alliance has over 300 member companies from around the world, and over 4,200 products have received Wi-Fi certification. The stated goal of the Wi-Fi Alliance is to enhance the user experience through product interoperability.

Even though they have many advantages, wireless networks also have limitations and disadvantages. For example, because 802.11b and 802.11g devices use the 2.4 GHz band, these devices can pick up interference from appliances such as microwave ovens and cordless telephones that use the same band. More important, wireless networks pose major security concerns because wireless transmissions are much more susceptible to interception and intrusion than wired networks. These issues are discussed in detail in Chapter 12, Managing Systems Support and Security.

In addition to Wi-Fi, another form of wireless transmission called **Bluetooth** is very popular for short-distance wireless communication that does not require high power. Examples of Bluetooth devices include wireless keyboards, mice, printers, cell phone headsets, and digital cameras, among others. People with Bluetooth-equipped phones or PDAs can even beam information to each other and exchange digital notes.

Although the expansion of Wi-Fi has been dramatic, future technology promises even greater wireless speed, range, and compatibility. For example, in addition to 802.11 protocols for LANs, IEEE is working on **802.16** standards, which are broadband wireless communications protocols for **MANs (metropolitan area networks)**. These specifications, which IEEE calls **WirelessMAN™**, or **WiMAX**, are expected to enable wireless multimedia applications with a range of up to 30 miles. Also, a new standard called 802.11r was recently approved by the IEEE. The **802.11r** standard enhances roaming in the 802.11 ESS topology by reducing the downtime involved when a client roams from one access point to another.

CASE IN POINT 10.4:   SPIDER IT SERVICES

Spider IT Services specializes in custom network design and installation. Firms hire Spider to do an overall analysis of their network needs, including a detailed cost-benefit study. Recently, a problem arose. One of Spider's clients complained that the relatively new network was too slow and lacked sufficient capacity. Reviewing the case, Spider's top management realized that the rapidly growing client had simply outgrown the network much earlier than anticipated.

Could this problem have been avoided? Note that IBM, in the article shown in Figure 10-19 on page 461, commented that performance can "degrade exponentially" in certain kinds of network situations. Consider the IBM article and other material in this chapter, and offer your views.

SYSTEMS DESIGN COMPLETION

System architecture marks the end of the systems design phase of the SDLC. Recall that back in the systems analysis phase, all functional primitives were identified and documented with process descriptions. The objective then was to identify the system's functions and determine *what* each logical module would do, without attempting to determine *how* that function would be carried out. Moving from analysis to design tasks, the development process continued with consideration of output and user interface design, data design, and system architecture issues. Now, based on a clear definition of system requirements and design, software applications can be developed, documented, and tested as part of the systems implementation phase of the SDLC, which is described in Chapter 11, Managing System Implementation.

Developers must also consider system management and support tools that can monitor system performance, deal with fault management, handle backup, and provide for disaster recovery. These topics are covered in detail in Chapter 12, Managing Systems Support and Security.

The final activities in the systems design phase are preparing a system design specification, obtaining user approval, and delivering a presentation to management.

System Design Specification

The **system design specification** is a document that presents the complete design for the new information system, along with detailed costs, staffing, and scheduling for completing the next SDLC phase — systems implementation.

The system design specification is the baseline against which the operational system will be measured. Unlike the system requirements document, which is written for users to understand, the system design specification is oriented toward the programmers who will use it to create the necessary programs. Some sections of the system requirements document are repeated in the system design specification, such as process descriptions, data dictionary entries, and data flow diagrams.

The system design specification varies in length, so you must organize it carefully and number all pages in sequence. You should include a cover page, a detailed table of contents, and an index. The contents of the system design specification depend on company standards and the complexity of the system. A typical system design specification typically includes the following sections.

1. *Executive Summary.* The management summary provides a brief overview of the project for company managers and executives. It outlines the development efforts to date, provides a current status report, summarizes current project costs and costs for the remaining phases, reviews the overall benefits of the new system, presents the systems development phase schedule, and highlights any issues that management will need to address.

2. *System Components.* This section contains the complete design for the new system, including the user interface, outputs, inputs, files, databases, and network specifications. You should include source documents, report and screen layouts, DFDs, and all other relevant documentation. You also should include the requirements for all support processing, such as backup and recovery, start-up processing, and file retention. If the purchase of a software package is part of the strategy, you must include any interface information required between the package and the system you are developing. If you use a CASE design tool, you can print design diagrams and most other documentation directly from the tool.

3. *System Environment.* This section describes the constraints, or conditions, affecting the system, including any requirements that involve operations, hardware, systems software, or security. Examples of operational constraints include transaction volumes that must be supported, data storage requirements, processing schedules, reporting deadlines, and online response times.

4. *Implementation Requirements.* In this section, you specify start-up processing, initial data entry or acquisition, user training requirements, and software test plans.

5. *Time and Cost Estimates.* This section provides detailed schedules, cost estimates, and staffing requirements for the systems development phase and revised projections for the remainder of the SDLC. You also present total costs-to-date for the project and compare those costs with your prior estimates.

6. *Appendices.* Supplemental material can be included in appendices at the end of the system design specification. In this section, you might include copies of documents from the first three phases if they would provide a helpful reference for readers.

User Approval

Users must review and approve the interface design, report and menu designs, data entry screens, source documents, and other areas of the system that affect them. The review and approval process continues throughout the systems design phase. When you complete the design for a report, you should meet with users to review the prototype, adjust the design if necessary, and obtain written approval. Chapter 8 contains guidelines and suggestions that will help you design attractive, effective reports.

Securing approvals from users throughout the design phase is very important. That approach ensures that you do not have a major task of obtaining approvals at the end, it keeps the users involved with the system's development, and it gives you feedback about whether or not you are on target. Some sections of the system design specification might not interest users, but anything that does affect them should be approved as early as possible.

Other IT department members also need to review the system design specification. IT management will be concerned with staffing, costs, hardware and systems software requirements, network impact, and the effect on the operating environment when the new system is added. The programming team will want to get ready for its role, and the operations group will be interested in processing support, report distribution, network loads, integration with other systems, and any hardware or software issues for which they need to prepare. You must be a good communicator to keep people up to date, obtain their input and suggestions, and obtain necessary approvals.

When the system design specification is complete, you distribute the document to a target group of users, IT department personnel, and company management. You should distribute the document at least one week before your presentation to allow the recipients enough time to review the material.

TOOLKIT TIME

The Communication Tools in Part 1 of the Systems Analyst's Toolkit can help you develop better reports and presentations. To learn more about these tools, turn to Part 1 of the four-part Toolkit that follows Chapter 12.

Presentations

Usually, you will give several presentations at the end of the systems design phase. The presentations give you an opportunity to explain the system, answer questions, consider comments, and secure final approval. Part 1 of the Systems Analyst's Toolkit can provide valuable guidelines and tips about oral presentations.

The first presentation is to the systems analysts, programmers, and technical support staff members who will be involved in future project phases or operational support for the system. Because of the audience, the presentation is technically oriented.

Your next presentation is to department managers and users from departments affected by the system. As in the first presentation, your primary objective is to obtain support and approval for the systems design. This is not a technical presentation; it is aimed at user interaction with the system and management's interest in budgets, schedules, staffing, and impact on the production environment.

The final presentation is delivered to management. By the time you give this presentation, you should have obtained all necessary approvals from prior presentations, and you should have the support of users and the IT department. Just like the management presentation at the end of the systems analysis phase, this presentation has a key objective: to obtain management's approval and support for the next development step — systems implementation — including a solid commitment for financial and other resources needed.

Based on the presentation and the data you submitted, management might reach one of three decisions: proceed with systems development, perform additional work on the systems design phase, or terminate the project. The next chapter discusses systems implementation, which is the fourth SDLC phase.

A QUESTION OF ETHICS

The new accounting system is operational, but feedback from users has been negative. The most common complaint is that the system is not user-friendly. Some people in the IT department think that more user training would solve the problem. However, Sam, the IT manager, is opposed to a fresh round of training. "Let's just set up the network to monitor the users' keystrokes and mouse clicks, and see what the patterns are," he suggested. "We can analyze the data and come up with tips and suggestions that would make the system easier to use."

Your initial reaction is that Sam is wrong, for two reasons. First, you believe that monitoring would not be an effective method to learn what users really want. In your view, that should have been done in the system requirements phase. Second, you are bothered by an ethical question: Even though the proposed monitoring would involve company business, the company network, and company time, you feel that many users would resent the unannounced monitoring, and might feel that their performance or other computing activities were being appraised without their knowledge.

Sam has asked to you to write up a recommendation. What will you say about the ethical question that troubles you?

CHAPTER SUMMARY

An information system combines hardware, software, data, procedures, and people into a system architecture. The architecture translates the system's logical design into a physical structure that includes hardware, software, and processing methods. The software consists of application programs, also called applications, that handle the input, manage the processing logic, and provide the required output.

Before selecting an architecture, the analyst must consider enterprise resource planning, initial cost and TCO, scalability, Web integration, legacy interface requirements, processing options, and security issues.

Enterprise resource planning (ERP) establishes an enterprise-wide strategy for IT resources and specific standards for data, processing, network, and user interface design. Companies can extend ERP systems to suppliers and customers in a process called supply chain management. A systems analyst must assess initial cost and TCO and ensure that the design is scalable. Scalability means that a system can be expanded, modified, or downsized easily to meet business needs. The analyst also must consider if the system will be Web-centric and follow Internet design protocols, and if it must interface with existing systems, called legacy systems. System security is an important concern throughout the design process, especially for e-commerce applications that involve credit card and personal data. Processing options affect system design and resources required.

An architecture requires servers and clients. Servers are computers that supply data, processing services, or other support to one or more computers called clients. In mainframe architecture, the server performs all processing, and terminals communicate with the centralized system. Clients can be connected in distributed systems to form local area networks (LANs) or wide area networks (WANs). A typical LAN design involves file server design, where the client requests a copy of a data file and the server responds by transmitting the entire file to the client.

Client/server architecture divides processing between one or more clients and a central server. In a typical client/server system, the client handles the entire user interface, including data entry, data query, and screen presentation logic. The server stores the data and provides data access and database management functions. Application logic is divided in some manner between the server and the clients. In a typical client/server interaction, the client submits a request for information from the server, which carries out the operation and responds to the client. Compared to file server designs, client/server systems are more scalable and flexible.

A fat, or thick, client design places all or most of the application processing logic at the client. A thin client design places all or most of the processing logic at the server. Thin client designs provide better performance, because program code resides on the server, near the data. In contrast, a fat client handles more of the processing, and must access and update the data more often. Compared with maintaining a central server, fat client TCO also is higher, because of initial hardware and software requirements and the ongoing expense of maintaining and updating remote client computers. The fat client design is simpler to develop, because the architecture resembles traditional file server designs where all processing is performed at the client.

Client/server designs can be two-tier or three-tier (also called n-tier). In a two-tier design, the user interface resides on the client, all data resides on the server, and the application logic can run either on the server or on the client, or be divided between the client and the server. In a three-tier design, the user interface runs on the client and the data is stored on the server, just as with a two-tier design. A three-tier design also has a middle layer between the client and server that processes the client requests and translates them into data access commands that can be understood and carried out by the server. The middle layer is called an application server, because it provides the application logic, or business logic. Middleware is software that connects dissimilar applications and enables them to communicate and pass data. In planning the system design, a systems analyst also must consider cost-benefit and performance issues.

The Internet has had an enormous impact on system architecture. In implementing a design, an analyst should consider e-commerce strategies, the availability of packaged solutions, and corporate portals, which are entrances to a multifunction Web site. The analyst also should understand the concepts of cloud computing and Web 2.0, which may shape the future of Internet computing.

The primary processing methods are online and batch processing. Users interact directly with online systems that continuously process their transactions when and where they occur and continuously update files and databases. In contrast, batch systems process transactions in groups and execute them on a predetermined schedule. Many online systems also use batch processing to perform routine tasks, such as handling reports and accounting entries.

Networks allow the sharing of hardware, software, and data resources in order to reduce expenses and provide more capability to users. The network is represented by a seven-layer logical model called the OSI (Open Systems Interconnection) model. Various OSI layers handle specific functions as data flows down from the sending computer and up into the receiving computer.

The way a network is configured is called the network topology. Networks typically are arranged in four patterns: hierarchical, bus, ring, and star. A single mainframe computer usually controls a hierarchical network, a bus network connects workstations in a single-line communication path, a ring network connects workstations in a circular communication path, and a star network connects workstations to a central computer or networking device called a switch. Wireless networks, or WLANs, based on IEEE 802.11 standards, have seen explosive growth, especially in situations where the flexibility of wireless is important. The new IEEE 802.11n standard is expected to enhance the continued growth of wireless networking. WLANs have three major topologies: BSS, ESS, and ISS. Although wireless networks are very popular, they do have some limitations and disadvantages, including interference and security concerns.

The system design specification presents the complete systems design for an information system and is the basis for the presentations that complete the systems design phase. Following the presentations, the project either progresses to the systems development phase, requires additional systems design work, or is terminated.

Key Terms and Phrases

802.11 *476*
802.11a *477*
802.11b *477*
802.11g *477*
802.11n *477*
802.11r *479*
802.16 *479*
access point *477*
amendment *476*
application logic *458*
application server *458*
applications *451*
bandwidth *466*
Basic Service Set (BSS) *477*
batch processing *469*
Bluetooth *479*
bus network *473*
business logic *458*
client/server architecture *455*
clients *453*
cloud computing *465*
data frames *474*
data processing center *453*
distributed database management system (DDBMS) *461*
distributed systems *454*
e-marketplaces *451*
enterprise resource planning (ERP) *448*
environment *448*
Extended Service Set (ESS) *477*
extensibility *450*
fat client *458*
file server *455*
file sharing architecture *455*
File Transfer Protocol (FTP) *475*
gateway *475*
hierarchical network *472*
hub *474*
Independent Service Set (ISS) *478*
infrastructure mode *477*
Institute of Electrical and Electronics Engineers (IEEE) *476*
Internet operating system *467*
legacy data *456*
legacy systems *451*
local area network (LAN) *454*

mainframe architecture *453*
MAN (metropolitan area network) *479*
Mbps (megabits per second) *476*
middleware *459*
multiple input/multiple output (MIMO) *477*
Multistation Access Unit (MAU) *474*
network topology *471*
n-tier *458*
online system *468*
OSI (Open Systems Interconnection) model *471*
peer-to-peer mode *478*
platform *448*
point-of-sale (POS) *469*
portal *464*
protocol *475*
proxy server *475*
ring network *473*
roaming *477*
router *475*
scalability *450*
scaling on demand *465*
server *453*
stand-alone *454*
star network *474*
supply chain management *448*
switch *474*
system architecture *446*
system design specification *479*
terminal *453*
thick client *458*
thin client *458*
three-tier *458*
Transmission Control Protocol/Internet Protocol (TCP/IP) *475*
transparent *454*
two-tier *458*
Web 2.0 *467*
Web-centric *451*
Wi-Fi Alliance *478*
Wi-Fi (wireless fidelity) *478*
wide area network (WAN) *454*
wiki *467*
wireless local area network (WLAN) *476*
WirelessMAN™ *479*
WiMAX *479*

Learn It Online

Instructions: To complete the Learn It Online exercises, start your browser, click the Address bar, and then enter the Web address **scsite.com/sad8e/learn**. When the Systems Analysis and Design Learn It Online page is displayed, follow the instructions in the exercises below. Each exercise has instructions for saving your results, either for your own records or for submission to your instructor.

1 Chapter Reinforcement

TF, MC, and SA

Below SAD Chapter 10, click one of the Chapter Reinforcement links for Multiple Choice, True/False, or Short Answer. Answer each question and submit to your instructor.

2 Flash Cards

Below SAD Chapter 10, click the Flash Cards link and read the instructions. Type 20 (or a number specified by your instructor) in the Number of playing cards text box, type your name in the Enter your Name text box, and then click the Flip Card button. When the flash card is displayed, read the question and then click the ANSWER box arrow to select an answer. Flip through the Flash Cards. If your score is 15 (75%) correct or greater, click Print on the File menu to print your results. If your score is less than 15 (75%) correct, then redo this exercise by clicking the Replay button.

3 Practice Test

Below SAD Chapter 10, click the Practice Test link. Answer each question, enter your first and last name at the bottom of the page, and then click the Grade Test button. When the graded practice test is displayed on your screen, click Print on the File menu to print a hard copy. Continue to take practice tests until you score 80% or better.

4 Who Wants To Be a Computer Genius?

Below SAD Chapter 10, click the Computer Genius link. Read the instructions, enter your first and last name at the bottom of the page, and then click the Play button. When your score is displayed, click the PRINT RESULTS link to print a hard copy.

5 Wheel of Terms

Below SAD Chapter 10, click the Wheel of Terms link. Read the instructions, and then enter your first and last name and your school name. Click the PLAY button. When your score is displayed on the screen, right-click the score and then click Print on the shortcut menu to print a hard copy.

6 Crossword Puzzle Challenge

Below SAD Chapter 10, click the Crossword Puzzle Challenge link. Read the instructions, and then click the Continue button. Work the crossword puzzle. When you are finished, click the Submit button. When the crossword puzzle is redisplayed, submit it to your instructor.

Case-Sim: SCR Associates

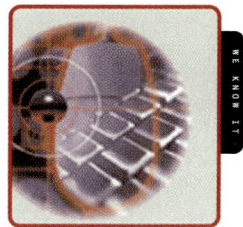

Background

SCR Associates is a consulting firm that offers IT solutions and training. SCR needs an information system to manage operations at its new training center. The new system will be called TIMS (Training Information Management System). As a newly hired systems analyst, you will report to Jesse Baker, systems group manager. You will work on various tasks and practice the skills you learned in this chapter.

Using the Case

The SCR Associates case study is a Web-based simulation that allows you to practice your skills in a real-world environment. The case study transports you to SCR's company intranet, where you can complete 12 work sessions, each aligning with a chapter. As you work on the case, you will receive e-mail and voice mail messages, obtain information from SCR's online libraries, and perform various tasks. The first time you enter the SCR Case, you should go to the starting page at **scsite.com/sad8e/scr** for detailed instructions.

Preview: Session 10

Your supervisor, Jesse Baker, wants you to be familiar with the main issues that a systems analyst should consider when selecting an architecture, including enterprise resource planning, initial costs and TCO, scalability, Web integration, legacy interface requirements, processing options, and security issues.

To start a work session, you log on to the SCR intranet. When you enter your name and password, an opening screen displays links to the work sessions, and you select Session 10. At this point, you check your e-mail and voice mail messages carefully. Then you begin working on your task list, which includes the following items:

Tasks: System Architecture Tasks

1. Jesse wants me to recommend a vendor who offers an ERP strategy. I need to review the SAP and Oracle Web sites, and at least two others that offer ERP solutions, and reply to her with the results and the reasons for my recommendations.

2. Visit SCR's data library to review SCR's network configuration and then send Jesse a recommendation for the TIMS system architecture. She wants me to suggest an overall client/server design, number of tiers, and network topology. She also asked me to comment on these issues: legacy data, Web-centricity, scalability, security, and batch processing that might be needed. Jesse said it was OK to make reasonable assumptions in my proposal to her.

3. Perform research on the Internet to learn more about TCO, and develop a TCO checklist that includes the five most important elements of TCO, because of their magnitude or potential impact on TIMS.

4. Prepare a system design specification as Jesse requested.

FIGURE 10-44   Task list: Session 10.

Chapter Exercises

Review Questions

1. Define the term system architecture. Define the term scalability, and explain why it is important to consider scalability in system design.
2. When selecting an architecture, what items should a systems analyst consider as part of the overall design checklist?
3. What is enterprise resource planning (ERP)? What is supply chain management?
4. Explain the term server and provide an example of server-based processing; explain the term client and provide an example of client-based processing.
5. Describe client/server architecture, including fat and thin clients, client/server tiers, and middleware.
6. Describe the impact of the Internet on system architecture. Include examples.
7. Explain the difference between online processing and batch processing and provide an example of each type.
8. Explain the difference between a LAN and a WAN, define the term topology, and draw a sketch of each wired and wireless network model. Also describe four IEEE 802.11 amendments.
9. Explain the differences between the BSS, ESS, and ISS wireless topologies. Which is rarely used in business? To what kind of network do the 802.16 standards apply?
10. List the sections of a system design specification, and describe the contents.

Discussion Topics

1. Information technology has advanced dramatically in recent years. At the same time, enormous changes in the business world have occurred as companies reflect global competition and more pressure for quality, speed, and customer service. Did the new technology inspire the business changes, or was it the other way around?
2. Internet-based sales have shown explosive growth in recent years. How does B2B interaction differ from consumer-based Internet marketing, and why is it growing so rapidly?
3. This chapter described seven guidelines that a systems analyst might use when considering an architecture. In your view, are all the items of equal weight and importance, or should some be ranked higher? Justify your position.
4. One manager states, "When a new system is proposed, I want a written report, not an oral presentation, which is like a sales pitch. I only want to see the facts about costs, benefits, and schedules." Do you agree with that point of view?

Projects

1. Visit the IT department at your school or a local company to determine what type of network it is using. Draw a sketch of the network configuration.
2. Prepare a 10-minute talk explaining Web 2.0 and cloud computing to a college class. Using the text and your own Internet research, briefly describe the five most important points you will include in your presentation.
3. Perform research on the Internet to identify an ASP that offers Web-based business solutions, and write a brief memo describing the firm and its services.
4. Perform research on the Internet to learn about trends in wireless networking, and typical costs involved in the installation of a wireless LAN.

Apply Your Knowledge

The Apply Your Knowledge section contains four mini-cases. Each case describes a situation, explains your role in the case, and asks you to respond to questions. You can answer the questions by applying knowledge you learned in the chapter.

1 Digital Dynamics

Situation:

After three years as a successful Web design firm in Southern California, Digital Dynamics has decided to add two new business ventures: a group that specializes in supplying qualified employees to high-tech firms and a training division that offers online courses in advanced Web design and e-commerce skills. As a senior systems analyst, you have been asked to study the situation and make recommendations.

1. Should Digital Dynamics adopt ERP? What specific advantages would ERP offer?
2. How could the concept of supply chain management apply to a company's service-based division? Provide some specific suggestions.
3. Should Digital Dynamics use separate portals for employees, customers, and suppliers? If you follow the example of Oracle Corporation in the textbook, what type of portal design would you suggest?
4. Is the experience of other companies relevant? Use the Internet to locate examples of Web-based firms that offer personnel services and technical training. How would you evaluate the Web sites you visited? What specific features impressed you favorably or unfavorably?

2 R/Way Trucking

Situation:

As you learned earlier in this chapter, R/Way is a small but rapidly growing trucking company headquartered in Cleveland, Ohio. R/Way's information system currently consists of a file server and three workstations where freight clerks enter data, track shipments, and prepare freight bills. To perform their work, the clerks obtain data from a file server and use database and spreadsheet programs stored on stand-alone PCs to process the data. At your meeting yesterday, R/Way's president approved your recommendation to create a relational database to handle R/Way operations and provide links for R/Way shippers and customers.

1. Review the concept of supply chain management. Although R/Way offers services rather than products, could that concept apply to the design of R/Way's new system? If so, how?
2. What would be the advantages of selecting an Internet-based architecture for R/Way's system?
3. Should R/Way's new system be based on file-server or client/server architecture? Why?
4. What would be the pros and cons of selecting in-house development versus a packaged solution for the R/Way system?

3 Nothing But Net

Situation:

Nothing But Net is an IT consulting firm that specializes in e-commerce solutions. As a newly hired systems analyst, you have been asked to research some current topics.

1. Obtain at least two estimates of how much consumers are projected to spend on Internet purchases during the next three years. Are the estimates similar? If not, which forecast do you believe, and why?

2. Many of Nothing But Net's customers are start-up firms that must fight hard to attract investment capital, and many traditional lending institutions are skeptical of new Web-based firms. Perform research to determine the mortality rate of new e-commerce firms that use the Web as their primary marketing channel, and write a brief memo that describes the results of your research.

3. Some IT professionals predict that traditional companies will increase their Internet marketing efforts, making it even harder for new Web-based firms to compete. Perform research to find out more about the topic and share your results with the class.

4. Suppose you were asked to draft a sales brochure for Nothing But Net. List all the services in which potential customers might be interested.

4 Aunt Ann's Kitchen

Situation:

Aunt Ann's Kitchen offers a line of specialty food products to institutional customers and restaurant chains. The firm prides itself on using only the finest ingredients and preparation methods. The owner, Ann Rose, hired you as an IT consultant to help her plan the system architecture for a new WLAN that will connect employee computers to the main (wired) network. She asked you to start with the following questions:

1. What possible IEEE 802.11 amendments could be used for the new system? What are the pros and cons of each amendment?

2. Choose an amendment to implement and explain your choice.

3. Suppose that microwave ovens and cordless telephones are used extensively in some parts of the facility. Would that affect your IEEE 802.11 amendment choice? If so, why?

4. Suppose that the new WLAN will also provide roaming services for employees with portable notebook computers. Which wireless topology will be required?

Case Studies

Case studies allow you to practice specific skills learned in the chapter. Each chapter contains several case studies that continue throughout the textbook, and a chapter capstone case.

NEW CENTURY HEALTH CLINIC

New Century Health Clinic offers preventive medicine and traditional medical care. In your role as an IT consultant, you will help New Century develop a new information system.

Background

The New Century clinic associates accepted your interface, output, input, and data designs and your recommendation to install a server and four personal computers as clients on a local area network. The network will include a tape backup unit and Internet access via a modem that can exchange data with insurance companies. A high-speed laser printer and an impact printer for multipart forms will be accessible by any of the four PCs. Now you will determine system architecture for the New Century system.

When you created ERDs and record designs for New Century during the data design process in Chapter 9, you considered whether to use a file-processing or database approach. As you know, each strategy has advantages and disadvantages, depending on the specific hardware and software environment and business requirements. At this point, you must decide which way to proceed, and Dr. Jones will accept your recommendation (with your instructor's approval). You should start by reviewing the DFDs and object-oriented diagrams that you prepared in the systems analysis phase, and the ERDs and table designs that you created in the systems design phase. Then, review the system architecture checklist at the beginning of this chapter.

Assignments

1. What would be the advantages of selecting an Internet-based architecture for the New Century system?
2. Should the New Century system be based on file-server or client/server architecture? Why?
3. Could the New Century system use both online and batch processing? How?
4. Prepare an outline for a system design specification and describe the contents of each section.

PERSONAL TRAINER, INC.

Personal Trainer, Inc., owns and operates fitness centers in a dozen Midwestern cities. The centers have done well, and the company is planning an international expansion by opening a new "supercenter" in the Toronto area. Personal Trainer's president, Cassia Umi, hired an IT consultant, Susan Park, to help develop an information system for the new facility. During the project, Susan will work closely with Gray Lewis, who will manage the new operation.

Background

Susan and Gray finished their work on the user interface, input, and output design. They developed a user-centered design that would be easy to learn and use. Now Susan turned her attention to the architecture for the new system.

Susan wanted to consider enterprise resource planning, total cost of ownership, scalability, Web integration, legacy systems, processing methods, and security issues. She also needed to select a network plan, or topology, that would dictate the physical cabling and network connections, or consider a wireless network. When all these tasks were completed, she would submit a system design specification for approval.

Assignments

1. Would an ERP strategy work well for Personal Trainer? Investigate ERP strategies and products available from Internet vendors and submit a recommendation based on your research.

2. If Susan chooses a client/server architecture, what issues must she consider? Prepare a checklist for her that includes the main topics and issues she should consider.

3. What would be the benefits of using a wireless network? What would be the drawbacks?

4. Prepare an outline for a system design specification and describe the contents of each section.

CHAPTER CAPSTONE CASE: SoftWear, Limited

SWL

SoftWear, Limited (SWL), is a continuing case study that illustrates the knowledge and skills described in each chapter. In this case study, the student acts as a member of the SWL systems development team and performs various tasks.

Background

Jane Rossman, manager of applications, and Rick Williams, systems analyst, had several meetings with True Blue Systems, the consulting firm hired to assist SWL in implementing the new ESIP system. Michael Jeremy, SWL's finance vice president, requested that True Blue also make recommendations about a possible SWL intranet to link all SWL locations and support client/server architecture.

The initial report from True Blue indicated that the new ESIP system should be designed as a DBMS so it could interface with the new mainframe payroll package. True Blue suggested that the ESIP system be implemented on a server in the payroll department and developed as a Microsoft Access application. They felt that this approach provided a relational database environment, client/server capability, and SQL command output to communicate with the mainframe. Figure 10-45 shows the proposed design of the system.

Jane Rossman met with Ann Hon to review True Blue's report and get her approval for training IT staff members. Jane had some prior experience in Access application development, but Rick had none, so Jane suggested that Rick and Becky Evans, another systems analyst, should attend a one-week workshop. Ann agreed.

Ann, Jane, and Rick met with Michael Jeremy to get his approval before proceeding further. He asked them to develop a specific budget and timetable including all necessary hardware, software, and training costs. They had most of the information, but they needed some help from True Blue to estimate the cost of network implementation, installation, and physical cabling.

The first phase of the project would use a local area network to link the various headquarter departments to the mainframe. A second phase, proposed by True Blue, would connect all SWL locations to a wide area network, with the possibility that employees could access their individual ESIP accounts over the internal network or from outside the company using the Internet.

A week later, Ann received a memo from Michael Jeremy that said he had approved the project and that she should start work immediately.

System Architecture

The ESIP development team included Jane, Rick, and Becky. The group discovered that the Microsoft Access application consists of interactive objects such as tables, queries, forms, reports, macros, and code modules.

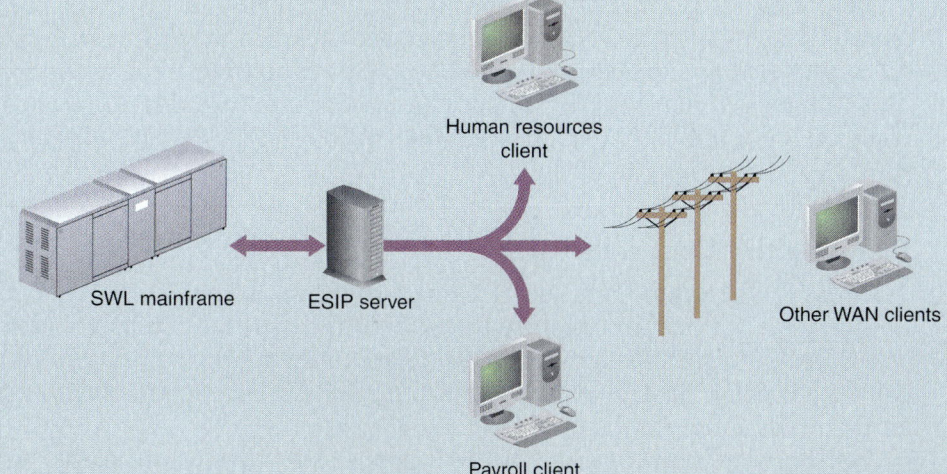

FIGURE 10-45 Diagram of the proposed ESIP system.

They decided to begin by reviewing the entity-relationship diagrams they prepared previously to determine the overall structure of the DBMS design. Then they identified the tables, reviewed the relationships among them, and analyzed the record designs they had developed.

CHAPTER CAPSTONE CASE: SoftWear, Limited (continued)   SWL

They also reviewed output requirements, input screen designs, processing considerations, backup and recovery procedures, and controls that must be built into the new system.

As recommended by True Blue Systems, the new ESIP system would be implemented as a client/server design, with the data stored on the payroll department server, which would be linked to clients in the payroll and human resources department.

Planning the System

In their first meeting, Jane asked the team to define all the tasks that the new system would perform, including a list of all reports and other required output. Jane explained that ESIP data would be stored on the server, but the application logic and objects such as forms, queries, and reports would be located on client workstations. Separating the objects from the data would provide better security and reduce the network loads, she explained.

Security Issues

In their next planning session, Jane asked the group to consider all security issues affecting the new system design. Because the system contains payroll data, it is important to control user access and updates. The team decided to use the security features in Access for control, including passwords and user and workgroup accounts for employees authorized to use the ESIP application. Each user would be assigned a permission level that grants access only to certain objects.

Jane explained her plan for system security by saying, "We'll create a comprehensive security plan later that will cover all the operational security issues for the new system. Meanwhile, let's go back to the department heads, Michael Jeremy and Tina Pham, to get their input on what the security levels should be." She added that "Users will be allowed to create and modify certain forms and reports, but most other actions will be permitted only by authorized IT department members."

Creating the Database Objects

Before creating the database objects, the team reviewed the ERDs and verified that the records designs were in 3NF. In addition, they verified that the new payroll system permitted cross-platform access because the ESIP system required data from the payroll master file. They discovered that the payroll package used a standard data format called open database connectivity (ODBC) that supported links to the Access database. After planning the system, they started creating the objects.

Planning the User Interface

From earlier interviews, the IT team knew that users in the payroll and human resources departments wanted an interface that would be easy to learn and simple to operate. Jane asked Becky to start designing a main form, or switchboard, that would display automatically when the ESIP application started. All ESIP screen forms would use buttons, menus, and icons as shown in Figure 10-46.

Becky created a prototype of the input screens to show to users. After securing user approval, the screen designs were added to the system specification document for the ESIP system.

CHAPTER CAPSTONE CASE: SoftWear, Limited (continued)

Using Visual Basic and Macros

Because she was a programmer before her assignment as a systems analyst, Becky wanted to know if they would be using Visual Basic as a program development language. Jane told her that they would write many of the procedures in Visual Basic because it allows more powerful data manipulation than macros and makes it easier to customize error messages. They would use macros, however, to speed up the development process and handle simple tasks such as opening and closing forms and running reports.

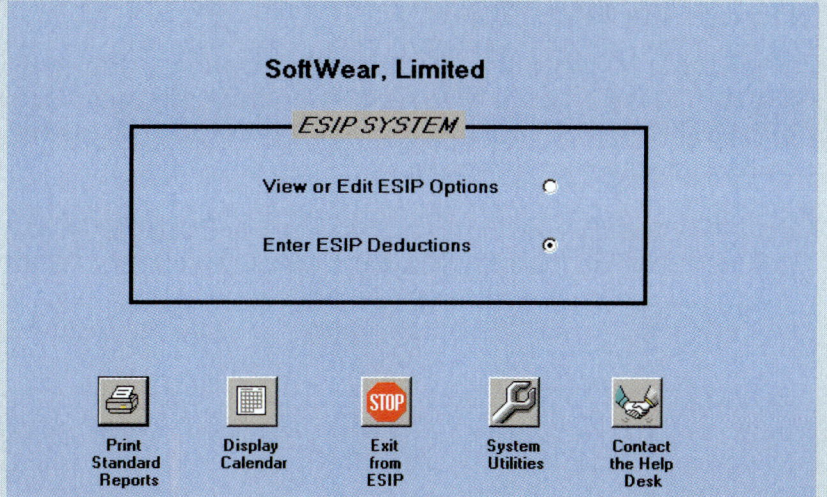

FIGURE 10-46 Sample of a main switchboard that displays when the ESIP system starts.

Completing the Systems Design Phase

The IT team completed the systems design phase by writing the documentation and designing backup and recovery, file retention, and start-up procedures. The final step was to develop a system design specification for the ESIP system.

Rick and Becky showed a draft of the system design specification to Jane and Ann. After incorporating their changes, they e-mailed copies to a distribution list of users, IT staff members, and managers. Their presentations to users and IT staff members went well, and at the management presentation, they received final approval to implement the ESIP system.

SWL Team Tasks

1. Rick Williams asked you to suggest software products that can provide network management features suitable for a network with 25–50 users. You can use the Internet to research the topic.

2. In the Background section of the SWL case, you learned that True Blue Systems had recommended a local area network to link the various headquarters departments to the SWL mainframe. Should a wireless network have been considered? Your task is to research wireless network products and submit a recommendation for a WLAN installation, supported by an explanation of wireless advantages and possible limitations. Be sure to consider flexibility, scalability, and security issues.

3. Assume that the recommendation you made in the preceding task was accepted. Now you have been asked to develop a specific budget for a WLAN that would consist of 50 workstations. You need to research the TCO of this project, including the cost of hardware, software, installation, and maintenance. You can make reasonable assumptions where you might not have specific facts, but you should state those assumptions clearly.

4. Michael Jeremy, finance vice president, has been reading about cloud computing in his favorite IT magazine, and he is considering the possibility of using the concept at SWL. He asked you to answer the following questions: What would be the benefits of a cloud computing architecture? What would be the disadvantages? What are examples of cloud computing services currently available? Perform research on the Internet and prepare a brief report for him.

CHAPTER CAPSTONE CASE: SoftWear, Limited (continued)

Manage the SWL Project

You have been asked to manage SWL's new information system project. One of your most important activities will be to identify project tasks and determine when they will be performed. Before you begin, you should review the SWL case in this chapter. Then list and analyze the tasks, as follows:

LIST THE TASKS Start by listing and numbering at least 10 tasks that the SWL team needs to perform to fulfill the objectives of this chapter. Your list can include SWL Team Tasks and any other tasks that are described in this chapter. For example, Task 3 might be to Review the logical design, and Task 6 might be to Begin the physical design process.

ANALYZE THE TASKS Now study the tasks to determine the order in which they should be performed. First identify all concurrent tasks, which are not dependent on other tasks. In the example shown in Figure 10-47, Tasks 1, 2, 3, 4, and 5 are concurrent tasks, and could begin at the same time if resources were available.

Other tasks are called dependent tasks, because they cannot be performed until one or more earlier tasks have been completed. For each dependent task, you must identify specific tasks that need to be completed before this task can begin. For example, you would want to review the logical design before you could begin the physical design process, so Task 6 cannot begin until Task 3 is completed, as Figure 10-47 shows.

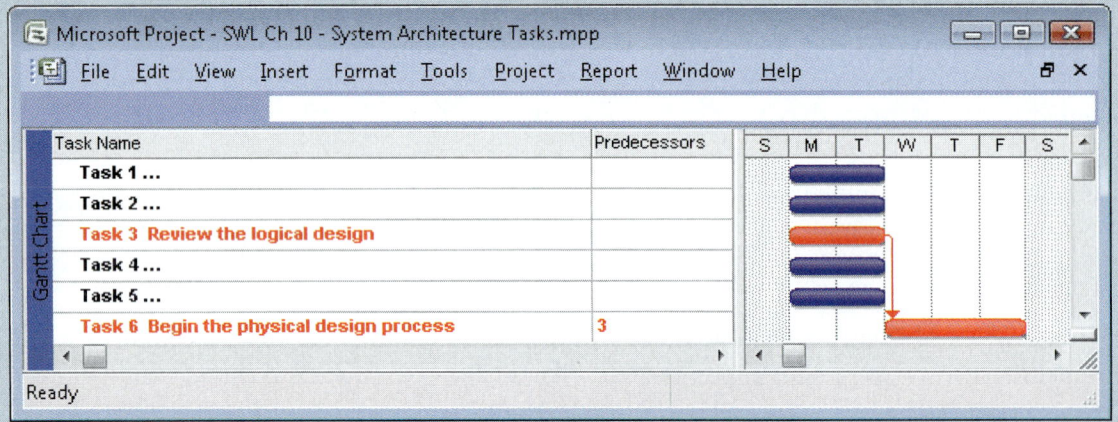

FIGURE 10-47 Tasks 1, 2, 3, 4, and 5 are concurrent tasks that could be performed at the same time. Task 6 is a dependent task that cannot be performed until Task 3 has been completed.

Chapter 3 describes project management tools, techniques, and software. To learn more, you can visit the Features section on your Student Study Tool CD-ROM, or the project management resources library at **scsite.com/sad8e/project**. On the Web, Microsoft offers demo versions, training, and tips for using Project 2007. You also can visit the OpenWorkbench.org site to learn more about this free, open-source software.

PHASE 4 SYSTEMS IMPLEMENTATION

As the Dilbert cartoon suggests, successful systems implementation requires effective methods, a capable team, and management support. You will learn more about these topics in the systems implementation phase.

Systems implementation is the fourth of five phases in the systems development life cycle. In the previous phase, systems design, you created a physical model of the system. Now you will implement that design. Your tasks will include application development, documentation, testing, training, data conversion, and system changeover. The deliverable for this phase is a completely functioning information system.

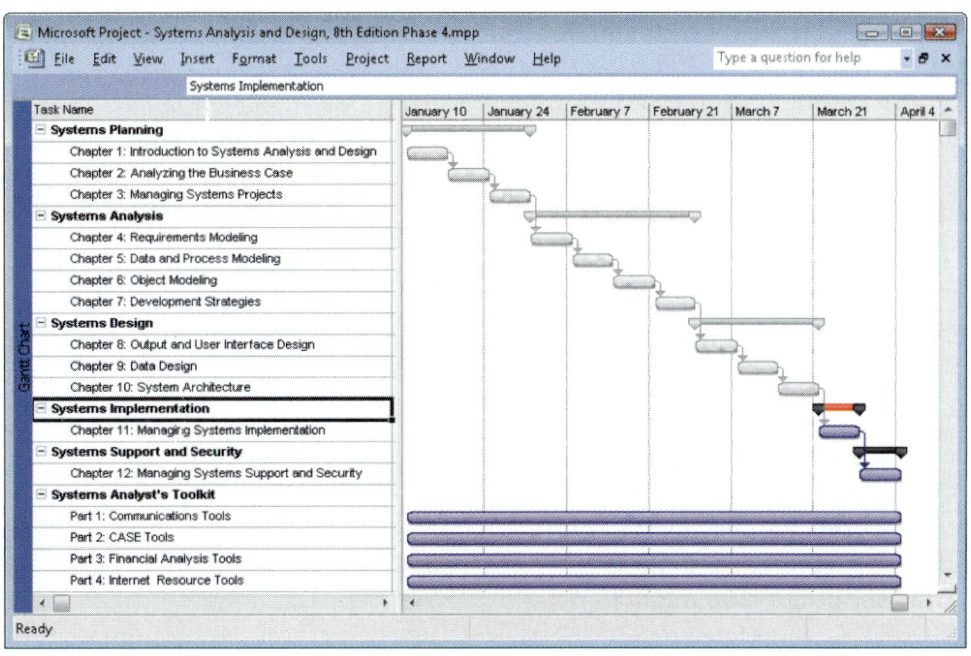

CHAPTER 11

Managing Systems Implementation

Chapter 11 describes the systems implementation phase of the SDLC. This chapter describes application development, installation, and evaluation.

INTRODUCTION

OBJECTIVES

When you finish this chapter, you will be able to:

- Explain the importance of software quality assurance and software engineering

- Describe the application development process for structured, object-oriented, and agile methods

- Draw a structure chart showing top-down design, modular design, cohesion, and coupling

- Explain the coding process

- Explain unit, integration, and system testing

- Differentiate between program, system, operations, and user documentation

- List the main steps in system installation and evaluation

- Develop a training plan for each group of participants, compare in-house and outside training, and describe effective training techniques

- Describe data conversion and changeover methods

- Explain post-implementation evaluation and the final report to management

Managing systems implementation involves application development, testing, documentation, training, data conversion, system changeover, and post-implementation evaluation of the results.

During systems implementation, the system design specification serves as a blueprint for constructing the new system. The initial task is application development, which requires systems analysts and programmers to work together to construct the necessary programs and code modules. Before a changeover can occur, the system must be tested and documented carefully, users must be trained, and existing data must be converted. After the new system is operational, a formal evaluation of the results takes place as part of a final report to management.

CHAPTER INTRODUCTION CASE: Mountain View College Bookstore

Background: Wendy Lee, manager of college services at Mountain View College, wants a new information system that will improve efficiency and customer service at the three college bookstores.

In this part of the case, Florence Fullerton (systems analyst) and Harry Boston (student intern) are talking about implementation tasks for the new system.

Participants:	Wendy, Florence, and Harry
Location:	Wendy Lee's office, Monday morning, February 8, 2010
Project status:	The system design specification was approved, and Florence and Harry are ready to implement the new bookstore information system.
Discussion topics:	Implementation tasks, including quality assurance, structure charts, testing, training, data conversion process, system changeover, and post-implementation evaluation

Florence:	Good morning, Wendy. We're ready to start the implementation process, and I'd like to go over our plans. Harry will be assisting me, so I asked him to join us.
Wendy:	*I'm glad you did. I met Harry during the interviews several months ago.*
Harry:	Hi, Wendy, good to see you again. What's next?
Florence:	Let's talk about quality assurance. We'll also discuss various implementation options, including agile methods, but we'll continue with a structured approach for now.
Wendy:	*Sounds good. What are the major tasks on your list?*
Florence:	Well, the biggest task is to translate the design into program code and produce a functioning system. We'll develop structure charts that the programmers can use as blueprints, and Harry will help me coordinate with the programmers.
Harry:	It will be great to see all the design work finally turn into a functioning system.
Florence:	It sure will. Anyway, we'll proceed to do several types of testing, and we'll document everything we do. When we're ready, we'll put the new system into what's called a test environment until we're ready to go online with the operational environment.
Wendy:	*What about training?*
Florence:	We'll consider several kinds of training — from vendors or we might do our own.
Wendy:	*Then what?*
Florence:	The final steps will be data conversion and system changeover. After the new system is up and running, we'll schedule a formal evaluation and submit a final report. Here's a task list to get us started:

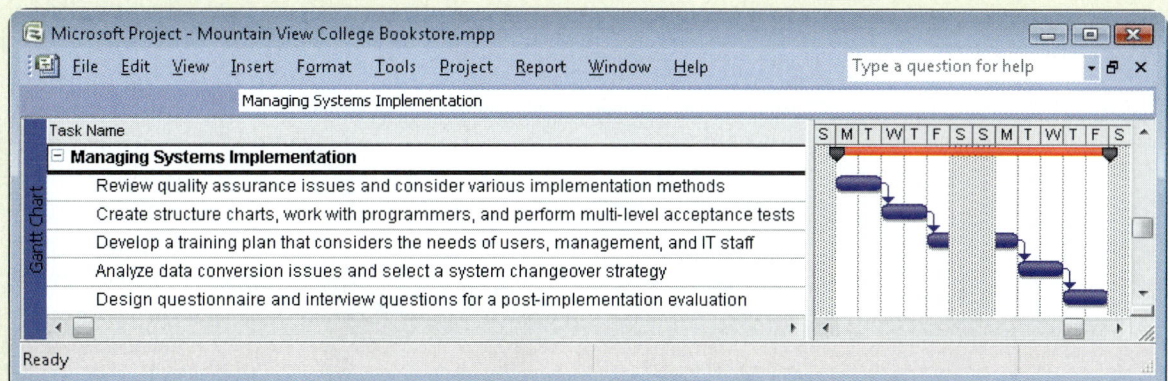

FIGURE 11-1   Typical systems implementation task list.

For more information about software engineering, visit **scsite.com/sad8e/more**, locate Chapter 11, and then click the Software Engineering link.

SOFTWARE QUALITY ASSURANCE

In today's competitive business environment, companies are intensely concerned with the quality of their products and services. A successful organization must improve quality in every area, including its information systems. Top management must provide the leadership, encouragement, and support needed for high-quality IT resources.

No matter how carefully a system is designed and implemented, problems can occur, especially in a complex system. Rigorous testing catches errors in the implementation stage, but it is much less expensive to correct mistakes earlier in the development process. The main objective of **quality assurance** is to avoid problems or to detect them as soon as possible. Poor quality can result from inaccurate requirements, design problems, coding errors, faulty documentation, and ineffective testing.

In an effort to achieve high standards of quality, software systems developers should consider software engineering concepts, internationally recognized quality standards, and careful project management techniques.

Software Engineering

Because quality is so important, you can use an approach called software engineering to manage and improve the quality of the finished system. **Software engineering** is a software development process that stresses solid design, accurate documentation, and careful testing.

The Web site for the Software Engineering Institute (SEI) at Carnegie Mellon University is shown in Figure 11-2. SEI is a leader in software engineering and provides quality standards and suggested procedures for software developers and systems analysts. SEI's primary objective is to find better, faster, and less-expensive methods of software development. To achieve that goal, SEI designed a set of software development standards called the **Capability Maturity Model (CMM)**®, which has been used successfully by thousands of organizations around the globe. The purpose of the model, which was introduced in 1991, was to improve software quality, reduce development time, and cut costs. More recently, SEI established a new model, called **Capability Maturity Model Integration (CMMI)**®, that integrates software and systems development into a much larger framework called **process improvement**. The

FIGURE 11-2 The Software Engineering Institute (SEI) provides leadership in software quality improvement. Notice that their vision is to lead the world to a software-enriched society.

CMMI® regards software as part of a larger quality improvement process, rather than as an end in itself. The CMMI® tracks an organization's processes, using five maturity levels, from Level 1, which is referred to as unpredictable, poorly controlled, and reactive, to Level 5, in which the optimal result is process improvement. The five maturity levels are shown in Figure 11-3.

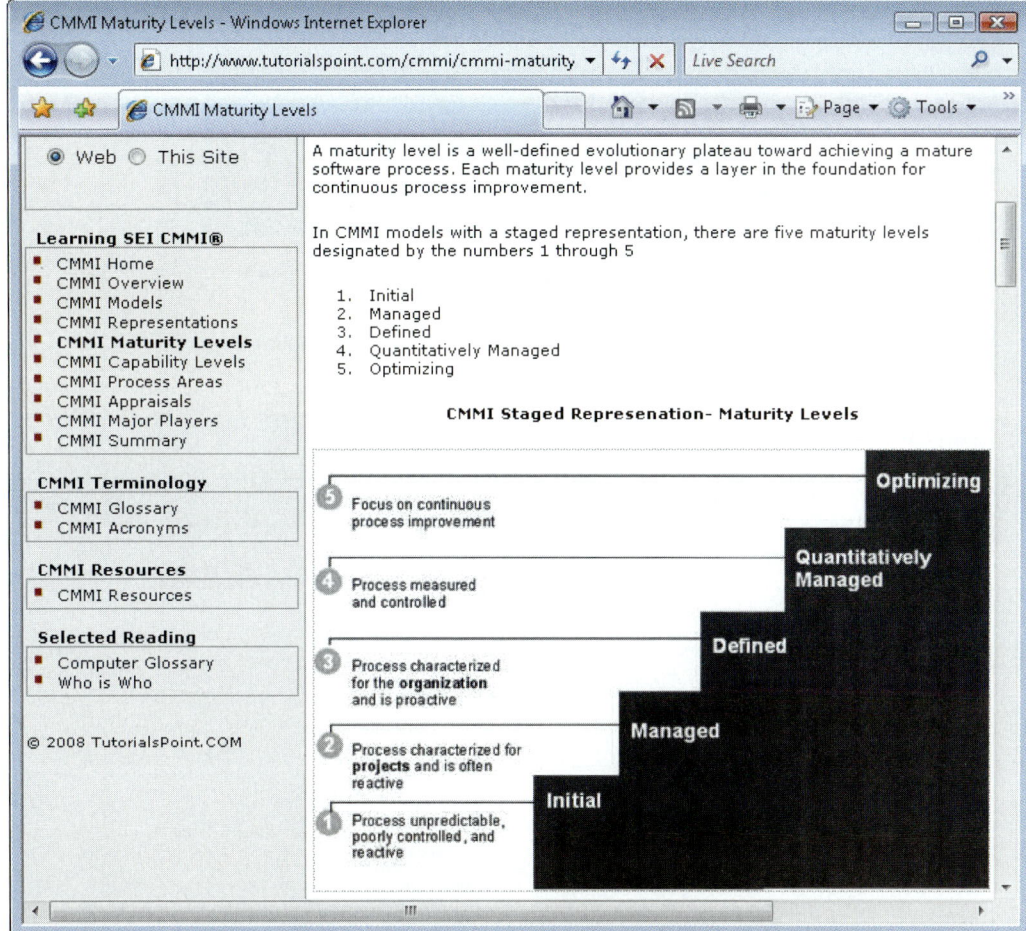

FIGURE 11-3   The CMMI® includes five maturity levels, from Level 1, which is referred to as unpredictable, poorly controlled, and reactive, to Level 5, in which the optimal result is process improvement.

International Organization for Standardization (ISO)

You learned in Chapter 9 that the International Organization for Standardization (ISO) is a worldwide body that establishes quality standards for products and services, as shown in Figure 11-4 on the next page. ISO standards include everything from internationally recognized symbols, such as those shown in Figure 11-5, to the ISBN numbering system that identifies this textbook. In addition, ISO seeks to offer a global consensus of what constitutes good management practices — practices that can help firms deliver consistently high-quality products and services.

Because software is so important to a company's success, many firms seek assurance that software systems, either purchased or developed in-house, will meet rigid quality standards. In 1991, ISO established a set of guidelines called ISO 9000-3, which provided a quality assurance framework for developing and maintaining software.

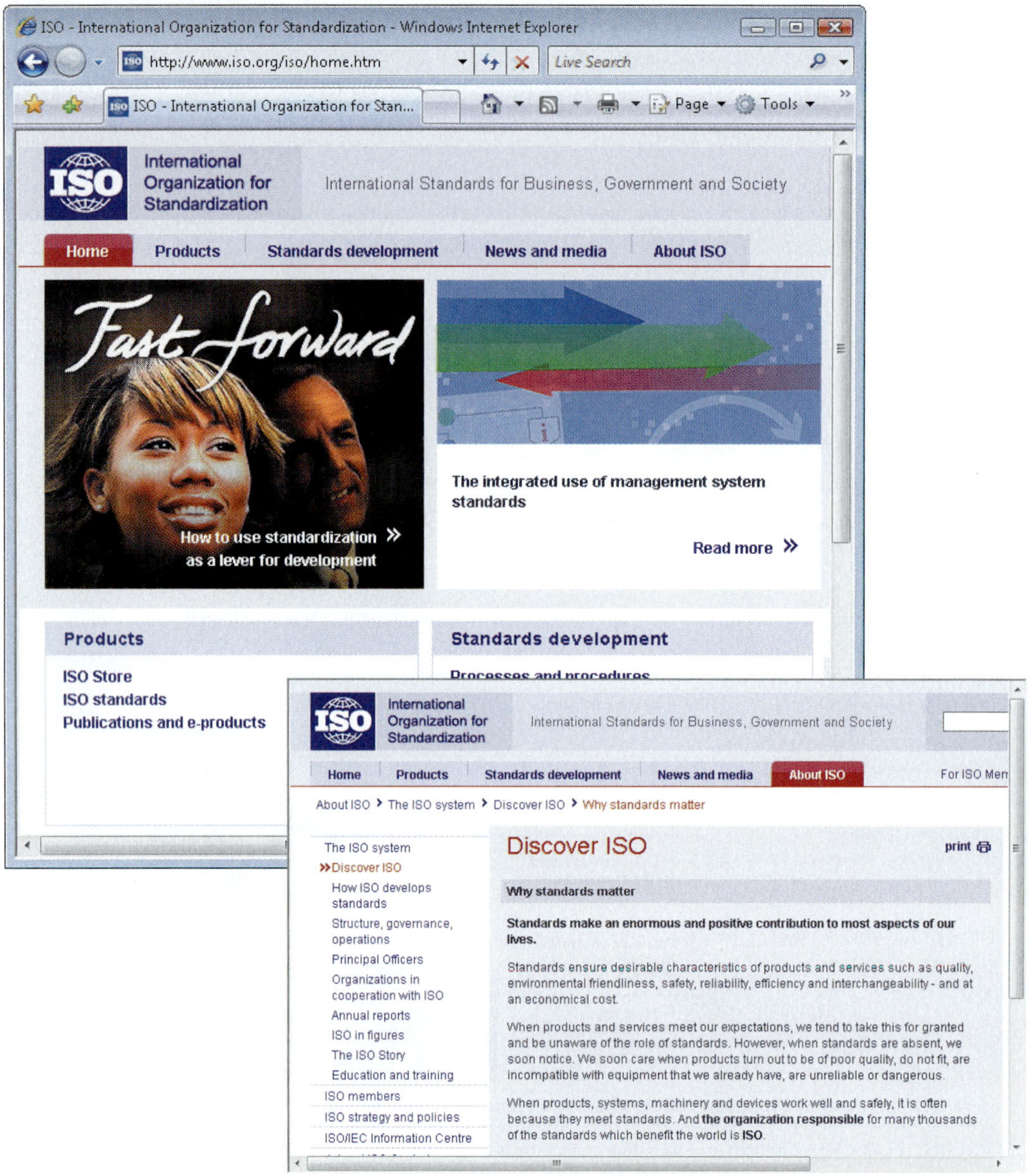

FIGURE 11-4   The International Organization for Standardization (ISO) is an international body that establishes standards for many products and services, including software development. ISO states that standards, which provide product quality, compatibility, and safety, often are taken for granted, and noticed only when they are absent.

FIGURE 11-5   ISO standards include internationally recognized symbols.

A company can specify ISO 9000-3 standards when it purchases software from a supplier or use ISO guidelines for in-house software development to ensure that the final result measures up to ISO standards. ISO requires a specific development plan, which outlines a step-by-step process for transforming user requirements into a finished product. ISO standards can be quite detailed. For example, ISO requires that a software supplier document all testing and maintain records of test results. If problems are found, they must be resolved, and any modules affected must be retested. Additionally, software and hardware specifications of all test equipment must be documented and included in the test records.

OVERVIEW OF APPLICATION DEVELOPMENT

Application development is the process of constructing the programs and code modules that serve as the building blocks of the information system. In Chapter 1, you learned that structured analysis, object-oriented (O-O) analysis, and agile methods are three popular development options. Regardless of the method, the objective is to translate the design into program and code modules that will function properly. Because systems implementation usually is very labor-intensive, developers often use project management tools and techniques to control schedules and budgets.

ON THE WEB

For more information about application development, visit **scsite.com/sad8e/ more**, locate Chapter 11, and then click the Application Development link.

Review the System Design

At this point, it might be helpful to review the tasks involved in the creation of the system design.

- In Chapter 4, you learned about requirements modeling and how to use functional decomposition diagrams (FDDs) to break complex business operations down into smaller units, or functions.

- In Chapter 5, you learned about structured data and process modeling, and you created data flow diagrams (DFDs). You also developed process descriptions for functional primitive processes that documented the business logic and processing requirements.

- In Chapter 6, you developed an object-oriented model of the new system that included use case diagrams, class diagrams, sequence diagrams, state transition diagrams, and activity diagrams.

- In Chapter 7, you selected a development strategy.

- In Chapter 8, you planned the system's output and designed the user interface.

- In Chapter 9, you worked with data design issues, analyzed relationships between system entities, and constructed entity-relationship diagrams (ERDs).

- In Chapter 10, you considered an overall system architecture.

Taken together, this set of tasks produced an overall design and a plan for physical implementation.

Application Development Tasks

If you used traditional structured or object-oriented (O-O) methods, you now are ready to translate the design into a functioning application. If you selected an agile development method, you will plan the project, lay the groundwork, assemble the team, and prepare to interact with the customers.

TRADITIONAL METHODS Building a new system requires careful planning. After an overall strategy is established, individual modules must be designed, coded, tested, and documented. A **module** consists of related program code organized into small units that are easy to understand and maintain. After the modules are developed and tested individually, more testing takes place, along with thorough documentation of the entire system, as shown in Figure 11-6.

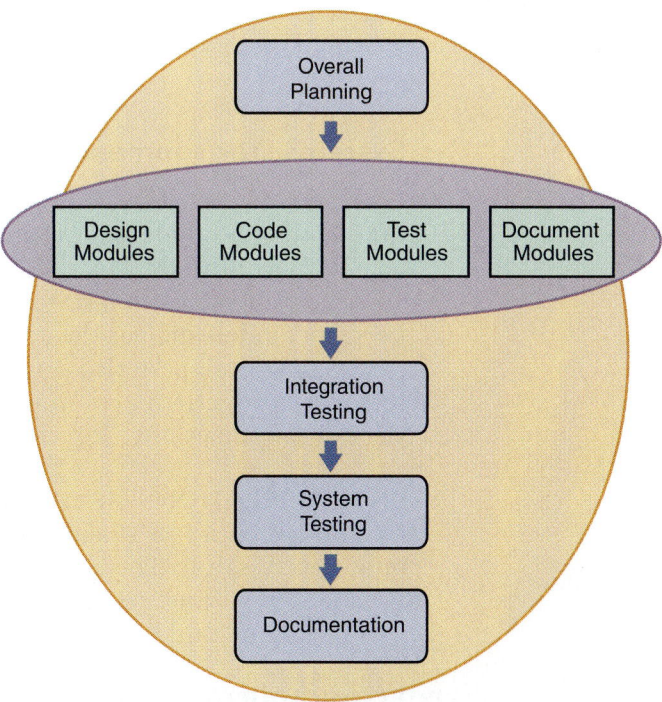

FIGURE 11-6 The main steps in application development.

When you create program modules using structured or object-oriented methods, you start by reviewing documentation from prior SDLC phases and creating a set of program designs. If you built a documentation file early in the development process and updated it regularly, you now have a valuable repository of information. The centerpiece of your documentation is the system design specification, accompanied by diagrams, source documents, screen layouts, report designs, data dictionary entries, and user comments. If you used a CASE tool during the systems analysis and design process, your job will be much easier. At this point, coding and testing tasks begin. Although programmers typically perform the actual coding, IT managers usually assign systems analysts to work with them as a team.

AGILE METHODS If you decided to use an agile approach, intense communication and collaboration will now begin between the IT team and the users or customers. The objective is to create the system through an iterative process of planning, designing, coding, and testing. Agile projects use various models, including the spiral model shown in Figure 1-30 on page 23, or the Extreme Programming (XP) example shown in Figure 11-7. Agile development and XP are discussed in detail later in this chapter.

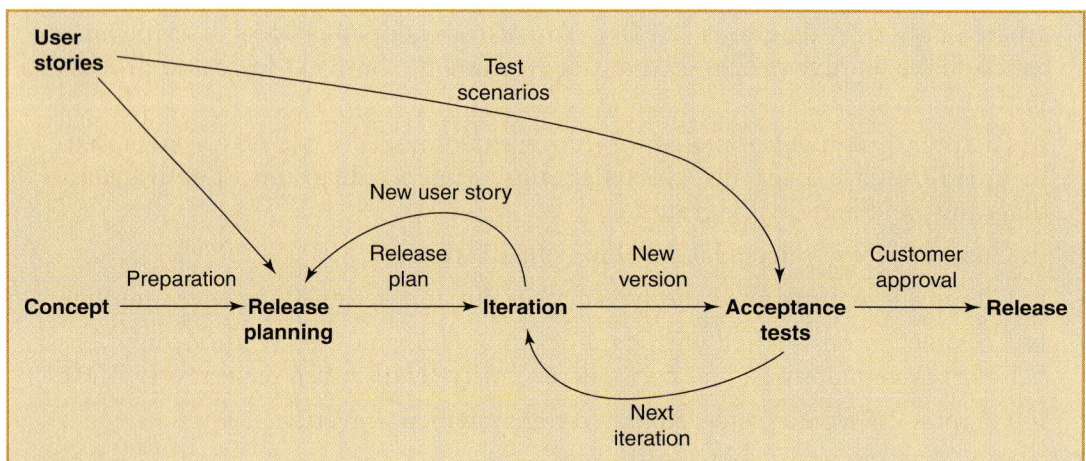

FIGURE 11-7 Simplified model of an Extreme Programming (XP) project. Note the emphasis on iteration and testing.

Systems Development Tools

Each systems development approach has its own set of tools that has worked well for that method. For example, structured development relies heavily on DFDs and structure charts; object-oriented methods use a variety of diagrams, including use case, class, sequence, and transition state diagrams; and agile methods tend to use spiral or other iterative models such as the example in Figure 11-7.

System developers also can use multipurpose tools to help them translate the system logic into properly functioning program modules. These generic tools include entity-relationship diagrams, flowcharts, pseudocode, decision tables, and decision trees.

ENTITY-RELATIONSHIP DIAGRAMS During data design, in Chapter 9, you learned how to use entity-relationship diagrams to show the interaction among system entities and objects. An ERD is a useful tool regardless of which methodology you are using, because the various relationships (one-to-one, one-to-many, and many-to-many) must be understood and implemented in the application development process.

FLOWCHARTS As you learned in Chapter 5, flowcharts can be used to describe program logic, and are very useful in visualizing a modular design. A **flowchart** represents logical rules and interaction graphically, using a series of symbols connected by arrows. Using flow-charts, programmers can break large systems into subsystems and modules that are easier to understand and code.

PSEUDOCODE Pseudocode is a technique for representing program logic. Pseudocode is similar to structured English, which was explained in Chapter 5. Pseudocode is not language-specific, so you can use it to describe a software module in plain English without requiring strict syntax rules. Using pseudocode, a systems analyst or a programmer can describe program actions that can be implemented in any programming language. Figure 11-8 illustrates an example of pseudocode that documents a sales promotion policy.

ON THE WEB

For more information about pseudocode, visit **scsite.com/ sad8e/more**, locate Chapter 11, and then click the Pseudocode link.

SAMPLE OF A SALES PROMOTION POLICY:

- Preferred customers who order more than $1,000 are entitled to a 5% discount, and an additional 5% discount if they used our charge card.

- Preferred customers who do not order more than $1,000 receive a $25 bonus coupon.

- All other customers receive a $5 bonus coupon.

PSEUDOCODE VERSION OF THE SALES PROMOTION POLICY:

```
IF customer is a preferred customer, and
        IF customer orders more than $1,000 then
                Apply a 5% discount, and
                IF customer uses our charge card, then
                        Apply an additional 5% discount
        ELSE
                        Award a $25 bonus coupon
ELSE
        Award a $5 bonus coupon
```

FIGURE 11-8 Sample of a sales promotion policy with logical rules, and a pseudocode version of the policy. Notice the alignment and indentation of the logic statements.

DECISION TABLES AND DECISION TREES As you learned in Chapter 5, decision tables and decision trees can be used to model business logic for an information system. In addition to being used as modeling tools, analysts and programmers can use decision tables and decision trees during system development, as they develop code modules that implement the logical rules. Figure 11-9 shows an example of a decision tree that documents the sales promotion policy shown in Figure 11-8. Notice that the decision tree accurately reflects the sales promotion policy, which has three separate conditions, and four possible outcomes.

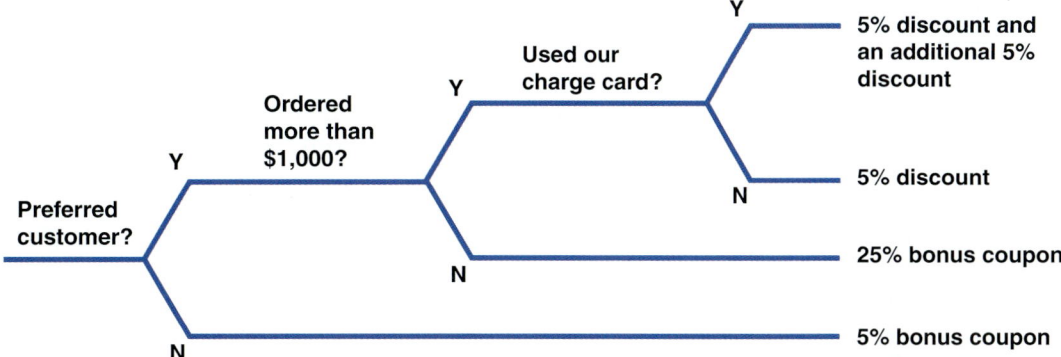

FIGURE 11-9 Sample decision tree that reflects the sales promotion policy in Figure 11-8. Like a decision table, a decision tree shows the action to be taken based on certain conditions.

Project Management

Regardless of whether structured analysis, object-oriented design, or agile methods are used, even a modest-sized project might have hundreds or even thousands of modules. For this reason, application development can become quite complex and difficult to manage. At this stage, project management is especially important. Users and managers are looking forward to the new system, and it is very important to set realistic schedules, meet project deadlines, control costs, and maintain quality. To achieve these goals, the systems analyst or project manager should use project management tools and techniques similar to those described in Chapter 3 to monitor and control the development effort.

The following sections describe the application development process. Structured development techniques and tools are discussed first, followed by object-oriented and agile development methods.

STRUCTURED APPLICATION DEVELOPMENT

Structured application development usually involves a **top-down approach**, which proceeds from a general design to a detailed structure. After a systems analyst documents the system's requirements, he or she breaks the system down into subsystems and modules in a process called **partitioning**. This approach also is called **modular design** and is similar to constructing a leveled set of DFDs. By assigning modules to different programmers, several development areas can proceed at the same time. As explained in Chapter 3, you can use project management software to monitor work on each module, forecast overall development time, estimate required human and technical resources, and calculate a critical path for the project.

Because all the modules must work together properly, an analyst must proceed carefully, with constant input from programmers and IT management to achieve a sound, well-integrated structure. The analyst also must ensure that integration capability is built into each design and thoroughly tested.

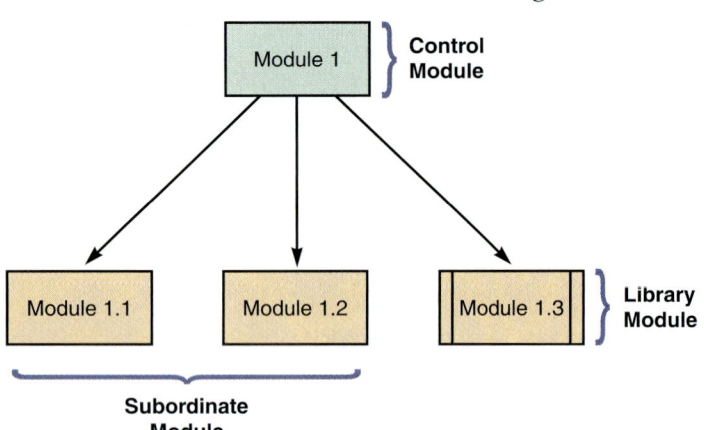

FIGURE 11-10 An example of structure chart modules.

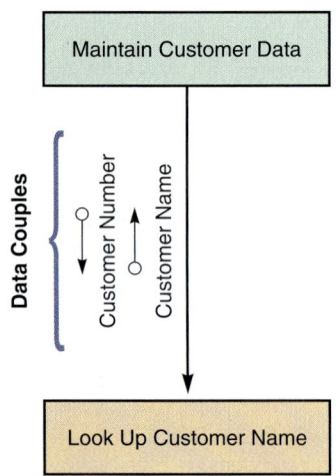

FIGURE 11-11 An example of a structure chart data.

Structure Charts

Structure charts show the program modules and the relationships among them. A **structure chart** consists of rectangles that represent the program modules, with arrows and other symbols that provide additional information. Typically, a higher-level module, called a **control module**, directs lower-level modules, called **subordinate modules**. In a structure chart, symbols represent various actions or conditions. Structure chart symbols represent modules, data couples, control couples, conditions, and loops.

MODULE A rectangle represents a module, as shown in Figure 11-10. Vertical lines at the edges of a rectangle indicate that module 1.3 is a library module. A **library module** is reusable code and can be invoked from more than one point in the chart.

DATA COUPLE An arrow with an empty circle represents a data couple. A **data couple** shows data that one module passes to another. In the data couple example shown in Figure 11-11, the *Look Up Customer Name* module exchanges data with the *Maintain Customer Data* module.

CONTROL COUPLE An arrow with a filled circle represents a control couple. A **control couple** shows a message, also called a **status flag**, which one module sends to another. In the example shown in Figure 11-12, the *Update Customer File* module sends an *Account Overdue* flag back to the *Maintain Customer Data* module. A module uses a flag to signal a specific condition or action to another module.

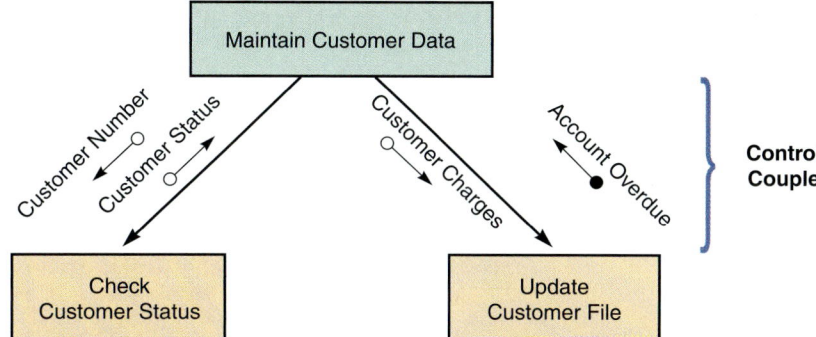

FIGURE 11-12 An example of a structure chart control couple.

CONDITION A line with a diamond on one end represents a condition. A **condition** line indicates that a control module determines which subordinate modules will be invoked, depending on a specific condition. In the example shown in Figure 11-13, *Sort Inventory Parts* is a control module with a condition line that triggers one of the three subordinate modules.

LOOP A curved arrow represents a loop. A **loop** indicates that one or more modules are repeated. In the example shown in Figure 11-14, the *Get Student Grades* and *Calculate GPA* modules are repeated.

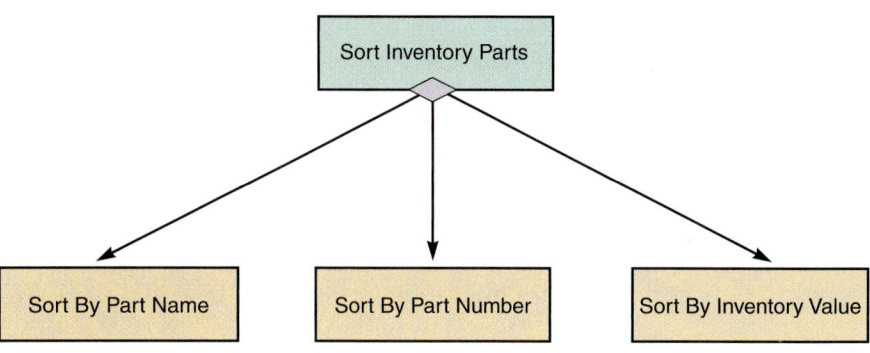

FIGURE 11-13 The diagram shows a control module that triggers three subordinate modules.

Cohesion and Coupling

Cohesion and coupling are important tools for evaluating the overall design. As explained in the following sections, it is desirable to have modules that are highly cohesive and loosely coupled.

Cohesion measures a module's scope and processing characteristics. A module that performs a single function or task has a high degree of cohesion, which is desirable. Because it focuses on a single task, a cohesive module is much easier to code and reuse. For example, a module named *Verify Customer Number*

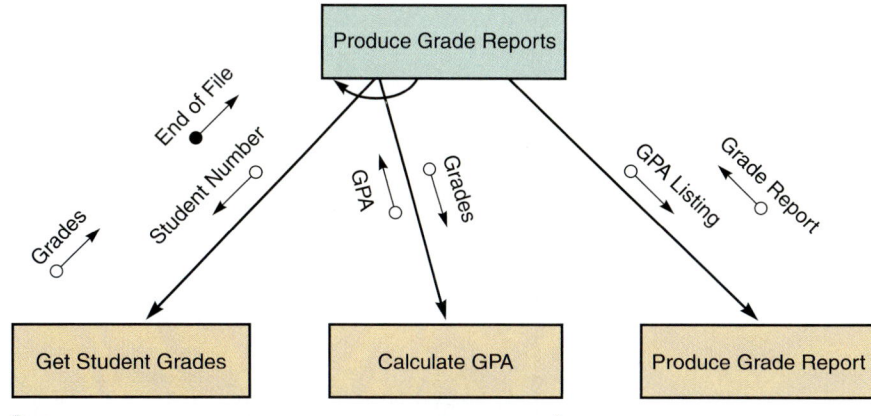

The curved arrow indicates that these modules are repeated.

FIGURE 11-14 The diagram shows a structure chart loop with two repeating modules.

is more cohesive than a module named *Calculate and Print Statements*. If you notice the word *and* in a module name, you know that more than one task is involved.

If a module must perform multiple tasks, more complex coding is required and the module will be more difficult to create and maintain. If you need to make a module more cohesive, you can split it into separate units, each with a single function. For example, by splitting the module *Check Customer Number and Credit Limit* in Figure 11-15 on the next page into two separate modules, *Check Customer Number* and *Check Customer Credit Limit*, cohesion is greatly improved.

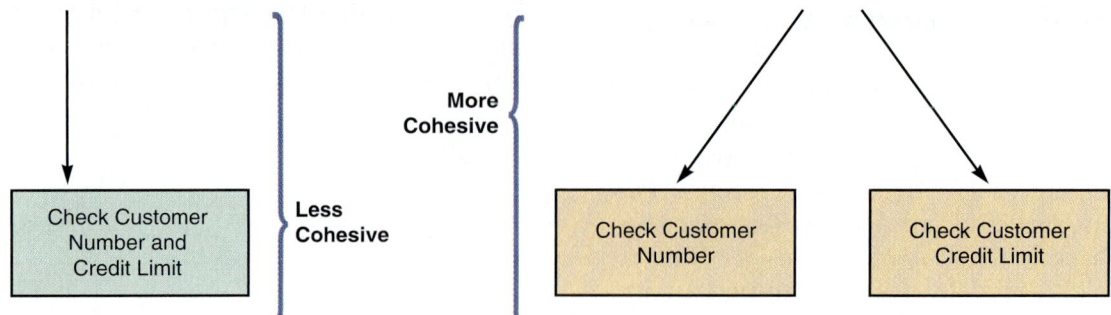

FIGURE 11-15 Two examples of cohesion. Notice that the single module on the left is less cohesive than the two modules on the right.

Coupling describes the degree of interdependence among modules. Modules that are independent are **loosely coupled**, which is desirable. Loosely coupled modules are easier to maintain and modify, because the logic in one module does not affect other modules. If a programmer needs to update a loosely coupled module, he or she can accomplish the task in a single location. If modules are **tightly coupled**, one module is linked to internal logic contained in another module. For example, Module A might refer to an internal variable contained in Module B. In that case, a logic error in the Module B will affect the processing in Module A. For that reason, passing a status flag down as a message from a control module is generally regarded as poor design. It is better to have subordinate modules handle processing tasks as independently as possible, to avoid a cascade effect of logic errors in the control module.

In Figure 11-16, the tightly coupled example on the left shows that the subordinate module *Calculate Current Charges* depends on a status flag sent down from the control module *Update Customer Balance*. It would be preferable to have the modules loosely coupled and logically independent. In the example on the right, a status flag is not needed because the subordinate module *Apply Discount* handles discount processing independently. Any logic errors are confined to a single location: the *Apply Discount* module.

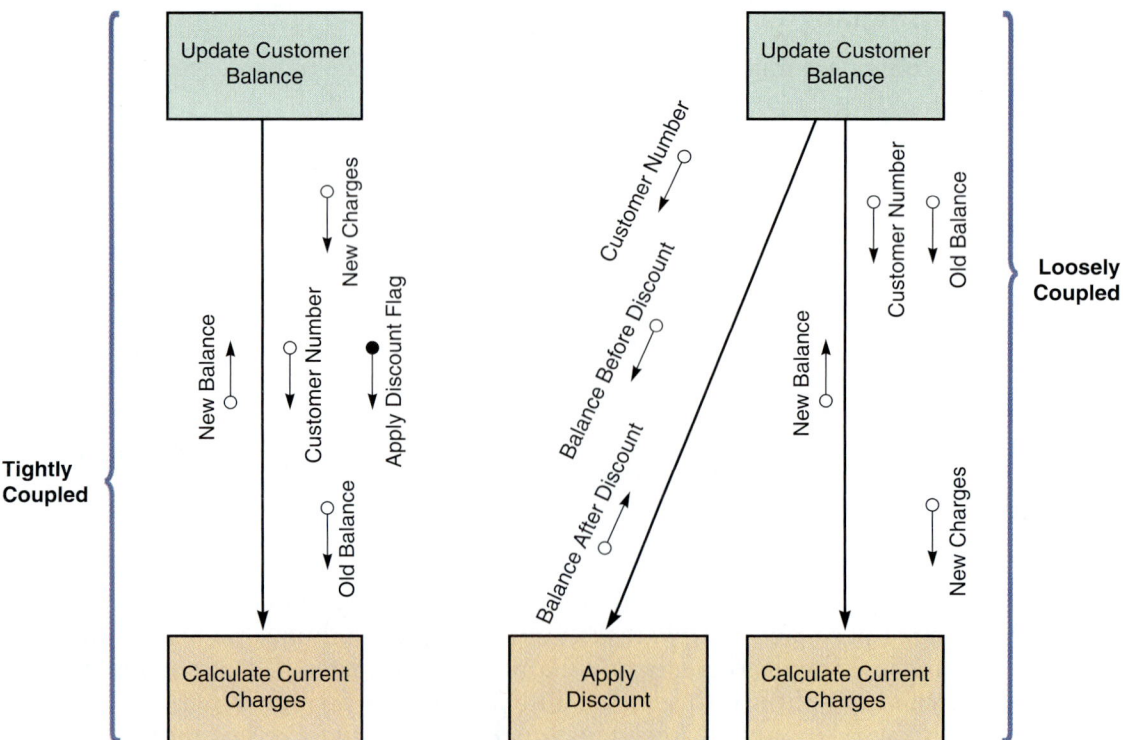

FIGURE 11-16 An example of tightly coupled and loosely coupled structure charts.

Drawing a Structure Chart

If you used a structured analysis method, your structure charts will be based on the DFDs you created during data and process modeling.

Typically, you follow four steps when you create a structure chart. You review DFDs to identify the processes and methods, identify the program modules and determine control-subordinate relationships, add symbols for couples and loops, and analyze the structure chart to ensure that it is consistent with your system documentation.

STEP 1: REVIEW THE DFDS Your first step is to review all DFDs for accuracy and completeness, especially if changes have occurred since the systems analysis phase. If object models also were developed, you should analyze them to identify the objects, the methods that each object must perform, and the relationships among the objects. A method is similar to a functional primitive, and requires code to implement the necessary actions.

STEP 2: IDENTIFY MODULES AND RELATIONSHIPS Working from the logical model, you transform functional primitives or object methods into program modules. When analyzing a set of DFDs, remember that each DFD level represents a processing level. If you are using DFDs, you would work your way down from the context diagram to the lower-level diagrams, identifying control modules and subordinate modules, until you reach functional primitives. If more cohesion is desired, you can divide processes into smaller modules that handle a single task. Figure 11-17 shows a structure chart based on the order system shown in Figures 5-16, 5-17, and 5-18 on pages 210 – 212. Notice how the three-level structure chart relates to the three DFD levels.

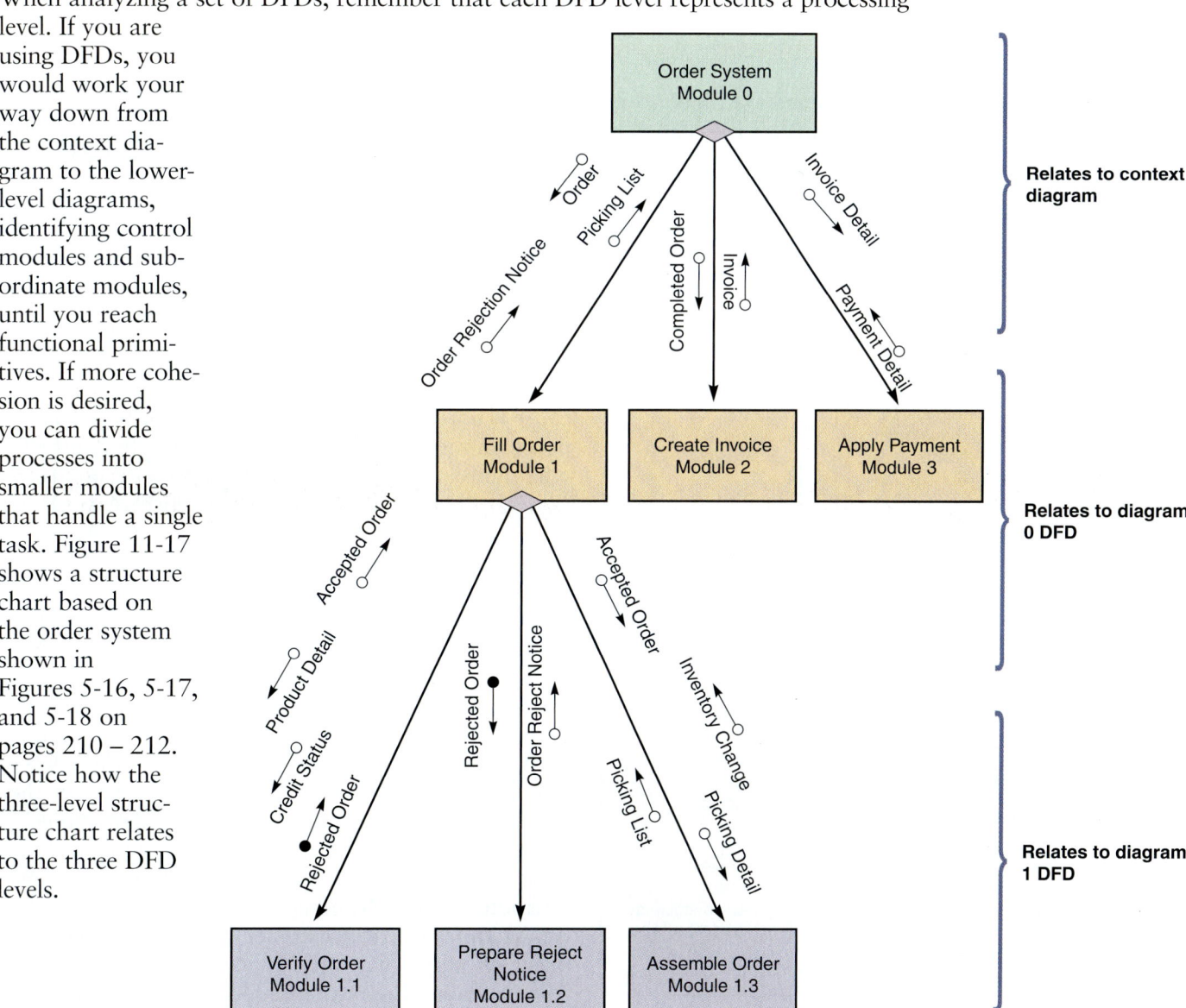

FIGURE 11-17 A structure chart based on the order system DFDs on pages 210 - 212. The three-level structure chart relates to the three DFD levels.

STEP 3: ADD COUPLES, LOOPS, AND CONDITIONS Next, you add couples, loops, and conditions to the structure chart. If you are working with DFDs, you can review the data flows and the data dictionary to identify the data elements that pass from one module to another. In addition to adding the data couples, you add control couples where a module is sending a control parameter, or flag, to another module. You also add loops and condition lines that indicate repetitive or alternative processing steps, as shown in Figure 11-17 on the previous page. If you also developed an object model, you can review the class diagrams and object relationship diagrams to be sure that you understand the interaction among the objects.

STEP 4: ANALYZE THE STRUCTURE CHART AND THE DATA DICTIONARY At this point, the structure chart is ready for careful analysis. You should check each process, data element, or object method to ensure that the chart reflects all previous documentation and that the logic is correct. You also should determine that modules are strongly cohesive and loosely coupled. Often, you must draw several versions of the chart. Some CASE tools can help you analyze the chart and identify problem areas.

OBJECT-ORIENTED APPLICATION DEVELOPMENT

When you studied the object-oriented methods described in Chapter 6, you learned that O-O analysis makes it easier to translate an object model directly into an object-oriented programming language. This process is called **object-oriented development**, or OOD. Although many structured design concepts also apply to object-oriented methodology, there are some differences.

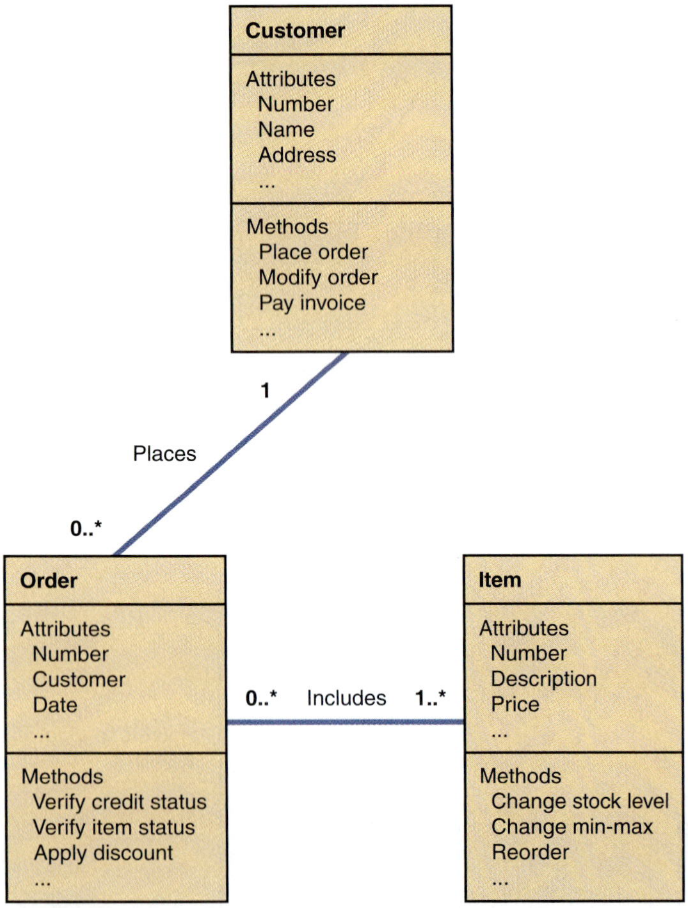

Characteristics of Object-Oriented Application Development

When implementing a structured design, a structure chart is used to describe the interaction between program modules, as explained earlier. In contrast, when implementing an object-oriented design, relationships between objects already exist. Because object interaction is defined during the O-O analysis process, the application's structure is represented by the object model itself.

As Chapter 6 explains, objects contain both data and program logic, called methods. Individual object instances belong to classes of objects with similar characteristics. The relationship and interaction among classes are described using a class diagram, such as the one shown in Figure 11-18. A class diagram includes the class **attributes**, which describe the characteristics of objects in the class, and **methods**, which represent program logic. For example, the Customer class describes customer objects. Customer attributes include Number, Name, Address, and so on. Methods for the Customer class include Place order, Modify order, and Pay invoice, among others. The Customer class can exchange messages with the Order class.

FIGURE 11-18 A simplified class diagram for a customer order processing system.

In addition to class diagrams, programmers get an overview of object interaction by using object relationship diagrams that were developed during the O-O analysis process. For example, Figure 11-19 shows an object relationship diagram for a fitness center. Notice the model shows the objects and how they interact to perform business functions and transactions.

Properly implemented, object-oriented development can speed up projects, reduce costs, and improve overall quality. However, these results are not always achieved. Organizations sometimes have unrealistic expectations, and do not spend enough time learning about, preparing for, and implementing the OOD process. For example, no one would build a bridge without a analysis of needs, supporting data, and a detailed blueprint — and the bridge would not be opened for traffic until it had been carefully inspected and checked to ensure that all specifications were met. O-O software developers sometimes forget that the basic rules of architecture also apply to their projects.

In summary, to secure the potential benefits of object-oriented development, systems analysts must carefully analyze, design, implement, test, and document their O-O projects.

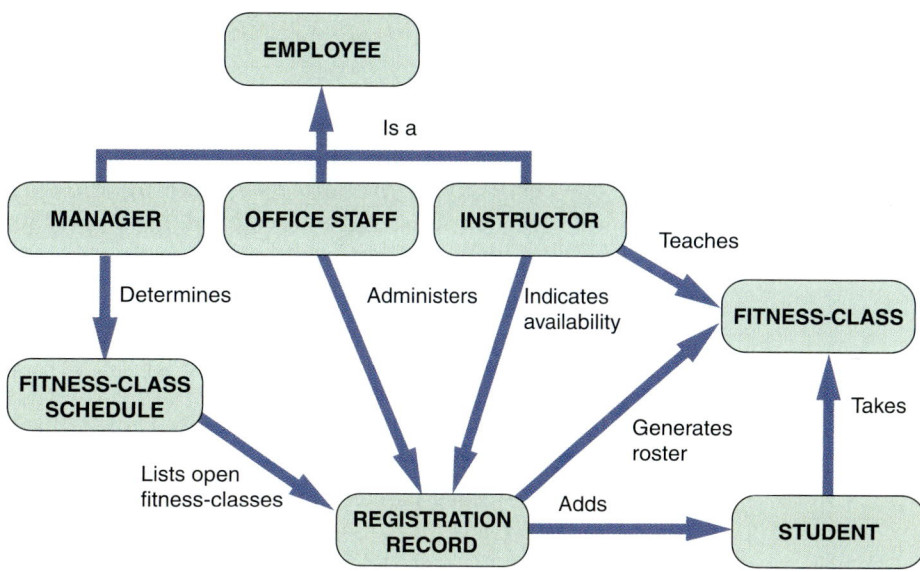

FIGURE 11-19   An object-relationship diagram for a fitness center.

Implementation of Object-Oriented Designs

When a programmer translates an object-oriented design into an application, he or she analyzes the classes, attributes, methods, and messages that are documented in the object model. During this process, the programmer makes necessary revisions and updates to class diagrams, sequence diagrams, state transition diagrams, and activity diagrams.

The programmer's main objective is to translate object methods into program code modules and determine what event or message will trigger the execution of each module. To accomplish the task, the programmer analyzes sequence diagrams and state transition diagrams that show the events and messages that trigger changes to an object. O-O applications are called event-driven, because each event, transaction, or message triggers a corresponding action. The programmer can represent the program steps in pseudocode initially, or use CASE tools and code generators to create object-oriented code directly from the object model.

Object-Oriented Cohesion and Coupling

The principles of cohesion and coupling also apply to object-oriented application development. Classes should be as loosely coupled (independent of other classes) as possible. In addition, an object's methods also should be loosely coupled (independent of other methods) and highly cohesive (perform closely related actions). By following these principles, classes and objects are easier to understand and edit. O-O programmers who ignore cohesion and coupling concepts may end up creating a web of code that is difficult to maintain. When code is scattered in various places, editing becomes complicated and expensive.

AGILE APPLICATION DEVELOPMENT

As you learned in Chapter 1, agile development is a distinctly different systems development method. It shares many of the steps found in traditional development, but uses a highly iterative process. The development team is in constant communication with the **customer** or primary user, shaping and forming the system to match the customer's specifications. Agile development is aptly named because it is based on a quick and nimble development process that easily adapts to change. Agile development focuses on small teams, intense communication, and rapid development iterations.

You also learned in Chapter 1 about Extreme Programming (XP), which is one of the newest agile methods. As shown in Figure 11-20, XP emphasizes customer satisfaction, teamwork, speed, and simplicity. The jigsaw puzzle symbolizes the many small pieces that combine to form a full picture. The following sections describe a typical XP project.

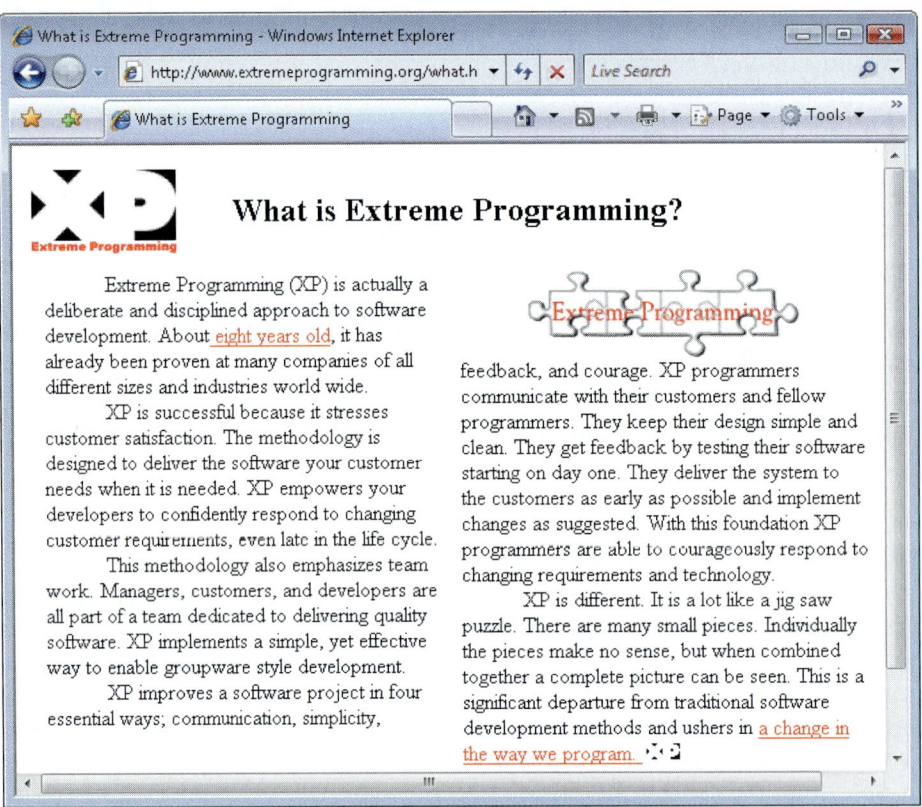

FIGURE 11-20 Supporters believe that XP will change how systems programmers work. Notice the jigsaw puzzle, which symbolizes many small pieces that make no sense individually, but combine to form a complete picture.

An Extreme Programming (XP) Example

ON THE WEB

For more information about extreme programming (XP), visit **scsite.com/ sad8e/more**, locate Chapter 11, and then click the Extreme Programming (XP) link.

Suppose that a customer has requested a sales tracking system. The first step in the XP process, like any other development method, would be to define the system requirements. The customer begins by meeting with programmers and providing user stories. A **user story** is a short, simple requirements definition. Programmers use user stories to determine the project's requirements, priorities, and scope.

In our example, suppose we have the following user stories:

- *As the sales manager, I want to identify fast or slow moving items so I can manage our inventory more effectively.*

- *As a store manager, I need enough lead time to replenish my stock so I don't run out of hot items.*

- *As a sales representative, I want to offer the best selection of fast selling items and clear out the old stock that is not moving.*

User stories do not deal with technical details and are so short that they are often written on index cards. Each user story is given a priority by the customer, so the requirements can be ranked. In addition, programmers assign a score to each user story that indicates the estimated difficulty of implementation. This information helps the team form a plan and assign its resources. Figure 11-21 shows a sample user story on an index card, with a priority notation and an estimated difficulty score. Projects are often composed of many user stories, from which programmers can estimate the scope, time requirements, and difficulty of the project. In addition to the user stories, frequent face-to-face meetings with customers provide a higher level of detail as the project progresses.

ID: 2 Inventory Management

As sales manager, I want to identify fast or slow moving items so I can manage our inventory more effectively.

Priority: 3
Difficulty: 5

FIGURE 11-21 An example of a user story written on an index card, with a priority notation and an estimated difficulty score.

The team must also develop a **release plan,** which specifies when user stories will be implemented and the timing of the releases. Releases are relatively frequent, and each system release is like a prototype that can be tested and modified as needed.

User stories are implemented in a series of iteration cycles. An **iteration cycle** includes planning, designing, coding, and testing of one or more features based on user stories. At the beginning of each iteration cycle, which is often two weeks long, the team holds an **iteration planning meeting** to break down the user stories into specific tasks that are assigned to team members. As new user stories or features are added, the team reviews and modifies the release plan.

As with any development process, success is determined by the customer's approval. The programming team regularly meets with the customer, who tests prototype releases as they become available. This process usually results in additional user stories, and changes are implemented in the next iteration cycle. As the project's code changes during each iteration, obsolete code is removed and remaining code is restructured to keep the system up to date. The iteration cycles continue until all user stories have been implemented, tested, and accepted.

Extreme Programming uses an interesting concept called parallel programming. In **parallel programming,** two programmers work on the same task on the same computer; one drives (programs) while the other navigates (watches). The onlooker examines the code strategically to see the *forest* while the driver is concerned with the individual *trees* immediately in front of him or her. The two discuss their ideas continuously throughout the process.

Another important concept in XP is that unit tests are designed *before* code is written. This **test-driven design** focuses on end results from the beginning and prevents programmers from straying from their goals. Because of the magnitude and intensity of the multi-cycle process, agile testing relies heavily on automated testing methods and software.

Programmers can use popular agile-friendly languages such as Python, Ruby, and Perl. However, agile methods do not require a specific programming language, and programmers also use various object-oriented languages such as Java, C++, and C#.

The Future of Agile Development

Agile methodology is becoming very popular for software projects. Its supporters boast that it speeds up software development and delivers precisely what the customer wants, when the customer wants it, while fostering teamwork and empowering employees.

ON THE WEB

For more information about the future of agile development, visit **scsite.com/ sad8e/more**, locate Chapter 11, and then click the Future of Agile Development link.

However, there are drawbacks to this adaptive rather than predictive method. Critics of agile development often claim that because it focuses on quick iterations and fast releases, it lacks discipline and produces systems of questionable quality. In addition, agile methodology may not work as well for larger projects because of their complexity and the lack of focus on a well-defined end product.

Before implementing agile development, the proposed system and development methods should be examined carefully. As experienced IT professionals know, a one-size-fits-all solution does not exist. For more information on agile methods, refer to the discussion of systems development methods that begins on page 18 in Chapter 1.

CODING

Coding is the process of turning program logic into specific instructions that the computer system can execute. Working from a specific design, a programmer uses a programming language to transform program logic into code statements. An individual programmer might create a small program, while larger programs typically are divided into modules that several individuals or groups can work on simultaneously.

Programming Environments

TOOLKIT TIME

The CASE tools in Part 2 of the Systems Analyst's Toolkit can help you understand IDEs. To learn more about these tools, turn to Part 2 of the four-part Toolkit that follows Chapter 12.

Each IT department has its own programming environment and standards. Visual Basic, Python, Ruby, and SQL are examples of commonly used programming languages, and many commercial packages use a proprietary set of commands. As the trend toward Internet-based applications continues, HTML/XML, Java, and other Web-centric languages will be used extensively.

To simplify the integration of system components and reduce code development time, many programmers use an **integrated development environment (IDE)**. IDEs can make it easier to program interactive software products by providing built-in tools and advanced features, such as real-time error detection, syntax hints, highlighted code, class browsers, and version control. As you learned in Chapter 7, IBM WebSphere and Microsoft .NET are popular IDEs. In addition to these commercial packages, programmers can use open-source IDEs such as Java-based NetBeans IDE and Eclipse. You can learn more about IDEs in Part 2 of the Systems Analyst's Toolkit.

Generating Code

You learned in earlier chapters that systems analysts use application generators, report writers, screen generators, fourth-generation languages, and other CASE tools that produce code directly from program design specifications. Some commercial applications can generate editable program code directly from macros, keystrokes, or mouse actions. Figure 11-22 shows a very simple example of a Visual Basic code module in Microsoft Access that opens a customer order form and produces a

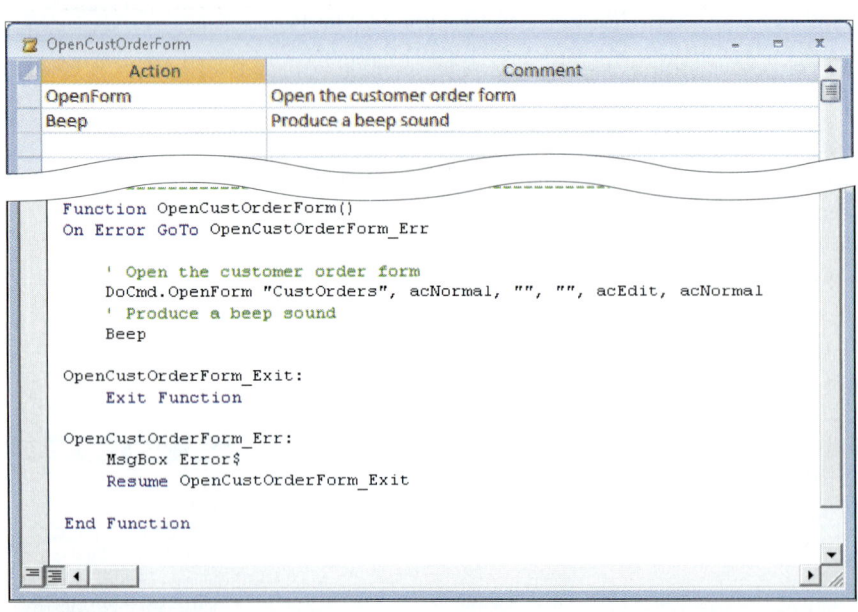

FIGURE 11-22 First, a programmer uses keystrokes and mouse actions to create the Microsoft Access macro shown in the top screen. The macro opens a customer order form and produces a beep sound. Access converts the macro to the editable code shown in the bottom screen — complete with commands, comments, and error-handling procedures.

beep sound. A macro automatically generated the code, and the macro itself was created by a series of keystrokes and mouse actions. Notice that the code module shown in Figure 11-22 includes program commands, comments, and error-handling procedures.

TESTING THE SYSTEM

After coding, a programmer must test each program to make sure it functions correctly. Later, programs are tested in groups, and finally the development team must test the entire system. The first step is to compile the program using a CASE tool or a language compiler. This process detects **syntax errors**, which are language grammar errors. The programmer corrects the errors until the program executes properly.

Next, the programmer desk checks the program. **Desk checking** is the process of reviewing the program code to spot **logic errors**, which produce incorrect results. This process can be performed by the person who wrote the program or by other programmers. Many organizations require a more formal type of desk checking called a **structured walkthrough**, or **code review**.

Typically, a group of three to five IT staff members participate in code review. The group usually consists of project team members and might include other programmers and analysts who did not work on the project. The objective is to have a peer group identify errors, apply quality standards, and verify that the program meets the requirements of the system design specification. Errors found during a structured walkthrough are easier to fix while coding is still in the developmental stages.

In addition to analyzing logic and program code, the project team usually holds a session with users called a **design walkthrough**, to review the interface with a cross-section of people who will work with the new system and ensure that all necessary features have been included. This is a continuation of the modeling and prototyping effort that began early in the systems development process.

The next step in application development is to initiate a sequence of unit testing, integration testing, and system testing, as shown in Figure 11-23.

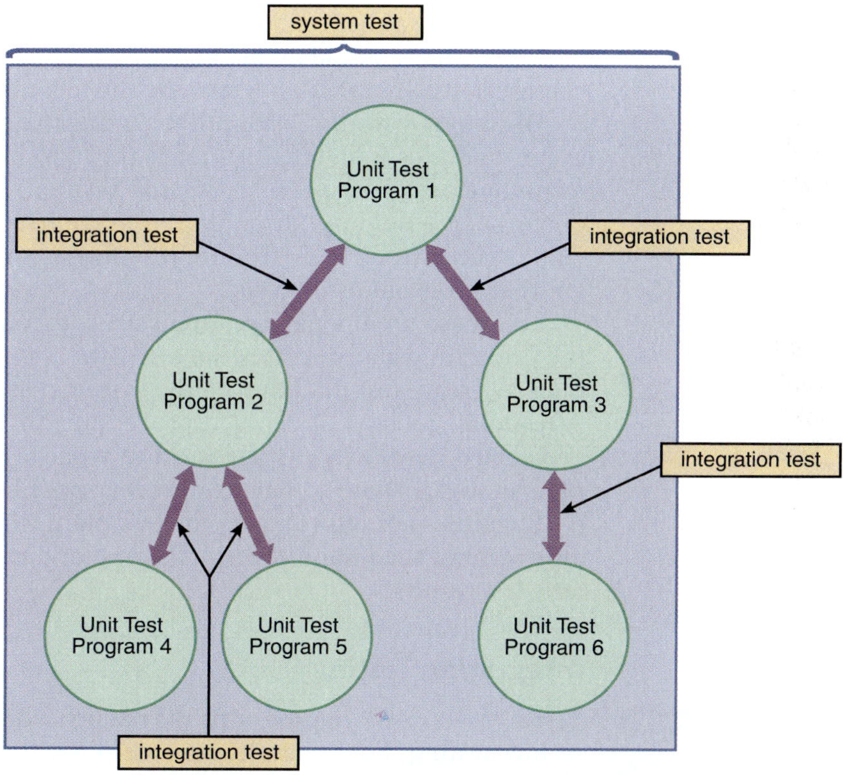

FIGURE 11-23   The first step in testing is unit testing, followed by integration testing, and then system testing.

Unit Testing

The testing of an individual program or module is called **unit testing**. The objective is to identify and eliminate execution errors that could cause the program to terminate abnormally, and logic errors that could have been missed during desk checking.

CASE IN POINT 11.1: YOUR MOVE, INC.

You work for April Olivia, the IT manager at Your Move, Inc., a large retailer specializing in games of all kinds. The company is in the final stages of developing a new inventory management system, and April wants you to handle the testing.

"Be sure you put lots of errors into the test data," she said. "Users are bound to make mistakes, and we need to design built-in safeguards that will catch the mistakes, and either fix them automatically, or alert the user to the problem."

Of course, April's comment makes a lot of sense, but you've never done this before and you wonder how to proceed. Should you try to invent every possible data error? How will you know that you've thought of every situation that could occur? Consider the problem, develop an approach, and write up your plan in a brief memo.

Test data should contain both correct data and erroneous data and should test all possible situations that could occur. For example, for a field that allows a range of numeric values, the test data should contain minimum values, maximum values, values outside the acceptable range, and alphanumeric characters. During testing, programmers can use software tools to determine the location and potential causes of program errors.

During unit testing, programmers must test programs that interact with other programs and files individually, before they are integrated into the system. This requires a technique called stub testing. In **stub testing**, the programmer simulates each program outcome or result and displays a message to indicate whether or not the program executed successfully. Each stub represents an entry or exit point that will be linked later to another program or data file.

To obtain an independent analysis, someone other than the programmer who wrote the program usually creates the test data and reviews the results. Systems analysts frequently create test data during the systems design phase as part of an overall test plan. A **test plan** consists of detailed procedures that specify how and when the testing will be performed, who will participate, and what test data will be used. A comprehensive test plan should include scenarios for every possible situation the program could encounter.

Regardless of who creates the test plan, the project manager or a designated analyst also reviews the final test results. Some organizations also require users to approve final unit test results.

Integration Testing

ON THE WEB

For more information about system testing, visit **scsite.com/ sad8e/more**, locate Chapter 11, and then click the System Testing link.

Testing two or more programs that depend on each other is called **integration testing**, or **link testing**. For example, consider an information system with a program that checks and validates customer credit status, and a separate program that updates data in the customer master file. The output from the validation program becomes input to the master file update program. Testing the programs independently does not guarantee that the data passed between them is correct. Only by performing integration testing for this pair of programs can you make sure that the programs work together properly. Figure 11-23 on the previous page shows integration testing for several groups of programs. Notice that a program can have membership in two or more groups.

Systems analysts usually develop the data they use in integration testing. As is the case with all forms of testing, integration test data must consider both normal and unusual situations. For example, integration testing might include passing typical records between two programs, followed by blank records, to simulate an unusual event or an operational problem. You should use test data that simulates actual conditions because you are testing the interface that links the programs. A testing sequence should not move to the integration test stage unless it has performed properly in all unit tests.

System Testing

After completing integration testing, you must perform **system testing**, which involves the entire information system, as shown in Figure 11-23. A system test includes all typical processing situations and is intended to assure users, developers, and managers that the program meets all specifications and that all necessary features have been included.

During a system test, users enter data, including samples of actual, or live, data, perform queries, and produce reports to simulate actual operating conditions. All processing options and outputs are verified by users and the IT project development team to ensure that the system functions correctly. Commercial software packages must undergo system testing similar to that of in-house developed systems, although unit and integration testing usually are not performed. Regardless of how the system was developed, system testing has the following major objectives:

- Perform a final test of all programs

- Verify that the system will handle all input data properly, both valid and invalid

- Ensure that the IT staff has the documentation and instructions needed to operate the system properly and that backup and restart capabilities of the system are adequate (the details of creating this sort of documentation are discussed later in this chapter)

- Demonstrate that users can interact with the system successfully

- Verify that all system components are integrated properly and that actual processing situations will be handled correctly

- Confirm that the information system can handle predicted volumes of data in a timely and efficient manner

Successful completion of system testing is the key to user and management approval, which is why system tests sometimes are called **acceptance tests**. Final acceptance tests, however, are performed during systems installation and evaluation, which is described later in this chapter.

How much testing is necessary? The answer depends on the situation and requires good judgment and input from other IT staff members, users, and management, as shown in Figure 11-24. Unfortunately, IT project managers often are pressured to finish testing quickly and hand the system over to users. Common reasons for premature or rushed testing are demands from users, tight systems development budgets, and demands from top management to finish projects early. Those pressures hinder the testing process and often have detrimental effects on the final product.

You should regard thorough testing as a cost-effective means of providing a quality product. Every error caught during testing eliminates potential expenses and operational problems. No system, however, is 100% error-free. Often, errors go undetected until the system becomes operational. Errors that affect the integrity or accuracy of data must be corrected immediately. Minor errors, such as typographical errors in screen titles, can be corrected later.

Some users want a system that is a completely finished product, while others realize that minor changes can be treated as maintenance items after the system is operational. In the final analysis, you must decide whether or not to postpone system installation if problems are discovered. If conflicting views exist, management will decide whether or not to install the system after a full discussion of the options.

FIGURE 11-24 System testing requires good judgment and input from other IT staff members, users, and IT management.

CASE IN POINT 11.2: WEBTEST, INC.

As a new systems analyst, you suspect that testing Web-based systems probably involves a different set of tools and techniques, compared to testing traditional LAN-based systems. Because you've always wanted to run your own IT company, you have decided to launch a start-up firm called WebTest, Inc., that would offer consulting services specifically aimed at testing the performance, integrity, efficiency, and security of Internet-based systems.

Your idea is to identify and purchase various Web site testing tools that currently are available, then use these tools as a Web site testing consultant. No one in your area offers this type of consulting service, so you have high hopes.

Now, you need to perform Internet research to learn more about Web testing software that is available. Review the Internet Resources Tools section, which is Part 4 of the Systems Analyst's Toolkit that follows Chapter 12, and use a search engine to develop a list of at least four products that you might want to use. For each product, write up a brief description of its overall purpose, its features and benefits, its cost, and how it would fit into your business game plan.

DOCUMENTATION

Documentation describes an information system and helps the users, managers, and IT staff who must interact with it. Accurate documentation can reduce system downtime, cut costs, and speed up maintenance tasks. Figure 11-25 shows an example of software that can automate the documentation process and help software developers generate accurate, comprehensive reference material.

Documentation is essential for successful system operation and maintenance. In addition to supporting a system's users, accurate documentation is essential for IT staff members who must modify the system, add a new feature, or perform maintenance. Documentation includes program documentation, system documentation, operations documentation, and user documentation.

FIGURE 11-25 In addition to CASE tools, software such as Imagix can automate the task of software documentation.

Program Documentation

Program documentation describes the inputs, outputs, and processing logic for all program modules. The program documentation process starts in the systems analysis phase and continues during systems implementation. Systems analysts prepare overall documentation, such as process descriptions and report layouts, early in the SDLC. This documentation guides programmers, who construct modules that are well supported by internal and external comments and descriptions that can be understood and maintained easily. A systems analyst usually verifies that program documentation is complete and accurate.

System developers also use **defect tracking software**, sometimes called **bug tracking software**, to document and track program defects, code changes, and replacement code, called **patches**. One popular example is Bugzilla, shown in Figure 11-26. According to its Web site, Bugzilla is a free, open-source program that can track bugs and manage software quality assurance.

FIGURE 11-26 Bugzilla is an example of a defect tracking program that can track bugs and manage software quality assurance.

System Documentation

System documentation describes the system's functions and how they are implemented. System documentation includes data dictionary entries, data flow diagrams, object models, screen layouts, source documents, and the systems request that initiated the project. System documentation is necessary reference material for the programmers and analysts who must support and maintain the system.

Most of the system documentation is prepared during the systems analysis and systems design phases. During the systems implementation phase, an analyst must review prior documentation to verify that it is complete, accurate, and up to date, including any changes made during the implementation process. For example, if a screen or report has been modified, the analyst must update the documentation. Updates to the system documentation should be made in a timely manner to prevent oversights.

ON THE WEB

For more information about documentation, visit **scsite.com/ sad8e/more**, locate Chapter 11, and then click the Documentation link.

Operations Documentation

If the information system environment involves a minicomputer, a mainframe, or centralized servers, the analyst must prepare documentation for the IT group that supports centralized operations. A mainframe installation might require the scheduling of batch jobs and the distribution of printed reports. In this type of environment, the IT operations staff serves as the first point of contact when users experience problems with the system.

Operations documentation contains all the information needed for processing and distributing online and printed output. Typical operations documentation includes the following information:

- Program, systems analyst, programmer, and system identification
- Scheduling information for printed output, such as report run frequency and deadlines

- Input files and where they originate; and output files and destinations
- E-mail and report distribution lists
- Special forms required, including online forms
- Error and informational messages to operators and restart procedures
- Special instructions, such as security requirements

Operations documentation should be clear, concise, and available online if possible. If the IT department has an operations group, you should review the documentation with them, early and often, to identify any problems. If you keep the operations group informed at every phase of the SDLC, you can develop operations documentation as you go along.

User Documentation

User documentation consists of instructions and information to users who will interact with the system and includes user manuals, Help screens, and tutorials. Programmers or systems analysts usually create program documentation and system documentation. To produce effective and clear user documentation — and hence have a successful project — you need someone with expert skills in this area doing the development, just as you need someone with expert skills developing the software. The skill set required to develop documentation usually is not the same as that to develop a system. This is particularly true as you move into the world of online documentation, which needs to coordinate with print documentation and intranet and Internet information. Technical writing requires specialized skills, and competent technical writers are valuable members of the IT team.

Just as you cannot throw a system together in several days, you cannot add documentation at the end. That is a common misconception and often proves fatal to a project. While that has always been true of software user documentation, this is an even more critical issue now that online Help and context-sensitive Help so often are needed. Context-sensitive Help is part of the program. You must put coded callouts in the text that link to the correct page of information in the documentation. To try to go back and add this after the fact would take a great deal of time; depending on the project size, it could take months! Additionally, it could introduce other coding errors — and it all has to be tested as well.

Systems analysts usually are responsible for preparing documentation to help users learn the system. In larger companies, a technical support team that includes technical writers might assist in the preparation of user documentation and training materials. Regardless of the delivery method, user documentation must be clear, understandable, and readily accessible to users at all levels.

User documentation includes the following:

- A system overview that clearly describes all major system features, capabilities, and limitations
- Description of source document content, preparation, processing, and samples
- Overview of menu and data entry screen options, contents, and processing instructions
- Examples of reports that are produced regularly or available at the user's request, including samples
- Security and audit trail information
- Explanation of responsibility for specific input, output, or processing requirements
- Procedures for requesting changes and reporting problems

- Examples of exceptions and error situations

- Frequently asked questions (FAQs)

- Explanation of how to get help and procedures for updating the user manual

Most users prefer **online documentation,** which provides immediate Help when they have questions or encounter problems. Many users are accustomed to context-sensitive Help screens, hints and tips, hypertext, on-screen demos, and other user-friendly features commonly found in popular software packages; they expect the same kind of support for in-house developed software.

If the system will include online documentation, that fact needs to be identified as one of the system requirements. If the documentation will be created by someone other than the analysts who are developing the system, that person or group needs to be involved as early as possible to become familiar with the software and begin developing the required documentation and support material. In addition, system developers must determine whether the documentation will be available from within the program, or as a separate entity in the form of a tutorial, slide presentation, reference manual, or Web site.

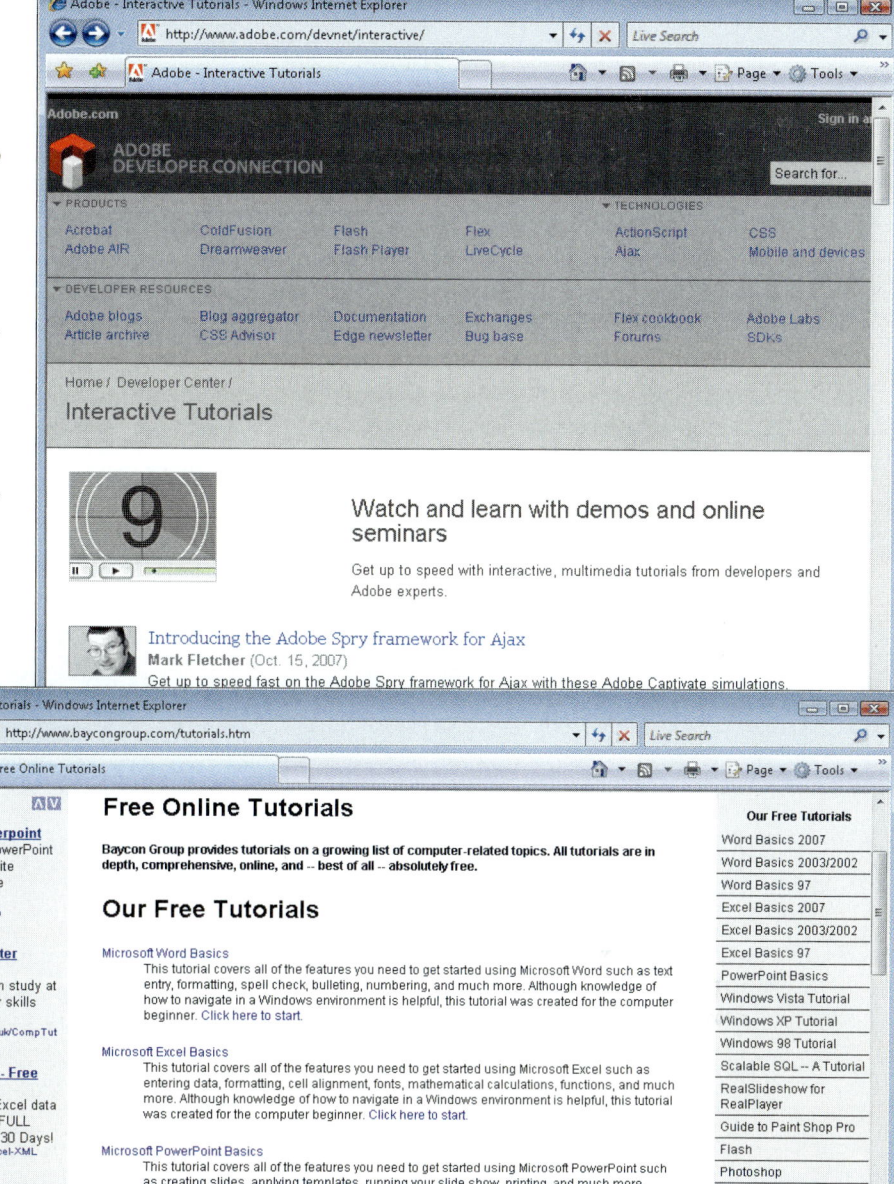

FIGURE 11-27 Adobe offers interactive tutorials, demos, and seminars for its products, while Baycon Group provides a variety of free tutorials.

If necessary, links should be created within the program that will take the user to the appropriate documentation.

Effective online documentation is an important productivity tool because it empowers users and reduces the time that IT staff members must spend in providing telephone, e-mail, or face-to-face assistance. Interactive tutorials are especially popular with users who like to learn by doing. Many software packages include tutorials, and additional tutorials are available online, as shown in Figure 11-27.

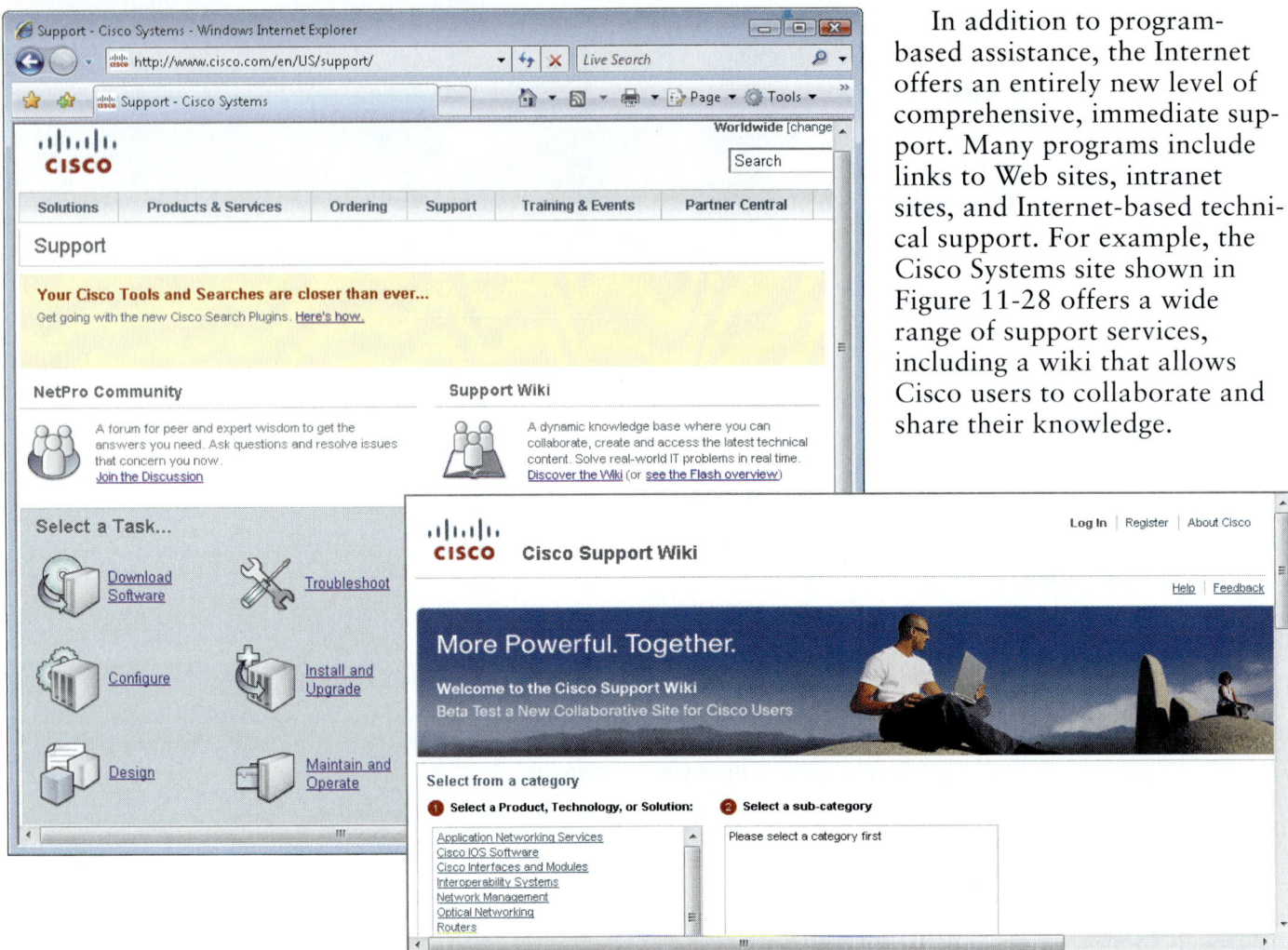

In addition to program-based assistance, the Internet offers an entirely new level of comprehensive, immediate support. Many programs include links to Web sites, intranet sites, and Internet-based technical support. For example, the Cisco Systems site shown in Figure 11-28 offers a wide range of support services, including a wiki that allows Cisco users to collaborate and share their knowledge.

FIGURE 11-28 In addition to traditional types of technical support, the Cisco Systems Web site features a support wiki.

Although online documentation is essential, written documentation material also is valuable, especially in training users and for reference purposes. A sample page from a user manual is shown in Figure 11-29. Systems analysts or technical writers usually prepare the manual, but many companies invite users to review the material and participate in the development process.

No matter what form of user documentation your system will require, you must keep in mind that it can take a good deal of time to develop. The time between finishing software coding and the time when a complete package — including documentation — can be released to users is entirely dependent on how well the documentation is thought out in advance. If the completion of your project includes providing user documentation, this issue needs to be addressed from the very beginning of the project. Determining what the user documentation requirements are and ascertaining who will complete the documents is critical to a timely release of the project.

Neglecting user documentation issues until after all the program is complete often leads to one of two things: (1) The documentation will be thrown together quickly just to get it out the door on time, and it more than likely will be inadequate; or (2) it will be done correctly, and the product release will be delayed considerably.

User training typically is scheduled when the system is installed; the training sessions offer an ideal opportunity to distribute the user manual and explain the procedures for updating it in the future. Training for users, managers, and IT staff is described later in this chapter.

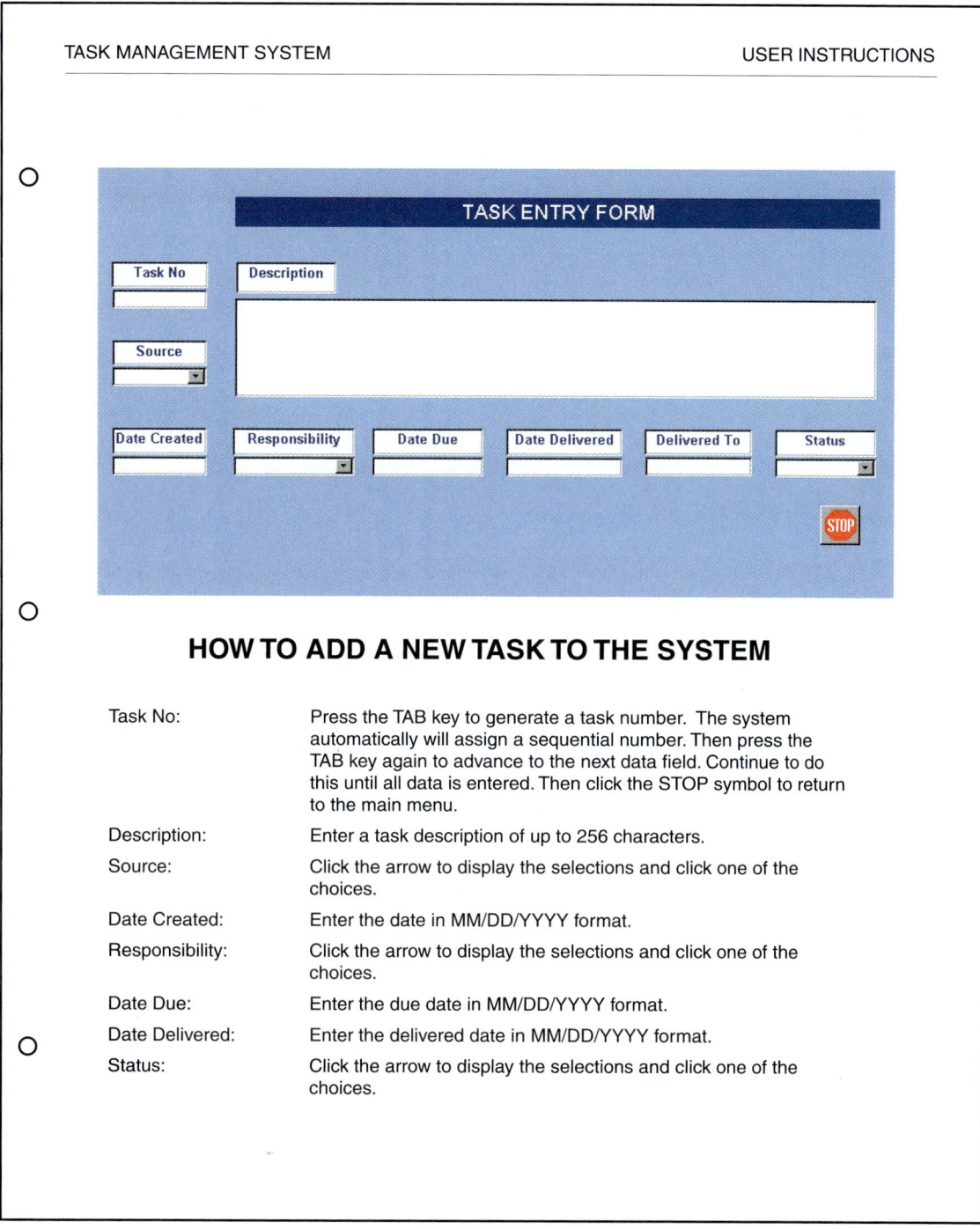

TASK MANAGEMENT SYSTEM   USER INSTRUCTIONS

TASK ENTRY FORM

Task No   Description

Source

Date Created   Responsibility   Date Due   Date Delivered   Delivered To   Status

STOP

HOW TO ADD A NEW TASK TO THE SYSTEM

Task No: Press the TAB key to generate a task number. The system automatically will assign a sequential number. Then press the TAB key again to advance to the next data field. Continue to do this until all data is entered. Then click the STOP symbol to return to the main menu.

Description: Enter a task description of up to 256 characters.

Source: Click the arrow to display the selections and click one of the choices.

Date Created: Enter the date in MM/DD/YYYY format.

Responsibility: Click the arrow to display the selections and click one of the choices.

Date Due: Enter the due date in MM/DD/YYYY format.

Date Delivered: Enter the delivered date in MM/DD/YYYY format.

Status: Click the arrow to display the selections and click one of the choices.

FIGURE 11-29   A sample page from a user manual. The instructions explain how to add a new task to the system.

MANAGEMENT APPROVAL

After system testing is complete, you present the results to management. You should describe the test results, update the status of all required documentation, and summarize input from users who participated in system testing. You also must provide detailed time schedules, cost estimates, and staffing requirements for making the system fully operational. If system testing produced no technical, economical, or operational problems, management determines a schedule for system installation and evaluation.

SYSTEM INSTALLATION AND EVALUATION

The following sections describe system installation and evaluation tasks that are performed for every information systems project, whether you develop the application in-house or purchase it as a commercial package.

The new system now is ready to go to work. Your earlier design activities produced the overall architecture and processing strategy, and you consulted users at every stage of development. You developed and tested programs individually, in groups, and as a complete system. You prepared the necessary documentation and checked it for accuracy, including support material for IT staff and users. Now, you will carry out the remaining steps in systems implementation:

- Prepare a separate operational and test environment
- Provide training for users, managers, and IT staff
- Perform data conversion and system changeover
- Carry out a post-implementation evaluation of the system
- Present a final report to management

OPERATIONAL AND TEST ENVIRONMENTS

You learned earlier that an environment, or platform, is a specific combination of hardware and software. The environment for the actual system operation is called the **operational environment** or **production environment**. The environment that analysts and programmers use to develop and maintain programs is called the **test environment**. A separate test environment is necessary to maintain system security and integrity and protect the operational environment. Typically, the test environment resides on a limited-access workstation or server located in the IT department.

Access to the operational environment is limited to users and must strictly be controlled. Systems analysts and programmers should not have access to the operational environment except to correct a system problem or to make authorized modifications or enhancements. Otherwise, IT department members have no reason to access the day-to-day operational system.

The test environment for an information system contains copies of all programs, procedures, and test data files. Before making any changes to an operational system, you must verify them in the test environment and obtain user approval. Figure 11-30 shows the differences between test environments and operational environments.

An effective testing process is essential, whether you are examining an information system or a batch of computer chips. Every experienced systems analyst can tell you a story about an apparently innocent program change that was introduced without being tested properly. In those stories, the innocent change invariably ends up causing some unexpected and unwanted changes to the program. After any modification, you should repeat the same acceptance tests you ran when the system was developed. By

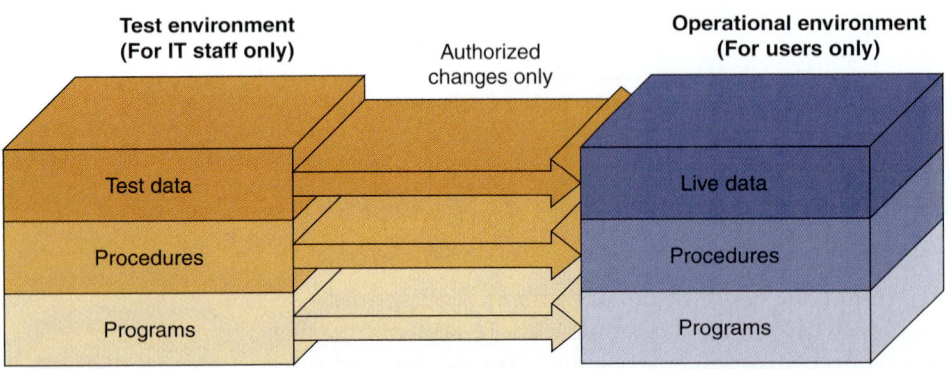

FIGURE 11-30 The test environment versus the operational environment. Notice that access to the test environment is limited to IT staff, while the operational environment is restricted to users.

restricting access to the operational area and performing all tests in a separate environment, you can protect the system and avoid problems that could damage data or interrupt operations.

The operational environment includes hardware and software configurations and settings, system utilities, telecommunications resources, and any other components that might affect system performance. Because network capability is critically important in a client/server environment, you must verify connectivity, specifications, and performance before installing any applications. You should check all communications features in the test environment carefully, and then check them again after loading the applications into the operational environment. Your documentation should identify all network specifications and settings, including technical and operational requirements for communications hardware and software. If you have to build or upgrade network resources to support the new system, you must test the platform rigorously before system installation begins.

TRAINING

No system can be successful without proper training, whether it involves software, hardware, or manufacturing, as shown in Figure 11-31. A successful information system requires training for users, managers, and IT staff members. The entire systems development effort can depend on whether or not people understand the system and know how to use it effectively.

FIGURE 11-31 In any situation, training must fit the needs of users and help them carry out their job functions.

Training Plan

You should start to consider a **training plan** early in the systems development process. As you create documentation, you should think about how to use the material in future training sessions. When you implement the system, it is essential to provide the right training for the right people at the right time. The first step is to identify who should receive training and what training is needed. You must look carefully at the organization, how the system will support business operations, and who will be involved or affected. Figure 11-32 shows specific training topics for users, managers, and IT staff. Notice that each group needs a mix of general background and detailed information to understand and use the system.

As shown in Figure 11-32 on the next page, the three main groups for training are users, managers, and IT staff. A manager does not need to understand every submenu or feature, but he or she does need a system overview to ensure that users are being trained properly and are using the system correctly. Similarly, users need to know how to perform their day-to-day job functions, but do not need to know how the company allocates system operational charges among user departments. IT staff people probably need the most information. To support the new system, they must have a clear understanding of how the system functions, how it supports business requirements, and the skills that users need to operate the system and perform their tasks.

After you identify the objectives, you must determine how the company will provide training. The main choices are to obtain training from vendors, outside training firms, or use IT staff and other in-house resources.

ON THE WEB

For more information about training, visit **scsite.com/ sad8e/ more**, locate Chapter 11, and then click the Training link.

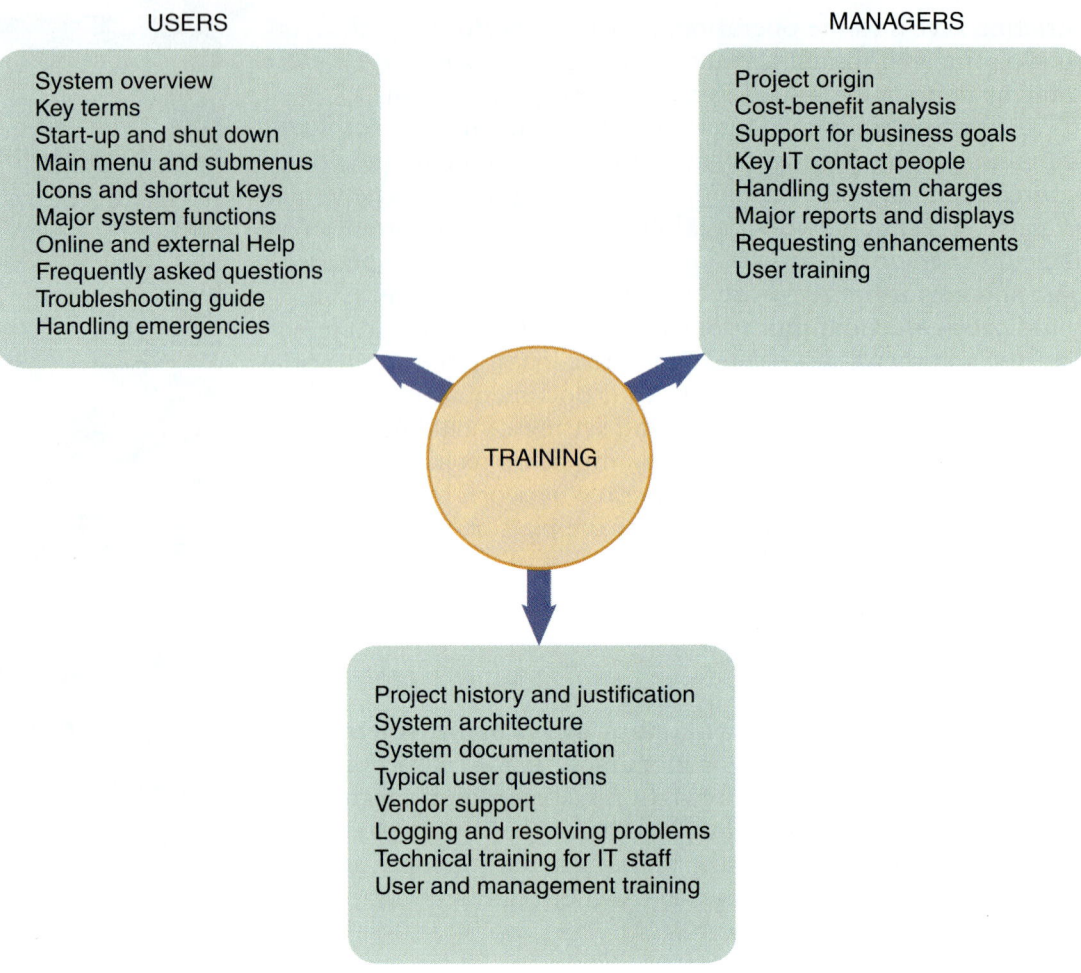

USERS

- System overview
- Key terms
- Start-up and shut down
- Main menu and submenus
- Icons and shortcut keys
- Major system functions
- Online and external Help
- Frequently asked questions
- Troubleshooting guide
- Handling emergencies

MANAGERS

- Project origin
- Cost-benefit analysis
- Support for business goals
- Key IT contact people
- Handling system charges
- Major reports and displays
- Requesting enhancements
- User training

TRAINING

IT STAFF

- Project history and justification
- System architecture
- System documentation
- Typical user questions
- Vendor support
- Logging and resolving problems
- Technical training for IT staff
- User and management training

FIGURE 11-32 Examples of training topics for three different groups. Users, managers, and IT staff members have different training needs.

Vendor Training

If the system includes the purchase of software or hardware, then vendor-supplied training is one of the features you should include in the RFPs (requests for proposal) and RFQs (requests for quotation) that you send to potential vendors.

Many hardware and software vendors offer training programs free or at a nominal cost for the products they sell. In other cases, the company might negotiate the price for training, depending on their relationship with the vendor and the prospect of future purchases. The training usually is conducted at the vendor's site by experienced trainers who provide valuable hands-on experience. If a large number of people need training, you might be able to arrange classes at your location.

Vendor training often gives the best return on your training dollars because it is focused on products that the vendor developed. The scope of vendor training, however, usually is limited to a standard version of the vendor's software or hardware. You might have to supplement the training in-house, especially if your IT staff customized the package.

Webinars, Podcasts, and Tutorials

Many vendors offer Web-based training options, including Webinars, podcasts, and tutorials. Figure 11-33 shows a Webinar and a podcast. A **Webinar**, which combines the words Web and seminar, is an Internet-based training session that provides an interactive experience. Most Webinars are scheduled events with a group of preregistered users and an online presenter or instructor. A pre-recorded Webinar session also can be delivered as a **Webcast**, which is a one-way transmission, whenever a user wants or needs training support.

A **podcast** refers to a Web-based broadcast that allows a user to receive audio or multimedia files using music player software such as iTunes®, and listen to them on a PC or download them to an iPod® or other portable MP3 player. Podcasts can be prescheduled, made available on demand, or delivered as automatic updates, depending on a user's preference. An advantage of a podcast is that **subscribers** can listen to the recorded material anywhere, any-time. As technology continues to advance, other wireless devices such as PDAs and cell phones will be able to receive podcasts.

A **tutorial** is a series of online interactive lessons that present mate-rial and provide a dialog with users. Tutorials can be developed by software vendors, or by a company's IT team. A tutorial example is included in the in-house training section.

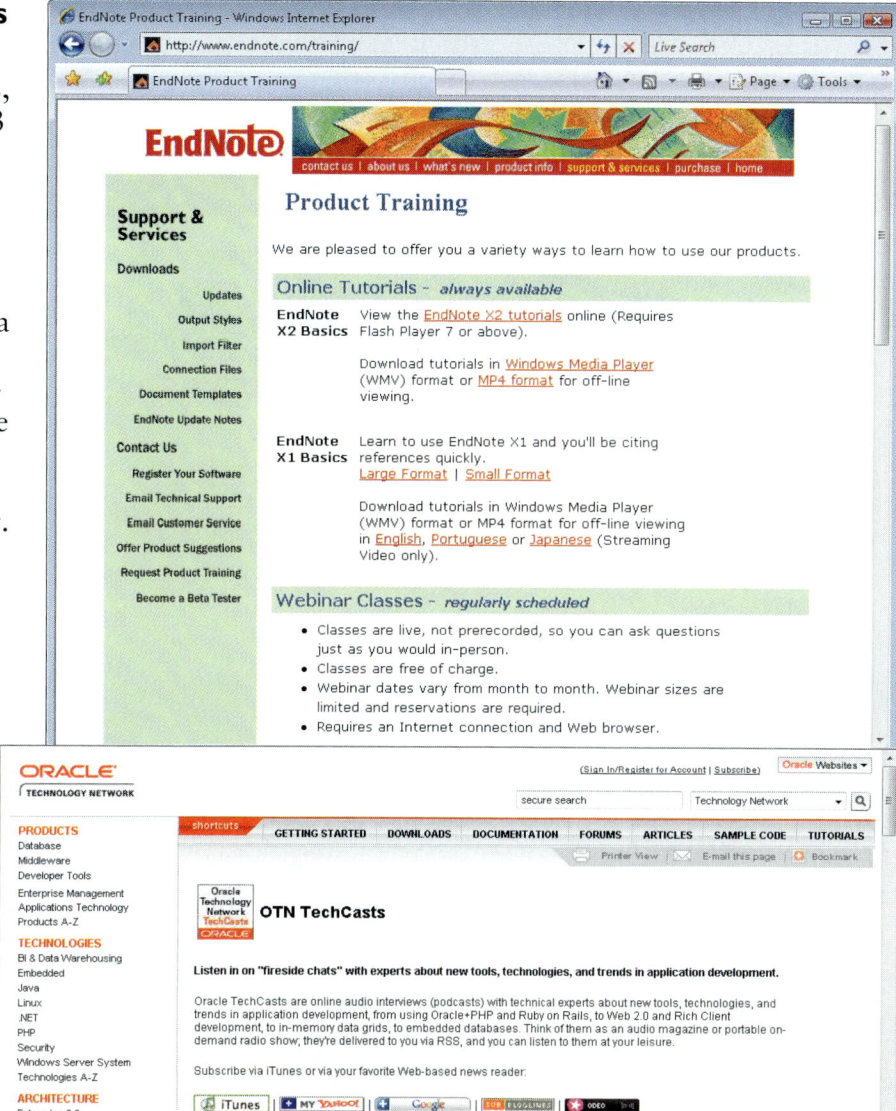

FIGURE 11-33 Vendor training and support can include online tutorials, Webinars, and podcasts, which Oracle calls *Techcasts*.

Outside Training Resources

You also can look into an independent training firm to provide in-house hardware or software training. If vendor training is not practical and your organization does not have the internal resources to perform the training, you might find that outside training consultants are a desirable alternative.

The rapid expansion of information technology has produced tremendous growth in the computer-training field. Many training consultants, institutes, and firms are avail-able that provide either standardized or customized training packages. IT industry lead-ers, such as Hewlett-Packard and IBM, offer a wide variety of training solutions, as shown in Figure 11-34 on the next page.

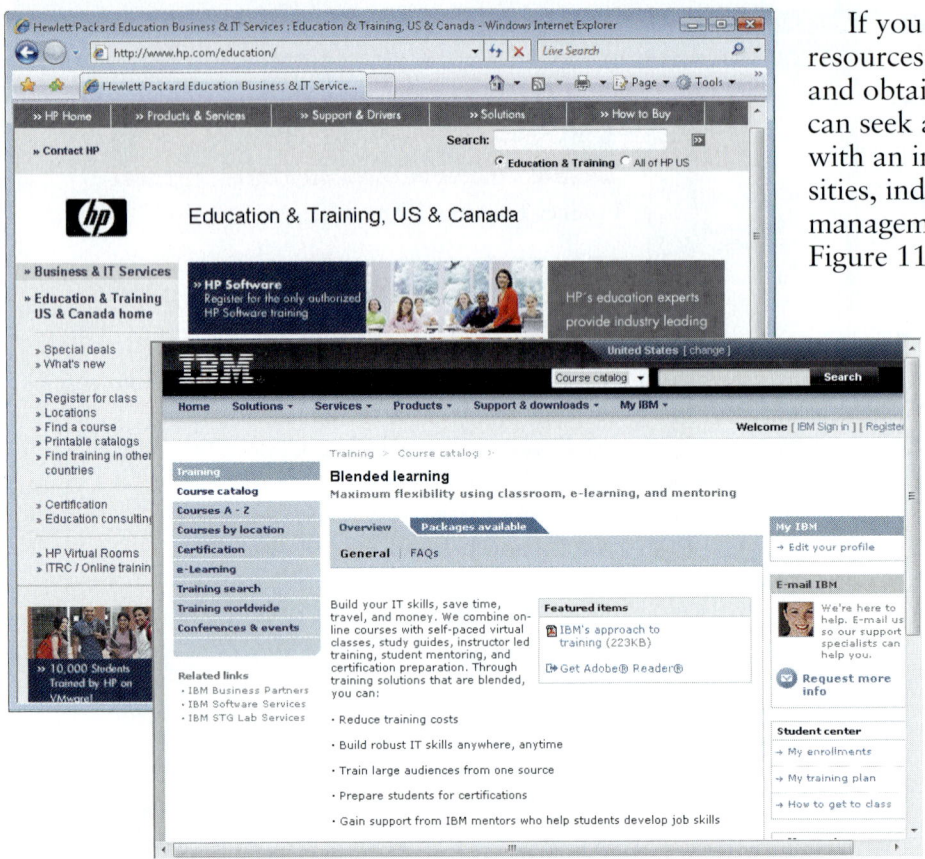

FIGURE 11-34 IT industry leaders, such as Hewlett-Packard and IBM, offer a wide variety of training solutions.

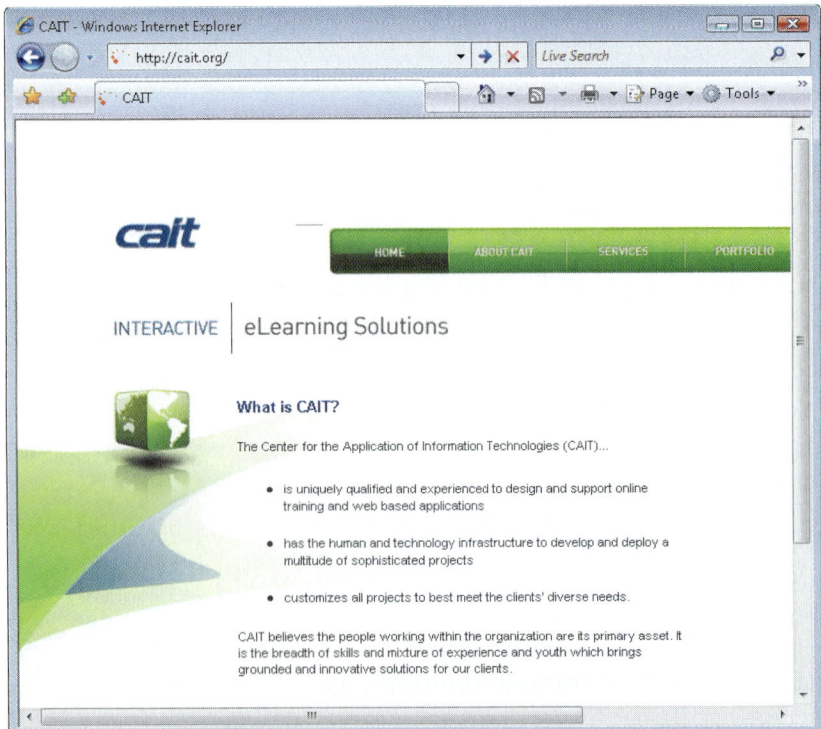

FIGURE 11-35 Western Illinois University's Center for the Application of Information Technologies (CAIT) helps organizations design, develop, and deploy information technology education and training.

If you decide to investigate outside training resources, you can contact a training provider and obtain references from clients. You also can seek assistance from nonprofit sources with an interest in training, including universities, industry associations, and information management organizations. For example, Figure 11-35 shows the Web site for Western Illinois University's Center for the Application of Information Technologies (CAIT), which describes a variety of IT education and training resources.

In-House Training

The IT staff and user departments often share responsibility for developing and conducting training programs for internally developed software. If your organization has a help desk, the staff might be able to handle user training.

Multimedia is an effective training method. Presentation software, such as Microsoft PowerPoint, OpenOffice Impress, or Corel Presentations, allows you to design training sessions that combine slides, animation, and sound. You also can use programs that capture actual keystrokes and mouse actions, and then replay the screens as a demonstration for users. If your firm has a media or graphic arts group, they can help you prepare training aids such as videotapes, charts, and other instructional materials. When developing a training program, you should keep the following guidelines in mind.

Train people in groups, with separate training programs for distinct groups. Group training, as shown in Figure 11-36, makes the most efficient use of time and training facilities. In addition, if the group is small, trainees can learn from the questions and problems of others. A training program must address the job interests and skills of a wide range of participants. For example, IT staff personnel and users require very different information. Problems often arise when some participants have technical backgrounds and others do not. A single program will not meet everyone's needs.

Select the most effective place to conduct the training. Training employees at your company's location offers several advantages. Employees incur no travel expense, they can respond to local emergencies that require immediate attention, and training can take place in the actual environment where the system will operate. You can encounter some disadvantages, however. Employees who are distracted by telephone calls and other duties will not get the full benefit of the training. In addition, using the organization's computer facilities for training can disrupt normal operations and limit the amount of actual hands-on training.

Provide for learning by hearing, seeing, and doing. Some people learn best from lectures, discussions, and question-and-answer sessions. Others learn best from viewing demonstrations or from reading documentation and other material. Most people learn best from hands-on experience. You should provide training that supports each type of learning.

FIGURE 11-36 Users must be trained on the new system. Training sessions might be one-on-one or group situations such as the one shown here. Many vendors provide product training as part of an overall service to customers.

Prepare effective training materials, including interactive tutorials. User-friendly training materials contribute to training effectiveness and provide a valuable resource for users. You can prepare the material in various forms, including traditional training manuals, printed handouts, or online material. The main goal is to deliver user-friendly, cost-effective training. Regardless of the instructional method, the training lessons should include step-by-step instructions for using all the features of the information system.

Most people learn best when they participate actively in the training process. Figure 11-37 shows a sample tutorial lesson for a sales prospect management system. In Lesson 1, the user learns how to enter and exit the system. In Lesson 2, which is shown in Figure 11-38 on the next page, the user learns how to add a sales prospect and return to the main menu.

A training package also should include a reference section that summarizes all options and commands, and a listing of all error messages and what action the user should take when a problem occurs. More sophisticated tutorials might offer interactive sessions where users practice various tasks and get feedback on their progress.

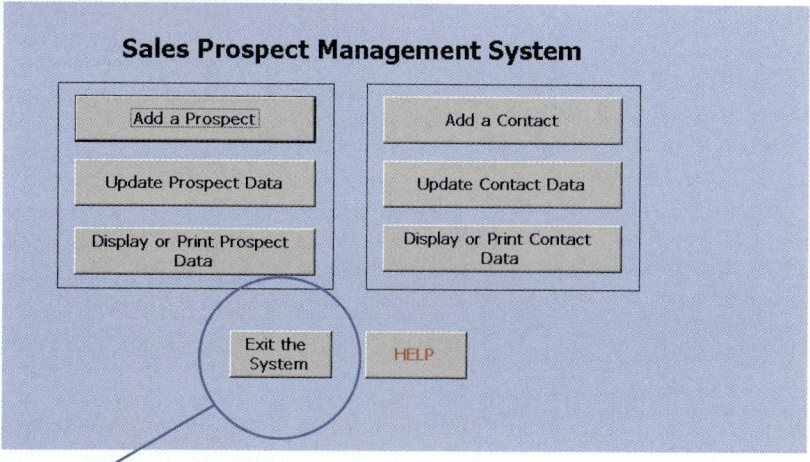

FIGURE 11-37 A sample tutorial for a sales prospect management system. In the first lesson, the user learns how to enter and exit the system.

Sales Prospect Management System
Lesson 2: Adding a Sales Prospect

- Start your computer and enter the Sales Prospect Management System as explained in Lesson 1. Click the Add a Prospect button on the main switchboard. The blank form shown below displays.

Add a Sales Prospect

First Name		Prospect ID		Dial Phone Number
Last Name		Title		Contact Via E-mail
Company		Work Phone		Generate Memo
Dear		Work Extension		
Address		Fax Number		Add Another Prospect
City				Return to Main Menu
State/Province				HELP
Postal Code				
Country				
E-mail Address				
Referred By				
Notes				

- Enter the data into the fields as shown on the screen below. The system will add the Prospect ID automatically. Press the ENTER key to move to the next field.

Add a Sales Prospect

First Name	Samuel	Prospect ID	10343	Dial Phone Number
Last Name	Rose	Title	Tech Support Manager	Contact Via E-mail
Company	Last Resort Systems, Inc.	Work Phone	(555) 123-4567	Generate Memo
Dear	Sam	Work Extension	2219	
Address	1303 Van Buren Drive	Fax Number	(555) 123-9999	Add Another Prospect
City	Annapolis			Return to Main Menu
State/Province	MD			HELP
Postal Code	21403-9999			
Country	US			
E-mail Address	sam_rose@lrs.com			
Referred By	Emma Nell			
Notes	Sam's company might be merging with SuperSystems. This would be a great time to get our foot in the door.			

- When you have entered all the data, compare your screen to the one shown above. If it matches, you have entered a sales prospect successfully. Click the Return to Main Menu button. You now are ready for Lesson 3, Updating Prospect Data.

FIGURE 11-38 A second sample lesson in the online tutorial.

Even if you lack the resources to develop interactive tutorials, you can design a series of dialog boxes that respond with Help information and suggestions when users select various menu topics.

Rely on previous trainees. After one group of users has been trained, they can assist others. Users often learn more quickly from coworkers who share common experience and job responsibilities. Using a **train-the-trainer** strategy, you can select knowledgeable users who then conduct sessions for others. When utilizing train-the-trainer techniques, the initial training must include not only the use of the application or system, but some instruction on how to present the materials effectively.

When training is complete, many organizations conduct a full-scale test, or **simulation**, which is a dress rehearsal for users and IT support staff. Organizations include all procedures, such as those that they execute only at the end of a month, quarter, or year, in the simulation. As questions or problems arise, the participants consult the system documentation, Help screens, or each other to determine appropriate answers or actions. This full-scale test provides valuable experience and builds confidence for everyone involved with the new system.

DATA CONVERSION

Data conversion is an important part of the system installation process. During **data conversion**, existing data is loaded into the new system. Depending on the system, data conversion can be done before, during, or after the operational environment is complete. You should develop a data conversion plan as early as possible, and the conversion process should be tested when the test environment is developed.

ON THE WEB

For more information about data conversion, visit **scsite.com/sad8e/ more**, locate Chapter 11, and then click the Data Conversion link.

Data Conversion Strategies

When a new system replaces an existing system, you should automate the data conversion process, if possible. The old system might be capable of **exporting** data in an acceptable format for the new system or in a standard format, such as ASCII or ODBC. **ODBC (Open Database Connectivity)** is an industry-standard protocol that allows DBMSs from various vendors to interact and exchange data. Most database vendors provide ODBC drivers, which are a form of middleware. As you learned in Chapter 10, middleware connects dissimilar applications and enables them to communicate.

If a standard format is not available, you must develop a program to extract the data and convert it to an acceptable format. Data conversion is more difficult when the new system replaces a manual system, because all data must be entered manually unless it can be scanned. Even when you can automate data conversion, a new system often requires additional data items, which might require manual entry.

Data Conversion Security and Controls

You should maintain strict input controls during the conversion process, when data is extremely vulnerable. You must ensure that all system control measures are in place and operational to protect data from unauthorized access and to help prevent erroneous input.

Even with careful data conversion and input controls, some errors will occur. For example, duplicate customer records or inconsistent part numbers might have been tolerated by the old system, but will cause the new system to crash. Most organizations require that users verify all data, correct all errors, and supply every missing data item during conversion. Although the process can be time-consuming and expensive, it is essential that the new system be loaded with accurate, error-free data.

SYSTEM CHANGEOVER

System changeover is the process of putting the new information system online and retiring the old system. Changeover can be rapid or slow, depending on the method. The four changeover methods are direct cutover, parallel operation, pilot operation, and phased operation. Direct cutover is similar to throwing a switch that instantly changes over from

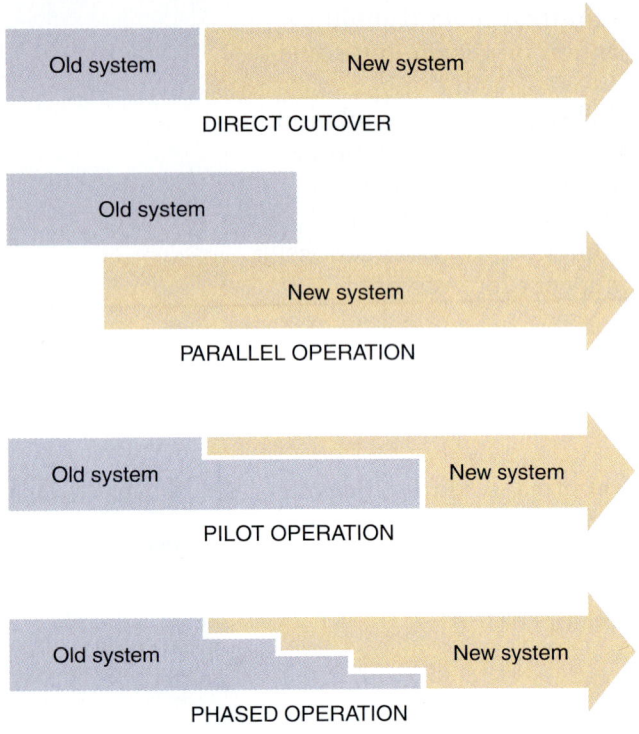

FIGURE 11-39 The four system changeover methods.

the old system to the new. Parallel operation requires that both systems run simultaneously for a specified period, which is the slowest method. The other methods, pilot and phased operation, fall somewhere between direct cutover and parallel operation. Figure 11-39 illustrates the four system changeover methods.

Direct Cutover

The **direct cutover** approach causes the changeover from the old system to the new system to occur immediately when the new system becomes operational. Direct cutover usually is the least expensive changeover method because the IT group has to operate and maintain only one system at a time.

Direct cutover, however, involves more risk than other changeover methods. Regardless of how thoroughly and carefully you conduct testing and training, some difficulties can arise when the system goes into operation. Problems can result from data situations that were not tested or anticipated or from errors caused by users or operators. A system also can encounter difficulties because live data typically occurs in much larger volumes than test data.

Although initial implementation problems are a concern with all four changeover methods, they are most significant when the direct cutover approach is used. Detecting minor errors also is more difficult with direct cutover because users cannot verify current output by comparing it to output from the old system. Major errors can cause a system process to terminate abnormally, and with the direct cutover method, you cannot revert to the old system as a backup option.

Companies often choose the direct cutover method for implementing commercial software packages because they feel that commercial packages involve less risk of total system failure. Commercial software is certainly not risk-free, but the software vendor usually maintains an extensive knowledge base and can supply reliable, prompt fixes for most problems.

For systems developed in-house, most organizations use direct cutover only for noncritical situations. Direct cutover might be the only choice, however, if the operating environment cannot support both the old and new systems or if the old and new systems are incompatible.

Timing is very important when using a direct cutover strategy. Most systems operate on weekly, monthly, quarterly, and yearly cycles. For example, consider a payroll system that produces output on a weekly basis. Some employees are paid twice a month, however, so the system also operates semimonthly. Monthly, quarterly, and annual reports also require the system to produce output at the end of every month, quarter, and year. When a cyclical information system is implemented in the middle of any cycle, complete processing for the full cycle requires information from both the old and the new systems. To minimize the need to require information from two different systems, cyclical information systems usually are converted using the direct cutover method at the beginning of a quarter, calendar year, or fiscal year.

Parallel Operation

The **parallel operation** changeover method requires that both the old and the new information systems operate fully for a specified period. Data is input into both systems, and output generated by the new system is compared with the equivalent output from the old system. When users, management, and the IT group are satisfied that the new system operates correctly, the old system is terminated.

The most obvious advantage of parallel operation is lower risk. If the new system does not work correctly, the company can use the old system as a backup until appropriate changes are made. It is much easier to verify that the new system is working properly under parallel operation than under direct cutover, because the output from both systems is compared and verified during parallel operation.

Parallel operation, however, does have some disadvantages. First, it is the most costly changeover method. Because both the old and the new systems are in full operation, the company pays for both systems during the parallel period. Users must work in both systems and the company might need temporary employees to handle the extra workload. In addition, running both systems might place a burden on the operating environment and cause processing delays.

Parallel operation is not practical if the old and new systems are incompatible technically, or if the operating environment cannot support both systems. Parallel operation also is inappropriate when the two systems perform different functions or if the new system involves a new method of business operations. For example, until a company installs data scanners in a factory, it is impractical to launch a new production tracking system that requires such technology.

Pilot Operation

The **pilot operation** changeover method involves implementing the complete new system at a selected location of the company. A new sales reporting system, for instance, might be implemented in only one branch office, or a new payroll system might be installed in only one department. In these examples, the group that uses the new system first is called the **pilot site**. During pilot operation, the old system continues to operate for the entire organization, including the pilot site. After the system proves successful at the pilot site, it is implemented in the rest of the organization, usually using the direct cutover method. Therefore, pilot operation is a combination of parallel operation and direct cutover methods.

Restricting the implementation to a pilot site reduces the risk of system failure, compared with a direct cutover method. Operating both systems for only the pilot site is less expensive than a parallel operation for the entire company. In addition, if you later use a parallel approach to complete the implementation, the changeover period can be much shorter if the system proves successful at the pilot site.

Phased Operation

The **phased operation** changeover method allows you to implement the new system in stages, or modules. For example, instead of implementing a new manufacturing system all at once, you first might install the materials management subsystem, then the production control subsystem, then the job cost subsystem, and so on. You can implement each subsystem by using any of the other three changeover methods.

Analysts sometimes confuse phased and pilot operation methods. Both methods combine direct cutover and parallel operation to reduce risks and costs. With phased operation, however, you give a part of the system to all users, while pilot operation provides the entire system, but to only some users.

One advantage of a phased approach is that the risk of errors or failures is limited to the implemented module only. For instance, if a new production control subsystem fails to operate properly, that failure might not affect the new purchasing subsystem or the existing shop floor control subsystem.

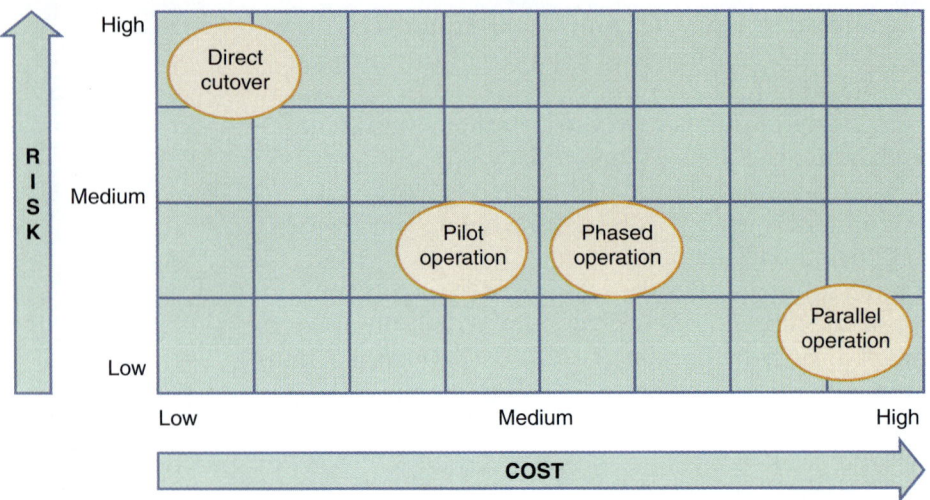

FIGURE 11-40 Relative risk and cost characteristics of the four changeover methods.

Phased operation is less expensive than full parallel operation because you have to work with only one part of the system at a time. A phased approach is not possible, however, if the system cannot be separated easily into logical modules or segments. In addition, if the system involves a large number of separate phases, phased operation can cost more than a pilot approach.

Figure 11-40 shows that each changeover method has risk and cost factors. As a systems analyst, you must weigh the advantages and disadvantages of each method and recommend the best choice in a given situation. The final changeover decision will be based on input from the IT staff, users, and management — and the choice must reflect the nature of the business and the degree of acceptable risk.

CASE IN POINT 11.3: GLOBAL COOLING

You are a systems analyst at Global Cooling, a leading manufacturer of air conditioning units. You are leading a team that is developing a new production scheduling system. The project is now in the application development stage. Unit testing has been completed, and you are in the final stages of integration testing. Your supervisor, Ella Pham, is eager to implement the new application ahead of schedule and asked if you could trim system testing from two weeks to three days, and use a direct cutover method instead of the parallel changeover method that originally was planned. Write a brief memo expressing your views.

POST-IMPLEMENTATION TASKS

Once the new system is operational, you must perform two additional tasks: Prepare a post-implementation evaluation and deliver a final report to management.

Post-Implementation Evaluation

A **post-implementation evaluation** assesses the overall quality of the information system. The evaluation verifies that the new system meets specified requirements, complies with user objectives, and produces the anticipated benefits. In addition, by providing feedback to the development team, the evaluation also helps improve IT development practices for future projects.

A post-implementation evaluation should examine all aspects of the development effort and the end product — the developed information system. A typical evaluation includes feedback for the following areas:

- Accuracy, completeness, and timeliness of information system output
- User satisfaction

- System reliability and maintainability
- Adequacy of system controls and security measures
- Hardware efficiency and platform performance
- Effectiveness of database implementation
- Performance of the IT team
- Completeness and quality of documentation
- Quality and effectiveness of training
- Accuracy of cost-benefit estimates and development schedules

You can apply the same fact-finding techniques in a post-implementation evaluation that you used to determine the system requirements during the systems analysis phase. When evaluating a system, you should:

- Interview members of management and key users
- Observe users and computer operations personnel actually working with the new information system
- Read all documentation and training materials
- Examine all source documents, output reports, and screen displays
- Use questionnaires to gather information and opinions from a large number of users
- Analyze maintenance and help desk logs

Figure 11-41 on the next page shows the first page of a sample user evaluation form for the new information system where users evaluate 18 separate elements on a numerical scale, so the results can be tabulated easily. Following that section, the form provides space for open-ended comments and suggestions.

Whenever possible, people who were not directly involved in developing the system should conduct the post-implementation evaluation. IT staff and users usually perform the evaluation, although some firms use an internal audit group or independent auditors to ensure the accuracy and completeness of the evaluation.

When should post-implementation evaluation occur? Is it better to wait until the new system has been in operation for one month, six months, one year, or longer? Users can forget details of the developmental effort if too much time elapses before the evaluation. After several months or a year, for instance, users might not remember whether they learned a procedure through training, from user documentation, or by experimenting with the system on their own.

Users also might forget their impressions of IT team members over time. An important purpose of the post-implementation evaluation is to improve the quality of IT department functions, including interaction with users, training, and documentation. Consequently, the evaluation team should perform the assessment while users are able to recall specific incidents, successes, and problems so they can offer suggestions for improvement. Post-implementation evaluation primarily is concerned with assessing the quality of the new system. If the team performs the evaluation too soon after implementation, users will not have enough time to learn the new system and appreciate its strengths and weaknesses. Although many IT professionals recommend conducting the evaluation after at least six months of system operation, pressure to finish the project sooner usually results in an earlier evaluation in order to allow the IT department to move on to other tasks.

Ideally, conducting a post-implementation evaluation should be standard practice for all information systems projects. Sometimes, evaluations are skipped because users are eager to work with the new system, or because IT staff members have more pressing priorities. In some organizations, management might not recognize the importance and benefits of a post-implementation evaluation. The evaluations are extremely important, however, because they enable the development team and the IT department to learn what worked and what did not work. Otherwise, developers might commit the same errors in another system.

USER EVALUATION FORM

System: Evaluator: Date:

Please evaluate the information system project by circling the one number for each factor that best represents your assessment.

	Unsatisfactory		Acceptable		Excellent

SYSTEM OUTPUT

1. Accuracy of information 1	2	3	4	5	6
2. Completeness of information 1	2	3	4	5	6
3. Ease of use. 1	2	3	4	5	6
4. Timeliness of information 1	2	3	4	5	6

USER INTERFACE

5. Clarity of instructions 1	2	3	4	5	6
6. Quality of Help messages. 1	2	3	4	5	6
7. Ease of use. 1	2	3	4	5	6
8. Appropriateness of options. 1	2	3	4	5	6
9. Clarity of error messages 1	2	3	4	5	6
10. Prevention of input errors 1	2	3	4	5	6

INFORMATION TECHNOLOGY STAFF

11. Cooperation 1	2	3	4	5	6
12. Availability. 1	2	3	4	5	6
13. Knowledge 1	2	3	4	5	6
14. Reporting of progress. 1	2	3	4	5	6
15. Communication skills 1	2	3	4	5	6

TRAINING

16. Completeness. 1	2	3	4	5	6
17. Appropriateness 1	2	3	4	5	6
18. Schedule. 1	2	3	4	5	6

FIGURE 11-41   Sample user evaluation form. The numerical scale allows easy tabulation of results. Following this section, the form provides space for open-ended comments and suggestions.

CASE IN POINT 11.4: YORKTOWN INDUSTRIES

Cindy Winslow liked her new job as lead systems analyst at Yorktown Industries. She was pleased that her development team completed the new human resources system ahead of schedule and under budget. Cindy looked forward to receiving the post-implementation evaluation because she was confident that both the system and the development team would receive high marks from users and managers.

After the system operated for one month, Cindy received a call from her supervisor, Ted Haines. Ted told her that she would have to handle the evaluation, even though she headed the development effort. Cindy told Ted that she did not feel comfortable evaluating her own team's work. She explained that someone who was not involved in its development should do an independent evaluation. Ted responded that he had full confidence in Cindy's ability to be objective. He explained that no one else was available and he needed the evaluation quickly so he could move forward with the next stage in the corporate development plan.

Cindy was troubled about the situation and she called you, a professional acquaintance, for your advice. What would you tell her and why?

Final Report to Management

At the end of each SDLC phase, you submit a report to management, and the systems implementation phase is no exception. Your report should include the following:

- Final versions of all system documentation
- Planned modifications and enhancements to the system that have been identified
- Recap of all systems development costs and schedules
- Comparison of actual costs and schedules to the original estimates
- Post-implementation evaluation, if it has been performed

The final report to management marks the end of systems development work. In the next chapter, you will study the role of a systems analyst during systems operation, security, and support, which is the final phase of the SDLC.

TOOLKIT TIME

The Communication tools in Part I of the Systems Analyst's Toolkit can help you develop better reports and presentations. To learn more about these tools, turn to Part I of the four-part Toolkit that follows Chapter 12.

A QUESTION OF ETHICS

Your friend Jill is handling the testing for the new accounting system, and right now she is very upset about the most recent results. "It seems like every time we fix one thing, another issue pops up! After ten days of testing and adjusting, we are meeting over 90% of the goals and benchmarks. If we're looking for perfection, we'll never make the implementation deadline for the new system, and the users will be all over us. Not to mention top management's reaction to a delay. I'm sure we can resolve some of these issues after the system becomes operational."

How would you respond to Jill? Are ethical issues involved? What are your responsibilities, as an employee, as an IT professional, and as a friend?

CHAPTER SUMMARY

The systems implementation phase consists of application development, testing, installation, and evaluation of the new system. During application development, analysts determine the overall design strategy and work with programmers to complete design, coding, testing, and documentation. Quality assurance is essential during the implementation phase. Many companies utilize software engineering concepts and quality standards established by the International Organization for Standardization (ISO).

Each systems development approach has its own set of tools. For example, structured development relies heavily on DFDs and structure charts. A structure chart consists of symbols that represent program modules, data couples, control couples, conditions, and loops. Object-oriented methods use a variety of diagrams, including use case, class, sequence, and transition state diagrams. Agile methods tend to use a spiral or other iterative model.

System developers also can use more generic tools to help them translate the system logic into properly functioning program modules. These tools include entity-relationship diagrams, flowcharts, pseudocode, decision tables, and decision trees.

If an agile development approach is used, then the customer creates user stories that describe required features and priority levels. In agile methodology, new system releases are made after many iterations and each is test-driven carefully by the customer.

Cohesion measures a module's scope and processing characteristics. A module that performs a single function or task has a high degree of cohesion, which is desirable. Coupling measures relationships and interdependence among modules. Modules that are relatively independent are loosely coupled, which is desirable. Cohesion and coupling concepts are used in structured development, but also are applicable to object-oriented development.

Typically, you follow four steps when you create a structure chart. You review DFDs and object models to identify the processes and methods, identify the program modules and determine control-subordinate relationships, add symbols for couples and loops, and analyze the structure chart to ensure that it is consistent with your system documentation.

Programmers perform desk checking, code review, and unit testing tasks during application development. Systems analysts design the initial test plans, which include test steps and test data for integration testing and system testing. Integration testing is necessary for programs that interact. The final step is system testing for the completed system. System testing includes users in the testing process.

In addition to system documentation, analysts and technical writers also prepare operations documentation and user documentation. Operations documentation provides instructions and information to the IT operations group. User documentation consists of instructions and information for users who interact with the system and includes user manuals, Help screens, and tutorials.

During the installation process, you establish an operational, or production, environment for the new information system that is completely separate from the test environment. The operational environment contains live data and is accessible only by authorized users. All future changes to the system must be verified in the test environment before they are applied to the operational environment.

Everyone who interacts with the new information system should receive training appropriate to his or her role and skills. The IT department usually is responsible for training. Software or hardware vendors or professional training organizations also can provide training. When you develop a training program, remember the following guidelines: Train people in groups; utilize people already trained to help train others; develop separate programs for distinct employee groups; and provide for learning by using discussions, demonstrations, documentation, training manuals, interactive tutorials, Webinars, and podcasts.

Data conversion often is necessary when installing a new information system. When a new system replaces a computerized system, you should automate the data conversion process if possible. The old system might be capable of exporting data in a format that the new system can use, or you might have to extract the data and convert it to an acceptable format. Data conversion from a manual system often requires labor-intensive data entry or scanning. Even when data conversion can be automated, a new system often requires additional data items, which might require manual entry. Strict input controls are important during the conversion process to protect data integrity and quality. Typically, data is verified, corrected, and updated during the conversion process.

System changeover is the process of putting the new system into operation. Four changeover methods exist: direct cutover, parallel operation, pilot operation, and phased operation.

With direct cutover, the old system stops and the new system starts simultaneously; direct cutover is the least expensive, but the riskiest changeover method. With parallel operation, users operate both the old and new information systems for some period of time; parallel operation is the most expensive and least risky of the changeover methods. Pilot operation and phased operation represent compromises between direct cutover and parallel operation; both methods are less risky than direct cutover and less costly than parallel operation. With pilot operation, a specified group within the organization uses the new system for a period of time, while the old system continues to operate for the rest of the users. After the system proves successful at the pilot site, it is implemented throughout the organization. With phased operation, you implement the system in the entire organization, but only one module at a time, until the entire system is operational.

A post-implementation evaluation assesses and reports on the quality of the new system and the work done by the project team. Although it is best if people who were not involved in the systems development effort perform the evaluation, that is not always possible. The evaluation should be conducted early so users have a fresh recollection of the development effort, but not before users have experience using the new system.

The final report to management includes the final system documentation, describes any future system enhancements that already have been identified, and details the project costs. The report represents the end of the development effort and the beginning of the new system's operational life.

Key Terms and Phrases

acceptance tests *517*
application development *503*
attributes *510*
bug tracking software *519*
Capability Maturity Model (CMM)® *500*
Capability Maturity Model Integration
   (CMMI)® *500*
code review *515*
coding *514*
cohesion *507*
condition *507*
control couple *507*
control module *506*
coupling *508*
customer *512*
data conversion *531*
data couple *506*
defect tracking software *519*
design walkthrough *515*
desk checking *515*
direct cutover *532*
documentation *518*
exporting *531*
flowchart *504*
integrated development environment (IDE) *514*
integration testing *516*
iteration cycle *513*
iteration planning meeting *513*
library module *506*
link testing *516*
logic errors *515*
loop *507*
loosely coupled *508*
methods *510*
modular design *506*
module *503*
object-oriented development (OOD) *510*
ODBC (Open Database Connectivity) *531*
online documentation *521*
operational environment *524*
operations documentation *519*

parallel operation *532*
parallel programming *513*
partitioning *506*
patches *519*
phased operation *533*
pilot operation *533*
pilot site *533*
podcast *527*
post-implementation evaluation *534*
process improvement *500*
production environment *524*
program documentation *518*
pseudocode *505*
quality assurance *500*
release plan *513*
simulation *531*
software engineering *500*
status flag *507*
structure chart *506*
structured walkthrough *515*
stub testing *516*
subordinate modules *506*
subscribers *527*
syntax errors *515*
system changeover *531*
system documentation *519*
system testing *517*
test data *516*
test-driven design *513*
test environment *524*
test plan *516*
tightly coupled *508*
top-down approach *506*
training plan *525*
train-the-trainer *530*
tutorial *527*
unit testing *515*
user documentation *520*
user story *512*
Webcast *527*
Webinar *527*

Learn It Online

Instructions: To complete the Learn It Online exercises, start your browser, click the Address bar, and then enter the Web address **scsite.com/sad8e/learn**. When the Systems Analysis and Design Learn It Online page is displayed, follow the instructions in the exercises below. Each exercise has instructions for saving your results, either for your own records or for submission to your instructor.

1   Chapter Reinforcement

TF, MC, and SA

Below SAD Chapter 11, click one of the Chapter Reinforcement links for Multiple Choice, True/False, or Short Answer. Answer each question and submit to your instructor.

2   Flash Cards

Below SAD Chapter 11, click the Flash Cards link and read the instructions. Type 20 (or a number specified by your instructor) in the Number of playing cards text box, type your name in the Enter your Name text box, and then click the Flip Card button. When the flash card is displayed, read the question and then click the ANSWER box arrow to select an answer. Flip through the Flash Cards. If your score is 15 (75%) correct or greater, click Print on the File menu to print your results. If your score is less than 15 (75%) correct, then redo this exercise by clicking the Replay button.

3   Practice Test

Below SAD Chapter 11, click the Practice Test link. Answer each question, enter your first and last name at the bottom of the page, and then click the Grade Test button. When the graded practice test is displayed on your screen, click Print on the File menu to print a hard copy. Continue to take practice tests until you score 80% or better.

4   Who Wants To Be a Computer Genius?

Below SAD Chapter 11, click the Computer Genius link. Read the instructions, enter your first and last name at the bottom of the page, and then click the Play button. When your score is displayed, click the PRINT RESULTS link to print a hard copy.

5   Wheel of Terms

Below SAD Chapter 11, click the Wheel of Terms link. Read the instructions, and then enter your first and last name and your school name. Click the PLAY button. When your score is displayed on the screen, right-click the score and then click Print on the shortcut menu to print a hard copy.

6   Crossword Puzzle Challenge

Below SAD Chapter 11, click the Crossword Puzzle Challenge link. Read the instructions, and then click the Continue button. Work the crossword puzzle. When you are finished, click the Submit button. When the crossword puzzle is redisplayed, submit it to your instructor.

Case-Sim: SCR Associates

Background

SCR Associates is a consulting firm that offers IT solutions and training. SCR needs an information system to manage operations at its new training center. The new system will be called TIMS (Training Information Management System). As a newly hired systems analyst, you will report to Jesse Baker, systems group manager. You will work on various tasks and practice the skills you learned in this chapter.

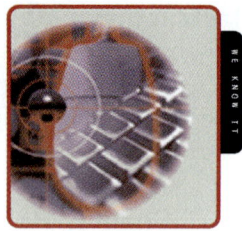

Using the Case

The SCR Associates case study is a Web-based simulation that allows you to practice your skills in a real-world environment. The case study transports you to SCR's company intranet, where you can complete 12 work sessions, each aligning with a chapter. As you work on the case, you will receive e-mail and voice mail messages, obtain information from SCR's online libraries, and perform various tasks. The first time you enter the SCR Case, you should go to the starting page at **scsite.com/sad8e/scr** for detailed instructions.

Preview: Session 11

You assisted your supervisor, Jesse Baker, in planning an architecture for the new TIMS system. Now she wants you to work on system implementation tasks and issues. Specifically, she wants you to develop a structure chart, testing and training plans, implementation guidelines, and a post-implementation review.

To start a work session, you log on to the SCR intranet. When you enter your name and password, an opening screen displays links to the work sessions, and you select Session 11. At this point, you check your e-mail and voice mail messages carefully. Then you begin working on your task list, which includes the following items:

Tasks: Systems Implementation

1. Jesse wants to see a structure chart. She said to use program modules based on the processes we identified earlier. She wants the modules to be cohesive and loosely coupled.

2. Need to develop a testing plan that includes unit testing, integration testing, and system testing as Jesse requested in her message. Also, draft a reminder to all IT staff members about the importance of careful documentation.

3. Jesse says it's important for SCR support staff and users to understand the difference between the test environment and the operational environment. Draft a message that explains the installation process and provides guidelines for all concerned. Also need to develop a training plan for TIMS, using Jesse's message as a guide. List the groups that should receive training, the topics that should be covered, and training methods we might use.

4. Jesse wants me to recommend a data conversion plan and a changeover method for TIMS. She also wants to see a plan for post-implementation review of TIMS. She wants suggestions on fact-gathering methods, a list of topics to cover, the best time to conduct the review, and who should perform it.

FIGURE 11-42 Task list: Session 11.

Chapter Exercises

Review Questions

1. Where does systems implementation fit in the SDLC, what tasks are performed during this phase, and why is quality assurance so important?
2. How are structured, object-oriented, and agile methods similar? How are they different?
3. Describe structure charts and symbols, and define cohesion and coupling.
4. Define unit testing, integration testing, and system testing.
5. What types of documentation does a systems analyst prepare, and what would be included in each type?
6. What is the purpose of an operational environment and a test environment?
7. Who must receive training before a new information system is implemented?
8. List and describe the four system changeover methods. Which one generally is the most expensive? Which is the riskiest? Explain your answers.
9. Who should be responsible for performing a post-implementation evaluation?
10. List the information usually included in the final report to management.

Discussion Topics

1. A supervisor states, "Integration testing is a waste of time. If each program is tested adequately, integration testing is not needed. Instead, we should move on to system testing as soon as possible. If modules don't interact properly, we'll handle it then." Do you agree or disagree with this comment? Justify your position.
2. Suppose you are a systems analyst developing a detailed test plan. Explain the testing strategies you will use in your plan. Will you use live or simulated data?
3. Using the Internet, locate an example of training for a software or hardware product. Write a brief summary of the training, the product, the type of training offered, and the cost of training (if available). Discuss your findings with the class.
4. Suppose that you designed a tutorial to train a person in the use of specific software or hardware, such as a Web browser. What specific information would you want to know about the recipient of the training? How would that information affect the design of the training material?

Projects

1. In this chapter, you learned about the importance of testing. Design a generic test plan that describes the testing for an imaginary system.
2. Design a generic post-implementation evaluation form. The form should consist of questions that you could use to evaluate any information system. The form should evaluate the training received and any problems associated with the program.
3. Create a one-page questionnaire to distribute to users in a post-implementation evaluation of a recent information system project. Include at least 10 questions that cover the important information you want to obtain.
4. Using the material in this chapter and your own Internet research, prepare a presentation on the pros and cons of agile development methods.

Apply Your Knowledge

The Apply Your Knowledge section contains four mini-cases. Each case describes a situation, explains your role in the case, and asks you to respond to questions. You can answer the questions by applying knowledge you learned in the chapter.

1 Sand and Surf Retailers

Situation:

Sand and Surf Retailers recently acquired several smaller companies to expand its chain of clothing outlets. To establish consistency for the current organization and future acquisitions, Sand and Surf decided to develop an in-house application called SPS (Standard Purchasing System). The SPS system would standardize purchasing practices for each Sand and Surf subsidiary and manage all purchasing information. System testing will be completed by the end of the week.

1. What types of documentation are needed for this application?
2. During application development, what steps should the IT staff follow to develop a structure chart?
3. What suggestions do you have for Help screens and online tutorials?
4. What types of testing should be performed? What types of test data should be used?

2 Albatross Airfreight

Situation:

Albatross Airfreight specializes in shipping cargo via air across North America. In an effort to modernize, the company has begun to computerize manual business processes. The new IT infrastructure includes a computer system that tracks cargo from departure point to destination. At this point, the new system is ready for implementation. Systems analysts are modularizing the system, and programmers are ready to start coding the first modules.

1. What issues should systems analysts and programmers discuss before they proceed with the project?
2. As a systems analyst on this project, how would you describe your primary responsibilities, and how could you contribute to the quality of the finished product?
3. As a programmer, how would you describe your primary responsibilities, and how could you contribute to the quality of the finished product?
4. Will the use of structure charts be beneficial during this stage of development? Discuss the advantages of structure charts compared with flowcharts and pseudocode.

3 Victorian Creations

Situation:

Victorian Creations is a growing business that specializes in the reproduction of furniture from the Victorian era. Since 2004, sales have increased steadily. The original accounting system was a package from Peachtree Software, which initially ran on a stand-alone PC and later on a LAN. Now, the firm is preparing to install a powerful, scalable accounting package that can support the company's current and future operations. You have been asked to develop a training plan for users.

1. Who should receive training on the new software, and what topics should the training cover?
2. Investigate an accounting package such as Peachtree to learn if the product can convert data from other accounting programs.
3. What changeover strategy would you suggest for the new accounting system? Explain your answer.
4. When should a post-implementation review be scheduled? Explain your answer.

4 Calico Prints

Situation:

Calico Prints creates a wide range of fabrics and wallpapers. Recently the company updated its payroll software as an in-house development project. Users and IT staff members have completed a comprehensive training curriculum. The system has been in the operational environment for approximately six weeks, and no major problems have occurred. You are responsible for the post-implementation evaluation.

1. What are some techniques you might use to obtain an accurate evaluation?
2. What are some specific questions you would include in a questionnaire? Who should receive the questionnaire?
3. Are there general guidelines that apply to the timing of a post-implementation evaluation? What are they?
4. In your view, should the company schedule a post-implementation evaluation for the payroll system at this time? Give specific reasons for your answer.

Case Studies

Case studies allow you to practice specific skills learned in the chapter. Each chapter contains several case studies that continue throughout the textbook, and a chapter capstone case.

NEW CENTURY HEALTH CLINIC

New Century Health Clinic offers preventive medicine and traditional medical care. In your role as an IT consultant, you will help New Century develop a new information system.

Background

You completed the systems design for the insurance system at New Century Health Clinic. The associates at the clinic have approved the design specification, and you hired two programmers, Bill Miller and Celia Goldring, to assist you with the programming and testing of the insurance system.

For Assignments 3 and 4, assume that a server and six client workstations have been purchased, installed, and networked in the clinic offices. You now are ready to begin installation and evaluation of the system.

Assignments

1. Plan the testing required for the system. You should consider unit, integration, and system testing in your test plan and determine who should participate in the testing. Also design the test data that you will use. Prepare a structure chart that shows the main program functions for the New Century system.

2. You have asked Anita Davenport, New Century's office manager, to contribute to the user manual for the insurance system. She suggested that you include a section of frequently asked questions (FAQs), which you also could include in the online documentation. Prepare 10 FAQs and answers for use in the printed user manual and context-sensitive Help screens. Also identify the specific people who will require training on the new system. Describe the type and level of training you recommend for each person or group.

3. Recommend a changeover method for New Century's system and justify your recommendation. If you suggest phased operation or pilot operation, specify the order in which you would implement the modules or how you would select a pilot workstation or location.

4. Should the associates perform a post-implementation evaluation? If an assessment is done, who should perform it? What options are available and which would you recommend?

PERSONAL TRAINER, INC.

Personal Trainer, Inc., owns and operates fitness centers in a dozen Midwestern cities. The centers have done well, and the company is planning an international expansion by opening a new "supercenter" in the Toronto area. Personal Trainer's president, Cassia Umi, hired an IT consultant, Susan Park, to help develop an information system for the new facility. During the project, Susan will work closely with Gray Lewis, who will manage the new operation.

Background

Susan finished work on system architecture issues, and her system design specification was approved. Now she is ready to address system implementation tasks, including quality assurance, structure charts, testing, training, data conversion, system changeover, and post-implementation evaluation.

Assignments

1. Identify the specific groups of people who need training on the new system. For each group, describe the type of training you would recommend and list the topics you would cover.

2. Suggest a changeover method for the new billing system and provide specific reasons to support your choice. If you recommend phased operation, specify the order in which you would implement the modules. If your recommendation is for pilot operation, specify the department or area you would select as the pilot site and justify your choice.

3. Develop a data conversion plan that specifies which data items must be entered, the order in which the data should be entered, and which data items are the most time-critical.

4. You decide to perform a post-implementation evaluation to assess the quality of the system. Who would you involve in the process? What investigative techniques would you use and why?

FANCIFUL CRYSTAL

Fanciful Crystal has produced fine crystal products for many years. The company once dominated the global market, but sales have declined recently. Last year, Fanciful Crystal rushed to implement a new Web-based system to boost sagging sales. Unfortunately, the online system was not tested thoroughly and experienced start-up problems. For example, customers complained about order mix-ups and overcharges for deliveries. You are a new system analyst with the company, and your supervisor asked you to investigate the problems.

1. Based on what you know about e-commerce, how would you have tested a new Web-based system?

2. Should ISO standards have been considered? Explain your answer.

3. What should Fanciful Crystal do in the future to avoid similar problems when developing new systems?

4. Three months after the system changeover, you perform a post-implementation evaluation. Prepare three evaluation forms for the new information system: one for users, one for managers, and one for the IT operations staff.

CHAPTER CAPSTONE CASE: SoftWear, Limited

SoftWear, Limited (SWL), is a continuing case study that illustrates the knowledge and skills described in each chapter. In this case study, the student acts as a member of the SWL systems development team and performs various tasks.

Background

The ESIP development team of Jane Rossman, Tom Adams, and Becky Evans started work on the ESIP system, which they would develop as a Microsoft Access application in a client/server environment. Jane and Tom scheduled additional meetings with the consulting firm, True Blue Systems, while Becky started designing the main switchboard and the data input screens.

The ESIP system design included a server to interact with various SWL client workstations and an interface to the new payroll package from Pacific Software. The payroll package was implemented successfully on SWL's mainframe, and several payroll cycles were completed without any processing problems.

When the ESIP development team met on Monday morning, the members studied the overview that True Blue submitted, shown in Figure 11-43. Jane said they would use a top-down design approach. Their first step was to partition the system and break it down into a set of modules on a structure chart. Each module would represent a program or function to be performed by one or more macros or Visual Basic procedures.

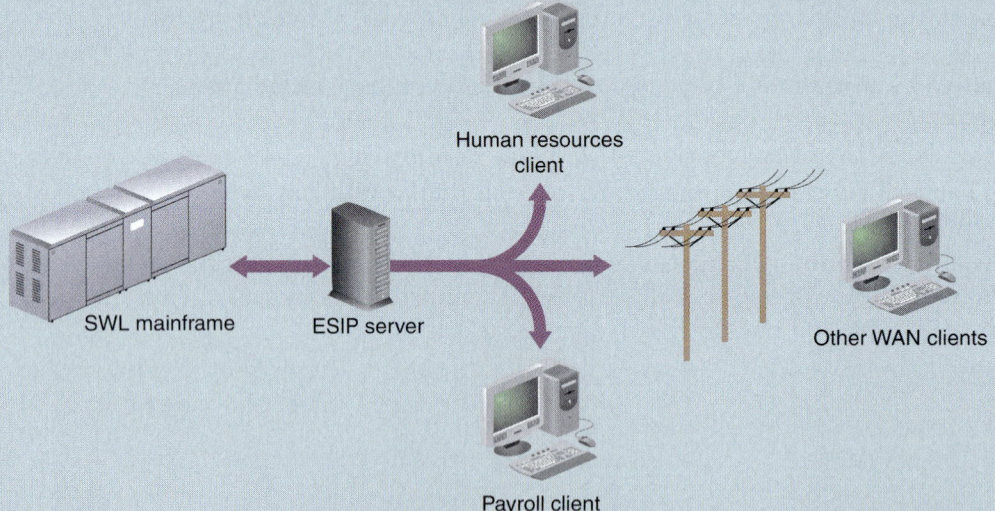

FIGURE 11-43   The ESIP system plan as submitted by True Blue Systems.

The team reviewed the documentation carefully, using the DFDs they prepared during the systems analysis phase. The team determined that the ESIP system needed to perform five main tasks: Extract the ESIP deductions during payroll processing; apply the extracted deductions to specific ESIP options; update employee deduction selections; update ESIP option choices; and handle fund transfers to internal and external ESIP entities. To accomplish those tasks, the system would need a variety of reports, controls, query and display capabilities, input screens, security provisions, and other features.

CHAPTER CAPSTONE CASE: SoftWear, Limited (continued)   **SWL**

Of the five main ESIP processes, only the extracting of payroll deductions would be done on SWL's mainframe. Jane said they would need to develop an interface program to control the extraction processing, but all the other functions would run on the ESIP server and clients. By afternoon, they had produced the structure chart shown in Figure 11-44.

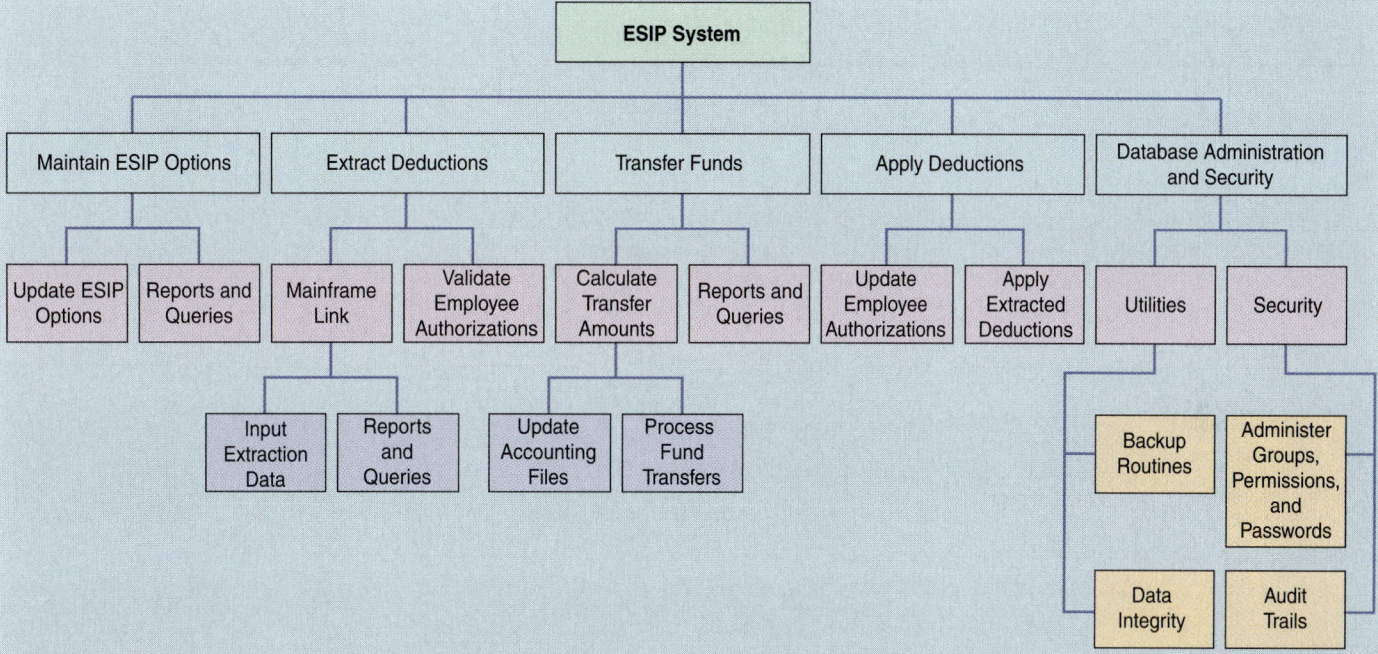

FIGURE 11-44 Structure chart for the ESIP system.

After studying the tasks and requirements, Jane estimated the time needed to design, code, unit test, and document each module. She also estimated the time needed for integration and system testing, completing the ESIP system documentation, and receiving management approval. Jane used Microsoft Project software to create a Gantt chart, which is explained in the Systems Analyst's Toolkit, to manage and track the project and show the individual assignments. The chart indicated that many tasks could be performed concurrently and displayed the critical path. The development team agreed that they would meet daily to review their progress and discuss any problems.

Mainframe Interface

Jane's next step was to meet with Rick Williams, who was familiar with the new payroll package, to discuss the ESIP deduction extraction program that executes when the mainframe runs the weekly payroll. Jane learned that the new payroll package was developed as a database application, and she could write the extract module in Visual Basic.

Working together, Jane and Becky prepared a design, wrote the commands, and unit tested the ESIP program modules with test files they created. They used stubs to indicate inputs and outputs to files with which the extraction program would interact. After verifying the results, Jane created a procedure for downloading the deduction file from the mainframe to the ESIP server in the payroll department. She tested the download procedure and updated the documentation.

CHAPTER CAPSTONE CASE: SoftWear, Limited (continued) — SWL

ESIP Server

The team started developing the Access database application that would handle the other ESIP functions. The plan was for Becky to finish the basic switchboard and screen designs, then add features that users had requested, including custom menus and icons for frequently used functions. Becky also would design all the reports documented in the data dictionary.

Meanwhile, Jane and Tom worked on the other modules. Their first task was to examine the ERDs developed in the systems design phase to ensure that the entities and normalized record designs still were valid. Then they would start creating individual objects, including tables, queries, macros, and code modules, using the application design tools in Access.

Jane and Tom defined the data tables, identified primary keys, and linked the tables into a relational structure, as shown in Figure 11-45. They used online Access Help to make sure that they were using the correct field types, sizes, and names, and that the tables would work properly with the input screens created by Becky. Jane and Tom used an agreed-upon naming convention for all objects to ensure consistency among the systems.

After they designed and loaded the tables with test data, Jane and Tom developed queries that would allow users to retrieve, display, update, and delete records. Some queries affected only individual records, such as ESIP options, while they designed other queries for specific reports, such as the ESIP Deduction Register query shown in Figure 11-46, which displays deductions first in date order, then by employee SSN. This query also produces SQL commands shown in the lower screen that will be transmitted to the ESIP server.

Jane and Tom also developed macros that performed specific actions when certain events occurred. They tested each macro by stepping through it to make

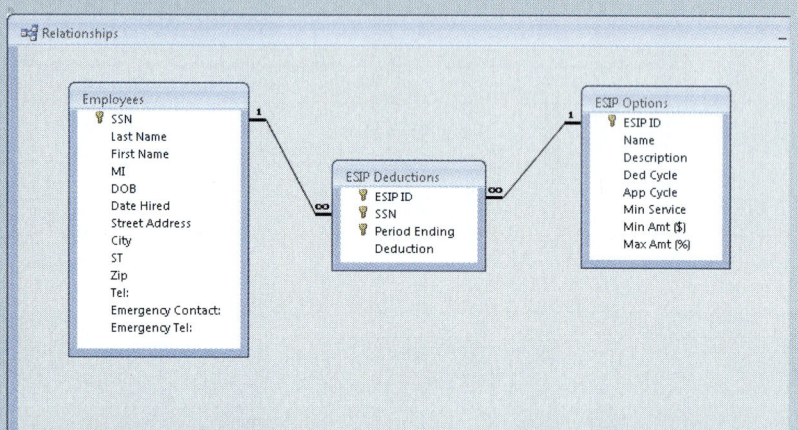

FIGURE 11-45 Relational structure for the ESIP system tables. Notice that the field or fields that comprise the primary key are shown in bold. Common fields link the tables, and referential integrity is indicated by the 1 (one) and ∞ (many) symbols.

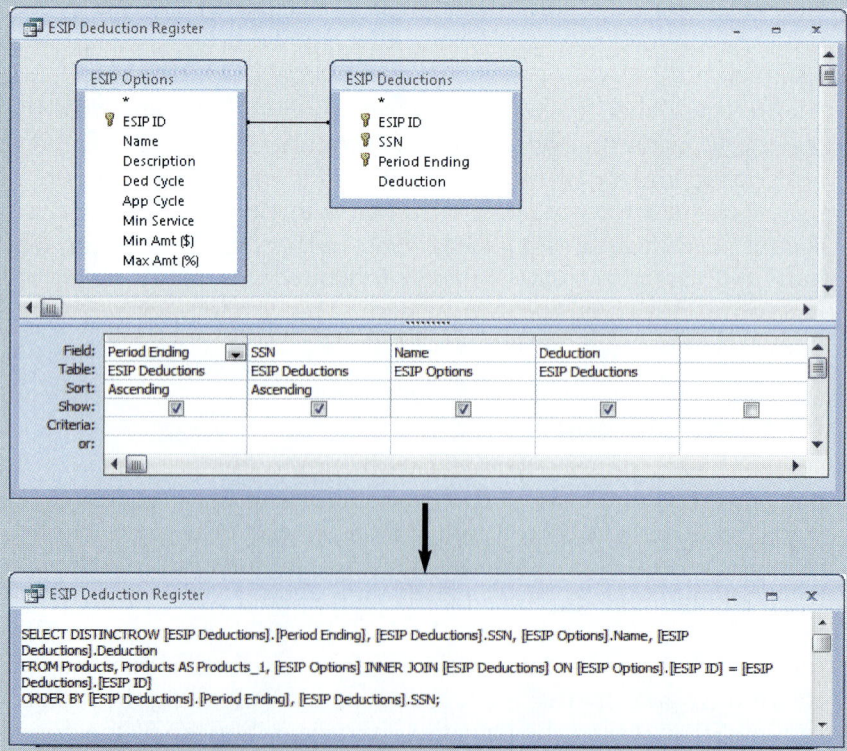

FIGURE 11-46 Example of a query that will provide data for the ESIP Deduction Register. Notice that the query sorts deductions by date, then by employee SSN. The query also produces SQL commands shown in the lower screen that will be transmitted to the ESIP server.

sure the commands executed properly. The macros later would be linked to various buttons and menus that Becky was designing into her switchboard and input screens. To save coding time, Tom converted several macros to Visual Basic in order to work on additional features and capabilities.

Jane and Tom also completed work on various security features, including password protection and several levels of permission required to view, modify, or delete specific objects. Later, when the system became operational, management would authorize specific permission levels, and Jane would designate a system administrator to maintain security and system policies.

Completing Application Development

In three weeks, the ESIP development team finished unit testing. Becky tested the switchboard, macros, queries, screen forms, menus, submenus, and code modules to ensure that they functioned correctly. Next, they linked the modules and performed integration testing. The testing ran smoothly, and they encountered no significant problems.

After integration testing, the analysts asked several principal users to participate in system testing. During this process, some minor screen changes were suggested and implemented after checking with users who would be affected by those changes. In addition, it turned out that one of the reports did not include a federally required Employer Identification Number (EIN). Because the reports had been designed using the Access report generator, the development team easily implemented this change.

The team members prepared user documentation as they completed each task. To produce a clear, understandable user manual, they decided to ask Amy Calico, SWL's payroll director, to review their notes and help them write a draft for current and future users. They wanted to explain the system in nontechnical terms, with adequate illustrations, screen shots, and a set of frequently asked questions. Jane said that the entire manual could be put online after SWL's intranet was developed.

Installation of the ESIP System

After a successful period of parallel operation, SWL fully implemented the payroll package purchased from Pacific Software and was ready to start installation of the ESIP system. In preparation, the IT development team of Jane Rossman, Tom Adams, and Becky Evans confirmed that SWL's existing network could handle the additional traffic generated by the new system. True Blue Systems, the outside consulting firm, noted in its report that the network might need to be upgraded in the future, especially if SWL expanded the number of networked applications and users.

Tom's first task was to install the ESIP application on the server in the payroll department and to verify that the system could communicate properly with SWL's mainframe. Then, he installed and tested a new high-speed tape cartridge backup system for the ESIP system.

Next, Tom loaded the ESIP application on a client PC in the human resources department. He checked all hardware and system software settings and used several test files to ensure that the client communicated with the ESIP server in the payroll department.

Meanwhile, Becky Evans and Rick Williams worked together on the interface between the ESIP system and the mainframe. They previously created a module called an extract program that directed the mainframe payroll system to capture the ESIP payroll deductions, store them in a file, and transmit the file back to the ESIP system. They already tested the interface using stubs to represent actual input and output files. Next, they would use a test data file with examples of every possible combination of permissible deductions and several improper deductions that testing should detect.

CHAPTER CAPSTONE CASE: SoftWear, Limited (continued)

As soon as Rick confirmed that the payroll package was ready for the interface test, Tom set up the ESIP server and sent the test file to the mainframe. Then, he ran the module that sent processing commands to the payroll system. Everyone was pleased to see that the mainframe handled the test data properly and generated an extracted deduction file, which it downloaded to the ESIP server.

Becky and Rick were ready to conduct hands-on training with the payroll group, so they arranged an early morning session with Amy Calico, Nelson White, Britton Ellis, and Debra Williams. Becky walked them through the steps, which were described clearly in the user manual, and then answered several questions. The payroll employees seemed pleased with the explanation and commented on how much easier the new system would be for them to use.

Next, Becky went to see Mike Feiner, director of human resources. Mike would be the only person allowed to add, change, or delete any of the ESIP options. Based on written authorization from Tina Pham, vice president of human resources, Mike would have a special password and permission level to allow him to perform those actions. Becky described to Mike how the system worked and then showed him how to enter, modify, and delete a test option she prepared. For security reasons, the special documentation for those functions would not be printed in the user manual itself, but would be retained in the IT department files.

Becky and Tom met again with the payroll group to show users how to enter the deduction authorizations for individual employees. Although the new payroll system was operational, the company still handled ESIP deductions manually. Using the ESIP server and another networked payroll PC, the payroll clerks were able to enter actual payroll data during a three-day test period. The built-in edit and validation features detected the errors that the team purposely inserted as test data and even identified several invalid authorizations that they had not noticed previously. The group produced printed reports and asked other payroll department members to review and verify the output.

Jane Rossman had a last-minute idea: Perhaps they should send a notice to all SWL employees describing the new ESIP system. The flyer also could remind employees how to select options and invite their questions or comments. Michael Jeremy, vice president of finance, thought that Jane's idea was excellent.

Up to that point, the company had made no final decision about the changeover method for the new system. Because the ESIP system replaced a series of manual processing steps, the main question was whether they should run the manual system in parallel operation for a specified period. Managers in the payroll, human resources, and accounting departments all wanted the new system operational as soon as possible, and Jane decided to use a direct cutover method, although not everyone agreed with the decision. In any case, the cutover was scheduled for Friday, May 7, 2010, when the first weekly payroll in May was processed.

Starting on Monday, April 26, the IT team met again with each of the users and reviewed a final checklist. No problems appeared, and the system was ready to interface with the mainframe and handle live data in an operational environment. On Friday morning, May 7, the payroll department ran the ESIP module that sent the processing commands to the payroll system. Later that morning, during the weekly processing cycle, the payroll package created a file with the extracted deductions and passed it back to the ESIP server.

With the ESIP system using real input data, IT department members visited each of the recently trained users to make sure they were experiencing no difficulties. They received good reports — users in the payroll and human resources departments seemed pleased with the new system. They were able to access the ESIP data, enter new deductions, and had no problems with screen output or printed reports.

The direct cutover to the ESIP system occurred without major problems. By the end of June, the system had completed nine weekly payroll cycles, produced all required reports

CHAPTER CAPSTONE CASE: SoftWear, Limited (continued) SWL

and outputs, and properly handled the monthly transfer of funds to the credit union and the SWL stock purchase plan.

During the first part of July, the IT department conducted a post-implementation evaluation with a team that consisted of two people: Katie Barnes, a systems analyst who had not been involved in the ESIP system development, and Ben Mancuso, a member of the finance department designated by Michael Jeremy. The evaluation team reviewed system operations, conducted interviews, and asked users to complete a brief questionnaire. The results were favorable, and it appeared that users were very satisfied with the new ESIP system. When the evaluation was completed in mid-July, Ann Hon sent the e-mail message shown in Figure 11-47

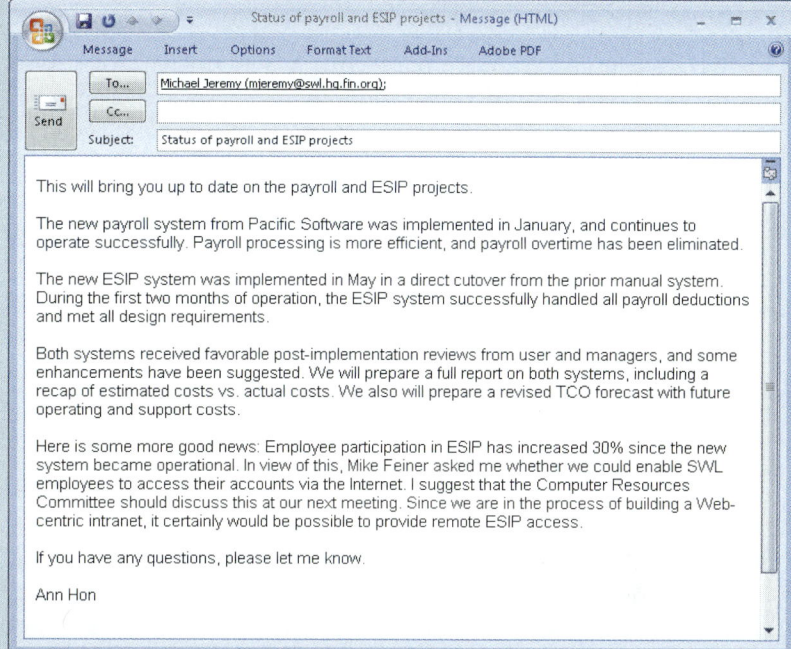

FIGURE 11-47 E-mail message from Ann Hon to Michael Jeremy regarding the payroll and ESIP systems.

to Michael Jeremy. The systems development effort for the ESIP system was completed successfully.

SWL Team Tasks

1. Rick has asked you to help him develop a recommendation for future software testing. One of his concerns is that in systems testing, it is virtually impossible to simulate every system transaction and function. Specifically, Rick wants you to suggest some guidelines that will produce the most reliable test results. He suggested that you should consider who should be involved in the testing, what types of transactions should be tested, and when the testing should be done. Based on what you already have learned about the ESIP system in earlier chapters, what would you recommend?

2. Rick also wants you to prepare a list of interview questions for users and design a brief questionnaire that measures the effectiveness of the new system. Develop the interview questions and questionnaire by following the guidelines suggested in this chapter and in Chapter 4, which discusses fact-finding techniques.

3. Rick is interested in using agile development methods for the next SWL project. He asked you to write a "Guide to Agile Methods" that he could distribute to other members of the IT team. The guide should be thorough, easy to understand, and should include your own research on the Internet.

4. You know that Jane was under pressure from SWL managers to get the ESIP system up and running. You also know that not everyone on the IT staff agreed with her decision to use a direct cutover method. What are some of the disadvantages of direct cutover? What other methods might have been used? What would you have done?

CHAPTER CAPSTONE CASE: SoftWear, Limited (continued)

Manage the SWL Project

You have been asked to manage SWL's new information system project. One of your most important activities will be to identify project tasks and determine when they will be performed. Before you begin, you should review the SWL case in this chapter. Then list and analyze the tasks, as follows:

LIST THE TASKS   Start by listing and numbering at least ten tasks that the SWL team needs to perform to fulfill the objectives of this chapter. Your list can include SWL Team Tasks and any other tasks that are described in this chapter. For example, Task 3 might be to Identify training needs, and Task 6 might be to Develop training solutions.

ANALYZE THE TASKS   Now study the tasks to determine the order in which they should be performed. First identify all concurrent tasks, which are not dependent on other tasks. In the example shown in Figure 11-48, Tasks 1, 2, 3, 4, and 5 are concurrent tasks, and could begin at the same time if resources were available.

Other tasks are called dependent tasks, because they cannot be performed until one or more earlier tasks have been completed. For each dependent task, you must identify specific tasks that need to be completed before this task can begin. For example, you would want to identify training needs before you could develop training solutions, so Task 6 cannot begin until Task 3 is completed, as Figure 11-48 shows.

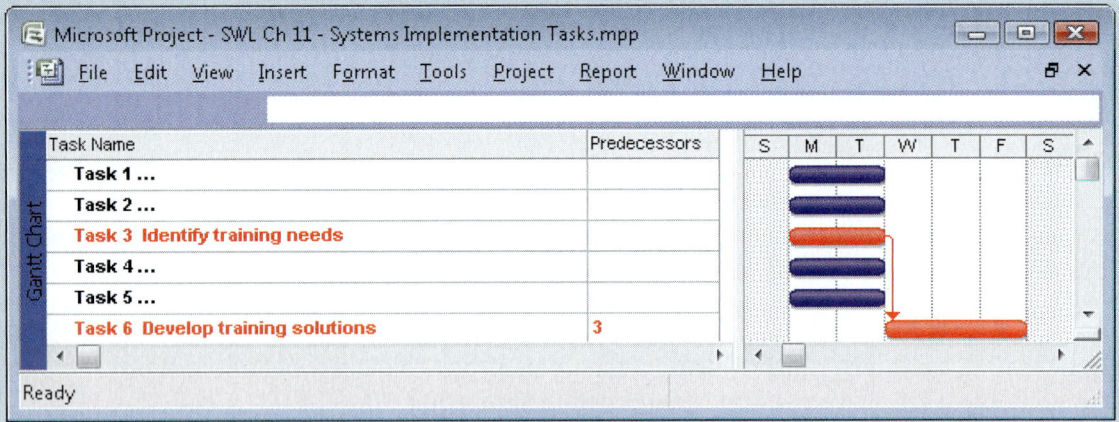

FIGURE 11-48 Tasks 1, 2, 3, 4, and 5 are concurrent tasks that could be performed at the same time. Task 6 is a dependent task that cannot be performed until Task 3 has been completed.

Chapter 3 describes project management tools, techniques, and software. To learn more, you can visit the Features section on your Student Study Tool CD-ROM, or the project management resources library at **scsite.com/sad8e/project**. On the Web, Microsoft offers demo versions, training, and tips for using Project 2007. You also can visit the OpenWorkbench.org site to learn more about this free, open-source software.

PHASE 5 SYSTEMS SUPPORT AND SECURITY

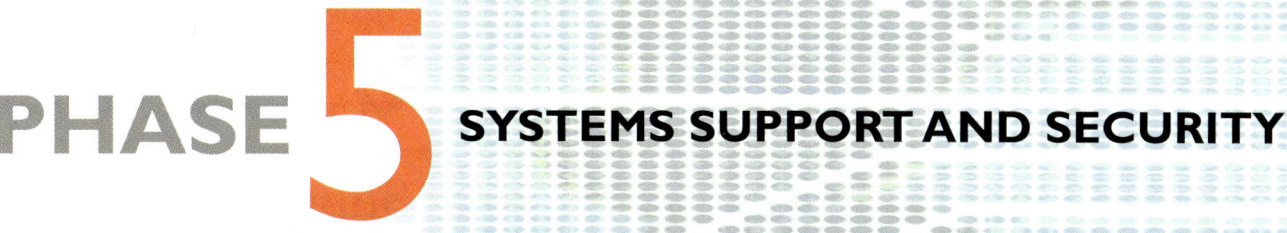

DELIVERABLE
An operational information system that is maintained, supported, and secured properly

TOOLKIT SUPPORT
Primary tools: Communications, CASE, and financial analysis tools
Other tools as required

As the Dilbert cartoon suggests, users expect prompt support, but they might not always receive it. Unhappy users do not bode well for system success. You will learn more about effective user support in this phase.

Systems support and security is the final phase in the systems development life cycle. In the previous phase, systems implementation, you delivered a functioning system. Now, you will support and maintain the system, handle security issues, protect the integrity of the system and its data, and be alert to any signs of obsolescence. The deliverable for this phase is an operational system that is properly maintained, supported, and secured.

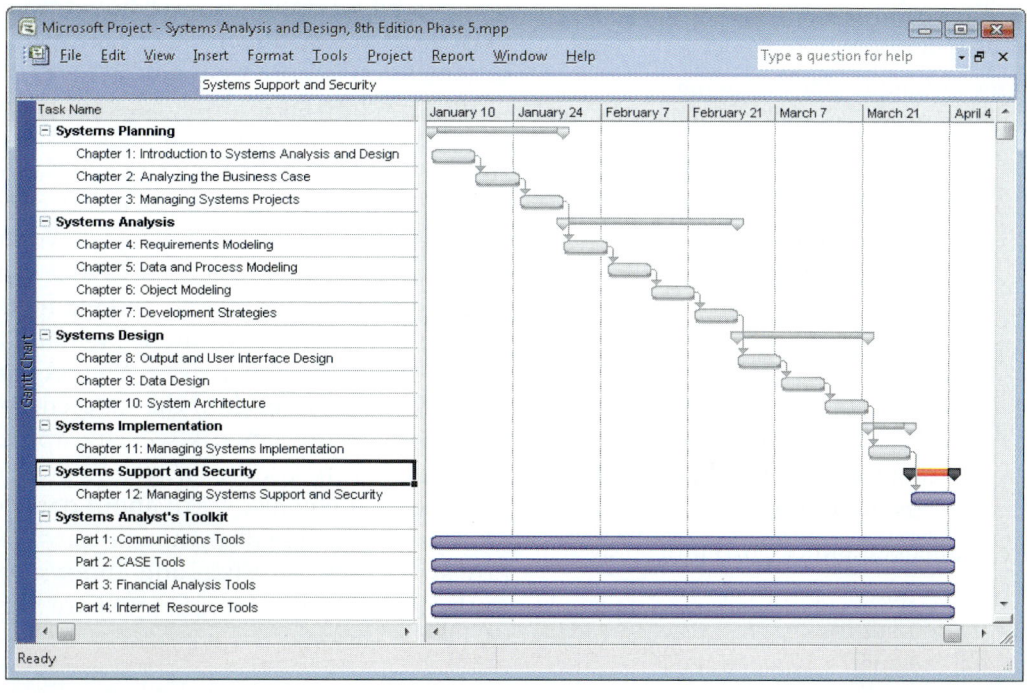

CHAPTER 12

Managing Systems Support and Security

Chapter 12 describes systems support and security tasks that continue throughout the useful life of the system. In addition to user support, this chapter discusses maintenance, security, backup and disaster recovery, performance measurement, and system obsolescence.

INTRODUCTION

OBJECTIVES

When you finish this chapter, you will be able to:

- Explain the systems support and security phase
- Describe user support activities, including user training and help desks
- Define the four types of maintenance
- Explain various techniques for managing systems maintenance and support
- Describe techniques for measuring, managing, and planning system performance
- Explain risk management concepts
- Assess system security at six levels: physical security, network security, application security, file security, user security, and procedural security
- Describe backup and disaster recovery
- List factors indicating that a system has reached the end of its useful life
- Assess future challenges and opportunities for IT professionals
- Develop a strategic plan for career advancement and strong IT credentials

Managing systems support and security involves three main concerns: user expectations, system performance, and security requirements.

A systems analyst is like an internal consultant who provides guidance, support, and training. Successful, robust systems often need the most support because users want to learn the features, try all the capabilities, and discover how the system can help them perform their business functions. In most organizations, more than half of all IT department effort goes into supporting existing systems.

This chapter begins with a discussion of systems support, including user training and help desks. You will study the four main types of maintenance: corrective, adaptive, perfective, and preventive. You also will learn how the IT group uses maintenance teams, configuration management, and maintenance releases, and you will examine system performance issues and maintenance tools. You will analyze the security system at each of the six security levels: physical security, network security, application security, file security, user security, and procedural security. You will also learn about data backup and recovery issues. Finally, you will learn how to recognize system obsolescence, and about some of the challenges and opportunities you are likely to face as an IT professional.

CHAPTER INTRODUCTION CASE: Mountain View College Bookstore

Background: Wendy Lee, manager of college services at Mountain View College, wants a new information system that will improve efficiency and customer service at the three college bookstores.

In this part of the case, Florence Fullerton (systems analyst) and Harry Boston (student intern) are talking about operation, support, and security issues for the new system.

Participants:	Florence and Harry
Location:	Florence's office, Friday afternoon, March 26, 2010
Project status:	Florence and Harry successfully have implemented the bookstore information system. Now they will discuss strategies for supporting and maintaining the new system.
Discussion topics:	Support activities, user training, system maintenance, and techniques for managing systems operation, enhancing system performance and security, and detecting factors indicating system obsolescence

Florence:	Well, we finally made it. The system is up and running and the users seem satisfied. Now we focus on supporting the system, ensuring that it delivers its full potential, and is properly secured and protected.
Harry:	*How do we do that?*
Florence:	First, we need to set up specific procedures for handling system support and maintenance. We'll set up a help desk that will offer user training, answer technical questions, and enhance user productivity.
Harry:	*Sounds good. I'll set up a training package for new users who missed the initial training sessions.*
Florence:	That's fine. You also should learn about the four types of maintenance. Users typically ask for help that requires corrective maintenance to fix problems or adaptive maintenance to add new features. As IT staff, we will be responsible for perfective maintenance, which makes the system more efficient, and preventive maintenance to avoid problems.
Harry:	*Anything else for us to do?*
Florence:	Yes, we'll need a system for managing maintenance requests from users. Also, we'll need to handle configuration management, maintenance releases, and version control. These tools will help us keep the system current and reduce unnecessary maintenance costs.
Harry:	*What about keeping tabs on system performance issues?*
Florence:	That's important, along with capacity planning to be sure the system can handle future growth.
Harry:	*What about the security of the system?*
Florence:	That's an excellent point. We'll look at physical security, network security, application security, file security, user security, and procedural security. We'll also look at backup and disaster recovery issues.
Harry:	*Sounds like we'll be busy for quite a while.*
Florence:	Well, that depends on the system itself and user expectations. Every system has a useful life, including this one. We'll try to get a good return on our investment, but we'll also watch for signs of obsolescence. Here are some tasks we can work on:

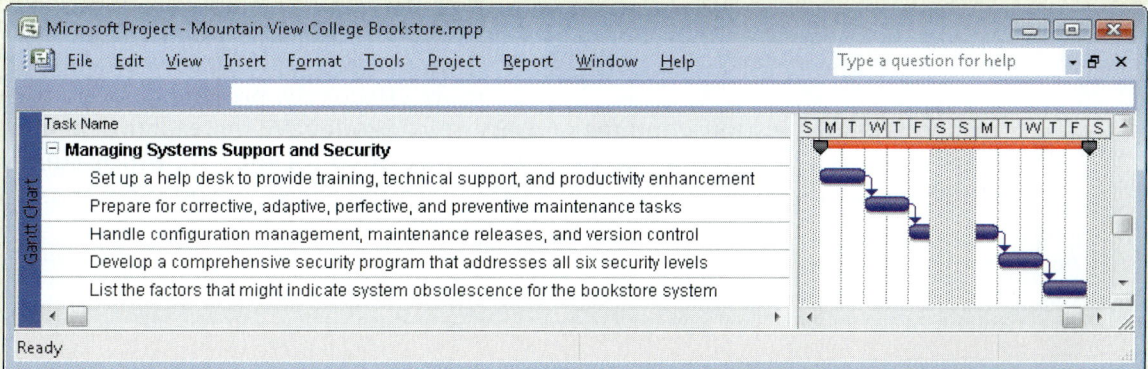

FIGURE 12-1 Typical systems support and security task list.

OVERVIEW

The systems support and security phase begins when a system becomes operational and continues until the system reaches the end of its useful life. Throughout the development process, the objective has been to create an information system that is efficient, easy to use, and affordable. After delivering the system, the IT team focuses on support and maintenance tasks.

The first part of this chapter covers four main topics. You will learn how to provide user support, maintain the system, manage the maintenance process, and handle system performance issues.

USER SUPPORT

Companies provide user support in many forms, including user training and a help desk to provide technical support and assistance.

User Training

In Chapter 11, you learned about initial training that is performed when a new system is introduced. Additionally, new employees must be trained on the company's information systems. For example, a firm that produces electronic assemblies must train its new employees, as shown in Figure 12-2.

If significant changes take place in the existing system or if a new version is released, the IT department might develop a **user training package**. Depending on the nature of the changes, the package could include online support via e-mail, a special Web site, a revision to the user guide, a training manual supplement, or formal training sessions. Training users about system changes is similar to initial training. The main objective is to show users how the system can help them perform their jobs.

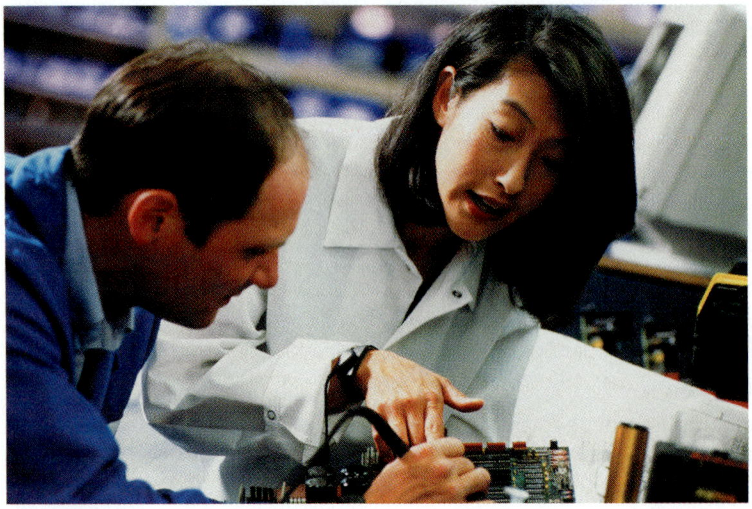

FIGURE 12-2 Whether a company is training manufacturing technicians, data entry personnel, or customer service representatives, employees need high-quality instruction to perform their jobs efficiently.

Help Desks

As systems and data structures become more complex, users need constant support and guidance. To make data more accessible and to empower users, many IT departments create help desks. A **help desk** is a centralized resource staffed by IT professionals who provide users with the support they need to do their jobs. A help desk has three main objectives: Show people how to use system resources more effectively, provide answers to technical or operational questions, and make users more productive by teaching them how to meet their own information needs. A help desk often is called an **information center** (**IC**) because it is the first place users turn when they need information or assistance.

ON THE WEB

For more information about help desks, visit **scsite.com/ sad8e/more**, locate Chapter 12, and then click the Help Desks link.

A help desk does not replace traditional IT maintenance and support activities. Instead, help desks enhance productivity and improve utilization of a company's information resources.

Help desk representatives need strong interpersonal and technical skills plus a solid understanding of the business, because they interact with users in many departments. A help desk should document carefully all inquiries, support tasks, and activity levels. The information can identify trends and common problems and can help build a technical support knowledge base.

During a typical day, the help desk staff member shown in Figure 12-3 might have to perform the following tasks:

- Show a user how to create a data query or report that displays specific business information

- Resolve network access or password problems

- Demonstrate an advanced feature of a system or a commercial package

- Help a user recover damaged data

- Offer tips for better operation

- Explain an undocumented software feature

- Show a user how to use Web conferencing

- Explain how to access the company's intranet or the Internet

- Assist a user in developing a simple database to track time spent on various projects

- Answer questions about software licensing and upgrades

- Provide information about system specifications and the cost of new hardware or software

- Recommend a system solution that integrates data from different locations to solve a business problem

- Provide hardware support by installing or reconfiguring devices such as scanners, printers, network cards, wireless devices, optical drives, backup devices, and multimedia systems

- Show users how to maintain data consistency and integrity among a desktop computer, a notebook computer, and a handheld computer

FIGURE 12-3 A help desk, also called an information center (IC), provides guidance and assistance to system users. When a user contacts a help desk, the response should be prompt and effective.

In addition to functioning as a valuable link between IT staff and users, the help desk is a central contact point for all IT maintenance activities. The help desk is where users report system problems, ask for maintenance, or submit new systems requests. A help desk can utilize many types of automated support, just as outside vendors do, including e-mail responses, on-demand fax capability, an online knowledge base, frequently asked questions (FAQs), discussion groups, bulletin boards, and automated voice mail. A help desk also can provide a live chat feature for users or external customers. For example, as shown in Figure 12-4 on the next page, Hewlett-Packard enables customers to chat interactively with a tech support person.

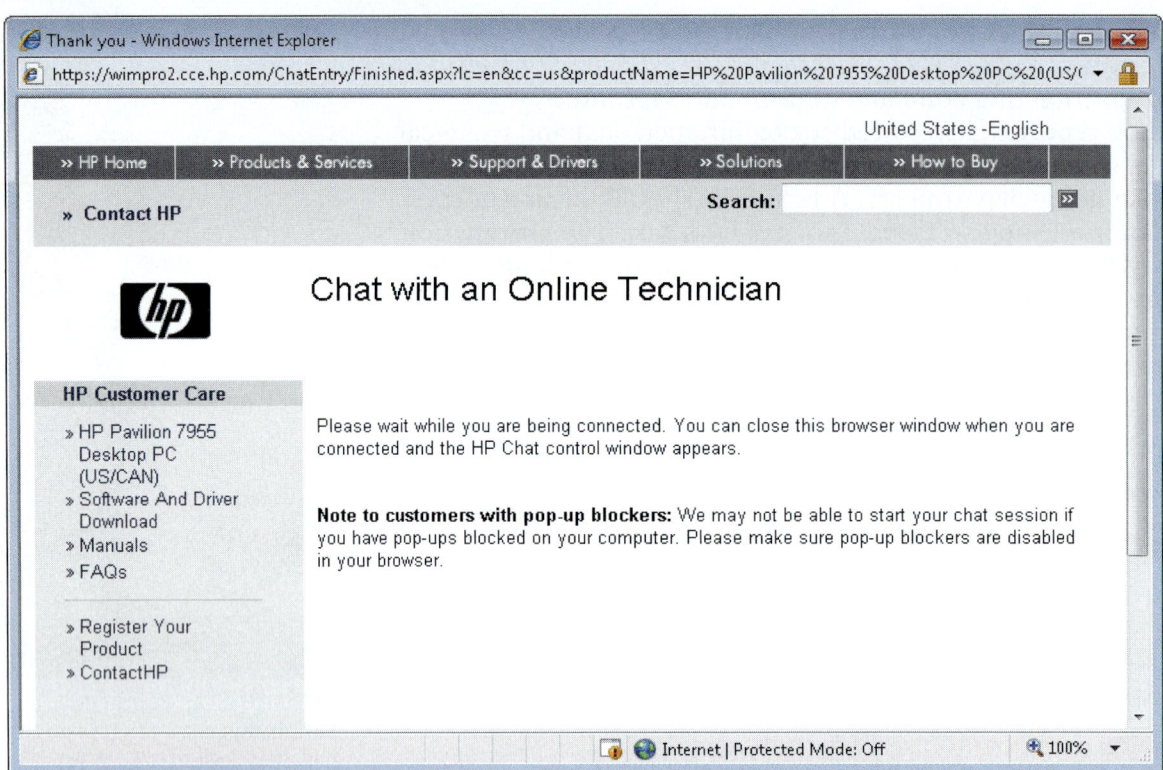

FIGURE 12-4 The HP support site enables a customer to chat with an online technician.

MAINTENANCE TASKS

The systems support and security phase is an important component of TCO (total cost of ownership) because ongoing maintenance expenses can determine the economic life of a system.

Figure 12-5 shows a typical pattern of operational and maintenance expenses during the useful life of a system. **Operational costs** include items such as supplies, equipment rental, and software leases. Notice that the lower area shown in Figure 12-5 represents fixed operational expenses, while the upper area represents maintenance expenses.

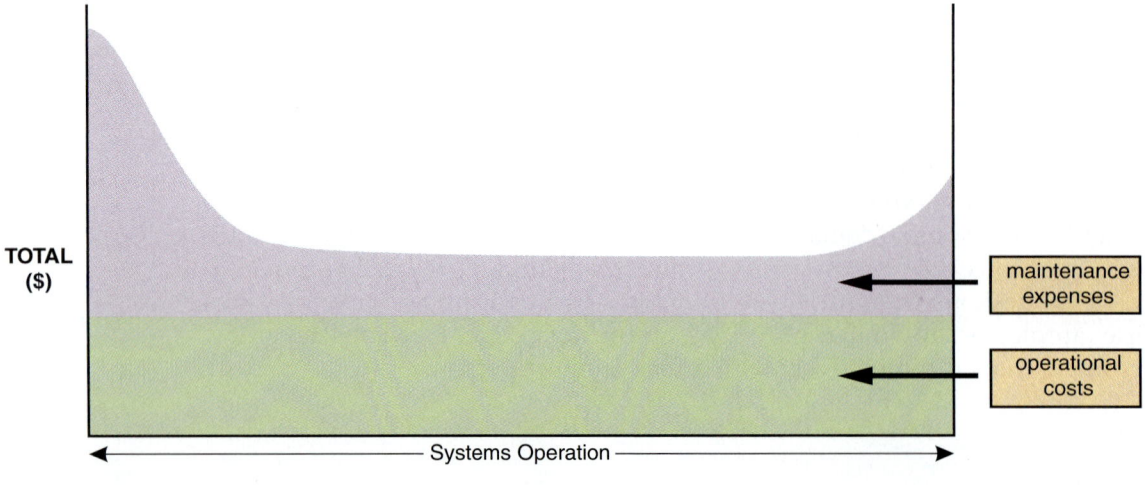

FIGURE 12-5 The total cost of operating an information system includes operational and maintenance costs. Operational costs (green) are relatively constant, while maintenance expenses (purple) vary over time.

Maintenance expenses vary significantly during the system's operational life and include spending to support maintenance activities. **Maintenance activities** include changing programs, procedures, or documentation to ensure correct system performance; adapting the system to changing requirements; and making the system operate more efficiently. Those needs are met by corrective, adaptive, perfective, and preventive maintenance.

Although some overlap exists, four types of maintenance tasks can be identified, as shown by the examples in Figure 12-6. **Corrective maintenance** is performed to fix errors, **adaptive maintenance** adds new capability and enhancements, **perfective maintenance** improves efficiency, and **preventive maintenance** reduces the possibility of future system failure. Some analysts use the term *maintenance* to describe only corrective maintenance that fixes problems. It is helpful, however, to view the maintenance concept more broadly and identify the different types of tasks.

Maintenance expenses usually are high when a system is implemented because problems must be detected, investigated, and resolved by corrective maintenance. Once the system becomes stable, costs usually remain low and involve minor adaptive maintenance. Eventually, both adaptive and perfective maintenance activities increase in a dynamic business environment.

Near the end of a system's useful life, adaptive and corrective maintenance expenses increase rapidly, but perfective maintenance typically decreases when it becomes clear that the company plans to replace the system. Figure 12-7 on the next page shows the typical patterns for each of the four classifications of maintenance activities over a system's life span.

Corrective Maintenance

Corrective maintenance diagnoses and corrects errors in an operational system. To avoid introducing new problems, all maintenance work requires careful analysis before making changes. The best maintenance approach is a scaled-down version of the SDLC itself, where investigation, analysis, design, and testing are performed before implementing any solution. Recall that in Chapter 11 you learned about the difference between a test environment and an operational environment. Any maintenance work that could affect the system must be performed first in the test environment, and then migrated to the operational system.

IT support staff respond to errors in various ways, depending on the nature and severity of the problem. Most organizations have standard procedures for minor errors, such as an incorrect report title or an improper format for a data element. In a typical procedure, a user submits a systems request that is evaluated, prioritized, and scheduled by the system administrator or the systems review committee. If the request is approved, the maintenance team designs, tests, documents, and implements a solution. As you learned in Chapter 2, many organizations use a standard online form for systems requests. In smaller firms, the process might be an informal e-mail message similar to the example shown in Figure 12-8 on page 563, which requests a relatively minor change in a monthly sales report.

TOOLKIT TIME

The Financial Analysis tools in Part 3 of the Systems Analyst's Toolkit can help you analyze and manage maintenance costs, and determine when a system is reaching the end of its useful life. To learn more about these tools, turn to Part 3 of the four-part Toolkit that follows Chapter 12.

Examples of Maintenance Tasks

Corrective Maintenance
- Diagnose and fix logic errors
- Replace defective network cabling
- Restore proper configuration settings
- Debug program code
- Update drivers
- Install software patch

Adaptive Maintenance
- Add online capability
- Create new reports
- Add new data entry field to input screen
- Install links to Web site
- Create employee portal

Perfective Maintenance
- Install additional memory
- Write macros to handle repetitive tasks
- Compress system files
- Optimize user desktop settings
- Develop library for code reuse
- Install more powerful network server

Preventive Maintenance
- Install new antivirus software
- Develop standard backup schedule
- Implement regular defragmentation process
- Analyze problem report for patterns
- Tighten all cable connections

FIGURE 12-6 Corrective maintenance fixes errors and problems. Adaptive maintenance provides enhancements to a system. Perfective maintenance improves a system's efficiency, reliability, or maintainability. Preventive maintenance avoids future problems.

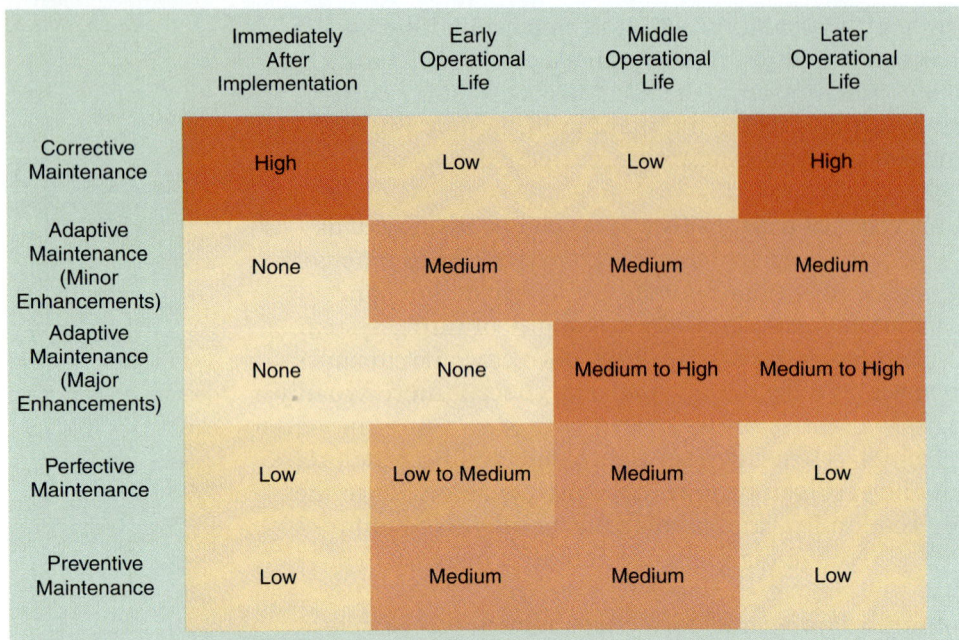

	Immediately After Implementation	Early Operational Life	Middle Operational Life	Later Operational Life
Corrective Maintenance	High	Low	Low	High
Adaptive Maintenance (Minor Enhancements)	None	Medium	Medium	Medium
Adaptive Maintenance (Major Enhancements)	None	None	Medium to High	Medium to High
Perfective Maintenance	Low	Low to Medium	Medium	Low
Preventive Maintenance	Low	Medium	Medium	Low

FIGURE 12-7 Information systems maintenance depends on two major factors: the type of maintenance and the age of the system.

For more serious situations, such as incorrect report totals or inconsistent data, a user submits a systems request with supporting evidence. Those requests receive a high priority and a maintenance team begins work on the problem immediately.

The worst-case situation is a system failure. If an emergency occurs, the maintenance team bypasses the initial steps and tries to correct the problem immediately. Meanwhile, a written systems request is prepared by a user or a member of the IT department and added to the maintenance log. When the system is operational again, the maintenance team determines the cause, analyzes the problem, and designs a permanent solution. The IT response team updates the test data files, thoroughly tests the system, and prepares full documentation.

The process of managing system support is described in more detail on page 564, including an overview of maintenance tasks and a procedural flowchart, which is shown in Figure 12-11 on page 567.

Adaptive Maintenance

Adaptive maintenance adds enhancements to an operational system and makes the system easier to use. An **enhancement** is a new feature or capability. The need for adaptive maintenance usually arises from business environment changes such as new products or services, new manufacturing technology, or support for a new Web-based operation.

The procedure for minor adaptive maintenance is similar to routine corrective maintenance. A user submits a systems request that is evaluated and prioritized by the systems review committee. A maintenance team then analyzes, designs, tests, and implements the enhancement. Although the procedures for the two types of maintenance are alike, adaptive maintenance requires more IT department resources than minor corrective maintenance.

A major adaptive maintenance project is like a small-scale SDLC project because the development procedure is similar. Adaptive maintenance can be more difficult than new systems development because the enhancements must work within the constraints of an existing system.

Perfective Maintenance

Perfective maintenance involves changing an operational system to make it more efficient, reliable, or maintainable. Requests for corrective and adaptive maintenance normally come from users, while the IT department usually initiates perfective maintenance.

During system operation, changes in user activity or data patterns can cause a decline in efficiency, and perfective maintenance might be needed to restore performance. When users are concerned about performance, you should determine if a perfective maintenance project could improve response time and system efficiency.

Perfective maintenance also can improve system reliability. For example, input problems might cause a program to terminate abnormally. By modifying the data entry process, you can highlight errors and notify the users that they must enter proper data. When a system is easier to maintain, support is less costly and less risky. In many cases, you can simplify a complex program to improve maintainability.

In many organizations, perfective maintenance is not performed frequently enough. Companies with limited resources often consider new systems development, adaptive maintenance, and corrective maintenance more important than perfective maintenance. Managers and users constantly request new projects, so few resources are available for perfective maintenance work. As a practical matter, perfective maintenance can be performed as part of another project. For example, if a new function must be added to a program, you can include perfective maintenance in the adaptive maintenance project.

Perfective maintenance usually is cost effective during the middle of the system's operational life. Early in systems operation, perfective maintenance usually is not needed. Later, perfective maintenance might be necessary, but have a high cost. Perfective maintenance is less important if the company plans to discontinue the system.

When performing perfective maintenance, analysts often use a technique called software reengineering. **Software reengineering** uses analytical techniques to identify potential quality and performance improvements in an information system. In that sense, software reengineering is similar to business process reengineering, which seeks to simplify operations, reduce costs, and improve quality — as you learned in Chapter 1.

Programs that need a large number of maintenance changes usually are good candidates for reengineering. The more a program changes, the more likely it is to become inefficient and difficult to maintain. Detailed records of maintenance work can identify systems with a history of frequent corrective, adaptive, or perfective maintenance.

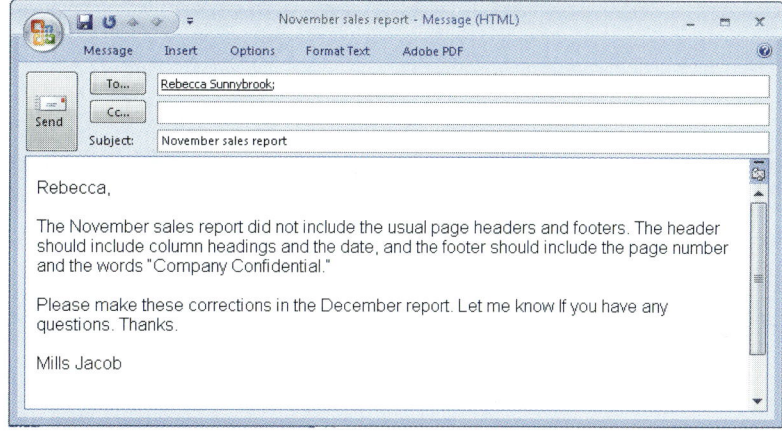

FIGURE 12-8 Example of an e-mail request for the correction of a formatting error in a report.

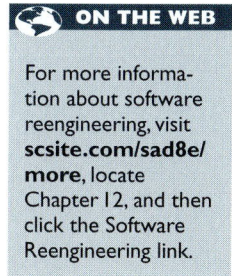

ON THE WEB

For more information about software reengineering, visit **scsite.com/sad8e/more**, locate Chapter 12, and then click the Software Reengineering link.

Preventive Maintenance

To avoid problems, preventive maintenance requires analysis of areas where trouble is likely to occur. Like perfective maintenance, the IT department normally initiates preventive maintenance. Preventive maintenance often results in increased user satisfaction, decreased downtime, and reduced TCO. Preventive maintenance competes for IT resources along with other projects, and sometimes does not receive the high priority that it deserves.

Regardless of the type of maintenance, computer systems must be supported by trained professionals, just as the carrier-based aircraft shown in Figure 12-9 must be serviced by skilled technicians. In both cases, the quality of the maintenance will directly affect the organization's success.

FIGURE 12-9 Regardless of the type of system, high-quality maintenance must be performed by trained professionals.

CASE IN POINT 12.1: OUTBACK OUTSOURCING, INC.

You are a systems analyst at Outback Outsourcing, a firm that handles payroll processing for many large companies. Outback Outsourcing uses a combination of payroll package programs and in-house developed software to deliver custom-made payroll solutions for its clients. Lately, users have flooded you with requests for more new features and Web-based capability to meet customer expectations. Your boss, the IT manager, comes to you with a question. She wants to know when to stop trying to enhance the old software and develop a totally new version better suited to the new marketplace. How would you answer her?

MAINTENANCE MANAGEMENT

System maintenance requires effective management, quality assurance, and cost control. To achieve these goals, companies use various strategies, such as a maintenance team, a maintenance management program, a configuration management process, and a maintenance release procedure. In addition, firms use version control and baselines to track system releases and analyze the system's life cycle. These concepts are described in the following sections.

The Maintenance Team

A **maintenance team** includes a system administrator and one or more systems analysts and programmers. The system administrator should have solid technical expertise, and experience in troubleshooting and configuring operating systems and hardware. Successful analysts need a strong IT background, solid analytical abilities, good communication skills, and an overall understanding of business operations.

SYSTEM ADMINISTRATOR   A **system administrator** manages computer and network systems. A system administrator must work well under pressure, have good organizational and communication skills, and be able to understand and resolve complex issues in a limited time frame. In most organizations, a system administrator has primary responsibility for the operation, configuration, and security of one or more systems. The system administrator is responsible for routine maintenance, and usually is authorized to take preventive action to avoid an immediate emergency, such as a server crash, network outage, security incident, or hardware failure.

Systems administration is a vital function, and various professional associations, such as SAGE, which is shown in Figure 12-10, offer a wide variety of technical information and support for system administrators. Notice that SAGE members subscribe to a code of ethics that includes professionalism, integrity, privacy, and social responsibility, among other topics.

SYSTEMS ANALYSTS   Systems analysts assigned to a maintenance team are like skilled detectives who investigate and rapidly locate the source of a problem by using analysis and synthesis skills. **Analysis** means examining the whole in order to learn about the individual elements, while **synthesis** involves studying the parts to understand the overall system. In addition to strong technical skills, an analyst must have a solid grasp of business operations and functions. Analysts also need effective interpersonal and communications skills and they must be creative, energetic, and eager for new knowledge.

PROGRAMMERS   In a small organization, a programmer might be expected to handle a wide variety of tasks, but in larger firms, programming work tends to be more

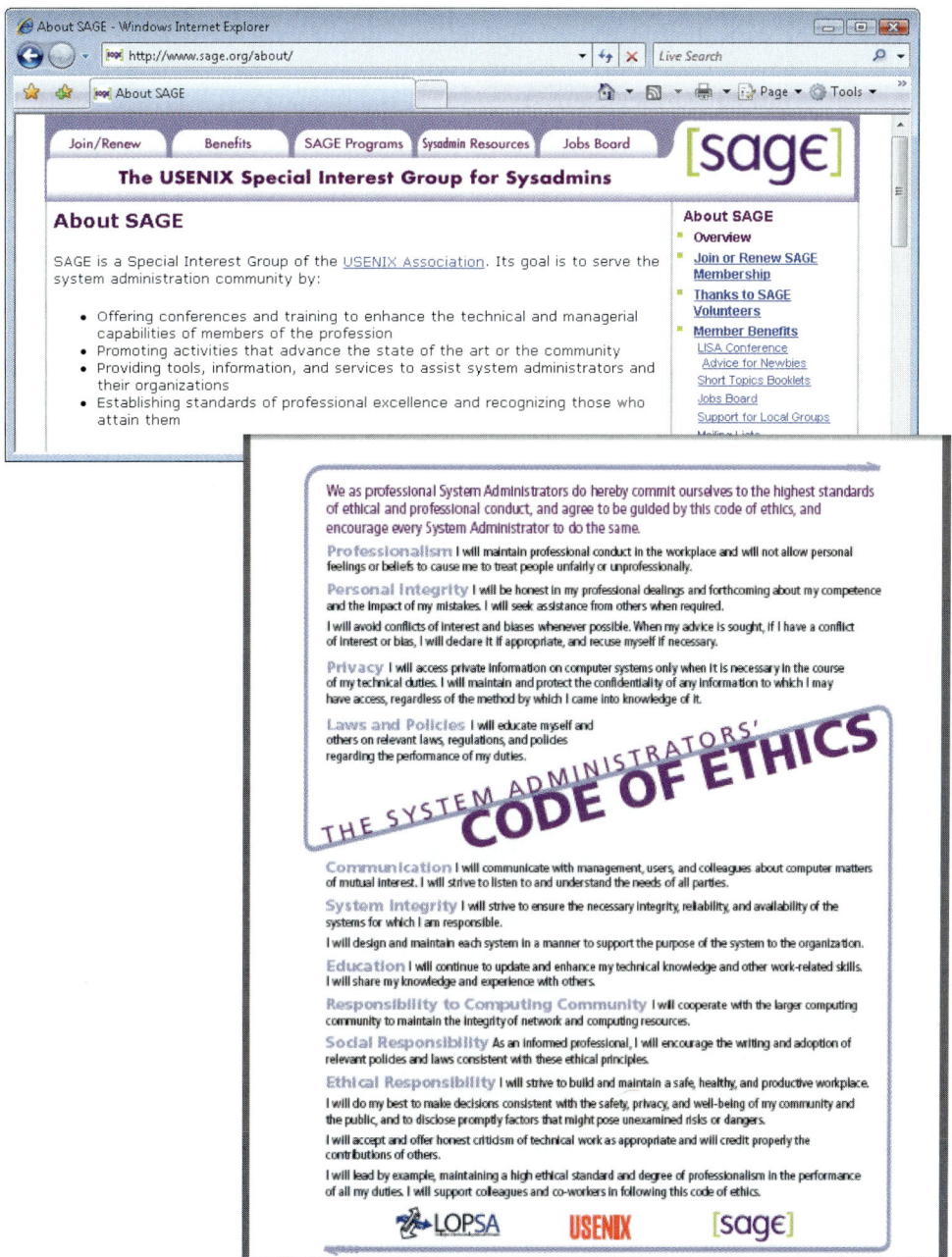

FIGURE 12-10 SAGE seeks to establish standards of professional excellence, improve the technical skills of its members, and promote a comprehensive code of ethics.

specialized. For example, typical job titles include an **applications programmer**, who works on new systems development and maintenance; a **systems programmer**, who concentrates on operating system software and utilities; and a **database programmer**, who focuses on creating and supporting large-scale database systems. Many IT departments also use a job title of **programmer/analyst** to designate positions that require a combination of systems analysis and programming skills.

ORGANIZATIONAL ISSUES IT managers often divide systems analysts and programmers into two groups: One group performs new system development, and the other group handles maintenance. Some organizations use a more flexible approach and assign IT staff members to various projects as they occur. By integrating development and support work, the people developing the system assume responsibility for maintaining it. Because the team is familiar with the project, additional training or expense is unnecessary, and members are likely to have a sense of ownership from the onset.

Unfortunately, many analysts feel that maintenance is less interesting and creative than developing new systems. In addition, an analyst might find it challenging to troubleshoot and support someone else's work that might have been poorly documented and organized.

Some organizations that have separate maintenance and new systems groups rotate people from one assignment to the other. When analysts learn different skills, the organization is more versatile and people can shift to meet changing business needs. For instance, systems analysts working on maintenance projects learn why it is important to design easily maintainable systems. Similarly, analysts working on new systems get a better appreciation of the development process and the design compromises necessary to meet business objectives.

One disadvantage of rotation is that it increases overhead because time is lost when people move from one job to another. When systems analysts constantly shift between maintenance and new development, they have less opportunity to become highly skilled at any one job.

Newly hired and recently promoted IT staff members often are assigned to maintenance projects because their managers believe that the opportunity to study existing systems and documentation is a valuable experience. In addition, the mini-SDLC used in many adaptive maintenance projects is good training for the full-scale systems development life cycle. For a new systems analyst, however, maintenance work might be more difficult than systems development, and it might make sense to assign a new person to a development team where experienced analysts are available to provide training and guidance.

CASE IN POINT 12.2: BRIGHTSIDE INSURANCE, INC.

As IT manager at Brightside Insurance Company, you organized your IT staff into two separate groups — one team for maintenance projects and the other team for new systems work. That arrangement worked well in your last position at another company. Brightside, however, previously made systems assignments with no particular pattern.

At first, the systems analysts in your group did not comment about the team approach. Now, several of your best analysts have indicated that they enjoyed the mix of work and would not want to be assigned to a maintenance team. Before a problem develops, you have decided to rethink your organizational strategy. Should you go back to the way things were done previously at Brightside? Why or why not? Do other options exist? What are they?

Maintenance Requests

Typically, maintenance requests involve a series of steps, as shown in Figure 12-11. After a user submits a request, a system administrator determines whether immediate action is needed and whether the request is under a prescribed cost limit. In nonemergency requests that exceed the cost limit, a systems review committee assesses the request and either approves it, with a priority, or rejects it. The system administrator notifies affected users of the outcome.

Users submit most requests for corrective and adaptive maintenance when the system is not performing properly, or if they want new features. IT staff members usually initiate requests for perfective and preventive maintenance. To keep a complete maintenance log, all work must be covered by a specific request that users submit in writing or by e-mail.

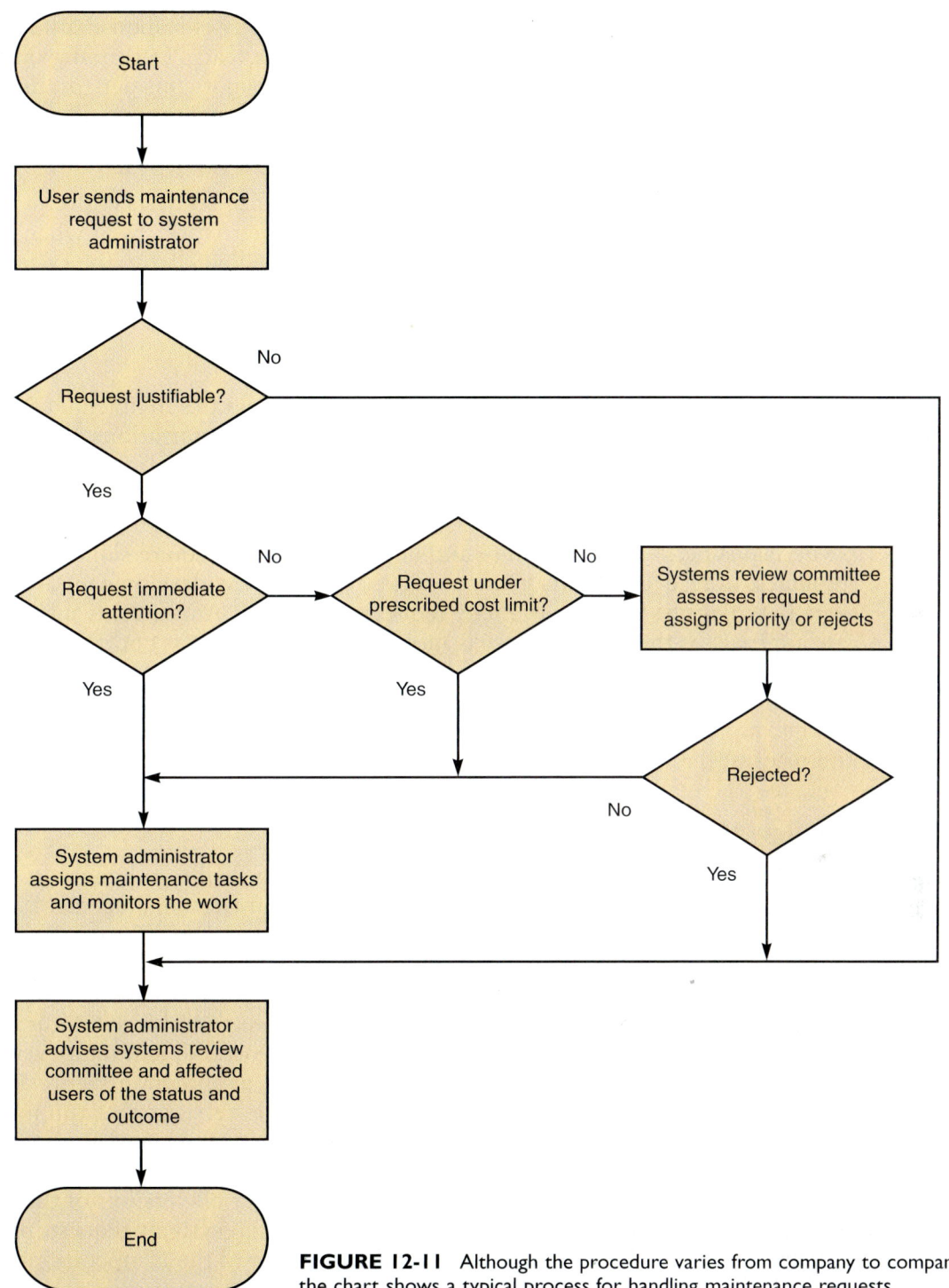

FIGURE 12-11 Although the procedure varies from company to company, the chart shows a typical process for handling maintenance requests.

INITIAL DETERMINATION When a user submits a maintenance request, the system administrator makes an initial determination. If the request is justifiable and involves a severe problem that requires immediate attention, the system administrator takes action at once. In justifiable, but noncritical, situations, the administrator determines whether the request can be performed within a preauthorized cost level. If so, he or she assigns the maintenance tasks and monitors the work.

THE SYSTEMS REVIEW COMMITTEE When a request exceeds a predetermined cost level or involves a major configuration change, the systems review committee either approves it and assigns a priority, or rejects it.

TASK COMPLETION The system administrator usually is responsible for assigning maintenance tasks to individuals or to a maintenance team. Depending on the situation and the company's policy, the system administrator might consider rotating assignments among the IT staff or limiting maintenance tasks to certain individuals or teams, as explained in the previous section.

USER NOTIFICATION Users who initiate maintenance requests expect a prompt response, especially if the situation directly affects their work. Even when corrective action cannot occur immediately, users appreciate feedback from the system administrator and should be kept informed of any decisions or actions that could affect them.

Establishing Priorities

In many companies, the systems review committee separates maintenance and new development requests when setting priorities. In other organizations, all requests are considered together, and the most important project gets top priority, whether it is maintenance or new development.

Some IT managers believe that evaluating all projects together leads to the best possible decisions because maintenance and new development require similar IT department resources. In IT departments where maintenance and new development are not integrated, it might be better to evaluate requests separately. Another advantage of a separate approach is that maintenance is more likely to receive a proportional share of IT department resources.

The most important objective is to have a procedure that balances new development and necessary maintenance work to provide the best support for business requirements and priorities.

ON THE WEB

For more information about configuration management, visit **scsite.com/sad8e/ more**, locate Chapter 12, and then click the Configuration Management link.

Configuration Management

Configuration management (CM) is a process for controlling changes in system requirements during software development. Configuration management also is an important tool for managing system changes and costs after a system becomes operational. Most companies establish a specific process that describes how system changes must be requested and documented.

As enterprise-wide information systems grow more complex, configuration management becomes critical. Industry standards have emerged, and many vendors offer configuration management software and techniques, as shown in Figure 12-12.

CM is especially important if a system has multiple versions that run in different hardware and software environments. Configuration management also helps to organize and handle documentation. An operational system has extensive documentation that covers development, modification, and maintenance for all versions of the installed system. Most documentation material, including the initial systems request, project management data, end-of-phase reports, data dictionary, and the IT operations and user manuals, is stored in the IT department.

Keeping track of all documentation and ensuring that updates are distributed properly are important aspects of configuration management.

Maintenance Releases

Keeping track of maintenance changes and updates can be difficult, especially for a complex system. When a **maintenance release methodology** is used, all noncritical changes are held until they can be implemented at the same time. Each change is documented and installed as a new version of the system called a **maintenance release**.

For an in-house developed system, the time between releases usually depends on the level of maintenance activity. A new release to correct a critical error, however, might be implemented immediately rather than saved for the next scheduled release.

When a release method is used, a numbering pattern distinguishes the different releases. In a typical system, the initial version of the system is 1.0, and the release that includes the first set of maintenance

FIGURE 12-12 CM Crossroads provides a source of information and resources for configuration management professionals.

changes is version 1.1. A change, for example, from version 1.4 to 1.5 indicates relatively minor enhancements, while whole number changes, such as from version 1.0 to 2.0 or from version 3.4 to 4.0, indicate a significant upgrade.

The release methodology offers several advantages, especially if two teams perform maintenance work on the same system. When a release methodology is used, all changes are tested together before a new system version is released. This approach results in fewer versions, less expense, and less interruption for users. Using a release methodology also reduces the documentation burden, because all changes are coordinated and become effective simultaneously.

A release methodology also has some potential disadvantages. Users expect a rapid response to their problems and requests, but with a release methodology, new features or upgrades are available less often. Even when changes would improve system efficiency or user productivity, the potential savings must wait until the next release, which might increase operational costs.

Commercial software suppliers also provide maintenance releases, often called **service packs,** similar to the Microsoft Service Pack for Windows Vista shown in Figure 12-13. As Microsoft explains, a service pack contains all the fixes and enhancements that have been made available since the last program version or service pack.

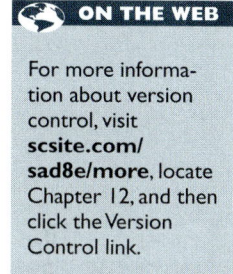

ON THE WEB

For more information about version control, visit **scsite.com/ sad8e/more,** locate Chapter 12, and then click the Version Control link.

Version Control

Version control is the process of tracking system releases, or versions. When a new version of a system is installed, the prior release is **archived,** or stored. If a new version causes a

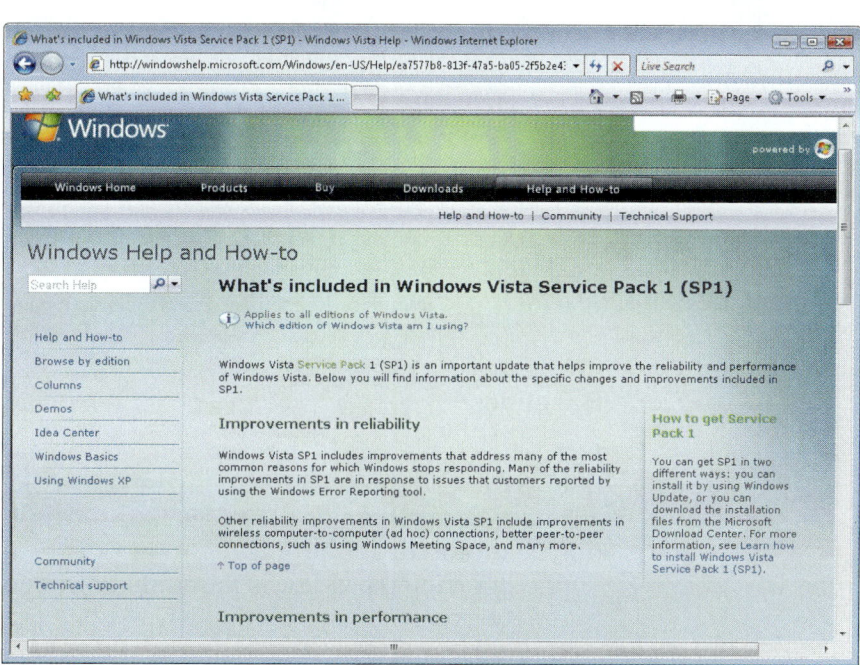

FIGURE 12-13 A Microsoft service pack provides access to up-to-date drivers, tools, security patches, and customer-requested product changes.

system to fail, a company can reinstall the prior version to restore operations. In addition to tracking system versions, the IT staff is responsible for configuring systems that have several modules at various release stages. For example, an accounting system might have a one-year old accounts receivable module that must interface with a brand-new payroll module.

As systems grow more complex, version control becomes an essential part of system documentation. In addition to in-house version control procedures, companies can purchase software from vendors such as Serena, as shown in Figure 12-14. Notice that Serena also offers software that can support agile development.

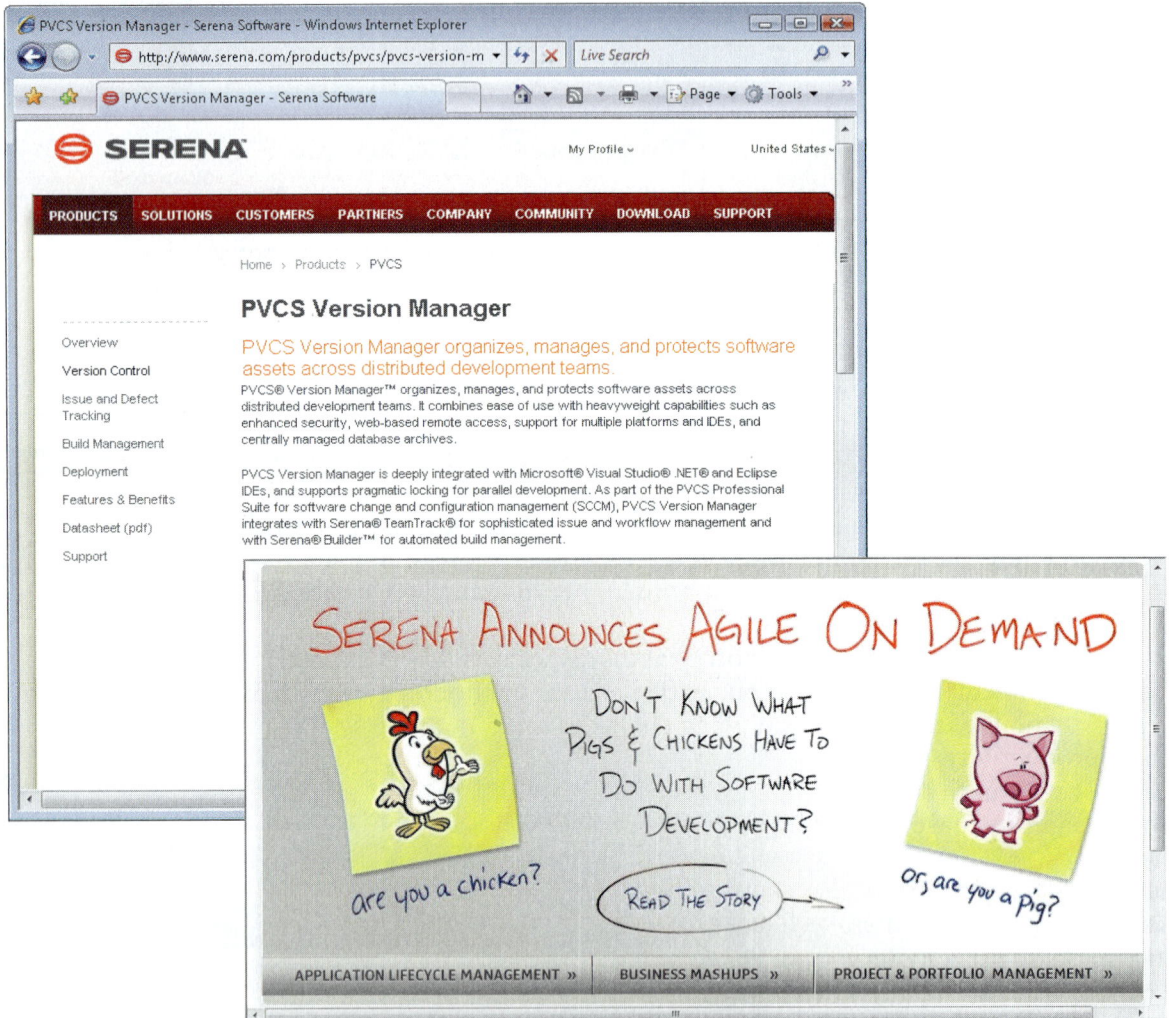

FIGURE 12-14 Serena offers version control software and tools that agile developers can use to support and manage projects.

Baselines

A **baseline** is a formal reference point that measures system characteristics at a specific time. Systems analysts use baselines as yardsticks to document features and performance during the systems development process. The three types of baselines are functional, allocated, and product.

The **functional baseline** is the configuration of the system documented at the beginning of the project. It consists of all the necessary system requirements and design constraints.

The **allocated baseline** documents the system at the end of the design phase and identifies any changes since the functional baseline. The allocated baseline includes testing and verification of all system requirements and features.

The **product baseline** describes the system at the beginning of system operation. The product baseline incorporates any changes made since the allocated baseline and includes the results of performance and acceptance tests for the operational system.

SYSTEM PERFORMANCE MANAGEMENT

Years ago, when most firms used a central computer for processing data, it was relatively simple to manage a system and measure its efficiency. Today, companies use complex networks and client/server systems to support business needs. A user at a client workstation often interacts with an information system that depends on other clients, servers, networks, and data located throughout the company. Rather than a single computer, it is the integration of all those components that determines the system's capability and performance. In many situations, IT managers use automated software and CASE tools to manage complex systems.

To ensure satisfactory support for business operations, the IT department must manage system faults and interruptions, measure system performance and workload, and anticipate future needs. The following sections discuss these topics.

Fault Management

No matter how well it is designed, every system will experience some problems, such as hardware failures, software errors, user mistakes, and power outages. A system administrator must detect and resolve operational problems as quickly as possible. That task, often called **fault management**, includes monitoring the system for signs of trouble, logging all system failures, diagnosing the problem, and applying corrective action.

The more complex the system, the more difficult it can be to analyze symptoms and isolate a cause. In addition to addressing the immediate problem, it is important to evaluate performance patterns and trends. Automated tools can assist system administrators in that task, such as the TeMIP software offered by HP, which is shown in Figure 12-15. In addition to automated notification, fault management software can identify underlying causes, speed up response time, and reduce service outages.

Although system administrators must deal with system faults and interruptions as they arise, the best strategy is to prevent problems by monitoring system performance and workload.

Performance and Workload Measurement

In e-business, slow performance can be as devastating as no performance at all. Network delays and application bottlenecks affect customer satisfaction, user productivity, and business results. In fact, many IT managers believe that network delays do more damage than actual stoppages, because they occur more frequently and are difficult to predict, detect, and prevent. Customers

FIGURE 12-15 HP claims that its TeMIP fault management software can reduce network downtime, improve service quality, and track resource usage.

expect reliable, fast response 24 hours a day, seven days a week. To support that level of service, companies use performance management software, such as the NetQoS product shown in Figure 12-16.

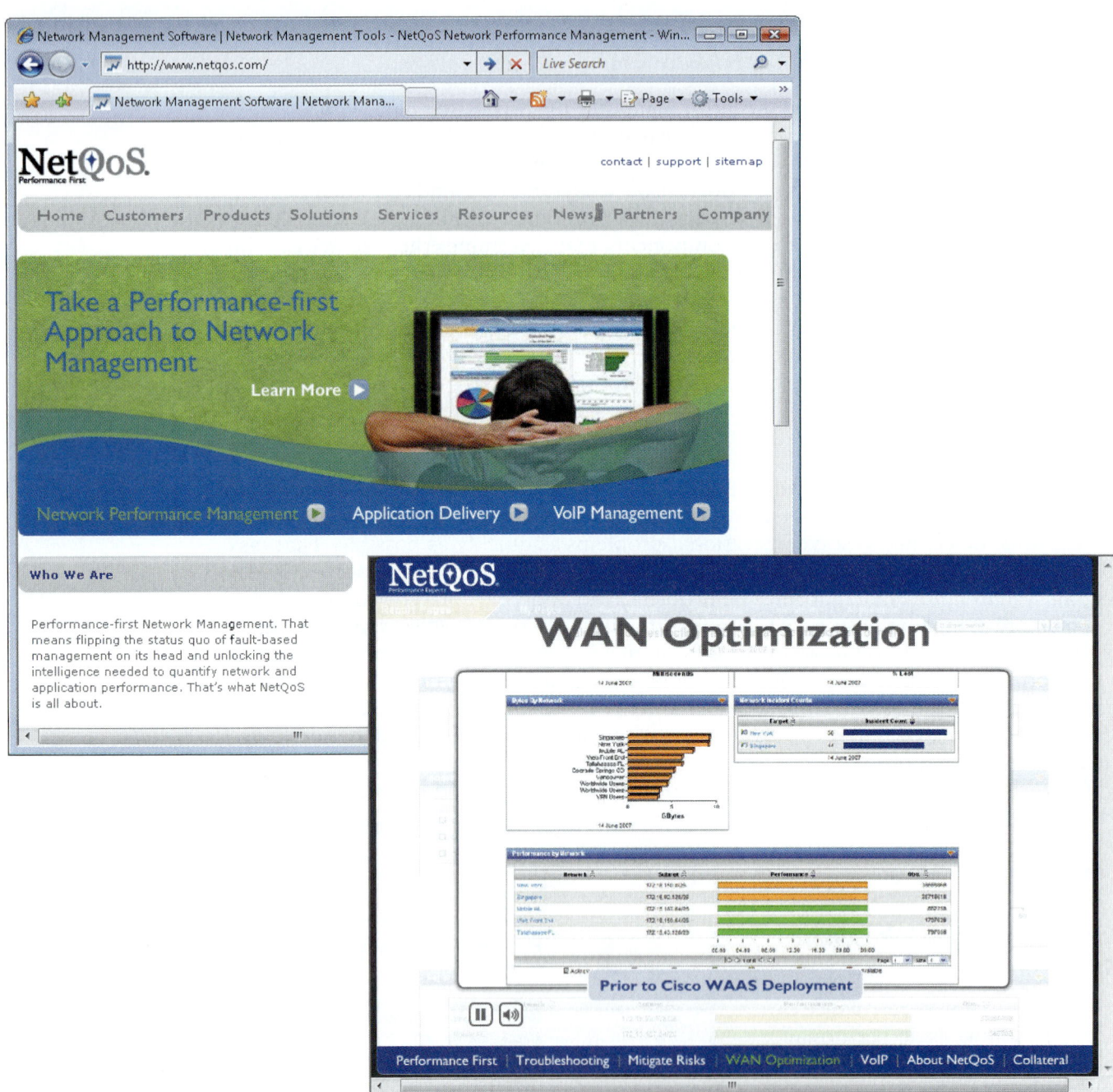

FIGURE 12-16 NetQos offers network performance management tools that feature graphical analysis and measurement of various performance metrics.

To measure system performance, many firms use **benchmark testing**, which uses a set of standard tests to evaluate system performance and capacity. In addition to benchmark testing, performance measurements, called **metrics**, can monitor the number of transactions processed in a given time period, the number of records accessed, and the volume of online data. Network performance metrics include response time, bandwidth, throughput, and turnaround time, among others.

RESPONSE TIME **Response time** is the overall time between a request for system activity and the delivery of the response. In the typical online environment, response time is measured from the instant the user presses the ENTER key or clicks a mouse button until the requested screen display appears or printed output is ready. Response time is affected by the system design, capabilities, and processing methods. If the request involves network or Internet access, response time is affected by data communication factors.

Online users expect an immediate response, and they are frustrated by any apparent lag or delay. Of all performance measurements, response time is the one that users notice and complain about most.

BANDWIDTH AND THROUGHPUT Bandwidth and throughput are closely related terms, and many analysts use them interchangeably. **Bandwidth** describes the amount of data that the system can transfer in a fixed time period. Bandwidth requirements are expressed in bits per second. Depending on the system, you might measure bandwidth in **Kbps (kilobits per second)**, **Mbps (megabits per second)**, or **Gbps (gigabits per second)**. Analyzing bandwidth is similar to forecasting the hourly number of vehicles that will use a highway in order to determine the number of lanes required.

Throughput measures actual system performance under specific circumstances and is affected by network loads and hardware efficiency. Throughput, like bandwidth, is expressed as a data transfer rate, such as Kbps, Mbps, or Gbps. Just as traffic jams delay highway traffic, throughput limitations can slow system performance and response time. That is especially true with graphics-intensive systems and Web-based systems that are subject to Internet-related conditions.

In addition to the performance metrics explained in the previous section, system administrators measure many other performance characteristics. Although no standard set of metrics exists, several typical examples are:

- Arrivals — The number of items that appear on a device during a given observation time.

- Busy — The time that a given resource is unavailable.

- Completions — The number of arrivals that are processed during a given observation period.

- Queue length — The number of requests pending for a service.

- Service time — The time it takes to process a given task once it reaches the front of the queue.

- Think time — The time it takes an application user to issue another request.

- Utilization — How much of a given resource was required to complete a task.

- Wait time — The time that requests must wait for a resource to become available.

The Computer Measurement Group (CMG®) maintains a site, shown in Figure 12-17, that provides support and assistance for IT professionals concerned with performance evaluation and capacity planning.

FIGURE 12-17 The Computer Measurement Group is a nonprofit organization that primarily is concerned with performance evaluation and capacity management.

TURNAROUND TIME   Turnaround time applies to centralized batch processing operations, such as customer billing or credit card statement processing. Turnaround time measures the time between submitting a request for information and the fulfillment of the request. Turnaround time also can be used to measure the quality of IT support or services by measuring the time from a user request for help to the resolution of the problem.

The IT department often measures response time, bandwidth, throughput, and turnaround time to evaluate system performance both before and after changes to the system or business information requirements. Performance data also is used for cost-benefit analyses of proposed maintenance and to evaluate systems that are nearing the end of their economically useful lives.

Finally, management uses current performance and workload data as input for the capacity planning process.

ON THE WEB

For more information about capacity planning, visit **scsite.com/sad8e/more**, locate Chapter 12, and then click the Capacity Planning link.

Capacity Planning

Capacity planning is a process that monitors current activity and performance levels, anticipates future activity, and forecasts the resources needed to provide desired levels of service.

As the first step in capacity planning, you develop a current model based on the system's present workload and performance specifications. Then you project demand and user requirements over a one- to three-year time period and analyze the model to see what is needed to maintain satisfactory performance and meet requirements. To assist you in the process, you can use a technique called what-if analysis.

What-if analysis allows you to vary one or more elements in a model in order to measure the effect on other elements. For example, you might use what-if analysis to answer questions such as: How will response time be affected if we add more PC workstations to the network? Will our client/server system be able to handle the growth in sales from the new Web site? What will be the effect on server throughput if we add more memory?

Powerful spreadsheet tools also can assist you in performing what-if analysis. For example, Microsoft Excel contains a feature called Goal Seek that determines what changes are necessary in one value to produce a specific result for another value. In the example shown in Figure 12-18, a capacity planning worksheet indicates that the system can handle 3,840 Web-based orders per day, at 22.5 seconds each. The user wants to know the effect on processing time if the number of transactions increases to 9,000. As the Goal Seek solution in the bottom figure shows, order processing will have to be performed in 9.6 seconds to achieve that goal.

When you plan capacity, you need detailed information about the number of transactions; the daily, weekly, or monthly transaction patterns; the number of queries; and the number, type, and size of all generated reports. If the system involves a LAN, you need to estimate network traffic levels to determine whether or not the existing hardware and software can handle the load. If the system uses a client/server design, you need to examine performance and connectivity specifications for each platform.

Most important, you need an accurate forecast of future business activities. If new business functions or requirements are predicted, you should develop contingency plans based on input from users and management. The main objective is to ensure that the system meets all future demands and provides effective support for business operations. Many firms offer capacity planning software and services, as shown in Figure 12-19.

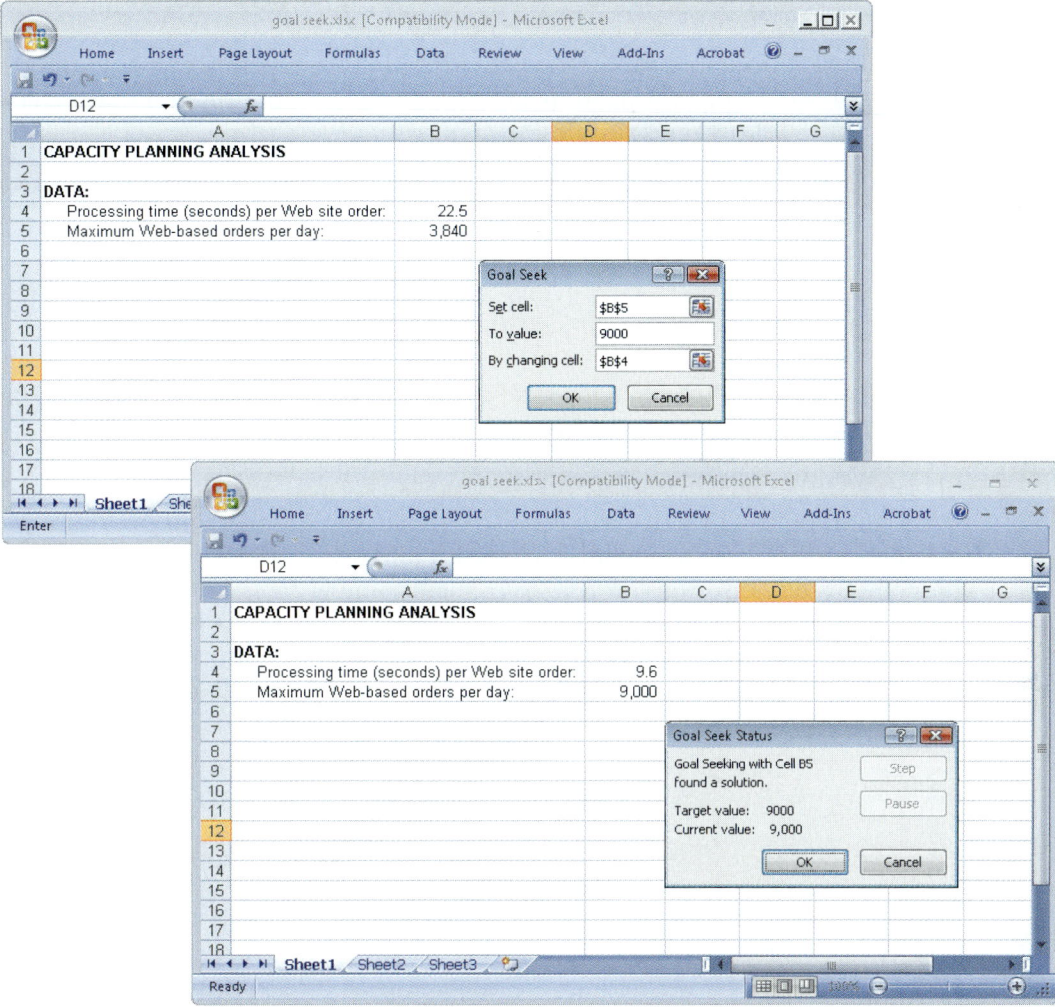

FIGURE 12-18 Microsoft Excel provides a Goal Seek feature that permits what-if analysis.

FIGURE 12-19 TeamQuest is an example of a firm that offers capacity planning tools and services.

System Maintenance Tools

You can use automated tools that provide valuable assistance during the operation and support phase. Many CASE tools include system evaluation and maintenance features, including the following examples:

- Performance monitor that provides data on program execution times
- Program analyzer that scans source code, provides data element cross-reference information, and helps evaluate the impact of a program change
- Interactive debugging analyzer that locates the source of a programming error
- Reengineering tools
- Automated documentation
- Network activity monitor
- Workload forecasting tool

In addition to CASE tools, you also can use spreadsheet and presentation software to calculate trends, perform what-if analyses, and create attractive charts and graphs to display the results. Information technology planning is an essential part of the business planning process, and you probably will deliver presentations to management. You can review Part 1 of the Systems Analyst's Toolkit for more information on using spreadsheet and presentation software to help you communicate effectively.

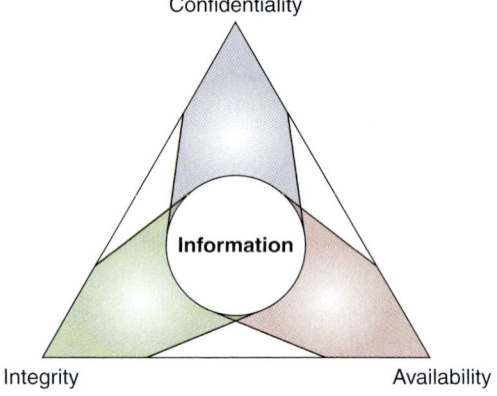

FIGURE 12-20 System security must provide information confidentiality, integrity, and availability.

FIGURE 12-21 Risk management requires continuous risk identification, assessment, and control.

SYSTEM SECURITY OVERVIEW

Security is a vital part of every information system. **Security** protects the system, and keeps it safe, free from danger, and reliable. In a global environment that includes many types of threats and attacks, security is more important than ever. This section includes a discussion of system security concepts, risk management, and common attacks against the system.

System Security Concepts

The **CIA triangle** in Figure 12-20 shows the three main elements of system security: confidentiality, integrity, and availability. **Confidentiality** protects information from unauthorized disclosure and safeguards privacy. **Integrity** prevents unauthorized users from creating, modifying, or deleting information. **Availability** ensures that authorized users have timely and reliable access to necessary information. The first step in managing IT security is to develop a **security policy** based on these three elements.

Risk Management

In the real world, *absolute* security is not a realistic goal. Instead, managers must balance the value of the assets being protected, potential risks to the organization, and security costs. For example, it might not be worth installing an expensive video camera monitoring system to protect an empty warehouse. To achieve the best results, most firms use a **risk management** approach that involves constant attention to three interactive tasks: risk identification, risk assessment, and risk control, as shown in Figure 12-21.

Risk identification analyzes the organization's assets, threats, and vulnerabilities. **Risk assessment** measures risk likelihood and impact. **Risk control** develops safeguards that reduce risks and their impact.

RISK IDENTIFICATION The first step in risk identification is to list and classify business assets. An **asset** might include company hardware, software, data, networks, people, or procedures. For each asset, a risk manager rates the impact of an attack and analyzes possible threats. A **threat** is an internal or external entity that could endanger an asset. For example, threat categories might include natural disasters, software attacks, or theft, as shown in Figure 12-22.

ON THE WEB

To learn more about risk management, visit **scsite.com/sad8e/ more**, locate Chapter 12, and then click the Risk Management link.

Threat Categories and Examples

THREAT	CATEGORY
Extortion	Hacker steals trade secrets and threatens to release them if not paid.
Hardware and software failures	Router stops functioning, or software causes the application server to crash.
Human error or failure	Employee accidentally deletes a file.
Natural disasters	Flood destroys company building and networked systems.
Service failure	Electricity is disrupted and brings the entire system down for hours.
Software attack	A group plants destructive software, a virus, or a worm into a company network.
Technical obsolescence	Outdated software is slow, difficult to use, and vulnerable to attacks.
Theft of physical or intellectual property	Physical server is stolen, intellectual property is stolen or used without permission; may be physical or electronic.
Trespass and espionage	Employee enters unlocked server room and views the payroll data on a forbidden system.
Vandalism	Attacker defaces Web site logo, or destroys CEO's hard drive physically or electronically.

FIGURE 12-22 System threats can be grouped into several broad categories. Note the examples provided for each category.

Next, the risk manager identifies vulnerabilities and how they might be exploited. A **vulnerability** is a security weakness or soft spot, and an **exploit** is an attack that takes advantage of a vulnerability. To identify vulnerabilities, a risk manager might ask questions like these: *Could hackers break through the proxy server? Could employees retrieve sensitive files without proper authorization? Could people enter the computer room and sabotage our servers?* Each vulnerability is rated and assigned a value. The output of risk identification is a list of assets, vulnerabilities, and ratings.

RISK ASSESSMENT In IT security terms, a **risk** is the impact of an attack multiplied by the likelihood of a vulnerability being exploited. For example, an impact value of 2 and a vulnerability rating of 10 would produce a risk of 20. On the other hand, an impact value of 5 and a vulnerability rating of 5 would produce a risk of 25. When risks are calculated and prioritized, **critical risks** will head the list. Although ratings can be subjective, the overall process provides a consistent approach and framework.

RISK CONTROL After risks are identified and assessed, they must be controlled. Control measures might include the following examples: *We could place a firewall on the proxy server; We could assign permissions to sensitive files; We could install biometric devices to guard the computer room.* Typically, management chooses one of four risk control strategies: avoidance, mitigation, transference, or acceptance. **Avoidance** eliminates the risk by adding protective safeguards. For example, to prevent unauthorized access to LAN computers, a secure firewall might be installed. **Mitigation** reduces the impact of a risk by careful planning and preparation. For example, a company can prepare a disaster recovery plan in case a natural disaster occurs. **Transference** shifts the risk to another asset or party, such as an insurance company. **Acceptance** means that nothing is done. Companies usually accept a risk only when the protection clearly is not worth the expense.

The risk management process is iterative — risks constantly are identified, assessed, and controlled. To be effective, risk managers need a combination of business knowledge, IT skills, and experience with security tools and techniques.

Attacker Profiles and Attacks

An **attack** is a hostile act that targets the system, or the company itself. Thus, an attack might be launched by a disgruntled employee, or a hacker who is 10,000 miles away. Attackers break into a system to cause damage, steal information, or gain recognition, among other reasons. Attackers can be grouped into categories, as shown in Figure 12-23, while Figure 12-24 describes some common types of attacks.

Attacker Characteristics

ATTACKER	DESCRIPTION	SKILL SET
Cyberterrorist	Attacks to advance political, social, or ideological goals.	High
Employee	Uses unauthorized information or privileges to break into computer systems, steal information, or cause damage.	Varies
Hacker	Uses advanced skills to attack computer systems with malicious intent (black hat) or to expose flaws and improve security (white hat).	High
Hacktivist	Attacks to further a social or political cause; often involves shutting down or defacing Web sites.	Varies
Script kiddie	Inexperienced or juvenile hacker who uses readily available malicious software to disrupt or damage computer systems, and gain recognition.	Low
Spy	Non-employee who breaks into computer systems to steal information and sell it.	High

FIGURE 12-23 IT security professionals have coined labels for various types of attackers.

Types of Attacks and Examples

ATTACK	EXAMPLES
Back door	Attacker finds vulnerability in software package and exploits it.
Denial of service or distributed denial of service	One or more computers send a stream of connection requests to disable a Web server.
Dumpster diving	Attacker scours the trash for valuable information that can be used to compromise the system.
Mail bombing	Enormous volumes of e-mail are sent to a target address.
Malicious code	Attacker sends infected e-mail to the target system. Attackers may use viruses, worms, Trojan horses, keystroke loggers, spyware, or scripts to destroy data, bog down systems, spy on users, or assume control of infected systems.
Man in the middle	The attacker intercepts traffic and poses as the recipient, sending the data to the legitimate recipient but only after reading the traffic or modifying it.
Password cracking	Hacker attempts to discover a password to gain entry into a secured system. This can be a dictionary attack, where numerous words are tried, or a brute force attack, where every combination of characters is attempted.
Privilege escalation	Employee tricks a computer into raising his or her account to the administrator level.
Sniffing	Network traffic is intercepted and scanned for valuable information.
Social engineering	An attacker calls the help desk posing as a legitimate user and requests that his or her password be changed.
Spam	Unwanted, useless e-mail is sent continuously to business e-mail accounts, wasting time and decreasing productivity.
Spoofing	IP address is forged to match a trusted host, and similar content may be displayed to simulate the real site for unlawful purposes.

FIGURE 12-24 Attacks can take many forms, as this table shows. IT security managers must be able to detect these attacks and respond with suitable countermeasures.

The following sections discuss how companies combat security threats and challenges by using a multilevel strategy.

SECURITY LEVELS

To provide system security, you must consider six separate but interrelated levels: physical security, network security, application security, file security, user security, and procedural security. The following sections describe these security levels, which are shown in Figure 12-25, and the issues that must be addressed. Top management often makes the final strategic and budget decisions regarding security, but systems analysts should understand the overall picture in order to make informed recommendations.

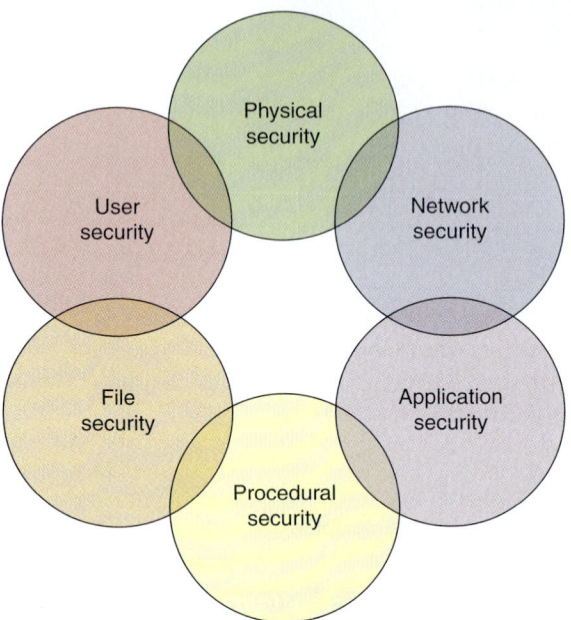

FIGURE 12-25 Each of the six related security levels has a specific focus.

Physical Security

The first level of system security concerns the physical environment, including IT resources and people throughout the company. Special attention must be paid to critical equipment located in operations centers, where servers, network hardware, and related equipment operate. Large companies usually have a dedicated room built specifically for IT operations. Smaller firms might use an office or storage area. Regardless of its size and shape, an operations center requires special protection from unwanted intrusion. In addition to centrally located equipment, all computers on the network must be secure, because each server or workstation can be a potential access point. Physical access to a computer represents an entry point into the system and must be controlled and protected.

OPERATIONS CENTER SECURITY Perimeter security is essential in any room or area where computer equipment is operated or maintained. Physical access must be controlled tightly, and each entrance must be equipped with a suitable security device. All access doors should have internal hinges and electromagnetic locks that are equipped with a battery backup system to provide standby power in the event of a power outage. When the battery power is exhausted, the doors should fail in a closed position, but it should be possible for someone locked inside the room to open the door with an emergency release.

To enhance security, many companies are installing **biometric scanning systems**, which map an individual's facial features, fingerprints, handprint, or eye characteristics, as shown in Figure 12-26. These hi-tech authentication systems replace magnetic identification badges, which can be lost, stolen, or altered.

Video cameras and motion sensors can be used to monitor computer room security and provide documentation of all physical activity in the area. A motion sensor uses infrared technology to detect movement, and can be configured to provide audible or silent alarms, and to send e-mail messages when it is triggered. Other types of sensors can monitor temperature and humidity in the computer room. Motion sensor alarms can be activated at times when there is no expected activity in the computer room, and authorized technicians should have codes to enable or disable the alarms.

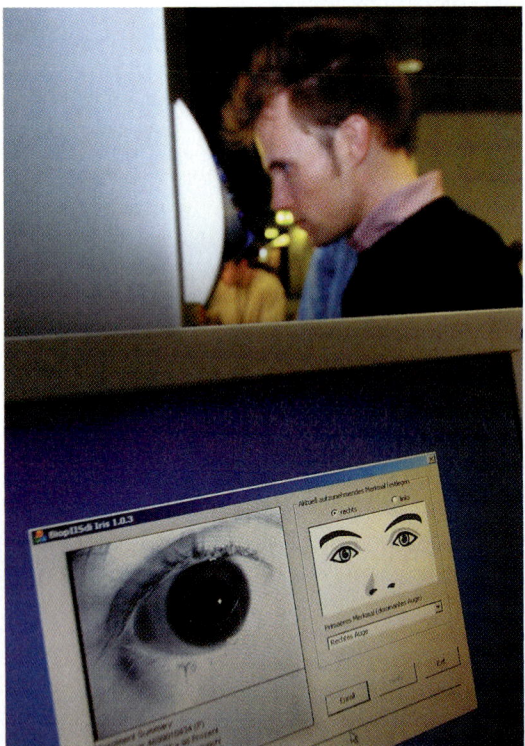

FIGURE 12-26 Companies use biometric scanning to analyze the features of the eye's iris, which has more than 200 points that can be measured and used for comparison.

SERVERS AND DESKTOP COMPUTERS If possible, server and desktop computer cases should be equipped with locks. This simple, but important, precaution might prevent an intruder from modifying the hardware configuration of a server, damaging the equipment, or removing a disk drive. Server racks should be locked, to avoid the unauthorized placement and retrieval of keystroke loggers. A **keystroke logger** is a device that can be inserted between a keyboard and a computer. Typically, the device resembles an ordinary cable plug, so it does not call attention to itself. The device can record everything that is typed into the keyboard, including passwords, while the system continues to function normally. Keystroke loggers can be used legitimately to monitor, back up, and restore a system, but if placed by an intruder, a keystroke logger represents a serious security threat.

In addition to hardware devices, keystroke logging software also exists. A keystroke logging program can be disguised as legitimate software and downloaded from the Internet or a company network. The program remains invisible to the user as it records keystrokes and uploads the information to whoever installed the program. Such malicious software can be removed by antivirus and antispyware software, discussed later in the Application Security section.

Tamper-evident cases should be used where possible. A tamper-evident case is designed to show any attempt to open or unlock the case. In the event that a computer case has been opened, an indicator LED remains lit until it is cleared with a password. Tamper-evident cases do not prevent intrusion, but a security breach is more likely to be noticed. Many servers now are offered with tamper-evident cases as part of their standard configuration.

Monitor screen savers that hide the screen and require special passwords to clear should be used on any server or workstation that is left unattended. Also, use a **BIOS-level password**, also called a **boot-level password** or a **power-on password**, that must be entered before the computer can be started. Also, a boot-level password prevents an unauthorized person from booting a computer by using a CD-ROM or USB device.

Finally, companies must consider electric power issues. In mission-critical systems, large-scale backup power sources are essential to continue business operations. In other cases, computer systems and network devices should be plugged into an **uninterruptible power supply** (UPS) that includes battery backup with suitable capacity. The UPS should be able to handle short-term operations in order to permit an orderly backup and system shutdown.

NOTEBOOK COMPUTERS When assessing physical security issues, be sure to consider additional security provisions for notebook, laptop, and tablet computers. Because of their small size and high value, these computers are tempting targets for thieves and industrial spies. Although the following suggestions are intended as a checklist for notebook computer security, many of them also apply to desktop workstations.

- Select an operating system, such as Windows Vista, that allows secure logons, BIOS-level passwords, and strong firewall protection. Also, log on and work with a user account that has limited privileges rather than an administrator account, and mask the administrator account by giving it a different name that would be hard for a casual intruder to guess.

- Mark or engrave the computer's case with the company name and address, or attach a tamper-proof asset ID tag. These measures might not discourage a professional thief, but might deter a casual thief, or at least make your computer relatively less desirable because it would be more difficult to use or resell. Security experts also recommend that you use a generic carrying case,

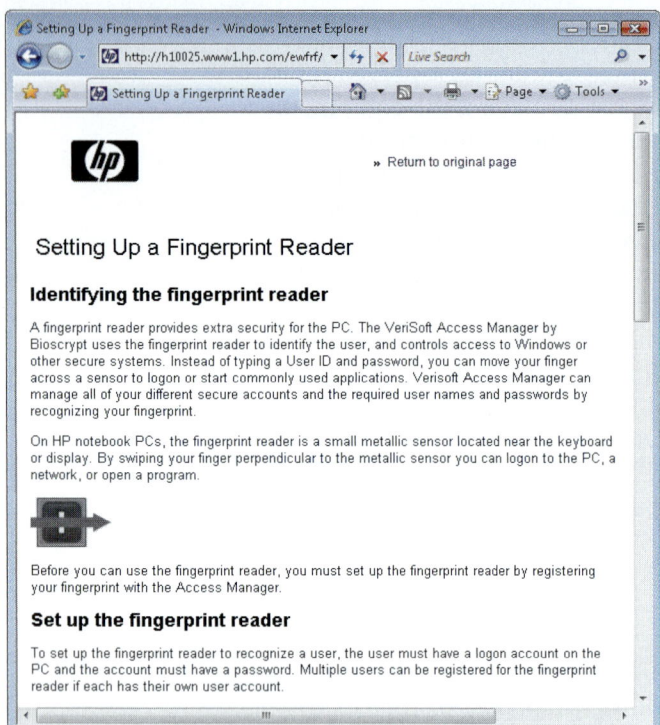

FIGURE 12-27 Some notebook computers feature a fingerprint reader, which is a small metallic sensor located near the keyboard or display.

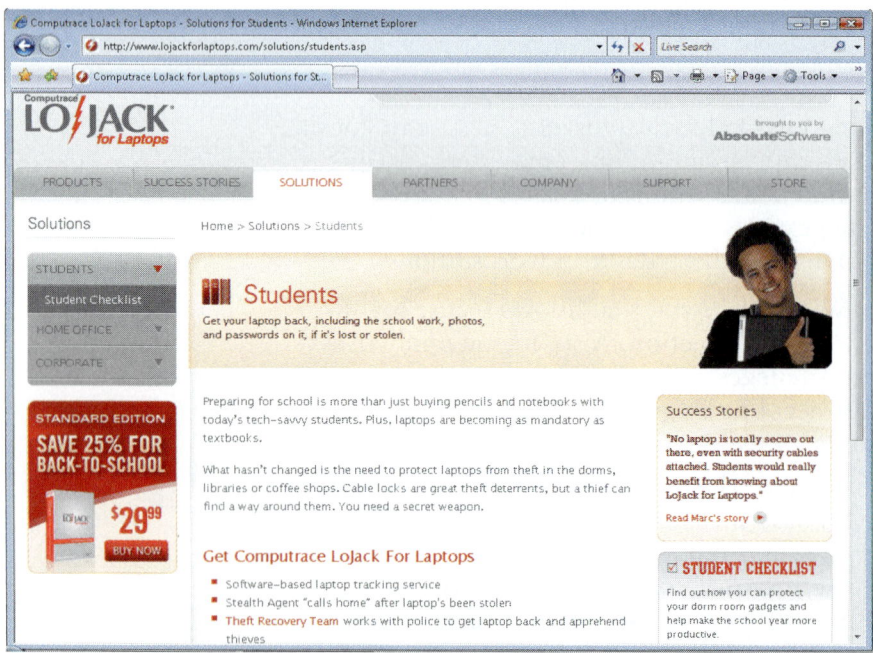

FIGURE 12-28 Many students use LoJack on their notebook computers. The product offers automated call-in identification and remote data erase capability.

such as an attaché case, rather than a custom carrying case that calls attention to itself and its contents. Also be sure to complete and submit all manufacturer registration cards.

- Consider notebook models that have a built-in fingerprint reader, as shown in Figure 12-27.

- Many notebook computers have a **Universal Security Slot (USS)** that can be fastened to a cable lock or laptop alarm. Again, while these precautions might not deter professional thieves, they might discourage and deter casual thieves.

- Back up all vital data before using the notebook computer outside the office. Also, instead of using your computer's hard drive, save and transport highly sensitive data on removable media, such as a flash memory device.

- Use tracking software that directs your laptop periodically to contact a security tracking center. If your notebook is stolen, the call-in identifies the computer and its physical location. Armed with this information, the security tracking center can alert law enforcement agencies and communications providers. As shown in Figure 12-28, Computrace sells a product called LoJack for Laptops, which offers call-in service, as well as a remote data erase capability. Some versions of the product even provide a payment of up to $1,000 if the firm does not recover your stolen laptop.

- While traveling, try to be alert to potential high-risk situations, where a thief, or thieves, might attempt to distract your attention and snatch your computer. These situations often occur in crowded, noisy places like airport baggage claim areas, rental car counters, and security checkpoints. Also, when traveling by car, store your computer in a trunk or lockable compartment where it will not be visible.

- Establish stringent password protection policies that require minimum length and complexity, and set a limit on how many times an invalid password can be entered before the system locks itself down. In some situations, you might want to establish file encryption policies to protect extremely sensitive files.

CASE IN POINT 12.3: OUTER BANKS COUNTY

Outer Banks County is a 200-square-mile area in coastal North Carolina, and you are the IT manager. The county has about a hundred office employees who perform clerical tasks in various departments. A recent budget crisis has resulted in a wage and hiring freeze, and morale has declined. The county manager has asked you to install some type of keystroke logger to monitor employees and determine whether they are fully productive. After your conversation, you wonder whether there might be some potential privacy and security issues involved.

For example, does an employer have a duty to notify its employees that it is monitoring them? Should the employer notify them even if not required to do so? From a human resources viewpoint, what would be the best way to approach this issue? Also, does a potential security issue exist? If an unauthorized person gained possession of the keystroke log, he or she might be able to uncover passwords and other sensitive data.

What are your conclusions? Are these issues important, and how would you respond to the county manager's recommendation? Before you answer, you should go on the Internet and learn more about keystroke loggers generally, and specific products that currently are available.

Network Security

A **network** is defined as two or more devices that are connected for the purpose of sending, receiving, and sharing data, which is called network traffic. In order to connect to a network, a computer must have a **network interface**, which is a combination of hardware and software that allows the computer to interact with the network. To provide security for network traffic, data can be **encrypted**, which refers to a process of encoding the data so it cannot be accessed without authorization.

ENCRYPTING NETWORK TRAFFIC Network traffic can be intercepted and possibly altered, redirected, or recorded. For example, if an **unencrypted**, or **plain text**, password or credit card number is transmitted over a network connection, it can be stolen. When the traffic is encrypted, it still is visible, but its content and purpose are masked.

Figure 12-29 on the next page shows an example of encrypted traffic compared to plain text traffic. In the upper screen, the user has logged on to the SCR Associates case study, using a password of *sad8e*. Notice that anyone who gains access to this data easily could learn the user's password. In the lower screen, the user has logged on to an online bank account and used a password, but the encryption process has made it impossible to decipher the keystrokes.

Two commonly used encryption techniques are private key encryption and public key encryption. **Private key encryption** is symmetric, because a single key is used to encrypt and decrypt information. While this method is simple and fast, it poses a fundamental problem. To use symmetric encryption, both the sender and receiver must possess the same key beforehand, or it must be sent along with the message, which increases the risk of interception and disclosure.

In contrast, **public key encryption (PKE)** is asymmetric, because each user has a pair of keys: a public key and a private key, as shown in Figure 12-30 on the next page. Public keys are used to encrypt messages. Users can share their public keys freely, while keeping their private keys tightly guarded. Any message encrypted with a user's public key can only be decrypted with that user's private key. This method is commonly used in secure online shopping systems.

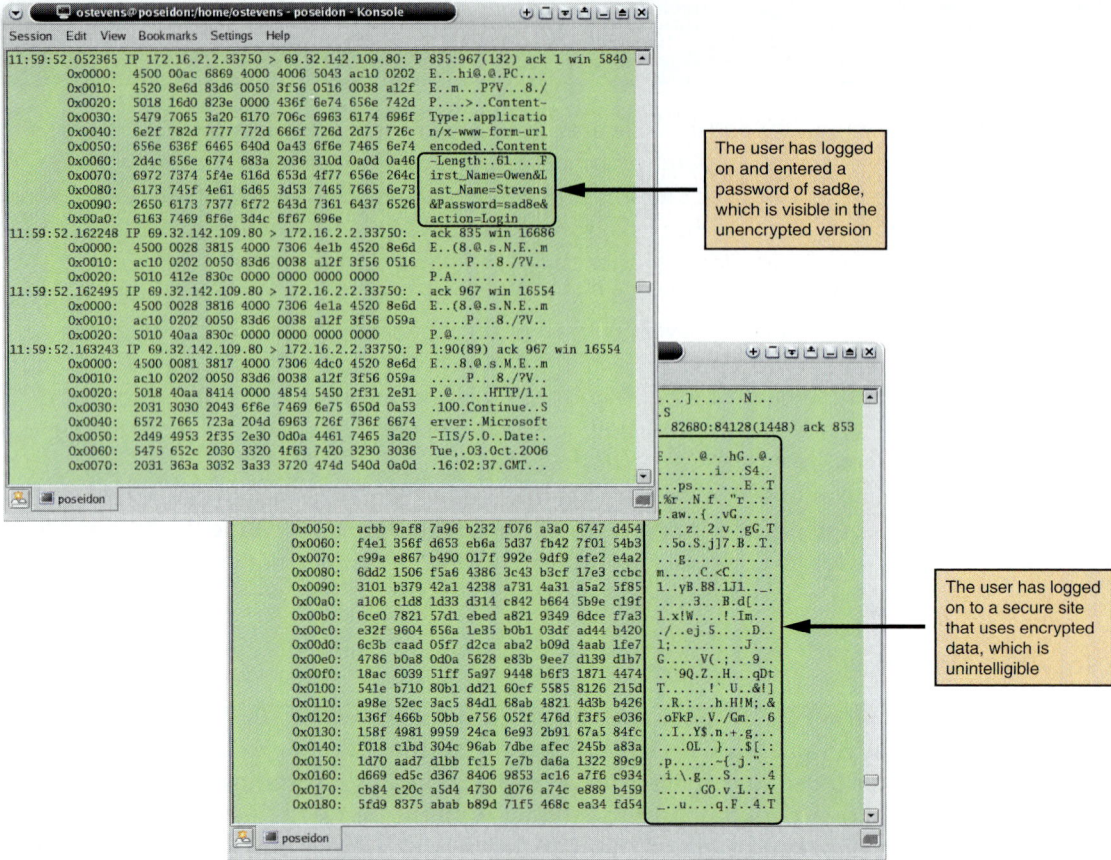

FIGURE 12-29 The upper screen shows an example of unencrypted text, which contains a visible password. In the lower screen, the encrypted text cannot be read.

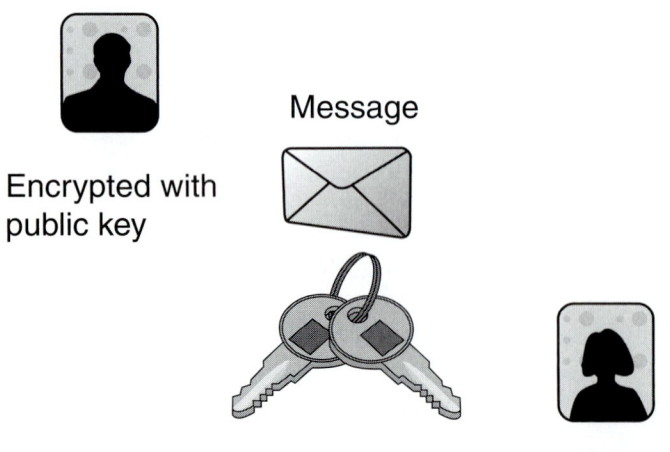

FIGURE 12-30 In a PKE environment, a message encrypted with a public key only can be decrypted with the matching private key.

A recent Wikipedia article uses an interesting analogy for public key encryption. The article suggests that PKE is similar to a locked mailbox with a mail slot that is accessible to the public. The mailbox's location (street address) represents the public key. Anyone knowing the street address can drop a message through the slot. However, only a person with a key can open the box and read the message.

WIRELESS NETWORKS As you learned in Chapter 10, wireless network security is a vital concern, because wireless transmission is much more vulnerable than traffic on a wired network. However, if wireless traffic is encrypted, any data that is intercepted by an unintended recipient will be useless to the intruder.

The earliest form of wireless security, called **Wired Equivalent Privacy (WEP)**, required each wireless client to use a special, preshared key. Although this method was used by many home and small office networks, it provided relatively weak protection.

WEP was replaced by **Wi-Fi Protected Access (WPA)**, which offered major security improvements based on protocols created by the Wi-Fi Alliance. The most recent wireless security enhancement, called **WPA2**, further strengthens the level of wireless protection. WPA2 is an extension of WPA based on a full implementation of the **IEEE 802.11i** standard. According to the WiFi Alliance, the WPA2 standard became mandatory for all new devices seeking Wi-Fi certification after March 2006. WPA2 is compatible with WPA, so companies easily can migrate to the new security standard.

PRIVATE NETWORKS   It is not always practical to secure all network traffic. Unfortunately, encrypting traffic increases the burden on a network, and can decrease network performance significantly. In situations where network speed is essential, such as a Web server linked to a database server, many firms use a private network to connect the computers. A **private network** is a dedicated connection, similar to a leased telephone line. Each computer on the private network must have a dedicated interface to the network, and no interface on the network should connect to any point outside the network. In this configuration, unencrypted traffic safely can be transmitted because it is not visible, and cannot be intercepted from outside the network.

VIRTUAL PRIVATE NETWORKS   Private networks work well with a limited number of computers, but if a company wants to establish secure connections for a larger group, it can create a virtual private network (VPN). A **virtual private network (VPN)** uses a public network, such as the Internet or a company intranet, to connect remote users securely. Instead of using a dedicated connection, a VPN allows remote clients to use a special key exchange that must be authenticated by the VPN. Once authentication is complete, a secure network connection, called a **tunnel**, is established between the client and the access point of the local intranet. All traffic is encrypted through the VPN tunnel, which provides an additional level of encryption and security.

PORTS AND SERVICES   A **port**, which is identified by a positive integer, is used for routing incoming traffic to the correct application on a computer. In TCP/IP networks, such as the Internet, all traffic received by a computer contains a destination port. Because the destination port determines where the traffic will be routed, the computer sorts the traffic by port number, which is included in the transmitted data. An analogy might be a large apartment building with multiple mailboxes. Each mailbox has the same street address, but a different box number.

A **service** is an application that monitors, or listens on, a particular port. For example, a typical e-mail application listens on port 25. Any traffic received by that port is routed to the e-mail application. Services play an important role in computer security, and they can be affected by port scans and denial-of-service attacks.

- Port scans. **Port scans** attempt to detect the services running on a computer by trying to connect to various ports and recording the ports on which a connection was accepted. For example, the result of an open port 25 would indicate that a mail server is running. Port scans can be used to draw an accurate map of a network, and pinpoint possible weaknesses.

- Denial of service. A **denial-of-service (DOS)** attack occurs when an attacking computer makes repeated requests to a service or services running on certain ports. Because the target computer has to respond to each request, it can become bogged

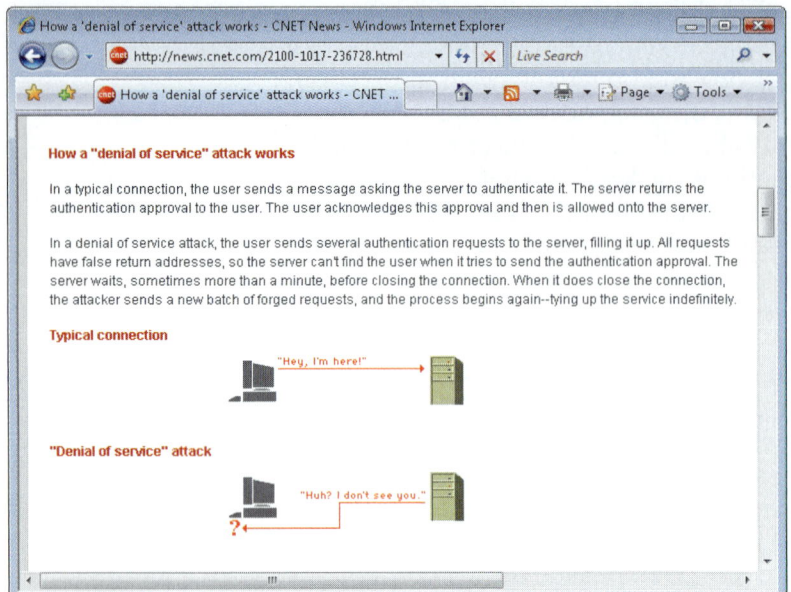

FIGURE 12-31 In a denial-of-service attack, a hacker sends numerous authentication requests with false return addresses. The server tries unsuccessfully to send authentication approval, and eventually is disabled by the flood of requests.

down and fail to respond to legitimate requests. A much more devastating attack based on this method is called a **distributed denial-of-service (DDOS)** attack. This attack involves multiple attacking computers that can synchronize DOS attacks on a server, as shown in Figure 12-31.

FIREWALLS A **firewall** is the main line of defense between a local network, or intranet, and the Internet. A firewall must have at least one network interface with the Internet, and at least one network interface with a local network or intranet. Firewall software examines all network traffic sent to and from each network interface. Preset rules establish certain conditions that determine whether the firewall will allow the traffic to pass. When a matching rule is found, the firewall automatically accepts, rejects, or drops the traffic. When a firewall rejects traffic, it sends a reply indicating that the traffic is not permissible. When a firewall drops traffic, no reply is sent. Firewalls can be configured to detect and respond to denial-of-service attacks, port scans, and other suspicious activity.

Figure 12-32 shows a basic set of firewall rules for a company that has a Web server and a mail server. In this example, the firewall would accept public Web server traffic only on ports 80 and 443, and public mail server traffic only on port 25. The firewall would allow private LAN traffic to any destination and port.

RULE	INTERFACE	SOURCE	DESTINATION	PORT	ACTION
1	Public	Any	Web Server	80	Accept
2	Public	Any	Web Server	443	Accept
3	Public	Any	Web Server	Any	Reject
4	Public	Any	Mail Server	25	Accept
5	Public	Any	Mail Server	Any	Reject
6	Public	Any	Any	Any	Drop
7	Private	LAN	Any	Any	Accept

FIGURE 12-32 Examples of rules that determine whether the firewall will allow traffic to pass.

NETWORK INTRUSION DETECTION Suppose an intruder attempts to gain access to the system. Obviously, an intrusion alarm should be sounded when certain activity or known attack patterns are detected. A **network intrusion detection system (NIDS)** is like a burglar alarm that goes off when it detects a configuration violation. The NIDS also can alert the administrator when it detects suspicious network traffic patterns. A NIDS requires fine-tuning to detect the difference between legitimate network traffic and an attack. It is also important that a NIDS be placed on a switch or other network device that can monitor all network traffic. Although a NIDS requires some administrative overhead, it can be very helpful in documenting the efforts of attackers and analyzing network performance.

Application Security

In addition to securing the computer room and shielding network traffic, it is necessary to protect all server-based applications. To do so, you must analyze the application's functions, identify possible security concerns, and carefully study all available documentation. Application security requires an understanding of services, hardening, application permissions, input validation techniques, software patches and updates, and software logs.

SERVICES   In the network security section, you learned that a service is an application that monitors, or listens, on a particular port. You can determine which services are running by using a port scan utility. If a particular application is not needed, it should be disabled. This will improve system security, performance, and reliability. An unnecessary or improperly configured service could create a vulnerability called a **security hole**. For example, if a loosely configured FTP (File Transfer Protocol) service is available to a hacker, he or she might be able to upload destructive code to the server.

HARDENING   The **hardening** process makes a system more secure by removing unnecessary accounts, services, and features. Hardening is necessary because the default configuration of some software packages might create a vulnerability. For example, initial software settings might include relatively weak account permissions or file sharing controls. Hardening can be done manually or by using a configuration template, which speeds up the process in a large organization.

Hardening also includes additional protection such as antivirus and antispyware software. These programs can detect and remove **malware**, which is hostile software designed to infiltrate, damage, or deny service to a computer system. Malware includes worms, Trojan horses, keystroke loggers, and spyware, among others.

APPLICATION PERMISSIONS   Typically, an application is configured to be run only by users who have specific rights. For example, an **administrator**, or **superuser** account, allows essentially unrestricted access. Other users might be allowed to enter data, but not to modify or delete existing data. To prevent unauthorized or destructive changes, the application should be configured so that nonprivileged users can access the program, but cannot make changes to built-in functions or configurations. **User rights**, also called **permissions**, are discussed in more detail in the file security section.

INPUT VALIDATION   As you learned in Chapter 8, when designing the user interface, input validation can safeguard data integrity and security. For example, if an application requires a number from *1* to *10*, what happens if an alphabetic character or the number *31* is entered? If the application is designed properly, it will respond with an appropriate error message. Chapter 8 also explained data entry and validation checks, which are important techniques that can improve data integrity and quality. Failure to validate input data can result in output errors, increased maintenance expense, and erratic system behavior.

PATCHES AND UPDATES   In an operational system, security holes or vulnerabilities might be discovered at any time. **Patches** are software modules that can repair these holes, reduce vulnerability, and update the system. Like any other new software, patches must be tested carefully. Before applying a patch, an effort should be made to determine the risks of *not* applying the patch, and the possibility that the patch might affect other areas of the system.

Many firms purchase software packages called **third-party software**. Patches released by third-party software vendors usually are safe, but any patch must be reviewed carefully before it is applied. Because researching and applying patches is time consuming and expensive, many software vendors offer an **automatic update service** that enables an application to contact the vendor's server and check for a needed patch or update. Depending on the configuration, available patches can be downloaded and installed without human intervention, or might require approval by IT managers. Although it is convenient, automatic updating carries substantial risks, and should be used only if changes can readily be undone if unexpected results or problems develop.

SOFTWARE LOGS   Operating systems and applications typically maintain a **log** that documents all events, including dates, times, and other specific information. Logs can be important in understanding past attacks and preventing future intrusions. For example, a pattern of login errors might reveal the details of an intrusion attempt. A log also can include system error messages, login histories, file manipulation, and other information that could help track down unauthorized use. Software logs should be monitored constantly to determine if misuse or wrongdoing has occurred. As explained in the network security section, a network intrusion detection system (NIDS) can alert a system administrator whenever suspicious events occur.

File Security

Computer configuration settings, users' personal information, and other sensitive data are stored in files. The safety and protection of these files is a vital element in any computer security program, and a systems analyst needs to consider the importance of permissions, which can be assigned to individual users or to user groups.

PERMISSIONS   File security is based on establishing a set of permissions, which describe the rights a user has to a particular file or directory on a server. The most common permissions are read, write, and execute. Typical examples of permissions include the following:

- Read a file — The user can read the contents of the file.
- Write a file — The user can change the contents of the file.
- Execute a file — The user can run the file, if it is a program.
- Read a directory — The user can list the contents of the directory.
- Write a directory — The user can add and remove files in the directory.

When assigning file permissions, a system administrator should ensure that each user has only the minimum permissions necessary to perform his or her work — not more. In some firms, the system administrator has broad discretion in assigning these levels; in other companies, an appropriate level of management approval is required for any permissions above a standard user level. In any case, a well-documented and enforced permissions policy is necessary to promote file security and reduce system vulnerability.

USER GROUPS   Individual users who need to collaborate and share files often request a higher level of permissions that would enable any of them to change file content. A better approach, from a system administrator's viewpoint, might be to create a user group, add specific users, and assign file permissions to the group, rather than to the individuals. Many firms use this approach, because it allows a user's rights to be determined by his or her work responsibilities, rather than by job title or rank. If a person is transferred, he or she leaves certain groups and joins others that reflect current job duties.

User Security

User security involves the identification of system users and consideration of user-related security issues. Regardless of other security precautions and features, security ultimately depends on system users and their habits, practices, and willingness to support security goals. Unfortunately, many system break-ins begin with a user account that is compromised in some way. Typically, an intruder accesses the system using the compromised account, and may attempt a **privilege escalation attack**, which is an unauthorized attempt to increase permission levels.

User security requires identity management, comprehensive password protection, defenses against social engineering, an effective means of overcoming user resistance, and consideration of new technologies. These topics are discussed in the following sections.

IDENTITY MANAGEMENT   **Identity management** refers to controls and procedures necessary to identify legitimate users and system components. An identity management strategy must balance technology, security, privacy, cost, and user productivity. Identity management is an evolving technology that is being pursued intensively by corporations, IT associations, and governments.

The Burton Group, a leading IT security consultant, has described identity management as a "set of electronic records that represent ... people, machines, devices, applications, and services." This definition suggests that not just users, but each component in a system, must have a verifiable identity that is based on unique characteristics. For example, user authentication might be based on a combination of a password, a Social Security number, an employee number, a job title, and a physical location.

Because of the devastating consequences of intrusion, IT managers are giving top priority to identity management strategies and solutions.

PASSWORD PROTECTION   As the section on physical security points out, a secure system must have a password policy that requires minimum length, complexity, and a limit on invalid login attempts. Although passwords are a key element in any security program, users often choose passwords that are easy to recall, and they sometimes resent having to remember complex passwords. Even so, IT managers should insist on passwords that have a minimum length, require a combination of case-sensitive letters and numbers, and must be changed periodically. Unfortunately, any password can be compromised if a user writes it down and stores it in an easily accessible location such as a desk, a bulletin board, or under the keyboard.

During the U.S. election campaign in 2008, a hacker made global headlines by gaining access to the e-mail account of a vice-presidential candidate. The intruder signed on as the candidate, requested a new password, guessed the answers to the security questions, and was able to enter the account. These actions were totally illegal, and constituted a serious felony under federal law.

SOCIAL ENGINEERING   Even if users are protecting and securing their passwords, an intruder might attempt to gain unauthorized access to a system using a tactic called **social engineering**. In a social engineering attack, an intruder uses social interaction to gain access to a computer system. For example, the intruder might pretend to be a new employee, an outside technician, or a journalist. Through a series of questions, the intruder tries to obtain the information that he or she needs to compromise the system. A common ploy is for the attacker to contact several people in the same organization, and use some information from one source to gain credibility and entry to another source.

An intruder also might contact a help desk and say: "Hi. This is Mark Tompkins from accounting. I seem to have forgotten my password. Could you give me a new one?" Although this request might be legitimate, it also might be an attacker trying to access the system. A password never should be given based solely on this telephone call. The user should be required to provide further information to validate his or her identity, such as a Social Security number, employee ID, telephone extension, and company e-mail address.

ON THE WEB

For more information about identity management, visit **scsite.com/ 8e/more**, locate Chapter 12, and then click the Identity Management link.

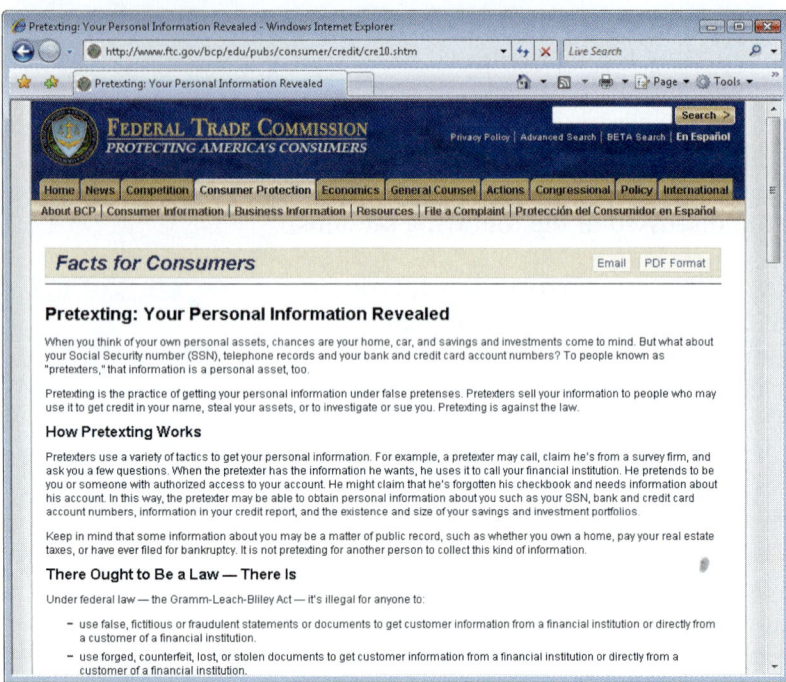

FIGURE 12-33 The Federal Trade Commission provides information about pretexting and how to guard against it.

One highly publicized form of social engineering is called **pretexting**, which is a method of obtaining personal information under false pretenses. Pretexting, which is described in the Federal Trade Commission statement shown in Figure 12-33, is a very real threat. The best way to combat social engineering attacks is with employee education, more training, and a high level of awareness during day-to-day operations.

USER RESISTANCE Many users, including some senior managers, dislike tight security measures because they can be inconvenient and time consuming. Systems analysts should remind users that the company owes the best possible security to its customers, who have entrusted personal information to the firm; to its employees, who also have personal information stored in company files; and to its shareholders, who expect the company to have a suitable, effective, and comprehensive security program that will safeguard company assets and resources. When users understand this overall commitment to security and feel that they are part of it, they are more likely to choose better passwords, be more alert to security issues, and contribute to the overall success of the company's security program.

NEW TECHNOLOGIES In addition to traditional measures and biometric devices, technology can enhance security and prevent unauthorized access. For example, a **security token** is a physical device that authenticates a legitimate user. The Wikipedia screen in Figure 12-34 shows several types of security tokens. Some firms provide employees with security tokens that generate a numeric validation code, which the employee enters in addition to his or her normal password.

FIGURE 12-34 Security tokens, which come is various forms, can provide an additional level of security.

Unfortunately, new technology sometimes creates new risks. For example, Google offers a desktop-based search engine, Google Desktop, with a powerful indexing feature that scans all the files, documents, e-mails, chats, and stored Web pages on a user's computer. Although the program provides a convenient way for users to locate and retrieve their data, it also can make it easier for an intruder to obtain private information, especially in a multiuser environment, because the program can recall and display almost anything stored on the computer. Also, if an intruder uses the term *password* in a search, the program might be able to find password reminders that are stored anywhere on the computer. According to Google, the search index resides on the user's computer and is never sent or made accessible to Google or anyone else without explicit consent. However, to maintain privacy for multiuser computers, Google strongly recommends that each user have a separate account, with individual usernames and passwords.

Google Desktop also offers a way for users to search across multiple computers, and this option has caused some concern among IT managers and privacy advocates. To perform a multi-computer search, it is necessary to store a user's data temporarily on Google's servers. Some observers feel that this makes the data more vulnerable and possibly subject to examination by third parties, including government agencies. Google states that if you choose to enable the *Search Across Computers* feature, your shared data is encrypted and treated as personal information, in accordance with the Google Privacy Policy. Software such as Google Desktop is powerful, convenient, and fun to use. However, you should understand the possible risks involved before installing this type of software on personal or business workstations.

Procedural Security

Procedural security, also called **operational security**, is concerned with managerial policies and controls that ensure secure operations. In fact, many IT professionals believe that security depends more on managerial issues than technology. Management must work to establish a corporate culture that stresses the importance of security to the firm and its people. Procedural security defines how particular tasks are to be performed, from large-scale data backups to everyday tasks such as storing e-mails or forms. Other procedures might spell out how to update firewall software or how security personnel should treat suspected attackers.

All employees should understand that they have a personal responsibility for security. For example, an employee handbook might require that users log out of their system accounts, clear their desks, and secure all documents before leaving for the day. These policies reduce the risk of **dumpster diving** attacks, as shown in Figure 12-35, in which an intruder raids desks or trash bins for valuable information. In addition, **paper shredders** should be used to destroy sensitive documents.

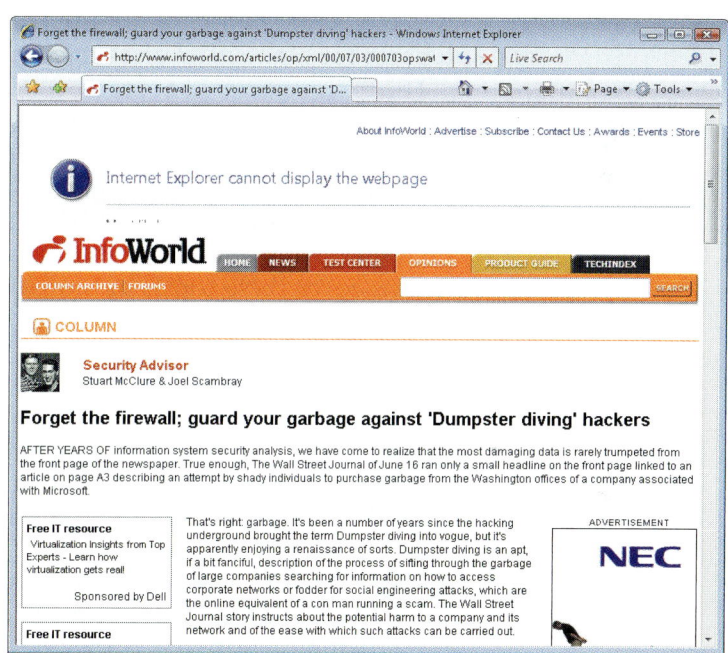

FIGURE 12-35 Dumpster diving might be a low-tech intrusion, but it represents a serious security threat.

CASE IN POINT 12.4: CHAIN LINK CONSULTING, INC.

Chain Link Consulting is an IT consulting firm that specializes in system security issues. The company's president has asked you to help her put together a presentation to a group of potential clients at a trade show meeting next month. First, she wants you to review system security issues, considering all six security levels. Then she wants you to come up with a list of ways that Chain Link could test a client's security practices, in order to get a real-world assessment of vulnerability.

To make matters more interesting, she told you it was OK to be creative in your recommendations, but not to propose any action that would be illegal or unethical. For example, it would be OK to pose as a job applicant with false references to see if they were being checked, but it would not be appropriate to pick a lock and enter the computer room.

Your report is due tomorrow. What will you suggest?

Procedural security also includes safeguarding certain procedures that would be valuable to an attacker. The most common approach is a *need-to-know* concept, where access is limited to employees who need the information to perform security-related tasks. Many firms also apply a set of classification levels for access to company documents. For example, highly sensitive technical documents might be available only to the IT support team, while user-related materials would be available to most company employees. If classification levels are used, they should be identified clearly and enforced consistently.

Procedural security must be supported by upper management and fully explained to all employees. The organization must provide training to explain the procedures and issue reminders from time to time that will make security issues a priority.

ON THE WEB

For more information about backup and disaster recovery, visit **scsite.com/ 8e/more**, locate Chapter 12, and then click the Backup and Disaster Recovery link.

BACKUP AND RECOVERY

Every system must provide for data backup and recovery. **Backup** refers to copying data at prescribed intervals, or continuously. **Recovery** involves restoring the data and restarting the system after an interruption. An overall backup and recovery plan that prepares for a potential disaster is called a **disaster recovery plan**.

The tragic events of September 11, 2001, and increased concern about global terrorism have led many companies to upgrade their backup and disaster recovery plans. Heightened focus on disaster recovery has spawned a whole new industry, which includes new tools and strategies. Many IT professionals feel that terrorism concerns have raised security awareness throughout the corporate world. Although they are separate topics, backup and disaster recovery issues usually are intertwined. The following sections cover these topics in more detail.

Backup Policies

The cornerstone of business data protection is a **backup policy**, which contains detailed instructions and procedures. An effective backup policy can help a firm continue business operations and survive a catastrophe. The backup policy should specify backup media, backup types, and retention periods.

BACKUP MEDIA Backup media can include tape, hard drives, optical storage, and online storage. Physical backups must be carefully identified and stored in a secure location. **Offsiting** refers to the practice of storing backup media away from the main business location, in order to mitigate the risk of a catastrophic disaster such as a flood, fire, or earthquake. Even if the operating system includes a backup utility, many system administrators prefer to use specialized third-party software that offers more options and better controls for large-scale operations.

In addition to on-site data storage, many companies use Web-based data backup and retrieval services offered by vendors such as Rocky Mountain Software and IBM. For a small- or medium-sized firm, this option can be cost effective and reliable.

BACKUP TYPES Backups can be full, differential, incremental, or continuous. A **full backup** is a complete backup of every file on the system. Frequent full backups are time consuming and redundant if most files are unchanged since the last full backup. Instead of performing a full backup, another option is to perform a **differential backup**, which is faster because it backs up *only* the files that are new or changed since the last full backup. To restore the data to its original state, you restore the last full backup, and then restore the last differential backup. Many IT managers believe that a combination of full and differential backups is the best option, because it uses the least amount of storage space and is simple.

The fastest method, called an **incremental backup**, only includes recent files that never have been backed up by any method. This approach, however, requires multiple steps to restore the data — one for each incremental backup.

Most large systems use **continuous backup,** which is a real-time streaming method that records all system activity as it occurs. This method requires expensive hardware, software, and substantial network capacity. However, system restoration is rapid and effective because data is being captured in real time, as it occurs. Continuous backup often uses a **RAID (redundant array of independent disks)** system that mirrors the data. RAID systems are called **fault-tolerant,** because a failure of any one disk does not disable the system. Compared to one large drive, a RAID design offers better performance, greater capacity, and improved reliability. When installed on a server, a RAID array of multiple drives appears to the computer as a single logical drive. Figure 12-36 shows a comparison of various backup methods.

Comparison of Backup Methods

BACKUP TYPE	CHARACTERISTICS	PROS AND CONS	TYPICAL FREQUENCY
Full	Backs up all files.	Slowest backup time and requires the most storage space. Rapid recovery because all files are restored in a single step.	Monthly or weekly.
Differential	Only backs up files that are new or changed since the last full backup.	Faster than a full backup and requires less storage space. All data can be restored in just two steps by using the last full backup and the last differential backup.	Weekly or daily.
Incremental	Only backs up files that are new or changed since the last backup of any kind.	Fastest backup and requires the least storage space because it only saves files that have never been backed up. However, requires many restore steps — one for each incremental backup.	Daily or more often.
Continuous	Real-time, streaming method that records all system activity.	Very expensive hardware, software, and network capacity. Recovery is very fast because system can be restored to just before an interruption.	Usually only used by large firms and network-based systems.

FIGURE 12-36   Comparison of full, differential, incremental, and continuous backup methods.

RETENTION PERIODS   Backups are stored for a specific **retention period** after which they are either destroyed or the backup media is reused. Retention periods can be a specific number of months or years, depending on legal requirements and company policy. Stored media must be secured, protected, and inventoried periodically.

Business Continuity Issues

Global concern about terrorism has raised awareness levels and increased top management support for a business continuity strategy in the event of an emergency. A disaster recovery plan describes actions to be taken, specifies key individuals and rescue authorities to be notified, and spells out the role of employees in evacuation, mitigation, and recovery efforts. The disaster recovery plan should be accompanied by a **test plan**, which can simulate various levels of emergencies and record the responses, which can be analyzed and improved as necessary.

After personnel are safe, damage to company assets should be mitigated. The plan might require shutting down systems to prevent further data loss, or moving physical assets to a secure location. Afterward, the plan should focus on resuming business operations, including the salvaging or replacement of equipment and the recovery of backup data. The main objective of a disaster recovery plan is to restore business operations to pre-disaster levels.

Disaster recovery plans are often part of a larger **business continuity plan (BCP)**, which goes beyond a recovery plan, and defines how critical business functions can continue in the event of a major disruption. Some BCPs specify the use of a hot site. A **hot site** is an alternate IT location, anywhere in the world, that can support critical systems in the event of a power outage, system crash, or physical catastrophe. A hot site requires **data replication**, which means that any transaction on the primary system must be mirrored on the hot site. If the primary system becomes unavailable, the hot site will have the latest data and can function seamlessly, with no downtime.

Although hot sites are attractive backup solutions, they are very expensive. However, a hot site provides the best insurance against major business interruptions. In addition to hot sites, business insurance can be important in a worst-case scenario. Although expensive, **business insurance** can offset the financial impact of system failure and business interruption.

SYSTEM OBSOLESCENCE

At some point, every system becomes obsolete. For example, you might not remember the punch cards shown in Figure 12-37, but they represented the cutting edge of data management back in the 1960s. Data was stored by punching holes at various positions, and was retrieved by machines that could sense the presence or absence of a punched hole. Most full-size cards stored only 80 characters, or bytes, so more than 12,000 cards would be needed to store a megabyte. It is interesting to note that almost 50 years later, punch card technology still survives in voting systems used in some states, but this method rapidly is being replaced by newer technology.

Constantly changing technology means that every system has a limited economic life span. Analysts and managers can anticipate system obsolescence in several ways and it never should come as a complete surprise.

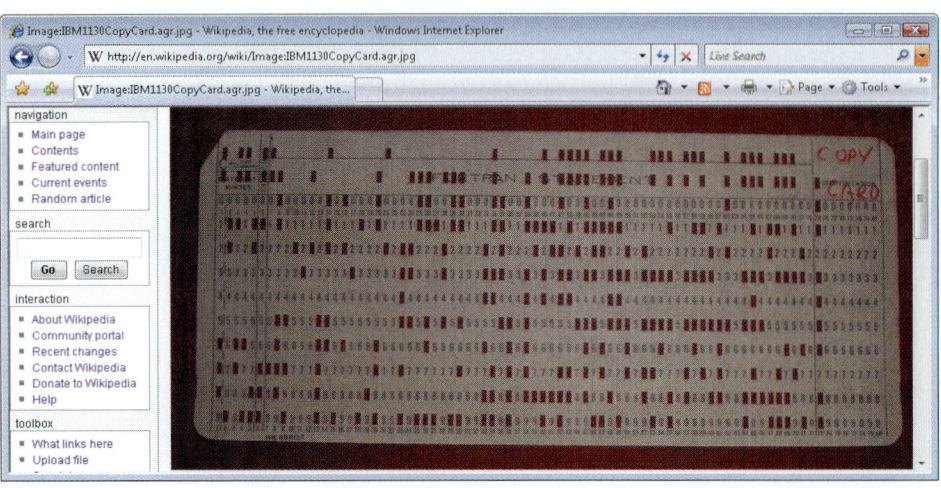

FIGURE 12-37 In the 1960s, punch cards represented the state-of-the-art in data management.

A system becomes **obsolete** when it no longer supports user needs, or when the platform becomes outmoded. The most common reason for discontinuing a system is that it has reached the end of its economically useful life, as indicated by the following signs:

- The system's maintenance history indicates that adaptive and corrective maintenance are increasing steadily.

- Operational costs or execution times are increasing rapidly, and routine perfective maintenance does not reverse or slow the trend.

- A software package is available that provides the same or additional services faster, better, and less expensively than the current system.

- New technology offers a way to perform the same or additional functions more efficiently.

- Maintenance changes or additions are difficult and expensive to perform.

- Users request significant new features to support business requirements.

Systems operation and support continues until a replacement system is installed. Toward the end of a system's operational life, users are unlikely to submit new requests for adaptive maintenance because they are looking forward to the new release. Similarly, the IT staff usually does not perform much perfective or preventive maintenance because the system will not be around long enough to justify the cost. A system in its final stages requires corrective maintenance only to keep the system operational.

User satisfaction typically determines the life span of a system. The critical success factor for any system is whether or not it helps users achieve their operational and business goals. As an IT staff member, you should expect to receive input from users and managers throughout the systems development process. You should investigate and document all negative feedback, because it can be the first signal of system obsolescence.

At some point in a system's operational life, maintenance costs start to increase, users begin to ask for more features and capability, new systems requests are submitted, and the SDLC begins again.

FUTURE CHALLENGES AND OPPORTUNITIES

The only thing that is certain about the future is continuous change. Change itself is neither good nor bad — the real issue is how people and companies deal with the challenges and opportunities that are bound to occur.

No one would start a complex journey without a map and a plan. To navigate the future of information technology, companies require strategic plans, which were discussed in Chapter 2. An individual also needs a plan to reach to a specific goal or destination. This section discusses some predictions and stresses the importance of personal planning and development, including the acquisition of professional credentials.

Predictions

Although no one can foresee the future, it is safe to assume that companies will face intense competition and economic, social, and political uncertainty. Many IT experts believe that in this environment, the highest priorities will be the safety and security of corporate operations, environmental concerns, and bottom-line TCO.

Gartner Inc. is a leading IT consulting firm that is famous for forecasting industry trends. Here is a summary of predictions that Gartner published in 2008:

- By 2009, more than 30% of IT organizations will use environmental criteria when purchasing hardware and software. By 2010, 75% of organizations will use total energy and CO_2 footprint measurement as mandatory hardware buying criteria.

- By 2010, end users will shape 50% of software and hardware acquisitions. Web-based computing has empowered users, and IT organizations will incorporate user preferences into future business strategy.

- By 2011, Apple will double its U.S. and Western Europe unit market share with software that provides ease of use and flexibility, more innovation, and a focus on interoperability across multiple devices.

- By 2011, many firms will purchase 40% of their IT infrastructure as a service. Increased high-speed bandwidth will encourage firms to use hosted sites that produce no reduction in response times. As service-oriented architecture (SOA) becomes common, cloud computing will take off, and applications will not require a specific infrastructure.

- By 2012, 50% of traveling workers will use lighter, smaller Internet-centric devices rather than notebook computers. Also, a new class of portable applications will enable users to re-create their work environment across multiple locations or systems.

- By 2012, 80% of commercial software will include open-source components. Many open-source applications are mature, stable, and well supported, and represent value opportunities for vendors as well as customers.

- By 2012, at least one-third of business software spending will be for Software as a Service (SaaS).

Gartner also predicted that by 2011, large enterprises will require suppliers to certify their green credentials and sourcing policies. One issue might relate to the explosion of data storage and server farms, such as the one shown in Figure 12-38. In his 2008 book, *Planet Google*, author Randall Stross notes that the enormous amount of energy needed to drive cloud computing, including Google's servers, has raised serious environmental concerns. The author also states that "By 2006, data centers already consumed more power in the United States than did television sets." It seems clear that this issue will have to be addressed as technology moves forward.

FIGURE 12-38 The rapid growth of data centers and server farms has increased energy consumption significantly and raised environmental concerns.

Strategic Planning for IT Professionals

An IT professional should think of himself or herself as a business corporation that has certain assets, potential liabilities, and specific goals. Individuals, like companies, must have a strategic plan. The starting point is to formulate an answer to the following career planning question: What do I want to be doing three, five, or ten years from now?

Working backwards from your long-term goals, you can develop intermediate milestones and begin to manage your career just as you would manage an IT project. You can even use the project management tools described in Chapter 3 to construct a Gantt chart or a PERT/CPM chart using months (or years) as time units. Once the plan is developed, you would monitor it regularly to see whether you were still on schedule.

Planning a career is not unlike planting a tree that takes several years to reach a certain height. Once you know the desired height and the annual growth rate, you can determine when you must plant the tree. Similarly, if you want to possess a particular educational credential two years from now, and the credential takes two years to earn, then you need to start on it immediately if you want to adhere to stay on track.

IT Credentials and Certification

In recent years, technical credentials and certification have become extremely important to IT employers and employees. In a broad sense, **credentials** include formal degrees, diplomas, or certificates granted by learning institutions to show that a certain level of education has been achieved successfully. The term **certification** also has a special meaning that relates to specific hardware and software skills that can be measured and verified by examination. For example, a person might have a two- or four-year degree in Information Systems and possess an A+ certification, which attests to the person's computer hardware knowledge and skills.

In addition to Microsoft, as shown in Figure 12-39, many other IT industry leaders offer certification, including Cisco, Novell, Oracle, and Sun Microsystems. To learn more about certification generally, you can visit the Web site maintained by the ICCP (Institute for Certification of Computing Professionals), which is shown in Figure 12-40.

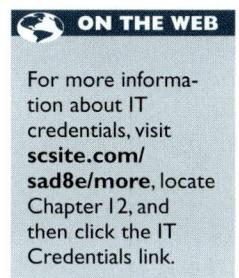

ON THE WEB

For more information about IT credentials, visit **scsite.com/ sad8e/more**, locate Chapter 12, and then click the IT Credentials link.

FIGURE 12-39 To earn the MCP (Microsoft Certified Professional) credential, a person must demonstrate a high level of IT knowledge and skill.

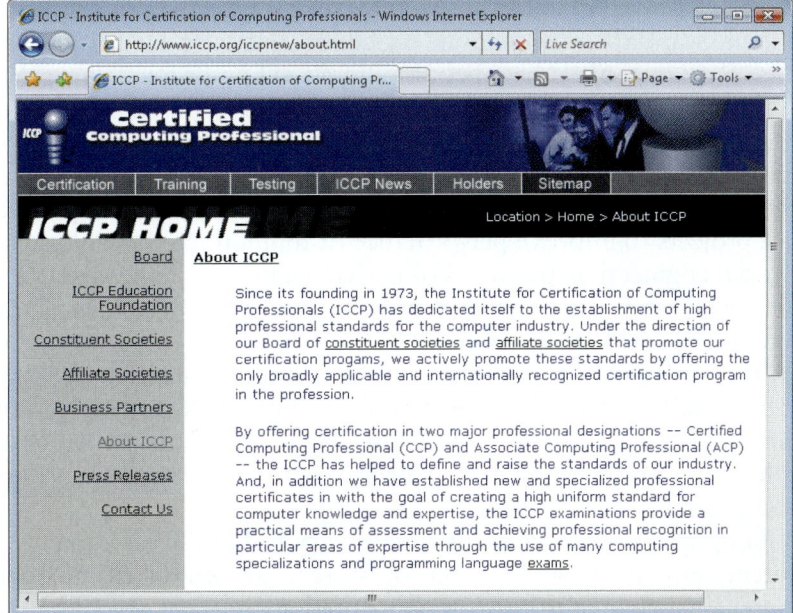

FIGURE 12-40 ICCP (Institute for Certification of Computing Professionals) maintains a Web site that provides information about certification opportunities.

A QUESTION OF ETHICS

Jamie just completed a routine security audit on the company's information systems, and she found several areas of vulnerability. For example, file permissions have not been updated in some time, no comprehensive password policy exists, and network traffic is not fully encrypted. She noted these areas, among others, in a report to Tamika, her supervisor. The report included specific recommendations to fix the problems.

Tamika responded by saying that budgets are tight right now, and she could not approve Jamie's requests to resolve these issues. As an IT professional, Jamie is very uncomfortable with the risk level, but she has been unable to sway Tamika. When Jamie discussed the situation with her friend, Ethan, he said, "Why worry about it? If it's good enough for Tamika, it should be good enough for you."

What do you think of Ethan's advice, and why? Is this an ethical question? If Jamie still is uncomfortable, what are her options?

CHAPTER SUMMARY

Systems support and security covers the period from the implementation of an information system until the system no longer is used. A systems analyst's primary involvement with an operational system is to manage and solve user support requests.

Corrective maintenance includes changes to correct errors. Adaptive maintenance satisfies new systems requirements, and perfective maintenance makes the system more efficient. Adaptive and perfective maintenance changes often are called enhancements. Preventive maintenance is performed to avoid future problems.

The typical maintenance process resembles a miniature version of the systems development life cycle. A systems request for maintenance work is submitted and evaluated. If it is accepted, the request is prioritized and scheduled for the IT group. The maintenance team then follows a logical progression of investigation, analysis, design, development, testing, and implementation.

Corrective maintenance projects occur when a user or an IT staff member reports a problem. Standard maintenance procedures usually are followed for relatively minor errors, but work often begins immediately when users report significant errors.

In contrast to corrective maintenance, adaptive, perfective, and preventive maintenance projects always follow the organization's standard maintenance procedures. Adaptive maintenance projects occur in response to user requests for improvements to meet changes in the business or operating environments. The IT staff usually initiates perfective maintenance projects to improve performance or maintainability. Automated program restructuring and reengineering are forms of perfective maintenance. In order to avoid future problems, IT staff performs preventive maintenance, which involves analysis of areas where trouble is likely to occur.

A maintenance team consists of one or more systems analysts and programmers. Systems analysts need the same talents and abilities for maintenance work as they use when developing a new system. Many IT departments are organized into separate new development and maintenance groups where staff members are rotated from one group to the other.

Configuration management is necessary to handle maintenance requests, to manage different versions of the information system, and to distribute documentation changes. Maintenance changes can be implemented as they are completed or a release methodology can be used in which all noncritical maintenance changes are collected and implemented simultaneously. A release methodology usually is cost effective and advantageous for users because they do not have to work with a constantly changing system. Systems analysts use functional, allocated, and product baselines as formal reference points to measure system characteristics at a specific time.

System performance measurements include response time, bandwidth, throughput, and turnaround time. Capacity management uses those measurements to forecast what is needed to provide future levels of service and support. Also, CASE tools that include system evaluation and maintenance features can be used during the systems operation, security, and support phase.

Security is a vital part of every computer system. System security is dependent upon a comprehensive security policy that defines how organizational assets are to be protected and how attacks are to be responded to.

Risk management creates a workable security policy by identifying, analyzing, anticipating, and reducing risks to an acceptable level. Because information systems face a wide array of threats and attacks, six separate but interrelated security levels should be analyzed: physical security, network security, application security, file security, user security, and procedural security. Physical security concerns the physical environment, including critical equipment located in a computer room, as well as safeguards for servers and desktops throughout the company. Network security involves encryption techniques, as well as private networks and other protective measures, especially where wireless transmissions are concerned. Application security requires an understanding of services, hardening, application permissions, input validation techniques, software patches and updates, and software logs. File security involves the use of permissions, which can be assigned to individual users or to user groups. User security involves identity management techniques, a comprehensive password protection policy, an awareness of social engineering risks, and an effective means of overcoming user resistance. Procedural security involves managerial controls and policies that ensure secure operations.

Data backup and recovery issues include backup media, backup schedules, and retention periods, as well as backup designs such as RAID and Web-based backups.

All information systems eventually become obsolete. The end of a system's economic life usually is signaled by rapidly increasing maintenance or operating costs, the availability of new software or hardware, or new requirements that cannot be achieved easily by the existing system. When a certain point is reached, an information system must be replaced, and the entire systems development life cycle begins again.

Many IT experts predict intense competition in the future, along with economic, political, and social uncertainty. Facing these challenges, top IT priorities will be the safety and security of corporate operations, environmental concerns, and bottom-line TCO.

An IT professional should have a strategic career plan that includes long-term goals and intermediate milestones. An important element of a personal strategic plan is the acquisition of IT credentials and certifications that document specific knowledge and skills. Many IT industry leaders offer certification, and the ICCP (Institute for Certification of Computing Professionals) can provide additional information about certification questions and issues.

Key Terms and Phrases

acceptance 578

adaptive maintenance 561

administrator 587

allocated baseline 570

analysis 564

applications programmer 565

archived 569

asset 577

automatic update service 587

availability 576

avoidance 578

backup 592

backup media 592

backup policy 592

bandwidth 573

baseline 570

benchmark testing 572

biometric scanning systems 580

BIOS-level password 581

boot-level password 581

business continuity plan (BCP) 594

business insurance 594

capacity planning 574

certification 597

CIA triangle 576

confidentiality 576

configuration management (CM) 568

continuous backup 593

corrective maintenance 561

credentials 597

critical risk 578

database programmer 565

data replication 594

denial of service (DOS) 585

differential backup 592

disaster recovery plan 592

distributed denial of service (DDOS) 586

dumpster diving 591

encrypted 583

enhancement 562

exploit 577

fault management 571

fault tolerant 593

firewall 586

full backup 592

functional baseline 570

Gbps (gigabits per second) 573

hardening 587

help desk 558

hot site 594

identity management 589

IEEE 802.11i 585

incremental backup 592

information center (IC) 558

integrity 576

Kbps (kilobits per second) 573

keystroke logger 581

log 588

maintenance activities 561

maintenance expenses 561

maintenance release 568

maintenance release methodology 568

maintenance team 564

malware 587

Mbps (megabits per second) 573

metrics 572

mitigation 578

network 583

network interface 583

network intrusion detection system (NIDS) 586

obsolete 595

offsiting 592

operational costs 560

operational security 591

paper shredder 591

patches 587

perfective maintenance 561

permissions 587

plain text 583

port 585

port scan 585

power-on password 581

pretexting 590

preventive maintenance 561

private key encryption 583

private network 585

privilege escalation attack 589

procedural security 591

product baseline 571

programmer/analyst 565

public key encryption (PKE) 583

RAID (redundant array of independent disks) 593

recovery 592

response time 573

retention period 593

risk 577

risk assessment 577

risk control 577

risk identification 577

risk management 576

security 576

security hole 587

security policy 576

security token 590

service 585

service packs 569

social engineering 589

software reengineering 563

superuser 587

synthesis 564

system administrator 564

systems programmer 565

tamper-evident cases 581

test plan 594

third-party software 587

threat 577

throughput 573

transference 578

tunnel 585

turnaround time 574

unencrypted 583

uninterruptible power supply (UPS) 581

Universal Security Slot (USS) 582

user rights 587

user training package 558

version control 569

virtual private network (VPN) 585

vulnerability 577

what-if analysis 574

Wi-Fi Protected Access (WPA) 585

Wired Equivalent Privacy (WEP) 585

WPA2 585

Learn It Online

Instructions: To complete the Learn It Online exercises, start your browser, click the Address bar, and then enter the Web address **scsite.com/sad8e/learn**. When the Systems Analysis and Design Learn It Online page is displayed, follow the instructions in the exercises below. Each exercise has instructions for saving your results, either for your own records or for submission to your instructor.

1 Chapter Reinforcement

TF, MC, and SA

Below SAD Chapter 12, click one of the Chapter Reinforcement links for Multiple Choice, True/False, or Short Answer. Answer each question and submit to your instructor.

2 Flash Cards

Below SAD Chapter 12, click the Flash Cards link and read the instructions. Type 20 (or a number specified by your instructor) in the Number of playing cards text box, type your name in the Enter your Name text box, and then click the Flip Card button. When the flash card is displayed, read the question and then click the ANSWER box arrow to select an answer. Flip through the Flash Cards. If your score is 15 (75%) correct or greater, click Print on the File menu to print your results. If your score is less than 15 (75%) correct, then redo this exercise by clicking the Replay button.

3 Practice Test

Below SAD Chapter 12, click the Practice Test link. Answer each question, enter your first and last name at the bottom of the page, and then click the Grade Test button. When the graded practice test is displayed on your screen, click Print on the File menu to print a hard copy. Continue to take practice tests until you score 80% or better.

4 Who Wants To Be a Computer Genius?

Below SAD Chapter 12, click the Computer Genius link. Read the instructions, enter your first and last name at the bottom of the page, and then click the Play button. When your score is displayed, click the PRINT RESULTS link to print a hard copy.

5 Wheel of Terms

Below SAD Chapter 12, click the Wheel of Terms link. Read the instructions, and then enter your first and last name and your school name. Click the PLAY button. When your score is displayed on the screen, right-click the score and then click Print on the shortcut menu to print a hard copy.

6 Crossword Puzzle Challenge

Below SAD Chapter 12, click the Crossword Puzzle Challenge link. Read the instructions, and then click the Continue button. Work the crossword puzzle. When you are finished, click the Submit button. When the crossword puzzle is redisplayed, submit it to your instructor.

Case-Sim: SCR Associates

Background

SCR Associates is a consulting firm that offers IT solutions and training. SCR needs an information system to manage operations at its new training center. The new system will be called TIMS (Training Information Management System). As a newly hired systems analyst, you will report to Jesse Baker, systems group manager. You will work on various tasks and practice the skills you learned in this chapter.

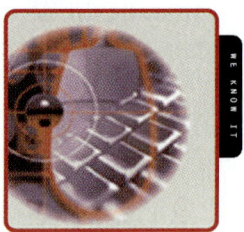

Using the Case

The SCR Associates case study is a Web-based simulation that allows you to practice your skills in a real-world environment. The case study transports you to SCR's company intranet, where you can complete 12 work sessions, each aligning with a chapter. As you work on the case, you will receive e-mail and voice mail messages, obtain information from SCR's online libraries, and perform various tasks. The first time you enter the SCR Case, you should go to the starting page at **scsite.com/sad8e/scr** for detailed instructions.

Preview: Session 12

You assisted your supervisor, Jesse Baker, in various implementation tasks, and the TIMS system is up and running. Now she wants you to focus on system operation, support, and security tasks. Specifically, she wants you to work on a help desk, version control, configuration management, capacity planning, and system security issues. She also wants you to create a checklist that will help SCR know when the TIMS system is reaching the end of its useful life.

To start a work session, you log on to the SCR intranet. When you enter your name and password, an opening screen displays links to the work sessions, and you select Session 12. At this point, you check your e-mail and voice mail messages carefully. Then you begin working on your task list, which includes the following items:

Tasks: Managing Systems Support and Security

1. Jesse wants a recommendation about creating an SCR help desk. She said that I can find lots of information about help desks on the Internet.

2. At our meeting, Jesse asked me how SCR should manage the TIMS system in the future. I need to search the Internet to learn more about version control, configuration management, and capacity planning, and send her the results of my research.

3. Another important issue: Security! Jesse wants my thoughts on how SCR should manage IT security. She wants me to consider all six levels, and prepare an outline for a corporate security policy.

4. Jesse says that no one likes surprises or problems. She wants me to draft a checklist that SCR can use to detect TIMS obsolescence as early as possible. She also said that I might be receiving some interesting news very soon. Wonder what that's about?

FIGURE 12-41 Task list: Session 12.

Chapter Exercises

Review Questions

1. Describe the four classifications of maintenance and provide an example of each type.
2. Why are newly hired systems analysts often assigned to maintenance projects?
3. What is configuration management and why is it important?
4. What is the purpose of capacity planning? How is what-if analysis used in capacity planning?
5. What is a release methodology and what are the pros and cons of this approach? What is the purpose of version control?
6. Define the following terms: response time, bandwidth, throughput, and turnaround time. How are the terms related?
7. What are some key issues that you must address when considering data backup and recovery?
8. Explain the concept of risk management, including risk identification, assessment, and control.
9. What are the six security levels? Name at least three specific issues that apply to each level. Also provide three examples of threat categories, attacker profiles, and types of attacks.
10. List six indications that an information system is approaching obsolescence.

Discussion Topics

1. Assume that your company uses a release methodology for its sales system. The current version is 4.5. Decide whether each of the following changes would justify a version 5.0 release, or be included in a version 4.6 update: (a) Add a new report, (b) add a Web interface, (c) add data validation checks, (d) add an interface to the marketing system, and (e) change the user interface.
2. The four types of IT system maintenance also apply to other industries. Suppose you were in charge of aircraft maintenance for a small airline. What would be an example of each type of maintenance — corrective, adaptive, perfective, and preventive?
3. An IT manager assigns programmers and systems analysts to maintenance projects if they have less than two years of experience or if they received an average or lower rating in their last performance evaluation. Do you agree with this practice?
4. What are the most important security issues facing companies today? Have these changed in the last five years, and will they continue to change? How should companies prepare themselves for security threats and problems in the future?

Projects

1. Using the Internet, locate a software package designed to automate version control. List the key features and describe your findings in a brief memo.
2. Develop a process for managing change requests and design a form to handle a generic change request. The process should include a contingency plan for changes that must be resolved immediately.
3. Visit the IT department at your school or at a local company and find out whether performance measurements are used. Write a brief report describing your findings.
4. Explain how to use the Goal Seek feature in Microsoft Excel, and create a worksheet that demonstrates this feature.

Apply Your Knowledge

The Apply Your Knowledge section contains four mini-cases. Each case describes a situation, explains your role in the case, and asks you to respond to questions. You can answer the questions by applying knowledge you learned in the chapter.

1 Premium Publishers

Situation:

Premium Publishers is a small publishing firm that specializes in reprinting classic literature. A year ago the IT staff developed a Web-based order entry system. The system has performed well, but the company would like to add more features and improve performance. So far, most of the maintenance has involved correcting minor errors.

1. What types of maintenance have the IT staff performed? What types of maintenance will they perform if the existing system is retained?
2. If new features are added, what methodology should the IT staff use to add new functions and enhancements?
3. What IT security measures should the firm adopt? Prepare a security checklist, and be sure to consider all six security levels.
4. Even though the new system is only a year old, e-commerce changes constantly. At what point should Premium Publishers consider replacing the Web-based system with a new system, and why?

2 Oceanside Furniture

Situation:

Oceanside Furniture produces indoor and outdoor wicker furniture. The company grew from one store in 2004 to eight locations today. Two years ago, the company's IT department developed an inventory control system to keep track of products and reorder out-of-stock items. The new system was well received by users, and inventory problems have decreased significantly. Since the inventory system became operational, however, users steadily have requested increased functionality and changes in screen forms and reports.

1. Should Oceanside have a specific process to manage future changes and enhancements? What should it be?
2. What about version control? Should Oceanside institute a maintenance release methodology? Why or why not?
3. Suppose that you had to assign specific IT staff members to maintain the inventory control system. How would you accomplish the task? Describe your strategy in a brief memo.
4. What should Oceanside watch for to detect possible obsolescence in the future? Develop a checklist with specific examples that Oceanside management could use.

3   Robin Hood Associates

Situation:

Robin Hood Associates is an IT consulting firm that develops new systems and maintains older systems for its clients. Robin Hood recently was awarded a contract to correct problems with an existing system. The system is three years old, and the consulting firm that initially designed the system did a poor job of documentation. The data dictionary, user manuals, and other reference material never have been updated, and no process exists for version control.

1. As one of the Robin Hood team members, how should you proceed? What steps would you take, and what would be your priorities?
2. Are CASE tools available that you could use on this assignment? What are they?
3. What advice would you give to the client regarding capacity planning for the future?
4. What steps should the client take to ensure that the system is secure? Prepare a checklist with at least 15 security items that the client should evaluate and monitor. Be sure to consider all six security levels.

4   Economy Travel

Situation:

Economy Travel specializes in personalized travel packages at popular prices, and the firm operates 12 offices in major U.S. cities. A key selling point is the firm's client management database, which includes preferences such as airline seating choices and favorite hotels. Economy Travel purchased the client management software as an off-the-shelf vendor package and modified the program to meet the company's needs. The package has been operational for one year and has performed well. Economy Travel, however, is in the process of expanding its operation to include six additional locations. You have been called in as a consultant to help the company make some decisions about IT support.

1. What performance and workload measurement issues should the company consider at the present time?
2. What capacity planning issues should the company consider at the present time?
3. Should the company establish a system baseline before the integration of the six new sites? Explain your answer.
4. As an IT consultant, you must understand the client's business. From that perspective, consider the impact of the Internet on the travel agency business. Investigate this topic using the Internet and other sources of information, and decide what issues to discuss with Economy Travel.

Case Studies

Case studies allow you to practice specific skills learned in the chapter. Each chapter contains several case studies that continue throughout the textbook, and a chapter capstone case.

NEW CENTURY HEALTH CLINIC

New Century Health Clinic offers preventive medicine and traditional medical care. In your role as an IT consultant, you will help New Century develop a new information system.

Background

You implemented the new system at New Century Health Clinic successfully, and the staff has used the system for nearly four months. New Century is pleased with the improvements in efficiency, office productivity, and patient satisfaction.

Some problems have surfaced, however. The office staff members call you almost daily to request assistance and suggest changes in certain reports and forms. You try to be helpful, but now you are busy with a major project for a local distributor of exercise equipment. Actually, your contract with New Century required you to provide support only during the first three months of operation. Anita Davenport, New Century's office manager, reported that the system seems to slow down at certain times during the day, making it difficult for the staff to keep up with its workload. Also, you increasingly are concerned about system security. A recent article in the local newspaper described an incident where a disgruntled former employee was about to break into the computer system and destroy or alter data.

Assignments

1. You are willing to charge a lower rate for ongoing support services because you designed the system. You want New Century to use a specific procedure for requesting assistance and changes, however, so that you can plan your activities efficiently. Prepare a complete, written procedure for New Century Health Clinic maintenance change requests. Include appropriate forms with your procedure.

2. What could be causing the periodic slowdowns at New Century? If a problem does exist, which performance and workload measures would you monitor to pinpoint the problem?

3. At the end of the systems analysis phase, you studied the economic feasibility of the system and estimated the future costs and benefits. Now that the system is operational, should those costs and benefits be monitored? Why or why not?

4. You decide to prepare a security checklist for New Century. Prepare a list of security issues that the firm should evaluate and monitor. Be sure to organize the items into categories that match the six security levels.

PERSONAL TRAINER, INC.

Personal Trainer, Inc., owns and operates fitness centers in a dozen Midwestern cities. The centers have done well, and the company is planning an international expansion by opening a new "supercenter" in the Toronto area. Personal Trainer's president, Cassia Umi, hired an IT consultant, Susan Park, to help develop an information system for the new facility. During the project, Susan will work closely with Gray Lewis, who will manage the new operation.

Background

System changeover and data conversion were successful for the new Personal Trainer system. The post-implementation evaluation indicated that users were pleased with the system. The evaluation also confirmed that the system was operating properly. Several users commented, however, that system response seemed slow. Susan Park, the project consultant, wants to meet with you to discuss operation, maintenance, and security issues affecting the new system.

Assignments

1. What might be causing the slow response time? Prepare a brief memo explaining system performance and workload measurement, using nontechnical language that Personal Trainer users can understand easily.

2. Personal Trainer's top management asked you to provide ongoing maintenance for the new system. In order to avoid any misunderstanding, you want to provide a brief description of the various types of maintenance. Prepare a brief memo that does this, and include at least two realistic examples of each type of maintenance.

3. Although the system has been operational for a short time, users already have submitted several requests for enhancements and noncritical changes. Should Personal Trainer use a maintenance release methodology to handle the requests? Why or why not?

4. What are the main security issues that Personal Trainer should address? Prepare a memo that lists the primary concerns and offers a specific recommendation for dealing with each issue.

TARHEEL INDUSTRIES

Tarheel Industries is a medium-sized sporting goods manufacturer located in North Carolina. Tarheel's online production support system was developed in-house and was implemented two months ago. The system runs 24 hours a day in Tarheel's three manufacturing facilities.

Background

Last Monday morning, the production support system developed a problem. When a screen display for certain parts was requested, the displayed values were garbled.

When she was alerted to the situation, Marsha Stryker, Tarheel's IT manager, immediately assigned a systems analyst to investigate the problem. Marsha instructed the analyst, Eric Wu, to resolve the problem and get the system up and running as soon as possible. Eric previously worked on two small maintenance projects for the production control system, so he was somewhat familiar with the application.

Eric worked all day on the problem, and by 6:30 p.m., he developed and implemented a fix. After verifying that the production support system was capable of producing correct part displays, Eric went home. Early the following morning, Marsha called Eric and two other members of the applications maintenance group to a meeting in her office, where she briefed them on a new adaptive maintenance project for another high-priority system. She asked them to begin work on the new project immediately.

Several nights later, the production control system crashed shortly after midnight. Every time the system was reactivated, it crashed again. Finally, around 2:30 a.m., all production lines were shut down and third-shift production workers were sent home. The production support system finally was corrected and full production was restored the following day, but by that time, Tarheel Industries had incurred thousands of dollars in lost production costs. The cause of the production support system crash was identified as a side effect of the fix that Eric made to the system.

Assignments

1. Is the second production support system failure entirely unexpected?
2. Who is most to blame for the second system failure?
3. What might Marsha have done differently to avoid the situation? What might Eric have done differently?
4. Outline a new set of maintenance procedures that will help Tarheel Industries avoid such problems in the future.

MILLS IMPORTS

Mills Imports is a successful importer of gourmet coffees, cheeses, and specialty foods from around the world. Mills Imports recently developed and implemented an online sales information system.

Background

Using a client/server design, the PCs in each of the firm's 12 retail stores were networked with a server located in the sales support center at the main office. Salespeople in the retail stores use the customer sales information system to record sales transactions; to open, close, or query customer accounts; and to print sales receipts, daily sales reports by salesperson, and daily sales reports by merchandise code. The sales support staff uses the system to query customer accounts and print various daily, weekly, and monthly reports.

When the customer sales system was implemented, the IT department conducted extensive training for the salespeople and the sales support center staff. One member of the systems development team also prepared a user manual, but users are familiar with the system so the manual rarely is used.

Two weeks ago, Mills opened two additional stores and hired six new sales representatives. A manager gave the user manual to the new sales representatives and asked them to read it and experiment with the system. Now, salespeople in both new stores are having major problems using the sales system. When a representative from the main office visited the stores to investigate the problem, she discovered that the new people could not understand the user manual. When she asked for examples of confusing instructions, several salespeople pointed to the following examples:

- *Obtaining the authorization of the store manager on Form RBK-23 is required before the system can activate a customer charge account.*

- *Care should be exercised to ensure that the BACKSPACE key is not pressed when the key on the numeric keypad with a left-facing arrow is the appropriate choice to accomplish nondestructive backspacing.*

- *To prevent report generation interruption, the existence of sufficient paper stock should be verified before any option that requires printing is selected. If not, the option must be reselected.*

- *The F2 key should be pressed in the event that a display of valid merchandise codes is required. That same key terminates the display.*

Assignments

1. What could Mills Imports have done to avoid the situation?
2. Should the sales support staff ask the IT department to rewrite the user manual as a maintenance project, or should they request a training session for the new salespeople? Can you offer any other suggestions?
3. Rewrite the user manual instructions so they are clear and understandable for new users. What steps might you take to ensure the accuracy of the new user manual instructions?
4. In the process of rewriting the user manual instructions, you discover that some of the instructions were not changed to reflect system maintenance and upgrade activities. A request form on the firm's intranet, for example, has replaced Form RBK-23. Mills also has phased out printed reports in favor of online reports, which users can view by entering a username and password. Rewrite the user manual instructions to reflect the changes.

CHAPTER CAPSTONE CASE: SoftWear, Limited

SoftWear, Limited (SWL), is a continuing case study that illustrates the knowledge and skills described in each chapter. In this case study, the student acts as a member of the SWL systems development team and performs various tasks.

Background

In mid-December 2010, five months after the post-implementation evaluation, the payroll package and the ESIP system were operating successfully and users seemed satisfied with both systems.

During that time, users requested minor changes in reports and screen displays, which the IT staff handled easily. Jane Rossman, manager of applications, continued to assign a mixture of new systems and maintenance tasks to the IT team, and the members indicated that they enjoyed the variety and challenge of both types of work.

Debra Williams, the payroll clerk who prints the ESIP checks, reported the only operational problem. She could not load and align the special check stock in the printer correctly. Becky Evans visited Debra to study the situation and then wrote a specific procedure to solve the problem.

No overtime had been paid in the payroll department since the new system was implemented, and errors in payroll deductions had stopped. Michael Jeremy, SWL's vice president of finance, who initiated the payroll and ESIP projects, is very pleased with the system's operation and output. He recently visited an IT department staff meeting to congratulate the entire group personally.

Some requests for enhancements also occurred. Mike Feiner recently submitted a systems request for the ESIP system to produce an annual employee benefits statement with the current value of all savings plan deductions, plus information on insurance coverage and other benefits data. Mike also indicated that the company would offer several new ESIP choices, including various mutual funds.

In mid-December, Pacific Software announced the latest release of its payroll package. The new version supported full integration of all payroll and human resources functions and data. Ann Hon, director of information systems, was interested in the announcement because she knew that Tina Pham, SWL's vice president of human resources, wanted a human resources information system (HRIS) to support SWL's long-term needs. At Ann's request, Jane Rossman assigned Becky Evans to analyze the new payroll package to determine if SWL could implement the latest version as a company-wide client/server application.

Becky began the preliminary investigation by reviewing the current system and meeting with Mike Feiner to learn more about the new ESIP options. Next, she met with Marty Hoctor, a representative from Pacific Software, to review the features of the new release. After describing the new software, Marty mentioned that a large Midwestern retail chain recently implemented the package, and he invited Becky to contact Sean Valine, director of IT at that company, to discuss the new release. Becky spoke with Sean, and he agreed to e-mail her a summary of comments that users had made about the new software.

Becky completed her preliminary investigation, including a cost-benefit analysis, and worked with Jane Rossman and Ann Hon to prepare a report and presentation to SWL's newly formed systems review committee, which was created at Ann's suggestion. In their presentation, the IT team recommended that SWL upgrade to the new release of the payroll package and build a client/server application for all of SWL's payroll and personnel functions, including the ESIP system. They also suggested that a team of IT and human resources people get together to study preliminary economic, technical, and operational factors involved in a human resources information system and report back to the systems review committee. They pointed out that if the project was approved, the same team could handle the systems development using JAD or RAD techniques. After the presentation, the committee approved the

CHAPTER CAPSTONE CASE: SoftWear, Limited (continued)

request and Ann called an IT department staff meeting for the next morning to start planning the systems analysis phase.

During the meeting, Ann and Jane thanked the entire department for its efforts on the payroll and ESIP projects. Ann pointed out that although the payroll package and the ESIP system support SWL's current needs, the business environment changes rapidly and a successful, growing company must investigate new information management technology constantly. At this point, the systems development life cycle for SWL begins again.

SWL Team Tasks

1. Now that the new ESIP system is operational, Jane Rossman wants you to track system performance using various measurements. At a minimum, she expects you to monitor operational costs, maintenance frequency, technical issues, and user satisfaction. You can add other items if you choose. Write a proposal for Jane that lists each factor you will measure, and make sure that you explain why the item is important and how you plan to obtain the information.

2. Jane assigned you to the SWL team that will study the feasibility of a human resources information system (HRIS). Using the Internet, identify several commercial packages and the names of firms or consultants who specialize in HRIS implementation. Write a brief memo to Jane with your findings.

3. Jane wants you to prepare a security audit procedure for SWL. Specifically, she wants you to prepare a checklist of security issues that need to be evaluated and rated. She said to consider all six security levels, and to include as many specific items as possible that should be assessed.

4. As Ann Hon pointed out in the last meeting, the business environment changes rapidly and a successful, growing company like SWL must investigate new information management technology constantly. Ann has asked you to describe trends in software and hardware that might affect SWL's future IT plans. Perform research on the Internet to identify several technology issues that might represent potential problems or opportunities for SWL, and present the results in a memo to Ann.

Manage the SWL Project

You have been asked to manage SWL's new information system project. One of your most important activities will be to identify project tasks and determine when they will be performed. Before you begin, you should review the SWL case in this chapter. Then list and analyze the tasks, as follows:

LIST THE TASKS Start by listing and numbering at least 10 tasks that the SWL team needs to perform to fulfill the objectives of this chapter. Your list can include SWL Team Tasks and any other tasks that are described in this chapter. For example, Task 3 might be to Perform necessary corrective maintenance and Task 6 might be to Identify perfective maintenance tasks.

ANALYZE THE TASKS Now study the tasks to determine the order in which they should be performed. First identify all concurrent tasks, which are not dependent on other tasks. In the example shown in Figure 12-42, Tasks 1, 2, 3, 4, and 5 are concurrent tasks, and could begin at the same time if resources were available.

CHAPTER CAPSTONE CASE: SoftWear, Limited (continued)

Several nights later, the production control system crashed shortly after midnight. Every time the system was reactivated, it crashed again. Finally, around 2:30 a.m., all production lines were shut down and third-shift production workers were sent home. The production support system finally was corrected and full production was restored the following day, but by that time,

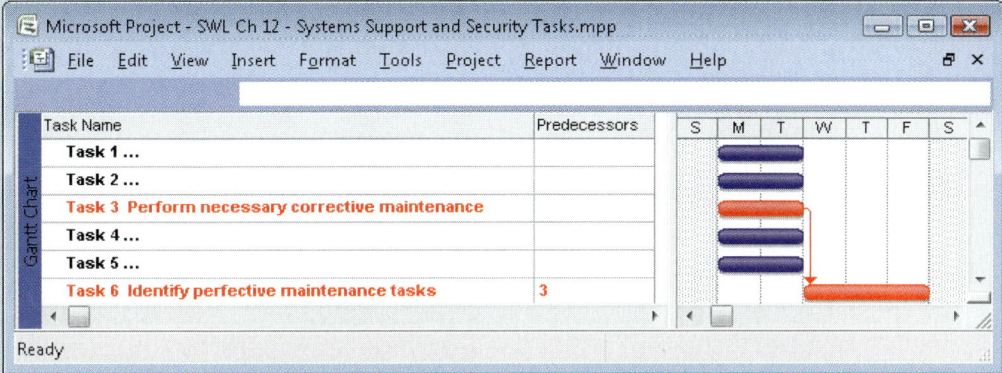

FIGURE 12-42 SWL Tasks 1, 2, 3, 4, and 5 are concurrent tasks that could be performed at the same time. Task 6 is a dependent task that cannot be performed until Task 3 has been completed.

Tarheel Industries had incurred thousands of dollars in lost production costs. The cause of the production support system crash was identified as a side effect of the fix that Eric made to the system.

Assignments

1. Is the second production support system failure entirely unexpected?
2. Who is most to blame for the second system failure?

THE SYSTEMS ANALYST'S TOOLKIT

The Systems Analyst's Toolkit presents a valuable set of cross-phase skills and knowledge that you can use throughout the systems development process. Part 1 discusses communication tools that can help you write clearly, speak effectively, and deliver powerful presentations. Part 2 describes CASE tools that you can use to design, construct, and document an information system. Part 3 demonstrates financial analysis tools you can use to measure project feasibility, develop accurate cost-benefit estimates, and make sound decisions. Part 4 describes Internet resource tools that you can use to locate information, obtain reference material, and monitor IT trends and developments.

PART

Communication Tools

In **Part I** of the Systems Analyst's Toolkit, you will learn about written and oral communication skills that you can use as a systems analyst.

INTRODUCTION

OBJECTIVES

When you finish this part of the Toolkit, you will be able to:

- List overall guidelines for successful communications
- Write effective letters, memos, and e-mail messages
- Measure the readability of written material
- Organize and prepare written reports that are required during systems development
- Follow guidelines for effective oral communication
- Plan, develop, and deliver a successful presentation
- Use effective speaking techniques to achieve your objectives
- Manage and strengthen your communication skills

A successful systems analyst must have good written and oral communication skills to perform his or her job effectively. Never underestimate the importance of effective communications, whether you are using a memo, e-mail, or an oral presentation to convey your ideas. The following guidelines will help you prepare and deliver effective presentations. Remember, however, that nothing increases your ability to communicate better than practicing these skills.

TOOLKIT INTRODUCTION CASE: Mountain View College Bookstore

Background: Wendy Lee, manager of college services at Mountain View College, wants a new information system that will improve efficiency and customer service at the three college bookstores.

In this part of the case, Florence Fullerton (systems analyst) and Harry Boston (student intern) are talking about communication tools and techniques.

Participants:	Florence and Harry
Location:	Mountain View College Cafeteria, at the beginning of the bookstore information system project
Discussion topics:	Guidelines for successful oral and written communications

Florence:	Harry, before we start the project for Wendy, let's talk about written and oral communication tools and techniques. We'll be using these throughout the systems development process.
Harry:	*Fine with me. I've always enjoyed my English and writing courses.*
Florence:	Well, everything you learned in school certainly applies in the workplace. The basic principles of good communications apply everywhere. But in a business environment there are some additional issues to keep in mind.
Harry:	*Such as?*
Florence:	For example, you will be preparing documents, presentations, and training material for users. In each situation, you have to consider the audience very carefully, including their technical knowledge, organizational level, and experience. You'll need to adjust your approach based on those factors, and be aware of the readability of your written work. Also, you'll be using e-mail as a primary communications tool.
Harry:	*I understand. Are there any special ground rules for using e-mail?*
Florence:	Well, e-mail can be more casual than typical business writing — but not *too* casual. We'll talk about that, along with ways to maximize your online effectiveness using software such as Microsoft Outlook. Also, we'll go over some guidelines for preparing presentations.
Harry:	*Sounds good to me. When do we start?*
Florence:	Right now. Here's a task list to get us under way:

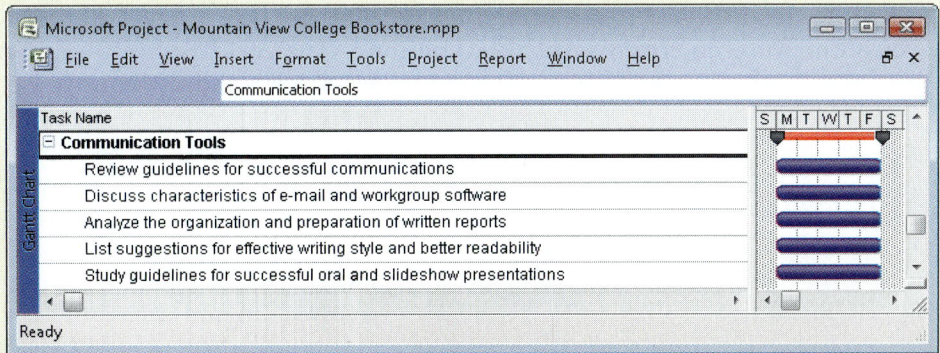

FIGURE TK 1-1 Typical communication tasks.

SUCCESSFUL COMMUNICATION STRATEGIES

Successful communication does not just happen. Usually, it is the result of a specific strategy that includes careful planning, hard work, and practice. To be a successful communicator, you must consider five related questions about yourself, your audience, and your objectives: why, who, what, when, and how. You also must consider the cultural context of your communication. Above all, you must know your subject and have confidence in yourself.

Why, Who, What, When, and How

The **why, who, what, when, and how of communications** are important questions that you must answer before you communicate. These five questions are described in the following section.

WHY   Know *why* you are communicating, and what you want to accomplish. Ask yourself the question, "Is this communication necessary, and what specific results am I seeking?" Your entire communication strategy depends on the results that you need.

WHO   Know *who* your targets are. Chapter 1 describes how information needs of users depend on their organizational and knowledge levels. When communicating with management, for example, sometimes a fine line exists between saying enough and saying too much. Each situation is different, so you must use good judgment and be alert for input from your audience.

WHAT   Know *what* is expected of you and when to go into detail. This is directly related to knowing who your targets are and the organizational and knowledge levels of your audience. For example, a vice president might expect less detail and more focus on how a project supports the company's strategic business goals. You must design your communications just as carefully as your systems project. For example, will the recipients expect you to address a specific issue or topic? Will they expect cost estimates or charts? Design your communications based on the answers to those questions.

WHEN   Know *when* to speak and *when* to remain silent and let others continue the discussion. To be an effective speaker, you must be a good listener — and use audience feedback to adjust your presentation. Good timing is an essential part of every presentation. Your delivery must be paced properly — too fast and you will lose your audience; too slow and they might become bored.

HOW   Know *how* to communicate effectively. You can strengthen your communication skills by using Toolkit suggestions, reflecting upon your own experiences, and observing successful and unsuccessful techniques used by others.

Cultural Context

Communication strategy is affected by the cultural context in which the communication takes place, as shown in Figure TK 1-2. Cultural factors can include geography, background, educational level, and societal

FIGURE TK 1-2 Every communication takes place within an overall cultural context.

differences, among others. These differences must be considered when asking and answering the *why, who, what, when,* and *how* questions.

In addition to these factors, you learned in Chapter 1 that corporate culture is very important. A **corporate culture** includes the beliefs, rules, traditions, values, and attitudes that define a company and influence its way of doing business. To be successful, a systems analyst must understand, and work within this culture. For example, if you speak to a group in a company that encourages a highly participative style, you might want to solicit feedback, invite audience comments, or conduct a poll during your presentation. Similarly, if the organization or group is very formal, or very informal, you might want to adjust your style accordingly.

Know Your Subject

No matter how well you plan your communication, you must know your subject inside and out. Your credibility and effectiveness will depend on whether others believe you and support your views. No one can know everything, so it is important to adopt a specific preparation strategy. For example, before a presentation, consider what others expect you to know and what questions they will ask. No matter how well you prepare, however, you will not have an answer for every question. Remember that it is better to say, "I don't know, but I'll find out and get back to you," rather than to guess.

WRITTEN COMMUNICATIONS

Good writing is important because others often judge you by your writing. If you make a mistake while speaking, your audience probably will forget it. Your written errors, however, might stay around for a long time. Grammatical, typographical, and spelling errors distract readers from your message. Your written communications will include e-mail messages, memos, letters, workgroup communications, and formal reports.

ON THE WEB

For more information about effective written communications, visit **scsite.com/ sad8e/more**, locate Toolkit Part 1, and then click the Effective Written Communications link.

Writing Style and Readability

If you have not taken a writing course, you should consider doing so. If you have a choice of courses, select one that focuses on business writing or technical writing. Any writing class, however, is worth the effort. Most bookstores and libraries have excellent books on effective communications, and many Internet sites offer writing guidelines, tips, and grammar rules. As you prepare written documents, keep in mind the following suggestions:

ON THE WEB

For more information about grammar checkers, visit **scsite.com/sad8e/ more**, locate Toolkit Part 1, and then click the Grammar Checkers link.

1. Know your audience. If you are writing for nontechnical readers, use terms that readers will understand.
2. Use the **active voice** whenever possible. For example, the active voice sentence "Tom designed the system," is better than, "The system was designed by Tom," which is an example of the **passive voice**.
3. Keep your writing clear, concise, and well-organized. Each paragraph should present a single topic or idea.
4. Use an appropriate style. For example, use a conversational tone in informal documents and a business tone in formal documents.
5. Use lists. If a topic has many subtopics, a list can organize the material and make it easier to understand.
6. Use short, easy-to-understand words. Your objective is not to impress your audience with the size of your vocabulary.
7. Avoid repeating the same word too often. Use a thesaurus to locate synonyms for frequently repeated words. Many word processing programs include a thesaurus and other tools to help you write better.

8. Check your spelling. You can use the spell checker in your word processing program to check your spelling, but remember that a **spell checker** is a tool that identifies only words that do not appear in the program's dictionary. For example, a spell checker will not identify instances when you use the word *their*, instead of the word *there*.

9. Check your grammar. Most word processing programs include a **grammar checker**, which is a tool that can detect usage problems and offer suggestions. When you use a grammar checker, you can set various options to match the level and style of the writing and to highlight or ignore certain types of usage. For example, you can set the grammar checker in Microsoft Word to check grammar rules only, or you can configure it to check writing style, as shown in Figure TK 1-3, including gender-specific words, sentence fragments, and passive sentences.

10. Review your work carefully. Then double-check it for spelling, grammatical, and typographical mistakes. If possible, ask a colleague to proofread your work and suggest improvements.

All writers must consider **readability**, which analyzes ease of comprehension by measuring specific characteristics of syllables, words, and sentences. Two popular readability tools are the Flesch Reading Ease Score and the Flesch-Kincaid Grade Level Score.

The **Flesch Reading Ease score** measures the average sentence length and the average number of syllables per word and rates the text on a 100-point scale using the formula shown in Figure TK 1-4. With this tool, the higher the score, the easier it is to understand. Notice that Microsoft suggests that for most standard documents, you should aim for a score of 60-70.

The **Flesch-Kincaid Grade Level score** uses the same variables, but in a different formula that produces a rating keyed to a U.S. grade-school level. For example, a score of 8.0 would indicate material easily understood by a person at an eighth-grade reading level. With this tool, Microsoft suggests that for most standard documents, you should aim for a score of 7.0 to 8.0.

E-Mail, Memos, and Letters

Because e-mail will be your primary tool for written communication, it is important to use it properly and effectively. E-mail usually is less formal than other written correspondence, but you still must follow the rules of good grammar, correct spelling, and clear writing.

FIGURE TK 1-3 You can set the grammar checker in Microsoft Word to check grammar rules only, or you can configure it to check your writing style.

Although many authors use a more conversational style for e-mail, you should remember that e-mail messages often are forwarded to other recipients or groups, and so you must consider the users to whom it might be distributed. If you regularly exchange messages with a specific group of users, most e-mail programs allow you to create a distribution list that includes the members and their e-mail addresses. For example, Figure TK 1-5 shows how to use Microsoft Outlook to send an e-mail to a six-person systems development team. Now that e-mail has become the standard method of business communication, it is important that all users know how to use e-mail properly, professionally, and courteously. This topic is discussed in the following section.

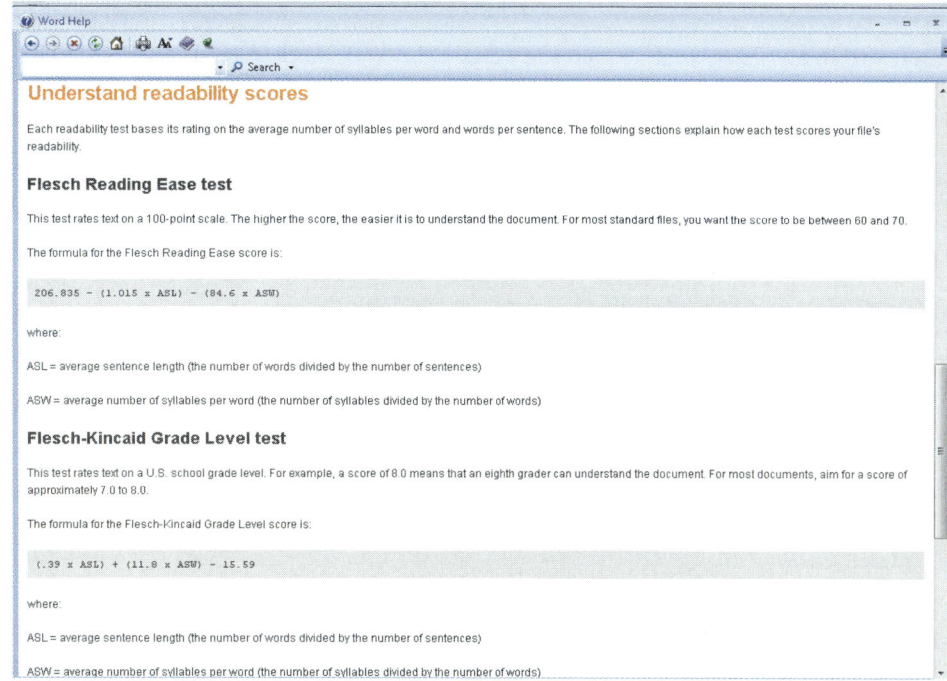

FIGURE TK 1-4 Two popular readability measurement tools are the Flesch Reading Ease Score and the Flesch-Kincaid Grade Level Score.

Although e-mail is the main form of internal communication, internal memos and announcements still are important, and external communications often require letters printed on company letterhead. Most companies use a standard format, or template, for internal memos and external letters. If your company stores those on a network, you can download and use the templates. If you want to create your own designs, you can use a word processor to create templates with specific layouts, fonts, and margin settings. A **template** gives your work a consistent look and makes your job easier. Most word processing programs also provide a feature that allows you to design your memos as forms and fill in the blanks as you work.

ON THE WEB

For more information about readability, visit **scsite.com/sad8e/more**, locate Toolkit Part 1, and then click the Readability link.

Netiquette

Netiquette is a term that combines the words *Internet* and *etiquette*. On the Web, you can find many sources of information about netiquette. One example is the site shown in Figure TK 1-6 on the next page, which offers an excellent source of netiquette guidelines, tips, and links.

All e-mail users should be aware of some simple rules, most of which are nothing more than good manners and common sense. For example, an excellent starting point

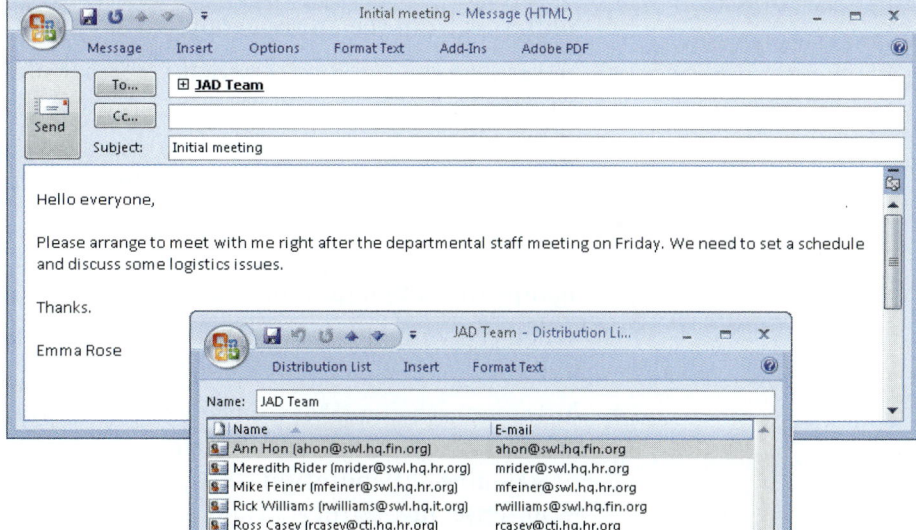

FIGURE TK 1-5 Microsoft Outlook allows users to create distribution lists for sending e-mail messages.

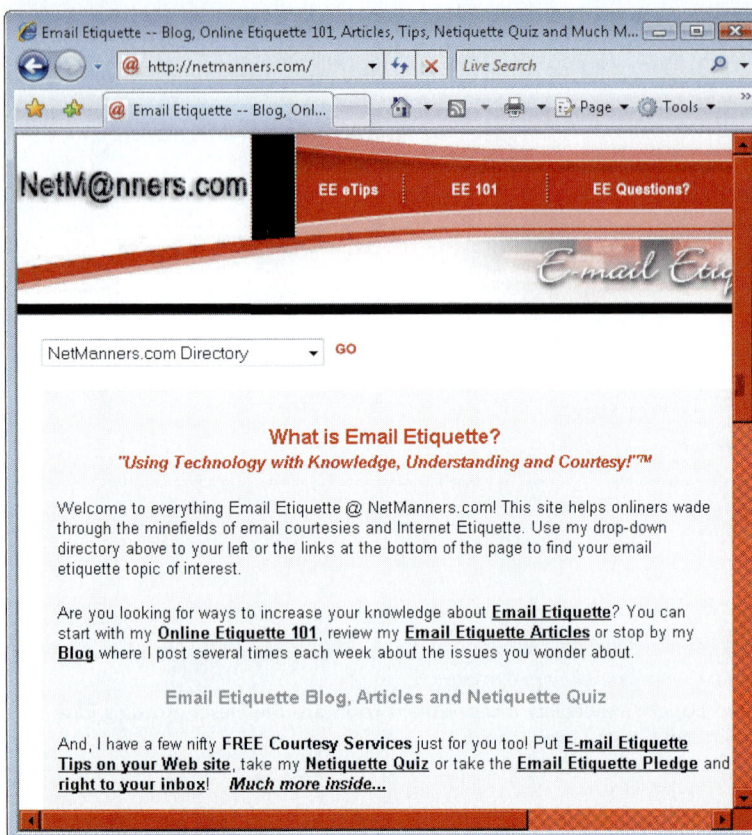

FIGURE TK 1-6   This site offers netiquette guidelines, tips, and links.

is to avoid sending material that is personal or confidential, because your messages might be forwarded by others and distributed more widely than you intended. Another important rule is never to send or reply to an e-mail when you are tired or upset. Instead, you can write a draft if you want to, but save the unsent message so you can review it later.

Here are some common rules and tips:

- Always fill in the subject field with a brief description of the contents.

- Be brief — in most cases, less is more.

- Be professional. Remember, if it has your name on it, it reflects on you personally and that often is how people will view your messages.

- Be sure to check your spelling.

- Don't forward jokes or chain letters without the permission of the recipient.

- Don't overuse humor or sarcasm that might work in a face-to-face situation, but not in an e-mail context.

- Don't type in all caps — it is like YELLING! It is also hard to read.

- Don't use colored fonts, background, or images in business e-mail messages.

- Don't use the return receipt request feature unless there is a valid business reason to do so.

- If you have large attachment files, try to zip or compress them before sending.

- If you send a message to a group of people, especially if they don't know each other, use a blind copy (Bcc) for all of the recipients in order to shield the addresses from the entire group.

- Never give out personal contact information of others without their specific permission to do so.

- Never include personal information unless you are 100% sure of your recipient and no other means of communication would provide better privacy and security.

- Remember that there are copyright laws. You do not have an unrestricted right to do whatever you please with someone else's e-mail message to you. Laws against discrimination and defamation can also apply to e-mail messages.

- When replying, don't include all the earlier messages unless there is a reason to do so.

- **Instant messaging** (IM) and cell-phone **texting** are popular because they allow informal, interactive, and immediate communication. While IM and texting can be valuable collaboration tools, users should exercise good judgment and common sense, just as they would in any form of business conversation.

In addition to these guidelines, it is important to follow company policy regarding communications at work. Many firms restrict personal communications that involve company time or equipment, and courts have upheld an employer's right to limit or monitor such communications.

Workgroup Software

Many companies use **workgroup software**, often called **groupware**, because it enhances employee productivity and teamwork. In addition to basic e-mail, workgroup software enables users to manage and share their calendars, task lists, schedules, contact lists, and documents. Popular examples of workgroup software include Microsoft Outlook and Novell's GroupWise, which is shown in Figure TK 1-7.

Google Docs, which is shown in Figure TK 1-8, offers free, Web-based collaboration. Using this application, a team can work on centrally stored documents instead of e-mailing drafts back and forth. Teams also can use powerful multiauthoring software, such as Adobe Acrobat, to add revisions, notes, and comments to PDF documents.

Reports

You must prepare many reports during systems development, including the preliminary investigation report, the system requirements document at the end of the systems analysis phase, the system design specification at the end of the system design phase, and the final report to management when the system goes into operation. You also might submit other reports, such as status reports, activity reports, proposals, and departmental business plans. You will create your reports as electronic documents, so you can attach them to e-mails and print them, if desired.

In some cases, you must present reports more formally. The example in Figure TK 1-9 on the next page shows a binder for a system requirements document, which includes an introduction, an executive summary, findings, recommendations, time and cost estimates, expected benefits, and an appendix.

You can use a **cover memo**, or an e-mail message similar to the one shown in Figure TK 1-9, when you send a report, and you can set a date, time, and place for an oral presentation. You also can request that the recipients read the report in advance of the presentation.

The **introduction** usually includes a title page, table of contents, and brief description of the proposal. The title page should be clean and neat and contain the name of the proposal, the subject, the date, and the names of the

FIGURE TK 1-7 Workgroup software, such as Novell's GroupWise, allows a user to collaborate with others by sharing documents and folders.

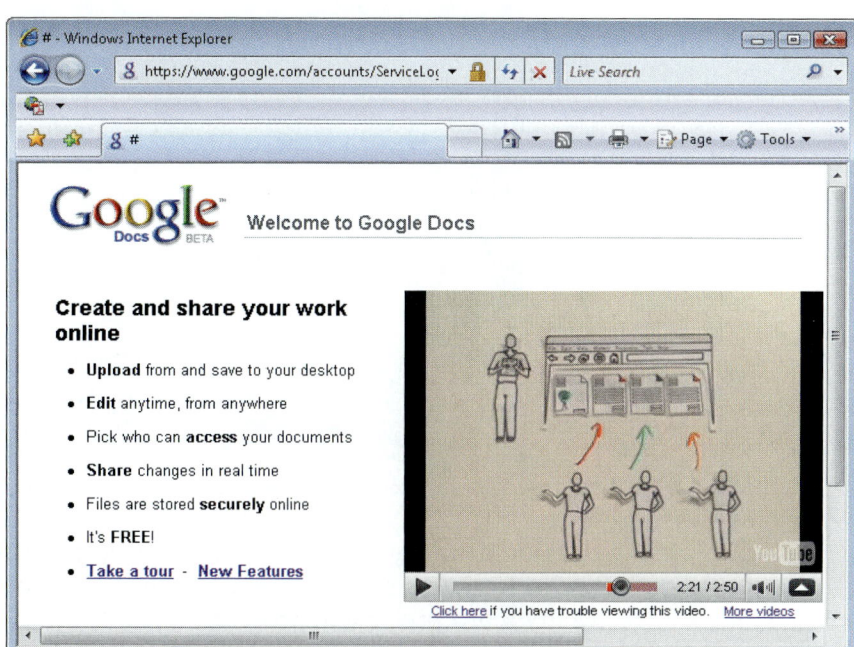

FIGURE TK 1-8 An employee team can use Google Docs to work on centrally stored documents.

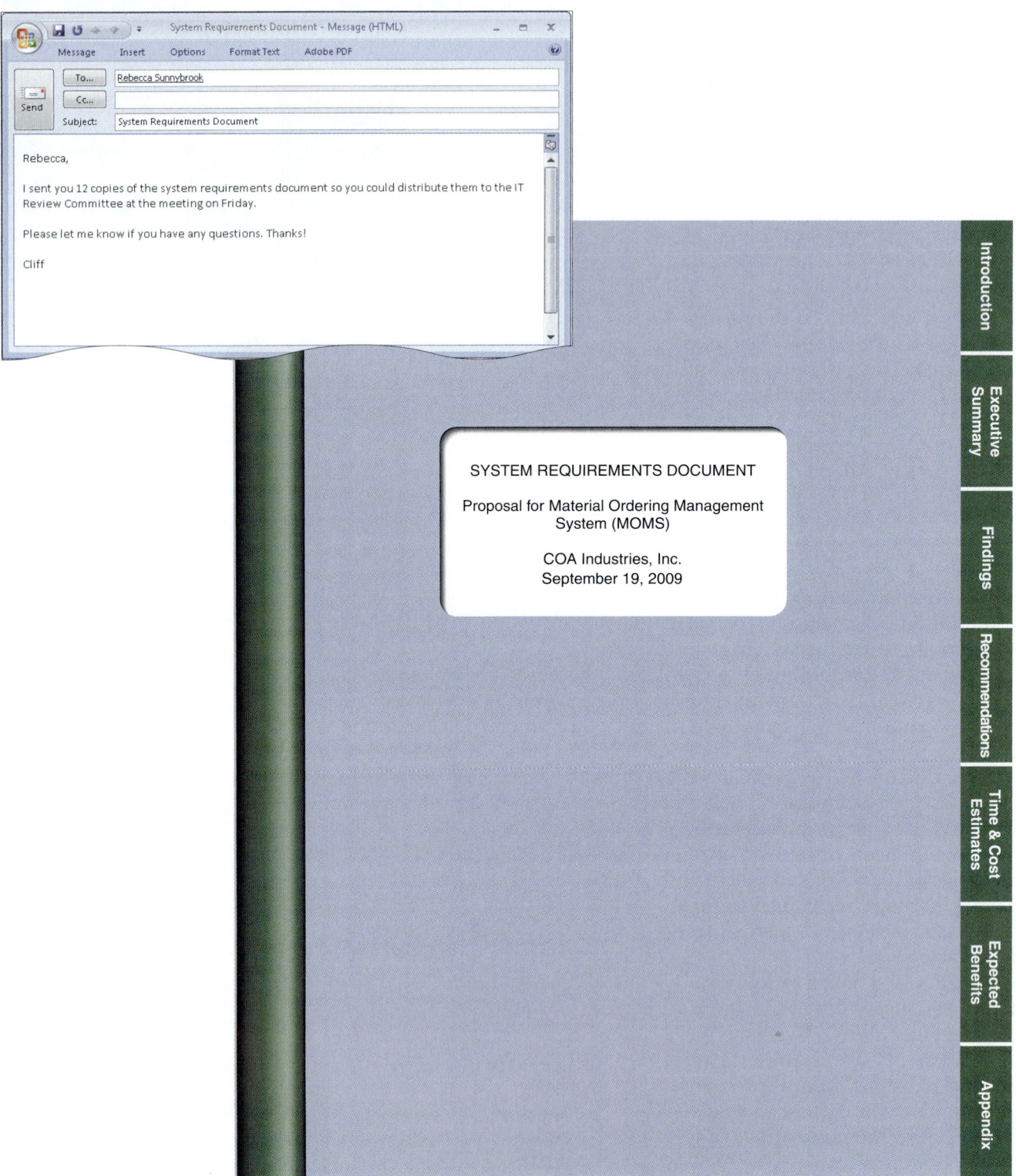

FIGURE TK 1-9   Typical binder for a system requirements document, with an explanatory e-mail message.

development team members. If the project already has a recognized name or acronym, use it. Include a **table of contents** when the report is long or includes many exhibits. Many word processing programs include a tool that can generate a table of contents automatically.

The **executive summary** is used to summarize the entire project, including your recommendations, in several paragraphs. Generally, the executive summary should not exceed 200 words or one page.

Use the **findings section** to describe the major conclusions that you or the team reached during the systems analysis phase. You can make the findings section detailed or summarized, depending on the project. You must explain the logical design of the new system in a way that nontechnical managers can understand clearly. With a management audience, the most important task is to explain how the proposed system supports the company's business needs.

The **recommendations section** presents the best system alternative, with a brief explanation that does not disparage anyone who favors a different alternative. Your recommendation should mention the essential factors of economic, technical, and operation feasibility.

The **alternatives section** can be a separate section or part of the recommendations section. The alternatives section identifies various strategies and alternatives, as discussed in Chapter 7. In the alternatives section, you should list the advantages and disadvantages of each major system alternative. In this section, you should include the cost-benefit results, with a clear description of the economic analysis techniques that were used. You can use tables or graphs to support and clarify your alternatives when necessary.

When you have a large number of supporting documents such as questionnaires or sampling results, you should put those items in an **appendix** located at the end of the document. Make sure you include only relevant information, and provide references for interested readers.

ORAL COMMUNICATIONS

An **oral presentation** is required at the end of the preliminary investigation and again at the conclusion of the systems analysis phase. You might need to give more than one presentation in some situations to present technical material to members of the IT department or to present an overview for top managers. When preparing an oral presentation, you should perform six important tasks: Define the audience, define the objectives for your presentation, organize the presentation, define any technical terms you will use, prepare your presentation aids, and practice your delivery.

ON THE WEB

For more information about effective presentations, visit **scsite.com/sad8e/ more**, locate Toolkit Part I, and then click the Effective Presentations link.

Define the Audience

Before you develop a detailed plan for a management presentation, you must define the audience. Senior managers often prefer an executive summary rather than a detailed presentation, but that is not always the case, especially in smaller companies where top management is more involved in day-to-day activities. If you consider the expectations of your audience and design your presentation accordingly, you will improve your chances of success.

Define the Objectives

When you communicate, you should focus on your objectives. In the management presentation for the systems analysis phase, your goals are the following:

- Inform management of the status of the current system
- Describe your findings concerning the current system problems
- Explain the alternative solutions that you developed
- Provide detailed cost and time estimates for the alternative solutions
- Recommend the best alternative and explain the reasons for your selection

Organize the Presentation

Plan your presentation in three stages: the introduction, the information, and the summary. First, you should introduce yourself and describe your objectives. During the presentation, make sure that you discuss topics in a logical order. You should be as specific as possible when presenting facts — your listeners want to hear your views about what is wrong, how it can be fixed, how much it will cost, and when the objectives can be accomplished. In your summary, briefly review the main points, and then ask for questions.

Define Any Technical Terms

You should avoid specialized or technical terminology whenever possible. If your audience might be unfamiliar with a term that you plan to use, either define the term or find another way to say it so they will understand your material.

Prepare Presentation Aids

Much of what people learn is acquired visually, so you should use helpful, appropriate visual aids to help the audience follow the logic of your presentation and hold their attention. Visual aids also can direct audience attention away from you, which is helpful if you are nervous when you give the presentation. You can use a visual aid with an outline of topics that will help you stay on track. You can enhance the effect of your presentation with visual aids that use various media and software, as explained in the following sections.

VISUAL AIDS Visual aids can help you display a graphical summary of performance trends, a series of cost-benefit examples, or a bulleted list of important points. You can use white-boards, flip charts, overhead transparencies, slides, films, and videotapes to enhance your presentation. When preparing your visual aids, make sure that the content is clear, readable, and easy to understand. Verify ahead of time that the audience can see the visual material from anywhere in the room. Remember that equipment can fail unexpectedly, so be prepared with an alternate plan.

PRESENTATION SOFTWARE With a computer and a projection system, you can use **presentation software**, such as Microsoft PowerPoint, to create slides with sounds, animation, and graphics. As shown in Figure TK 1-10, Microsoft offers free Webcasts and podcasts with advanced tips and techniques that can improve your presentations.

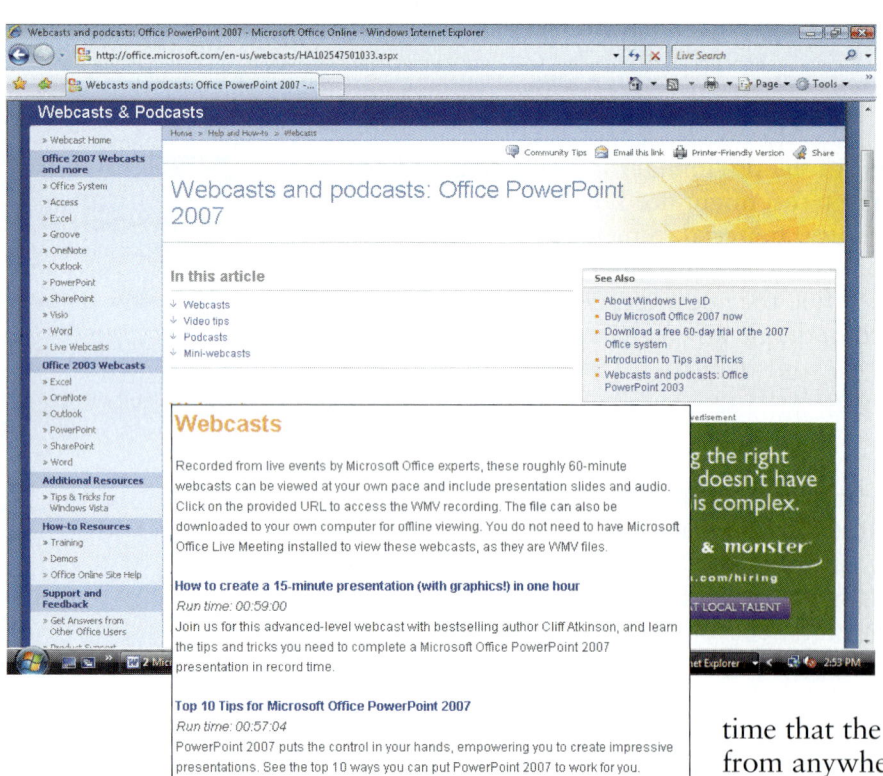

FIGURE TK 1-10 Microsoft offers free, self-paced Webcasts and podcasts.

Preparing an effective slide presentation requires time and effort, and personal experience is the best teacher. There is no universal agreement about how to prepare a slide show, and many sources of information exist. Some overall guidelines include the following:

- Your first step (and perhaps the most important) is to prepare an overall outline that will be the foundation of your presentation. You should focus on the content and structure of your presentation before you consider visual issues.

- Remember that a fine line exists between providing too little information and too much.

- Display one topic per slide, and try to follow the rule often called the **7 by 7 rule**: no more than seven items per slide, and no more than seven words per item. Some presenters believe that a **6 by 6 rule** is even more effective.

- When displaying a list of items, consider using a series of slides to add each point sequentially, especially if you want to focus attention on the item being discussed.

- Use bullets rather than numbers, unless you are showing a specific sequence or order.

- Choose easily readable fonts. Use sans serif styles, such as Arial, for all body text. If you do use a serif style (such as Times Roman), apply it only in titles.

- Use appropriate point sizes for titles and body text. Your goal is to prepare slides that are readable and visually attractive. Although point size selection depends on individual judgment and experience, here are some suggestions to get you started: For titles, try either 40- or 36-point fonts; for body text, 32- or 24-point fonts usually work well.

- Select special effects carefully — too many graphics, colors, sounds, or other special effects will distract your audience.

- You can include tables or graphics, but keep them simple and easy to understand. Also, you can use a special effect, such as boldface, italic, underlining, or a different color, to highlight an important word or phrase.

- Strive for a consistent look and feel among your slides, and position visual elements in the same place on each slide. You should use a master template to ensure uniformity and conform to company-wide standards that might apply, such as a copyright notice, a confidentiality statement, or placement of the company name and logo. Choose colors carefully, and keep them consistent. Usually, light letters on a dark background are easiest to read. Presentation software normally has predefined color palettes that provide background and text colors that ensure readability. Use these palettes as a guideline for selecting colors when possible.

- Be sure to check spelling and grammar!

- During the presentation, do *not* read your slides to the audience! They can read the slides on their own. Your slide presentation is an outline that provides structure and emphasis — it is not the presentation itself.

- It is important to deliver a presentation that can be viewed easily from anywhere in the room. When setting up, consider the size of the room, the number of people attending, the size and location of your visual aids, and the characteristics of any projection equipment you will be using.

Practice

The most important part of your preparation is practice. You should rehearse several times to ensure that the presentation flows smoothly and the timing is correct. Practicing will make you more comfortable and build your confidence.

Do not be tempted to write a script. If you read your presentation, you will be unable to interact with your audience and adjust your content based on their reactions. Instead, prepare an outline of your presentation and practice from the outline. Then, when you deliver the actual presentation, you will not have to struggle to remember the exact words you planned to say, and you will be able to establish a good rapport with your audience.

The Presentation

When you deliver your presentation, the following pointers will help you succeed:

SELL YOURSELF AND YOUR CREDIBILITY   As a presenter, you must sell yourself and your credibility. A brilliant presentation will not convince top managers to approve the system if they are not sold on the person who gave the presentation. On the other hand, projects often are approved on the basis of the presenter's knowledge, commitment, and enthusiasm.

Your presentation must show confidence about the subject and your recommendations. You should avoid any conflicts with the people attending the presentation. If you encounter criticism or hostility, remain calm and stay focused on the issues — not the person making the comments. You will have a successful presentation only if you know the material thoroughly, prepare properly, and sell yourself and your credibility.

CONTROL THE PRESENTATION   During the presentation, you must control the discussion, maintain the pace of the presentation, and stay focused on the agenda — especially when answering questions. Although you might be more familiar with the subject material, you should not display a superior attitude toward your listeners. Maintain eye contact with the audience and use some humor, but do not make a joke at someone else's expense.

ANSWER QUESTIONS APPROPRIATELY   Let your audience know whether you would prefer to take questions as you go along or have a question-and-answer session at the end. Sometimes the questions can be quite difficult. You must listen carefully and respond with a straightforward answer. Try to anticipate the questions your audience will ask so you can prepare your responses ahead of time.

When answering a difficult or confusing question, repeat the question in your own words to make sure that you understand it. For example, you can say, "If I understand your question, you are asking …" This will help avoid confusion and give you a moment to think on your feet. To make sure that you gave a clear answer, you can say, "Have I answered your question?" Allow follow-up questions when necessary.

USE EFFECTIVE SPEAKING TECHNIQUES   The delivery of your presentation is just as important as its content. You can strengthen your delivery by speaking clearly and confidently and projecting a relaxed approach. You also must control the pace of your delivery. If you speak too fast, you will lose the audience, and if the pace is too slow, people will lose their concentration and the presentation will not be effective.

Many speakers are nervous when facing an audience. If this is a problem for you, keep the following suggestions in mind:

- Control your environment. If you are most nervous when the audience is looking at you, use visual aids to direct their attention away from you. If your hands are shaking, do not hold your notes. If you are delivering a computer-based presentation, it is a good idea to use a handheld wireless device to control the slides. Concentrate on using a strong, clear voice. If your nervousness distracts you, take a deep breath and remind yourself that you really do know your subject.

- Turn your nervousness to your advantage. Many people do their best work when they are under a little stress. Think of your nervousness as normal pressure.

- Avoid meaningless filler words and phrases. Using words and phrases such as *okay, all right, you know, like, um*, and *ah* are distracting and serve no purpose.

- Practice! Practice! Practice! Some people are naturally gifted speakers, but most people need lots of practice. You must work hard at practicing your presentation and building your confidence. Many schools offer speech or public speaking courses that are an excellent way of practicing your skills. It also can be advantageous to preview your presentation with one or more people and ask for input.

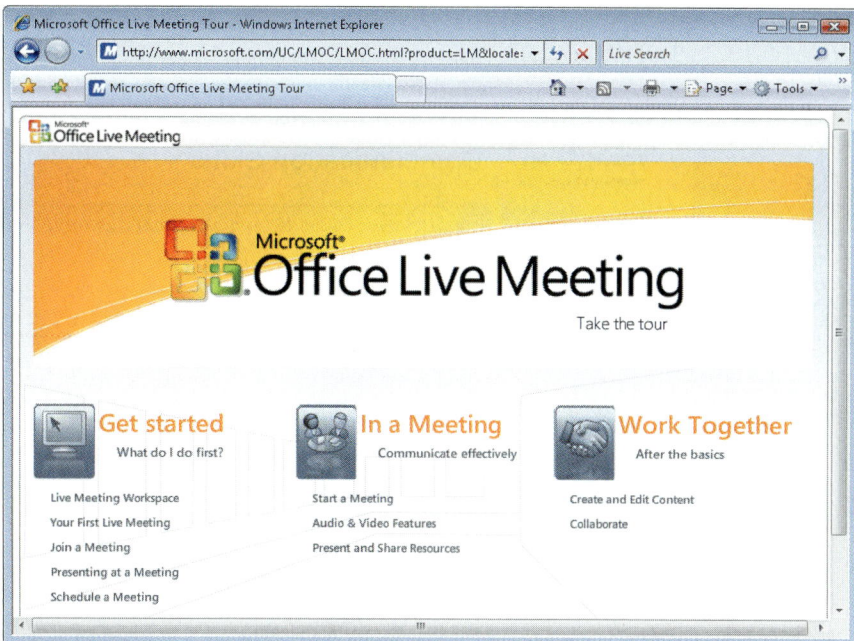

FIGURE TK 1-11 Microsoft Office Live Meeting is a powerful Web-conferencing application that can deliver PowerPoint presentations.

Delivering Online Presentations

In addition to face-to-face presentations, you might be asked to deliver an online presentation, possibly with two-way communication between you and the audience. Past versions of Microsoft PowerPoint allowed you to broadcast PowerPoint presentations to small groups. The presentation could be live, with two-way questions and answers, or stored on a server to be downloaded later.

Microsoft PowerPoint 2007, however, uses a new platform, called **Microsoft Office Live Meeting**, to accomplish the same results, as shown in Figure TK 1-11. In addition to delivering presentations, the Live Meeting application offers Web conferencing, with powerful audio and video capability that allows you to host or participate in group meetings and collaborate with others.

MANAGING YOUR COMMUNICATION SKILLS

More than ever, employees must rely on their personal skills and experience. In an uncertain world and a turbulent economy, individuals should think of themselves as profit-making companies, complete with assets, liabilities, strengths, and areas for development. In Chapter 2, you learned that a company must have a strategic plan, and the same is true for an individual. Armed with a plan to improve your communication skills, you are much more likely to reach your full potential.

Communicating is like any other activity — the more you practice, the better you become. Many resources are available for students and IT professionals who want to improve their written and oral communication skills. For example, the Vocational Information Center site shown in Figure TK 1-12 on the next page offers a wide range of free resources and links that can help you become a better writer, presenter, and public speaker. The Association for Computing Machinery (ACM) also offers many online courses and tutorials for members, including students and IT professionals.

Some people find it difficult to stand in front of a group and deliver a presentation or report. For many years, membership in Toastmasters, International has been a popular way to gain confidence, overcome stage fright, and develop public speaking skills.

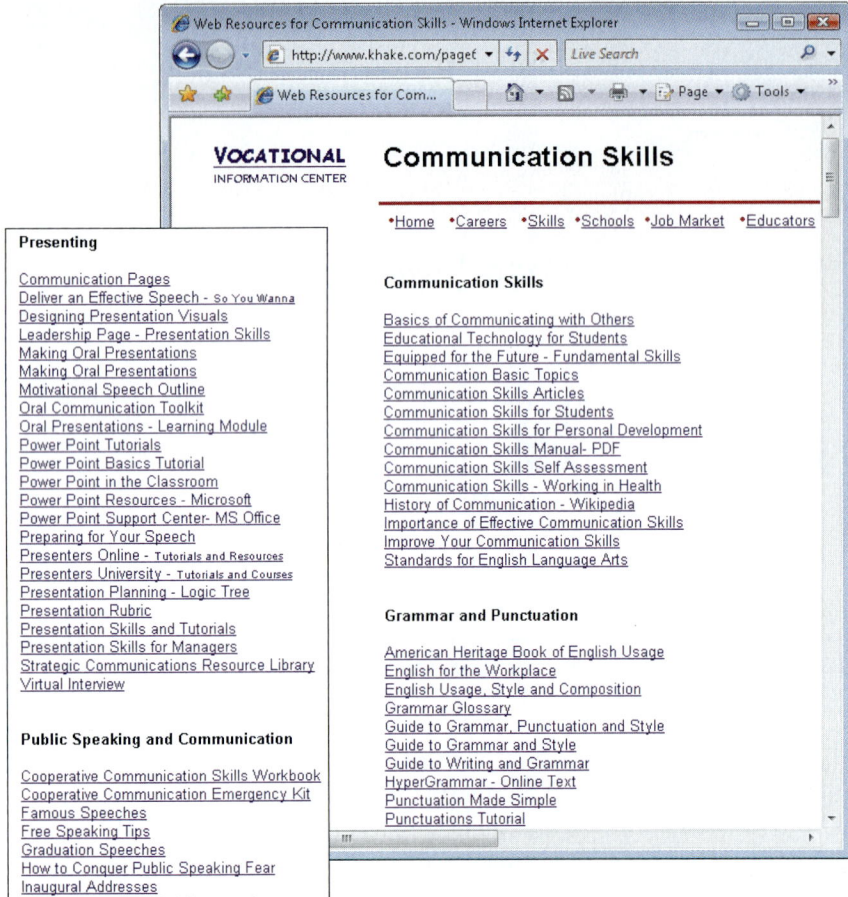

As shown in Figure TK 1-13, Toastmasters offers a friendly environment where members critique each speech in a positive manner, note the strengths, and offer suggestions about what might be improved. With more than 200,000 members in 92 countries, this organization offers an excellent way to develop better public speaking skills.

FIGURE TK 1-12   The Vocational Information Center offers many online courses and tutorials on communications skills.

FIGURE TK 1-13   Toastmasters International is famous for helping people become better public speakers.

TOOLKIT SUMMARY

Your success as a systems analyst depends on your ability to communicate effectively. You must know why you are communicating, what you want to accomplish, who your targets are, what is expected of you, and when to go into detail. You must know your subject and how to use good written and oral communications techniques.

You will be judged by your written work, so it must be free of grammatical, spelling, and punctuation errors. You should write e-mail, letters, and memos clearly, and the writing style should match the situation. Many firms have standard formats for letters and memos, and you can use templates to achieve consistency.

Your writing must be clear and understandable. You can use readability measurement tools such as the Flesch Reading Ease score and the Flesch-Kincaid Grade Level score.

You will prepare various reports during systems development, and the format will vary depending on the nature of the report. Your reports should have a cover memo and might include an introduction, an executive summary, findings, recommendations, time and cost estimates, expected benefits, and an appendix.

In addition to written communications, you must communicate effectively in person. You might have to deliver several presentations to different audiences at different times during the SDLC. Presentations are an important form of oral communication, and you should follow specific guidelines in preparing your presentation. You prepare by defining your audience, identifying your objectives, and organizing the presentation itself. You also need to define technical terms and prepare visual aids to help your audience understand the material. Most important, you must practice your delivery to gain confidence and strengthen your presentation skills.

When you develop slide presentations, you should follow the 6 by 6 rule or 7 by 7 rule and other guidelines that will make your slides easy to read and understand. You should select fonts and point sizes carefully, and strive for a consistent look and feel throughout the presentation. Special effects can be interesting, but do not overuse them.

When you give the presentation, you are selling your ideas and your credibility. You must control the discussion, build a good rapport with the audience, answer all questions clearly and directly, and try to use good speaking techniques. Again, the best way to become a better speaker is to practice.

Every IT professional should have a strategic plan to manage and improve written and oral communication skills. Many online resources offer courses, tutorials, and support to help you develop the skills you will need in the workplace.

Key Terms and Phrases

6 by 6 rule *625*

7 by 7 rule *625*

active voice *617*

alternatives section *623*

appendix *623*

corporate culture *617*

cover memo *621*

executive summary *622*

findings section *623*

Flesch-Kincaid Grade
   Level score *618*

Flesch Reading Ease score *618*

grammar checker *618*

groupware *621*

instant messaging (IM) *620*

introduction *621*

Microsoft Office Live
   Meeting *627*

netiquette *619*

oral presentation *623*

passive voice *617*

presentation software *624*

readability *618*

recommendations
   section *623*

spell checker *618*

table of contents *622*

template *619*

texting *620*

visual aids *624*

why, who, what, when,
   and how of
   communications *616*

workgroup software *621*

Toolkit Exercises

Review Questions

1. Describe the why, who, what, when, and how of communications. Explain each term and give an example. Also, what is a corporate culture and why is it important?
2. Mention five specific techniques you can use to improve your written documents.
3. What techniques can help to improve your e-mail communications?
4. Describe the main sections of a written report to management.
5. When preparing an oral presentation, what six tasks should you perform?
6. When you organize the presentation, what three stages do you plan?
7. Why are visual aids important? Give at least three examples of different types of visual aids, and explain how you would use each type in a presentation.
8. What can you do during your presentation to improve its success?
9. Name three specific strategies you can use if you get nervous during a presentation.
10. Why is practice so important when preparing a presentation?

Discussion Topics

1. Most people agree that business e-mail can be more conversational than formal written documents, but even e-mail has its limits. As a manager, what guidance would you give people regarding e-mail style and usage?
2. Is it possible to overcommunicate? For example, in Chapter 4 you learned to avoid leading questions, which might suggest an answer. Can you think of other examples, like newspaper headlines, where "less is more"?
3. Many articles stress the importance of body language. Think of examples where you noticed a person's body language. Did it relate to something they were trying to communicate — or something they were trying *not* to communicate?
4. Should e-mail monitoring by an employer always be permissible, never permissible, or does the answer depend on specific factors? If so, what are they?

Projects

1. *The Elements of Style* by William Strunk, Jr. and E. B. White is a popular reference manual for proper English usage. The book identifies many words and phrases that are commonly misused, including *between* and *among*, *affect* and *effect*, *different from* and *different than*, *like* and *as*, and *infer* and *imply*. Review *The Elements of Style* or another source, and explain how these words should be used.
2. Using Microsoft PowerPoint or another program, prepare a presentation on "How to Prepare an Effective Slide Presentation." Assume that your audience is familiar with presentation software, but never had any formal training.
3. As a training manual writer, choose a simple hardware or software task and write a two- or three-paragraph description of how to perform the task. Then check the readability statistics. Try to keep the Flesch Reading Ease score above 60 and the Flesch-Kincaid Grade Level score to 8.0 or less.
4. View at least three examples of public speaking. You can investigate TV network news broadcasts, C-SPAN, or any other source. Describe each speaker's gestures, expressions, voice levels, inflections, timing, eye contact, and effectiveness.

PART 2 CASE Tools

In **Part 2** of the Systems Analyst's Toolkit, you will learn how CASE tools can help you perform systems development and maintenance tasks.

INTRODUCTION

OBJECTIVES

When you finish this part of the Toolkit, you will be able to:

- Explain CASE tools and the concept of a CASE environment

- Trace the history of CASE tools and their role in a fourth-generation environment

- Define CASE terms and concepts, including a repository, modeling tools, documentation tools, engineering tools, and construction tools

- Explain an integrated development environment

- Provide examples of CASE tool features

- Describe CASE tool trends, and how they relate to object-oriented analysis and agile methods

Computer-aided systems engineering (CASE), also called **computer-aided software engineering**, is a technique that uses powerful software, called **CASE tools**, to help system developers design and construct information systems. In this part of the Systems Analyst's Toolkit, you will learn about the history, characteristics, and features of CASE tools. You will see specific examples of CASE tools and how they are used in various development tasks. In addition, you will learn about integrated software development environments.

TOOLKIT INTRODUCTION CASE: Mountain View College Bookstore

Background: Wendy Lee, manager of college services at Mountain View College, wants a new information system that will improve efficiency and customer service at the three college bookstores.

In this part of the case, Florence Fullerton (systems analyst) and Harry Boston (student intern) are talking about CASE tools and concepts.

Participants:	Florence and Harry
Location:	Florence's office, early in the systems planning phase
Discussion topics:	CASE tools, integrated development environments, CASE terms and concepts

Florence: Harry, before we get too far into the project, let's talk about CASE tools and concepts and how we can use them in the development process.

Harry: *Sure. I saw examples of CASE tools in my systems analysis course. CASE stands for computer-aided systems engineering, right?*

Florence: Yes. Some people say CASE stands for computer-aided *software* engineering instead of computer-aided *systems* engineering. There's really no difference. Either way, CASE tools are an integral part of the systems development process, and can be used with structured, object-oriented, or agile development methods.

Harry: *Will we use CASE tools on the bookstore project?*

Florence: Yes. We'll start by identifying and modeling business functions and processes. That information will become part of a central repository, which is a database that stores all the characteristics of the information system.

Harry: *Then what?*

Florence: We'll use CASE tools to draw various diagrams, including DFDs, UML diagrams, functional decomposition diagrams, and business process diagrams. You'll learn about these as we go along. We might also use CASE tools to generate program code, screens, and reports. Also, depending on the software environment we select, we might use an integrated development environment.

Harry: *What's that?*

Florence: An integrated development environment is like a built-in CASE tool that a vendor integrates into a software product.

Harry: *Anything else?*

Florence: Well, before we decide on a CASE tool, we'll take a look at several examples. Here's a task list to get us started:

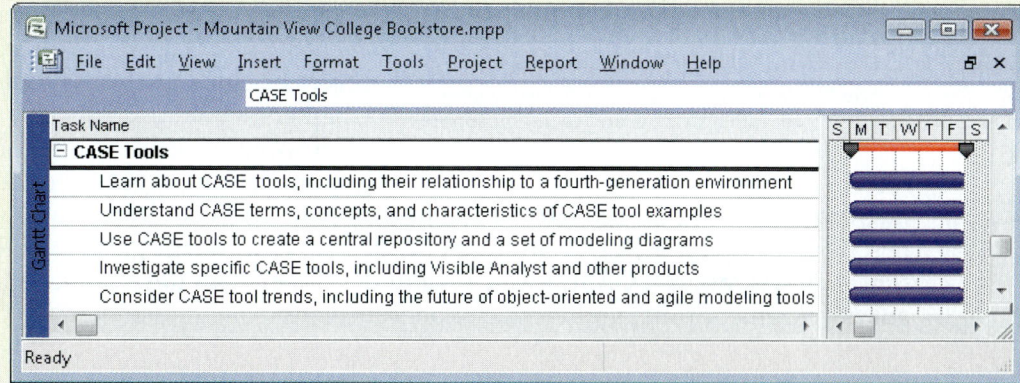

FIGURE TK 2-1 Typical CASE tool tasks.

OVERVIEW OF CASE TOOLS

If you ask a carpenter for an example of a tool, the response might be a hammer, drill, or screwdriver. Put the same question to a chef, and the answer might be a measuring cup, knife, or spatula. Every type of work requires specific tools to do the job properly, and system development is no different. CASE tools can reduce costs, speed up development, and provide comprehensive documentation that can be used for future maintenance or enhancements. System developers also use CASE tools to plan business-driven systems starting with the company's operational needs and future requirements.

The Carnegie Mellon Software Engineering Institute, a leader in software standards and quality management, suggests that a **CASE environment** is more than a set of CASE tools — it includes any use of computer-based support in the software development process. As Figure TK 2-2 shows, the Carnegie Mellon definition includes managerial, administrative, and technical support for a software project.

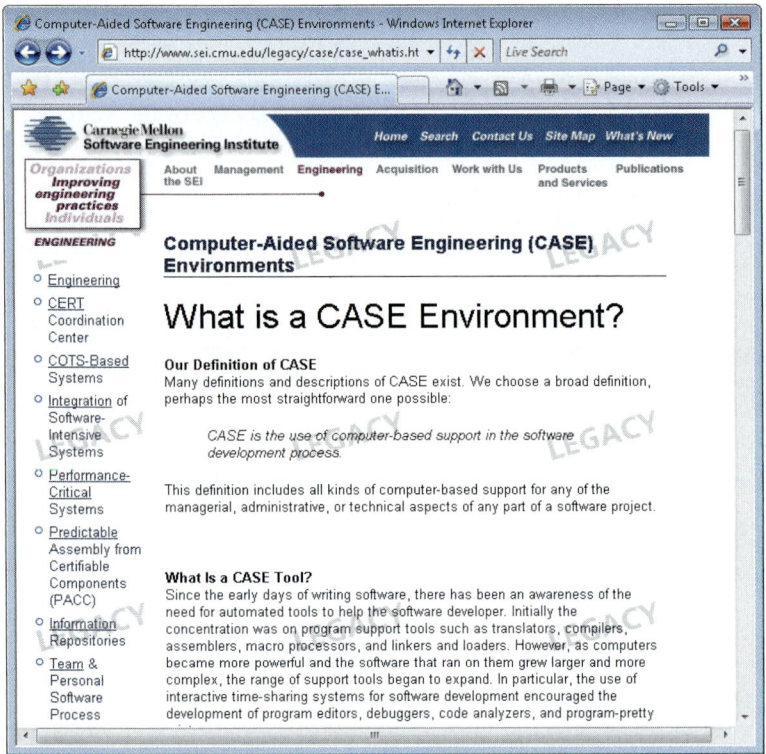

FIGURE TK 2-2 The Carnegie Mellon definition of a CASE environment includes any software-based managerial, administrative, and technical support for a software project.

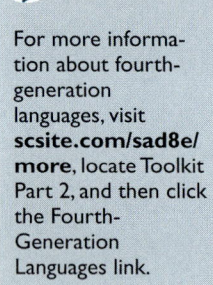

ON THE WEB

For more information about fourth-generation languages, visit **scsite.com/sad8e/more**, locate Toolkit Part 2, and then click the Fourth-Generation Languages link.

CASE Tools History

As early as the 1960s, programmers used tools such as editors and code debuggers to write programs for mainframe computers. Today, in our software-driven world, CASE tools have evolved into powerful resources that systems analysts need to build and maintain complex information systems.

Forty years ago, virtually all code was written in **procedural** languages such as COBOL, which required a programmer to create code statements for each processing step. In contrast, modern languages such as Visual Basic or Java are called **non-procedural,** or **event-driven,** languages because instead of writing a series of sequential instructions, a programmer defines the actions that the program must perform when certain events occur. Because non-procedural languages are **object-oriented programming languages** (OOPL), they are especially valuable in implementing an object-oriented system design. You learned about object-oriented analysis and design concepts in Chapter 6.

Another trend involves powerful programming languages called **fourth-generation languages** (4GLs) that are part of the **fourth-generation environment,** which was described in Chapter 7, Development Strategies. In a fourth-generation environment that includes modern CASE tools, system developers can develop accurate prototypes, cut development time, and reduce expense.

The Marketplace for CASE Tools

The CASE tool marketplace includes a wide variety of vendors and products, and no one tool dominates the market. You can use a site such as the one shown in Figure TK 2-3 to locate CASE tool products and vendors. Depending on their features, some CASE tools can cost thousands of dollars, while others are available as shareware, or even as freeware, as shown in Figure TK 2-4.

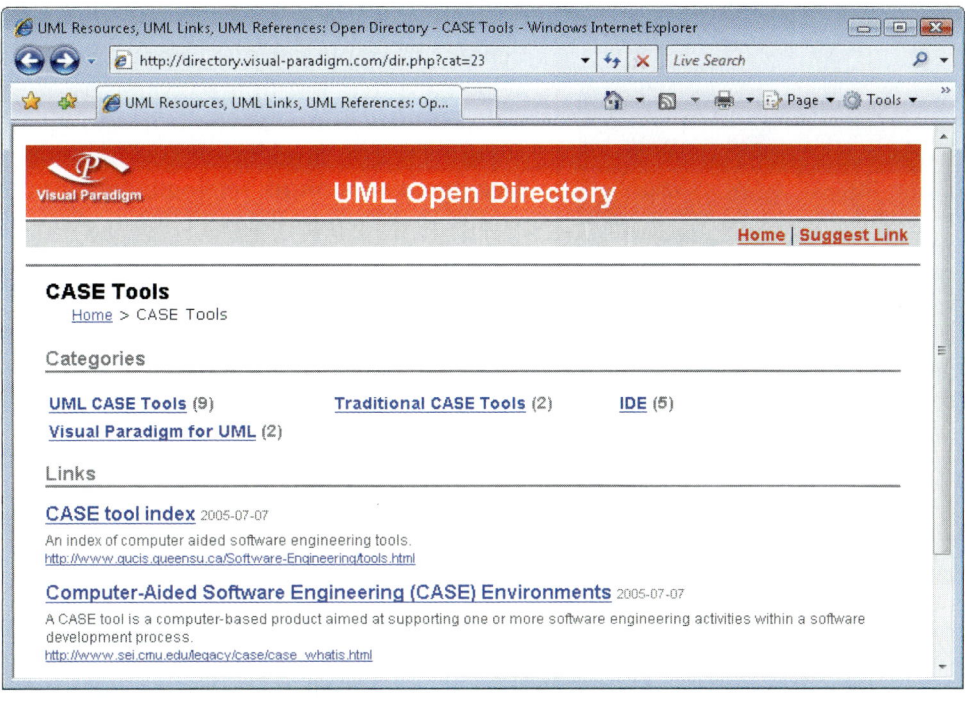

FIGURE TK 2-3   Visual Paradigm's Web site offers information about various types of CASE tools.

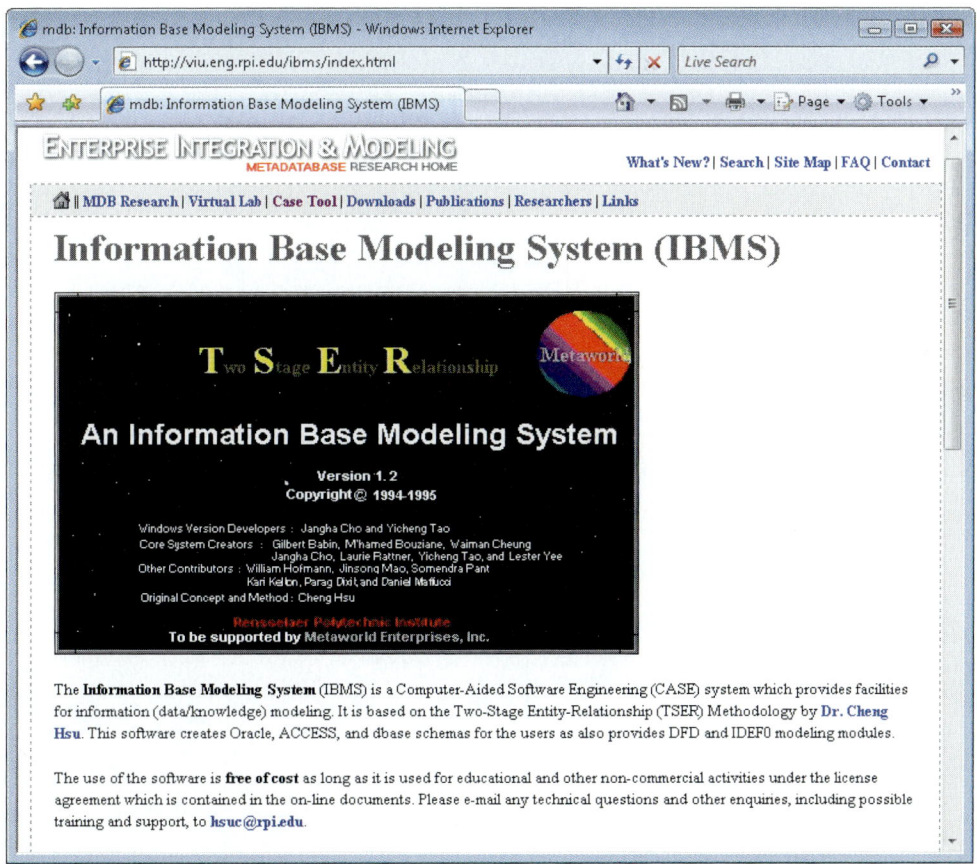

FIGURE TK 2-4   Information Base Modeling System (IBMS) is a freeware CASE tool that can be downloaded and used for educational purposes.

ON THE WEB

For more information about the CASE tool marketplace, visit **scsite.com/sad8e/more**, locate Toolkit Part 2, and then click The CASE Tool Marketplace link.

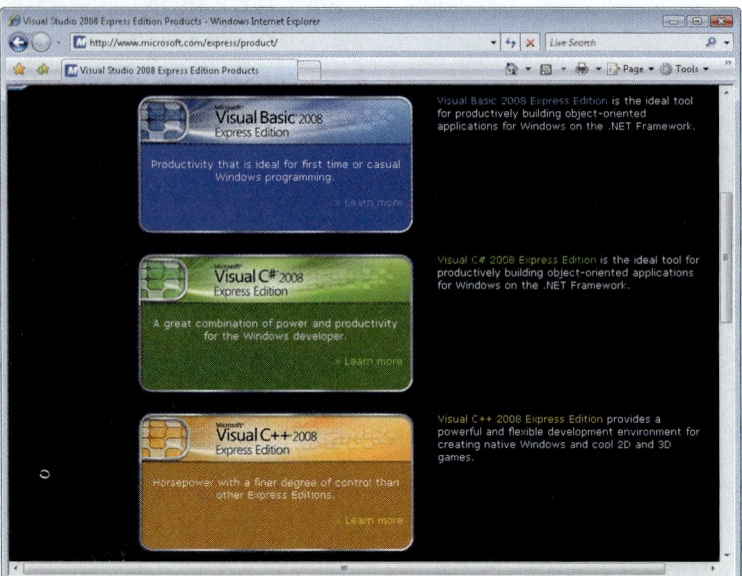

FIGURE TK 2-5 Microsoft offers Express Editions, which are free, but limited, versions of various software development tools.

You also can visit Microsoft's download center to sample various **Express Editions**, which are free, but limited, versions of various software development tools, as shown in Figure TK 2-5. Students also can download a free demo version of Telelogic Tau Modeler, which is shown in Figure TK 2-6. According to Telelogic, the program is a UML-based tool with a model-driven approach that can enhance productivity and software quality. Telelogic, an IBM company, also offers System Architect, which is another powerful CASE tool.

How do you select a CASE tool? The answer depends on the type of project, its size and scope, possible budgetary and time constraints, and the preferences and experience of the system development team. After you study the terms, concepts, and examples in this part of the Toolkit, you will be able to evaluate various products and make an informed decision. The first step in learning about CASE tools is to understand basic CASE terms and concepts.

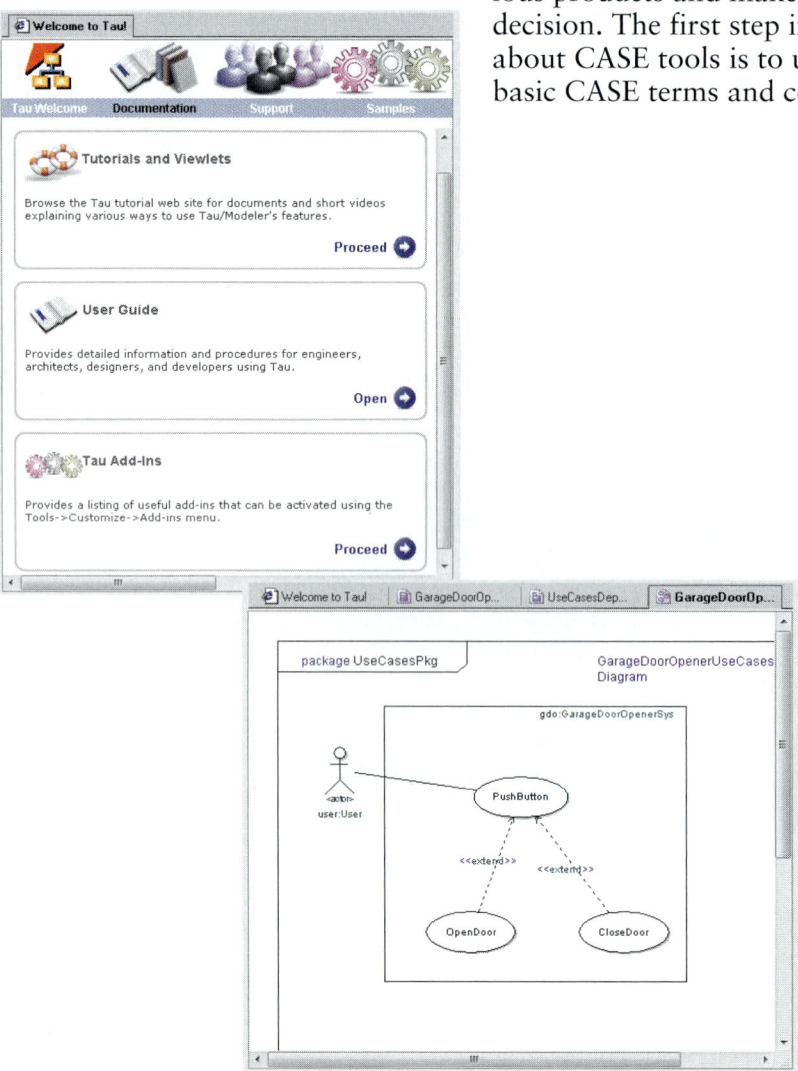

FIGURE TK 2-6 Telelogic Tau Modeler is a free, UML-based tool with a model-driven approach.

CASE Terms and Concepts

A typical CASE tool is actually a set of individual tools that share a repository of information. The important terms and concepts are explained in the following sections.

Repository

A **repository** is a database that serves as a central storage location for all information about the system being developed. Once a data element has been defined in the repository, it can be accessed and used by processes and other information systems. For example, your sales processing, accounts receivable, and shipping systems all might require data about customers. After the CUSTOMER data element is entered in the repository, all three systems can share a consistent, up-to-date definition.

When you define a data element in the repository, you can assign a data type and format, a range of acceptable values, and one or more aliases. An **alias** is an alternative name for a data element. The repository can be searched, and all instances of the data element will be listed. For example, Figure TK 2-7 shows a Visible Analyst search for the data element named CUSTOMER NUMBER. As the screens show, you can search the entire repository, or among specific types of diagrams, and the results will show all instances of the data element.

Individual Tools

An integrated set of CASE tools can be used to model, document, engineer, and construct the information system, as explained in the following sections.

MODELING TOOLS Throughout the SDLC, system developers use modeling tools and diagrams to represent the system graphically. The textbook describes many examples, including Unified Modeling Language diagrams and functional decomposition diagrams (Chapter 4), data flow diagrams (Chapter 5), various object diagrams (Chapter 6), entity-relationship diagrams (Chapter 9), and structure charts (Chapter 10). Most popular CASE products offer these modeling tools, among others. One of the most important benefits of a CASE environment is that it provides an overall framework that allows a developer to create a series of graphical models based on data that has already been entered into a central repository.

DOCUMENTATION TOOLS The main source of system documentation is the repository, which was explained in the previous section. In most CASE software, the repository automatically identifies the new entries and adds them to the database. In addition to the repository itself, many CASE products provide tools that

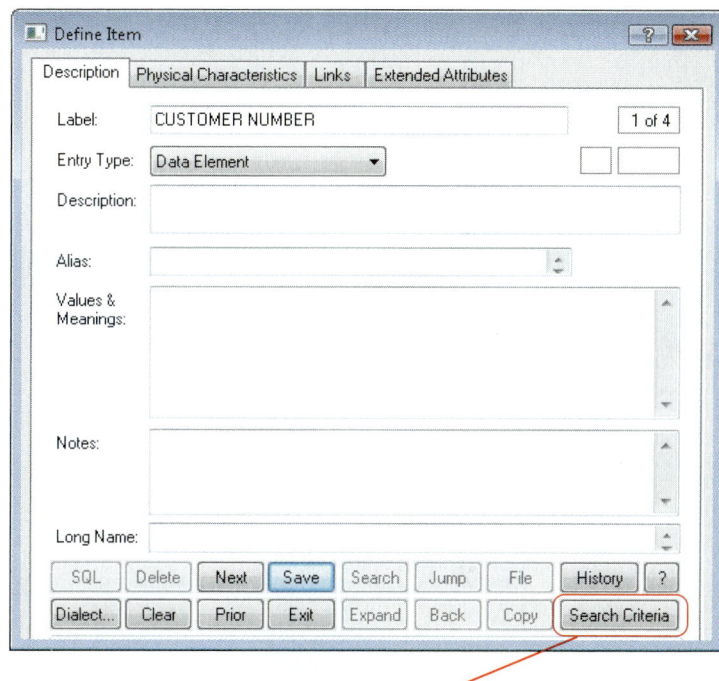

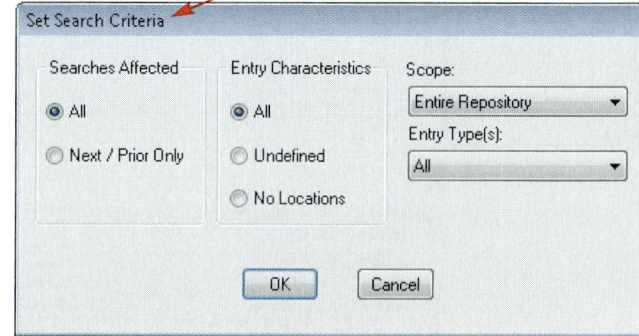

FIGURE TK 2-7 A Visible Analyst repository search for the data element named CUSTOMER NUMBER. The results will show all instances of the data element.

check automatically for inconsistent or incomplete information in forms, reports, and diagrams. This is especially important in large, complex systems.

ENGINEERING TOOLS Engineering tools include forward engineering and reverse engineering tools. **Forward engineering** means translating business processes and functions into applications. Some CASE tools allow you to build the system either by editing objects and code directly, or by modifying graphical representations such as DFDs and UML diagrams. As you learned in Chapter 1, CASE tools such as System Architect and Visible Analyst allow you to develop a business model that can be translated into information system components. **Reverse engineering** allows you to examine an existing application and break it down into a series of diagrams, structure charts, and source code. Using a reverse engineering CASE tool, an analyst can transform existing application source code into a working model of the system. This can be especially important when integrating new systems with legacy systems or systems that were developed in different environments. Figure TK 2-8 shows two examples of reverse engineering tools.

CONSTRUCTION TOOLS A full-featured CASE tool can handle many program development tasks, such as generating application code, screens, and reports.

- An **application generator**, also called a **code generator**, allows you to develop computer programs rapidly by translating a logical model directly into code. As shown in Figure TK 2-9, Telelogic's Tau Modeler uses information stored in the data repository to generate applications in languages such as C, C++, Java, and AgileC.

- A **screen generator**, or **form painter**, is an interactive tool that helps you design a custom interface, create screen forms, and handle data entry format and procedures. The screen generator allows you to control how the screen will display captions, data fields, data, and other visual attributes. Modern CASE tools usually include a screen generator that interacts with the data dictionary. As shown in Figure TK 2-10, Gillani Software's FourGen® CASE tool set includes a form painter and a code generator, along with many other powerful features.

- A **report generator**, also called a **report writer**, is a tool for designing formatted reports rapidly. Using a report generator, you can modify a report easily at any stage of the design process. When you are satisfied with the report layout, the report writer creates a report definition and program code that actually produces the report. You also can input sample field values to create a **mock-up report** for users to review and approve.

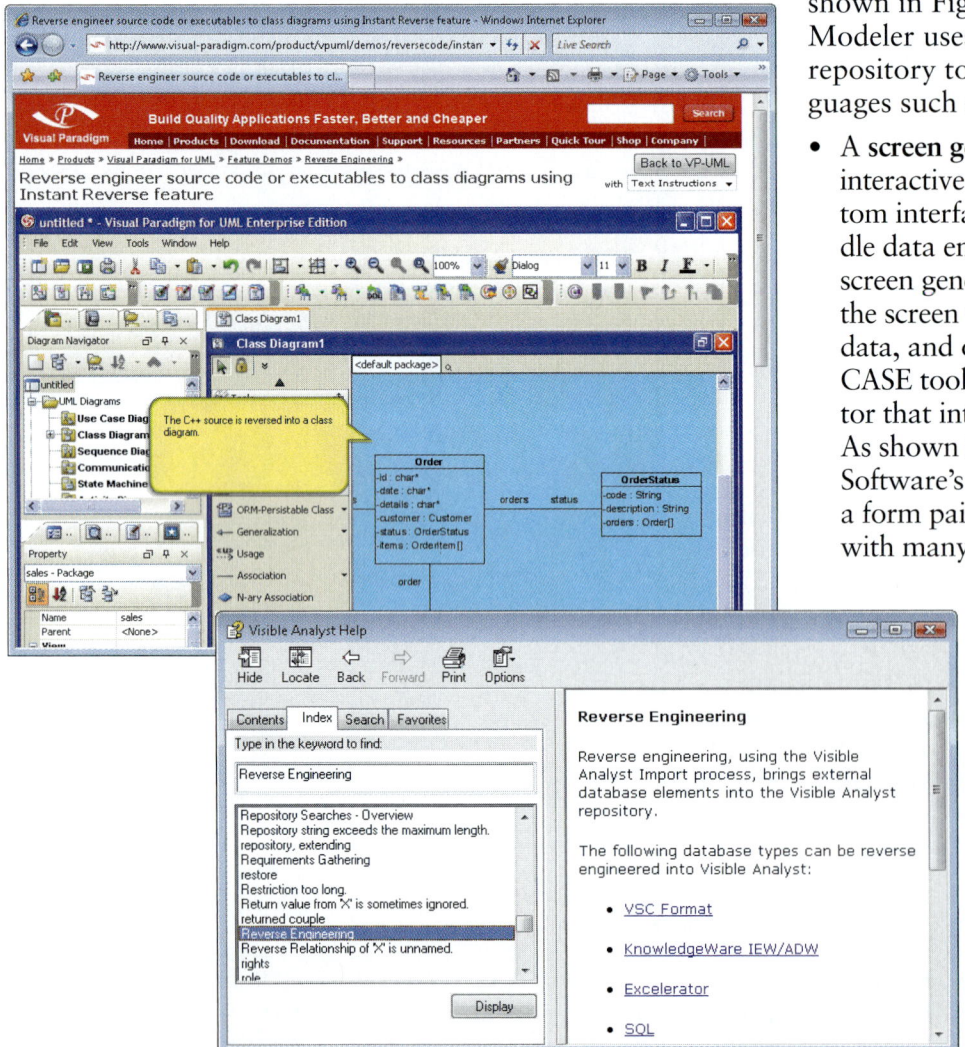

FIGURE TK 2-8 Examples of CASE tools that offer reverse engineering that allows an analyst to study an application or database from a design viewpoint.

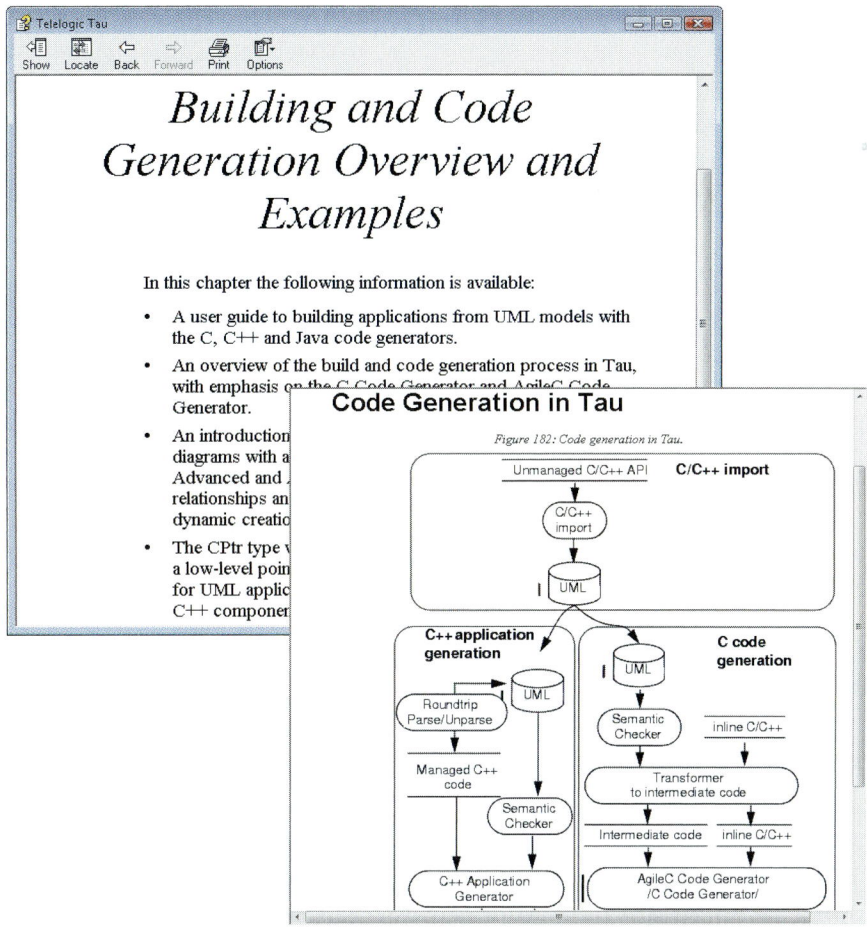

FIGURE TK 2-9 Telelogic Tau Modeler can generate application code in languages such as C, C++, Java, and Agile C.

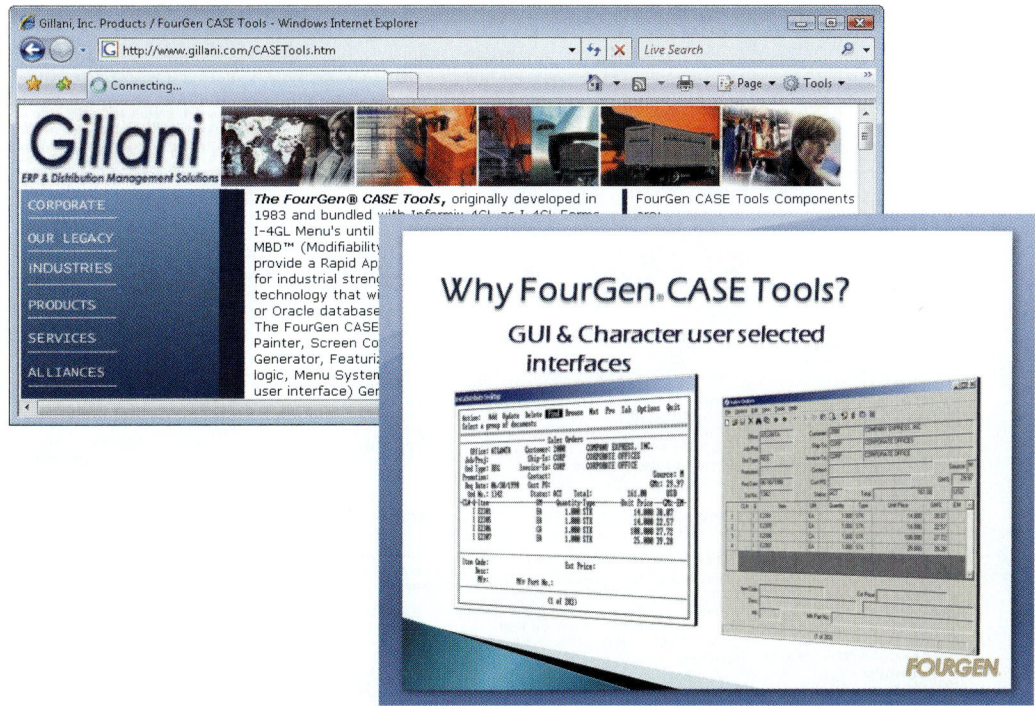

FIGURE TK 2-10 Gillani's FourGen® CASE tools include a form painter and a screen code generator.

INTEGRATED DEVELOPMENT ENVIRONMENTS

ON THE WEB

For more information about integrated development environments, visit **scsite.com/sad8e/ more**, locate Toolkit Part 2, and then click the Integrated Development Environments link.

An **integrated development environment (IDE)** uses a built-in CASE tool that a software vendor includes to make it easier to plan, construct, and maintain a specific software product. Many large firms such as Microsoft, Oracle, and IBM offer IDEs that relate to their family of products. The following sections explain how these tools are used.

Examples of Integrated Development Environments

Although generic CASE tools can be used to plan and design any type of information system, it usually is easier to use an integrated tool that the vendor has provided. For example, as shown in Figure TK 2-11, Oracle provides **Oracle Designer**, which is packaged with Oracle's application software. According to the company, Oracle Designer models business processes, data entities, and relationships — and can transform these models into designs that generate complete applications. Other leading firms, such as SAP and Sybase, also offer powerful development tools.

Figure TK 2-12 shows Microsoft's **Visual Studio 2008**, which is another example of an integrated development environment. Visual Studio 2008 includes various application development tools that are specifically designed to support Microsoft's Web-based application development strategy. In addition to these commercial packages, programmers can use open-source software such as Java-based NetBeans IDE and Eclipse.

Pros and Cons of Integrated Development Tools

In a specific software environment, an integrated development tool is highly effective because it is an integral part of the vendor's software package. The only possible disadvantage is that each IDE is different, and requires a learning curve and skills that might or might not be readily transferable. In contrast, non-specific CASE tools such as Visible Analyst or System Architect can be used in any development environment. Given the dynamic changes in IT, a systems analyst should seek to learn as many development and CASE tools as possible.

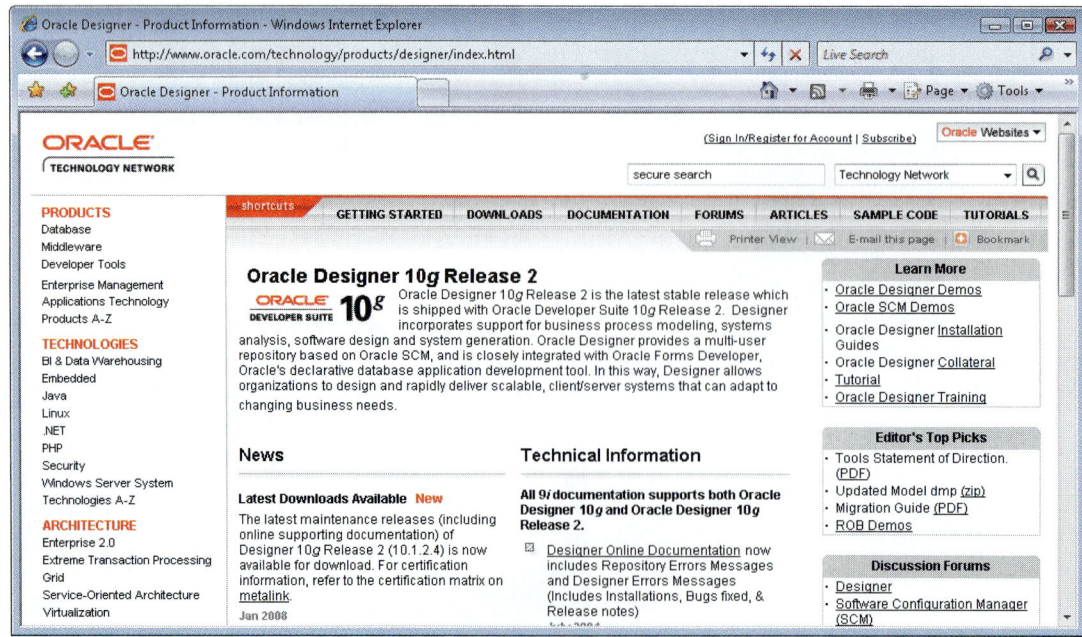

FIGURE TK 2-11   Oracle Designer is a modeling and application development tool packaged with Oracle's application software.

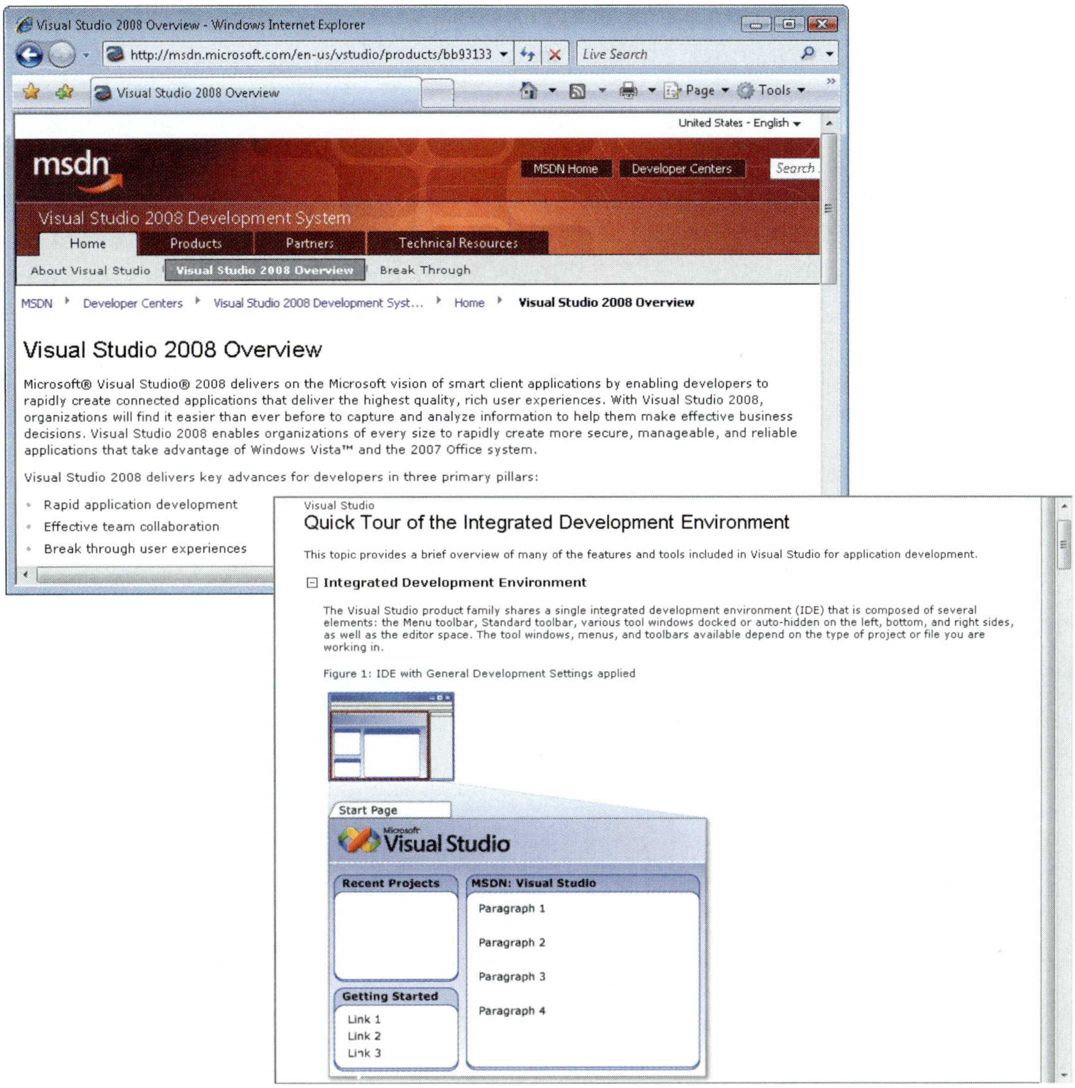

Visual Studio 2008 Overview - Windows Internet Explorer

http://msdn.microsoft.com/en-us/vstudio/products/bb93133   Live Search

Visual Studio 2008 Overview   Page ▾   Tools ▾

FIGURE TK 2-12 Visual Studio 2008 is an integrated development environment that supports Microsoft's family of applications.

CASE TOOL EXAMPLES

You can choose from dozens of vendors and CASE tools that offer a wide range of functions, characteristics, and appearance. The following sections illustrate several features of products offered by three leading CASE tool suppliers: Visible Systems Corporation, Telelogic Software, and Rational Software.

Visible Analyst

Visible Systems Corporation is an important player in the software development market. Visible offers tools for data and application modeling, code generation, and software configuration management.

ON THE WEB

For more information about the Visible Analyst CASE tool, visit **scsite.com/ sad8e/more**, locate Toolkit Part 2, and then click the Visible Analyst CASE Tool link.

The **Visible Analyst**® CASE tool can generate many types of models and diagrams. Figure TK 2-13 shows three sample diagrams for a library system: an entity-relationship diagram, a data flow diagram, and a structure chart. All three examples are integrated with a central data repository.

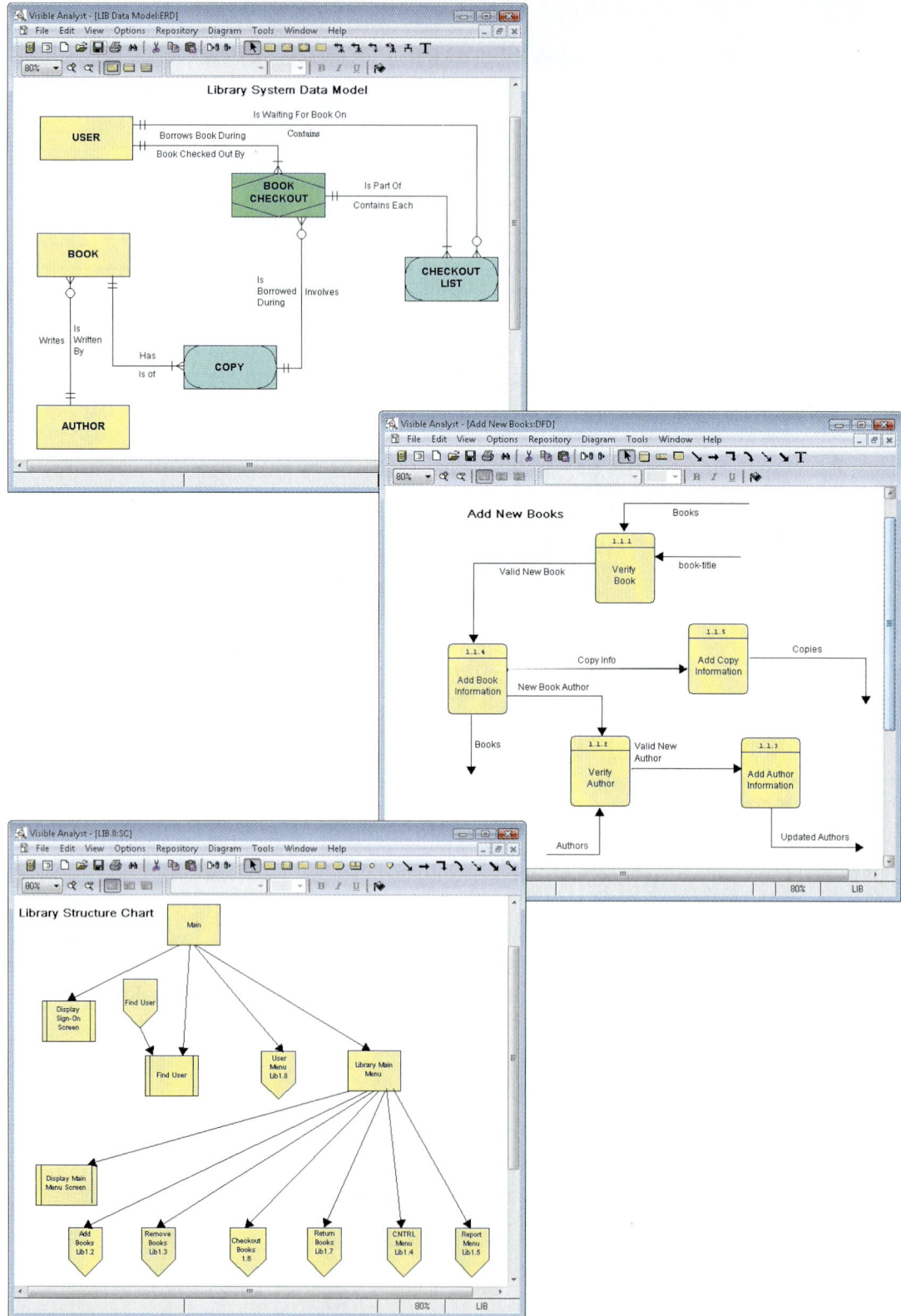

FIGURE TK 2-13 Three sample Visible Analyst diagrams are integrated with the central data repository for a library system.

Visible Analyst also provides a full range of Help features, including the error message analysis screen shown in Figure TK 2-14. When a user clicks an alphabetic letter, he or she can learn more about a specific error message that the program has displayed.

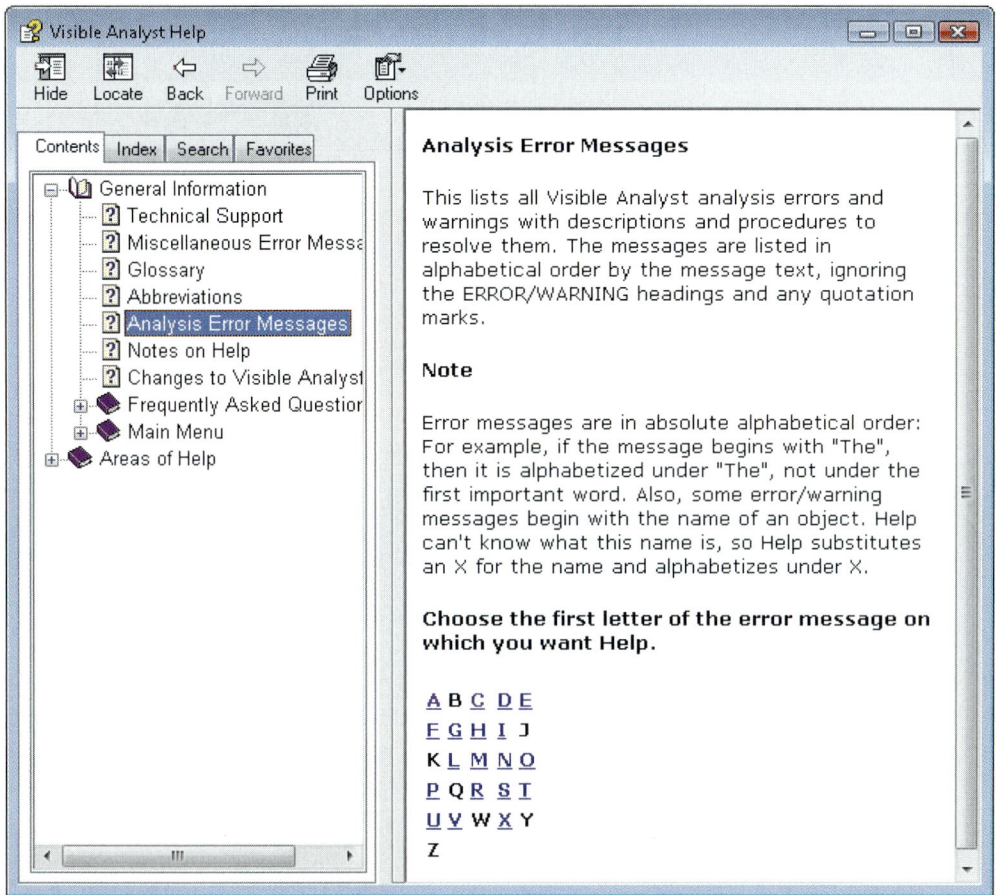

FIGURE TK 2-14 Visible Analyst provides error message analysis. When a user clicks an alphabetic letter, he or she can learn more about a specific error message.

System Architect

On its Web site, Telelogic claims that **System Architect**® leads the market for modeling and design tools that can build enterprise systems in full alignment with business requirements. Telelogic states that System Architect is the first fully integrated modeling tool to unify the enterprise with complete business and systems information management.

Telelogic's emphasis on business modeling is apparent. In addition to the framework concept, which is described in the next section, Telelogic offers numerous diagrams and definitions, all keyed to a particular view of the system. For example, as shown in Figure TK 2-15, when a user wants to model the system from a specific viewpoint, he or she can create a new diagram or definition from an extensive list.

Telelogic also offers extensive tutorials with its System Architect product. For example, as shown in Figure TK 2-15 on the next page, a user can learn about business process analysis and study a sample project to learn the features of the program.

ON THE WEB

For more information about the System Architect CASE tool, visit **scsite.com/ sad8e/more**, locate Toolkit Part 2, and then click the System Architect CASE Tool link.

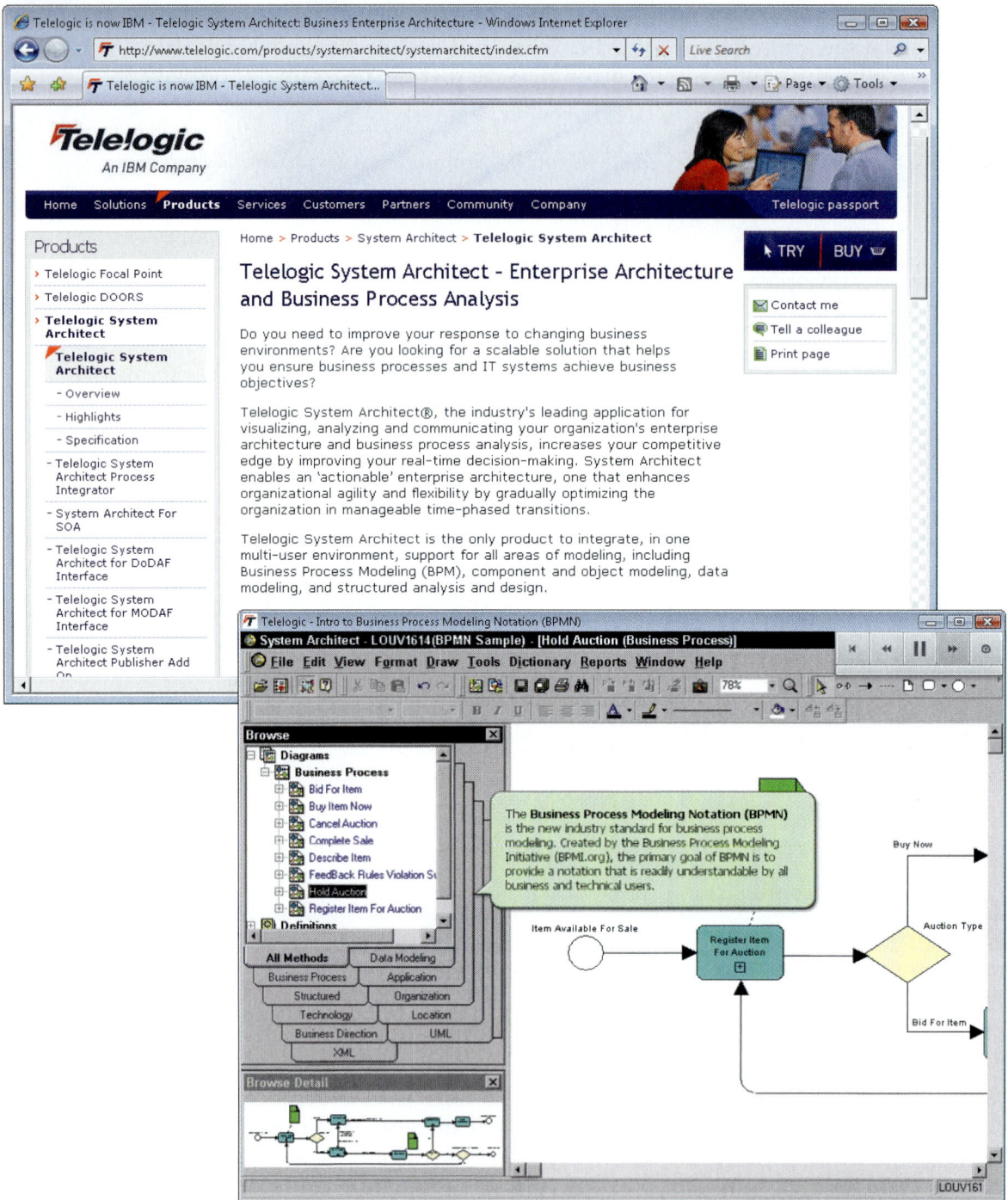

FIGURE TK 2-15 System Architect tutorials inlcude a sample project showing business process modeling notation, which graphically represents system functions.

Rational Software

IBM's Rational Software offers a wide range of systems development and modeling products, including a powerful tool called **Rational Software Architect**. As the upper screen in Figure TK 2-16 shows, Rational Software Architect uses model-driven development and the UML to produce applications and services. By visiting the Rational Web site, you can view and download an online publication called the **Rational Edge**, which is shown in the lower screen in Figure TK 2-16.

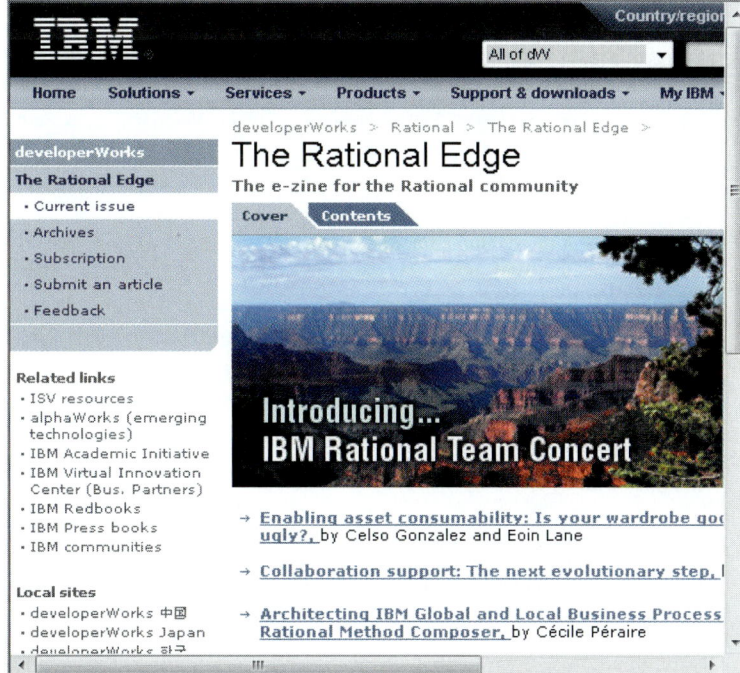

FIGURE TK 2-16 The Rational Edge is an online publication that offers news and information about Rational Software products.

CASE TOOL TRENDS

Traditional CASE software evolved from simple utilities, such as code editors, to powerful multipurpose tools that can help you envision, plan, and build an entire information system. The evolution of CASE software will continue, as developers seek even more powerful tools that can model complex business processes and integrate with customer and supplier systems.

Just as modern spacecraft could not have been built without specialized, high-technology tools, future software will be planned, constructed, and maintained with a new generation of CASE tools. The following sections discuss CASE tool trends and method-specific tools.

New Products and Features

CASE tool vendors constantly offer more features and greater flexibility. One example is a framework to help transform business processes into an information system. A **framework** organizes and documents system development tasks. For example, the **Zachman Framework** shown in Chapter 4 on page 154 arranges traditional fact-finding questions into a useful matrix.

Another example is Telelogic Software's **Framework Manager,** which is shown in Figure TK 2-17. According to Telelogic, the software enables a company to create a business model, where each cell represents a specific business operation or process. Many vendors offer framework models that are industry-specific, such as the U.S. Department of Defense Architecture Framework.

FIGURE TK 2-17 The Framework Manager feature in Telelogic's System Architect allows organizations to build integrated business models.

Also, as software becomes more powerful and complex, the lines between traditional CASE tools and other modeling tools continue to blur. For example, **Microsoft Visio** generally is regarded as a charting tool, but also can model networks, business processes, and many types of special diagrams. Visio offers a variety of online tutorials and training sessions, some of which include self-assessment, as shown in Figure TK 2-18. Visio training starts with the simple examples and progresses to more powerful tools, such as the cross-functional flowcharts shown in Figure TK 2-19 on the next page.

FIGURE TK 2-18 Microsoft Visio offers self-paced online tutorials, complete with a self-assessment feature.

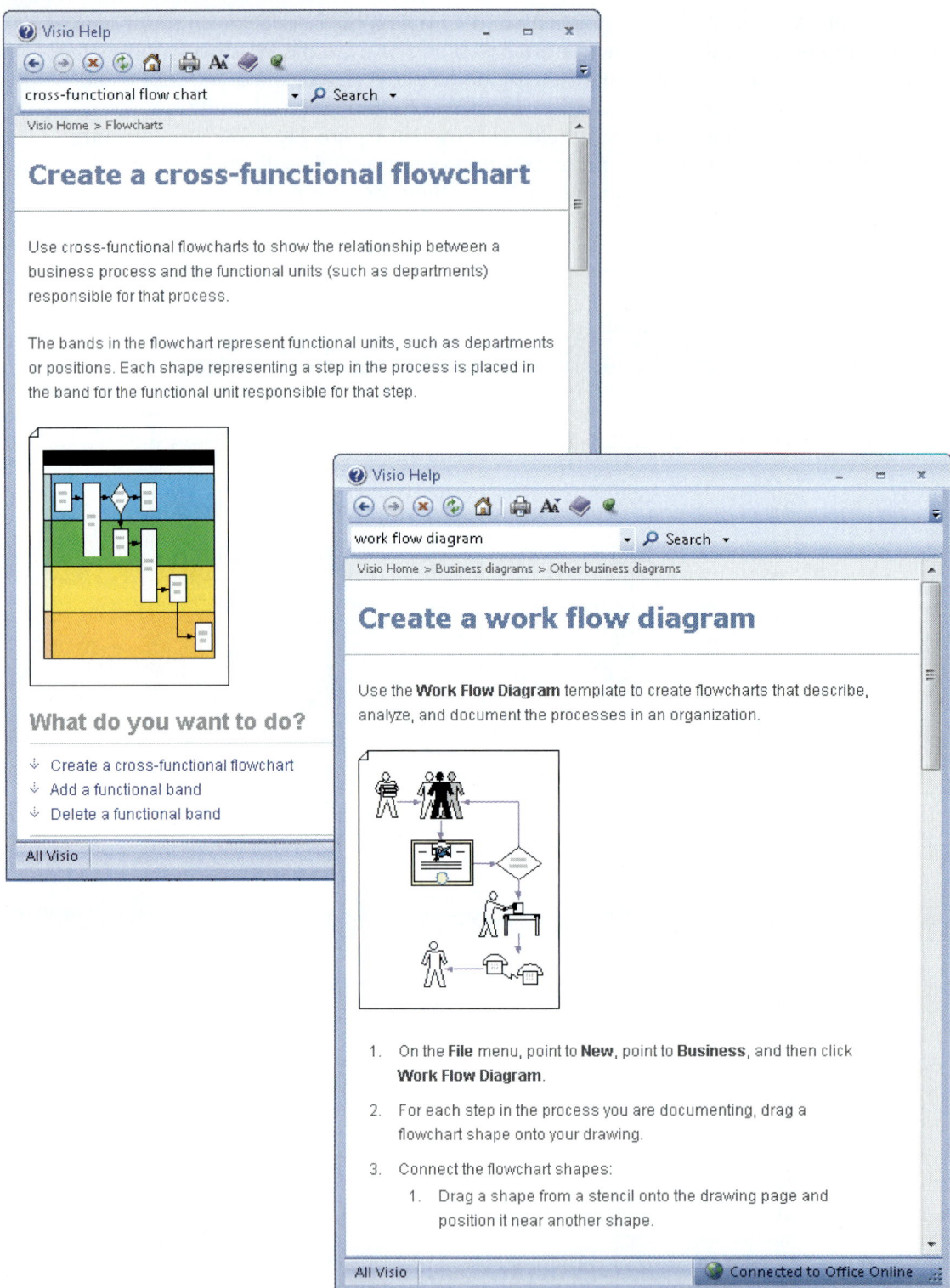

FIGURE TK 2-19 Microsoft Visio offers a wide selection of diagrams, including cross-functional flowcharts and work flow diagrams.

Another trend is the increasing use of integrated development environments. Chapter 11 explains how an IDE can simplify the integration of system components and reduce code development time. An IDE typically includes built-in tools such as real-time error detection, syntax hints, highlighted code, class browsers, and version control. In addition to IBM's

WebSphere and Microsoft's .NET, programmers can use open-source IDEs such as Java-based NetBeans IDE and Eclipse. Microsoft's Visual Studio series, which was discussed in an earlier section, is an IDE that focuses on Web-based development.

Method-Specific CASE Tools

As Chapter 11 explains, each systems development approach has a set of tools that has worked especially well for that method. For example, structured development relies heavily on DFDs and structure charts. Object-oriented methods use a variety of diagrams, such as use case, class, sequence, and transition state diagrams. Agile methods tend to use spiral or other iterative models. In Chapter 1, Figure 1-23 on page 18 lists several method-specific modeling tools. System developers also use multipurpose tools to help them translate the system logic into properly functioning program modules. These generic tools include entity-relationship diagrams, flowcharts, pseudocode, decision tables, and decision trees.

Structured analysis is a traditional approach that is time-tested and easy to understand. Structured modeling tools are described in detail in Chapter 5, Data and Process Modeling. However, as Chapter 1 points out, **object-oriented analysis and design (OOAD)** is very popular. Widespread use of object-oriented languages has spurred interest in O-O CASE and UML-based modeling tools, which provide seamless development from planning to actual coding. Other O-O features include modular design and reusable code, which can reduce costs and speed up development. Object-oriented analysis and design tools and techniques are described in Chapter 6, Object Modeling.

The most recent trend is the popularity of agile methods. Chapter 11 includes a detailed description of an agile project, including the iterative cycles and the intense contact between developers and users. According to Scott W. Ambler, a well-known IT consultant, agile developers use a wide range of modeling tools, including CASE tools. However, many agile teams find that simple whiteboard sketching works best for them, as shown in Figure TK 2-20.

Although it is difficult to predict the future, it seems clear that CASE tools will continue to evolve and become more powerful. At the same time, system developers will sometimes choose simpler, low-tech methods and techniques as modeling tools.

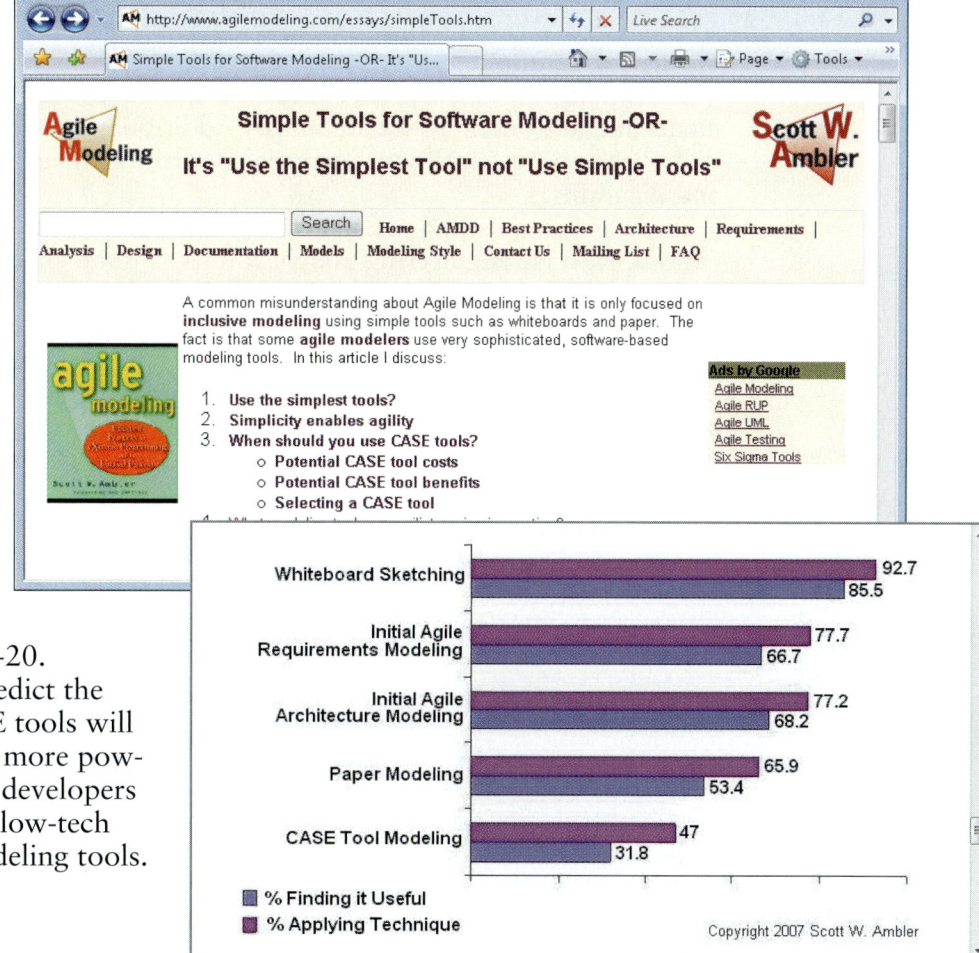

FIGURE TK 2-20 Agile teams use a wide range of modeling techniques, from simple whiteboards to CASE tools.

TOOLKIT SUMMARY

CASE stands for computer-aided systems engineering. CASE tools are software programs that system developers use to help them design and construct information systems. CASE tools can reduce costs, speed up development, and provide comprehensive documentation that can be used for future maintenance or enhancements.

Older systems used program code that was written in procedural languages such as COBOL, which required a programmer to create code statements for each processing step. Modern languages such as C++ and Java are non-procedural, or event-driven, languages because a programmer defines the actions that the program must perform when certain events occur.

Non-procedural languages, called 4GLs, are object-oriented programming languages (OOPL). 4GLs are important components of a fourth-generation environment that allows system developers to develop accurate prototypes, cut development time, and reduce expense.

A repository is a database that serves as a central storage location for all information about the system being developed. Once a data element has been defined in the repository, it can be accessed and used by processes and other information systems. An alias is an alternative name for a data element. The repository can be searched, and all instances of the data element will be listed.

An integrated set of CASE tools can be used to model, document, engineer, and construct the information system. Modeling tools represent the system graphically by using various types of diagrams, including data flow diagrams (DFDs), Unified Modeling Language (UML) diagrams, functional decomposition diagrams, structure charts, and network diagrams.

The main source of system documentation is the repository, which identifies new elements and adds them to the database. Additional documentation is provided by tools that check automatically for inconsistent or incomplete information in forms, reports, and diagrams.

Forward engineering means translating business processes and functions into applications. Reverse engineering allows you to examine an existing application and break it down into a series of diagrams, structure charts, and, in some cases, source code.

A CASE tool can handle many program development tasks, such as generating application code, screens, and reports.

An integrated development environment (IDE) uses a built-in CASE tool that a software vendor includes to make it easier to plan, construct, and maintain a specific software product. Examples of IDEs include Oracle Designer and Microsoft's Visual Studio 2008.

Two trends seem clear: CASE tool vendors will continue to include powerful new features, and the popularity of object-oriented tools will continue to grow.

Key Terms and Phrases

alias 637

application generator 638

CASE environment 634

CASE tools 632

code generator 638

computer-aided software engineering 632

computer-aided systems engineering (CASE) 632

event-driven programming language 634

Express Editions 636

form painter 638

forward engineering 638

fourth-generation environment 634

fourth-generation languages (4GLs) 634

framework 646

Framework Manager 646

integrated development environment (IDE) 640

Microsoft Visio 647

mock-up report 638

non-procedural programming language 634

object-oriented analysis and design (OOAD) 649

object-oriented programming languages (OOPL) 634

Oracle Designer 640

procedural 634

Rational Edge 644

Rational Software Architect 644

report generator 638

report writer 638

repository 637

reverse engineering 638

screen generator 638

structured analysis 649

System Architect® 643

Visible Analyst® 642

Visual Studio 2008 640

Zachman framework 646

Toolkit Exercises

Review Questions

1. Define CASE, CASE tools, and a CASE environment.
2. Explain the difference between procedural and non-procedural languages.
3. Describe 4GLs and their characteristics.
4. Define a repository, and explain its role in the systems development process.
5. What are forward and reverse engineering tools, and how are they used?
6. Provide an example of an application generator and a screen generator.
7. How is a report generator used, and what is a mock-up report?
8. Explain the concept of an integrated development environment and provide two examples of IDEs.
9. What are some features of Telelogic's Tau Modeler and Microsoft's Visio?
10. What is the emerging role of object-oriented analysis and design methods? Agile methods?

Discussion Topics

1. Would a systems analyst be better off in a position where he or she works with an IDE, or where generic CASE tools are used? Explain your answer.
2. Visit the Web sites for System Architect and Visible Analyst. If you could choose only one of these products, which one would you select, and why?
3. If you were a programmer, would you prefer to work with procedural or non-procedural languages? Explain your reasons.
4. Review the Dilbert© cartoon on page 137. Although the example might be far-fetched, perhaps future software will be able to identify business opportunities and requirements. Meanwhile, if the trend toward more powerful CASE tools continues, many of the tedious program development tasks might be performed automatically. Is there a limit to the capabilities of future CASE software? Could a complete information system be designed by describing a business operation and specifying certain inputs and outputs? Explain your answer.

Projects

1. Go to the site shown in Figure TK 2-3 on page 635 and choose a CASE tool. Visit the vendor's site and learn all you can about the product. Write a brief report that describes your experience.
2. Visit the Telelogic Web site and navigate to the area that describes Telelogic Tau Modeler. Check to be sure that the free demo software is available. Download and install the program. Experiment with the software, and write a brief report that describes your experience.
3. Search the Internet and locate an example of a screen generator. Visit the vendor's site and learn all you can about the product. Write a brief report that describes your experience.
4. Go to the site shown in Figure TK 2-10 on page 639 and learn more about Gillani's FourGen CASE Tools. Write a brief report summarizing your findings.

PART 3 Financial Analysis Tools

In **Part 3** of the Systems Analyst's Toolkit, you will learn how to use financial analysis tools during the planning, analysis, design, implementation, support, and securing of an information system.

INTRODUCTION

OBJECTIVES

When you finish this part of the Toolkit, you will be able to:

- Define economic feasibility
- Classify costs and benefits into various categories, including tangible or intangible, direct or indirect, fixed or variable, and developmental or operational
- Understand chargeback methods and how they are used
- Use payback analysis to calculate the length of time that it takes for a project to pay for itself
- Use return on investment analysis to measure a project's profitability
- Use present value analysis to determine the value of a future project measured in current dollars

Part 3 of the Systems Analyst's Toolkit shows you how to use various tools to calculate a project's costs and benefits. As a systems analyst, you need to know how to calculate costs and benefits when you conduct preliminary investigations, evaluate IT projects, and make recommendations to management.

Financial analysis tools are important throughout the systems development life cycle. For example, in Chapter 2 you learn that economic feasibility depends on a comparison of costs and benefits. A project is economically feasible if the future benefits outweigh the estimated costs of developing or acquiring the new system. In Chapter 7, when you analyze development strategies, you apply financial analysis tools and techniques as you examine various options. Then, as Chapter 12 explains, you use these tools again to recognize the end of a system's useful life.

TOOLKIT INTRODUCTION CASE: Mountain View College Bookstore

Background: Wendy Lee, manager of college services at Mountain View College, wants a new information system that will improve efficiency and customer service at the three college bookstores.

In this part of the case, Florence Fullerton (systems analyst) and Harry Boston (student intern) are talking about financial analysis tools and techniques.

Participants:	Florence and Harry
Location:	Mountain View College Cafeteria, during the systems planning phase
Discussion topics:	Economic feasibility, chargeback methods, cost-benefit classification, payback analysis, return on investment analysis, and net present value analysis

Florence: Hi, Harry. Before we go any further with the preliminary investigation, I wanted to meet with you to discuss financial analysis tools and techniques, and how we will apply them.

Harry: *Fine with me. Where do we start?*

Florence: Well, because economic feasibility depends on a project's costs and benefits, the first step is to classify those costs and benefits into specific categories. For example, costs can be tangible or intangible, direct or indirect, fixed or variable, and developmental or operational.

Harry: *Okay, I understand. What do we do after we classify everything?*

Florence: Then we use one or more financial analysis tools: payback analysis, return on investment analysis, and net present value analysis. We'll learn how to use them to evaluate the system and compare alternatives. We'll also talk about chargeback methods.

Harry: *What are they?*

Florence: A chargeback method is just a way to allocate costs for an IT system. We'll have to talk to Wendy about that. Meanwhile, here's a task list we can work on:

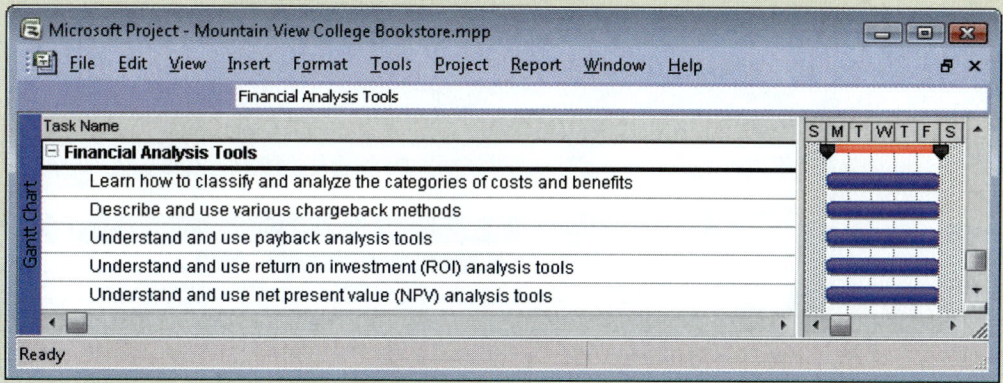

FIGURE TK 3-1 Typical financial analysis tasks.

DESCRIBING COSTS AND BENEFITS

As a systems analyst, you must review a project's costs and benefits at the end of each SDLC phase so management can decide whether or not to continue the project. Before you can use the economic analysis tools described in this section of the Toolkit, you must learn how to identify and classify all costs and benefits.

As you learned in Chapter 2, **economic feasibility** means that the projected benefits of the proposed system outweigh the projected costs. When you determine economic feasibility, you must consider the project's benefits compared to the project's **total cost of ownership (TCO)**, which includes ongoing support and maintenance costs, as well as acquisition costs.

Figure TK 3-2 shows an online TCO analysis tool provided by HP Services. HP stresses the importance of TCO analysis, and points to studies showing that the majority of total IT costs occur *after* the purchase, and that nearly half the costs lie outside the IT

FIGURE TK 3-2 HP Services offers an online TCO analysis tool.

department's budget. HP also cites a study that shows a staggering TCO of $21,000 for a $2,000 PC when all costs are considered over a five-year period. HP noted that the most significant cost factor is user support, including peer-to-peer assistance that rarely is documented or measured.

Cost-benefit analysis tools also are available from various vendors and organizations, including the site shown in Figure TK 3-3, which is supported by the European Union, and is intended to provide various tools that encourage technology innovation and management.

FIGURE TK 3-3   Example of cost-benefit analysis software available on the Web.

Cost-benefit analysis is used by business and government as a decision-making tool. For example, Figure TK 3-4 on the next page shows the Web site of the Center for Information Technology (CIT), which is a support division for the National Institutes of Health. Notice that the agency describes cost-benefit analysis as a vital management tool that is used throughout the systems development process.

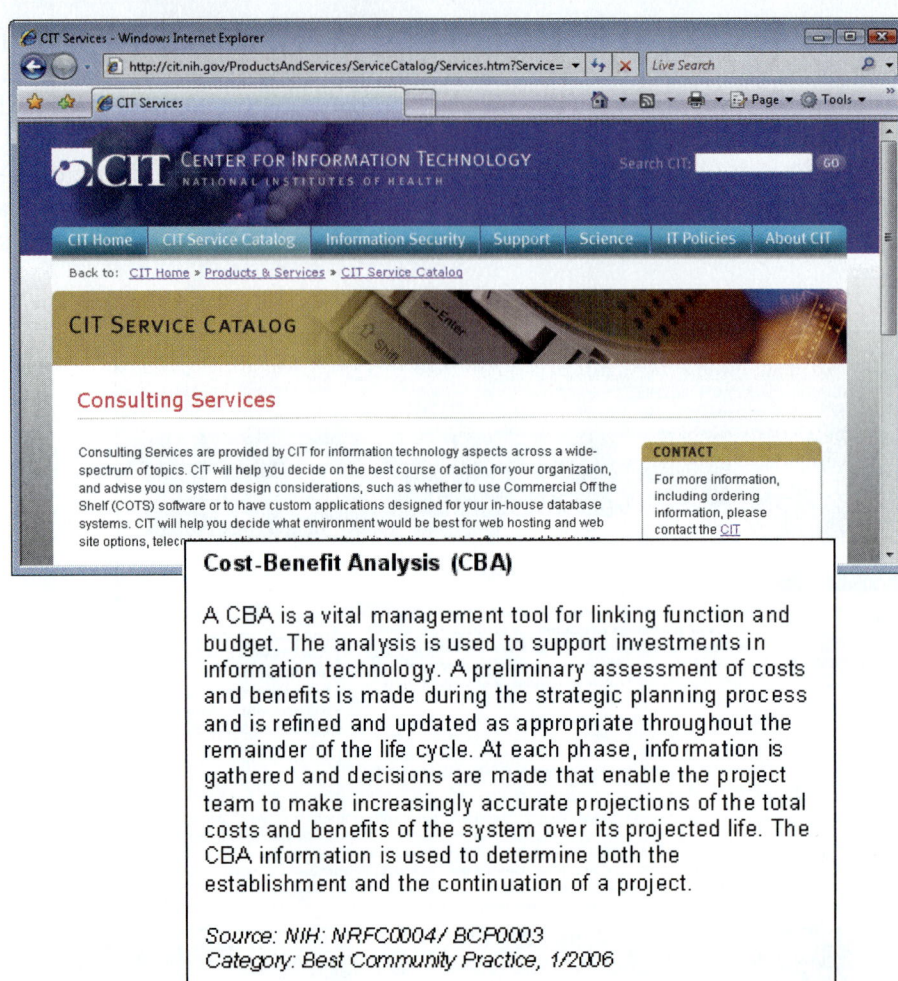

FIGURE TK 3-4 The Center for Information Technology (CIT) provides IT support for the National Institutes of Health, including cost-benefit analysis.

Cost Classifications

Costs can be classified as tangible or intangible, direct or indirect, fixed or variable, and developmental or operational. **Tangible costs** are costs for which you can assign a specific dollar value. Examples of tangible costs include employee salaries, hardware and software purchases, and office supplies. Tangible costs also include the interest charges that firms must pay when they need to borrow money for working capital or to finance new projects. In times of high interest rates, these costs can be significant and must be considered.

In contrast, **intangible costs** are costs whose dollar value cannot be calculated easily. The cost of customer dissatisfaction, lower employee morale, and reduced information availability are examples of intangible costs.

If the analyst examines an intangible item carefully, however, it sometimes is possible to estimate a dollar value. For example, users might dislike a system because it is difficult to learn. Their dissatisfaction is an intangible cost, but if it translates into an increase in errors that must be corrected, you probably could assign a tangible dollar cost. You should try to work with tangible costs whenever possible.

Direct costs are costs that can be associated with the development of a specific system. Examples of direct costs include the salaries of project team members and the purchase of hardware that is used only for the new system. In contrast, **indirect costs**, or

overhead expenses, cannot be attributed to the development of a particular information system. The salaries of network administrators, copy machine rentals, and insurance expenses are examples of indirect costs.

Fixed costs are costs that are relatively constant and do not depend on a level of activity or effort. Many fixed costs recur regularly, such as salaries and hardware rental charges. **Variable costs** are costs that vary depending on the level of activity. The costs of printer paper, supplies, and telephone line charges are examples of variable costs.

Developmental costs are incurred only once, at the time the system is developed or acquired. Those costs might include salaries of people involved in systems development, software purchases, initial user training, and the purchase of necessary hardware or furniture. **Operational costs** are incurred after the system is implemented and continue while the system is in use. Examples of operational costs include system maintenance, ongoing training, annual software license fees, and communications expense.

Some costs apply to more than one category of expenses. For example, overtime pay for clerical staff during the systems analysis phase would be classified as developmental, variable, and direct. A monthly fee for maintaining the company's Web site would be regarded as operational, fixed, and indirect.

Managing Information Systems Costs and Charges

Management wants to know how much an information system costs, so it is important for the systems analyst to understand direct costs, indirect costs, and methods of allocating IT charges within the company.

Direct costs usually are easier to identify and predict than indirect costs. For example, the salaries of project team members and the purchase of hardware, software, and supplies for the new system are direct costs. After a new information system goes into operation, other direct costs might include the lease of system-specific hardware or software.

Many IT department costs cannot be attributed directly to a specific information system or user group. Those indirect costs can include general hardware and software acquisition expenses; facility maintenance, air conditioning, security, rent, insurance, and general supplies; and the salaries of operations, technical support, and information center personnel.

A **chargeback method** is a technique that uses accounting entries to allocate the indirect costs of running the IT department. Most organizations adopt one of four chargeback methods: no charge, a fixed charge, a variable charge based on resource usage, or a variable charge based on volume.

1. **No charge method.** Some organizations treat information systems department indirect expenses as a necessary cost of doing business, and IT services are seen as benefiting the entire company. Thus, indirect IT department costs are treated as general organizational costs and are not charged to other departments. In this case, the information systems department is called a **cost center,** because it generates accounting charges with no offsetting credits for IT services.

2. **Fixed charge method.** With this method, the indirect IT costs are divided among all the other departments in the form of a fixed monthly charge. The monthly charge might be the same for all departments or based on a relatively constant factor such as department size or number of workstations. By using a fixed charge approach, all indirect costs are charged to other departments, and the IT group is regarded as a profit center. A **profit center** is a department that is expected to break even or show a profit. Under the profit center concept, company departments purchase services from the IT department and receive accounting charges that represent the cost of providing the services.

3. **Variable charge method based on resource usage. Resource allocation** is the charging of indirect costs based on the resources used by an information system. The allocation might be based on connect time, server processing time, network resources required, printer use, or a combination of similar factors. **Connect time** is the total time that a user is connected actively to a remote server — some Internet service providers use this as a basis for charges. In a client/server system, **server processing time** is the time that the server actually responds to client requests for processing. The amount a particular department is charged will vary from month to month, depending not only on that department's resource usage, but also on the total resource usage. The IT department is considered a profit center when an organization uses the resource allocation method.

4. **Variable charge method based on volume.** The indirect IT department costs are allocated to other departments based on user-oriented activity, such as the number of transactions or printing volume. As with the resource allocation method, a department's share of the costs varies from month to month, depending on the level of activity. In this case, the IT department is considered a profit center.

Benefit Classifications

In addition to classifying costs, you must classify the benefits that the company expects from a project. Like costs, benefits can be classified as tangible or intangible, fixed or variable, and direct or indirect. Another useful benefit classification relates to the nature of the benefit: positive benefits versus cost-avoidance benefits. **Positive benefits** increase revenues, improve services, or otherwise contribute to the organization as a direct result of the new information system. Examples of positive benefits include improved information availability, greater flexibility, faster service to customers, higher employee morale, and better inventory management.

In contrast, **cost-avoidance benefits** refer to expenses that would be necessary if the new system were not installed. Examples of cost-avoidance benefits include handling the work with current staff instead of hiring additional people, not having to replace existing hardware or software, and avoiding problems that otherwise would be faced with the current system. Cost-avoidance benefits are just as important as positive benefits, and you must consider both types when performing cost-benefit analysis.

COST-BENEFIT ANALYSIS

ON THE WEB

For more information about payback analysis, visit **scsite.com/ sad8e/more**, locate Toolkit Part 3, and then click the Payback Analysis link.

Cost-benefit analysis is the process of comparing the anticipated costs of an information system to the anticipated benefits. Cost-benefit analysis is performed throughout the SDLC to determine the economic feasibility of an information system project and to compare alternative solutions. Many cost-benefit analysis techniques exist. This section covers discussion of only the three most common methods: payback analysis, return on investment analysis, and present value analysis. Each of the approaches analyzes cost-benefit figures differently, but the objective is the same: to provide reliable information for making decisions.

Payback Analysis

Payback analysis is the process of determining how long it takes an information system to pay for itself. The time it takes to recover the system's cost is called the **payback period**. To perform a payback analysis, you carry out the following steps:

1. Determine the initial development cost of the system.

2. Estimate annual benefits.

3. Determine annual operating costs.

4. Find the payback period by comparing total development and operating costs to the accumulated value of the benefits produced by the system.

When you plot the system costs over the potential life of the system, you typically see a curve such as the one shown in Figure TK 3-5. After the system is operational, costs decrease rapidly and remain relatively low for a period of time. Eventually, as the system requires more maintenance, costs begin to increase. The period between the beginning of systems operation and the point when operational costs are rapidly increasing is called the **economically useful life** of the system.

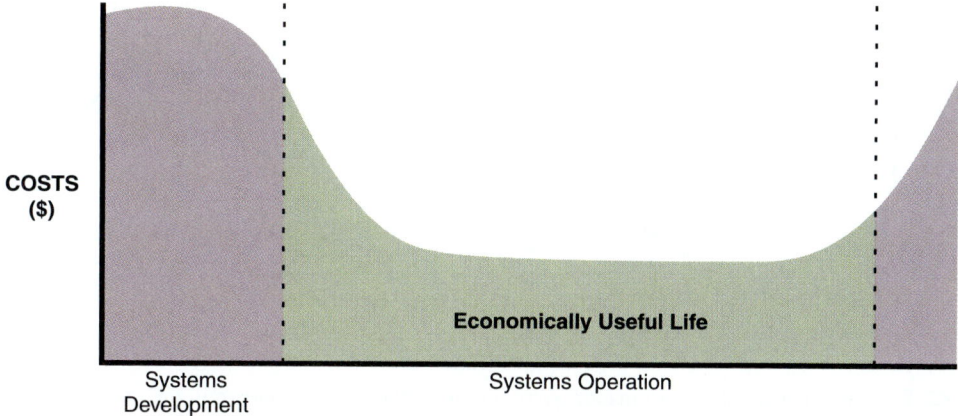

FIGURE TK 3-5 The costs of a typical system vary over time. At the beginning, system costs are high due to initial development expense. Costs then drop during systems operation. Maintenance costs begin to increase until the system reaches the end of its economically useful life. The area between the two dashed lines shows the economically useful life of this system.

When you plot the benefits provided by an information system against time, the resulting curve usually resembles the one shown in the upper graph in Figure TK 3-6 on the next page. Benefits start to appear when the system becomes operational, might increase for a time, and then level off or start to decline.

When conducting a payback analysis, you calculate the time it takes for the accumulated benefits of an information system to equal the accumulated costs of developing and operating the system.

In the lower graph in Figure TK 3-6, the cost and benefit curves are plotted together. The dashed line indicates the payback period. Notice that the payback period is not the point when current benefits equal current costs, where the two lines cross. Instead, the payback period compares accumulated costs and benefits. If you graph current costs and benefits, the payback period corresponds to the time at which the areas under the two curves are equal.

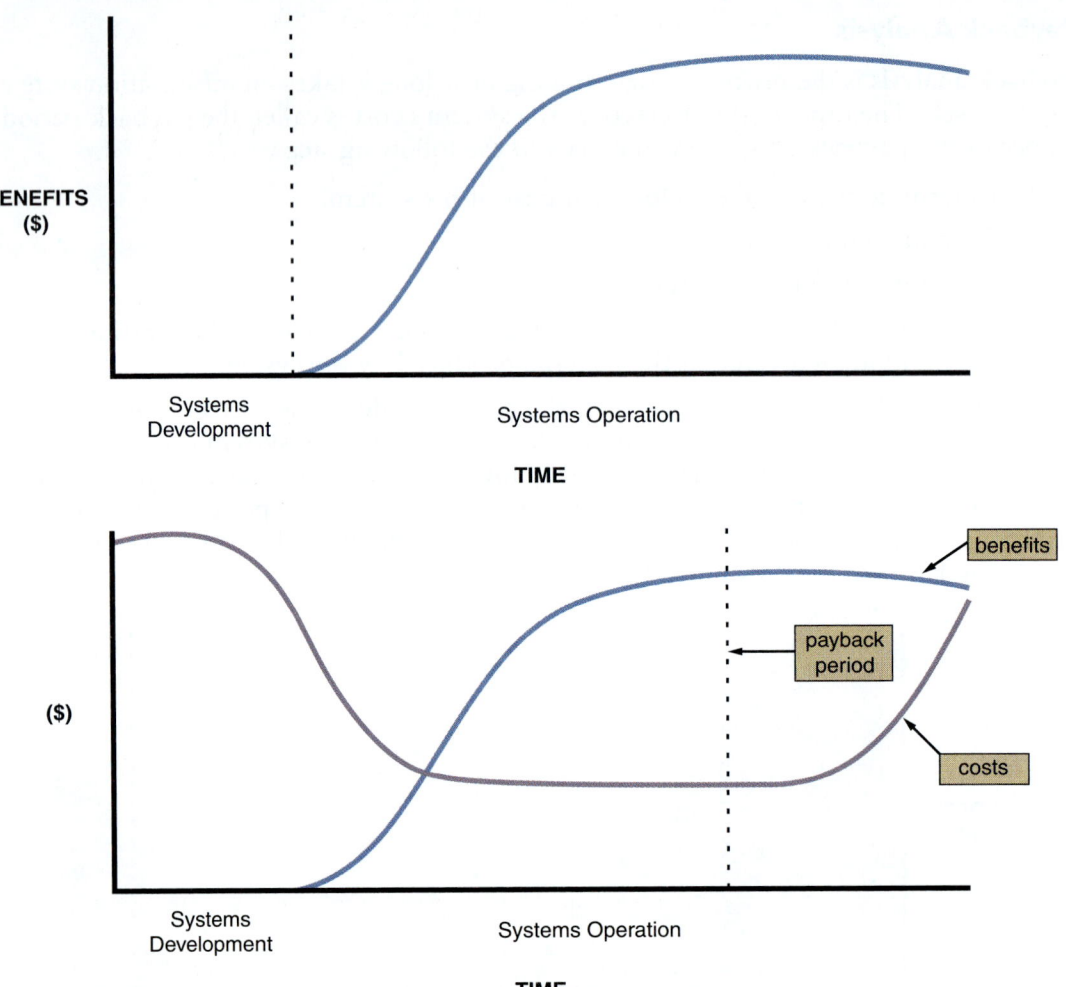

FIGURE TK 3-6 Benefits of an information system change over time, as shown in the upper graph. The lower graph shows costs and benefits plotted on the same graph. The dashed line indicates the payback period, when accumulated benefits equal accumulated costs.

Figure TK 3-7 contains two cost-benefit tables. The tables show the anticipated annual costs, cumulative costs, annual benefits, and cumulative benefits for two information systems projects. Year 0 (zero) corresponds to the year in which systems development begins. The development of Project A takes less than one year, so some benefits are realized in Year 0. Systems development for Project B requires more than one year, so the benefits do not begin until some time in Year 1.

In Project A, by the end of Year 4, the cumulative costs are $135,700, which slightly exceeds the $132,000 cumulative benefits. By the end of Year 5, however, the cumulative benefits of $171,000 far exceed the cumulative costs, which are $157,700. Therefore, at some point in time during Year 5, the accumulated costs and benefits are equal, and the payback period is established. In Project B, a similar situation exists. By the end of Year 4, Project B's cumulative costs are $191,000, which is greater than the cumulative benefits of $156,000. At some point during Year 5, cumulative benefits will exceed cumulative costs, and the system will have paid for itself.

If more specific information is available regarding the timing of costs and benefits during a year, you can calculate the payback period more precisely. Another approach is to create a chart that shows the exact point when cumulative benefits exceed cumulative costs, which is explained in the following section.

PAYBACK ANALYSIS EXAMPLES

PROJECT A:

YEAR	COSTS	CUMULATIVE COSTS	BENEFITS	CUMULATIVE BENEFITS
0	60,000	60,000	3,000	3,000
1	17,000	77,000	28,000	31,000
2	18,500	95,500	31,000	62,000
3	19,200	114,700	34,000	96,000
4	21,000	135,700	36,000	132,000
5	22,000	157,700	39,000	171,000
6	23,300	181,000	42,000	213,000

PROJECT B:

YEAR	COSTS	CUMULATIVE COSTS	BENEFITS	CUMULATIVE BENEFITS
0	80,000	80,000	——	——
1	40,000	120,000	6,000	6,000
2	25,000	145,000	26,000	32,000
3	22,000	167,000	54,000	86,000
4	24,000	191,000	70,000	156,000
5	26,500	217,500	82,000	238,000
6	30,000	247,500	92,000	330,000

FIGURE TK 3-7 Payback analysis data for two information systems proposals: Project A and Project B.

Some managers are critical of payback analysis because it places all the emphasis on early costs and benefits and ignores the benefits received after the payback period. Even if the benefits for Project B in Year 6 soared as high as $500,000, the payback period for that project still occurs during the fifth year of operation. In defense of payback analysis, the earlier cost and benefit predictions usually are more certain. In general, the further out in time that you extend your projections, the more unsure your forecast will be. Thus, payback analysis uses the most reliable of your cost and benefit estimates.

Payback analysis rarely is used to compare or rank projects because later benefits are ignored. You would never decide that Project A is better than Project B simply because the payback period for A is less than that for B; considering all the costs and all the benefits when comparing projects makes more sense.

Even with its drawbacks, payback analysis is popular. Many business organizations establish a minimum payback period for approved projects. If company policy requires a project to begin paying for itself within three years, then neither project in Figure TK 3-7 would be approved, though both are economically feasible because total benefits exceed total costs.

Using a Spreadsheet to Compute Payback Analysis

You can use a spreadsheet to record and calculate accumulated costs and benefits, as shown in Figure TK 3-8 on the next page. The first step is to design the worksheet and label the rows and columns. After entering the cost and benefit data for each year, you enter the formulas. For payback analysis, you will need a formula to display cumulative totals, year by year. For example, the first year in the cumulative costs column is the same as Year 0 costs, so the formula in cell C6 is =B6. The cumulative cost total for the second year is Year 0 cumulative total + Year 1 costs, so the formula for cell C7 is =C6+B7, and so on. The first worksheet shows the initial layout and the second worksheet shows the finished spreadsheet.

After you verify that the spreadsheet operates properly, you can create a line chart that displays the cumulative costs, benefits, and payback period, which is identified by the intersection of the cost and benefit lines, as shown in Figure TK 3-9.

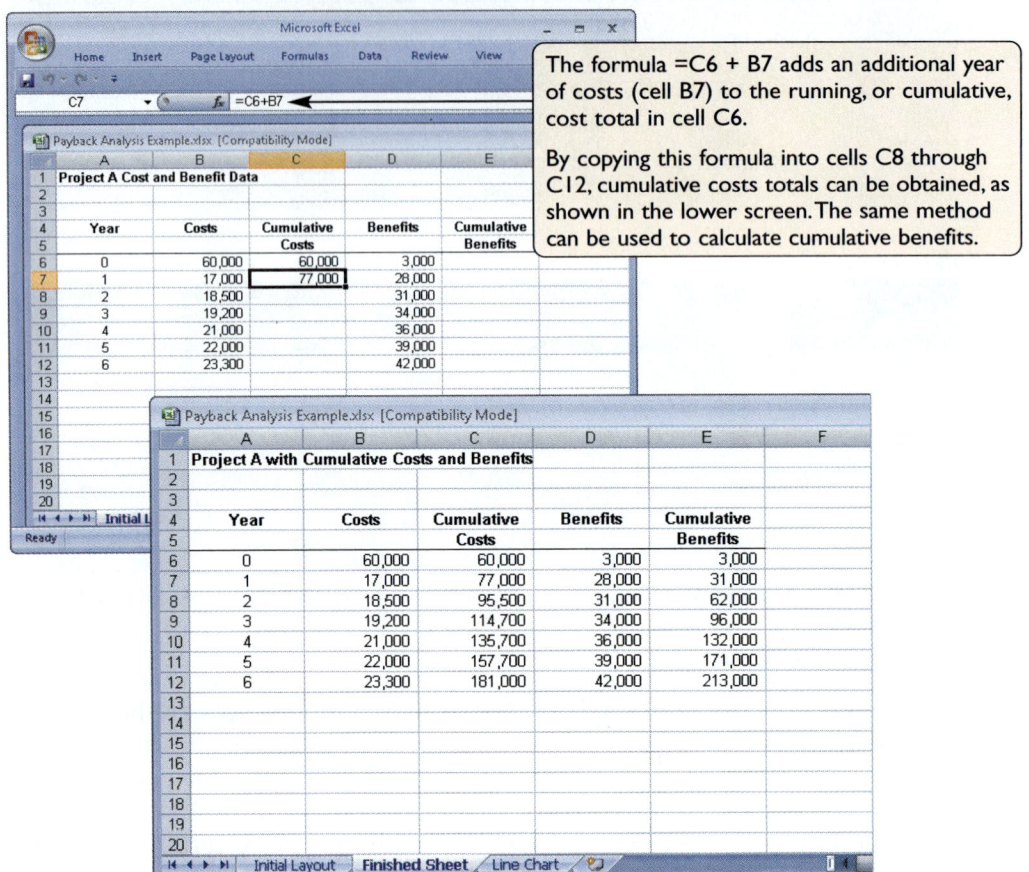

FIGURE TK 3-8 A Microsoft Excel worksheet displays payback analysis data for Project A in the upper screen. When cumulative cost and benefit formulas are entered, the finished worksheet in the lower screen appears.

Return on Investment Analysis

Return on investment (ROI) is a percentage rate that measures profitability by comparing the total net benefits (the return) received from a project to the total costs (the investment) of the project. ROI is calculated as follows:

ROI = (total benefits – total costs) / total costs

Return on investment analysis considers costs and benefits over a longer time span than payback analysis. ROI calculations usually are based on total costs and benefits for a period of five to seven years. For example, Figure TK 3-10 shows the ROI calculations for Project A and Project B. The ROI for Project A is 17.7%, and the ROI for Project B is 33.3%.

In many organizations, projects must meet or exceed a minimum ROI. This minimum ROI can be an estimate of the return the organization would receive from investing its money in other investment opportunities such as treasury bonds, or it can be a higher rate that the company requires for all new projects. If a company requires a minimum ROI of 15%, for example, then both Projects A and B would meet the criterion.

You also can use ROI for ranking projects. If Projects A and B represent two different proposed solutions for a single information systems project, then the solution represented by Project B is better than the Project A solution. If Projects A and B represent two

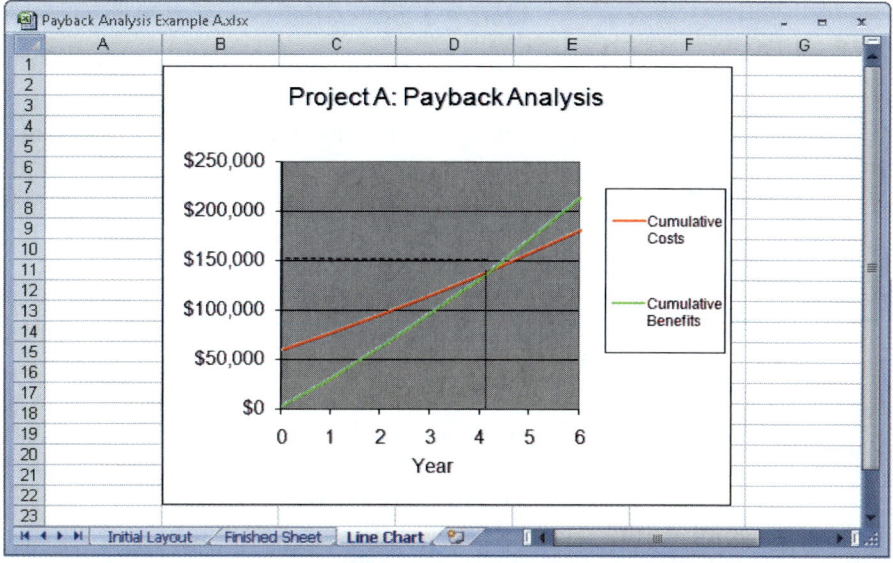

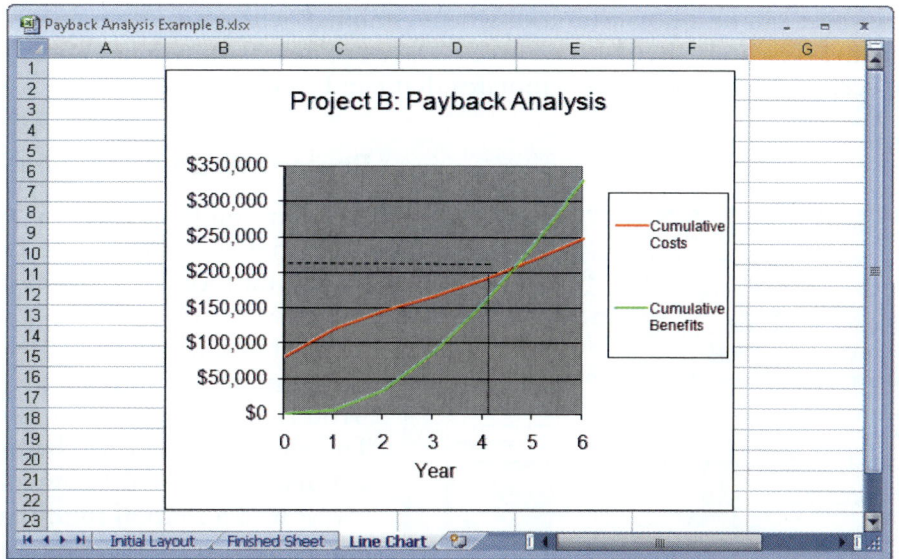

FIGURE TK 3-9 Microsoft Excel can be used to show the payback period by creating a chart of cumulative costs and benefits. Note that Project A has a shorter payback period than Project B.

ON THE WEB

For more information about return on investment analysis, visit **scsite.com/ sad8e/more**, locate Toolkit Part 3, and then click the Return on Investment Analysis link.

different information systems projects, and if the organization has sufficient resources to pursue only one of the two projects, then Project B is the better choice.

Critics of return on investment analysis raise two points. First, ROI measures the overall rate of return for the total period, and annual return rates can vary considerably. Two projects with the same ROI might not be equally desirable if the benefits of one project occur significantly earlier than the benefits of the other project. The second criticism is that the ROI technique ignores the timing of the costs and benefits. This concept is called the time value of money, and is explained in the section on the present value analysis method.

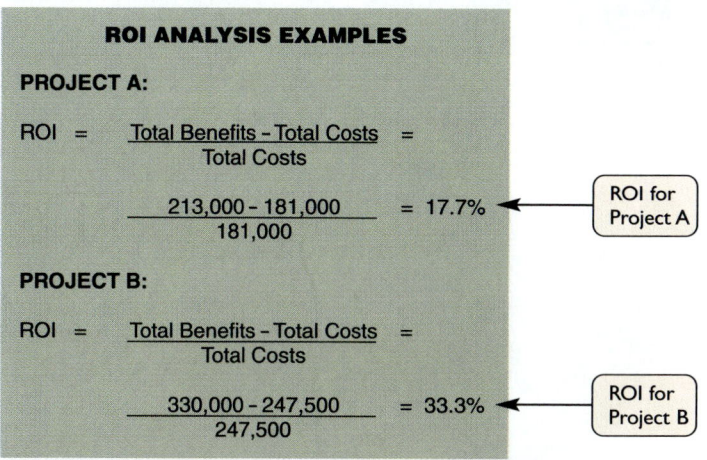

ROI ANALYSIS EXAMPLES

PROJECT A:

$$ROI = \frac{Total\ Benefits - Total\ Costs}{Total\ Costs} =$$

$$\frac{213,000 - 181,000}{181,000} = 17.7\%$$ ← ROI for Project A

PROJECT B:

$$ROI = \frac{Total\ Benefits - Total\ Costs}{Total\ Costs} =$$

$$\frac{330,000 - 247,500}{247,500} = 33.3\%$$ ← ROI for Project B

FIGURE TK 3-10 Return on investment analysis for Project A and Project B shown in Figure TK 3-7 on page 663.

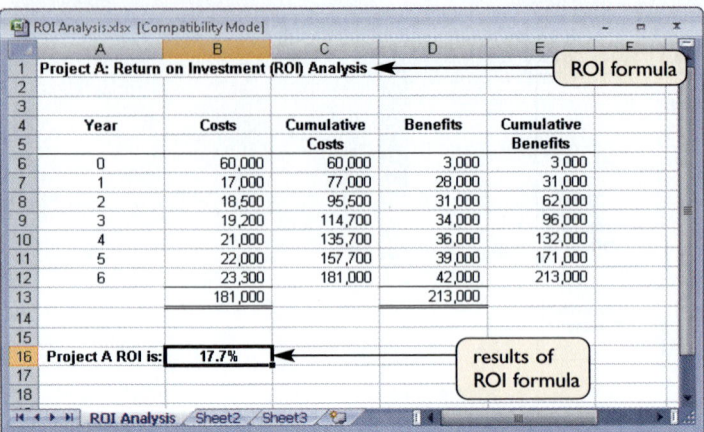

FIGURE TK 3-11   The worksheet for ROI analysis for Project A.

Using a Spreadsheet to Compute ROI

You also can use spreadsheet programs to calculate the ROI. To do so for Project A, first set up the worksheet and enter the cost and benefit data. You can use cumulative columns (as you did in payback analysis) but you also will need two overall totals (one for costs and one for benefits), as shown in Figure TK 3-11.

The last step is to add a formula to calculate the ROI percentage rate, which is displayed in cell B16 in Figure TK 3-11. As stated previously, the ROI calculation is total benefits minus total costs, divided by total costs. Therefore, the formula that displays the ROI percentage in cell B16 is =(D13 − B13)/B13.

A major advantage of using a spreadsheet is if your data changes, you can modify your worksheet and calculate a new result instantly.

A spreadsheet can be a powerful tool when combined with an ROI template. Hall Consulting and Research offers several free ROI templates, as shown in Figure TK 3-12. The lower screen shows one of the templates that can be downloaded.

Present Value Analysis

A dollar you have today is worth more than a dollar you do not receive until one year from today. If you have the dollar now, you can invest it and it will grow in value. For example, would you rather have $100 right now or a year from now? The answer should be obvious. If you receive the $100 now, you can invest it in a mutual fund that has an annual return of 8%. One year from now, you will have $108 instead of $100.

You might decide to approach ROI from a different direction. For example, instead of asking, "How much will my $100 be worth a year from now?" you can ask, "How much do I need to invest today, at 8%, in order to have $100 a year from now?" This concept is known as the **time value of money**, as shown in Figure TK 3-13, and it is the basis of the technique called **present value analysis**.

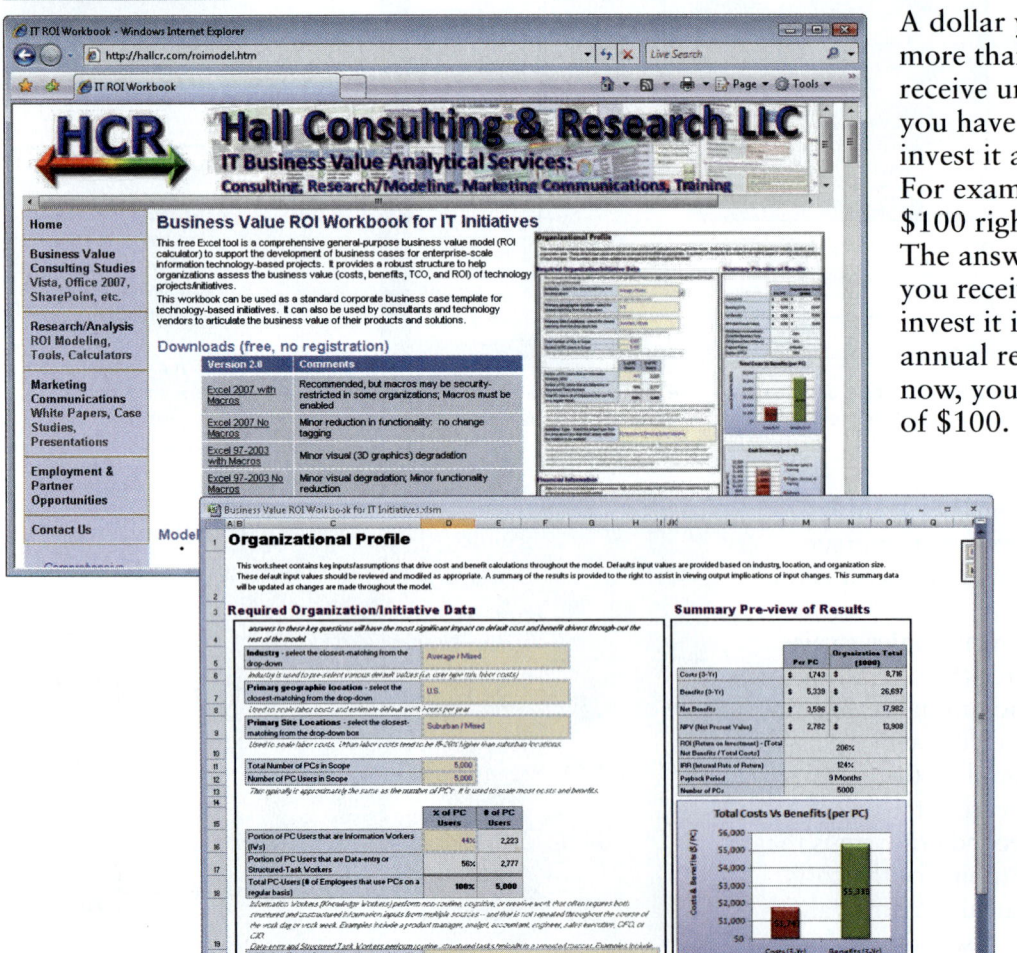

FIGURE TK 3-12   ROI information and templates can be viewed and downloaded from this site.

The **present value** of a future dollar is the amount of money that, when invested today at a specified interest rate, grows to exactly one dollar at a certain point in the future. The specified interest rate is called the discount rate. In present value analysis, a company uses a discount rate that represents the rate of return if the money is put into relatively risk-free investments, such as bonds, instead of being invested in the project.

Most companies require a rate of return that is higher than the discount rate because of the degree of risk in any project compared with investing in a bond. Companies often reject projects that seem attractive because the risk is not worth the potential reward.

To help you perform present value analysis, adjustment factors for various interest rates and numbers of years are calculated and printed in tables called **present value tables**. Figure TK 3-14 shows a portion of a present value table, including values for 10 years at various discount rates.

To use a present value table, you locate the value in the column with the appropriate discount rate and the row for the appropriate number of years. For example, to calculate the present value of $1 at 12% for five years, you look down the 12% column in Figure TK 3-14 until you reach the row representing five years. The table value is 0.567. To determine what the present value of $3,000 will be in five years with a discount rate of 12%, multiply the present value factor from the table by the dollar amount; that is, PV = $3,000 × 0.567 = $1,701.

Many finance and accounting books contain comprehensive present value tables, or you can obtain this information on the Internet, as shown in Figure TK 3-15.

To perform present value analysis, you must time-adjust the cost and benefit figures. First, you multiply each of the projected benefits and costs by the proper present value factor, which depends on when the cost will be incurred or the benefit will be received. The second step is to sum all the time-adjusted benefits and time-adjusted

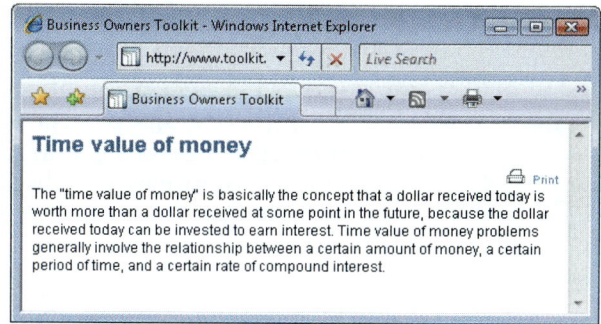

FIGURE TK 3-13 Another definition of the *time value of money*.

PERIODS	6%	8%	10%	12%	14%
1	0.943	0.926	0.909	0.893	0.877
2	0.890	0.857	0.826	0.797	0.769
3	0.840	0.794	0.751	0.712	0.675
4	0.792	0.735	0.683	0.636	0.592
5	0.747	0.681	0.621	0.567	0.519
6	0.705	0.630	0.564	0.507	0.456
7	0.665	0.583	0.513	0.452	0.400
8	0.627	0.540	0.467	0.404	0.351
9	0.592	0.500	0.424	0.361	0.308
10	0.558	0.463	0.386	0.322	0.270

FIGURE TK 3-14 Portion of a present value table showing adjustment factors for various time periods and discount rates. Values in the table are calculated using the formula shown in the text. Notice how the factors decrease as time and percentages increase.

FIGURE TK 3-15 The CCH Web site provides a variety of information to the business community, including present value tables.

ON THE WEB

For more information about the time value of money, visit **scsite.com/sad8e/more**, locate Toolkit Part 3, and then click The Time Value of Money link.

costs. Then, you calculate the **net present value (NPV)** of the project, which is the total present value of the benefits minus the total present value of the costs. Figure TK 3-16 shows the calculation of net present value for two sample projects.

In theory, any project with a positive NPV is economically feasible because the project will produce a larger return than would be achieved by investing the same amount of money in a discount rate investment. Remember that risks are associated with any project, however, and management typically insists on a substantially higher return for high-risk projects. For example, both projects in Figure TK 3-16 have positive net present values and appear economically worthwhile.

Suppose, however, that you knew one of the projects had a 90% probability of achieving its goals, while the other project had only a 70% chance. To be attractive, the project with the higher risk would have to offer a corresponding higher reward. Chapter 3 explains how project managers evaluate risks. Figure 3-20 on page 112 shows a matrix that includes various combinations of risk probability and impact.

Net present value also can be used to compare and rank projects. All things being equal, the project with the highest net present value is the best investment. Figure TK 3-16 shows that Project B is a better investment than Project A because it has a higher net present value.

NET PRESENT VALUE EXAMPLES

PROJECT A:

	Year 0	Year 1	Year 2	Year 3	Year 4	Year 5	Year 6	Total
Benefits:	3,000	28,000	31,000	34,000	36,000	39,000	42,000	
Present Value Factor (12%):	1.000	0.893	0.797	0.712	0.636	0.567	0.507	
Present Value:	3,000	25,004	24,707	24,208	22,896	22,113	21,294	143,222
Costs:	60,000	17,000	18,500	19,200	21,000	22,000	23,300	
Present Value Factor (12%):	1.000	0.893	0.797	0.712	0.636	0.567	0.507	
Present Value:	60,000	15,181	14,745	13,670	13,356	12,474	11,813	141,239
Net Present Value:						net present value of Project A		➤ 1,983

PROJECT B:

	Year 0	Year 1	Year 2	Year 3	Year 4	Year 5	Year 6	Total
Benefits:	——	6,000	26,000	54,000	70,000	82,000	92,000	
Present Value Factor (12%):	——	0.893	0.797	0.712	0.636	0.567	0.507	
Present Value:	——	5,358	20,722	38,448	44,520	46,494	46,644	202,186
Costs:	80,000	40,000	25,000	22,000	24,000	26,500	30,000	
Present Value Factor (12%):	1.000	0.893	0.797	0.712	0.636	0.567	0.507	
Present Value:	80,000	35,720	19,925	15,664	15,264	15,026	15,210	196,809
Net Present Value:						net present value of Project B		➤ 5,377

FIGURE TK 3-16 Net present value analysis for Project A and Project B.

Present value analysis provides solutions to the shortcomings of payback analysis and return on investment analysis. Unlike payback analysis, present value analysis considers all the costs and benefits, and not just the earlier values. In addition, present value analysis takes into account the timing of costs and benefits, so their values can be adjusted by the discount rate that provides a common yardstick and recognizes the time value of money. Even so, companies often use all three methods to get more input for making decisions. Sometimes a project will score higher on one method of analysis and lower on another.

Using a Spreadsheet to Calculate Present Value

You can use a worksheet such as the one shown in Figure TK 3-17 to calculate the present value based on the data for Project A. You begin by entering the unadjusted cost and benefit values and the discount factors for each year. Next, enter a formula to produce an adjusted value for each cost and benefit entry. To produce an adjusted value, you multiply the cost or benefit value by the discount factor. You start with cell B12 by entering the formula =B10*B11. Because the factor is 1.000, the 3,000 amount remains unchanged.

The formula =I12 - I16 subtracts the present value of all costs from the present value of all benefits, and produces the net present value (NPV) in cell D18.

D18   fx =I12-I16

NPV Analysis.xlsx

	A	B	C	D	E	F	G	H	I
1	Project A: Net Present Value (NPV) Analysis								
2									
3	*Note:*								
4	*Cells B12:H12 and B16:H16 contain formulas to multiply discount factors * benefit and cost values*								
5	*Cells I12 and I16 contain adjusted cost and benefit totals*								
6	*Cell D18 contains the calculation for net present value (NPV)*								
7									
8		Year	Year	Year	Year	Year	Year	Year	Total
9		0	1	2	3	4	5	6	
10	Benefits	3,000	28,000	31,000	34,000	36,000	39,000	42,000	
11	Factor	1.000	0.893	0.797	0.712	0.636	0.567	0.507	
12	PV of Benefits	3,000	25,004	24,707	24,208	22,896	22,113	21,294	143,222
13									
14	Costs	60,000	17,000	18,500	19,200	21,000	22,000	23,300	
15	Factor	1.000	0.893	0.797	0.712	0.636	0.567	0.507	
16	PV of Costs	60,000	15,181	14,745	13,670	13,356	12,474	11,813	141,239
17									
18	Net Present Value of Project A:			1,983					
19									
20									

Ready   100%

FIGURE TK 3-17 A Microsoft Excel worksheet can be used to calculate net present value (NPV).

Now, you copy the formula from cell B12 to cells C12 through H12, and the adjusted values will display. You total the adjusted benefits in cell I12 with the formula =SUM(B12:H12), and then use the same method for the cost figures. Your final step is to calculate the net present value in cell D18 by subtracting the adjusted costs in cell I16 from the adjusted benefits in cell I12.

There is another way to use a worksheet in present value analysis. Most spreadsheet programs include a built-in present value function that calculates present value and other time-adjusted variable factors. The program inputs the formula and then you input the investment amount, discount rate, and number of time periods. You can use your spreadsheet's Help feature to find out more about using these functions.

TOOLKIT SUMMARY

As a systems analyst, you must be concerned with economic feasibility throughout the SDLC, and especially during the systems planning and systems analysis phases. A project is economically feasible if the anticipated benefits exceed the expected costs. When you review a project, you work with various feasibility and cost analysis tools.

You must classify project costs as tangible or intangible, direct or indirect, fixed or variable, and developmental or operational. Tangible costs are those that have a specific dollar value, whereas intangible costs involve items that are difficult to measure in dollar terms, such as employee dissatisfaction. Direct costs can be associated with a particular information system, while indirect costs refer to overhead expenses that cannot be allocated to a specific project. Fixed costs remain the same regardless of activity levels, while variable costs are affected by the degree of system activity. Developmental costs are one-time systems development expenses, while operational costs continue during the systems operation and use phase.

Every company must decide how to charge or allocate information systems costs and the chargeback method. Common chargeback approaches are no charge, a fixed charge, a variable charge based on resource usage, or a variable charge based on volume.

Some companies use a no charge approach because IT services benefit the overall organization. This method treats the IT group for accounting purposes as a cost center that offers services without charge. In contrast, if management imposes charges on other departments, the IT department is regarded as a profit center that sells services that otherwise would have to be purchased from outside the company.

You also must classify system benefits. Many benefit categories are similar to costs: tangible or intangible, fixed or variable, and direct or indirect. Benefits also can be classified as positive benefits that result in direct dollar savings or cost-avoidance benefits that allow the firm to avoid costs that they would otherwise have incurred.

Cost-benefit analysis involves three common approaches: payback analysis, return on investment (ROI) analysis, and present value analysis. You can use spreadsheet programs to help you work with those tools.

Payback analysis determines the time it takes for a system to pay for itself, which is called the payback period. In payback analysis, you compare total development and operating costs to total benefits. The payback period is the point at which accumulated benefits equal accumulated costs. A disadvantage of this method is that payback analysis analyzes only costs and benefits incurred at the beginning of a system's useful life.

Return on investment (ROI) analysis measures a system by comparing total net benefits (the return) to total costs (the investment). The result is a percentage figure that represents a rate of return that the system offers as a potential investment. Many organizations set a minimum ROI that all projects must match or exceed and use ROI to rank several projects. Although ROI provides additional information compared with payback analysis, ROI expresses only an overall average rate of return that might not be accurate for a given time period. Also, ROI does not recognize the time value of money.

Present value analysis adjusts the value of future costs and benefits to account for the time value of money. By measuring all future costs and benefits in current dollars, you can compare systems more accurately and consistently. Present value analysis uses mathematical factors that you can derive or look up in published tables. You also can use a spreadsheet function to calculate present value. Many companies use present value analysis to evaluate and rank projects.

Key Terms and Phrases

chargeback method *659*

connect time *660*

cost center *659*

cost-avoidance benefits *660*

cost-benefit analysis *660*

developmental costs *659*

direct costs *658*

economic feasibility *656*

economically useful life *661*

fixed charge method *659*

fixed costs *659*

indirect costs *658*

intangible costs *658*

net present value (NPV) *668*

no charge method *659*

operational costs *659*

overhead expenses *659*

payback analysis *661*

payback period *661*

positive benefits *660*

present value *667*

present value analysis *666*

present value tables *667*

profit center *659*

resource allocation *660*

return on investment (ROI) *664*

server processing time *660*

tangible costs *658*

time value of money *666*

total cost of ownership (TCO) *656*

variable charge method based on resource usage *660*

variable charge method based on volume *660*

variable costs *659*

Toolkit Exercises

Review Questions

1. What is economic feasibility? How do you know if a project is economically feasible?
2. How can you classify costs? Describe each cost classification, and provide a typical example for each category.
3. What is a chargeback method? What are four common chargeback approaches?
4. How can you classify benefits? Describe each benefit classification, and provide a typical example for each category.
5. What is payback analysis, and what does it measure? What is a payback period, and what is the formula to calculate the payback period?
6. What is return on investment (ROI) analysis, and what does it measure? What is the formula to calculate ROI?
7. What is present value analysis, and what does it measure?
8. What is the meaning of the phrase, *time value of money*?
9. Why is it difficult to assign a dollar figure to an intangible cost? When and how can it be done? Provide an example with your explanation.
10. What is a system's economically useful life, and how is it measured?

Discussion Topics

1. Suppose your supervisor asks you to inflate the benefit figures for an IT proposal, in order to raise the priority of his or her favorite project. Would this be ethical or not? Does internal cost-benefit analysis affect company shareholders? Why or why not?
2. In this Toolkit Part, you learned how to use payback analysis, ROI, and NPV to assess IT projects. Could these tools also be used in your personal life? Give an example of how you might use each one to help you make a financial decision.
3. Is there a role for intuition in the decision-making process, or should all judgments be made strictly on the numbers? Explain your answer.
4. The time value of money is an important factor when analyzing a project's NPV. Is the time value of money more important, less important, or of the same importance in periods of low inflation compared with periods of high inflation? Explain your answer.

Projects

1. Suppose you are studying two hardware lease proposals. Option 1 costs $4,000, but requires that the entire amount be paid in advance. Option 2 costs $5,000, but the payments can be made $1,000 now and $1,000 per year for the next four years. If you do an NPV analysis assuming a 14% discount rate, which proposal is less expensive? What happens if you use an 8% rate?
2. Assume the following facts:
   A project will cost $45,000 to develop. When the system becomes operational, after a one-year development period, operational costs will be $9,000 during each year of the system's five-year useful life. The system will produce benefits of $30,000 in the first year of operation, and this figure will increase by a compound 10% each year. What is the payback period for this project?
3. Using the same facts as in Project 2, what is the ROI for this project?
4. Using the same facts as in Project 2, what is the NPV for this project?

PART 4 Internet Resource Tools

In **Part 4** of the Systems Analyst's Toolkit, you will learn about Internet resource tools that can help you perform your duties and achieve your professional goals.

INTRODUCTION

OBJECTIVES

When you finish this part of the Toolkit, you will be able to:

- Describe the characteristics of the Internet and the World Wide Web

- Plan an Internet search strategy, review your information requirements, use the proper search tools and techniques, evaluate the results, and consider copyright and data integrity issues

- Use search engines, subject directories, and the invisible Web to locate the information you require

- Demonstrate advanced search techniques, including Boolean logic and Venn diagrams

- Describe Internet communication channels including newsgroups, newsletters, blogs, podcasts, RSS feeds, Webinars, mailing lists, Web-based discussion groups, chat rooms, instant messaging, and text messaging

- Provide examples of IT community resources of value to a systems analyst

- Explain the benefits and disadvantages of online learning opportunities

The Internet offers a wealth of information about every conceivable subject. Without a good road map, however, it is easy to become overwhelmed by the sheer volume of material. Part 4 of the Systems Analyst's Toolkit will describe various Internet resources, assist you in formulating an effective information gathering strategy, and explain Internet resource tools and techniques that you can use to access the information you need.

TOOLKIT INTRODUCTION CASE: Mountain View College Bookstore

Background: Wendy Lee, manager of college services at Mountain View College, wants a new information system that will improve efficiency and customer service at the three college bookstores.

In this part of the case, Florence Fullerton (systems analyst) and Harry Boston (student intern) are talking about Internet resource tools.

Participants:	Florence and Harry
Location:	Mountain View College Cafeteria, near the end of the systems analysis phase
Discussion topics:	Internet resource tools

Florence: Hi, Harry. Did you get a chance to try that new search engine?

Harry: *I sure did. First, I formulated an Internet search strategy, as you suggested, and then I did a search using the logical operators that we talked about. I liked the way the search engine organized the results for me. I also followed your suggestions about checking out the quality of the results.*

Florence: Sounds good. But remember — sometimes it might be better not to start your search with a search engine.

Harry: *Why not?*

Florence: Well, the Internet is a huge place. In some situations, you might want to get a broad overview of a topic before plunging into a specific search. If so, you might want to use a subject directory like Yahoo! or the Librarians' Index to the Internet. Another important resource is called the invisible Web, which is a collection of searchable databases that usually are not accessed by search engines.

Harry: *Okay. What about using other communication channels like newsgroups, newsletters, blogs, podcasts, RSS feeds, Webinars, mailing lists, Web-based discussion groups, chat rooms, instant messaging, and text messaging?*

Florence: They all are important ways to get online information. Also remember that the IT community has many resources that can help you locate information, keep up with IT developments, and advance your career with knowledge and training.

Harry: *It seems like the more I know about the Internet, the more I have to learn.*

Florence: I feel the same way. Let's see what else we can learn about Internet resource tools. Here's a task list to get us started:

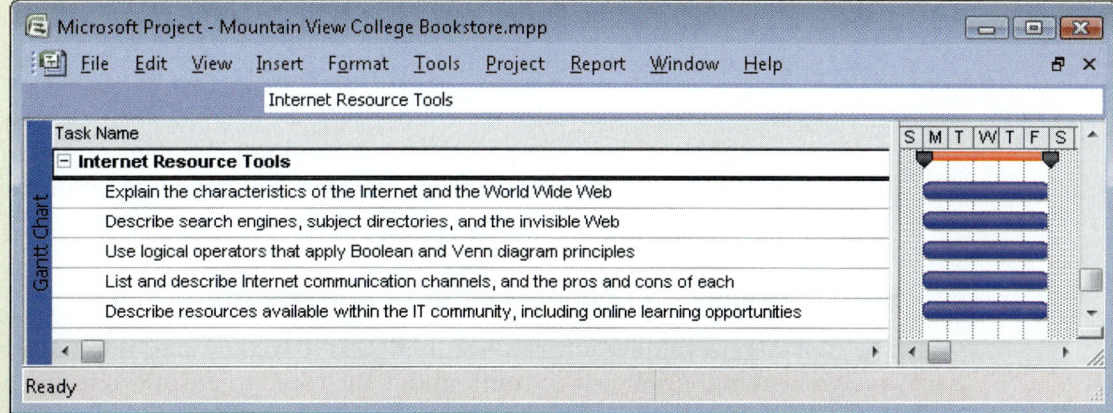

FIGURE TK 4-1   Typical Internet resource tasks.

OVERVIEW

As a systems analyst, you rely on your knowledge, skills, and experience as you respond to a variety of short- and long-term challenges. Fortunately, you have access to the Internet, where an enormous storehouse of information is available to you at little or no cost. The **Internet** is a worldwide network that integrates many thousands of other networks, which in turn link millions of government, business, educational, and personal users around the globe. The Internet can assist you in your daily work by helping you solve technical problems and can help you advance your career by offering access to training, education, and communication with other IT professionals.

The Internet allows you to visit the **World Wide Web**, usually referred to as the **Web**, which contains billions of text and multimedia documents called **Web pages**. A collection of related Web pages is called a **Web site** and is stored on a computer called a **Web server**. A **Web browser**, or **browser**, is a software program that allows you to access and display Web pages that are delivered to you by a Web server. Microsoft Internet Explorer, Mozilla Firefox, and Apple Safari are popular browsers that offer powerful graphic interfaces to help you navigate the Web.

This Toolkit Part begins with a step-by-step plan for Internet research, followed by a summary of search basics and a detailed discussion of search engines, subject directories, and the invisible Web. Internet communications tools are covered next, including newsgroups, newsletters, blogs, podcasts, RSS feeds, Webinars, mailing lists, Web-based discussion groups, chat rooms, and instant messaging. The last section presents valuable online resources available within the IT community.

ON THE WEB

For more information about the World Wide Web, visit **scsite.com/ sad8e/ more**, locate Toolkit Part 4, and then click the World Wide Web link.

PLANNING AN INTERNET RESEARCH STRATEGY

An Internet research strategy is necessary to avoid frustration and wasted time. A pilot or ship captain would not begin a journey without knowing the destination. Similarly, you can use a four-step plan to navigate efficiently and confidently toward your objectives:

Step 1. Review your information requirements.

Step 2. Use the proper search tools and techniques.

Step 3. Evaluate the results.

Step 4. Consider copyright and data integrity issues.

Over time, you will gain experience and develop your own preferences for using the Internet. You should remember that each research situation is unique, and several tools and techniques might be necessary to achieve the results you seek.

Step 1. Review Your Information Requirements

The first step to finding information online is to make sure you really understand what you are seeking. You need to think about the topic to ensure that you are casting an appropriate net. For example, a supervisor might ask you to help decide between two specific CASE products. Your initial inclination might be to find a review of various CASE applications. Upon reflection, however, you realize it would be more useful to understand CASE tools in a general sense before comparing specific products. Therefore, you decide to start with a more generalized search instead of going directly to vendor sites.

Step 2. Use the Proper Search Tools and Techniques

Once you feel that you understand the information required, it is time to pick an initial tool. At this point, you face some choices. Should you use search engines or subject directories, or should you try to access the invisible Web? Should you seek commercial sites, IT publications, professional associations, newsgroups, or other areas to explore?

As you gain experience, you will be able to handle a wide range of Internet tools and resources. As with most skills, the more you use them, the more expertise you acquire. In time, you probably will develop your own list of favorite tools and resources.

Step 3. Evaluate the Results

By definition, the Internet is essentially open and unregulated. On the plus side, a huge diversity of information is available. The quality of content, however, varies greatly. Unlike published journals or textbooks, almost anyone can post **content**, or material, on the Web. This means that the searcher must review the information very carefully. Questions to ask when accessing content include the following:

SOURCE Is the author identifiable? Does the author have expertise on the subject? You might need to trace back through Web site addresses or URLs to find biographical information, or do a separate search on the author's name.

ACCURACY Does the information come from a commercial source that is offering its own solution? Is it from an association with an inherent bias? Often, it is very difficult to find completely objective information. Identifying biases and finding information from a variety of sources is a way to address this problem.

SCOPE Is the information specific enough? If not, you should narrow and refine your search and seek additional resources until you locate the information you need. When you use a search engine, one way to do this is to perform a **subsearch** using the results of the initial search as a starting point.

CURRENCY How old is the information? Is the topic static or dynamic? In the IT world, technology changes very quickly. If you locate information that appears to be out of date, you might seek more recent data to ensure that your results are valid.

LOOK AND FEEL Is the information easy to access and navigate? If the site is designed in a logical manner and offers value-added links to worthwhile pages and resources, do not be overly concerned with style — some excellent material is created by authors and producers who focus on content, not design.

If you find the content useful, be sure to credit the source properly when you use it. You must observe legal and ethical standards when you deal with Internet material. To learn more about proper citation and to view specific examples of how to cite electronic material, you can visit the Library of Congress site at *http://memory.loc.gov/learn/start/*.

Step 4. Consider Copyright and Data Integrity Issues

Before you copy or download your search results, you must ensure that you legally can use the material, and that the content is safe and free of threats.

The first issue involves copyright law. Many people regard the Web as a public domain, but in reality, it is more like a book or a CD. In other words, you might own a CD, but you do not own the material on that CD — you only possess a license to use the content in certain ways. On the Web, you should look for copyright notices and restrictions. If in doubt, you might have to contact the copyright holder to seek permission.

FIGURE TK 4-2 McAffee's site shows how many computers and files are infected at any given time.

The second issue involves data integrity. It is important to protect your network and computer system from any unwanted viruses or **malware,** which is malicious software that might jeopardize your security or privacy. Unlike intellectual content, which is easy to evaluate, it is impossible for you to determine the integrity and validity of the internal file structure and format without a virus detection tool. As Figure TK 4-2 shows, many thousands of viruses are identified each day. Viruses and other intrusions cost businesses many millions of dollars in lost data and additional effort. Without proper protection, you run the risk of not only corrupting your own files or hard drive, but bringing your entire company network down.

If the information is legally usable and safe, you can save it to your hard drive or network, depending on the content and purpose. The information you find may be in one of many formats, including word-processing documents, spreadsheets, and databases; Adobe PDF files; and multimedia material with file extensions such as .jpg, .mp3, and .mov.

You can download files directly to your hard drive from many Web sites. If this option is not offered explicitly, you can try right-clicking a link or embedded object to display a list of choices that includes downloading the selected information.

The following sections discuss search engines, subject directories, the invisible Web, and other Internet tools in more detail.

SEARCH BASICS

As you journey on the Internet, you will use various navigation tools and techniques. To reach your destination, you must know how to use search engines, subject directories, and a collection of searchable databases called the invisible Web.

A **search engine** is an application that uses keywords and phrases to locate information on the Internet. **Meta-search engines** are tools that can apply multiple search engines simultaneously. For most people, search engines are the workhorses of information gathering. Search engines employ a variety of approaches to gathering information, and although they are extremely valuable, users should be aware of potential problems. For example, search results can be affected if the search engine permits commercial users to achieve higher priority based on payment of fees. Also, search engines access only a portion of the Internet.

A **subject directory** or **topic directory** is a Web site that allows you to find information by using a hierarchy, starting with a general heading and proceeding to more specific topics. A subject directory is an excellent starting point when you want an overview of a topic before proceeding to specific Web sites. Typically, a subject directory is created by an editorial staff that visits, evaluates, and organizes the sites into various categories and subcategories.

The **invisible Web**, also called the **deep Web** or **hidden Web**, refers to a vast collection of documents, databases, and Web pages that are usually not detected by search engines, but can be accessed using other tools and techniques. The invisible Web is a huge information storehouse, many times larger than the searchable Web, and includes thousands of university, scientific, and government libraries.

The following sections cover search engines, subject directories, and the invisible Web in more detail.

SEARCH ENGINES

A search engine often is the best starting point for gathering information. A well-planned search will narrow the range of content to a manageable level and will allow you to explore the choices or execute a subsearch within the focused results. As with any tool, it is important to understand the intended use and limitations of a search engine before applying it to a task.

Search Engine Concepts

Search engines use a specialized computer program called a **spider** or **crawler** that travels from site to site **indexing**, or cataloging, the contents of the pages based on **keywords**. The results are compiled into a database, so what you are searching is not the Web itself, but the contents of the search engine's database.

No single search engine can catalog the shifting contents of the Web, and even the most powerful engines cover a fraction of known Web content. If a particular site is not widely linked, or its author does not submit it to major search engines, then the material is invisible to them. Also, any site that requires a visitor to type in data, such as a name, cannot be accessed by search engines.

Although search engine indexes are incomplete and often dated, they are capable of delivering an overwhelming number of results, or **hits**. The real issue is quality versus quantity. When comparing search engines, it is important to know the company's policy toward allowing commercial sites to boost their ranking in a **pay for performance** arrangement. Links that are subsidized by companies are called **sponsored links**.

Not all search engines work the same way. By understanding the underlying algorithms, or specific rules, that drive these information engines, it is possible to better target your search. For example, some sites, like Google, rank their pages by analyzing the number of other sites that link to that page. Other search engines organize results differently. For example, Ask.com uses an interesting approach called ExpertRank™ technology, which ranks a site based on the number of authoritative sites that reference it, not just based on general popularity.

You usually can determine an engine's approach by clicking the *About* tab or link on the search site. The following section describes several examples of search engines, which are shown in Figure TK 4-3 on the next page.

ON THE WEB

For more information about search engines, visit **scsite.com/ sad8e/more**, locate Toolkit Part 4, and then click the Search Engines link.

INDEXED SEARCH ENGINES Google, like most other popular research tools, is an **indexed search engine** that organizes and ranks the results of a search. Although they have much in common, each tool has its own search algorithms, features, and user interface. Other examples of indexed search engines include Yahoo! and AltaVista, among others.

META-SEARCH ENGINES Meta-search engines can apply multiple search engines simultaneously. A meta-search engine examines the indexed results of several search engines to provide broader coverage. In addition to Ixquick, other examples of meta-search engines include Metacrawler and Dogpile, which claims to use meta-search technology to remove duplicates and bring the most relevant sites to the top of the list.

Search Techniques

Consider the following suggestions when you begin a search:

- Refine your topic. Unless you limit the scope of your search, you might be overwhelmed by the number of results. If you are looking for general information on a broad topic, consider a subject directory site.

- Translate your question into an effective search query. Searches are executed on keywords. You will improve your success if you pick the proper keywords. Try to find unique words or phrases and avoid those with multiple uses. For example, a search for the term *hard drive* might produce information about a computer hardware device or a difficult auto trip. Also consider using advanced search techniques, which are described in the following section.

FIGURE TK 4-3 Google is an example of an indexed search engine, Ixquick is an example of a meta-search tool, and Ask is a combination engine that includes ranked pages, suggestions to refine or narrow the search, and an expert-compiled resources section.

- Review the search results and evaluate the quality of the results. If the search needs refinement or additional material, you can either use the site's advanced search techniques or select a different Internet resource altogether.

- It is important to organize the results of your search, so you can recognize and revisit important sites. Some search engines offer a personalized search history, which you can review and edit for this purpose. Many people find that the easiest solution is to create favorites or bookmarks in their browser for sites visited in important searches, using a set of folders and subfolders. If you do this, you can wait until you start the search, or you can create your filing system ahead of time.

To be effective, you should understand the mechanics of the search engine, use proper spelling, find unique phrases, and experiment with a variety of approaches. If you are consistently returning too many results, try using topic-specific terms and advanced search techniques. Conversely, if too few results are returned, eliminate the least important terms or concepts, broaden your subject, or use more general vocabulary when you select terms. An excellent tutorial, which includes a glossary and a comparison of several search engines, can be found at the University of California at Berkeley Library site, as shown in Figure TK 4-4.

Advanced Search Techniques

Many search engines offer powerful features that allow you to refine and control the type of information returned from searches. These features can include the option to search within returned results and the ability to search within specific areas, such as newsgroups. Perhaps the most powerful advanced feature is the option to use Boolean logic.

Boolean logic is a system named after British mathematician George Boole and refers to the relationships among search terms. You can use various combinations of the **logical operators** OR, AND, and NOT to improve your search success greatly. Figure TK 4-5 on the next page graphically illustrates the use of the operators with search terms. The circles shown in the figure are called Venn diagrams. A **Venn diagram** uses circular symbols to illustrate Boolean logic. Venn diagrams are named after John Venn, a nineteenth-century scholar who devised a scheme for visualizing logical relationships. In the sample diagrams, the shaded area indicates the results of the search.

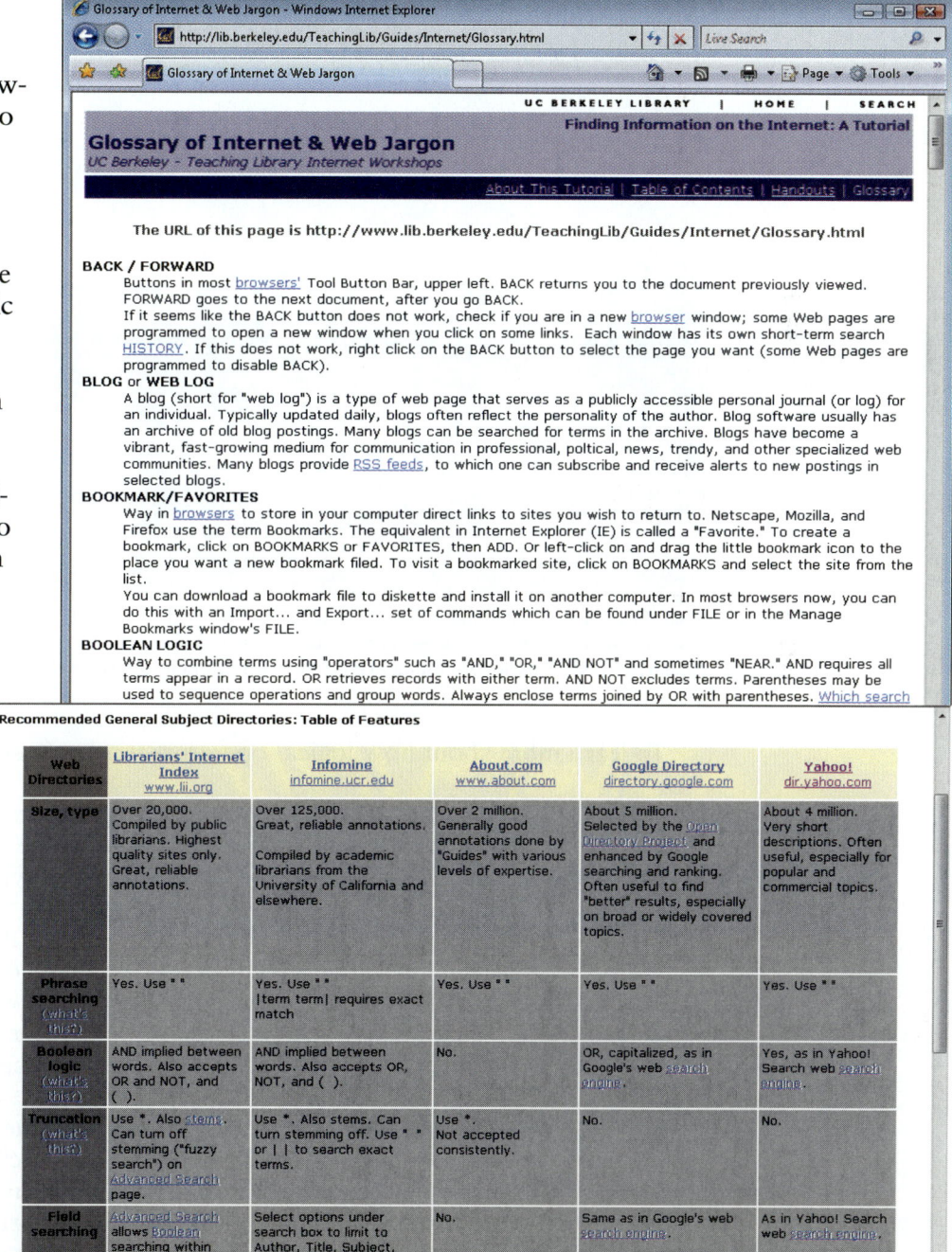

FIGURE TK 4-4 The University of California at Berkeley offers search-related information, tips, and resources.

LOGICAL OPERATORS	EXAMPLE	VENN DIAGRAM (SHADED AREA INDICATES RETURNED RESULTS)		
A **or** B	Baseball **or** Football	Baseball		Football
A **and** B	Baseball **and** Football	Baseball		Football
A **not** B	Internet **not** Web	Internet		Web
A **or** B **or** C	Colorado **or** Mining **or** Gold	Colorado		Mining / Gold
A **and** B **and** C	Colorado **and** Mining **and** Gold	Colorado		Mining / Gold
A **and** B **not** C	Colorado **and** Mining **not** Gold	Colorado		Mining / Gold

FIGURE TK 4-5 Examples of logical operators OR, AND, and NOT. The shaded area represents the returned results. OR is the most inclusive term, returning results if any of the terms appear; AND requires all keywords to appear; and NOT excludes results even if they are found in the same document.

To learn more about how logical operators work, consider the following diagrams:

OR The first diagram at the top of Figure TK 4-5 shows that the **OR** term will retrieve all results containing either term. Notice that the shaded area includes both circles. The OR operator can be used when you need a wide search net.

AND The second diagram in Figure TK 4-5 shows that the **AND** term will retrieve only those results in which all terms linked by the AND operator are present. Notice that the shaded area includes only the overlapping portion of the two circles. The more terms or concepts combined in an AND search, the fewer pages will be returned. The AND operator often is used to narrow a set of search results.

NOT The third diagram in Figure TK 4-5 shows how the NOT operator can be used to exclude certain records. In this diagram, consider the closely related terms, Internet and Web. It is likely that these terms both appear in many Web pages. The **NOT** operator will strip out the results containing the unwanted term. For example the search string, Internet NOT Web, will return only those pages with the term, Internet. You should use the NOT operator carefully, because the term you want to exclude may be intertwined with the term you seek in many documents that would be useful to you.

The last three diagrams in Figure TK 4-5 illustrate other combinations of logical operators where three search terms are involved.

Figure TK 4-6 shows an excellent tutorial published by New York University Libraries. The tutorial explains Boolean logic, and features an interactive search for information about UFOs.

USING PHRASES Suppose you want to find sites that sell board games that require players to use strategy, such as Monopoly®. In your search, you could specify both terms, *strategy* AND *game*, but your results probably would include many documents that describe game strategies used in various sports, which is not what you are seeking. A better approach might be to search using the phrase "strategy game" enclosed in quotes. A **phrase** is more specific than an AND operator, because it specifies an exact placement of terms. In this example, the phrase "strategy game" will not retrieve any documents unless they contain that exact phrase.

The implementation of Boolean logic varies by search engine. Some engines require the use of full Boolean searching using the complete operators (OR, AND, NOT) in the search window. Others use implied Boolean logic with keyword searching. In **implied Boolean logic**, symbols are used to represent Boolean operators, such as a plus sign (+) for AND, and a minus sign (-) for NOT. If two search terms are entered in the search window with a space between them, some search engines may assume an OR; others assume an AND. You need to consult the site's Help files to understand the underlying rules.

FIGURE TK 4-6 The New York University Libraries tutorial explains Boolean logic, and features an interactive search for information about UFOs.

USING FILL-IN FORMS Most search engines provide an **advanced search** feature that offers a fill-in form similar to the example shown in Figure TK 4-7 on the next page. Notice that a user can narrow the search by entering search terms in the upper part of the form, and by specifying dates and various other conditions.

Search Checklist

Many people find it helpful to prepare for an Internet search by using a checklist similar to the following.

- Does the topic have any unique words, phrases, or acronyms? If so, use these terms in the search.

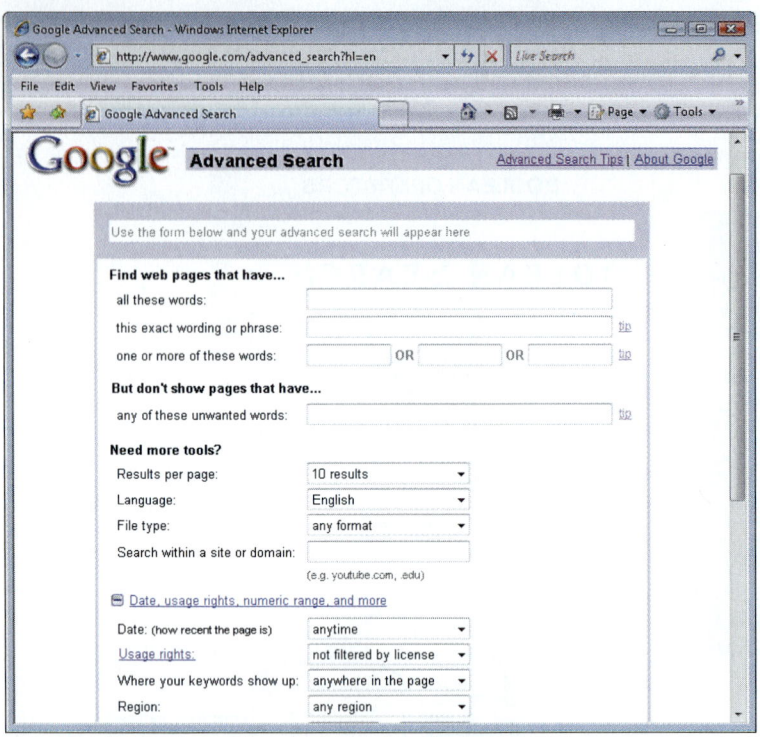

FIGURE TK 4-7 For users unfamiliar with Boolean terms, many search engines offer an advanced search feature with fill-in forms. This example shows that *all these words* is like the AND operator, *this exact wording or phrase* is like putting quotes around the search terms, *one or more of these words* is like the OR operator, and *any of these unwanted words* is like the NOT operator.

- Do any of the search terms have other spellings or names? If so, include these with an OR operator in the search.

- Are certain additional words or phrases likely to appear in any Web document? If so, consider adding an AND operator to narrow the search.

- Is there any unrelated material that my search terms might pick up? If so, consider using the NOT operator to exclude these documents.

- Are any organizations, publications, or institutions likely to have an interest in my topic? If so, try to locate their Web sites and then conduct a further search using the site indexes and databases available on the site.

- Is the search returning results too numerous to examine? If so, keep adding additional terms to narrow the search and reduce the number of hits until a reasonable number is achieved.

SUBJECT DIRECTORIES

A subject directory collects and organizes Web sites in a top-down format, based on subjects and topics. An analogy might be a corporate organization chart, where you could go to the top person for an overview, then visit with lower-level employees to obtain specific information about their areas.

A subject directory is an excellent tool when you want general information about a topic before plunging into an array of specific Web sites. Yahoo! is a popular directory site that serves as a **portal**, or entrance to other Internet resources. Other academic and professional directories target the specific needs of researchers and users who concentrate on particular subjects. Many subject directories are reviewed by human experts, rather than computer robots, to ensure relevance and quality of links.

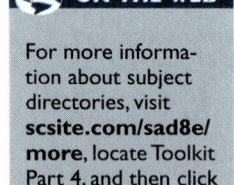

ON THE WEB

For more information about subject directories, visit **scsite.com/sad8e/more**, locate Toolkit Part 4, and then click the Subject Directories link.

A Subject Directory Example

Subject directories organize information into various categories and provide an overall framework for finding information. For example, suppose you wanted to use the Yahoo! Directory to learn more about IT security magazines. You could visit the Yahoo! site, and start with a list of main topics, one of which is called *Computers and Internet*. When you click this item, you would see a display similar to the upper screen in Figure TK 4-8, which shows various subtopics, including *Security and Encryption*. As you continue, you would click the *Magazines* link, and you would see a list of seven publications that you could explore.

In addition to Yahoo!, other popular subject directories include About.com and the Librarians' Index to the Internet, which maintains a site at *lii.org*.

Advantages and Disadvantages of Subject Directories

The main advantage of a subject directory is that it provides an overview when you are not sure of the size and scope of your topic. Later, when you have a better understanding of your subject, you can use a search engine to seek additional information and examples.

Subject directories also have shortcomings. Many subject directories use human expertise to formulate the subject organization and determine the placement of links. This process involves subjective decisions that might affect the quality of search results. Some subject directories are updated continually; others might not be current. Also, unlike a search engine, a subject directory forces you to work your way through a series of levels, rather than using specific words and phrases to locate directly the material you seek.

THE INVISIBLE WEB

Everyone is familiar with what is called the **visible Web**, which refers to Web sites that are indexed by the major search engines and are publicly accessible. As discussed earlier, much more information is available on the Internet that is not indexed. This valuable information source includes numerous text, graphics, and data files stored in collections that are unreachable by search engines.

The invisible Web includes searchable databases that contain an enormous amount of information in university and government libraries, as well as thousands of specialized databases that are maintained by institutions and organizations around the world.

Much of the invisible Web is open to the public, but some databases are password protected. Many sites allow guest access, but some areas can be accessed only by members of a specific group.

In an article in the *Journal of Electronic Publishing*, author Michael K. Bergman compared Internet searching to dragging a net across the surface of the ocean. He stated that "While a great deal may be caught in the net, there is still a wealth of information that is deep, and therefore, missed. The reason is simple: Most of the Web's information is buried far down on dynamically generated sites, and standard search engines never find it." Mr. Bergman pointed out that traditional search engines create their indices by spidering, or crawling, through many millions of Web pages. He pointed out that "To be discovered, the page must be static and linked to other pages.... Because traditional search engine crawlers cannot probe beneath the surface, the deep Web has heretofore been hidden."

FIGURE TK 4-8 In the example shown, the user has visited the Yahoo! Directory, clicked the *Computers and Internet* topic, then clicked the *Security and Encryption* link, and the *Magazines* link, which displays a list of IT security magazines.

ON THE WEB

For more information about the invisible Web, visit **scsite.com/ sad8e/more**, locate Toolkit Part 4, and then click the Invisible Web link.

Invisible Web Examples

A few examples of information on the invisible Web include the following:

- Specialized topic databases: subject-specific collections of information, such as corporate financial filings and reports, genealogy records, or Ellis Island immigration data

- Hardware and software vendors: searchable technical support databases for large sites such as Microsoft's or Oracle's knowledge bases

- Publications: databases of published and archived articles

- Libraries: searchable catalogs for thousands of libraries, including the Library of Congress and numerous university and institutional libraries

- Government databases: census data, statutes, patents, copyrights, and trademarks

- Auction sites: searchable listings of items, bidders, and sellers

- Locators: telephone numbers, addresses, and e-mail addresses

- Career opportunities: job listings and résumé postings

Navigation Tools for the Invisible Web

You can access the invisible Web in several ways. One approach is to use a search engine to locate a portal, or entrance, to a searchable database by including the word, database, as a required search term. For example, if you are searching for information about printer drivers, you could specify "printer drivers" AND database. The additional term will narrow the search results and increase the likelihood of finding searchable collections of printer drivers.

You also can access the invisible Web by using specialized portals that list and organize searchable databases. An example of an invisible Web portal is CompletePlanet.com, which is shown in Figure TK 4-9. CompletePlanet.com claims to access over 70,000 invisible Web databases that are frequently overlooked by traditional searching, and also offers a special search tool called the Deep Query Manager™.

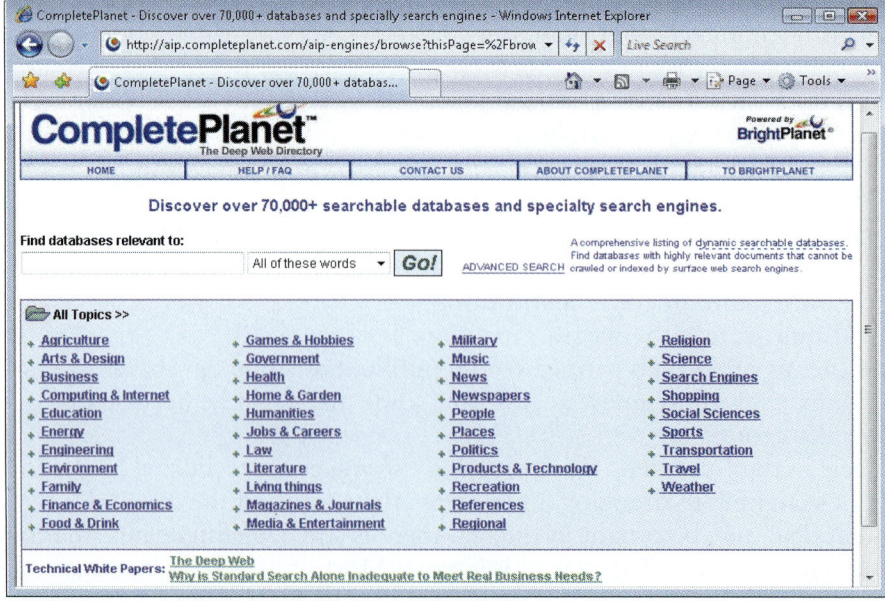

FIGURE TK 4-9 The CompletePlanet site is organized by topics, and has links to many searchable databases whose contents cannot be indexed by traditional search engines.

If you spend a significant amount of time searching the invisible Web, you might want to use special navigation software to assist you. For example, EndNote is an application offered by Thomson Reuters, as shown in Figure TK 4-10.

EndNote uses a special information transfer method called the **Z39.50 protocol** that allows a user to connect directly to hundreds of government and university databases. The top screen in Figure TK 4-10 shows a sample of available databases. In the middle screen, the user has selected the Library of Congress catalog and is ready to enter specific search terms and logical operators. The bottom screen shows that the user has connected to the remote database.

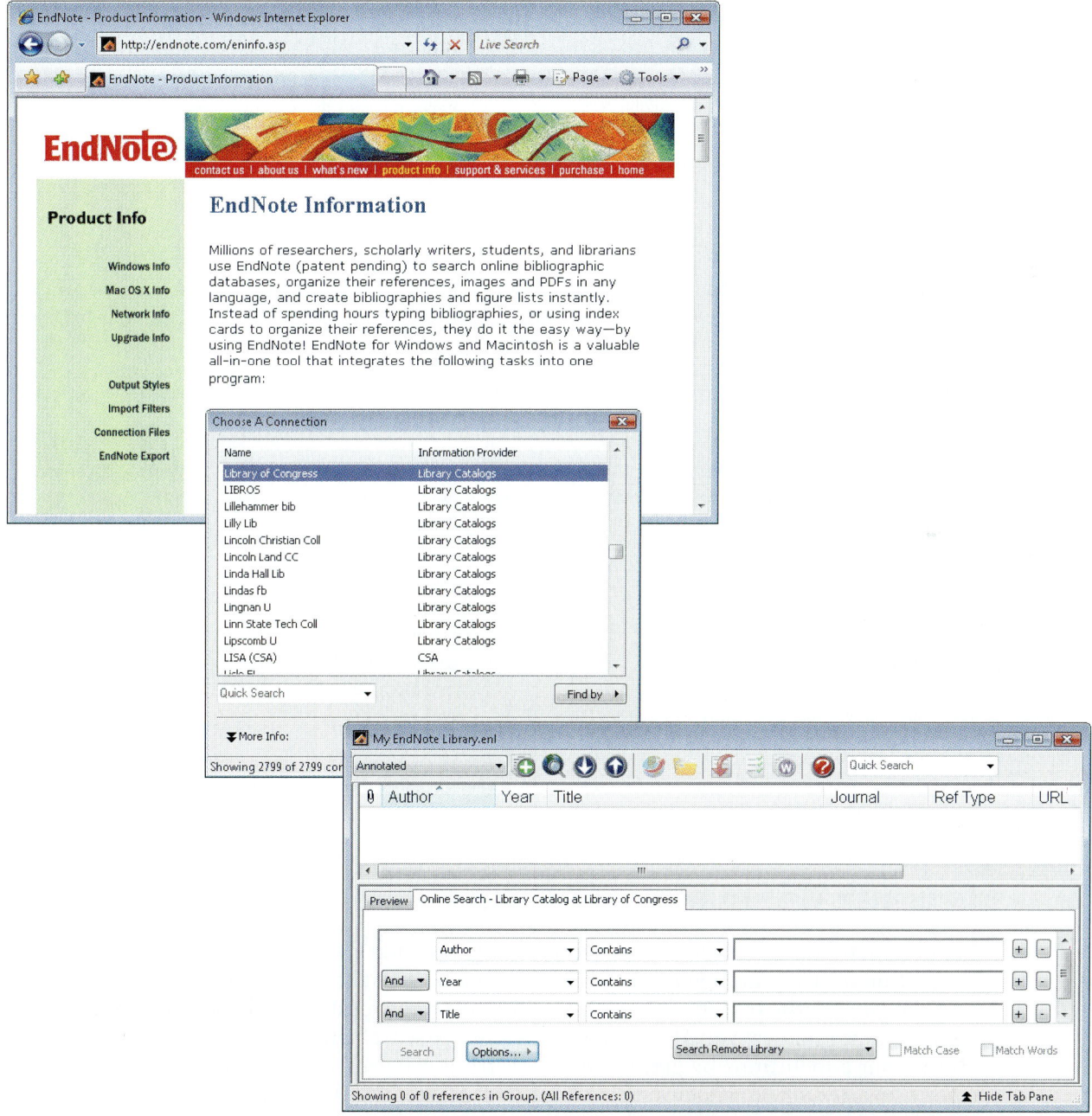

FIGURE TK 4-10 EndNote is an application that uses a special information transfer method called the Z39.50 protocol that allows you to connect directly to hundreds of government and university databases. In the example shown, the user has selected the Library of Congress catalog and is ready to enter specific search terms and logical operators. The bottom screen shows that the user has connected to the remote database.

Figure TK 4-11 shows a recap of the Internet tools and resources: search engines, subject directories, and the invisible Web. Notice that each has advantages and disadvantages.

Recap of Internet Search Resources

RESOURCE	POSSIBLE USES	ADVANTAGES	DISADVANTAGES	NOTES
Search Engines	Good initial starting point, especially if you have an overall understanding of the topic. Search engines can lead you to important government, professional, or commercial sites.	Flexibility. You can choose from many different engines with various features. Some allow newsgroup access as well. Meta-search engines can return and rank results from multiple sources. Advanced search techniques can be used.	Frequently produces information overload. Can return many irrelevant or out-of-date links. Without a refined search, it is impossible to examine results carefully. Some sites allow companies to improve their hits by "pay for performance."	You can improve search effectiveness significantly by using advanced search techniques based on logical operators.
Subject Directories	Good way to get a broad overview of a topic before accessing specific sites.	With a subject directory, you can maintain broader focus and perspective, and work from the general to the specific without getting lost in a maze of Web sites.	Material is organized by human intervention; quality, currency, and accuracy might vary.	Once you work your way through the various levels, you might be able to bookmark the resource for more direct future access.
Invisible Web	Information from nonindexed databases and searchable directories such as company financial reports, library holdings, industry reports, and government information.	Very diverse resource. Many more pages on the Web are non-indexed rather than indexed.	Can be difficult to access unless you know where to look. Navigating a searchable database can be more difficult than using a traditional search engine, because no common interface exists.	Portal sites are available to help you navigate the invisible Web. You can use a general search engine to locate searchable databases by searching a subject term and the word *database*.

FIGURE TK 4-11 A recap of the three main Internet search resources: search engines, subject directories, and the invisible Web. Notice that each option has advantages and disadvantages.

INTERNET COMMUNICATION CHANNELS

Suppose that you are asked to analyze your organization's virus protection requirements. As part of your research you would want to learn about relevant news, developments, and the latest virus threats. You also might want to suggest several specific products. Assume that you performed your research using a search engine. Now you want to check your conclusions by getting feedback from other IT professionals. You can consider using newsgroups, newsletters, blogs, podcasts, RSS feeds, Webinars, mailing lists, Web-based discussion groups, chat rooms, and instant messaging.

Newsgroups

Most people are familiar with bulletin boards they see at school, at work, and in their communities. Using thumb tacks or tape, people post information and read what others have posted. A **newsgroup** is the electronic equivalent of the everyday bulletin board. Newsgroups offer online discussion forums that address every conceivable subject and interest area. A newsgroup can put you in touch with the knowledge, experience, and opinions of a large online community.

Newsgroups are part of the **Usenet**, which is a discussion system that consists of Internet-based bulletin boards. The Usenet was designed to allow users to communicate quickly and effectively in the early days of the Internet. Many newsgroups are open to the public; others are closed and can be accessed only with a username and password.

Articles or messages are posted to the appropriate newsgroups by users, and posted topics are broken down into **threads**, or subtopics. To view and post messages to a newsgroup, you need a program called a **newsreader**. You can use a stand-alone newsreader, such as Forté software's popular Agent program, or you can work with newsreaders that are built into browsers such as Internet Explorer and Firefox.

ON THE WEB

For more information about newsgroups, visit **scsite.com/sad8e/more**, locate Toolkit Part 4, and then click the Newsgroups link.

Some search engines and subject directories allow you to conduct a specific search among newsgroups. Figure TK 4-12 shows some of the many computer-related newsgroups that can be searched by the Google search engine. This portal provides a convenient entry point for a systems analyst who wants to explore and participate in a wide range of computer-related discussions.

To understand how newsgroups work, consider the following example. In your research on virus protection, which was mentioned in the previous section, assume that you have narrowed the product choices down to two. You are having trouble, however, differentiating between them and would like feedback from current users. You might want to visit individual product sites for user testimonials, but you would be unlikely to find a negative opinion on a vendor's site. You also could poll your professional colleagues, but you want to survey a variety of users that goes beyond your limited circle.

At this point, you might decide to tap into an appropriate newsgroup and see if there have been any postings that are relevant to you. Some newsgroups are moderated, in which articles are sent to a person who approves them before they are posted for the group. Before you post, you should read the FAQ files associated with each newsgroup. The term **FAQs** stands for **frequently asked questions**. FAQs are a common method of providing guidance on questions that users are likely to ask. In many cases, FAQs describe the particular **netiquette**, or Web guidelines for protocol and courtesy, that exist on a particular newsgroup or site. You can learn more about netiquette in Part 1 of the Systems Analyst's Toolkit.

As you gain experience, you might decide to subscribe to newsgroups that address topics of interest to you, including some of the ones listed in Figure TK 4-12.

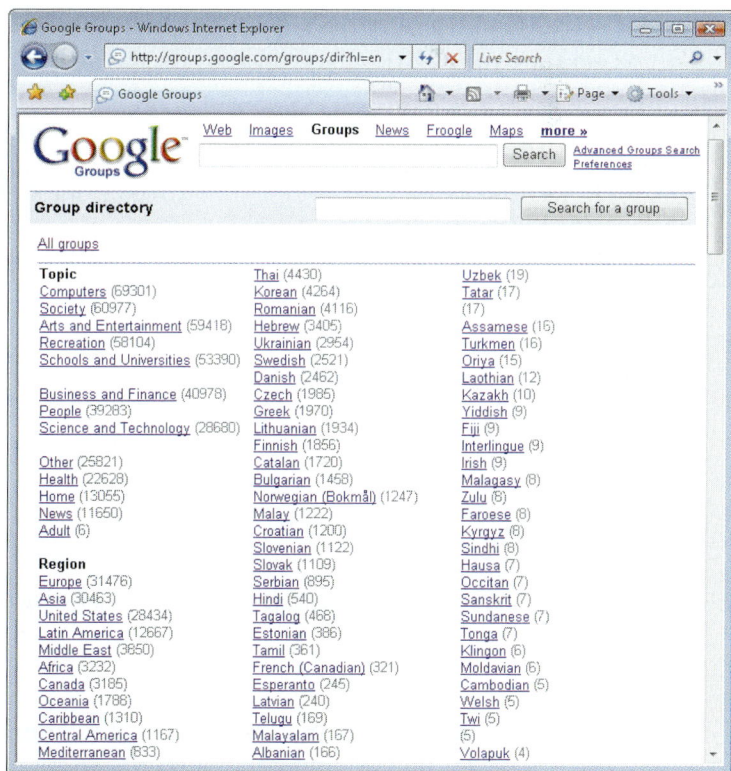

FIGURE TK 4-12 Google provides easy access to discussion groups. In this example, the user is browsing for computer-related newsgroups.

Newsletters, Blogs, and Podcasts

Newsletters are a convenient way to keep current on topics of interest. Many online magazines and other groups offer free e-mail newsletters to subscribers interested in specific topics. For example, as shown Figure TK 4-13, *InfoWorld* readers can request updates on various subjects.

A **blog** is a Web-based log, or journal. Computer-related blogs can provide valuable information for a systems analyst. Also, many vendors offer Web-based training options, including podcasts.

A **podcast** refers to a Web-based broadcast that allows a user to receive audio or multimedia files using music player software such as iTunes®, and either listen to them on a computer or download them to an iPod® or other portable player. Podcasts can be prescheduled, made available on demand, or delivered as automatic updates, depending on a user's preference. An advantage of a podcast is that users, called subscribers, can listen to the recorded material anywhere, anytime. As technology continues to advance, other wireless devices such as PDAs and cell phones will be able to receive podcasts directly.

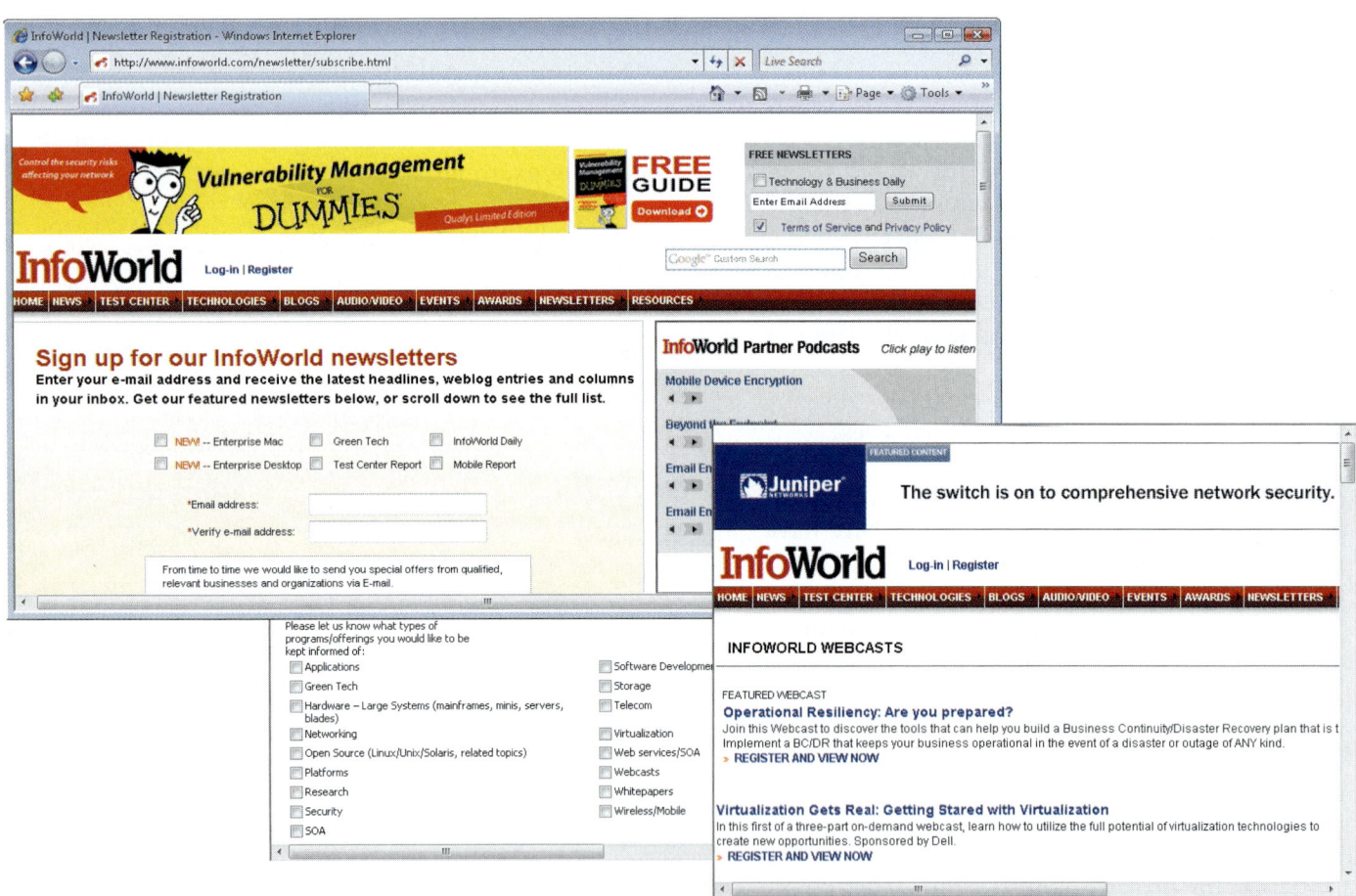

FIGURE TK 4-13 Subscribers to *InfoWorld* newsletters can request updates on various subjects.

RSS Feeds

The term **RSS** stands for **Really Simple Syndication**. RSS is a format for publishing frequently updated content to users who subscribe to an RSS download, also called a **feed**, an **RSS feed**, or a **Web feed**. Web publishers such as Yahoo!, Google, CNN, MSNBC, and many other newspapers, magazines, vendors, and blogs use RSS feeds to distribute news and updates to subscribers, who can read the content with software called an **RSS reader**, a **feed reader**, or an **aggregator**.

Figure TK 4-14 shows how Yahoo! uses RSS feeds to publish news items, and a sample of the wide variety of available topics. Many main headings, such as *Technology*, also contain a lengthy list of subtopics. In addition, Yahoo! allows you to create a custom RSS feed by typing a term or phrase into a search box, which is shown at the bottom of Figure TK 4-14. For example, if you were especially interested in news updates about Cisco routers, you could create a Yahoo! feed that would capture and download all articles on that topic. Yahoo! feeds make it easy for an IT professional to stay well informed on topics of interest. However, notice the important caution regarding the use of the content. Yahoo! states that the feeds are free for individuals and nonprofit organizations for non-commercial use, but that proper attribution is required.

ON THE WEB

For more information about RSS feeds, visit **scsite.com/ sad8e/more**, locate Toolkit Part 4, and then click the RSS Feeds link.

Webinars

A **Webinar**, which combines the words *Web* and *seminar*, is an Internet-based training session that provides an interactive experience. Most Webinars are scheduled events with a group of pre-registered users and an online presenter or instructor. A prerecorded Webinar session also can be delivered as a **Webcast**, which is a one-way transmission, whenever a user wants or needs training support.

Mailing Lists

A **mailing list**, also called a **listserv**, is similar to a newsgroup in that it provides a forum for people who want to exchange information about specific topics. Like a newsgroup, users can post messages and view postings made by others. Instead of a bulletin board approach, however, a mailing list uses e-mail to communicate with users. A computer called a **list server** directs e-mail to people who subscribe to, or join, the mailing list.

When a person subscribes to a list, he or she can receive e-mail messages as they are posted. Subscribers also can access a collection of messages called a **digest**. Many mailing lists maintain Web sites where users can search message archives.

A systems analyst would be interested in mailing lists that focus on information technology. To locate IT-related mailing lists, you can visit the Web sites of professional organizations, and you can try adding the phrase *mailing list or listserv* to your search topic. Also, as shown in Figure TK 4-15 on the next page, you can visit Web sites such as lsoft.com and tile.net that organize mailing lists by name and subject.

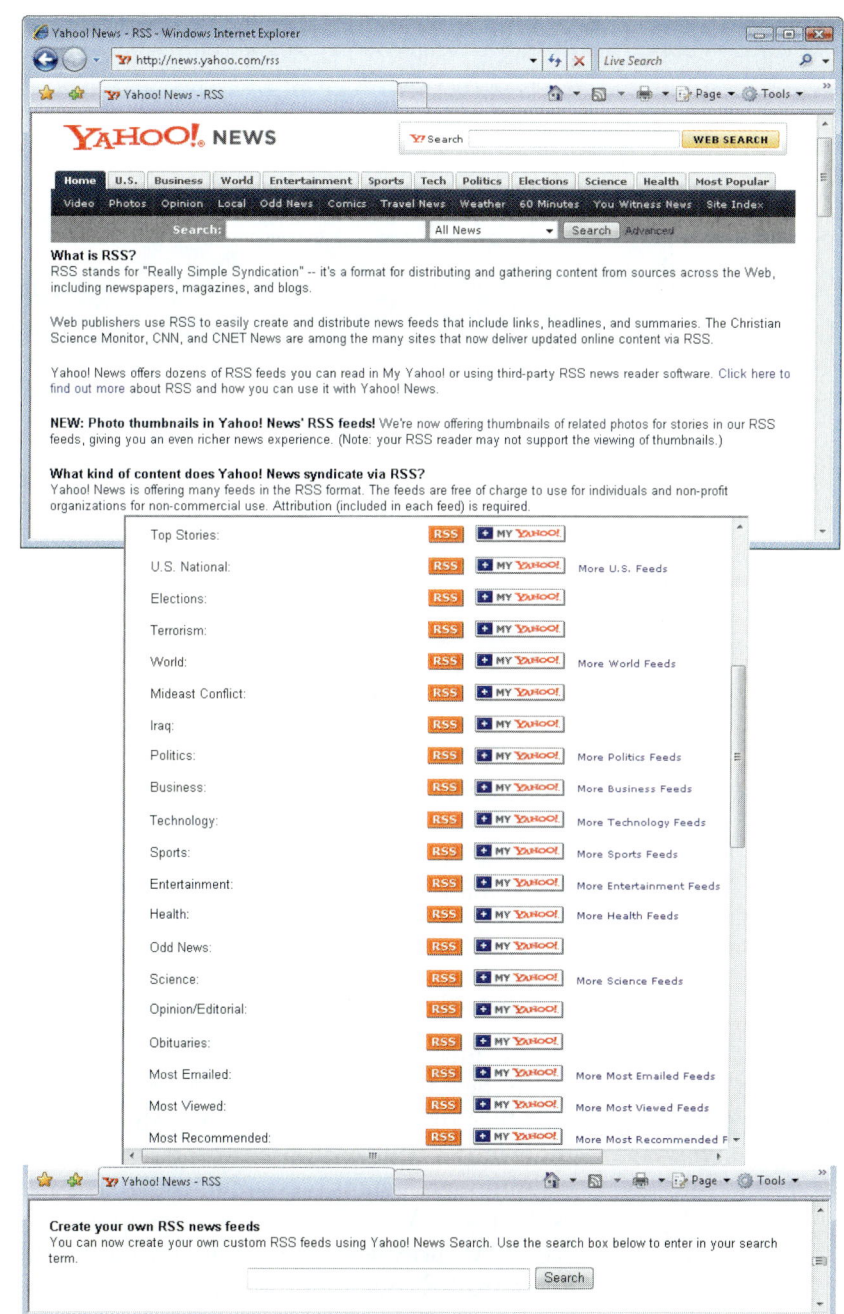

FIGURE TK 4-14 Yahoo! uses RSS feeds to publish news about a wide variety of topics. You also can create a custom feed by typing a term or phrase into the search box at the bottom of the Web page.

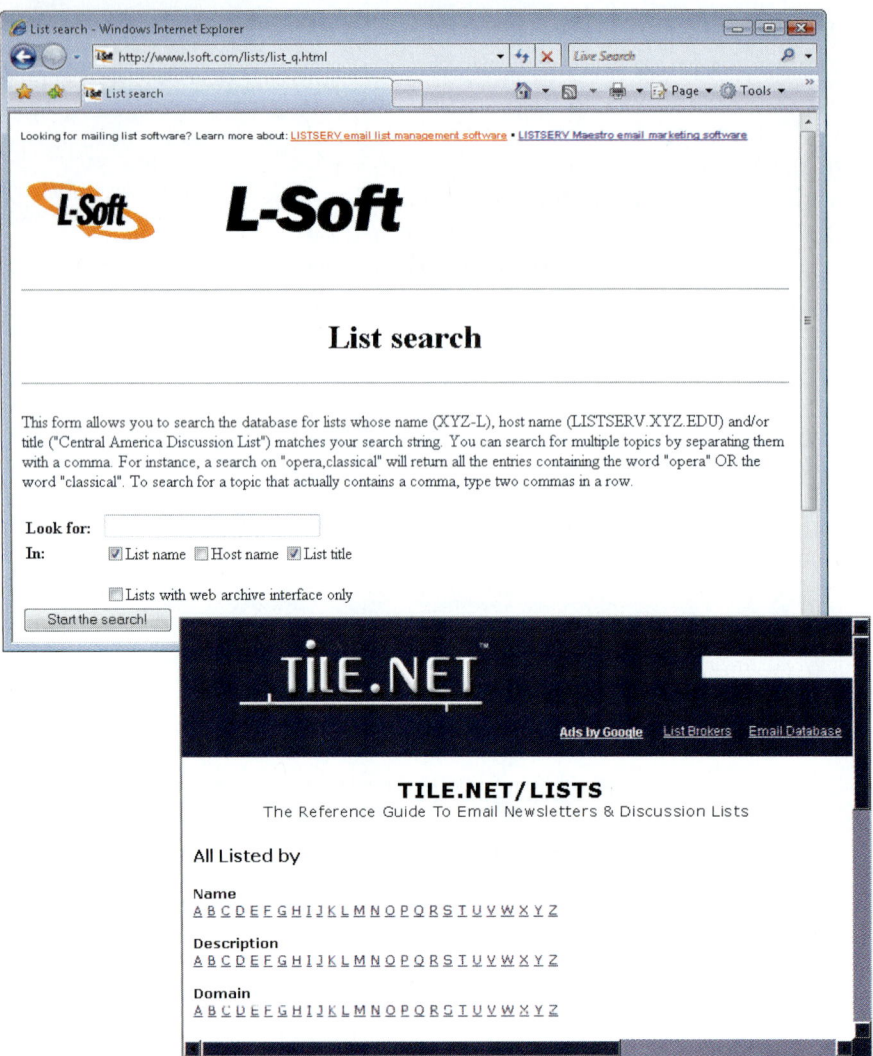

FIGURE TK 4-15 Lsoft.com and tile.net are examples of Web sites that organize mailing lists by name and subject.

Web-Based Discussion Groups

Another online communication resource is called a **Web-based discussion group**, which is an online community that combines features of mailing lists and newsgroups. Web-based discussion groups can be accessed through portals, such as Yahoo! or Google, or by visiting vendor sites, such as Cisco, IBM, or Apple. Group members can receive messages either individually or in digest form like a mailing list, and they can browse messages stored on the group's Web site. In addition to text messages, group Web sites often include membership lists and contact information, photo galleries, and links of interest to members.

Web-based discussion groups are popular because they offer a convenient and free meeting place with a graphical interface that makes it easy for users to access information and exchange messages.

Chat Rooms

A **chat room**, also called a **channel,** is an online meeting place where users can interact and converse in real time. The chat room concept originated with **IRC**, which stands for **Internet Relay Chat**. IRC is a multichannel system supported by servers that enable conversations, group or individual, on a worldwide basis. You can learn more about IRC at the Web site *www.irc.org*, and you can follow links to see a list of available IRC channels that might be of interest to you.

Various companies also provide chat rooms that are available to IT professionals interested in specific topics. For example, Microsoft offers a wide variety of technical chat rooms, as shown in Figure TK 4-16.

Instant Messaging and Text Messaging

Instant messaging (IM) allows online users to exchange messages immediately, even while they are working in another program or application. Users are alerted that other members of their group are available online, and users can send and receive messages or enter into a chat with other users.

Although instant messaging began as a popular feature in home-oriented services such as AOL and Yahoo!, it has become an important business communications tool. Corporate use of IM, however, raises serious security and privacy concerns because it is relatively uncontrolled. Also, certain industries such as banking and health care must observe legal regulations that govern all their communications, including IM, which must be logged and documented. As shown in Figure TK 4-17, firms such as Blue Coat offer software that can manage instant messaging and provide necessary security and controls.

ON THE WEB

For more information about IM and texting abbreviations, visit **scsite.com/sad8e/more**, locate Toolkit Part 4, and then click the IM and Texting Abbreviations link.

Many people use **text messaging,** or **texting,** to send brief written messages from one mobile phone or wireless device to another. Users can also send text messages from a computer to a handheld device. The popularity of IM and texting has given rise to numerous abbreviations that reduce message size and speed up the communication process. Some well-known examples are *BTW (by the way), JMO (just my opinion),* and *TIA (thanks in advance).*

Figure TK 4-18 on the next page shows a recap of online IT channels that can assist a

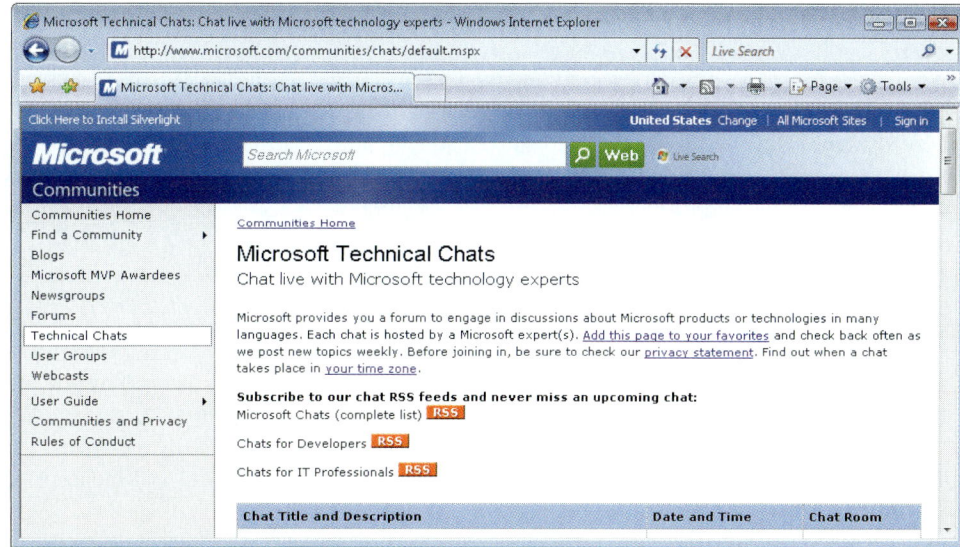

FIGURE TK 4-16 Microsoft offers a wide variety of technical chat rooms where users can interact and converse in real time.

systems analyst in online research and communication: newsgroups, newsletters, mailing lists, RSS feeds, Web-based discussion groups, chat rooms, and instant messaging. Notice that each resource has advantages and disadvantages.

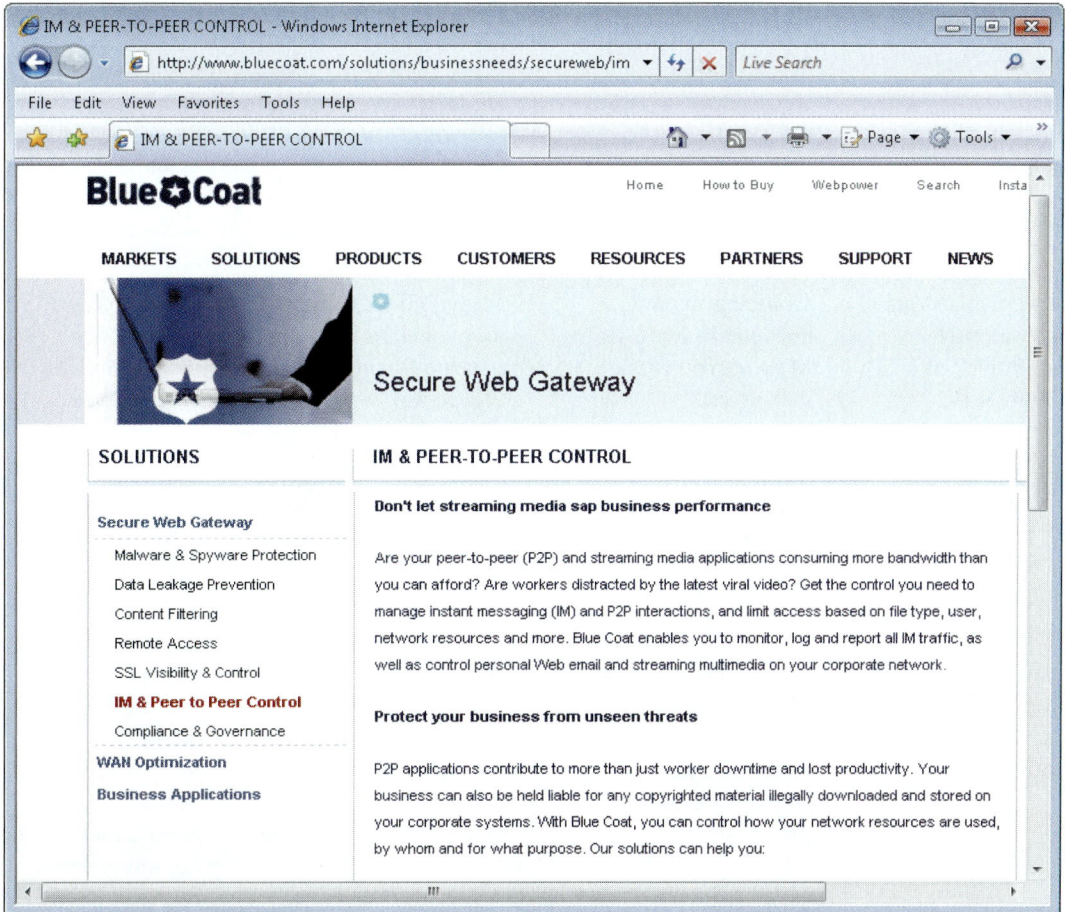

FIGURE TK 4-17 Firms such as Blue Coat offer software that can manage instant messaging and provide necessary security and controls.

Recap of Internet Communication Channels

RESOURCE	POSSIBLE USES	ADVANTAGES	DISADVANTAGES	NOTES
Newsgroups	Answers to technical questions, advice, and support.	Can find information on every conceivable subject — good place to exchange views with other analysts.	Cumbersome to search through message threads. Need to assess quality of information carefully.	Some major search engines allow newsgroup access.
Newsletters, Blogs, and Podcasts	Good way to follow trends and developments regarding specific IT topics.	Most newsletters, blogs, and podcasts are free to users who can subscribe to specific topics.	Not interactive, and the amount of information can be overwhelming. Difficult to filter irrelevant items.	Some newsletters, blogs, and podcasts are published daily; others weekly or monthly.
RSS Feeds	Provide news and updates for readers of online newspapers, magazines, and blogs.	Convenient way to keep posted on virtually any topic of interest.	Unless topics are specific, the volume of information might be difficult to sift through.	Rapid growth is expected to continue, as more users discover this valuable resource.
Webinars	Internet-based training that can provide an interactive experience.	Users can plan, schedule, and pre-register for online sessions.	Only available when scheduled — otherwise not interactive.	Very popular technique because of convenience and cost-effectiveness.
Mailing Lists	Members can exchange information with others about specific topic of interest.	Messages arrive by e-mail, rather than in the form of bulletin board postings.	Amount of material might be difficult to read and analyze.	Some mailing lists, or listservs, allow members to search archived messages.
Web Discussion Groups	Members form a Web-based community regarding topics of common interest.	Combine many convenient features of newsgroups and mailing lists.	Although free, these groups usually must be accessed through a Web portal.	Web discussion groups offer a mix of features and convenience.
Chat Rooms	Online meeting places where users can interact and converse.	Many IT chat rooms attract professionals who are willing to help each other solve problems.	Discussions take place in real time, which might not be convenient. Dialog might be unfocused and irrelevant to your needs.	Many large vendors, such as Microsoft, offer technical chat rooms.
Instant Messaging and Text Messaging	Users can exchange messages immediately, either online or by cell phone.	Highly efficient means of real-time communication on topics of interest. Good way to collaborate on team projects.	Can be distracting to a busy user, and sheer volume of nonessential messages can be a problem.	IM and text messaging have moved beyond the personal desktop and are acceptable business communication tools.

FIGURE TK 4-18 A recap of Internet communication channels that can assist a systems analyst in online research and communication. Notice that each option has advantages and disadvantages.

INFORMATION TECHNOLOGY COMMUNITY RESOURCES

If you were asked to check a stock price or research the weather in a distant city for a business trip, you probably would not use a search engine. Instead, you would visit a favorite site you use regularly to access specific information. Similarly, when you require IT information, you can access a huge assortment of sites and resources that can be called the **information technology (IT) community**. This vast collection includes many sites that IT professionals can use to research specific questions or obtain background information. As a systems analyst, you are a member of this community. Like most communities, it offers you resources and support, including answers to technical questions, updates on new products and services, and information about training opportunities. The IT community includes numerous publications and online magazines, searchable databases, Web-based discussion groups, and mailing lists.

Four important components of the IT community are corporate resources, government resources, professional resources, and online learning resources, which are described in the following sections.

Corporate Resources

Corporate resources can provide general IT knowledge and background, as well as help solve specific business challenges. It is very important to evaluate corporate content carefully, because some sites are developed by companies with an interest in selling you a specific solution or product. Figure TK 4-19 shows an example of a site that offers product reviews and information.

If you are looking for help on a software application, it is a good idea to start by reviewing the software documentation. In many cases, technical support is included free of charge for a specific period of time. If you are working with an application with expired technical support, the software provider's Web site will describe support options that are available to you, including various fees and charges. Common problems often are addressed in the FAQ section.

An important corporate resource to systems analysts is their own internal company Web site or intranet. An intranet must be easy to access and provide access to valuable information. Companies increasingly are using intranets as a means of sharing information and working towards common solutions. Intranets can contain company policies and procedures, lessons-learned files, and financial information. They also enable employees to access and update their personal benefit information. In many organizations, the intranet is reducing the volume of paper memos and reports by serving as an enterprise-wide library and clearinghouse.

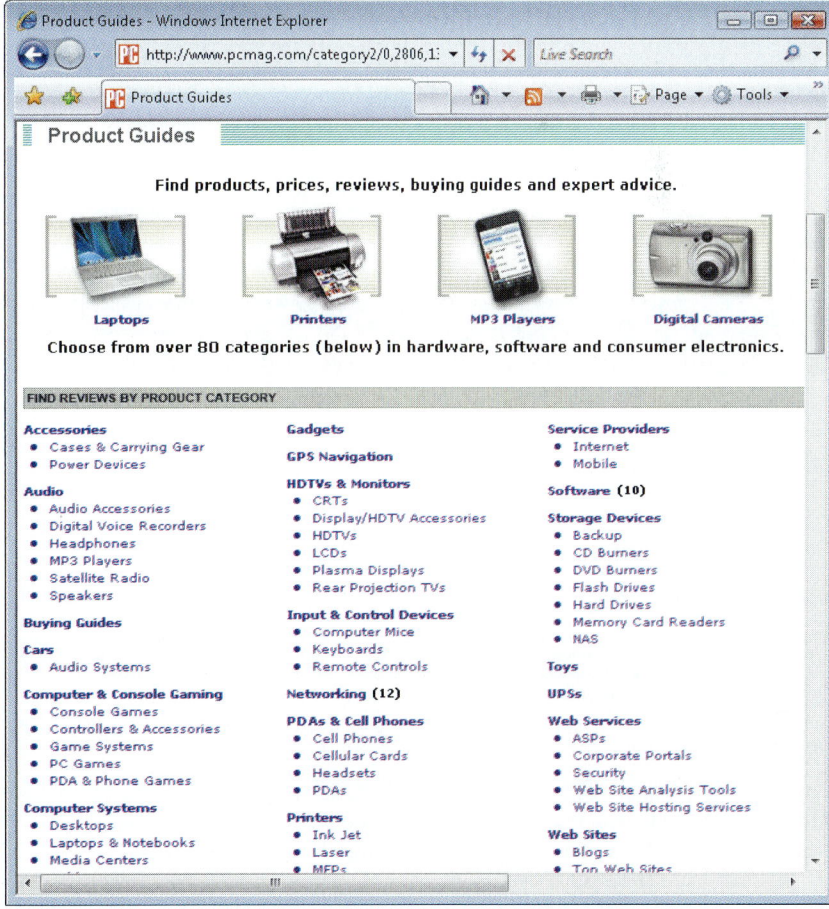

FIGURE TK 4-19 PC Magazine is an example of a site that offers product guides and reviews.

Information Technology Community Resources

Government Resources

The IT needs of the federal government are enormous. Not surprisingly, a number of excellent federal IT resources are available on the Internet. Many sites offer comprehensive, nonbiased information and valuable advice for IT professionals. For example, recent General Accounting Office (GAO) reports on the IT industry have covered everything from an analysis of the information security practices to a framework for assessing IT investments. Additionally, government sites can provide information on federal, state, and local business policies and regulations. The General Services Administration (GSA) site depicted in Figure TK 4-20 is a good source for federal policies and regulations, especially for firms that do business with the government.

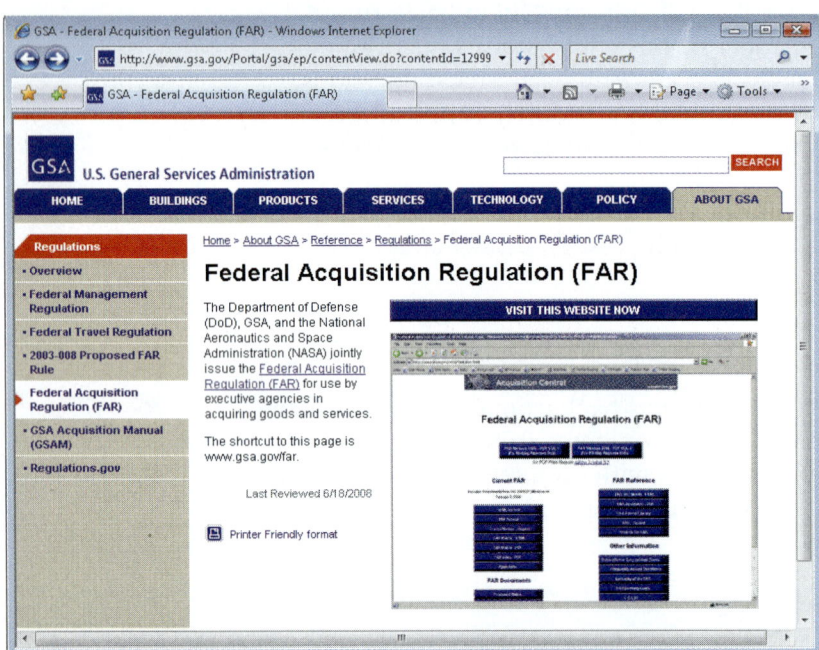

FIGURE TK 4-20 The GSA site contains information about federal IT policies, news, and related links. The screen shows a sample from the IT Regulations, Guidelines and Laws section.

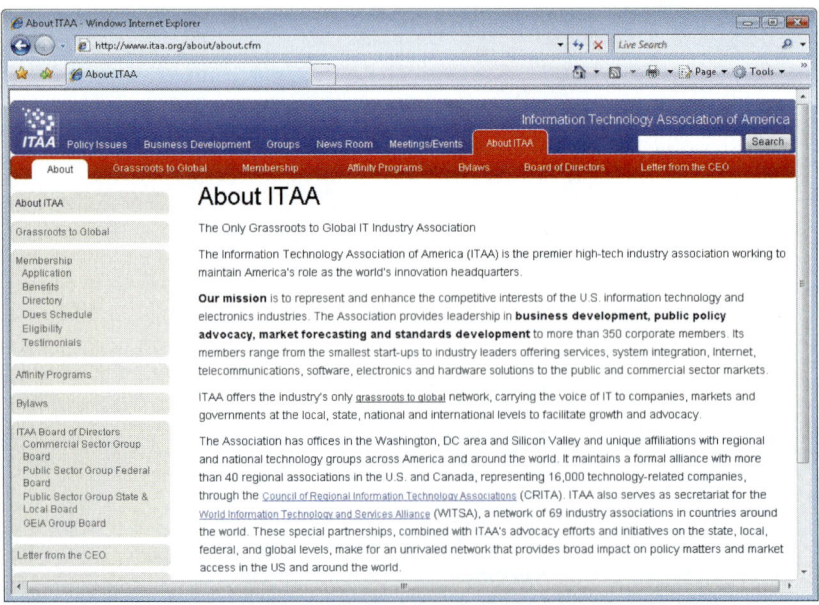

FIGURE TK 4-21 The ITAA site is one of many resources that a systems analyst can use to keep up with current issues, trends, and opportunities.

Professional Resources

The sites of professional associations are another resource worth consideration. Many associations focus on a specific topic such as project management, software engineering, or information security. Some of these associations are free, though most require you to provide basic information in order to access the complete site. The links provided by these sites often are quite useful. Professional organizations, such as the **Information Technology Association of America (ITAA)** site shown in Figure TK 4-21, also sponsor seminars and training. Many associations offer electronic newsletters that relate to your area of interest and are delivered to you by e-mail on a periodic basis.

Online Learning Resources

It is difficult to keep up with the constantly changing IT landscape. Targeted professional development is a way for IT workers to remain focused and current in their chosen areas. In the past, this goal often meant attending lengthy and expensive off-site training courses. Advances in bandwidth and processing power have made online learning an increasingly attractive option for many IT professionals.

Online learning, also referred to as **e-learning**, is a term that refers to the delivery of educational or training content over the public Internet or an intranet. You can locate learning opportunities by searching the Web or through various professional associations. Many schools and colleges have seen a tremendous increase in demand for online learning and have increased their course offerings accordingly. The GoLearn.gov Learning Center shown in Figure TK 4-22 is a government site with many online learning information and opportunities for federal employees.

Online learning can take many forms, ranging from individual self-paced instruction with little or no instructor involvement, to interactive, instructor-led groups with streaming audio and video capability. When choosing an online learning method, your learning goals and the quality of the content are the most important considerations. You need to think about your personal learning preferences. For example, you might learn better in a collaborative environment rather than working alone. If that is the case, consider options that include an interactive peer community. The following are some advantages and disadvantages that apply to the use of online resources:

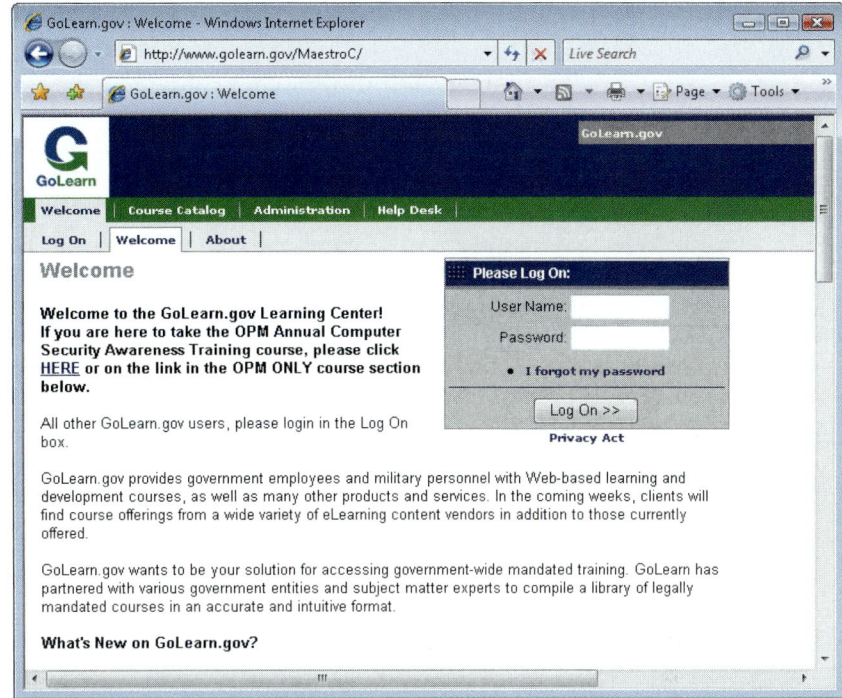

FIGURE TK 4-22 The GoLearn.gov Learning Center offers a wide range of online training for federal employees.

BENEFITS OF ONLINE LEARNING Benefits of online learning include the following:

- Convenient. You can participate in training when and where you want.

- Economical. Online learning options generally are less expensive than traditional face-to-face learning.

- Customizable. Generally, you can tailor the learning experience to your interests and needs.

DISADVANTAGES OF ONLINE LEARNING Disadvantages of online learning include the following:

- Interaction. Although online learning can be highly collaborative, it lacks the face-to-face component that some learners find necessary. Additionally, you are less likely to receive the focused feedback that you would get when participating in traditional classroom training.

- Interface. Although bandwidth and computing power have increased greatly in recent years, the interface in an online learning experience might be a limitation. For example, you might experience slower processing performance, especially if the training uses interactive video, audio, or high-resolution graphics.

- Suitability. Online learning might not be the best option, particularly if the content is complicated and unfamiliar to you.

Figure TK 4-23 on the next page shows a recap of online IT resources, including commercial resources, government and professional resources, and online learning opportunities. Notice that each resource has advantages and disadvantages.

Recap of Online IT Resources

RESOURCE	POSSIBLE USES	ADVANTAGES	DISADVANTAGES	NOTES
Corporate Resources	Specific technical hardware or software help, training opportunities.	First stop for troubleshooting proprietary software or hardware.	Very often an agenda is associated with the site — for example, advocating a particular product or service.	Vendor sites contain valuable specific product or solution information (check the site's FAQs). Many offer newsletters for interested visitors.
Government Resources	Information on IT regulations.	Wide variety of general topics, from congressional studies to industry-relevant government regulations.	Not all information can be accessed via the Web — sometimes sites refer the searcher to a document number for ordering. Information tends to be general and not always relevant to the commercial sector.	Suggested sites: Library of Congress, General Accounting Office, Government Computer News.
Professional Resources	Information on professional skills and standards, IT career advice, and job opportunities.	Very specific IT interest items. Sites often provide valuable links and information on training opportunities.	Quality of sites varies. Many very specialized sites.	These sites often are worth bookmarking and using as central IT portal sites.
Online Learning Resources	An IT professional needs to stay current in a constantly changing technology environment. Online learning can provide job-specific skills and support for career advancement.	Convenient, economical, and customizable. You can participate in training when and where you want, and online learning options are generally less expensive than traditional face-to-face learning. Also, you generally can tailor the learning experience to your interests and needs.	Online learning lacks the face-to-face component that some learners find necessary. Learners are less likely to receive focused feedback. Also, the interface in an online learning experience might be a limitation. Overall, online learning might not be the best option, particularly if the content is complicated and unfamiliar to you.	When considering online learning, you must know your personal learning style and what works best for you. You should examine the entire range of options, from individual self-paced instruction with little or no instructor involvement, to interactive, instructor-led classes.

FIGURE TK 4-23 A recap of online resources that a systems analyst can use to keep up with current issues and trends in IT. Notice that each option has advantages and disadvantages.

TOOLKIT SUMMARY

The Internet is a worldwide network that integrates many thousands of other networks, which in turn link millions of government, business, educational, and personal users around the globe. The Internet can assist you in your daily work by helping you solve technical problems and in the ongoing development of your career by providing access to training and professional education.

The Internet allows access to the World Wide Web, usually called the Web, which refers to the global collection of electronic documents stored on the Internet. These documents are referred to as Web pages, which are organized and stored on individual Web sites.

An Internet research strategy should use a four-step approach: review your information requirements, use the proper search tools and techniques, evaluate the results, and consider copyright and data integrity issues. Many people find it helpful to prepare for an Internet search by using a checklist to identify key terms, phrases, and characteristics of the topic.

The primary research tools include search engines, subject directories, and a collection of searchable database resources called the invisible Web. A search engine uses keywords and phrases to locate information on the Internet and list the results of the search. Users should be aware that results can be affected if the search engine permits commercial users to achieve higher priority based on payment of fees. Also, search engines access only a portion of the Internet. Meta-search engines are tools that can apply multiple search engines simultaneously.

A subject directory is a Web site that allows you to access topics by using a hierarchy, starting with general headings and proceeding to more specific topics. A subject directory is an excellent starting point when you want an overview of a particular topic before proceeding to specific Web sites.

The invisible Web, or hidden Web, describes numerous text, graphics, and data files stored in collections that are unreachable by search engines.

Other Internet tools that might be of value to a systems analyst are newsgroups, newsletters, blogs, podcasts, RSS feeds, Webinars, mailing lists, Web-based discussion groups, chat rooms, and instant messaging. Newsgroups, which are part of the Usenet, are online discussion groups that address every conceivable subject and interest area. Newsletters are published by numerous commercial and nonprofit groups that offer membership subscriptions to users who are interested in specific topics. A mailing list, or listserv, allows subscribing members to post and receive messages forwarded to them by a list server. A Web-based discussion group, usually accessed through a portal such as Yahoo!, combines features of newsgroups and mailing lists for its members. A chat room is an online meeting place where users can interact and converse in real time. Instant messaging allows online users to exchange messages immediately, even while they are working in another program or application.

When an IT professional needs to research a topic or seek background information, he or she can turn to an assortment of sites and resources called the information technology (IT) community. The IT community includes corporate, government, professional, and online learning resources.

Professional development through online learning is a way for IT workers to remain current in their chosen areas. Online learning refers to the delivery of educational or training content over the Internet or an intranet. Online learning is convenient, economical, and customizable. Some disadvantages, however, include a lack of face-to-face interaction, limitations of the interface, and the fact that not everyone works well with this type of training.

Key Terms and Phrases

Toolkit Exercises

Review Questions

1. Describe the size and characteristics of the Internet and the World Wide Web.
2. How do search engines differ from subject directories? Compare these approaches and describe their advantages and possible disadvantages.
3. What is the invisible Web, and how can it be accessed?
4. What steps should you follow when planning an Internet research strategy?
5. What questions should you ask when evaluating the quality of Internet research results?
6. What are sponsored links, and how can they affect the quality of your search results?
7. What is Boolean logic? Provide three examples.
8. How do Venn diagrams show the effect of the three logical operators?
9. Describe newsgroups, newsletters, blogs, podcasts, RSS feeds, Webinars, mailing lists, Web-based discussion groups, chat rooms, and instant messaging. How can these tools be used in Internet research?
10. What is the information technology community, and what resources does it offer for Internet research tasks?

Discussion Topics

1. The textbook explains that some companies pay to obtain a higher ranking when search results are displayed. Is this good, is it bad, or does it not matter to you as a user? Explain your position.
2. Some people rely heavily on instant messaging and texting to communicate with friends and business colleagues. Others find these methods distracting. Do you use instant messaging or texting? Why or why not?
3. Could Boolean logic and Venn diagrams be useful in everyday life? How might they be used?
4. The Internet has affected many aspects of our society. What are the most important benefits of the Internet, and what problems have been created by it?

Projects

1. Use a search engine and enter the following words: presidential candidates in 2016. Run the search and notice how many results appear. Now place quote marks around the phrase and run the search again. Explain the difference in the results.
2. Will the search phrase "*commercial television*" return the same results as the phrase "*television commercial*"? Experiment with a search engine, and explain the results you obtain.
3. Use the Yahoo! subject directory to identify two dictionaries of technical terms. Begin with the subject *Computers and Internet*, and then follow the appropriate links until you obtain results. Describe the results of your research. What other research strategies could you use for this task?
4. Perform research on the Web to learn more about RSS feeds. Then write a practical, step-by-step guide for users who want to set up RSS feeds at their workstations.

INDEX

member of an entity. For example, in a customer table the customer number is a unique primary key because no two customers can have the same customer number. That key also is minimal because it contains no information beyond what is needed to identify the customer. 398

private key encrpyption A common encryption technology called public key encryption (PKE). The private key is one of a pair of keys, and it decrypts data that has been encrypted with the second part of the pair, the public key. 583

private network A dedicated connection, similar to a leased telephone line. 585

privilege escalation attack An unauthorized attempt to increase permission levels. 589

probable-case estimate The most likely outcome is called a probable case estimate. 102

procedural A procedural language requires a programmer to create code statements for each processing step. 634

procedural security Also called operational security, is concerned with managerial policies and controls that ensure secure operations. 591

process Procedure or task that users, managers, and IT staff members perform. Also, the logical rules of a system that are applied to transform data into meaningful information. In data flow diagrams, a process receives input data and produces output that has a different content, form, or both. 7, 8, 138, 149, 198–199, 220

process 0 In a data flow diagram, process 0 (zero) represents the entire information system, but does not show the internal workings. 205

process control Process control allows users to send commands to a system. A process control screen (also known as a dialog screen) is part of the user interface, and enables a user to initiate or control system actions. 346

process description A documentation of a functional primitive's details, which represents a specific set of processing steps and business logic. 222–227

process improvement The framework used to integrate software and systems development by a new SEI model, Capability Maturity Model Integration (CMMI)®. 500

process model A process model describes system logic and processes that programmers use to develop necessary code modules. 16

process-centered An analytic approach that describes processes that transform data into useful information. 19

process control screens, 346
processes
    business, 8
    DFD symbol, 198
    documenting, 220
    in information systems, 7
    in requirements
      modeling, 138
    system requirements
      checklist, 149
processing
    batch, 469–470
    methods, 468–470
    options, 452–455, 482

product baseline The product baseline describes the system at the beginning of system operation. The product baseline incorporates any changes made since the allocated baseline and includes the results of performance and acceptance tests for the operational system. 571

production environment The environment for the actual system operation. It includes hardware and software configurations, system utilities, and communications resources. Also called

the operational environment. 524

productivity software Software such as word processing, spreadsheet, database management, and presentation graphics programs. 166

product-oriented Product-oriented firms manufacture computers, routers, or microchips. 9

profit center A department expected to break even, or show a profit. 601

program documentation Program documentation starts in the systems analysis phase and continues during systems implementation. Systems analysts prepare overall documentation, such as process descriptions and report layouts, early in the SDLC. Programmers provide documentation by constructing modules that are well-supported by internal and external comments and descriptions that can be understood and maintained easily. 518

Program Evaluation Review Technique (PERT) The Program Evaluation Review Technique (PERT) was developed by the U.S. Navy to manage very complex projects, such as the construction of nuclear submarines. At approximately the same time, the Critical Path Method (CPM) was developed by private industry to meet similar project management needs. The important distinctions between the two methods have disappeared over time, and today the technique is called either PERT, CPM, or PERT/ CPM. 105, 106–111, 596

programmer/analyst A designation for positions that require a combination of systems analysis and programming skills. 565

project coordinator The project coordinator handles administrative responsibilities for the development team and negotiates with users who might have

conflicting requirements or want changes that would require additional time or expense. 98, 119–120

project creep The process by which projects with very general scope definitions expand gradually, without specific authorization. 100

project leader The project leader, or project manager, usually is a senior systems analyst or an IT department manager if the project is large. An analyst or a programmer/analyst might manage smaller projects. 98

project management The process of planning, scheduling, monitoring, controlling, and reporting upon the development of an information system. 19, 121–123, 316
    and application
      development, 506
    overview, 98–99
    successful, 121

project management software Project management software can aid analysts in project planning, estimating, scheduling, monitoring, and reporting. Powerful project management packages offer many features, including PERT/CPM, Gantt charts, resource scheduling, project calendars, cost tracking, and cost-benefit analysis. 115–118, 123

project manager The project manager, or project leader, usually is a senior systems analyst or an IT department manager if the project is large. An analyst or a programmer/analyst might manage smaller projects. 98, 99–102, 113–115, 121, 123

project monitoring and controlling Project monitoring and controlling requires guiding, supervising, and coordinating the project team's workload. 99, 113–114

project planning Project planning includes identifying project tasks and estimating completion time and costs. 99

tuple A tuple (rhymes with couple), or record, is a set of related fields that describes one instance, or member of an entity, such as one customer, one order, or one product. A tuple might have one or dozens of fields, depending on what information is needed. 398

turnaround document Output document that is later entered back into the same or another information system. A telephone or utility bill, for example, might be a turnaround document printed by the company's billing system. When the bill is returned with payment, it is scanned into the company's accounts receivable system to record the payment accurately. 337

turnaround time Turnaround time applies to centralized batch processing operations, such as customer billing or credit card statement processing. Turnaround time measures the time between submitting a request for information and the fulfillment of the request. Turnaround time also can be used to measure the quality of IT support or services by measuring the time from a user request for help to the resolution of the problem. 574

tutorial A series of online interactive lessons that present material and provide a dialog with users. 527, 529–530

two-tier In a two-tier design, the user interface resides on the client, all data resides on the server, and the application logic can run either on the server or on the client, or be divided between the client and the server. 458

type In data dictionaries, type refers to whether a data element contains numeric, alphabetic, or character values. 217

unencrypted Data that is not encrypted. 583

Unicode Unicode is a relatively recent coding method that represents characters as integers. Unlike EBCDIC and ASCII, which use eight bits for each character, Unicode requires 16 bits per character, which allows it to represent more than 65,000 unique characters. 429

Unified Modeling Language (UML) a widely used method of visualizing and documenting software systems design. UML uses object-oriented design concepts, but it is independent of any specific programming language and can be used to describe business processes and requirements generally. 147

uninterruptible power supply (UPS) Battery-powered backup power source that enables operations to continue during short-term power outages and surges. 581

unit testing The testing of an individual program or module. The objective is to identify and eliminate execution errors that could cause the program to terminate abnormally, and logic errors that could have been missed during desk checking. 515–516

Universal Security Slot (USS) Can be fastened to a cable lock or laptop alarm. 582

unnormalized An unnormalized record is one that contains a repeating group, which means that a single record has multiple occurrences of a particular field, with each occurrence having different values. 407

unstructured brainstorming A group discussion where any participant can speak at any time. 165

usability metrics Data that interface designers can obtain by using software that can record and measure user interactions with the system. 352

usage model A service model that charges a variable fee for an application based on the volume of transactions or operations performed by the application. Also called a transaction model. 289

use case A use case represents the steps in a specific business function or process in UML (Unified Modeling Language). 257, 262, 266

use case description A description in UML (Unified Modeling Language) that documents the name of the use case, the actor, a description of the use case, a step-by-step list of the tasks required for successful completion, and other key descriptions and assumptions. 258

use case diagram A visual representation that represents the interaction between users and the information system in UML (Unified Modeling Language). 147–148, 259–260

Usenet A discussion system that consists of Internet-based bulletin boards. Newsgroups are part of the Usenet. 689

user application User applications utilize standard business software, such as Microsoft Office 2003, that has been configured in a specific manner to enhance user productivity. 295–296

user design phase In this phase, users interact with systems analysts and develop models and prototypes that represent all system processes, outputs, and inputs. 142

user documentation Instructions and information to users who will interact with the system. It includes user manuals, help screens, and tutorials. 520–523

user groups, security, 588

user ID A method of limiting access to files and databases to protect stored data. 430

user interface A user interface includes screens, commands, controls, and features that enable users to interact more effectively with an application. 346

user productivity systems Systems that provide employees of all levels a wide array of tools to improve job performance. Examples include e-mail,

word processing, graphics, and company intranets. 14

user security, 536–537

user rights User-specific privileges that determine the type of access a user has to a database, file, or directory. Also called permissions. 587

user story In agile development, a short, simple requirements definition provided by the customer. Programmers use user stories to determine a project's requirements, priorities, and scope. 512

user support A function typically performed by individuals within an IT department. User support provides users with technical information, training, and productivity support. 27, 558–560

user training package The main objective of a user training package is to show users how the system can help them perform their jobs. 558

user-centered A term that indicates the primary focus is upon the user. In a user-centered system, the distinction blurs between input, output, and the interface itself. 347

users Employees, customers, vendors, and others who interact with an information system. Sometimes referred to as end users. 7
  involvement in JAD, 140
  security resistance, 590
  systems design input, 311–312, 480–481
  training for, 525–526

user-selected Under the control of the system or application user. For example, user-selected help displays information when the user requests it. 355

validity check A type of data validation check that is used for data items that must have certain values. For example, if an inventory system has 20 valid item classes, then any input item that does not match one of the valid classes will fail the check. 365–366, 400

validity rules Rules that are applied to data elements when data is entered to

PHOTO CREDITS